EXPLORING THE UNKNOWN

ISBN 0-16-073135-6

90000

9 780160 731358

ISBN 0-16-073135-6

NASA SP-2004-4407

EXPLORING THE UNKNOWN

Selected Documents in the History of the U.S. Civil Space Program

Volume VI: Space and Earth Science

John M. Logsdon, General Editor
with Stephen J. Garber, Roger D. Launius,
and Ray A. Williamson

The NASA History Series

National Aeronautics and Space Administration
NASA History Office
Office of External Relations
Washington, DC   2004

Library of Congress Cataloguing-in-Publication Data

Exploring the Unknown: Selected Documents in the History of the U.S. Civil Space Program/ John M. Logsdon, editor ...[et al.]
p. cm.—(The NASA history series) (NASA SP: 4407)

     Includes bibliographical references and indexes.
     Contents: v. 6. Space and Earth Science
     1. Astronautics—United States—History.   I. Logsdon, John M., 1937–
           II. Series   III. Series   V. Series: NASA SP: 4407.
TL789.8.U5E87   1999   96-9066
387.8'0973-dc20   CIP

Dedicated to
William H. Pickering
Space Science Pioneer

and

Jonathan L. Friedman
Manager, Editor, and Friend

Contents

Chapter Two

Documents

Chapter Three

Documents

Acknowledgments

This volume is the sixth in a series that had its origins more than a decade ago. The individuals involved in initiating the series and producing the initial five volumes have been acknowledged in those volumes [Volume I—Organizing for Exploration (1995); Volume II—External Relationships (1996); Volume III—Using Space (1998); Volume IV—Accessing Space (1999); Volume V—Exploring the Cosmos (2001)]; those acknowledgments will not be repeated here.

We owe thanks to the individuals and organizations that have searched their files for potentially useful materials, and for the staffs at various archives and collections who have helped us locate documents. James Green of Goddard Space Flight Center, David DeVorkin of the National Air and Space Museum, and Joan Vernikos, formerly of NASA, in addition to authoring introductory essays, helped in the identification and acquisition of key documents. Graduate students Holly Carter Degn, Brian Dewhurst, Jonathan Krezel, Chirag Vyas, and Avery Sen also helped in the preparation of the volume.

My thanks go to all those mentioned above, and again to those who helped get this effort started almost a decade ago and who have been involved along the way.

John M. Logsdon, George Washington University

There are numerous people at NASA associated with historical study, technical information, and the mechanics of publishing who helped in myriad ways in the preparation of this documentary history. In the NASA History Division, M. Louise Alstork carefully edited the entire volume and prepared the index; Nadine J. Andreassen performed editorial and proofreading work on the project; Charles Brooks and Jennifer Troxell researched and wrote the entries for the biographical appendix; and Claire Rojstaczer expertly handled a great number of final editorial tasks. In addition, the staffs of the NASA Headquarters Library, the Scientific and Technical Information Program, and the NASA Document Services Center provided assistance in locating and preparing for publication the documentary materials in this work.

The NASA Headquarters Printing and Design Office developed the layout and handled printing. Specifically, we wish to acknowledge the work of Cindy Min, who expertly handled the layout of this book; Michelle Cheston, Lisa Jirousek, and Anne Marson, whose patient attention to detail is reflected in the excellent copyediting; and Jeffrey McLean and James Penny, who oversaw the printing of this volume.

Thanks are due to all these fine professionals.

Steven J. Dick
NASA Chief Historian

Introduction

One of the most important developments of the twentieth century has been the movement of humanity into space with machines and people. The underpinnings of that movement—why it took the shape it did; which individuals and organizations were involved; what factors drove a particular choice of scientific objectives and technologies to be used; and the political, economic, managerial, and international contexts in which the events of the space age unfolded—are all important ingredients of this epoch transition from an Earthbound to a spacefaring people. This desire to understand the development of spaceflight in the United States sparked this documentary history series.

The extension of human activity into outer space has been accompanied by a high degree of self-awareness of its historical significance. Few large-scale activities have been as extensively chronicled so closely to the time they actually occurred. Many of those who were directly involved were quite conscious that they were making history, and they kept full records of their activities. Because most of the activity in outer space was carried out under government sponsorship, it was accompanied by the documentary record required of public institutions, and there has been a spate of official and privately written histories of most major aspects of space achievement to date. When top leaders considered what course of action to pursue in space, their deliberations and decisions often were carefully put on the record. There is, accordingly, no lack of material for those who aspire to understand the origins and evolution of U.S. space policies and programs.

This reality forms the rationale for this series. Precisely because there is so much historical material available on space matters, the National Aeronautics and Space Administration (NASA) decided in 1988 that it would be extremely useful to have easily available to scholars and the interested public a selective collection of many of the seminal documents related to the evolution of the U.S. civilian space program. While recognizing that much space activity has taken place under the sponsorship of the Department of Defense and other national security organizations, the U.S. private sector, and in other countries around the world, NASA felt that there would be lasting value in a collection of documentary material primarily focused on the evolution of the U.S. government's civilian space program, most of which has been carried out since 1958 under the Agency's auspices. As a result, the NASA History Office contracted with the Space Policy Institute of George Washington University's Elliott School of International Affairs to prepare such a collection. This is the sixth volume in the documentary history series; two additional ones containing documents and introductory essays related to human space flight, including microgravity research in Earth orbit, will follow.

The documents collected during this research project were assembled from a diverse number of both public and private sources. A major repository of primary source materials relative to the history of the civil space program is the NASA Historical Reference Collection of the NASA History Office located at the Agency's Headquarters in Washington, DC. Project assistants combed this collection for the "cream" of the wealth of material housed there. Indeed, one purpose of this series from the start was to capture some of the highlights of the holdings at Headquarters. Historical materials housed at the other NASA installations, institutions of higher learning, and presidential libraries were other sources of documents considered for inclusion, as were papers in the archives of individuals and firms involved in opening up space for exploration.

Copies of the documents included in this volume in their original form will be deposited in the NASA Historical Reference Collection. Another complete set of project materials is located at the Space Policy Institute at George Washington University. These materials in their original forms are available for use by researchers seeking additional information about the evolution of the U.S. civil space program or wishing to consult the documents reprinted herein in their original form.

The documents selected for inclusion in this volume are presented in four major sections, each covering a particular aspect of the origins, evolution, and execution of the U.S. space and Earth science program. Section I deals with the scientific study of the Sun. Section II discusses the study of the physical characteristics of space, including both interactions between the Sun and Earth, and other areas of investigation. Section III deals with NASA's fundamental research in life sciences—space biology. (Issues associated with the study of the reactions of the human body to the space environment and the health of astronauts will be covered in the next two volumes.) Section IV discusses the most recent area of science to which space observations contribute—that intend to advance understanding of the Earth as a planetary system.

Volume I in this series covered the antecedents to the U.S. space program and the origins and evolution of U.S. space policy and of NASA as an institution. Volume II dealt with the relations between the civilian space program of the United States and the space activities of other countries; the relations between the U.S. civilian and national security space and military efforts; and NASA's relations with industry and academic institutions. Volume III provided documents on satellite communications, remote sensing, and the economics of space applications. Volume IV covered various forms of space transportation. Volume V covered the origins of NASA's space science program and its efforts in solar system exploration and astrophysics and astronomy. As noted above, two future volumes will cover human spaceflight (Volumes VII and VIII).

An overview essay introduces each section in the present volume. These essays are intended to introduce and complement the documents in the section, and to place them in a chronological and substantive context. Each essay contains references to the documents in the section it introduces, and may also contain references to documents in other sections of the collection. These introductory essays are the responsibility of their individual authors, and the views and conclusions contained therein do not necessarily represent the opinions of either George Washington University or NASA.

The project team in concert chose the documents included in each section with the essay writer from those assembled by the research staff for the overall project. The contents of this volume emphasize primary documents or long-out-of-print essays or articles and material from the private recollections of important actors in shaping space affairs. The contents of this volume thus do not comprise in themselves a comprehensive historical account; they must be supplemented by other sources, those both already available and to become available in the future. The documents included in each section are arranged chronologically, with the exception that closely related documents are grouped together. Each document is assigned its own number in terms of the section in which it is placed. Thus, the first document in the third section of this volume is designated "Document III-1." Each document or group of related documents is accompanied by a headnote setting out its context and providing a background narrative. These headnotes also provide specific information about people and events discussed. We have avoided the inclusion of explanatory notes in the documents themselves and have confined such material to the headnotes.

The editorial method we adopted for dealing with these documents seeks to preserve spelling, grammar, paragraphing, and use of language as in the original. We have sometimes changed punctuation where it enhances readability. We have used the designation [not included, or omitted] to note where sections of a document have not been included in this publication, and we have avoided including words and phrases that had been deleted in the original document unless they contribute to an understanding of what was going on in the mind of the writer in making the record. Marginal notations on the original documents are inserted into the text of the documents in brackets, each clearly marked as a marginal comment. Except insofar as illustrations and figures are necessary to understanding the text, those items have been omitted from this printed version. Page numbers in the original document are noted in brackets internal to the document text. Copies of all documents in their original form, however, are available for research by any interested person at the NASA History Office or the Space Policy Institute of George Washington University.

We recognize that there are certain to be quite significant documents left out of this compilation. No two individuals would totally agree on all documents to be included from the many we collected, and surely we have not been totally successful in locating all relevant records. As a result, this documentary history can raise an immediate question from its users: why were some documents included while others of seemingly equal importance were omitted? There never can be a fully satisfactory answer to this question. Our own criteria for choosing particular documents and omitting others rested on three interrelated factors:

- Is the document the best available, most expressive, most representative reflection of a particular event or development important to the evolution of the space program?

- Is the document not easily accessible except in one or a few locations, or is it included (for example, in published compilations of presidential statements) in reference sources that are widely available and thus not a candidate for inclusion in this collection?

- Is the document protected by copyright, security classification, or some other form of proprietary right and thus unavailable for publication?

As general editor of this volume, I was ultimately responsible for the decisions about which documents to include and for the accuracy of the headnotes accompanying them. It has been an occasionally frustrating but consistently exciting experience to be involved with this undertaking; my associates and I hope that those who consult it in the future find our efforts worthwhile.

John M. Logsdon
Director
Space Policy Institute
Elliott School of International Affairs
George Washington University

Biographies of Volume VI Editors

Stephen J. Garber is a historian with the National Aeronautics and Space Administration. He has worked in the NASA History Office since 1995 and served as the acting head of this office from July 2002 through October 2003. He has edited a book on the past and future of human spaceflight, and written on such aerospace history topics as the congressional cancellation of NASA's Search for Extraterrestrial Intelligence program, President Kennedy's attitudes toward space, the design of the Space Shuttle, and the Soviet Buran Space Shuttle.

Roger D. Launius is chairman of the Space History Department of the Smithsonian Institution's National Air and Space Museum and is the former NASA Chief Historian. He has produced several books and articles on aerospace history, including *Innovation and the Development of Flight* (Texas A&M University Press, 1999); *NASA & the Exploration of Space* (Stewart, Tabori, & Chang, 1998); *Frontiers of Space Exploration* (Greenwood Press, 1998); *Organizing for the Use of Space: Historical Perspectives on a Persistent Issue* (Univelt, Inc., AAS History Series, Volume 18, 1995), editor; *NASA: A History of the U.S. Civil Space Program* (Krieger Publishing Co., 1994); *History of Rocketry and Astronautics: Proceedings of the Fifteenth and Sixteenth History Symposia of the International Academy of Astronautics* (Univelt, Inc., AAS History Series, Volume 11, 1994), editor; *Apollo: A Retrospective Analysis* (Monographs in Aerospace History, Vol. 3, 1994); and *Apollo 11 at Twenty-Five*, electronic picture book issued on computer disk by the Space Telescope Science Institute, Baltimore, MD, 1994.

John M. Logsdon is Director of the Space Policy Institute of George Washington University's Elliott School of International Affairs, where he is also a professor of political science and international affairs. He holds a B.S. in physics from Xavier University and a Ph.D. in political science from New York University. He has been at George Washington University since 1970, and previously taught at The Catholic University of America. He is also a faculty member of the International Space University. He is an elected member of the International Academy of Astronautics and a member of the board of The Planetary Society. He is a former member of the NASA Advisory Council and served during 2003 on the Columbia Accident Investigation Board. Dr. Logsdon has lectured and spoken to a wide variety of audiences at professional meetings, colleges and universities, international conferences, and other settings, and has testified before Congress on numerous occasions. The electronic and print media frequently consult him for his views on various space issues. He has been a Fellow at the Woodrow Wilson International Center for Scholars and was the first holder of the Chair in Space History of the National Air and Space Museum. He is a Fellow of the American Association for the Advancement of Science and the American Institute of Aeronautics and Astronautics.

Ray A. Williamson is a Research Professor of Space Policy and International Affairs at George Washington University's Elliott School of International Affairs, focusing on the history, programs, and policy of Earth observations, space transportation, and space commercialization. He joined the Space Policy Institute in 1995. Previously, he was a Senior Associate and Project Director in the Office of Technology Assessment (OTA) of the U.S. Congress. He joined OTA in 1979. While at OTA, Dr. Williamson was Project Director for more than a dozen reports on space policy, including: *Russian Cooperation in Space* (1995), *Civilian Satellite Remote Sensing: A Strategic Approach* (1994), *Remotely Sensed Data: Technology, Management, and Markets* (1994), *Global Change Research and NASA's Earth Observing System* (1994), and *The Future of Remote Sensing from Space: Civilian Satellite Systems and Applications* (1993). He has written extensively about the U.S. space program. He holds a bachelor of arts degree in physics from Johns Hopkins University and a Ph.D. in astronomy from the University of Maryland. He spent two years on the faculty of the University of Hawaii studying diffuse emission nebulae and ten years on the faculty of St. John's College in Annapolis, Maryland. He is a member of the faculty of the International Space University and of the editorial board of *Space Policy*.

Acronyms

AAOE . . . . .Airborne Antarctic Experiment
AAS . . . . .American Astronomical Society
ACE . . . . .Advanced Composition Explorer
ACRIM . . . . .Active Cavity Radiometer Irradiance Monitor
ADCLS . . . . .Advanced Data Collection and Location System
ADEOS . . . . .Advanced Earth Observation Satellite
AEC . . . . .Atomic Energy Commission
AES . . . . .Apollo Extension Systems
AGU . . . . .American Geophysical Union
AIMP . . . . .Anchored, Interplanetary Monitoring Platforms
AIRS . . . . .Atmospheric Infrared Sounder
AIT . . . . .Asian Institute of Technology
AMOWG . . . . . .Astrophysics Management Operations Working Group
AMPTE . . . . .Active Magnetospheric Particle Tracer Experiment
AMS . . . . .American Meteorological Society
AO . . . . .Announcement of Opportunity
AOSO . . . . .Advanced Orbiting Solar Observatory
APA . . . . .Allowance for Program Adjustment
APACM . . . . .Atmospheric Physical and Chemical Monitor
ARC . . . . .Ames Research Center
ARISTOTELES . .Applications and Research Involving Space Technologies Observing
                     the Earth's Field from Low Earth Orbiting Satellites
ARPA . . . . .Advanced Research Projects Agency
ASI . . . . .Agencia Spatiale Italiano
ASTP . . . . .Apollo-Soyuz Test Project
ATLAS . . . . .Atmospheric Laboratory for Applications and Science
ATM . . . . .Apollo Telescope Mount
ATOM . . . . .Astronomical Telescope Orientation Mount
AU . . . . .Astronomical Unit
AVHRR . . . . .Advanced Very High Resolution Radiometer
AXAF . . . . .Advanced X-ray Astrophysics Facility

BAC . . . . .Bioscience Advisory Committee
BAC . . . . .British Aircraft Corporation
BASC . . . . .Board on Atmospheric Sciences and Climate
BMFT . . . . .Federal Ministry for Research and Technology (Germany)
BMRC . . . . .Bureau of Meteorology Research Center (Australia)

CCP . . . . .Cost Control Plan
CCRS . . . . .Canada Centre for Remote Sensing
CEES . . . . .Committee on Earth and Environmental Sciences
CELSS . . . . .Controlled Ecological Life Support Systems
CENR . . . . .Committee on Environment and Natural Resources

CEOS . . . . .Committee on Earth Observing
CERES . . . . .Clouds and Earth's Radiant Energy System
CES . . . . .Committee on Earth Sciences
CGED . . . . .Committee on Geophysical and Environmental Data
CIESIN . . . . .Consortium for International Earth Science Information Network
CLOS . . . . .Coordination on Land Observation Satellites
CME . . . . .Coronal Mass Ejection
CNES . . . . .Centre National d'Etudes Spatiales
COBE . . . . .Cosmic Background Explorer
COMETS . . . . . . .Communication and Broadcasting Engineering Test Satellite
CORSS . . . . .Coordination on Ocean Remote Sensing Satellites
COSPAR . . . . . . .Committee on Space Research
COSEPUP . . . . . .Committee on Science, Engineering, and Public Policy
COSTR . . . . .Collaborative Solar-Terrestrial Research
CRIE . . . . .Cosmic Ray Isotope Experiment
CRL . . . . .Communications Research Laboratory (Japan)
CRRES . . . . .Combined Release and Radiation Effects Satellite
CSA . . . . .Canadian Space Agency
CSAA . . . . .Committee on Space Astronomy and Astrophysics
CSBM . . . . .Committee on Space Biology and Medicine
CSIRO . . . . .Commonwealth Scientific and Industrial Research Organization
                       (Australia)
CSSP . . . . .Committee on Solar and Space Physics
CZCS . . . . .Coastal Zone Color Scanner

DAAC . . . . .Distributed Active Archive Centers
DARA . . . . .Deutsche Agentur fur Raumfahrtangelegenheiten
DE . . . . .Dynamics Explorer
DEM . . . . .Digital Elevation Model
DEP . . . . .Data Exchange Principles
DFRC . . . . .Dryden Flight Research Center
DMA . . . . .Defense Mapping Agency
DMSP . . . . .Defense Meteorological Satellite Program
DMWG . . . . .Data Management Working Group
DOC . . . . .U.S. Department of Commerce
DOD . . . . .U.S. Department of Defense
DOE . . . . .U.S. Department of Energy
DOI . . . . .U.S. Department of Interior
DOT . . . . .U.S. Department of Transportation
DSN . . . . .Deep Space Network
DTED . . . . .Digital Terrain Elevation Data

E-SPAN . . . . .European-Space Analysis Network
ECOM . . . . .EOS Communications
ELDO . . . . .European Launcher Development Organization

GOOS . . . . .Global Ocean Observing System
GOS . . . . .Global Observing System
GPM . . . . .Global Precipitation Mission
GPS . . . . .Global Positioning System
GRB . . . . .Geophysics Research Board
GRE . . . . .Gamma Ray Explorer
GRID . . . . .Global Resource Information Database
GRM . . . . .Geopotential Research Mission
GRM . . . . .Geopotential Radar Mapper
GSFC . . . . .Goddard Space Flight Center
GTOS . . . . .Global Terrestrial Observing System
GZ . . . . .Giacobini-Zinner

HAO . . . . .High Altitude Observatory
HCO . . . . .Harvard College Observatory
HEAO . . . . .High Energy Astronomical Observatory
HEDS . . . . .Human Exploration and Development of Space
HESP . . . . .High Energy Solar Platform
HEW . . . . .U.S. Department of Health, Education, and Welfare
HHS . . . . .U.S. Department of Health and Human Services
HIRDLS . . . . . . .High-Resolution Dynamics Limb Sounder
HIRIS . . . . .High-Resolution Imaging Spectrometer
HMMR . . . . .High-resolution Multifrequency Microwave Radiometer
HPCC . . . . .High-Performance Computing Centers
HRMS . . . . .High-Resolution Microwave Survey
HRTS . . . . .High-Resolution Telescope and Spectrograph
HRSO . . . . .High-Resolution Solar Observatory
HSO . . . . .Heliosynchronous Orbiter
HST . . . . .Hubble Space Telescope

IACG . . . . .Inter-Agency Consultative Group
IAF . . . . .International Astronautical Federation
ICCA . . . . .Interagency Coordinating Committee on Astronomy
ICE . . . . .International Cometary Explorer
ICS . . . . .Interface Control Specifications
ICSU . . . . .International Council of Scientific Unions
IEOS . . . . .International Earth Observing System
IFSAR . . . . .Interferometric Synthetic Aperture Radar
IGBP . . . . .International Geosphere-Biosphere Program
IGFA . . . . .International Group of Funding Agencies for Global Change Research
IGOS . . . . .International Global Observing Strategy
IGOSS . . . . .International Global Ocean Services System
IGY . . . . .International Geophysical Year
ILSSWG . . . . . . .International Life Sciences Strategic Working Group
IMAGE . . . . .Imager for Magnetopause-to-Aurora Global Exploration

IMP . . . . .Interplanetary Monitoring Platforms
IMS . . . . .International Magnetospheric Study
INPE . . . . .Instituto de Pesquisas Espacials
IOC . . . . .Intergovernmental Oceanographic Commission
IODE . . . . .International Ocean Data Exchange
IOSWG . . . . .Integrated Observing System Working Group
IPA . . . . .Intergovernmental Personnel Act
IPCC . . . . .Intergovernmental Panel on Climate Change
IRAS . . . . .Infra-Red Astronomy Satellite
IRS . . . . .Indian Remote Sensing Satellite
ISAS . . . . .Institute of Space and Astronautical Science
ISEE . . . . .International Sun-Earth Explorers
ISPM . . . . .International Solar Polar Mission
ISRO . . . . .Indian Space Research Organization
ISS . . . . .International Space Station
ISTPP . . . . .International Solar-Terrestrial Physics Program
ITHD . . . . .Interferometric Terrain Height Data
IUE . . . . .International Ultraviolet Explorer
IUS . . . . .Inertial Upper Stage
IV&V . . . . .Independent Verification and Validation
IWG . . . . .Investigators Working Group

JEA . . . . .Japanese Environment Agency
JEOS . . . . .Japanese Earth Observing System
JERS . . . . Japan Earth Remote Sensing Satellite
JMA . . . . .Japan Meteorological Agency
JPL . . . . .Jet Propulsion Laboratory
JSC . . . . .Johnson Space Center

KSC . . . . .Kennedy Space Center

LAGEOS . . . . . . .Laser Geodynamics Satellites
LaRC . . . . .Langley Research Center
LASA . . . . .Laser Atmospheric Sounder and Altimeter
LASCO . . . . .Light Spectrometric Coronagraph
LAWS . . . . .Laser Atmospheric Wind Sounder
LDEF . . . . .Long-Duration Exposure Facility
LERTS . . . . .Laboratoire d'Etudes et de Recherches en Télédétection Spatiale (France)
LFFP . . . . .Large Fine-Pointed Platform
LIDAR . . . . .Light Detection and Ranging Instrument
LIMS . . . . .Limb Infrared Monitor of the Stratosphere
LIS . . . . .Light Imaging Sensor
LSAC . . . . .Life Sciences Advisory Committee
ISCCP . . . . .International Satellite Cloud Climatology

LSLE . . . . .Life Sciences Laboratory Equipment
ISLSCP . . . . .International Satellite Land Surface Climatology Project
LSO . . . . .Large Solar Observatory
LSSPSC . . . . . . .Life Sciences Strategic Planning Study Committee
LWS . . . . .Living With a Star
LZP . . . . .Level Zero Processed Data

MC . . . . .Magnetospheric Constellation LS . . . Microwave Limb Sounder
MCC . . . . .Mission Control Center
McIDAS . . . . . . .Man-Computer Interactive Data Access System
MFE . . . . .Magnetic Field Explorer
MIMR . . . . .Multifrequency Imaging Microwave Radiometer
MIPAS . . . . .Michelson Interferometer for Passive Atmospheric Sounding
MIT . . . . .Massachusetts Institute of Technology
MITI . . . . .Ministry of International Trade and Industry (Japan)
MMS . . . . .Magnetospheric Multiscale
MO&DA . . . . . . .Mission Operations and Data Analysis
MODIS . . . . .Moderate-resolution Imaging Spectrometer
MORL . . . . .Manned Orbital Research Laboratory
MOS . . . . .Marine Observation Satellite
MOT . . . . .Manned Orbiting Telescope
MRI . . . . .Meteorological Research Institute (Japan)
MSC . . . . .Manned Spacecraft Center
MSF . . . . .Manned Space Flight
MSFC . . . . .Marshall Space Flight Center
MTE . . . . .Mesosphere-Thermosphere Explorer
MTPE . . . . .Mission To Planet Earth

N-ROSS . . . . . . .Navy Remote Ocean Sensing System
NAC . . . . .NASA Advisory Council
NACA . . . . .National Advisory Committee for Aeronautics
NAS . . . . .National Academy of Sciences
NASDA . . . . .National Space Development
NASM . . . . .National Air and Space Museum
NCAR . . . . .National Center for Atmospheric Research
NFLR . . . . .NASA Federal Laboratory Review task force
NIH . . . . .National Institutes of Health
NIMA . . . . .National Imagery and Mapping Agency
NITF . . . . .National Image Transfer Format
NOAA . . . . .National Oceanic and Atmospheric Administration
NOSS . . . . .National Oceanic Satellite System
NOZE . . . . .National Ozone Experiment
NPOESS . . . . . . .National Polar-Orbiting Environmental Satellite System
NRA . . . . .NASA Research Announcement
NRC . . . . .National Research Council

NREN . . . . .NASA Research and Education Network
NRL . . . . .Naval Research Laboratory
NSA . . . . .National Security Agency
NSBRI . . . . .National Space Biomedical Research Institute
NSC . . . . .National Space Council
NSCORT . . . . . . .National Specialized Centers of Research and Training
NSF . . . . .National Science Foundation
NSSDC . . . . .National Space Science Data Center
NSTC . . . . .National Science and Technology Council
NSTL . . . . .National Space Technology Laboratories

OAO . . . . .Orbiting Astronomical Observatory
OCI . . . . .Ocean Color Instrument
OMB . . . . .Office of Management and Budget
ONR . . . . .Office of Naval Research
OOE . . . . .Out-Of-the-Ecliptic missions
OPEN . . . . .Origins of Plasmas in the Earth's Neighborhood
OSB . . . . .Ocean Sciences Board
OSHA . . . . .Occupational Health and Safety Act of 1970
OSL . . . . .Orbital Solar Laboratory
OSO . . . . .Orbiting Solar Observatory
OSTDS . . . . .Office of Space Tracking and Data Systems
OSTP . . . . .Office of Science and Technology Policy

PAF . . . . .Payload Attach Fitting
PETA . . . . .People for the Ethical Treatment of Animals
PMC . . . . .Program Management Council
POEM . . . . .Polar Orbiting Earth observation Missions
POES . . . . .Polar-Orbiting Environmental Satellite
PR . . . . .Precipitation Radar
PSC . . . . .Physical Sciences Committee

QL . . . . .Quick Look Data

RA . . . . .Radar Altimeter
RCS . . . . .Reaction Control System
RF . . . . .Radio Frequency end-to-end test
RSA . . . . .Russian Space Agency
RSRP . . . . .Rocket and Satellite Research Panel

SACLS . . . . .Special Advisory Committee for Life Sciences
SAFIRE . . . . .Spectroscopy of the Atmosphere using Far Infrared Emission
SAGE . . . . .Stratospheric Aerosol and Gas Experiment
SAM . . . . .Sensing with Active Microwaves
SAM . . . . .Stratospheric Aerosol Measurement

SAMS . . . . .Stratospheric and Mesospheric Sounder
SAMPEX . . . . . . .Solar Anomalous and Magnetospheric Particle Explorer
SAR . . . . .Synthetic Aperture Radar
SBRC . . . . .Santa Barbara Research Center
SBUV/TOMS . . .Solar Backscatter Ultraviolet and Total Ozone Mapping Spectrometer
SCAPE . . . . .Self-Contained Atmosphere Protection Equipment
ScaRaB . . . . .Scanner Radiatsionnogo Balansa
SCIAMACHY . . . .Scanning Imaging Absorption Spectrometer for Atmospheric Chartography
SCOPE . . . . .Scientific Committee on Problems of the Environment
SCOR . . . . .Specialized Center of Research
SDI . . . . .Strategic Defense Initiative
SeaWiFs . . . . . . .Sea-viewing Wide Field-of-view Sensor
SEMMS . . . . .Solar Electric Multimission Spacecraft
SEP . . . . .Solar Electric Propulsion
SESAC . . . . .Space and Earth Science Advisory Committee
SETI . . . . .Search for Extraterrestrial Intelligence
SFFO . . . . .Solar Free-Flying Observatories
SFPP . . . . .Small Fine-Pointed Platform
SIR . . . . .Shuttle Imaging Radar
SIR-C/X-SAR . . . .Spaceborne Imaging Radar-C/X-Band Synthetic Aperture
SIRTF . . . . .Space Infrared Telescope Facility
SLS . . . . .Space Life Sciences
SMEX . . . . .Small Explorer
SMM . . . . .Solar Maximum Mission
SMMR . . . . .Scanning Multichannel Microwave Radiometer
SMRM . . . . .Solar Maximum Repair Mission
SNIP . . . . .Space Network Interoperability Panel
SOHO . . . . .Solar and Heliospheric Observatory
SOLSTICE . . . . . .Solar Stellar Irradiance Comparison Experiment
SOT . . . . .Solar Optical Telescope
SPAC . . . . .Space Program Advisory Council
SPAN . . . . .Space Physics Analysis Network
SPIS . . . . .Surface Imaging and Sounding Packages
SPO . . . . .Solar Polar Orbiter
SPOT . . . . .Systeme Pour l'observation de la Terre
SRTM . . . . .Shuttle Radar Topographic Mission
SSAC . . . . .Space Science Advisory Committee
SSB . . . . .Space Studies
SSC . . . . .Stennis Space Center
SSEC . . . . .Solar System Exploration Committee
SSMI . . . . .Special Sensor Microwave Imager
SSO . . . . .Sortie Solar Observatory
SSS . . . . .Solar Synoptic Satellite
SSWG . . . . .SETI Science Working Group

STA . . . . .Science and Technology Agency (Japan)
STC . . . . .Solar Telescope Cluster
STDN . . . . .Spaceflight Tracking and Data Network
STEREO . . . . . . .Solar-Terrestrial Relations Observatory
STG . . . . .Space Task Group
STIP . . . . .Study of Traveling Interplanetary Phenomena
STP . . . . .Solar-Terrestrial Probe
STS . . . . .Space Transportation System
STSP . . . . .Solar-Terrestrial Science Programme
SUNY . . . . .State University of New York
SUSIM . . . . .Solar Ultraviolet Spectral Irradiance Monitor

TDRS . . . . .Tracking and Data Relay Satellite
TDRSS . . . . .Tracking and Data Relay Satellite System
TFODM . . . . . . .Task Force on Observations and Data Management
THIR . . . . .Temperature Humidity Infrared Radiometer
TIMED . . . . .Thermosphere, Ionosphere, Mesosphere Energetics and Dynamics
TIROS . . . . .Television Infrared Observation Satellite
TMI . . . . .TRMM Microwave Imager
TOGA . . . . .Tropical Ocean Global Atmosphere Program
TOPEX . . . . .Ocean Topography Experiment
TOPS . . . . .Toward Other Planetary System
TOVS . . . . .TIROS Operational Vertical Sounder
TQM . . . . .Total Quality Management
TR&T . . . . .Targeted Research and Technology
TRACE . . . . .Transition Region And Coronal Explorer
TRMM . . . . .Tropical Rainfall Measuring Mission
TU . . . . .Technology Utilization

UARS . . . . .Upper Atmosphere Research Satellite
UCAR . . . . .University Corporation for Atmospheric Research
UCLA . . . . .University of California, Los Angeles
UFO . . . . .Unidentified Flying Object
UK . . . . .United Kingdom
UN . . . . .United Nations
UNEP . . . . .United Nations Environment Programme
UNESCO . . . . . . .United Nations Educational, Scientific, and Cultural Organization
UNEX . . . . .University class Explorers
UNWG . . . . .User Needs Working Group
US-AID . . . . .United States Agency for International Development
USDA . . . . .U.S. Department of Agriculture
USGCRP . . . . . . .U.S. Global Change Research Program
USGS . . . . .United States Geological Survey
USNC . . . . .United States National Committee
USPHS . . . . .United States Public Health Service

USSR ....Union of Soviet Socialist Republics
UV ....Ultraviolet Radiation

VIS ....Visible Infrared Scanner
VLA ....Very Large Array
VLBI ....Very Long Baseline Interferometry

WBDCS ....Wide Band Data Collection System
WCRP ....World Climate Research Program
WHO ....World Health Organization
WMO ....World Meteorological Organization
WOCE ....World Ocean Circulation Experiment
WWW ....World Weather Watch

XTE ....X-ray Timing Explorer
XUV ....Extreme Ultraviolet

Chapter One

Solar Physics from Space

by David H. DeVorkin

Introduction

The scientific study of the central body of the solar system—the Sun—has, of course, a long history. Studies of the Sun and allied research in solar-terrestrial relations (discussed in Chapter 2) predate access to space; as a result, there is a long history of attempts to circumvent the obstacles to observation posed by the atmosphere of Earth in examining both the Sun and its immediate environment, including Earth.[1] The problems critical to knowing the nature of the Sun and its influence on Earth include assessing the amount and character of its heat output, the origin and maintenance of solar energy, the question of the constancy of solar radiation (or its inconstancy), the magnetic properties of the Sun and its immediate environment, and the nature of solar radiation beyond the ultraviolet cutoff of Earth's atmosphere. The study of the solar constant and the means to harness solar energy have long been of interest for both theoretical and practical reasons.

This essay discusses the evolution of activities in the area of solar physics supported by the U.S. civilian space program. It is not primarily a history of the intellectual pursuits involved, but rather an examination of what areas of activity were considered to be candidates for support by NASA, based both upon advice from the scientific establishment and from program managers within NASA itself. The essay will explore the advice given to NASA by various bodies; it will look mainly at specific modes of investigation deemed to be most effective for the conduct of solar research and search out both scientific and nonscientific priorities that shaped the NASA research effort.

Solar physics has been the focus of much historical research, which has resulted in a critical mass of documentary narrative as well as a highly useful intellectual framework within which to examine changes in the field. Karl Hufbauer, in particular, has provided valuable insights into the nature of progress in the field in the modern era, including theory as well as ground-based and space-based observational trends, and he has set the stage to fully elucidate the efficacy of the priority-setting process set in place in 1958.[2] The essay will touch upon how the nature of this advice was influenced by the priorities

1. David H. DeVorkin, *Race to the Stratosphere: Manned Scientific Ballooning in America* (New York: Springer-Verlag, 1989); *Science with a Vengeance* (New York: Springer-Verlag, 1992); Homer E. Newell, *Beyond the Atmosphere: Early Years of Space Science* (Washington, DC: NASA SP-4211, 1980).

2. Karl Hufbauer, *Exploring the Sun: Solar Science Since Galileo* (Baltimore, MD: Johns Hopkins University Press, 1991); A. J. Meadows, *Science and Controversy: A Biography of Sir Norman Lockyer* (Cambridge, MA: Harvard University Press, 1972); A. J. Meadows, *Early Solar Physics* (Oxford: Oxford University Press, 1970). The historical literature on solar physics (to 1980) has been reviewed in David H. DeVorkin, *The History of Astronomy and Astrophysics, A Selected, Annotated Bibliography* (New York: Garland Press, 1982).

inherent in government patronage, especially those manifest in a highly politicized mission-oriented agency such as NASA (in contrast to the more research-oriented organizational culture of the National Science Foundation, for example).

Studying the Sun

This essay examines the period starting roughly in late 1957 and 1958 and extending to the present era, looking at how the scope, priorities, and goals of solar physics and, to some extent, solar-terrestrial relations changed in response to shifting scientific and programmatic needs in space research and to changes in technical capabilities, including launch weight, stabilization, and data retrieval and transmission. For instance, between 1958 and 1964, the United States, through its civilian space program and continuing programs in several branches of the military services (Navy and Air Force), developed the capability to continuously monitor from space the high-energy output of the Sun; roughly examine the large-scale features of solar activity as manifest in sunspot, flare, chromospheric, and coronal phenomena; and link those high-energy events directly to terrestrial ionospheric phenomena. The last achievement came both through suborbital and orbital programs and was primarily a result of progressively better systems for stabilization. In these first years, stabilization was one of the primary technological factors governing how future missions were planned. Especially in solar physics, this fact was reflected in the priorities established by scientific review panels as well as by program officers and panels within the civilian space agency.

Limits and Nature of Solar Physics

Solar physics in practice, though not in name, predates the rise of astrophysics; in many respects, astrophysics itself grew out of the physical study of the Sun. Given the enormous amount of light received from the Sun, it was the first celestial body to be studied to any great extent, its physical properties determined with any accuracy, and the amount and character of its heat radiation related to physical processes on Earth. Historians have identified five distinct eras for solar physics, starting with telescopic studies in the seventeenth century and progressing to spectroscopic studies in the nineteenth and early twentieth centuries; the establishment of a rational physical basis within which to interpret spectroscopic and photometric observations, approximately from 1910 to 1940; the transformation of solar physics by wartime and then Cold War imperatives from 1939 to 1957; and finally, solar physics since Sputnik.[3] Only the last era is covered here, even though its full appreciation requires some reference to the earlier periods. For one thing, unlike other disciplines, solar physics enjoyed a robust history in both military and civilian laboratories devoted to the use of rockets and balloons for research at high altitudes and near-space environments. These groups, formed in the wake of World War I in military research centers such as the Naval Research Laboratory (NRL) or at military-supported facilities such as the Applied Physics Laboratory of the Johns Hopkins University, were stimulated by the existence and availability of captured

---

3. Hufbauer, *Exploring the Sun.*

German V-2 missiles. The groups that formed to do science during V-2 tests defined their scientific interests in terms of the vehicle, in this case, a ballistic missile acting as a sounding rocket.[4]

By the late 1940s, several groups who pushed for improved means of stabilization and data retrieval were exploring the ultraviolet spectrum of the Sun. Two-axis stabilization was finally achieved in the early 1950s on Aerobee rockets, allowing observation of the far ultraviolet region of the spectrum through longer exposures. The primary driver for both solar physics and ionospheric physics was to determine which spectral features were responsible for the generation, maintenance, and variation of Earth's ionosphere, a critical element in long-range communication, command, and control of ballistic missiles. Throughout the early 1950s, few of these activities were sophisticated enough to be interesting to mainstream astrophysicists; thus, much of solar physics from rockets evolved along lines parallel to geophysical and military goals. By the mid-to-late 1950s, however, instrument stabilization, rocket reliability, and payload retrieval had improved to the point where astrophysically useful high-dispersion spectroscopic data were readily available to civilian groups, which brought astronomers back to consider the use of rockets, and, in the post-Sputnik era, satellites, as platforms for the study of the Sun.

Thus, in the wake of Sputnik, groups that had been devoted to scientific rocketry during the International Geophysical Year (IGY), as well as many mainstream scientists, began to think in detail about what could be done from satellite platforms. Within weeks of Sputnik 1, members of the Rocket and Satellite Research Panel (RSRP), the latest embodiment of the original V-2 Panel that had been created in 1946, met to consider the implications of research using satellites. Prior to Sputnik and independent of the meeting, but as a member of the panel, W. W. Kellogg of the RAND Corporation had prepared a report for the IGY Working Group on Internal Instrumentation of the Earth Satellite Program, entitled "Basic Objectives of a Continuing Program of Scientific Research in Outer Space." (See Document I-1.) Sputnik brought Kellogg's analysis to the attention of the National Academy IGY committees as well as the RSRP. He plotted out a scientific program assuming that development would be gradual and not revolutionary, that each stage of the program would help to design the next stage, and that human spaceflight would occur eventually but not immediately. In the immediate future, the primary vehicle for research would remain the sounding rocket, specifically for atmospheric studies, as well as solar physics and astrophysics; but, as the capability of building larger Earth satellites increased, the latter two areas would become the domain of orbiting vehicles. The bulk of Kellogg's attention was given over to terrestrial atmospheric studies, as well as to lunar and planetary studies, in two categories: Vanguard-type "lightweight satellite experiments" (50 to 75 pounds) and "advanced satellite experiments." Under the former, exploratory solar radiation studies using a variety of radiometers and bolometers to span the entire spectrum were highest on his list, mainly to better assess the radiation heat budget of Earth and its atmosphere. Time fluctuations of solar ultraviolet and x-ray radiation, monitoring solar activity, and observing its influence on Earth's atmosphere came next.

---

4. DeVorkin, *Science with a Vengeance*, pp. 344–45.

Kellogg's concept of an advanced satellite included two- and three-axis stabilization, larger power requirements, wider data transmission bandwidth, and the possibility of physical recovery. Satellites in this category that were devoted to solar studies would be capable of examining and photographing small regions of the Sun in a wide range of wavelengths, most specifically the ultraviolet. Pointing controls based upon those already operational on sounding rockets could be adapted to this purpose. Data retrieval could be physical, the preferred method at the time, or could be accomplished by scanning the data "photometrically" on board and then transmitting it electronically to Earth.[5]

In hindsight, the most interesting characteristic of Kellogg's assessment was its prudence. It shared this characteristic with other assessments of the same time period, as well as a strong conviction that the sounding rocket and balloon projects of the IGY should be extended beyond that period. This was the opinion of another stakeholder in space research, the National Advisory Committee for Aeronautics (NACA), which in February and March 1958 convened a series of working groups under the heading "Special Committee on Space Technology." One of the groups, on "Space Research Objectives," met in the spring of 1958 to deliberate over Kellogg's report and decided to extend it, with preliminary priorities and timetables.[6]

The goal of the NACA working group was to produce a more detailed position paper for the uses of three classes of Earth-orbiting satellites: 30-pound, 300-pound, and 3,000-pound. (See Document I-2.) The group envisioned one 30-pound satellite launch per month, starting immediately and lasting until the capability existed to launch 300-pound satellites (sometime in 1959, it was hoped). Larger satellites would be launched at the rate of one every two months. Satellites in the 3,000-pound class would begin to be launched in 1961 at a rate of one every four months. Although this plan was more explicit and ambitious than Kellogg's first assessment, the working group still advised that the number of "separate experiments" in each satellite in the 30-pound and 300-pound classes be limited, and that the 30-pound class continue to be launched well beyond 1959.[7] Solar and solar-terrestrial studies to be addressed with the 30-pound satellite included Kellogg's categories for a Vanguard class of launches: nondirectional monitoring including ultraviolet, x-ray, and gamma-ray detectors. The 300-pound class had the capability of stabilization for solar and stellar spectroscopy, as well as an imaging ultraviolet solar telescope. The 3,000-pound class could be capable of flying a 36-inch solar optical telescope able to perform a variety of observations; the specific character of these observations was left largely unspecified in the report.

Both the Kellogg study and the NACA Working Group endorsed human spaceflight and exploration "as proper objectives of a national program of space research."[8] The NACA report was otherwise silent on the subject, whereas Kellogg took care to elucidate human spaceflight as an inevitable goal, yet not currently justifiable on rational grounds.[9]

5. W. W. Kellogg, "Basic Objectives of a Continuing Program of Scientific Research in Outer Space," 9 December 1957, Dow Papers, Box 8.4. University of Michigan Library.
6. Working Group on Space Research Objectives, "Minutes of Meeting," 30 April 1958, Lyman Spitzer Papers, NACA file, Princeton University Library.
7. Ibid., p. 2.
8. Ibid., p. 5.
9. Kellogg, "Basic Objectives," p. 38.

Collective Advice: The Space Science Board

In June 1958, the Space Science Board (SSB) of the National Academy of Sciences was created as a vehicle to provide for "an orderly extension and continuation of the rocket and satellite work of the USNC/IGY."[10] (See Volume V, Chapter 1, for a discussion of the origins of the Space Science Board.) It was directed to gather information on space science activities and to assess priorities for space research. The Board met in New York City several times that summer to organize itself around a set of goals in space science and to become acquainted with projections of payload capability. Herbert F. York of the Advanced Research Projects Agency (ARPA) (NASA had not yet been created, and ARPA had responsibility for U.S. space activity) summarized future launch vehicle capabilities that could send 3,000 pounds into orbit by 1960, double that by 1962, over thirty times that amount by the mid-1960s, and some 50 tons into orbit by the late 1960s. Scientists, York implied, were to plan accordingly. However, Hugh Odishaw, Executive Director of USNC-IGY, warned that "the rocket and balloon potentialities for scientific research should not be neglected in the face of the prospective satellite program."[11] Richard W. Porter outlined the remaining space program of the IGY. Lloyd Berkner, using Porter's findings, summarized them in fifteen topical areas, including the continuing intensive study of solar corpuscular radiation. Berkner, Chairman of the Board, created twelve committees to prepare reports on specific fields and assigned astronomy and radio astronomy, which included solar physics, to a committee headed by renowned solar astrophysicist Leo Goldberg. He was charged to lead his committee in looking into ways to extend x-ray, ultraviolet, and corpuscular solar research into space.

At its second meeting in July, the Board discussed responses from the scientific community and prioritized specific problem areas for the next two years for satellites and rockets. Its selections were designed to supplement ongoing IGY projects and were chosen on the basis of scientific need and technical feasibility. Included in the short-range program for satellites were low-resolution optical scanning capabilities for the assessment of solar activity in the gamma, x-ray, and UV ranges.

The new National Aeronautics and Space Administration began operations on 1 October 1958. In December 1958, the new NASA Administrator, T. Keith Glennan, made it clear that the new space agency, not the Space Science Board, was to be the arbiter of space science policy and mission choices, setting priorities for scientific experiments and mission profiles. Up to that point, and indeed well into 1959, there were lingering tensions about who was in control. How Glennan managed to secure this role for NASA remains to be fully appreciated. (See Volume V, Chapter 1.) But for the purposes here, all the various deliberative bodies so far identified, and particularly the Space Science Board, had become advisory to NASA by the end of 1959.

In addition to the newly constituted review panels and institutions engaged in planning for space research, academics supported on military contracts were also encouraged to state their opinions on what kind of solar problems could be addressed in

---

10. Space Science Board, "Minutes," 27 June 1958, p. 2, NASA Historical Reference Collection, History Office, NASA Headquarters, Washington, DC.
    11. Ibid., p. 8.

space. This encouragement came from military as well as industrial patrons, all eager to stake a claim on outer space. In the late 1950s, few observatories and universities were engaged in space research and solar studies in a large way; only a few of the largest—the new Kitt Peak National Observatory, the University of Michigan, Princeton, and the Harvard-Smithsonian complex—were active. Some in the wings were interested; Jesse Greenstein, an astrophysicist at the California Institute of Technology who had attempted solar spectroscopy with the V-2 in the 1940s, was invited to address the Fourth Ballistic Missile and Space Technology Symposium at UCLA in August 1959 and looked forward to a "rapid evolution" of payload capability from 5 to 5,000 pounds.[12]

Missions

With NASA setting overall policy, the structure of the solar physics program evolved as NASA itself grew and reorganized, always promoting a mission-oriented approach to exploring space. As a result, in the fifteen years after Sputnik, space solar research became defined by the launch of a sequence of progressively larger spacecraft, each undertaking a series of flights and each with more sophisticated means of stabilization and pointing controls. As originally envisioned by York, Kellogg, and their colleagues, the program consisted of an evolutionary series of satellites defined by payload weight capabilities, rather than being based on order-of-magnitude increases in technological sophistication. The latter would come into play only by the 1980s and 1990s, allowing for far wider bandwidths, bigger data streams, and more complex batteries of simultaneously operated instruments. Between 1960 and 1975, Americans launched some fifty-three spacecraft that studied the Sun; forty-four contained up to three instruments; nine had more than four instruments. In comparison, the Soviet Union launched twenty payloads with three or fewer instruments each, and ten payloads were launched by all other nations combined. Accordingly, it has been claimed that "American leadership in studying the Sun from space was a testament both to the prominence of American scientists in the making of space policy and to the prowess of American engineers—especially in electronics and computing—in implementing that policy."[13] That policy included developing and maintaining a robust infrastructure that could produce a wide and varied array of new design prototypes, and both programmatic and developmental suborbital programs using balloons, aircraft, and sounding rockets. The mission-oriented approach, therefore, reflected launch vehicle capabilities as well as the acknowledged need to achieve specific levels of instrument performance. The advice given NASA reflected not only scientific interests and goals, but the range of choices NASA could make in how it would manage space science.

---

12. Jesse Greenstein, "Astrophysical Research in Space," p. 2. Air Force Office of Scientific Research, TN-59-907, 21 September 1959, Greenstein Papers, California Institute of Technology.
13. Hufbauer, *Exploring the Sun*, table 5.1, p. 167.

Explorer- and Observatory-Class Missions—The OSOs

So-called "Explorer-" and "Observatory-" class missions evolved more or less simultaneously. Both grew out of instrumentation requirements set in the sounding rocket era, but only the latter reached a level of sophistication that was attractive to mainstream solar astronomers and physicists. Thus, the most significant solar program that attracted the attention of astronomers in the 1960s and 1970s was the Orbiting Solar Observatory (OSO). After its first several flights in the early 1960s, it was enthusiastically supported by review bodies and by NASA advisory panels. How then did it come into being?

The first OSO was a partially spin-stabilized craft carrying thirteen scientific experiments. The 460-pound craft was launched by a Thor-Delta. During operation, the spin section rotated at some thirty rpm, and the biaxial pointing control, consisting of both coarse and fine adjustment modes, was capable of maintaining the solar-oriented instruments (x-ray spectrometers and gamma-ray monitors from Goddard Space Flight Center) to within one minute of arc of the center of the solar disk, or one-thirtieth of the angular diameter of the Sun.[14]

OSO was considerably more sophisticated than an Explorer-series craft, which typically would contain three or fewer instruments. Instruments in the OSO were clustered in two sections: an upper "sail" encrusted with solar panels that continually pointed toward the Sun, and a lower, nine-sided spin section that contained instruments that did not require two-axis stabilized pointing. But OSO was considerably smaller than the early ambitions of senior NASA space science managers like Homer E. Newell, who campaigned for a series of four stabilized orbiting astronomical observatories, including a fully dedicated solar observatory capable of handling a complex of instruments weighing some 500 to 700 pounds.[15] Arguments for and against the OSO program within NASA, compared to Explorer and to more advanced systems, were usually contained within briefing papers that described the program as complementary to the established Explorer series.

OSO's champion was an experimental physicist named John C. Lindsay. He had moved with Newell to NASA from the Naval Research Laboratory in 1958 and became a project manager in the Explorer program at the Goddard Space Flight Center (GSFC). According to Dr. Leo Goldberg, Lindsay was successful because "he was an experimenter himself, and a very good one."[16] In contrast to his boss's desire for huge orbiting observatories that fit Herbert York's optimistic predictions of future payload weights, Lindsay believed strongly that multiplexed instruments on the scale of those familiar to rocketsonde experimenters were a more practical route to a cost-effective program. The infrastructure for satellites on that scale would be a straightforward extension of rocketsonde technology and Explorer technology. Central to the effort was the existence of an established industrial capability: the founders of the Ball Brothers Research

---

14. NASA, *Orbiting Solar Observatory I* (Washington, DC: NASA SP-57, 1965), p. 219.
15. Dr. Leo Goldberg oral history, 17 May 1978, American Institute of Physics, College Park, MD, p. 119.
16. Ibid., p. 120.

Corporation of Boulder, Colorado, had over a decade of experience in building stabilized platforms for rocket spectrographs; monochromaters; and cameras flown on balloons, sounding rockets, and aircraft.[17]

Lindsay's campaign attracted Ball engineers, as well as influential astronomers and solar physicists. One such was Leo Goldberg of Michigan, who had initially agreed with Newell to instrument a large solar-oriented Orbiting Astronomical Observatory (OAO) but was relieved when NASA, for the moment, dropped the idea. The decision was made only after a review by the President's Science Advisory Committee concluded that the combined platform concept was impractical.[18] Part of this decision was due to Lindsay's persistence.

In May 1959, Lindsay organized an ad hoc "NASA Discussion Group on Orbiting Solar Observatory Project" to explore ways to perform solar research in orbit (see document I-3). Twenty-three astronomers from universities, military laboratories, and NASA Centers gathered to hear Lindsay describe what became OSO: a 300-pound payload capable of a pointing accuracy of one minute of arc in a sustained circular orbit of 300 to 500 miles altitude. Lindsay envisioned a gyroscopically stabilized spherical satellite with an operational lifetime of at least one month. He also invited others to describe possible payload instruments, calling upon astronomers from the Smithsonian Institution, the University of Michigan, the University of Colorado, the High Altitude Observatory, and the Naval Research Laboratory. Their responses made it plain that few were clear about the exact nature of their experiments, or whether the satellite would be a dedicated solar observer or a multifunctional solar and astronomical platform. Yet when Lindsay pressed the point about the need for a "smaller, less expensive solar observatory," he obtained both general endorsements and specific support from Michigan and NRL.[19]

In April 1959, measurements of the Sun from a pointed and stabilized platform were among the priorities identified as NASA's immediate flight objectives. By August, an Orbiting Solar Observatory was identified in an "Office of Space Sciences Ten-Year Program" document, and by October, the first contracts with Ball Brothers Research Corporation were signed to design and build the first two spacecraft and instrument systems. Some $250,000 was earmarked for OSO in the 1959 NASA budget, and this amount rose rapidly to $1.9 million in 1960 and $3.9 million in 1961. Small by space and military standards, OSO was still the most expensive solar physics project in history, and, of course, it would soon be dwarfed by NASA's other programs in stellar astronomy and human spaceflight. This level of funding seemed enormous to solar astronomers, who began to look with interest at the new program after the first successful flight.

The first OSO flew in March 1962. Later that year, the Solar Physics Subcommittee of NASA's Space Science Steering Committee approved the third through the fifth satellites in the OSO series. The subcommittee report reflected the opinion of a much broader group of scientists who had convened in June 1962 at the University of Iowa and who had called for better angular resolution, better access to a broader wavelength range, and the ability to isolate precise wavelength ranges for high-resolution spectral studies.

17. DeVorkin, *Science with a Vengeance*, chapter 12.
18. Goldberg oral history, p. 121.
19. NASA Discussion Group, "Minutes of Meeting," 8 June 1959, NASA Historical Reference Collection.

What came to be known as the Iowa Summer Study, convened by NASA program managers and funded by NASA, looked at all modes of space research. It was organized into a series of specialist "working groups," each reporting on a specific area of activity. A working group on astronomy was convened to comment on solar and stellar needs. For example, reflecting well-established priorities in solar physics and solar-terrestrial relations, the group agreed at the outset that it was more important to perform a complete reconnaissance of the ultraviolet and x-ray regions of the spectrum than the infrared, since the latter was partially observable from the ground.[20]

Even though the Iowa summer study endorsed the continuation of small Explorer-class satellites and OSO for studies of methods of solar flare prediction, it was clear that the top priority for solar physicists remained precision pointing accuracy to improve knowledge of the fine structure of the solar atmosphere in order to understand the physical mechanism responsible for producing solar flares. The first OSO provided for selective capability to continuously monitor solar activity. (The earlier NRL SOLRADS provided continuous monitoring but were not area-selective.) Scientists looked forward to the launch of the next OSO, which would have the added capability of scanning the solar disk with 1-arc-minute resolution, and endorsed a continuing program of two OSO launches per year by NASA.

The working group also wanted to see significant improvements in other related areas. Ground-based observations of the solar limb during the few moments of a solar eclipse had yielded information equivalent to a resolution as fine as 0.1 arc second, and balloon-based systems such as Stratoscope demonstrated that 0.5-arc-second resolution was possible at an altitude of 80,000 feet. There were many unknowns about the structure of the solar atmosphere and of the limitations of spacecraft design that made it impossible to predict what resolutions were needed and how they would be achieved, but the committee concluded that an angular resolution of 0.1 arc second was a meaningful goal for future solar observatories. The planned OAO series was set to meet that goal, and the committee concluded that there was no reason why true three-axis stabilization could not be adapted to a solar observatory.[21] Still, this requirement was several orders of magnitude beyond what had been possible with OSO and eventually led to calls for an advanced solar observatory series (see below).

The OSO program, however, soon came under scrutiny by scientists, NASA management, and Congress. Congressional attention led to a Government Accounting Office investigation in late 1963 and 1964; the result was allegations of serious cost overruns, up to some $800,000, attributed to mismanagement. The OSO budgets for those years were $10.0 and $20.0 million, respectively, which brought the motives for the investigation into question.

One factor may have been delays: OSO-2 fell far behind the two launches per year expected by the astronomical community. On 14 April 1964, a disastrous accident killed

---

20. Not stated explicitly at the time, but subject to future study, is the question of how the availability of infrared, as well as ultraviolet and x-ray, detector technology influenced this decision, since the former was still highly classified.

21. Space Science Board, National Academy of Sciences, Space Science Summer Study, 1962, pp. 2-6–2-7.

three men when OSO-B's third stage ignited on the test stand, deeply clouding the future of the program. Some of the parts of the craft were retrieved and refurbished, and a new satellite was built and flown successfully in February 1965. A history of the OSO-2, prepared at the Goddard Space Flight Center in April 1966, reported on the details of the fatal fire.[22] (See Document I-7.)

A "Fact-Finding Committee," composed mainly of Goddard employees, was quickly convened by the Goddard Space Flight Center directorate. Others were invited from Ball Brothers, Douglas Aircraft, the Air Force, the Army Ballistics Laboratory, NASA's Langley Research Center, and the Wallops Island launch facility. They reviewed the electrostatic environment of the components and concluded that the igniter squib was triggered by an electrostatic charge that had built up because the testing system and craft were not properly grounded. The parts were being handled under a large plastic shroud, which brushed up against the payload, creating the electrostatic charge.[23]

In contrast to the political ramifications of the 1967 Apollo 204 fire, this unfortunate incident did not catch the public eye.[24] Appropriate design changes were recommended, and the project resumed in June at Ball, using a rebuilt OBO-B prototype spacecraft, flight spares, undamaged components, and new procurements. Of the two pointed instruments, the NRL ultraviolet telescope and coronagraph needed little repair, but Harvard's ultraviolet spectrometer and spectroheliograph were ruined. A new instrument was built out of Harvard's flight spare unit and was delivered to Ball in late August 1964. Weight, balance, and other acceptance tests were performed throughout the fall, and the payload was shipped to Cape Kennedy in November. This time, it took only five months to produce a flyable payload. The short time was a testimony to the efficiencies of multicraft production. After routine tests at the Cape, the launch took place on 3 February 1965.

Everything went well until orbit fourteen, when the Harvard pointed experiment was turned on. It was quickly turned off due to anomalies in the readout, which raised the fear of internal electrical arcing. It was turned on again periodically during the next sixty orbits and then turned off until 11 May 1965. Although a protocol was established to turn the instrument on and off periodically, no useful data were gathered.[25] Goldberg, the Principal Investigator, who had been at Michigan but had moved to Harvard during the course of the project, painfully recalled that even though he and his team had taken every precaution to be sure that their instrument had properly outgassed, "one of the transistors in the output of the counter had just burned out, and that was the end of that experiment." He then added:

> Our hearts sank well below our boots. So it was back to the drawing board. We got some good advice from places like the Raytheon Corporation (who were building electronics for the Apollo mission), and we redesigned the circuitry so that nothing would be damaged, no matter how severe the arcing. We subjected it to very severe tests. We put it in a vacuum chamber and let it arc for 25 or 30

---

22. NASA, "Section 2—Goddard" in *History of Orbiting Solar Observatory* OSO-2, Report X-440-66-322, 1966.
23. The History of OSO-2 cited above mentions the shroud but does not explicitly state that it was the cause of the electrostatic buildup. That link was made in the Goldberg oral history interview, p. 123.
24. Homer Newell, *Beyond the Atmosphere*, p. 164.
25. *History of OSO-2*, pp. 5–27.

hours, and everything was still working. But that was already 1967. We'd missed OSO-1. We were now on OSO-4.[26]

Space research was not without considerable risk. Throughout its lifetime, the rest of OSO-2 maintained its pointing ability to within the specification set of ±1 arc minute of the center of the solar disk. Most of the other instruments worked nominally, and NASA deemed the flight a success.

The first OSOs were designed around a spinning wheel divided into nine chambers, with five reserved for instruments and the rest for flight systems. Their stabilized sail sections still depended upon the ability of the spin section to maintain azimuthal orientation. OSO-5 and OSO-6 were very much like the first four and were flown in 1969, but OSO-7 was a refined and more complex design. Instead of the deployable swinging ballast arms used on the first six satellites, OSO-7, launched in September 1971, used a simpler, kinematically more stable ballast system within a larger and heavier wheel. It also carried about twice the instrument payload weight, including an improved coronagraph.

OSO-7 was an intermediate step to an "advanced" or upgraded OSO series. Funding for OSO peaked in 1964 and remained high until 1966, when programmatic funding dropped to $10 million for 1967 and stayed in that range for about three years before rising again to the $20-million level in 1972 and 1973. Administrator Thomas Paine approved a follow-on program to OSO in January 1969, the larger OSO-I, -J, and -K series; by December 1970, a contract had been awarded to Hughes to build the spacecraft. During the authorization and appropriations cycle for the next year, however, NASA, expecting severe budget cuts, threatened to terminate the program. By March 1972, J and K were deferred, but work proceeded on I, even though Hughes simultaneously announced major cost overruns. NASA accordingly slipped OSO-I and cancelled J and K. OSO-8 (I) was launched successfully on 21 June 1975, and the program tapered off in the late 1970s, with funds drastically reduced to the $1-million level in 1977.

Each succeeding OSO was an incremental advance over previous craft. By the time of Goldberg's second flight on OSO-6, his team found that the spacecraft was stable enough to allow precise positioning and scanning over the entire solar disk. Goldberg's instrument was able to raster-scan small areas of the Sun "on command," which was very helpful in localizing and imaging short-lived solar events.[27]

The OSO program always had to compete with other NASA programs that emerged around it, and it suffered on a number of occasions from unfavorable comparisons and priority disputes between it and the far more costly Orbiting Astronomical Observatory series. (See Volume V, Chapter 3, for a discussion of this program.) In addition, the success of OSO revived some of the early dreams for an "advanced OSO" series, which had been promoted by NASA and endorsed by scientists at the 1962 Iowa Summer Study.

---

26. Goldberg oral history, p. 123.
27. Ibid., p. 124.

The Advanced Orbiting Solar Observatory (AOSO)

NASA representatives reported at the 1962 Iowa Summer Study that an advanced series of solar satellites was being planned that could accommodate instruments up to 10 feet long and 22 inches in diameter, with a pointing accuracy of better than ±5 arc seconds (the angular size of a golf ball at 1 mile, equivalent to about 2,000 miles of solar surface), and the stability of 1 arc second for at least 5 minutes, which theoretically would make it possible for some of the instruments to reach the desired 0.1-arc-second resolution. (See Document I-5.) These larger instruments would be comparable to ground-based solar instrumentation and could theoretically extend the spatial and spectral resolution of solar observations by several orders of magnitude. "The proposed spacecraft," the Summer Study concluded, "is therefore the next logical step needed to advance solar physics."[28]

Initial planning for an AOSO at NASA had begun in the spring of 1961 and led to a scientific requirements meeting held in June of that year at the High Altitude Observatory in Colorado. (See Document I-4.) But even by late 1962, John Lindsay, who had become the AOSO project scientist, could not be very specific about the instruments or even goals of the new program. The first of the series, originally to be called "Helios," was to provide the spatial and wavelength resolution necessary to better understand flare phenomena, with the capability of studying all forms of the energy released during these short-lived "transient" events. Helios was also to have a strong solar-terrestrial component, with experiments designed to better understand the mechanisms of energy transport between the Sun and Earth.

At the end of 1962, AOSO was still in its formative stage. Engineering studies were still being carried out, specifically to decide on options for pointing and stabilization. Mission definition continued for several years, and by the time of the Space Science Board's 1965 Summer Study, system design and the definition of the payload instruments were both well along. Solar telescopes for the first two flights were under development at Harvard under Dr. Leo Goldberg, at Goddard under Lindsay; at the High Altitude Observatory under Gordon Newkirk, and at the Naval Research Laboratory under J. D. Purcell, who was part of Richard Tousey's pioneering solar physics team. These instruments would examine all parts of the solar atmosphere, from the photosphere to the corona, in the visible, ultraviolet and x-ray portions of the spectrum. They had to fit into a cylindrical satellite about 3 meters long and 1.5 meters in diameter, with eight radial solar panels. AOSO was also intended to fly in a high-inclination, full sunlit orbit; far higher than the typical OSO orbit.

Faced with meeting this greater challenge, solar specialists realized that they also needed a broader base of support. The Working Group on Solar Astronomy, chaired by Goldberg with members from HAO, Sac Peak, Mount Wilson, Indiana, Minnesota, Hawaii, Michigan, California Institute of Technology (Caltech), and NRL, met with NASA program managers including Lindsay, John Naugle, Henry Smith and W. B. Taylor. They deliberated at Woods Hole over the various modes of research needed in solar astronomy and accordingly made a strong appeal for unity between solar astronomy and solar space

---

28. Space Science Board, National Academy of Sciences, Space Science Summer Study, 1962, pp. 2–8.

astronomy, especially with respect to how both were needed to meet the scientific challenge of answering current solar questions. (See Document I-6.) They also ratified the conclusions of the 1962 Iowa Summer Study on the need for fine pointing and triaxial stabilization. Arguing that all types of platforms were needed because they were complementary, they recommended in particular that the OSO program be augmented and that AOSO not be canceled in favor of programs employing human crews for solar research. The committee argued that "a satellite with AOSO specifications is an indispensable next step in NASA's solar program, and must be flown close to the coming solar maximum."[29]

Although the group report supported AOSO, it was clear that the group did not have unalloyed enthusiasm for the project. It was, as they implied in its recommendation, an interim step. AOSO was pinched between the group's strong promotion of an augmented OSO series and the fact that not everyone believed that AOSO would meet the 0.1-arc-second resolution requirement needed to address current solar questions. The group report also expressed concern that the 20-million-bit-per-orbit data capacity of AOSO was a "severe limitation on the performance of certain classes of observation in that spacecraft."[30] Many of the experiments could collect data at rates greater than any tape recorder could record. Reflecting the enthusiasm NASA held for human spaceflight, the group observed that "recovery of photographic data by return of the astronaut thus appears attractive as one way of breaking through the data barrier."[31] The group looked for alternatives appropriate for a robotic system: onboard data processing, automatic film return from an unpiloted AOSO (a capability demonstrated as early as 1960 by the CORONA reconnaissance satellites), or "real-time video-bandwidth telemetry" by relay to a high-altitude communications satellite.

Given these options, the group strongly considered doing solar observations from a crewed platform. Specific programs the group was asked to examine included ATOM, an Astronomical Telescope Orientation Mount that could be flown with some of the early Apollo Earth orbital missions in the 1965–70 period. They also examined Apollo Extension Systems (AES) for the 1970–75 period but argued in both cases that ATOM should not replace AOSO and that AES would be an appropriate follow-on to AOSO. Next came MOT, the Manned Orbiting Telescope, which would achieve the needed spatial resolution demanded by the problems in solar physics. One to 1.5 meter telescope apertures, up to 10 meters long; film recording; and physical recovery would meet the theoretical needs of the astronomers. However, the astronomers also knew that the presence of humans, although good for operating complex instruments in space and retrieving the data, would introduce unwanted shifting and vibration in the instruments; "even his breathing and involuntary muscular activity may be a major problem to the engineer developing an automatic stabilization system."[32]

AOSO also experienced budget and program pressure at just the time that the NASA budget stopped growing; this has been the usual reason given for its cancellation.

29. Space Science Board, National Academy of Sciences, *Space Research: Directions for the Future* (1965), p. 194.
30. Ibid., p. 200.
31. Ibid., p. 200.
32. Ibid., p. 199.

However, Leo Goldberg felt that it was also a "very marginal project" because "the pointing accuracy, the performance probably couldn't be achieved by that spacecraft. It just wasn't big enough . . . . We were after one second of arc pointing accuracy. It was supposed to be comparable to OAO—it was to be launched by a Thor-Agena, and that was the limiting factor."[33] Goldberg also felt that the positive report on the crewed option at Woods Hole in 1965 reflected the pressure felt from the human spaceflight side of NASA: "[NASA] wanted to operate in the Apollo extension program. George Mueller [head of the Office of Manned Space Flight] and his boys wanted to do scientific work, and solar work looked good. At first, there was a very crude concept, but it sure got to be sophisticated, and eventually led into the ATM, which was more than a match for AOSO."[34]

In August 1965, Homer Newell testified before Congress that AOSO would fly by 1969 and be in operation for a decade. In October 1965, Goddard signed a $58-million contract with Republic Aviation for the AOSO, and the Principal Investigators were directed to continue the development of their instruments. In December 1965, hardly two months later, the project was canceled "because of budgetary considerations." After the cancellation, Henry J. Smith, Chief of the Solar Physics Program in OSSA, lamented the decision, given the support the Woods Hole Summer Study expressed for AOSO. He noted that a cancellation because of "funds [that] were not available" wiped out any possibility of obtaining high-resolution observations during the forthcoming maximum in solar activity. He suggested various alternatives, such as adding a solar component to OAO to obtain continuous monitoring capabilities. But at least, he added, mission definition had proceeded to the point where the major instruments had been defined and partially funded, and support was forthcoming to continue their development at Harvard (building a scanning UV spectrometer), American Science and Engineering (imaging x-ray telescope), the High Altitude Observatory (white light coronagraph), and the Naval Research Laboratory (coronal and chromospheric spectroheliographs).[35] These were all to be general-purpose instruments, Smith added, which meant that they could be flown on an OAO, deployed from the Apollo Service Module, upgraded for the proposed Apollo Telescope Mount (ATM), or even operated from a lunar-based observatory. One way or the other, Smith drew the line; these instruments "represent the minimum scientific program necessary to carry on our investigations of the structure and behavior of the Sun."[36] By designing these instruments for any carrier, Smith suggested, solar physics was maximizing its chance to gain a flight opportunity.

The option that soon emerged was the ATM, which, for Smith and his astronomical colleagues, represented "a spaceborne equivalent of the equatorial solar telescope mounting at [ground-based] solar observatories." The ATM under design at that time

33. Dr. Leo Goldberg oral history, 22 February 1983, pp. 35–38, National Air and Space Museum.

34. Ibid., pp. 37–38

35. Henry J. Smith, "Solar Astronomy," in Homer Newell et al., *Astronomy in Space* (Washington, DC: NASA SP-127 1967), pp. 17–21.

36. Ibid., p. 22.

could handle instruments at least 3.7 meters long and provide stability of better than five arc seconds. But since they would be manually operated in low-Earth orbit, these instruments could not continuously monitor solar activity.

Smith outlined the various pros and cons of the ATM compared to AOSO in a document that was apparently part of a briefing package used sometime in 1966 to keep the robotic option alive. It argued that the Physics and Astronomy Program Office "must consider them complementary and not redundant flight programs." AOSO provided for continuous monitoring of the Sun, whereas the ATM could only observe intermittently. AOSO also provided for simultaneous instrument operations, whereas ATM was capable only of sequential operation of the various instruments. AOSO, on the other hand, required digital recording and was limited by bandwidth. The ATM utilized retrievable photographic recording with very high data capture rates, which would be needed to record all the details of transient events. The AOSO had to be programmed in advance, whereas the ATM would be comparatively flexible in its programming. The continuous operation of an AOSO was considered of paramount importance, however, because it stood the best chance to capture the highest intensity flare events. The briefing document therefore concluded that an ATM could not replace AOSO capabilities. Its "irreplaceable role in the Solar Physics Program" called for the "reinstitution of a high-resolution unmanned solar satellite to replace AOSO."[37]

The Apollo Telescope Mount

The history of the Apollo Telescope Mount and the steps taken to define the program have been reviewed extensively.[38] The focus in this discussion is on the nature of the advice given to NASA by scientists as the program was defined and their reactions as the program developed.

Before the cancellation of AOSO, the Solar Physics Program within OSSA, headed by Smith, looked forward to a three-pronged attack on the Sun: a Solar Explorer Satellite to be launched in late 1965 to study the continuous solar x-ray emissions during a quiet period of solar activity; the continuing OSO series, which had by then been extended to a total of eight launches through 1969; and, of course, the AOSO series, which was to have four launches starting in 1969 to examine the Sun during the active portion of its eleven-year cycle. More advanced capabilities, reaching to 0.025 arc second for a human-operated, Earth-orbiting, 100-inch diffraction limited telescope, capped the present program extrapolation. Smith also argued that smaller telescopes placed into lower solar orbits could achieve the same effective resolutions. These "solar probes" were thought to be best for the study of localized phenomena on the solar surface, not for large-scale synoptic observations.[39]

Before 1965, then, the possibility of an ATM-type mission was not prominent in the priorities of the Solar Physics Program. All this changed with the cancellation of AOSO,

37. "ATM vs. AOSO," Briefing, NASA Historical Reference Collection.

38. W. David Compton and Charles Benson, *Living and Working in Space: A History of Skylab* (Washington, DC: NASA SP-4208, 1983)

39. NASA, *Significant Achievement in Solar Physics 1958–1964* (Washington, DC: NASA SP-100, 1966), p. 86.

and the lobbying efforts for ATM by NASA human spaceflight personnel. George Mueller, Associate Administrator for Manned Space Flight, wrote passionately in March 1966 that "we are entering a new period in which it is becoming possible to place the astronomer in the space environment, near his instruments."[40] He was able to marshal support from prominent astronomers who had signed on to provide instruments for astronauts to use in the human spaceflight program. Earth-orbit flights would carry instruments for airglow photography, x-ray astronomy (Riccardo Giacconi), ultraviolet stellar astronomy (K. G. Henize), and UV and x-ray solar photography (Richard Tousey). "The Apollo hardware is capable of a wide variety of missions other than the manned lunar landing," Mueller claimed; larger payloads could replace fuel and oxidizer in intermediate Saturn stages.[41] He looked forward to using the instruments planned for AOSO on Apollo Applications flights, installing a telescope mount on the Apollo spacecraft for periods up to 14 days at a time. Deferring to Smith as well as to scientists for the details of the science, he felt this would be a valuable experience in assessing "the effectiveness of an astronaut in erecting, alining [sic], and operating relatively large astronomical instruments."[42]

By spring 1966, OSSA head Homer Newell had come to terms with George Mueller's vision of incorporating a major scientific project into the Apollo Applications Program, although the details still had to be worked out. In a March 1966 memorandum on the "Establishment of the Apollo Telescope Mount Project," Newell spelled out the "cluster concept" of nested experiments and how it would fit into the Apollo Applications Program in time to meet the 1969 solar maximum, which was the projected launch date for the new program.

Within a year, however, it became painfully clear to the scientists who had been recruited to build instruments for this crash program that there would be no flights during the next period of maximum solar activity. Some of the leaders in the community, notably Goldberg, began to express dissatisfaction with the pace of development, arguing that their instruments had been designed specifically for maximum solar activity, and would not be scientifically effective at any other time. Goldberg had always held prominent roles as an advisor on various committees, but in September 1967, he obtained an even greater role as the first chairman of NASA's Astronomy Missions Board, established by OSSA to advise on the design and conduct of astronomical experiments in space, and, most important, to achieve a greater degree of consensus formation between OSSA and academe. (See Volume V, Chapter 3, for a discussion of the creation of the Astronomy Missions Board.) The Board would also act as a new pressure point within NASA to further special interests of astronomers.

At its second and third meetings in late 1967 and mid-January 1968, the Astronomy Missions Board heard a series of extended briefings by NASA program officers on what had become by then the ATM-A mission. The Board ratified NASA's contention that the mission was a "logical and technically appropriate next step" for advancing knowledge, increasing the ability of humans to live and work in space; the mission would also provide

---

40. George E. Mueller, "Expanding Vistas in Astronomy," in Newell et al., *Astronomy in Space*, p. 49.
41. Ibid., p. 58.
42. Ibid., p. 62.

a means to gather knowledge about the Sun. Accommodating what was clearly NASA's mandate, the Board urged NASA to select appropriately trained astronauts to conduct the scientific observations and to improve expertise in "failure circumvention." Goldberg added in a cover letter, however, that the Board was concerned that the mission take place before the end of the coming solar maximum, expected to last until 1971.[43] (See Document I-8.)

Even though the Board acknowledged that the chief rationale for ATM was to demonstrate the utility of humans working in space, it still lobbied hard for a timely launch to maximize scientific return. At the least, it wanted to be assured that NASA would be sensitive to scientific needs in scheduling launches. Goldberg and the Board well knew that achieving the initial launch date of 1969 required serious compromises. In July 1967, Goldberg recalls, George Mueller visited his office, asking if there were some way that Harvard could have its instrument ready for a 1969 launch. One of Newell's primary conditions for OSSA's endorsement of ATM-A was that the Harvard experiment fly. Goldberg was willing to substitute a simpler instrument, a spectrum scanner optimized for the solar maximum and its expected high-energy solar flares, provided that Harvard's original experiment would still fly on ATM-B. Goldberg and Newell went ahead with this plan. But by May 1968, when the launch date had slipped into 1971, and looked like it would slip even further, Goldberg insisted that Harvard's original instrument, still under development, be reinstated. Harvard in fact had to threaten to withdraw from ATM entirely before NASA acquiesced.[44]

Not known to Goldberg in May 1968 was that slips for ATM launches were projected far beyond 1971, reaching to 1973 and even 1975. Newell, reacting to a series of inquiries from Goldberg, instructed John Naugle to respond, knowing full well that then-present "planning for the Saturn V workshops suggest[ed] that launches in 1973 and 1975 might be possible, but that these workshops [were] expected to be devoted primarily to the study of man himself." (See Document I-9.) Possible payloads, including ATM, as well as a large UV stellar package and a large x-ray and gamma-ray package, all competing for berths, could not possibly fly before 1975 and probably would not fly until much later. In an effort to meet Goldberg's demands, however, Newell instructed Naugle to search for "other equally effective means" to launch large-scale astronomical instruments. Options included extending the OAO program, converting the OSO and OGO programs from proprietary instruments to "guest observatory" status, expanding the Astronomy Explorer program, or modifying later OSOs to accommodate larger solar physics instruments and smaller stellar instruments. Above all, Newell wanted a "candid assessment of our space astronomy program, both in the unmanned and the manned spacecraft."[45]

Goldberg was, of course, distressed when he started hearing of the expected slips in the spring of 1968. Telegrams from NASA in April indicated at first a slip to 1972, which Goldberg argued would greatly diminish "the scientific importance of the payload." Then,

---

43. Astronomy Missions Board, "Resolution," 17–18 January 1968, attached to letter from Dr. Leo Goldberg to Dr. Homer Newell, 25 January 1968, NASA Historical Reference Collection.
44. Letter from Dr. Leo Goldberg to Montie Wright, 17 June 1980, NASA Historical Reference Collection.
45. Memorandum from Dr. Homer Newell to Dr. John Naugle, 9 April 1968, NASA Historical Reference Collection.

Naugle and Newell, despite their internal correspondence, somehow led Goldberg to believe that the slip would only be to June 1971, which hardly comforted the Harvard astronomer, who remained convinced that there would be further slips of "at least two to three months." There was a silver lining however, for Goldberg: this delay would give Harvard enough time to reinstate its original experiment. Goldberg urged Naugle to do this.[46] (See Documents I-10, I-11, and I-12.) Goldberg's frustrations were shared by the members of the Astronomy Missions Board, as well as by the Principal Investigators for the other major ATM instruments. A brief glance at the chronology of the definition of ATM, and eventually the Skylab mission, will reveal why this was so.

Skylab and the Apollo Telescope Mount

The Apollo Applications Program was born at Marshall Space Flight Center (MSFC) in September 1965, and the orbital workshop concept grew from that within the next month. In February 1966, MSFC submitted a proposal for an Apollo Telescope Mount based upon an engineering study by Ball Brothers. By that summer, plans called for building and launching no fewer than four ATMs involving some nineteen Saturn launches and twenty-six Saturn IB launches, with three Saturn-IVB wet workshops and four ATM payloads. The "wet" workshop concept (meaning that the stage would be launched with instruments mounted and its fuel tanks full, and unused fuel vented before the stage was used as a scientific laboratory) was the framework for the definition of the Apollo Applications Program.

In July 1966, Mueller's OMSF was given full responsibility for Apollo and Apollo Applications missions; Newell's OSSA would select experiments and analyze the data. Throughout the rest of that year, as OSSA identified major experiments, OMSF and NASA Headquarters personnel debated using a dry (launched with no fuel in the stage) workshop as a viable alternative. The dry workshop solved severe "habitability problems."

In January 1968, budget cuts reduced the Applications Program to one ATM flight, slated for April 1970, the first major slip, but well within the boundaries set by the Astronomy Missions Board. Further budget cuts, however, caused NASA to slip the first launch to November 1970. A major schedule shift occurred when NASA Administrator Thomas Paine finally approved the dry workshop in May 1969; this decision was announced on 22 July 1969 (two days after the Apollo 11 lunar landing). This was a good choice in terms of the design of the workshop, but inevitably it led to further slips; an 11 December 1969 press release stated that the change to a dry workshop would not cause any further slips but set the launch date at "mid-1972."[47] By August 1970, the Apollo Applications Program, now named Skylab, announced another slip to 1 November 1972, although internal planning dates were far more pessimistic. By April 1971, the launch was slipped to April 1973; the first Skylab was launched late that month.

---

46. D. L. Forsythe, Telex, 10 May 1968, NASA Historical Reference Collection.

47. NASA News Release 69-164, "Orbital Workshop Design Changes," NASA Historical Reference Collection.

48. Compton and Benson, *Living and Working in Space*.

The history of Skylab has been treated in some detail.[48] There were many problems that had to be overcome after the deployment of the laboratory, particularly the loss of one of the primary Skylab solar panels. In August 1973, NASA canceled plans for a second Skylab/ATM flight. Within a year, the program was closed down except for lingering support to process the data. The backup Skylab workshop, Multiple Docking Adapter, and Apollo Telescope Mount ended up at the National Air and Space Museum in Washington, D.C., as one of the most costly museum exhibits ever.

Looking back in 1980, however, astronomers like Goldberg expressed satisfaction with the Skylab program, once all the data had been retrieved and processing and analysis were well underway. Only then did Goldberg admit that the mission had evolved "into a marvel of engineering and scientific perfection (thanks in large measure to lengthy delays in the schedule)."[49] He and his colleagues recall their deep skepticism during the early years of the program, brought about by constant redefinition, budget cuts, and slips in almost every milestone. Nevertheless, Goldberg was among those who endorsed ATM/Skylab when called upon to do so by Homer Newell's office in preparation for budget briefings in November 1970. (See Document I-14.) As a quid pro quo, one month earlier, speaking for the Astronomy Missions Board, Goldberg urged that NASA improve support for ground-based facilities and analysis: "Full interpretation [of the ATM observations] will be possible only if ground-based observations are available and relevant laboratory studies have been carried out."[50]

Despite the Astronomy Missions Board's endorsements, making ATM a high priority for NASA was not supported by many in the astronomical community. It was neither discussed nor endorsed in the National Academy of Sciences's 1972 report *Astronomy and Astrophysics for the 1970s*. Headed by Caltech's Jesse Greenstein, the panel assessed all forms of astronomical practice, including space research. Even though Goldberg, by then Director of the Kitt Peak National Observatory, was on the central committee, he was not part of the study group that deliberated over priorities in solar research. There were deep fault lines in the community in the late 1960s and early 1970s, especially over spending priorities and maintaining the health of the disciplinary infrastructure, which was still perceived as optical and ground-based. As Greenstein noted in his introduction to the report, for the time period 1968 to 1971, NSF funds for basic research in astronomy (basic research grants alone) had flattened out at $6 million annually, whereas some 400 new Ph.D.s had entered the field in the same period, looking for support. Astronomy had made incredible advances in the 1970s but was heading for a period of retrenchment. He suggested that the field regain a balance over all areas of endeavor. Highest priority for the committee was that ground-based facilities not suffer due to overemphasis on space-based observatories.[51] Throughout the 1960s, total annual federal support for basic research in astronomy had been on the increase, averaging over $100 million from NASA (including all instrumentation) and between $10 and $20 million from the NSF. By far,

49. Goldberg to Wright.
50. Memorandum from Dr. Leo Goldberg to Dr. Homer Newell, 30 September 1970, NASA Historical Reference Collection.
51. Jesse Greenstein, "Introduction," in National Academy of Sciences, *Astronomy and Astrophysics for the 1970s* (1972), p. xii.
52. Ibid., Figures 3, 4, 5, and 6, pp. 61–64.

the largest single expenditure identified was for ATM, which reached the $70 million mark in the 1968–69 timeframe. By 1970, the funding was down to $30 million, which was still twice that of OSO and thirty times greater than levels of support for data analysis.[52]

In this fiscal state, the eleven members of the solar working group of the Greenstein committee, dominated by ground-based observers, endorsed continuing an ever-improving OSO program, with OSO-8 capable of one arc second of spatial resolution. "This program of continuous development and gradual improvement has made the OSO program among the most successful and productive of all astronomical satellite programs." They strongly recommended extending the OSO program to thirteen flights through the next solar maximum, expected to occur in the 1977–81 period. They called the OSOs and an upgraded sounding rocket program "the backbone of the solar space program."[53] Not included as a high priority for the coming decade, but definitely within the committee's sights, was a high-resolution solar telescope in space, of some 40-inch aperture, and capable of a guiding accuracy of 0.1 arc second.

The Space Science Board was hardly more sympathetic to scientific programs based on using astronauts as investigators. At its conference center at Woods Hole in July and August 1970, it articulated programs that would be possible at three levels of funding for the period 1971 to 1980. Among other projects, the working group on astronomy called for active design studies leading to a robotic solar observatory with capabilities equal to that of ATM-A: stability to one second of arc or better and payloads comparable to ATM. But the group also endorsed a crewed space station, for reasons reminiscent of those of the Working Group at the Iowa Summer Study in 1962: such operations offered a "great opportunity for solar research because of the high data rates inherent to solar observations."[54] More immediate priorities included a continuing OSO program for time-dependent studies of solar phenomena and the need to support allied solar-terrestrial programs, which depended upon continuing flights of the Navy's SOLRAD monitoring satellites, as well as the continued development of OGOs, the Atmospheric Explorer, IMP, and the Solar-Terrestrial Probe. Reflecting NASA's and national priorities, the astronomy working group highlighted the close relation between solar physics and Earth's environment.[55]

Given the complexity of the data-gathering requirements and perceived deficiencies in digital recording and transmission, coupled with the multiple goals of its advisory panels, NASA was always able to collect sufficient endorsements to demonstrate the general support of scientific advisory bodies for programs like ATM. By May 1971, for instance, NASA pointed to some seven different reports ranging from the 1965 Woods Hole study to a February 1971 Space Science Board summary in support of the program.[56] (See Document I-13.)

53. Ibid., pp. 92–93.

54. Space Science Board, National Academy of Sciences, *Priorities for Space Research*, 1971–1980 (1971), p. 78.

55. Ibid., p. 83.

56. J. Allen Crocker, "Advisory Group Comments on Skylab and ATM," 21 May 1971, NASA Historical Reference Collection.

The State of Solar Physics During the 1970s

After fifteen years of solar research from space, NASA's program officers and administrators could point to many valuable things learned about the Sun. The OSO era, 1962 to 1974, saw the first detailed studies of the sharp transition zone between the chromosphere and the corona and the first highly detailed images of the outer corona from rocket-borne and then satellite-borne coronagraphs. The extreme temperatures of flares were confirmed by OSO measurements, including studies of flare-related nuclear gamma-ray spectra. Coronal holes, the solar wind, and coronal transients were detected in a wide range of energies with resolutions capable of providing detailed information on energy transport.

The ATM and Skylab era saw highly detailed studies of chromospheric networks, prominences, and high-resolution x-ray imaging of solar photosphere and coronal structure. The post-Skylab era saw a return to AOSO-scale observatory-class programs using improved detectors and broadband data transmission, filling in the picture of the high-energy solar environment. During this period, NASA did begin to pay greater attention to solar-terrestrial relations, reflecting political pressures, societal trends, and institutional changes. Also of greater priority were joint international programs, which the United States initially embraced but, more than once, did not follow through to completion. During this period, priorities in NASA shifted from Apollo and Apollo Applications to supporting the Shuttle program.

The 1970s also saw significant changes in the relationship of specialists in solar physics to the general astronomical community, and this change may have been reflected in the advice given to NASA from its various boards and panels, which calls for comment here. Solar physicists were among the first specialists, along with planetary scientists, to feel that the mainstream American Astronomical Society was not able to meet their growing needs as a discipline. In the mid-1960s, a number of solar physicists and astronomers, including Henry J. Smith of NASA and Goldberg of Harvard, began to worry that fewer and fewer solar physicists attended AAS meetings. Smith in particular suggested that NASA-funded solar physicists meet periodically to air issues of mutual concern, whereas Goldberg, very much a leader of mainstream optical astronomy, preferred that these specialist meetings be held somehow under the aegis of the national society. Goldberg was, in fact, president of the AAS in the mid-1960s, and he was well aware of the concerns many of his colleagues had over the "Balkanization" of the society into specialist groups. Astronomy was one of the few disciplines in the physical sciences small enough to retain a unified national focus, and no one wanted that to change. For one thing, this unity gave the National Academy Decadal Surveys (the so-called Whitford and Greenstein Committees) significant political weight in Washington.

The result of this movement was the establishment of a Solar Physics Division (SPD) within the AAS in 1968, after several years of successful specialist meetings, rather than the creation of a new society. The planetary scientists, high-energy astrophysicists, dynamical astronomers, and even astronomy historians also established divisions, preserving to some extent the unity of the discipline under the parent society. The SPD sponsored many special sessions based upon space activities; in 1973, it convened a lengthy discussion of observations of the solar corona from Skylab. It also formed a conduit for interdisciplinary meetings with the American Geophysical Union and the American Meteorological Society in areas of strong NASA interest, such as solar-terrestrial relations, and became a forum for the preservation and improvement of critical ground-based facilities in solar astronomy.[57]

One of its most significant efforts came in the mid-1970s, when the Division sponsored an ad hoc committee on "Interaction Between Solar Physics and Astrophysics." (See Document I-16.) Headed by Andrea K. Dupree of the Harvard-Smithsonian Center for Astrophysics, what came to be called the "Dupree Committee" recommended ways to improve the relations between the two fields. Noting that "Communication and cross-fertilization among the subdisciplines of astrophysics has declined," the committee suggested ways to reestablish meaningful contact, partly through solar physics specialists' taking a more active role in advisory panels and boards. Reflecting concerns expressed by a similar committee convened by the National Academy, the effort was intended to highlight the continuing importance and relevance of solar physics to astrophysics by educating astronomers generally as to how knowledge of the workings of the Sun aided non-solar investigations.

Important too were solar-terrestrial relations, less an interest of the Dupree Committee than of its counterpart at the National Academy of Sciences, led by Eugene Parker. (See Document I-17.) The Parker committee's "Solar Physics Study" also explicitly identified the contributions of solar physics to mainstream astronomy: solar physics stimulated "new instruments, new diagnostic techniques, new interpretive insights, and new observational tests for existing theory."[58] The Parker committee was especially enthusiastic about a new AOSO-scale mission NASA was proposing—the Solar Maximum Mission (SMM)—to try once again to capture the Sun at the maximum point in its activity cycle. This was of great importance for solar-terrestrial relations, the Parker committee concluded, because the OSO and Skylab eras had established "exploratory observations that define the general nature of the complex atmosphere and activity of the sun" and provided the first detailed studies of the "inner working" of these high-energy phenomena. "The next stage is the detailed diagnostics," the committee added, "coordinating the necessary high-resolution observations to determine the precise physical nature of each phenomena." This was the job of SMM, considered timely and most critical for probing active regions, especially flare phenomena, and "a pivotal step in space research." SMM could lead to a series of Solar Synoptic Satellites (SSS), free-flyers that could monitor the evolution and life cycles of coronal structures and active regions, the drivers of solar-terrestrial phenomena. Although SMM was clearly the Parker Committee's top priority, it also mentioned a "Large Solar Observatory" (what was to become the Solar Optical Telescope (SOT)) and acknowledged that NASA was considering no fewer than five "facility-class" instruments to complement the NASA-ESA

---

57. See John H. Thomas, "The Solar Physics Division" in David DeVorkin, ed., *The American Astronomical Society's First Century* (College Park, MD: American Institute of Physics, 1999), pp. 238–44.

58. Andrea Dupree et al., "Report of the Ad Hoc Committee on Interaction Between Solar Physics and Astrophysics," Draft (18 June 1976), American Astronomical Society, Solar Physics Division; Hufbauer, *Exploring the Sun*, pp. 192–93; Space Science Board, National Academy of Sciences, *Report on Space Science*, 1975 (1976), chapter F.

Spacelab program. Ever mindful of NASA's priorities, the Parker Committee argued that the competing Spacelab series, consisting of 1-meter optical UV, XUV, soft x-ray, EUV, and hard x-ray telescopes, could not be considered as a substitute for SMM or SSS due to their short flight times.[59] This warning was very much in tune with priorities in the solar community and the old debate between AOSO and ATM.

Solar physicists were prompted to seek common ground with astrophysics because funding for both ground-based and space-based solar physics had peaked in the early 1970s and was steadily declining. The number of spacecraft used for solar science declined after the mid-1970s, due both to tightening budgets and to launch delays created by NASA's decision to use the Shuttle as its primary launch system. The decline was also due to greater competition from non-solar astrophysical programs, especially the emergence of NASA's "Great Observatories" program and its immediate precursors.[60] (See volume V, chapter 3, for a discussion of the Great Observatories.) In light of increased competition and lengthening lead times for developing new and larger missions, the 1975 Space Science Board "Report on Space Science" bluntly called into question the future of the space sciences and was uncharacteristically sharp in its commentary. Although it still gave highest priority to already-approved NASA programs such as the High Energy Astronomical Observatory (HEAO) series, to Pioneer Venus and Mariner Jupiter-Saturn, it also urged new starts in 1976 and 1977, including SMM, the Large Space Telescope, and the Gamma Ray Explorer. But the SSB also questioned the effectiveness of the "mission concept," feeling that it constrained science too much and fixed priorities too rigidly in an era of ever-dwindling support for science: "This trend, if not reversed, could lead to a national space program with minimal science, in contradiction to the stated objectives of the Space Act."[61]

The interdisciplinary nature of solar physics was also emphasized in 1975 when a "Solar Astronomy Task Force to the Ad Hoc Interagency Coordinating Committee on Astronomy" was convened under NASA solar physics program manager Henry J. Smith. Here, once again, the solar physics connection to solar-terrestrial relations was highlighted in order to broaden its support as far as possible among federal agencies. The many connections between solar physics and other problem areas, such as stellar and general astrophysics, magnetospheric physics, and plasma and nuclear physics, were all used to emphasize the importance of SMM, which the group saw as an "essential next step" in studying flares in high spatial and temporal resolution and coordinating observations across a wide spectrum. High time, spatial, and spectral resolution were all key to understanding the physical processes driving flares; SSM was "optimized for studying the dynamical processes associated with the energy buildup and release in flaring regions."[62]

During a period of deepening budget constraints and growing competition from other programs, both within astronomy and in space science generally, solar physicists found that their best course of action was to utilize the Solar Physics Division of the AAS

59. Space Science Board, *Report on Space Science* 1975, p. 149.

60. Hufbauer, *Exploring the Sun*, p. 195, table 5.2.

61. Space Science Board, *Report on Space Science* 1975, p. 6.

62. Federal Council for Science and Technology, "Report of the Solar Astronomy Task Force to the Ad Hoc Interagency Coordinating Committee on Astronomy," June 1975, p. II-63.

as an important platform from which to state their case. Solar physicists therefore remained a part of mainstream astrophysics under a parent society because they recognized that their primary audience still lay among astronomers, and astronomers offered, potentially, the most stable of audiences to support their goals.

The Shuttle Era

The Solar Maximum Mission was, in fact, the last significant solar physics satellite not to be launched by the Shuttle. When the Apollo Applications Program ended in the mid-1970s, NASA had already shifted much of its long-range planning to the Shuttle. Originally conceived as a means of establishing and supplying a space station, the Shuttle became an end unto itself when the space station was canceled. Throughout the 1970s and early 1980s, virtually all of NASA's scientific programs became redefined in terms of the Shuttle program, and solar physics was no exception. The lingering memory of ATM's failure to meet the previous solar maximum only heightened the need to monitor the active Sun at high spatial resolution during the next maximum. This time, meeting again at Woods Hole during the first two weeks of July 1973, the Space Science Board's solar physics panel made it clear that launching the SMM during a period of maximum solar activity was top priority. Drawing upon several recent studies emerging from NASA Shuttle workshops and European studies of the use of Spacelab for science, the various panels of the Board were convened to consider the scientific uses of the Space Shuttle. Most of the reports dealt with the various modes of doing science from the Shuttle. Notably at odds with this goal was the solar physics panel's view that, because of the "timing requirement imposed by the eleven-year solar cycle, we regard a free-flying satellite, with a carefully coordinated complement of instruments for the study of the next solar maximum, as the highest immediate priority item for solar physics."[63] The Shuttle could launch solar satellites but not operate them. The panel also noted that the chance of observing a truly major flare was rather slight, even during maximum activity, during the expected 7-day duration of a Shuttle flight. (See Document I-15.)

By 1973, Goddard Space Flight Center had already developed operational guidelines for SMM. It would be a free-flyer and was in fact suggested as a prototype for this class of Shuttle-related experiments. SMM itself would be launched on a Delta rocket and would carry some 500 kilograms of instruments in a stabilized cylindrical platform, not so different from the original AOSO concept but highly refined and far more feasible with the advances in electronic detector technology and higher bandwidth communications capabilities that had emerged in the intervening years. Its proposed capabilities were compared to the typical later OSO: it would be four times more precise in pointing than the projected OSO-I and would provide this precision over a greater area of the solar disk. It would also provide twice the power to the instruments, allowing them to be bigger and more robust than those limited by the OSO framework. The instruments themselves would be capable of detecting and imaging

---

63. Space Studies Board, National Academy of Sciences, *Scientific Uses of the Space Shuttle* (1974), pp. 138–39.

from the visible range down to the MEV x-ray and gamma-ray range, to catch the flood of high-energy particles and radiant flux expected from both the thermal and non-thermal phenomena associated with the biggest flares. The Woods Hole solar panel gave top priority to the SMM mission, provided that it was flown during the next solar maximum, 1977 to 1979; that the pointing accuracy be better than one second of arc over five minutes of time; and that the spacecraft be serviceable from the Shuttle "throughout the Shuttle era."[64]

Serviceability was a major design feature that defined all later Shuttle-based missions; it was to be demonstrated by SMM through the new mission design policy of building craft with modular units. Known as a "Multimission Modular Spacecraft," SMM was intended to act as a blueprint for a wide range of spacecraft missions. SMM inherited this role due to the downgrading and ultimate cancellation of the Gamma Ray Explorer (GRE), which had been in trouble for several years. Thus, alterations to the SMM program to take on this added technical capability had been under development for quite some.[65] As Noel Hinners described it in January 1976, the cancellation of GRE's Execution Phase Project Plan required that the System Definition and Execution Plan for SMM be altered to include, as its priority, the ability to "advance the concept of standardizing spacecraft subsystems." These standards would then be used "by many of NASA's future missions." (See Document I-18.) Hinners also confirmed that SMM would be launched by a Delta but had to be retrievable by the Shuttle, if necessary. "In-orbit servicing of the SMM," however, was "not a requirement."[66]

SMM and its relation to Shuttle was only one of the issues addressed by the 1975 Woods Hole solar study panel. It also examined the Shuttle-based "sortie" mode, using an astronaut as an observer, operator, or technician for performing solar research. Basic to the hardware would be the development of two general-purpose pointing platforms. One would be capable of handling telescopes over 2 meters in length and pointing them with an accuracy of one second of arc; the other would be able to handle instruments twice as long. The panel urged that the instrument integration design of the overall system be as transparent as possible, allowing the development of a wide range of instruments to proceed independently of the Shuttle program.

The Woods Hole panel used the experience of Skylab to discuss the role of the astronaut. Drawing upon the success of the second Skylab mission, the panel concluded that the role of astronaut as observer had been proven: suitably trained astronauts could make effective "real-time decisions" about what part of a flare to examine in detail; they could carry out "complicated observing sequences" when out of reach of ground control; and they could perform ad hoc procedures to correct mishaps, such as installing the thermal shield and cleaning dirty optics, fixing jammed film cameras, and correcting minor glitches in the instruments.[67] On the other hand, it was clear that the crew had contaminated the local space around Skylab, resulting in the need for cleaning the optics

---

64. Ibid., p. 140.

65. Memorandum from Anthony J. Calio to Director, Solar Terrestrial Programs, 19 December 1975, NASA Historical Reference Collection.

66. Memorandum from Noel Hinners to John F. Clark, 15 January 1976, NASA Historical Reference Collection.

67. Space Science Board, *Scientific Uses of the Space Shuttle*, p. 145.

in the first place. The panel therefore recommended that suitable precautions be taken for Shuttle and that the experience from the Skylab missions be used as a basis for improvement.

SMM was launched on 14 February 1980, carrying seven instruments designed and built by American and European consortia. Hufbauer[68] has reviewed in detail how some of the more significant instruments for SMM were chosen and how these choices were a reflection of critical issues being addressed by members of the solar community. One of the most significant problems was coming to closure on how much energy the Sun radiated into space and, accordingly, how much was being collected by Earth. Although error bars on this value had been localized to a few percent in the early twentieth century by the Smithsonian's Charles Greeley Abbot, far greater accuracy was required over long time periods to assess how constant this "solar constant" output of energy was, how and where it changed across the electromagnetic spectrum, and, finally, whether it oscillated periodically or not. Abbot had claimed to have detected major cyclic changes of several percent, and though few believed his statistics, the issue was very much alive when he died in 1972 at age 102.[69]

In late 1976, NASA approved flying multiple instruments to measure solar radiation characteristics, including solar irradiance, that would allow for an unequivocal measurement of the solar constant and some indication of the source of its variation, if it were verified to exist. An active cavity radiometer was included in SMM; this was the first instrument placed into orbit that had sufficient sensitivity to make the measurements. By the fall of 1980, SMM had detected minute changes in solar energy, amounting to only 10 degrees Celsius in the photosphere's average temperature of 5,700 degrees Celsius. The active cavity radiometer was capable of measuring changes in the bolometric flux as small as 0.001 percent and was an excellent example of the importation of new technical talent into solar physics, since it was promoted and built by a specialist in radiometric instrument development, not by a solar physicist.[70]

The data flow SMM was returning degraded and finally halted after some 300 days, when "Solar Max" lost its fine pointing capabilities on 11 December 1980. The problem was a set of fuses in the main electronics box that controlled stabilization and fine pointing. This failure, in fact, gave NASA an excellent reason to mount a Shuttle repair mission to demonstrate the utility of the human space program to science. (See Documents I-21 and I-22.) The repair mission was flown by *Challenger* in April 1984 and, among other tasks, replaced the main SMM electronics box for guidance and stabilization. This was far from a routine mission, however, since the grapples on the satellite turned out to be slightly different from those the astronauts carried in their maneuvering units. A contingency plan to use the large grappling arm succeeded in capturing the satellite, however, and the rest of the repair went smoothly, vindicating the whole modular concept and the utility of human-tended free-flyers.

68. Hufbauer, *Exploring the Sun*, pp. 197–98 and chapter 7.
69. D. H. DeVorkin, "Defending a Dream: The Abbot Years," *Journal for the History of Astronomy* 21 (1990): 121–36.
70. "NASA Satellite Detects Changes in Energy Output from Sun," NASA News Release 80-124, 6 August 1980.

In the Shuttle era, reflecting continually tightening budgets and increased costs, solar research with space satellites and probes dwindled further. A number of major programs were proposed, but all were either redefined or canceled outright. The Solar Max repair mission did not come cheaply, of course. Far more expensive than any ground-based projects, still it has been said to have cost less than flying a new mission. It was, moreover, funded primarily as a means to demonstrate the utility of NASA's space transportation system.

Competing Solar Programs in the 1980s and 1990s

The Shuttle era not only redefined the manner in which space solar physics could be done, but also created greater pressure on all of space science to search for more efficient ways to operate. Both the Carter and Reagan Administrations made it clear that the "golden age" for the space sciences (particularly planetary exploration), as the decade of the 1970s had been described by practitioners, was now over.[71] A clear NASA priority, therefore, was to find new ways to generate interest in scientific experiments that could be performed in orbit from the Shuttle and that would have appeal as part of a new international cooperative program between ESA and NASA called Spacelab. As early as 1975, however, Harold Glaser, Director of Solar-Terrestrial Programs, grew concerned that these programs would proceed without sufficient input from scientific experimenters. He petitioned the Director of Spacelab Programs in November of that year, noting that in order for NASA to "demonstrate to the public as early as possible the utility of the Shuttle/Spacelab as a valuable experimental facility," it was imperative that "experimenters and responsible payload personnel [were] involved in the trade offs which establish[ed] the payload constraints."[72] There were indeed many problems with a human-tended scientific experiment that required a high degree of stability and lack of environmental pollution. For instance, the acoustic and electrical environment of the Shuttle bay, as well as the influence of "shifting dynamic loads" (crew movements), all threatened to reduce the effectiveness of both solar and stellar investigations requiring sub-arc-second stability. Guarding against these sources of error vastly increased the expense and lead times of preparing suitable payloads.

The inevitable increase in costs, along with a general tightening of all budgets, amplified the difficulty of mounting major solar programs in the 1980s and 1990s. One of the most frustrating casualties was the International Solar Polar Mission (ISPM), put in place by a formal agreement between NASA and ESA in 1979.

The ISPM Saga

International cooperation was important for both political and economic reasons throughout the Cold War, but especially in the post-Apollo era. At first, in the 1960s, NASA

---

71. Robert Kraemer, *Beyond the Moon: A Golden Age of Planetary Exploration*, 1971–1978 (Washington, DC: Smithsonian Institution Press, 2000).

72. Harold Glaser to Director of Spacelab Program, 2 November 1975, NASA Historical Reference Collection.

had made launchers available to send British satellites like the Ariel series into orbit. The next phase of international cooperation grew out of a need to share costs and expertise: the International Ultraviolet Explorer (IUE), the Infra-Red Astronomy Satellite (IRAS), and even SMM were just a few of the many examples of this trend. Then, in March 1979, NASA and ESA agreed to develop a cooperative International Solar Polar Mission to make "coordinated observations of the interplanetary medium and the Sun simultaneously in the northern and southern hemispheres of the Sun."[73]

The key to the Solar-Polar mission was the ability to make stereoscopic, simultaneous viewing of the Sun possible in order to assess global solar behavior. Fifteen separate experiments from ESA and NASA Centers, JPL and Goddard, the University of Chicago, Bell Labs, and Los Alamos were proposed and partially developed. The rationale for the mission was that much solar phenomena was latitude-dependent, and thus far observations had been from the ecliptic plane only. For instance, the nature of the solar wind was only known within the narrow equatorial band, and scientists knew it was not wise to extrapolate to higher latitudes to assess the overall character of the solar wind. The technology was now available; by the mid-seventies, NASA felt that its ability to target probe trajectories to take advantage of gravity-assist maneuvers had advanced to the point where out-of-the-ecliptic missions were feasible. The swingbys of Jupiter by Pioneers 10 and 11 had provided ample proof of concept.[74] Two nearly identical SMM-type probes would be launched by the Shuttle and then inserted into a Jupiter encounter by an Inertial Upper Stage (IUS) system consisting of three or four stages. A February 1983 launch date was proposed to meet a narrow window for the Jupiter encounter.[75] (See Document I-20.)

The complex history of the cancellation of the American portion of the mission has been described as a result of congressional infighting and oversight, the incoming Reagan Administration's draconian reorganization of the budget process, and finally NASA's own shifting priorities. In December 1979, the Senate Appropriations Committee was worried that the IUS would not be adequate to launch both probes and that Shuttle development might be delayed, necessitating a slip in the launch schedule beyond the proposed Jupiter flyby window. The committee suggested slipping the mission until the next window. NASA also tried to slip the mission later that year when, during hearings on the 1980 Supplemental Appropriations Bill, the House side called for the termination of the mission and for rescinding funds already appropriated. This call was eventually reversed on a technicality, but it was evident throughout Fiscal Year (FY) 1979 and FY 1980 that NASA's priorities lay with funding the Shuttle over any one scientific mission. Throughout this phase, NASA kept ESA informed, and they both agreed to the delay even though the initial cuts were rescinded.

The incoming Reagan Administration overhauled the entire budgetary process, which resulted in deep cuts for NASA. NASA was directed by the White House to cut one

---

73. "Memorandum of Understanding Between the United States National Aeronautics and Space Administration and the European Space Agency for the International Solar Polar Mission," p. 1.
74. Jet Propulsion Laboratory, "Project Plan for International Solar Polar 1983 Mission," Document 628-1, pp. 1–3.
75. Ibid., pp. 1-3–6.

of its three large space science missions—the Large Space Telescope, Galileo, or ISPM. NASA reacted by canceling ISPM outright in the FY 1981 budget. "In just over one month, drastic cuts had been made, and made very quickly."[76] This time, ESA was informed after the fact, just hours before the Reagan Administration announced the cuts. (See Volume II, Chapter 1, for more details on this episode.) European officials were told that NASA would still launch the European spacecraft, but that the launch would be delayed until 1985 or 1986.

It came as no surprise to members of the Space Science Advisory Committee (SSAC), meeting at NASA Headquarters in early March 1981, that ESA quickly protested at the ambassadorial levels of its member countries, claiming that there was a "unilateral breach of the ISPM Memorandum of Understanding." ESA had approved the program over others specifically because of its international, transatlantic character.[77] To ESA observers, this event exposed an important difference between ESA and NASA priorities; it seemed as if NASA did not share ESA's sense of obligation to international agreements. To the SSAC's Leonard Fisk, NASA's decision did not bode well for any future collaborations, especially since a wholly domestic mission, Gamma Ray Observatory, was allowed to live. Answering his own question in the minutes of the SSAC meeting, he argued that the "magnitude of cut demanded that an existing [mission] must go. In addition, ISPM is slaved to solar cycle epoch and Hill anger over [delay and cost-growth] scenario." Andrew Stofan further explained that NASA's decision to save "near-term costs" at the expense of "run-out total" was not acceptable to Congress, "and NASA has been told so in no uncertain terms." David Morrison was recorded as feeling that the cut was not on scientific grounds, but on "programmatic" grounds, since NASA had not yet sunk much funding into the mission compared to what it had committed to the Space Telescope and Galileo.[78] (See Documents I-23 and I-24.)

ESA proposed several compromise solutions to NASA through the spring of 1981, some of which NASA sent on to Congress, the State Department, and OMB. A mix of positive but noncommittal reactions resulted, which gave ESA hope that it could reverse things. In late April, NASA petitioned the White House for additional funds, but the petition was rejected. NASA was left with the option of redirecting internal funds and priorities to keep ISPM a two-spacecraft mission. Deciding whether to follow this path became James Beggs's responsibility when he took office as NASA Administrator in late June 1981. By September, Beggs decided that NASA's priorities could not include the reduced two-spacecraft mission; no new funds would be requested for ISPM in the FY 1983 budget.

Though some observers hoped to dismiss the cancellation of ISPM because of its unusually complex nature, calling it an aberrant example of international collaboration, others pointed out that it was ESA's failure to appreciate the volatility of the "highly political US budget process."[79] This failure, however, was largely due to NASA's inability to

---

76. Joan Johnson-Freese, "Canceling the US Solar-Polar Spacecraft," *Space Policy* (February 1987): 24–37; quote is from p. 26.

77. Ibid., p. 28.

78. Space Science Advisory Committee, "Meeting Minutes," NASA Headquarters, 2 March 1981, NASA Historical Reference Collection.

79. Johnson-Freese, "Canceling the U.S. Solar-Polar Spacecraft," p. 32.

bring ESA into the budget process itself, before Reagan's public announcement, as well as NASA's failure at the time to "accept partners, rather than subordinate participants, and to treat them as such."[80] In 1989, looking back on the episode, NASA's chief of solar physics, J. David Bohlin, warned that "the real lesson, of course, is not just the facts, but that this episode colored our relations with the European Space Agency for many years (and still does)."[81]

The original ISPM concept was thus reduced to a single ESA probe, renamed *Ulysses*, which was launched by the Shuttle and a two-stage IUS on 6 October 1990, after several years of delay due to the 1986 *Challenger* disaster. *Ulysses* flew by Jupiter in February 1992 and has since reached both extremes of the northern and southern latitudes of the solar heliosphere. This first orbital cycle came during the solar minimum, but the program has been extended and has now lived long enough to execute another full orbit of the Sun at high latitudes during the most recent solar maximum.[82]

The Solar Polar mission, in the form of *Ulysses*, was conceived in the late 1970s, launched in 1990, and to date has provided some ten years of valuable heliospheric data. *Ulysses* is, however, only a moderate-sized craft, and the planetary program had already proved the programming involved. Other, larger programs in solar physics would not fare as well in the face of continuing budget stagnation, NASA's growing dependence on maintaining the Shuttle, and its focus on establishing a permanent presence in space.

The Solar Optical Telescope

A large, ultra-high-resolution solar telescope in space had long been suggested as a means of refining knowledge about the Sun. In 1979, Goetz Oertel, Deputy Chief of Solar Physics at NASA, reviewed his office's internal "Priorities in Instrumentation Development for Solar Astronomy." (See Document I-20.) The goals of solar physics as outlined by the Astronomy Missions Board in 1969 still created a serious challenge for spacecraft and instrument design: the ability to study structures on the Sun with angular sizes between 1 arc second and 0.1 arc second. This would require a 1.5-meter diffraction limited solar telescope in orbit.[83]

Called, among other titles, the "Large Solar Observatory" by SSB panels in the mid-1970s, what became the Solar Optical Telescope (SOT) was planned as a Shuttle-based, high-resolution solar observatory capable of a wide range of spectroscopic, photometric, and area imaging studies. The goals of SOT and its predecessors, however, were highly varied, and their evolution illustrates how "excessive planning can undermine the nation's efforts to achieve important scientific goals."[84]

---

80. Ibid., p. 37.

81. J. David Bohlin to Lennard Fisk et al., 21 June 1989, NASA Historical Reference Collection.

82. JPL Press Release, "Ulysses Studies the Sun's Polar Cap at Sunspot Maximum," 6 September 2000.

83. Memorandum from Goetz Oertel, Deputy Chief, Solar Physics, to Director, Physics and Astronomy Programs, 7 January 1970, NASA Historical Reference Collection; Astronomy Missions Board, 1969, p. 135.

84. "The Solar Telescope That Saw No Light," Appendix B in Commission on Geosciences, Environment and Resources, National Academy of Sciences, *A Space Physics Paradox: Why Has Increased Funding Been Accompanied by Decreased Effectiveness in the Conduct of Space Physics Research?*, 1994.

The origins of SOT have been traced to efforts in the mid-1960s by two solar physicists, Harold Zirin and Robert Howard, to build a 65-centimeter solar telescope for the second Skylab mission that would be capable of examining the fine structure of magnetically driven phenomena in the solar photosphere. They soon realized that a 1.5-meter telescope would be required to gain sufficient resolution, and when Skylab II was dropped, they decided to regroup and build a ground-based prototype. Meanwhile, NASA had a number of larger telescopes under study in the early 1970s. In 1976, Zirin and Howard jointly proposed with astronomers at Kitt Peak National Observatory to build and fly a larger Spacelab Orbital Telescope. NASA adopted the characteristics of the proposal, renaming it the Solar Optical Telescope, or SOT, and designating it as a NASA facility mission to be managed by the Goddard Space Flight Center.[85]

What started out as a $25-million mission for Spacelab was, by the end of 1985, estimated to cost $360 million. Study phases were complete, but design and construction were deferred, partly due to pressures caused by budget and management problems with the Large Space Telescope, by now known as the Hubble Space Telescope. But SOT's rising costs were also criticized by Congress, and in February 1986, in the wake of the *Challenger* disaster, OMB deleted the funds for SOT.

Goddard project managers retrenched, trying to downsize SOT to a $100-million program by reducing the size of the telescope to 1 meter, removing one of the focal plane instruments, and narrowing the wavelength range. They did not get even close, and even rejected an offer from the Naval Research Laboratory and Marshall to provide the complete system for $85 million. In the wake of *Challenger*, however, Goddard did accept the NRL's proposal for what became the High Resolution Solar Observatory, based upon the successful instrumentation designs of Richard Tousey's group at NRL.

The Shuttle flight of Spacelab 2 on 29 July 1985 carried the High Resolution Telescope and Spectrograph (HRTS) built at NRL. It verified and refined previous sounding rocket flights equipped with advanced stabilization systems called SPARCS that were able to resolve finely detailed structure at the point in the solar atmosphere where temperature is at a minimum. Getting at the fine structure of this region and its relation to upper regions was one of the central goals of SOT, shared by Richard Tousey and many other scientists like Zirin and Howard. Such observations might untangle the relative roles of "wavelike-matter" oscillations, magnetic fields, and random motions as agents promoting the transport of energy from the photosphere and chromosphere to the solar corona. In other words, SOT was directed at determining the source of heating of the million-degree corona, the first question posed about the solar corona when its extreme temperature was deduced in the late 1930s and early 1940s. Always thinking incrementally, in 1983, Tousey felt that more SPARCS-stabilized sounding rocket flights were needed to prepare this new technology for spaceflight. With the launch of Spacelab 2, he was happily reconciled to a human-tended platform, though he still held out hopes for SOT on the eve of its cancellation.[86]

85. Ibid., p. 92.
86. Richard Tousey, "Our Star—V2 to SOT," Transcript of Goddard Lecture, March 1983; "Solar Spectroscopy from Rowland to SOT," *Vistas in Astronomy* 29 (1986): 175–99.

Renamed the High Resolution Solar Observatory (HRSO), the program was revised as a space station payload (See Document I-25) but soon was changed into a robotic mission to save money. To restore its original capabilities, NASA looked for international support while it continued to support Tousey's experimentation. Even this retrenched position failed when OMB deleted all funds for the project in the FY 1988 budget.[87] (See Documents I-26 and I-27.) By then, HRSO had been redefined again, as Solar Physics was transferred out of NASA's Astrophysics Division and placed into a new Space Physics division. The new reincarnation, a free-flyer called the Orbiting Solar Laboratory (OSL), quickly ballooned into a $500-million and then an $811-million program. In 1986, it had been called a "truly marvelous and versatile laboratory" incorporating a wide range of instruments and facilities for simultaneous and synoptic studies of the Sun, in tune with the huge programs NASA was then campaigning for, such as the HST and the Earth Observing System (EOS). But as it grew larger, more complex, and costlier, OSL also began to lose support within NASA and, especially, among many solar physicists who either left the project in frustration or proposed different methods of gathering data. Finally, in the late summer of 1991, NASA slipped the OSL from 1993 to 1998, suggesting instead that the Space Physics Division think in terms of smaller missions.[88] There were a number of efforts to revive the program as a multifaceted assault on the Sun, but none of them materialized.

Shifting Support

Other factors influencing the lives of long-term projects included the 1987 introduction of a strategic planning process (See Volume V, Chapter 1) within NASA's Office of Space Science and Applications, which changed the character of its internal and external advisory structure. Influenced by the advice it received, NASA also changed its perception of disciplinary boundaries, which led to a number of internal reorganizations of OSSA offices.

Starting in the summer of 1984 and continuing through early 1986, the Space Science Board once again convened several task groups to explore the needs and directions of the various disciplines of space science. Initially, the SSB projected scientific goals between 1995 and 2015, but in the wake of *Challenger*, it revised its time frame. The Board also made a significant change in its advisory structure by linking solar physics to space physics, removing it completely from deliberations by the Astronomy and Astrophysics task group, and thereby foretelling the actual shift of the Solar Physics program out of the Astrophysics Branch at NASA. Accordingly, the astronomy and astrophysics task group called for the development of high-resolution optical interferometers and NASA's Great Observatories. But the task group's report made no mention of solar astronomy or any solar initiatives, whereas other related activities (such as gravitational wave physics) were covered.[89]

---

87. "Solar Telescope That Saw No Light," p. 94.
88. Ibid.
89. Space Studies Board, National Academy of Sciences, *Space Science in the Twenty-First Century—Imperatives for the Decades 1995 to 2015* (1988).

The advisory structure had changed significantly, and solar physics was in a new playing field, called "Solar System Space Physics" by the SSB coordinating panel but "Solar Space Physics" by members of the task group itself, since it was separate from planetary and lunar exploration. The structure of the Solar Space Physics Task Group's report identified the two major foci of the area as the physics of the Sun and "the processes that link solar variations to terrestrial phenomena."[90] (See Document I-28.) Emphasis in the first category was on the Sun's outer atmosphere and its influence on the magnetospheres and atmospheres of Earth and the other planets. Among prospective pre-1995 missions, only *Ulysses* was noted, along with *Galileo*, an Upper Atmosphere Research Satellite, and the International Solar-Terrestrial Physics Program. The recommended program post-1995 included new starts for large high-resolution UV and x-ray solar telescopes to examine small-scale transient structures in the solar photosphere, a solar probe to fly to within 2 million kilometers of the solar photosphere to investigate the sources of the solar wind, and continued development of advanced spacecraft for observing the Sun, Earth, and the solar-terrestrial environment.[91] The latter included reviving a solar polar mission as a follow-on to *Ulysses*, as well as the completion of SOT and its integration into an Advanced Solar Observatory on the Space Station.

The details of the Solar and Space Physics task group's report were fleshed out in an extensive appendix on the importance to astrophysics of studying small-scale features on the Sun, giving several examples of how small-scale processes could be of great importance generally. These included refining understanding of the nature of granulation and the relation of magnetic phenomena to turbulence, and gaining better knowledge of the physics behind the flow of matter and energy outward from the photosphere, chromosphere, corona, and solar wind. None of this was new, of course.

What distinguished the Solar Space Physics Task Group from its predecessors was its appeal to a new level of instrument and spacecraft technology. Instruments capable of making these observations from Earth orbit required ultra-high resolution and stability and the capability of working to their theoretical limits in the extreme ultraviolet and x-ray regions of the spectrum. The task group was able to point to only one instrument flown on Spacelab 2 in 1985 that had operational capabilities near to its instrumental limits, called SOUP, the Solar Optical Universal Polarimeter, a 30-centimeter-aperture visible light device. Significant advances in instrument development were therefore urged, including interferometric instruments in space capable of examining solar features beyond the 0.1-arc-second range and new technologies to achieve up to 0.001-arc-second accuracy in the x-ray and UV regions. Overall, however, the task group's priorities remained in line with those of its predecessors, calling for instruments capable of studying small-scale features on the solar surface and the details of its atmospheric structure.

In calling for these new technologies, the task group also called for improvements in spacecraft systems and propulsion. These included sustained low-thrust propulsion devices for the solar probes and possible interstellar probes (solar electric propulsion

---

90. Ibid., p. 27.
91. Ibid., p. 34.

systems); heat shield technologies capable of protecting craft flying to within 4 solar radii of the solar surface; high-reflectivity, multilayer coating technologies for extreme ultraviolet and x-ray imaging devices, allowing for normal-incidence reflectivity; and the development of a Lagrangian point platform for a wide array of studies.[92]

The task group expressed confidence that the interferometric and multilayer coating techniques that would make possible spatial resolutions better than 0.1 seconds of arc, over a wavelength range from 30 angstroms to the visible, "can be achieved with techniques that will be perfected in the twentieth century."[93] The task group then considered alternative sites for placing these devices, including near-Sun orbit, heliosynchronous orbit, lunar basing, solar orbit at one AU, Earth-orbit free-flyers, and crewed vehicles such as the Space Shuttle and International Space Station. The advantages of each were identified for specific missions. The lunar base was cited as being a particularly stable location for long-term studies. Earth-orbit free-flyers were preferred for similar reasons. Crewed vehicles, on the other hand, drew mixed comments. As before, the short duration and discontinuity of coverage of Shuttle flights was a continuing limitation on scientific return. Contamination was another lingering problem, especially the presence of thin polymerized hydrocarbons that block light in the extreme ultraviolet. Yet, the task group felt that the positive experience from Skylab indicated that it should not totally discount this mode of transport and operation. Thus, it looked kindly on the use of future space stations as platforms, while recognizing the problems inherent in an extended mechanical structure, which included lowered rigidity and an extended spectrum of vibrations and disturbances, making precise pointing and tracking difficult or impossible. Some form of isolated platform was suggested as a remedy, including co-orbiting platforms.[94]

Although the Solar Space Physics Task Group identified SOT as a desirable new mission, they did so only in passing and did not explore SOT priorities to any depth. Indeed, the implication might be that, in light of the advanced technologies proposed and the stated problems inherent with Shuttle-based or Space Station-based observatories, SOT had become obsolete, killed by its own long lead time and unrestricted growth.

The Role of Advice in Space Solar Physics

Further historical research needs to be done to be able to evaluate fully the process of consensus formation that produced the advice given to NASA by various elite task groups, panels, and commissions. NASA's influence in the process, especially, must be better understood, as well as other social and economic factors that evidently framed and informed the opinions of advisory groups. One especially fruitful path of new inquiry might be to explore more fully the relationship between scientific program and project managers within NASA and members of the various advisory bodies, especially in light of the historical fact that projects tend to balloon and expand themselves out of

92. Ibid., pp. 112–39.
93. Ibid., p. 133.
94. Space Studies Board, National Academy of Sciences, *Space Science in the Twenty-First Century: Imperatives for the Decades 1995 to 2015: Solar and Space Physics* (1988), pp. 133–35.

existence. What causes projects to balloon? Is there a correlation, for instance, between the size of a scientific mission or project and its ability to acquire internal and external support? Small projects may suffer from having only small constituencies among builders and users. Do larger projects have proportionally larger constituencies? And do very large projects suffer because they impinge on other constituencies? Is there an optimum point for project size, and to what extent does external scientific advice influence where this point lies? The importance of coalition building, so well illustrated in the case of the Hubble Space Telescope by Robert Smith,[95] may well turn out to have had a far different effect within solar space research, where a fluctuating set of disciplinary communities was involved.

This essay has only scratched the surface in assessing the degree to which NASA was able to follow external advice and how it interpreted and exploited the opinions of its advisory bodies in forming new programs and policies. Even at the superficial level examined here, it is evident that there were strong scientific motivations for achieving greater stability, pointing accuracy, and spectral resolution. These long-term goals are ongoing and are indeed seen in recent missions, such as Transition Region and Coronal Explorer (TRACE), one of the early products of NASA's Small Explorer (SMEX) program. Planned as the fourth mission in the "faster, better, cheaper" program championed by former NASA Administrator Dan Goldin, TRACE was developed in less than four years to refine knowledge of the relationship between solar magnetic fields and coronal heating. It was designed to observe the three-dimensional magnetic structure in the solar photosphere that determines the dynamics of the chromosphere and corona. TRACE, launched in April 1998, only a month behind schedule, aboard an air-launched Pegasus, is international and short-term, and it embodies many, if not all, of the characteristics desired by NASA's advisory groups. By 2000, TRACE had returned unprecedented high-resolution images of magnetic field phenomena in the lower corona.

From the standpoint of the nature of scientific advice reviewed here, TRACE and the SMEX program were appropriate NASA responses, given the economic and political pressures facing the space agency. But they are also a very positive reaction to earlier criticisms of NASA's mission-oriented policy: that the "emergence of a new generation of spacecraft does not . . . stem from demand by users."[96] Another indicator of NASA's ability to learn from past difficulties is its increased sensitivity to establishing and maintaining meaningful international partnerships. Possibly the most significant recent mission in solar physics is the Solar and Heliospheric Observatory, or SOHO, built by ESA, instrumented by teams in fourteen of its member nations, as well as by NASA, and launched aboard a U.S. Atlas booster in December 1995. NASA also is responsible for communications and daily operations of the craft.

The SOHO spacecraft is 4.3 meters long in the Sun-pointing direction and orbits a Lagrangian point in space between Earth and the Sun, allowing its battery of twelve instruments (nine from Europe and three from the United States) to maintain constant monitoring of solar activity. SOHO worked nominally in spite of various battery problems

---

95. Robert W. Smith, *The Space Telescope: A Study of NASA, Science, Technology, and Politics* (revised, New York: Cambridge University Press, 1993), Chapters 4 and 5.

96. André Lebeau, "The Astronaut and the Robot," *Space Policy* (February 1987): 211–12.

until June 1998, when attitude control and telemetry were completely lost. The 305-meter Arecibo radar dish in Puerto Rico was able to confirm SOHO's rotation rate and orbital position, which made it possible to direct commands to the craft to regain control. This was done by late September 1998; the satellite has been working well ever since, sending back critical information about the Sun that helps us to better appreciate its internal dynamic structure, as well as the onset of coronal bursts and mass ejections that would affect solar-terrestrial relations.

Recent programs like SOHO and TRACE bode well for the future health of space solar physics. They stand in contrast to the hugely ambitious programs like SOT-OSL, as well as those defined in terms of a human spaceflight program. In solar physics, the OSO series of missions has had the strongest continuing support from the scientific community and its elite deliberative panels. Least supported were those with the longest lead times and where mission control did not reside in the scientific community, such as SOT-OSL. Almost without exception, lead times were drastically lengthened as payloads had to be reconfigured to meet new NASA launch vehicle priorities. It is true that NASA's scientific advisors did respond positively to NASA's shifting priorities, but one would have to go well beyond the published literature and even the semipublished documents provided in connection with this essay to be able to assess how these reports were constructed and what other choices, if any, the advisory groups considered. It does seem that the advice given in later review panels and boards that were convened at the request of NASA, funded by NASA, and to some extent managed by NASA, though still under the aegis of the National Academy and other academic groups, was definitely constrained by NASA interests and priorities. One need go no farther than noting that many of the review panels were asked explicitly to provide scientific goals that could be met by NASA's continually changing priorities, starting with Explorer-class satellites, then Observatory-class, then Apollo and Apollo Applications, the Space Shuttle, and, at present, the International Space Station.

The extent to which these constraints have colored the advice given by the panels cannot be evaluated without extensive historical research using primary sources and oral history interviews. For example, one might speculate that the lessons learned from NASA's growing propensity for huge missions with long lead times, criticized widely by the scientific community, resulted in NASA's reworked "smaller, faster, cheaper" policy in the 1990s, which resulted in success stories like TRACE.[97] But this advice was couched in the context of a Shuttle-based and Shuttle-driven political economy, together with the spectacularly frustrating failures of many of the missions supposedly designed under this rubric. Because SMEX missions like TRACE and international missions like SOHO were not dependent upon the Shuttle, it remains to be seen whether this new policy will eventually result in a more cost-effective solar physics program that is responsive to the stated needs of the community it serves, and if, indeed, it is a policy reflecting a desire to promote scientific knowledge of the Sun.

---

97. See Howard McCurdy, *Faster, Better, Cheaper: Low-Cost Innovation in the U.S. Space Program* (Baltimore, MD: Johns Hopkins University Press, 2001) for an assessment of the "faster, better, cheaper" approach.

Document I-1

Document Title: W. W. Kellogg, RAND Corporation, "Basic Objectives of a Continuing Program of Scientific Research in Outer Space," 9 December 1957.

Source: Dow Papers, Box 84, University of Michigan Library, Ann Arbor, Michigan

This document represents one of the several general proposals for initial space science activities which were developed in the aftermath of the launches of Sputniks 1 and 2 in late 1957. Scientists who were interested in placing their instruments into orbit and beyond had clearly been giving a lot of thought on how best to proceed even before the initial satellite launches, and Kellogg's proposals reflect the ideas that sounding rockets would continue to be an important part of initial space science efforts and that human space flight was still some time in the future and not scientifically justified.

[cover letter]

9 December 1957

TO: Members Rocket and Satellite Research Panel

As agreed at our meeting last Friday I am sending a draft of a report prepared by me for the Working Group on Internal Instrumentation of the Earth Satellite Program entitled "Basic Objectives of a Continuing Program of Scientific Research in Outer Space." This was originally intended as a working paper for the National Academy of Sciences, but it can also serve the Executive Committee of our panel as a working paper and a back-up to our proposal.

If you have any comments I suggest you send them in duplicate to Van Allen and myself.

Sincerely yours,

[signature]
W. W. Kellogg
The RAND Corporation
1700 Main Street
Santa Monica, California

cc: W. W. Berning
L. A. Delsasso
W. G. Dow
K. Ehricke
C. F. Green
M. Greenberg
L. M. Jones
M. H. Nichols
M. Rosen
N. W. Spencer

K. Stehling
H. J. Stewart
W. G. Stroud
H. Strughold
E. Stuhlinger
J. W. Townsend
W. Von Braun
P. H. Wyckoff
M. Zelikoff

****

[no page number]

BASIC OBJECTIVES OF A CONTINUING PROGRAM
OF SCIENTIFIC RESEARCH IN OUTER SPACE
(Prepared for the National Academy of Sciences
National Research Council by the Technical Panel
for the Earth Satellite Program of the U.S.
National Committee for the I.G.Y. November, 1957)

1. Introduction

The International Geophysical Year marks the beginning of man's exploration of outer space. There have been previous rocket firings into the fringes of the earth's atmosphere, but the expanded rocket-sounding program on an international scale and the advent of artificial earth satellites represent by far the largest steps taken so far towards the scientific exploration and eventual habitation of outer space and the planets.

The interests of human progress and our national welfare now demand that a long term program of space exploration be formulated and pursued by the United States with the utmost energy. Although there will inevitably be benefits from such a program of a very practical nature, the basic goal of this exploration must be the quest of knowledge about our solar system and the universe beyond.

The scientific program proposed here has been formulated with the following ideas in mind:

- Technology of space flight will probably develop gradually. Therefore, the payloads and distances traveled will be relatively small at first, and the scientific experiments and observations will be correspondingly modest in the early stages.

- The scientific program should be designed to give information at each stage which will help in the planning of later flights.

[2] • Manned space flight will occur in the course of the program, but before this occurs certain crucial experiments must be performed which are aimed specifically at the design of a manned vehicle.

- In the quest for outer space we must not lose sight of the tremendous implications to life on earth which the occupation of space will have.

[3] 2.   Sounding Rockets

Sounding rockets have provided so much information about the upper atmosphere and its effects on incoming radiation of various kinds that they will continue to be useful in this area. A continuing program using such rockets should be aimed at determining the distribution in the vertical of such quantities as:

- Atmospheric composition.
- Atmospheric pressure, temperature, and density.
- Winds in the upper atmosphere.
- Atmospheric ionization.
- The absorption of electromagnetic radiation penetrating the atmosphere and the intensities of sources of such radiation in the atmospheric layers.
- The absorption of cosmic ray or solar particles, and the secondary effects of these particles.
- The geomagnetic field (also covered under satellites).
- Detection and location of electric current systems in the atmosphere.
- Experiments requiring recovery of packages (see below).

With a sufficiently intense program, it will be possible to detect latitudinal, diurnal, and seasonal changes of these quantities, and also the ways in which they are modified during periods of solar activity and magnetic storms.

Until the techniques for the recovery of packages from a satellite have been worked out in more detail and demonstrated, there will be a class of experiments requiring the return of various kinds of samples for which the vertical rocket is required. These may involve:

[4] • Film samples: Photographs, spectrographic data, cosmic ray packets, or data recordings where the quantity of information is too great to telemeter.
- Biological samples.

Experiments which will probably not be suitable for sounding rockets in the future, with the availability of earth satellites of progressively larger payloads, are solar or astrophysical observations, particularly those in which time changes are sought. Clearly, a satellite platform is superior for such observations.

[5] 3. Earth Satellites

An earth satellite is considered, for these purposes, to be a vehicle which is on an orbit controlled primarily by the earth's gravity. (This means, in effect, something less than 1,000,000 miles from the earth and with insufficient velocity to carry it further.) Even when the technology of space flight has progressed far beyond the ability to put satellites on orbit and vehicles are being directed on heliocentric and interplanetary missions, the earth satellite will surely continue to be a base for fruitful observations.

Fundamentally, a satellite well outside the earth's atmosphere can be used to observe only three kinds of things, namely: <u>photons, particles</u>, and <u>fields</u>.

The photons, since they represent electromagnetic radiation, may range from X-radiation and ultraviolet radiation to radio waves. In general, when dealing with photons coming from remote sources in the sun or beyond, the purpose of a satellite is to observe the wavelengths which do not penetrate the earth's atmosphere. This implies that the radiation of primary interest is at Wavelengths below the ozone cutoff in the ultra-violet (about .32 microns) and at wavelengths above the ionospheric cutoff in the radio wave region (about 30 m, or 10 Me). Most of the radiation inbetween these limits penetrates the atmosphere and can therefore be observed on the ground or from balloons, except for some important, but limited, regions in the infrared where water vapor, carbon dioxide, and ozone cause absorption.

In addition to observing these highly significant radiations from above, the satellite will be of great value in observing the earth, its [6] changing cloud patterns, its infrared radiation, etc. These are discussed further below.

The particles which can be observed from a satellite are solid meteoroids of various sizes and atomic nuclei with great energy emanating from the sun and beyond (auroral particles and cosmic rays). These are both of great significance to the development of manned space vehicles, since the solid particles constitute a hazard to the vehicle due to their ability to puncture its skin, and the atomic particles may be a hazard to the man inside.

The fields which are measurable from a satellite are the field of gravity and the magnetic field. The first, the field of gravity, is related to the masses and shapes of the earth and moon, and satellite observations promise to greatly improve the precision of our knowledge of these quantities. Magnetic field measurements not only tell about the magnetization of the earth and moon, but also tell about the electric current systems which flow in the vicinity of the earth.

Since a great deal has already been written about the uses of an artificial satellite, the following experiments are presented in outline rather than in detail. First are those which could be done in Vanguard-type satellites, assuming a growth potential in payload to 50 or 75 lbs and a wider choice of orbits than is available under the I.G.Y. program. With larger payloads and more advanced techniques there are some more elaborate experiments which could be done, experiments which require stable platforms, large transmission power and information bandwidth, recovery of packages, etc. Finally, there will be manned satellites.

[7] <u>Light Weight Satellite Experiments</u>
   a. <u>Creation of Visible Objects</u>
There are a number of reasons for wishing to have an easily visible satellite. In particular, precision orbit determinations will probably be done optically, and it is clearly desirable to have a satellite which reflects or emits a considerable amount of light. At night a flashing light with a brightness of 105 candlepower or more would be just visible at about 1000 mi range, provided the duration was about 0.1 sec or more. An alternative method is the creation of a large reflecting object such as a balloon or erectable corner reflector. Such an object to be seen optically or visually must be near the twilight zone of the earth, so that the observer can see the sunlit reflector against a darkened sky. Under such

conditions a 100 square foot diffuse reflector appears like a first magnitude star at about 200 mi (depending on the angle between the Sun and observer), and can still just be seen by the naked eye at about 2000 mi range. Naturally, with telescopes one can do much better, if one knows ahead of time where to look for the satellite.

With the sort of precision orbit determinations which can be obtained with optical tracking it is possible to do a number of important things, namely:

- Determination of air drag at high altitudes, from which atmospheric density can be derived.
- Geodetic measurements on the size and shape of the earth.
- Lunar mass, for orbits passing near the moon.
- Ion densities, when coupled with certain precision radio techniques.

[8] b. <u>Total Atmospheric Thermal and Solar Radiation Measurements</u>

A satellite is in an ideal position to measure the total flux of radiation in and out of the top of the atmosphere. The incoming radiation, being primarily from the Sun, is largely in the visible part of the spectrum, while the outgoing radiation from the atmosphere is infrared plus the solar radiation which is scattered and reflected upwards. These various fluxes can be sampled by a set of omnidirectional bolometers with coatings which are designed to absorb selectively a certain part of the spectrum. For example, a bolometer which is white in the visible but black in the infrared beyond about four or five microns, will respond to the thermal radiation from the earth and atmosphere, while one with the reverse spectral characteristics will measure the direct and reflected sunlight. Further, a directional detector of visible radiation pointed towards the Sun would, of course, monitor the incoming solar radiation alone. (Such a scheme is included in one of the I.G.Y. earth satellites.)

The purpose of this set of measurements is to determine the radiational heat budget of the earth and atmosphere. It is known that an excess of radiational energy is added to the atmosphere in low latitudes and that there is generally a net loss of energy from the Polar regions. An understanding of this energy imbalance is basic to an understanding of the general circulation of the atmosphere. Further, such a set of radiation measurements, provided that there were a reasonably fast response, would give a rough indication of the thermal inhomogeneity of the atmosphere and earth. It is likely that a measure of this inhomogeneity would provide an indication of the strength of the cyclonic and anticyclonic circulation. During periods of strong meridional transport of energy by the atmosphere there are rapid migrations north and south of warm and cold air masses, [9] and these could probably be distinguished by their thermal characteristics.

c. <u>Mapping the Cloud Cover</u>

On the sunlit side of the earth the contrast in the visible and near infrared between clouds and ground or open water is considerable, and it has been demonstrated dramatically by the use of rocket and balloon photography that the existing weather can be traced by the large area cloud patterns. These cloud patterns can be determined from a satellite by various means. A first approach, in which the scanning of the surface by photocells is performed by the uncontrolled rotation of the satellite, is being developed for the IGY program. In this case the reconstruction of the picture is complicated, however, and the data handling capacity of the telemetering link places an upper limit on the amount of coverage and degree of resolution.

The purpose of such an observation would be to show the cloud patterns over a large area of the earth with a degree of completeness unobtainable with present surface observation networks. For research in meteorology, this will throw new light on the way in which storm systems start and develop, on the broad pattern of flow, on the effects of mountain barriers, etc. If refined to the point where the observations can be made available to meteorologists immediately, it would represent one of the greatest advances ever made in meteorological data gathering, and would surely improve short-term forecasting-and hurricane predictions.

d. Time Fluctuations of Solar Ultraviolet and X-Radiation

Solar ultraviolet and X-ray intensities are quite variable, and appear to depend greatly on solar activity. Both X-rays and the ultra-violet are enhanced during a solar flare, in some wavelength regions by an order of [10] magnitude or more. These fluctuations have corresponding effects in the earth's atmosphere. Increased output of hard X-rays, for example, causes a pronounced D-layer and an associated interference with radio communications. An increase in the intensity of near ultraviolet solar light could contribute to the marked temperature excursions that have been noted in the ozone layer, and such temperate excursions undoubtedly interact with the surrounding wind patterns.

Since solar ultraviolet light and X-rays have such a pronounced effect on the atmosphere and since their fluctuations are associated with important related effects, it should be very fruitful to monitor these solar wavelengths over a long period of time, say for a year, for the purpose of correlating the ultraviolet and X-ray intensity-time curve with weather, radio, the ionosphere, airglow, winds, etc. Because these solar radiations are absorbed by the atmosphere, the logical place to monitor them is from above the appreciable atmosphere. This could be done in an artificial satellite orbiting entirely above 200 miles altitude. By using suitable windows and gas fillings, photon counters and ionization chambers can be constructed to respond only to radiation within a restricted band. (Such a photon counter, sensitive to Lyman-alpha radiation, is being flown on an early I.G.Y. satellite.) With such detectors, various bands from the near ultraviolet down to the hard X-rays could be monitored. Payloads on the order of fifty pounds should be adequate to permit coverage of a number of important wavelength bands in a single installation having indefinite duration of operations.

e. Distribution of Hydrogen in Space

The hydrogen population of interplanetary and interstellar space has [11] been a subject of much interest and speculation. On the basis of astrophysical observations, the current estimate is about 1000 atoms per cc in interplanetary space and about 1 atom per cc in inter-stellar space, but the basis for this is uncertain.

The density of hydrogen in space could be determined by observing the hydrogen Lyman-alpha radiation received from space and comparing it with solar Lyman-alpha radiation. Hydrogen ions in space would emit a more or less steady background of Lyman-alpha as they captured electrons. Hydrogen atoms would fluoresce under irradiation by solar Lyman-alpha, and this fluorescence would fluctuate directly with the solar curve. By analyzing the total Lyman-alpha intensity into the steady and solar-dependent components, one could then determine the relative densities of hydrogen ions and atoms. With suitable calibration the absolute densities could be determined.

The ionization chambers to be used to study solar Lyman-alpha radiation from an IGY satellite could also be used as the detectors for the hydrogen density experiment.

f. Extragalactic Light

Among the many radiations which strike the top of the earth's atmosphere, the light frown sources beyond our own galaxy is one of the most interesting, insofar as it contributes to the profound understanding of the astrophysical nature of the universe. The intensity of this extragalactic is already known to be quite weak in comparison with the light from our own galaxy, and its spectral character is known to be heavily shifted to the red. These facts alone are subject to an immediate cosmological interpretation viz. the expanding nature of the universe.

[12] The expanding universe hypothesis can be submitted to more specific test by detailed. measurements of the spectrum of extragalactic light and by the distribution of its intensity with respect to galactic latitude.

Such observations are impossible with ground-based or balloon-borne apparatus due to the great overburden of other radiations originating in the earth's upper atmosphere. They might be thought possible with vertically fired rockets which surmount the major emitting layers of the atmosphere; but the intensity is judged to be so weak that the several minutes of a rocket flight provide an inadequate period of time for significant measurement. The long time duration of a satellite's flight appears to be necessary in order to accumulate significant data.

The proposed apparatus consists of several high sensitivity, photoelectric telescopes equipped with a variety of spectral filters – all operating in the visible region of the spectrum. This experiment seems properly classified as an exploratory one. Results are not assured, but if they are obtained they will be of very far-reaching and profound significance.

g. Cosmic Ray Observations

The objectives of a cosmic ray experiment would be: (a) to make comprehensive observations on the total intensity of the cosmic radiation as a function of latitude, longitude, altitude, and time; (b) to investigate the presence of the nuclei of lithium, beryllium, and boron in the primary cosmic ray beam, and if present, to measure their intensities; and (c) to study, as in (a), the intensity of the heavy nuclei separately from the total intensity. Interpretation of the results of (a) and of (c) should yield a crucial test of the theory of the deflection of charged cosmic ray [13] particles approaching the earth through the geomagnetic field and should yield new information on the nature and importance of interplanetary magnetic fields. The data of (b) should settle one of the leading questions on the astrophysical origin of cosmic rays and on their propagation to the earth. The data from (a) and (c) should provide a greatly improved understanding of the systematic and sporadic fluctuations of the primary radiation, their astrophysical causes and their consequences on the rate of secondary cosmic ray phenomena within the atmosphere. A special question is whether the solar sources of cosmic rays yield the same distribution of nuclear species as that of the usual primary beam.

h. Primary Auroral Particles

The polar aurorae ("northern and southern lights") are caused by the interaction of energetic charged particles with the upper atmosphere. Due to their charges, they are deflected by the earth's magnetic field and are focused on the polar regions. It has been established that the intensity of these streams of auroral particles change rapidly, apparently due to changes in the Sun.

In order to observe these particles it would be necessary to have a satellite on a high inclination orbit, since the flux is concentrated towards the poles. By means of simple satellite-borne detectors it will be possible to map out the impact zones of the primary auroral particles on the top of the earth's atmosphere and to observe their changes locally and worldwide with time to a degree not ever likely to be approached by ground observatories. It will be possible to rapidly compare the northern and southern zones of incidence and to efficiently study the ways in which the position and configuration of these zones are influenced by and correlated with geomagnetic [14] field disturbances.

The temporal variations of the incidence of auroral radiations can be comprehensively correlated with observable activity on the Sun to an extent not presently conceivable by any other method. In addition, the nature of the primary auroral radiations (e.g. protons, electrons, heavy particles, etc.) can be comprehensively studied, as can their intensities and energy spectra. A comparison of these data with those from ground observatories should be very fruitful in establishing the physical processes which are induced in the earth's atmosphere.

These auroral observations are closely related to observations of the geomagnetic field. Indeed it would be desirable, for mutual support, to have two satellites aloft simultaneously – one carrying a magnetometer and the other carrying auroral radiation detectors. Eventually it may be possible to have a single satellite carry both types of apparatus.

i. Micrometeorites

There are various estimates of the number of micrometeorites striking the earth's atmosphere, but few actual measurements. For the IGY it is planned to count such particles in one or two satellites. The limited instrumentation and limited time of operation of the equipment will, however, leave unanswered such questions as: What is the mass spectrum? What is the energy spectrum? What are the fluctuations in total intensity? How are these particles related to visible meteor showers? In a satellite capable of operating over a period of a year most of these questions could be answered using calibrated microphones, thin diaphragms with photocells to observe punctures, electrostatic analyzers, and the like.

[15] j. Magnetic Field

The earth's magnetic field is mainly due to the magnetization of the Earth itself, a property which can be quite accurately measured by ground level surveys. However, the variations in this main field of external origin, amounting to as much as 7 per cent, say, are due to a variety of current systems in the ionosphere and above. (There are current systems induced in the earth also, but these are presumably secondary effects due to the phenomena at great altitudes.) A major source of geomagnetic variations are the direct current systems in the lower part of the E-region, which are below the satellite altitudes. However, at much greater distances, perhaps an Earth radius or more, there may be another highly variable current system known as the "ring current."

With a satellite borne magnetometer flying over a monitoring magnetometer on the ground making a simultaneous measurement of the magnetic field, it is possible to determine the horizontal flow of current between the ground and the satellite. The same technique can be used with two satellite magnetometers as they pass over each other. Thus, it is possible to map the electric current systems throughout the region of the ring current.

The use of vertical rockets to do this same thing has already been mentioned. In some ways a rocket is superior to a satellite for magnetic measurements, since it can make a vertical profile from the ground up and thereby determine where the electric currents lie. However, these currents are highly variable, and a satellite permits a determination of how they vary in time, how they are related to solar activity, and how they may vary in the horizontal. The ideal approach would be to use rockets and satellites in combination, thereby obtaining a more complete map of the [16] geomagnetic field in three dimensions and in time.

k. Ionospheric Observations

The ionized layers of the ionosphere (D, E, F^1, F^2, G) generally lie between 80 and 300 or 400 km. They are therefore mostly below the level of the satellite. A number of effective methods have been suggested for measuring the total free electron density between the satellite and the ground, one being a measure of the difference between the angle of incidence of the radio tracking signal and the optical line-of-sight as the satellite passes over a tracking station. This requires no additional experimental equipment beyond the minitrack and optical networks set up for the I.G.Y. program. Another observation yielding total electron densities is the rotation of the plane of polarization of the radio wave due to the Faraday effect. Such an observation requires a high gain antenna with a dipole to sense the plane of polarization.

Another class of satellite radio experiments would make use of the satellite as a known source of radiation to measure certain aspects of the fine structure of the ionosphere. It is observed that radio stars fluctuate, and these fluctuations are in part due to ionospheric inhomogeneities of various sorts, some of which are in the F-region. A satellite would permit a mapping of these horizontal inhomogeneities (sometimes known as "ionospheric lenses"), both in the horizontal and in the vertical, since the satellite may at times be below the F-region. An especially interesting aspect of the irregularities in ionization of the upper atmosphere is the pattern of the auroral clouds, streamers, draperies, etc. These patterns are marked by visual radiation, as is well known, but they are also regions of intense local ionization. The radio signal from a satellite in the auroral zone [17] would be influenced by the auroral ionization, and presumably a study of the fluctuations would tell a great deal about the character and distribution of the ionization in this region.

It should be borne in mind that the gross structure of the ionized layers can be measured from the ground continuously with ionospheric recorders, and that the general features of the ionosphere are already quite well understood. Furthermore, it was pointed out above that the fine structure of the ionospheric layers can probably best be obtained by a rocket which penetrates rapidly through the ionosphere, recording successive changes in "radio depth" as it goes. Until such additional experiments have been made it will be difficult to design satellite experiments specifically for ionospheric studies; it is certain, however, that the experiments described above can be invaluable by-products of any satellite experiment which provides a more or less steady signal with stable frequency and known polarization.

To date, no experiment has been proposed which can measure the free electron distribution above the top of the ionosphere from a single satellite without serious difficulties due to the dominant effects of inhomogeneities in the ionosphere itself, which tend to mask any second order effects at the satellite altitude. However, the distribution of free electrons above the ionosphere would be of great significance. The use of two satellites, with a two-frequency transmission link between them, offers an apparently feasible solution.

l. Biological Experiments

Biological experiments should be instituted at the earliest opportunity in the satellite program, since they will be crucial to the eventual attainment of manned space flight. There appear to be two main areas of [18] concern: The biological effects of prolonged exposure to the radiation in space, ranging from cosmic rays to the various solar emissions; and the subtle and complicated effects of prolonged weightlessness. With regard to the first, a program of exposure of biological samples and live animals to cosmic radiation at high altitude by balloons has been underway for some time, and at the altitudes atainable [sic] by balloons (over 100,000 feet) the cosmic radiation is essentially the same as at satellite altitudes. There are other kinds of radiation, such as solar ultra-violet and X-rays, which do not penetrate to balloon altitudes, but these can be reproduced conveniently in a laboratory. Thus the use of a satellite for the study of radiation effects on biological specimens does not appear to be too rewarding. However, for the study of prolonged weightlessness there is no known substitute for a vehicle floating freely in space. Biological specimens and live animals have been successfully flown and recovered from high-altitude rockets, having been exposed to a few minutes of weightlessness. The second Soviet satellite carried a dog, thereby lengthening the duration of the period of weightlessness ad mortuum. The U.S.I.G.Y. satellite program includes a biological sample (yeast). These first attempts to study weightlessness will have to be greatly expended in the future.

Advanced Satellite Experiments
a. Selective and Directional Thermal Radiation Measurements

Since certain constituents of the atmosphere, such as water vapor, ozone, and carbon dioxide, have strong absorption lines in the infrared region of the spectrum, a detector looking downward which is sensitive only in these regions does not "see" the earth's surface. Instead, it detects the radiation emitted upward from the upper levels of the constituent, the [19] radiation from the layers below having been absorbed by the atmosphere. Thus, for example, a detector looking down at around 9.6 microns (in a strong ozone band) would receive the thermal emission from the top of the ozone region at about 10 to 30 km altitude; a detector looking down at around 6 microns (in a strong water vapor band) would receive the emission from the top of the troposphere at 8 to 10 km, above which there is relatively little water vapor. A quantitative measurement of the thermal radiation in one of these narrow spectral intervals gives a measure of the temperature (and, to a second order, density) of the emitting layer. A more detailed analysis of the variation of this emission with zenith angle can give the vertical distribution of temperature in the emitting layer. This experiment would require a considerable degree of orientation control, particularly the measure of the "limb darkening" just described. To be most meaningful, the record for an entire satellite circuit should be complete, probably requiring data storage and retransmission over a telemetering station.

The purpose of such a set of measurements would be to map the effective temperature of various layers high in the atmosphere. Some of these layers are inaccessible to conventional sounding balloons, and even those which are accessible can only be sampled at a few points. As meteorologists have obtained progressively more information about the synoptic conditions in the upper atmosphere (using balloons and occasional rockets to date), they have gained more insight into the behavior of the atmosphere, and weather forecasting ability has gradually improved. However, balloons cannot penetrate the part of the atmosphere which is affected by solar ultraviolet radiation below about 0.3μ (the ozone cutoff). It seems [20] reasonably certain now that short-term changes in solar radiation have an immediate effect on parts of the upper atmosphere, and that these effects propagate slowly downward in a complicated and as yet unexplained way. A synoptic satellite observation of the kind described would probably provide a direct measurement of the immediate effects of a solar disturbance on the thermal structure of the atmosphere. It would, therefore, be a key to the development of a physical basis for long-range weather prediction.

b. Astronomical Spectrograms

A spectrograph mounted in an artificial satellite would be able to photograph the Sun, planets, and stars completely free from interference by the atmosphere, thus extending the sensitivity far into the ultraviolet end of the spectrum and permitting a much more detailed study of these bodies than is now possible.

Spectrographs to do this job are in essence available. Suitable light collectors would have to be designed. A pointing control would be necessary. Such a control could probably be worked out much along the lines of those now used in rockets, and would have a total weight less than thirty pounds. To retrieve the film, it would be necessary to work out techniques for recovery of a capsule from the satellite orbit (or of the satellite itself); however, such techniques have already been proposed and are considered to be feasible within the expected weight limitations. A less desirable alternative would be to analyze the data photometrically and transmit it back via the radiotelemetry link.

c. Ultraviolet Photographs of the Sun

Much of the photochemical and dynamical activity in the Sun is associated with the emission of ultraviolet radiation. Photographs of the [21] Sun in various regions of the ultraviolet should permit localization of regions associated with the respective wavelength emissions, and would be an important aid to understanding solar activity.

Suitable filters and U.V. sensitized films are available for making such photographs. If necessary, pointing controls similar to those already used in rockets could be constructed for directing a camera at the Sun. It would probably be desirable (but not necessarily essential) to recover the film after the pictures had been taken; however, as indicated in experiment b, it is believed that suitable techniques could be developed.

d. Planetary Spectrograms

A variation of experiment b. would measure the spectra of the various planets in the ultraviolet and infrared. All of the central. planets have visible atmospheres, but the composition of these atmospheres is difficult to observe spectrographically from the ground due to the presence of the, same or similar gases (in differing proportions) in our own atmosphere. For example, the solar ultraviolet radiation reflected from these planets is completely absorbed by our atmospheric ozone, and large segments of the infrared radiation which is emitted are absorbed by water vapor, carbon dioxide, and ozone, plus other trace constituents such as methane, nitric oxide, etc. A satellite would have a clear view of these planets.

The radiation from than is very weak, however, and would require quite accurate positioning of the spectrograph in order to provide long exposures with limited angular fields (in order to minimize the cosmic and stellar background). Moreover, it would probably be most desirable to recover the spectra in the form of exposed plates, though it is possible to telemeter the information to the ground

[22] e. An Experimental Test of the General Theory of Relativity

One of the predictions of the general theory of relativity is that the fundamental time scale of atomic phenomena (e.g. frequency of emitted spectral lines) is influenced by the gravitational potential in which the emitting system is located. This prediction has received thus far only a very few observational verifications and even these remain in a somewhat controversial state. It is conceivable to mount a so-called caesium or thalium "clock" in a satellite and a similar one at a ground station and intercompare the rates of these two clocks over an extended period of time. By the general relativistic theory, it is expected that there would be a systematic difference in the rate of running of these two "atomic clocks" due to the known difference of gravitational potential to which they are subjected.

The effect is a small one and it appears that accumulated observation over a period of the order of a month may be required to surmount reasonable experimental errors in location of the position of the satellite and in ionospheric conditions. (Both effects, of course, influence the transit time of the transmitted intercomparison signal from the satellite to the ground station.)

A proposal is known to be currently under consideration for a similar intercomparison between clocks, one of which is located on a mountain and the other in a neighboring valley. However, if the technical problems can be adequately solved, it may be desirable to utilize a satellite for a more sensitive test of this very profound theoretical hypothesis under different conditions.

f. Solar (Cosmic) Radio Noise in the HF and LF Spectrum

High frequency radio waves below about 5 Mc cannot penetrate the [23] ionosphere, and even radio waves at 20 Mc are sometimes totally absorbed. Thus, it is not possible to observe from the ground the lower frequency end of the radio noise which comes from the Sun and beyond.

A satellite would, of course, not suffer from ionospheric absorption, but the signal levels in this region are low and the antennas required to obtain much gain have to be large. However, by using large erectable reflectors or lenses to concentrate the signals and to obtain directionality measurements could be made.

g. Collection of Micrometeoritic Sales

If techniques can be worked out for recovery of the satellite or of small capsules from the satellite, a long-period collection of micro-meteorite particles could be obtained. These samples could be collected in containers filled with something like silicone grease, which could be opened while the satellite is on orbit and then closed just before the recovery operation was begun.

****

[38] 6. Manned Space Flight

Although it is impossible to predict how quickly man himself will follow his exploring instruments into outer space, the inevitable culmination of his efforts will be manned space flight and his landing on the nearer planets. It is clear that he can develop the ability to do this, and it is hard to conceive of mankind stopping short when such a tempting goal is within reach.

The attainment of manned space flight, however, cannot now be very clearly justified on purely rational grounds. It is possible, at least in principle, to design equipment which will do all the sensing needed to explore space and the planets. Mobile vehicles could be designed to land and crawl across the face of each of these distant worlds, measuring, touching, looking, listening, and reporting back to earth all the impressions gained. They could be remotely controlled, and so could act like hands, eyes, and ears for the operator on earth. Moreover, such robots could be abandoned without a qualm when they ran out of fuel or broke down.

Though all this could be done in <u>principle</u>, there may be a point at which the complexity of the machine to do the job becomes intolerable, and a man is found to be more efficient, more reliable, and above all more resourceful when unexpected obstacles arise. It is, in a sense, an article of faith that man will indeed be required to do the job of cosmic exploration personally– and, furthermore, that he will <u>want</u> to do the job himself, whether required to or not.

With man's first venture into outer space a new program of research and exploration will begin. The program described above will therefore be the dramatic prelude to the even more dramatic conquest to follow.

Document I-2

Document Title: "Report of the Working Group on Space Research Objectives, Special Committee on Space Technology," 14 November 1958.

Source: NASA Historical Reference Collection, History Office, NASA Headquarters, Washington, D.C.

The National Advisory Committee on Aeronautics created a Special Committee on Space Technology in the aftermath of Sputnik, and that committee continued its work even as NACA became the core of the new space agency, NASA, which came into existence on 1 October 1958. The committee expanded on Kellogg's report and other suggestions for initial steps in space science; its thinking formed an important basis for NASA's initial space science program.

[Stamped: Official Use Only]
[no page number]

Report of the
WORKING GROUP ON SPACE RESEARCH OBJECTIVES
Special Committee on Space Technology

The Working Group's recommendations to the Special Committee on Space Technology with regard to the scientific objects of a National Civil Space Research Program are concerned with the following categories of space flight:

I. Vertical Atmospheric Probes
II. Earth Satellites
III. Lunar and Solar System Probes
IV. Manned Space Flight and Exploration
and are summarized in the following paragraphs.

I. Vertical Atmospheric Probes

There are apparently no firm plans for continuing the experiments on the upper atmosphere beyond the expiration of the International Geophysical Year program. It is the recommendation of the Working Group that the IGY effort in probing the earth's atmosphere by vertically fired rockets and by balloon flights be extended beyond the IGY period at substantially the present level of activity. The Working Group further recommends that immediate attention be given to maintaining and operating the IGY research facilities at Ft. Churchill, Manitoba, and to providing rocket launching facilities in the Antarctic and from suitable islands.

II. Earth Satellites*

A program of scientific research with earth satellites is naturally tied closely to the practical matter of vehicle and propulsion system capabilities. From consideration of the possibilities it appears convenient to classify the recommended experimental objectives and research equipment in terms of compatibility with satellite payloads in three order-of-magnitude categories, 30 pounds, 300 pounds, and 3,000 pounds, as follows:
[2]
A. 30-pound (or smaller) satellite payloads

1. Non-directional monitoring of radio-frequency radiations. Of particular interest are those wavelengths absorbed or reflected by the atmosphere.
2. Observations in the ultra-violet region of the spectrum – wavelengths 1,000 to 3,000 angsgroms [sic] – in which the astrophysically important lines lie. This region of the spectrum is inaccessible to observation below an altitude of about 100 kilometers due to atmospheric absorption.
3. Exploratory ultra-violet observations in the far ultra-violet region of the spectrum – wavelengths less than 1,000 angstroms.
4. Studies of auroral radiations and of the interplanetary plasma.
5. Cosmic ray exploration.
6. X-ray exploration.
7. Gamma ray exploration.
8. Extra galactic radiation.
9. Observations on meteors, particularly meteor showers.
10. Magnetic field measurements (scalar magnitude).

---

*"Research in Outer Space - The Basic Objectives of a Continuing Program of Satellite Research". U.S. National Committee for the International Geophysical year, SCIENCE, Vol. 127, No. 3302, 11 April 1958, presents more complete discussions pertinent to most of the investigations suggested in this section.

11. Measurements of radiation energy balance of the earth.
12. Observations of cloud cover.
13. Measurements of atmospheric density.
14. Measurements of refraction of radio waves by the ionosphere.
15. Experiments with powered communications repeaters (10 kc/sec band width).
16. Biological experiments.

It is the opinion of the Working Group that a firing rate of one satellite per month in the 30-pound-payload class would represent a proper level of national effort for the above-listed [3] research program, at least until such time as 300-pound satellites become available. It is the further recommendation that the number of separate experiments per satellite flight be kept to a minimum in the interest of simplicity and reliability. It follows from this recommendation that it may be wise to continue the use of small satellites even after larger ones become available.

B. 300-pound satellite payloads

1. Ultra-violet stellar spectrometer (star tracking required).
2. Directional and selective cosmic ray experiments.
3. Solar ultra-violet telescope.
4. Mass spectrometry of low-energy particles.
5. Vector magnetic field measurements.
6. Meteorological reconnaissance satellite for observations and facsimile transmission of cloud patterns, etc.
7. Passive communication sphere (100-foot diameter reflective balloon).
8. Relativistic red-shift experiment (precision required is 1 part in 10^{11}, which may be possible with atomic clocks).
9. Biological experiments with small animals.
10. Navigational radio beacon.

The Working Group suggests that a firing rate of satellites in the 300-pound class of approximately one every two months, beginning in 1959, would provide the proper level of support for the program outlined.

C. 3,000-pound satellite payloads

1. Large solar telescope (of the order of 36-inch diameter).
2. Large stellar telescope (36-inch diameter).
3. Long-wave (λ >10 meters) radio telescope or interferometer (telescope diameter in the order of 300 feet or larger).
[4] 4. Gamma ray telescope and spectrometer.
5. X-ray telescope and spectrometer.
6. Manned vehicles – prelude to space flight.
7. Meteorological satellites for routine, continuing operations.
8. Broad-band communication repeater (6 mc/sec band width).

A firing rate of satellites in the 3,000-pound class of approximately one every four months, beginning in 1961, is considered to represent a proper level of effort in support of the research program outlined.

III. Lunar and Solar System Probes

1. Radio telescopic tracking beacons for vehicles on and in the vicinity of the moon.
2. Seismic measurements on the moon's surface by means of explosions and suitable detectors.
3. Magnetometer or plasma measurements for lunar magnetic field studies.
4. Measurement of density and composition of lunar atmosphere.
5. Study of interplanetary plasmas and interplanetary magnetic fields.

Experiments of, the kind listed above appear to be possible with the propulsion systems capable of putting large satellites in orbit. For example, it is estimated that a vehicle capable of launching a 3,000-pound earth satellite can be used alternatively to accomplish the following: Send a payload of. about 700 pounds to a hard landing on the moon; send a 500-pound payload to the vicinity of Mars or Venus; send a 100-pound payload to a soft landing on the moon. From a longer range viewpoint, the following broad exploratory objectives with respect to the moon and with respect to other planets may be listed:

6. Planetary and-satellite surfaces.
7. Planetography.
8. Atmospheric circulation – clouds.
[5] 9. Chemistry – rocks, erosion.
10. Magnetic fields
11. Ionosphere.
12. Biology, ecology, paleontology.

IV. Manned Space Flight

The Working Group endorses manned space flight and exploration as proper objectives of a national program of space research.

SUPPLEMENTARY. RECOMMENDATIONS AND REMARKS

A. Component development.

In connection with the implementation of the space research objectives, the Working Group calls attention to the vital need for reliable, long-lived, and efficient instrumentation system components, and strongly recommends an early beginning of a sustained program for their development. An expenditure of approximately 10 percent of available space research funds for component development is suggested as a reasonable level of support.

The following list compiles some of the important items in need of early development effort:

Equipment, Components, and Techniques:

(With emphasis on low weight, high reliability, operation over wide temperature ranges, mechanical ruggedness, and long life).

1. Electrical power supplies for on-board instrumentation.

    a. Storage batteries with increased reliability and cycling efficiency with specific outputs of greater than 40 watt-hours per pound.

    b. Photovoltaic cells with high efficiency and low weight.

    c. Solar heat engines.

    d. Power supplies using radioactive isotopes.

    e. Nuclear reactors in the range from 5 to 10 kilowatts.

[6] 2. Communications and electronic components.

    a. Long-lived and reliable microwave tube with low power consumption (minimum life 10 hours).

    b. Information storage devices.

    c. Photoelectric image tubes with long integration time (of order 1 hour) and long life.

    d. All passive components, which include resistors, capacitors, insulation, etc.

    e. Vacuum tubes.

    f. Transistors, diodes and other semiconductor devices.

3. Equipment for control of orientation of apparatus and entire vehicles.

4. Devices for control of temperature of on-board apparatus.

B. Operational Control, Data Reception and Processing.

The Working Group recommends that there should be an adequate worldwide network of tracking and data receiving stations. In addition, it is proposed that there should be established a central laboratory or station for direction and coordination of flights, reception and processing of telemetered data, computation of orbits, communications, and the like.

C. Launching Sites and Nature of Orbits Required.

The Working Group calls attention to the vital importance of high inclination orbits for the proper execution of many of the programs of research observations listed above. For other of the programs equatorial orbits are more desirable. Hence the establishment of new launching sites suitable for these purposes is recommended as an early undertaking.

A perigee altitude of approximately 250 miles will likely be adequate for most of the experimental satellite programs for the next several years. For certain experiments (e.g., magnetic field and auroral observations) orbits of high eccentricity are desired. For others (e.g., communications relays) low eccentricity orbits are preferable.

[7] D. Use of Obsolescent Military Vehicles.

In many cases large economies can be effected by using obsolescent military vehicles or combinations of such vehicles. It is recommended that close liaison be maintained with the military establishment in this connection.

E. Recovery.

Physical recovery of payloads is of special importance in biological and medical experiments and as a preparation for manned space flight. It will also be of value in certain physical experiments; though it is not as essential there

F. Grants and Contracts.

The Working Group recommends a greatly expanded program of grants and contracts for the accomplishment of many of the objectives listed herein.

G. Biological and Medical Research.

The Working Group on Space Research Objectives has given relatively cursory consideration to biological and medical research in view of two facts:
(a) The Special Committee on Space Technology has another Working Group on Human Factors and Training.
(b) The National Academy of Sciences, the National Science Foundation, and the American Institute of Biological Sciences will hold a four-day symposium (14-17 May 1958) on these specific topics.*

H. Classification and Security.

It is recommended that the national space research program be conducted on an unclassified level insofar as feasible without risk to the national security in order that the scientific talent of the country may participate to the fullest extent and in order that fruitful international cooperation may be developed.

Document I-3

Document Title: NASA Discussion Group on Orbiting Solar Observatory Project, "Minutes of Meeting," 23 May 1959.

Source: NASA Historical Reference Collection, History Office, NASA Headquarters, Washington, D.C.

---

*The Proceedings of the Symposium on the Possible Uses of Earth Satellites in Life Sciences Experiments will be published by the University of Michigan Press, probably in mid-1959.

In 1958, John Lindsay, along with several of his colleagues, moved from the Naval Research Laboratory to the new NASA to form the core of NASA's initial space science staff. In May 1959, Lindsay assembled a "discussion group" of those who were interested in solar research to discuss an initial NASA project in the area. Out of these discussions emerged what became known as the Orbiting Solar Observatory project.

[1]

8 June 1959

<u>MINUTES OF MEETING</u>
NASA Discussion Group
on
Orbiting Solar Observatory Project
NASA Headquarters
Washington, D. C.
23 May 1959

<u>SUMMARY OF GROUP ACTION</u>

1. The initial meeting of a group interested in the NASA Orbiting Solar Observatory Project was held at 9:00 a.m. in the 9th Floor Conference Room of the NASA Technical Building, Washington 25 [sic], D. C.

2. There was no action that required the attention of the Administrator.

3. The purpose of the meeting was to discuss the technical, scientific, and engineering possibilities of placing a solar observatory in orbit.

4. The first Working Group meeting will be arranged by the Chairman; no date nor location was suggested.

[2] In attendance were:

John C. Lindsay, NASA, Office of Space Sciences - Chairman
Edward T. Byram, Naval Research Laboratory
Talbot A. Chubb, Naval Research Laboratory
Robert M. Crane, NASA, Ames Research Center
Robert Davis, Smithsonian Astrophysical Observatory
Richard B. Dunn (for Walter O. Roberts), High Altitude Observatory
Laurence Dunkleman, NASA, Office of Space Sciences
Warren Gillespie, NASA, Langley Research Center
Harry Goett, NASA, Ames Research Center
Leo Goldberg, University of Michigan
Fred T. Haddock, University of Michigan
Robert Jones, NASA, Ames Research Center

James E. Kupperian, NASA, Goddard Space Flight Center
William Liner, University of Michigan
James Milligan, NASA, Goddard Space Flight Center
Roger C. Moore, NASA, Office of Space Sciences
Joseph J. Nemecek, Naval Research Laboratory
William A. Rense, University of Colorado
Nancy G. Roman, NASA, Office of Space Sciences
Gerhard F. Schilling, NASA, Office of Space Sciences
Morton J. Stoller, NASA, Office of Space Sciences
William C. Triplett, NASA, Ames Research Center
William A. White, NASA, Goddard Space Flight Center

After a short introduction by G. F. Schilling in which the purpose of the meeting – to discuss the technical, scientific and engineering possibilities of placing a solar observatory in orbit – was stated, the meeting was turned over to J. C. Lindsay.

Lindsay opened the discussion by presenting some of the background material and requirements for the project. An instrument will be used to point a satellite at the Sun. Three satellites are being considered. The launching vehicle in each case is expected to be the Thor-Delta which has a total payload capacity of the order of 300 pounds. The pointing control system with a desired pointing accuracy of 1° of arc will approximate about half the payload weight, leaving about 150 pounds for the experiment and its power supply. A circular orbit of 300-500 miles is contemplated and the satellite should have a minimum lifetime of one month. The [3] package, probably spherical, should withstand a temperature range of 15°-115° F and vibration testing of 20g.

Two types of pointing control instruments are being considered. One system essentially is a gyroscope spinning 2 rps with the spin axis perpendicular to the radius vector to the Sun. Attitude is accomplished by using the satellite's precession motion. Equipment can be mounted on the pointing control platform and in the rim of the spinning fly wheel. The second pointing control system removes the initial angular momentum by transferring it to a set of throw-away fly wheels. This system is more complex than the above scheme which is basically a one-step increase over rocket controls now in use.

Many questions were raised regarding optical packaging and equipment arrangements within the satellite. It was pointed out that since a fixed time schedule and known vehicle were involved elaborate ideas should be kept to a minimum. Since no final decisions have been made, however, the only set requirements are those originally stated regarding weight, etc.

The expected operational lifetime of one month was questioned and raised some concern. The lifetime of the first scheme depends on the helium needed to run the jets. Of course, a larger supply of helium at the expense of equipment for experiments would result in longer life. The lifetime of the second scheme, also one. month, depends on the batteries to run the stabilization fly wheels. Solar batteries would increase the expected lifetime, but the system-would be less economical and require a longer time scale for development. No firm answer was given regarding a desired lifetime for the observatory although it was-mentioned that six months to a year would be desirable. On the other hand, it was suggested that a six-month life was important if individual solar regions could be studied. The group agreed that the lifetime problem should be investigated very thoroughly.

Possible Experiments

Robert Davis – Smithsonian – was interested in a tag-along experiment on the solar camera pointing at something other than the Sun or in a slit spectrograph pointed at Sirius. This experiment, it was felt, was more suited to a rocket or X-15 airplane. Further discussion raised the possibility of need for an interim vehicle to point at stars.

[4] Leo Goldberg – University of Michigan – was interested primarily in the large orbiting astronomical observatory and would like to use this vehicle as a preliminary experiment. They Would install one of their three spectrographs, perhaps in the 600-1500 Å range. The scheme requires 3-axis stabilization but might be workable with 2-axis. Details of the experiment would depend on the answers to the questions raised regarding equipment arrangements and expected lifetime. Goldberg remarked he would like to know where the instrument is pointing exactly and suggested scanning the Sun in a fixed manner, say along a solar diameter with repeated operations after moving a small amount. He also requested a copy of the artist's conception of the pointer and mentioned that he would like to fly radio equipment in the same vehicle.

Fred T. Haddock – University of Michigan – Radio equipment would not be appropriate without optical equipment. Equipment required would be a sweep frequency receiver with its beam directed toward the Sun, a magnetic recorder, and 30 pounds of batteries, a total of 50 spinning pounds. The receiver would be in the 5-30 Me range and would require 1-2 kilowatt hours for steady operation. A short antenna, either a 3' diameter loop or a 30' whip erected after launching, would be needed. Ground station operations also were necessary. Haddock suggested that the listed weights could be lowered, which lead Lindsay to read typical STL (Space Technology Laboratories) package weights. The group requested copies of this published information.

William Liller – University of Michigan – would like to scan the Sun in the light of one line. This raised the scanning issue. The group preferred to wait on their answer, but indicated that they would require probably only a slow scan of 1' of arc in 1m of time.

William A. Reuse – University of Colorado – Using an Eschelle spectrograph with high resolving power, Colorado wants to monitor the Lyman Alpha profile as a function of time for neutral cloud detection, a portion of the Lyman continuum in the 890-910 Å range to determine cloud organization, and the 303 Å radiation. No total weight estimates were given although the optical parts would be about 45 pounds.

William A. Rense for High Altitude Observatory – With a coronograph mounted in, the satellite, HAO would hope to get spectra of the corona in the near or possibly even the far ultraviolet. [5] The project, including pointing control, would cost approximately $80,000.

Richard B. Dune (for Walter O. Roberts) – High Altitude Observatory – Using the 313-1216 Å range the Lyman continuum would be monitored continuously for meteorological effects.

Talbot A. Chubb – Naval Research Laboratory – was interested in a general purpose instrument with a long life. Chubb was in favor of placing all pointing controls in the wheel of the satellite, leaving the stabilized platform free for experiment. He would mount a telescope or spectroscope on the platform with separate azimuth and elevation drives in the wheel. Separate power supplies for the wheel and platform also were desirable. Since he basically was interested in making measurements from one place on earth the satellite's orbit would have to be a very long one. This experiment requires an appreciable angular momentum although Chubb felt it might be done without.

<u>General Remarks</u>

Lindsay asked the group if they felt the smaller, less expensive solar observatory would be useful even after the large astronomical observatory had become available. There was general agreement from the group and specific agreement from Goldberg and Chubb. It seems desirable, therefore, to consider a scheme for orienting the optics of the instrument after the satellite has been launched.

Lindsay also wondered to what extent the group was interested in correlated ground based programs. There was no clear agreement on this question although people seemed reluctant to commit their institutions to full-time space work.

Roman wondered what people would be willing to sacrifice experimentally to gain extended satellite life and 3-axis stabilization. Goldberg suggested a table of weights vs. gain to answer this question.

Goldberg was interested in a summary of the pointing control schemes available and the group wanted the specifications of the pointing control finally selected for the experiment.

Schilling suggested that interested parties in the group send advance proposals containing budgets for the next 6 to 12 months and additional requirements to Lindsay. NASA connected groups should submit memos. Formal proposals should be sent to NASA.

The first Working Group meeting for this project will be arranged by Lindsay.

The meeting adjourned at 1:00 p.m., 23 May 1959.

Document I-4

Document Title: Letter from Walter Orr Roberts, President, University Corporation for Atmospheric Research, to Dr. Leo Goldberg, Harvard College Observatory, 26 June 1961.

Source: Archives, National Academy of Sciences, Washington, D.C.

Once the Orbiting Solar Observatory project was well under way, NASA began planning for a follow-on project. An initial meeting to define the scientific objectives of the project was held at the new National Center for Atmospheric Research (NCAR) in Boulder, Colorado, which was operated by the University Corporation for Atmospheric Research (UCAR). In this letter, UCAR President Walter Orr Roberts communicates his thinking on research objectives to Dr. Leo Goldberg, a Harvard astronomer with a particular interest in solar research. Dr. Goldberg was chair of one of the committees of the Space Science Board working to define space science efforts for the 1960s and beyond. The "HAO" referred to in Roberts' letter is the High Altitude Observatory, also operated by UCAR.

[no pagination]

UNIVERSITY CORPORATION FOR ATMOSPHERIC RESEARCH

26 June 1961
Office of the Director
Boulder, Colorado

Dr. Leo Goldberg
Chairman, SSB Committee 2
Harvard College Observatory
Cambridge 38, Massachusetts

Dear Leo:

You no doubt have seen Ned Dyer's memo of 2 June to members of SSB Committee 2, asking for projects and proposals for space research within the framework of the COSPAR recommendations of last April. Many of the topics listed in the recommendations fall outside the scope of NCAR and HAO. However, I so want to comment on several items in which we are involved to some degree. The reference numbers are those of the list of resolutions in the memo.

12) Solar Spectrum.

We have no specific plans for taking spectra of the solar disk or corona from beyond the atmosphere of earth. We have had some interest in a far uv coronagraph — but have not moved beyond very general thinking and are clearly not at a proposal stage yet. I feel, however, that it is most important to maintain reliable and essentially continuous photometric standardization between surface and space measurements of spectra, especially of the corona. New data taken in space will lose much of their meaning and usefulness if they cannot be related reliably to the accumulation of solar spectral data from surface observatories. Because of this consideration, it is disquieting to note that the efforts to maintain continuous and high quality standardized coronal observatories at various stations have been deteriorating in recent years.

HAO and Sac Peak continue to calibrate all coronal spectra with reference to the solar disk as the photometric standard. Zirin and Firor are currently developing an application of the photoelectric image tube to the Climax spectrograph, based on some preliminary work last year. We expect that this technique will result in standardization of improved accuracy, as well as a more detailed analysis of the coronal emissions. However, coverage is not all it should be. And the ground back-up for evaluating space spectra may be inadequate unless this aspect of a solar activity service is not pushed.

We are currently renovating the HAO 5-inch coronagraph system at Climax, improving the optical system and making some changes for greater flexibility of operation. We are also in the midst of construction of a 16-inch coronagraph and spectrograph system, which we expect to have in operation sometime next year. This large system will be used however, primarily for detailed investigation of limited features and processes, rather than for general patrol observations.

13) Solar Constant; 14) Spectroheliograms.

No program planned here, but I would like to emphasize the comment under (12), on the importance of adequate standards to provide ground-based back-up for space photometry.

15) Solar Coronagraph.

Note comments under (12). In addition, HAO is actively interested in observation of the K-corona, having developed and operated, during IGY, an instrument for mapping the K-corona by polarization measurements at the surface. This instrument will continue to be available at Climax. We are also interested, tentatively, in adapting the K-coronameter for balloon or satellite use. Space observatories appear indispensable for continuity in observations of the corona; but I believe that the balloon may offer advantages in accuracy for selected supplementary observations, and may permit observations out to substantial coronal heights.

16) Interplanetary Medium.

Again, I believe that accurate correlation of surface and space observations is of highest importance, and need supplementation. HAO's part, so far, has been to operate a radio interferometer in the 8-50 Mc/s range, under Warwick's direction, principally on the Sun, radio stars, and Jupiter. This instrument is capable of localizing radio noise sources on the Sun, and thus of developing correlations between radio emissions and flares or other optical events. Such ground-based radio and optical observations of eruptive solar events can be of great value in coordination with space measurements in the interplanetary medium. Solar events can be expected to produce variations in the density, speed, and temperature of the interplanetary gas, in the magnetic field, and possibly other parameters of the medium. Again, the significance of such measurements will be greatly enhanced if they can be accurately related to surface observations.

19) Solar Events and Cosmic Rays.

Coordination of cosmic-ray and solar observations, particularly during the IGY, has greatly enriched our understanding of these phenomena. We feel that IQSY requires improvement, not relaxation, of this coordinated effort. In particular, the flare-patrol network needs to be improved particularly as to resolutions (both space and time), to record small flares and other minute or faint disk features that are of relatively minor importance during solar maximum phase. Intensified observations of the K-corona, zodiacal light from space, and magnetic observations of the disk are especially important in connection with the problem of M-region disturbances.

In addition to the solar observing activities already mentioned, HAO will continue to assist the University of Chicago and the University of Maryland by providing a location and routine maintenance at Climax for two cosmic-rays recording systems. These instruments already have produced valuable evidence of relationships between solar activity and cosmic-ray inputs to the earth.

22) Meteorological Rocket Network.

NCAR is actively interested in the development of meteorological rocket observations. Because of the early stage of development of our organization, the extent and direction of our interest is not yet clearly formulated. However, NCAR is sponsoring a conference next month of representatives of the participants in the meteorological rocket network in this country. The purpose of this meeting is primarily to evaluate alternative measurement techniques for temperature and density above 100,000 feet from the standpoint of accuracy and reliability.

26) Formation of Panel on Synoptic Rocket Soundings.

Again, because of the early stage of NCAR's development, we can only express interest in this area in a general way. While our research interests obviously may develop in various directions, we are particularly interested in experiments and observations that will help clarify the dynamics of the mesosphere. Consequently, we heartily approve the formation of the Panel, and will be interested in considering possible means of cooperation with it as our activities develop.

These are my feelings—developed rather hastily—as to the ways in which definite or tentative programs and activities at HAO and NCAR can relate to the space objectives outlined in the COSPAR resolutions. Undoubtedly modifications in these prospects will develop; I hope some of these can be in the direction of still more effective participation in a closely coordinated surface-space program of investigations.

With best regards.

Cordially,

[signature]
Walter Orr Roberts

Document I-5

Document Title: Space Science Board, National Academy of Sciences—National Research Council, A Review of Space Research, 1962.

Source: NASA Historical Reference Collection, History Office, NASA Headquarters, Washington, D.C.

The 1962 summer study on space research, organized by the Space Science Board at the University of Iowa, was a seminal event in defining future U.S. space science efforts. Participants in the study included those from all disciplines who were interested in space science and NASA staff. This is the excerpt from the report of the summer study which deals with solar research.

[2-1]

Chapter Two
ASTRONOMY*

I. Introduction

Since 1947, when the first ultraviolet spectrograms of the Sun were obtained from above the atmosphere with sounding rockets, astronomy has been on the threshold of a long-awaited era. With the launching of NRL's Solar Radiation Satellite I and, now, NASA's stabilized Orbiting Solar Observatory I, the narrow bounds which the terrestrial atmosphere has always imposed on the exploration of the full astronomical spectrum are breached still further. The spectrum opens out, not for just the few precious seconds when a sounding rocket or the X-15 climbs the apex of its quick trajectory, but for weeks and even months at a time.

The Orbiting Solar Observatory (OSO), with its complement of various detectors, marks the first in a carefully planned series of orbiting observatories, some to observe the Sun and some the stars. Successive members of the series will grow in power and versatility; the detailed designs are well advanced for the next three or four. Launchings of all spacecraft planned at present will occur within the next four years, if the existing schedule can be maintained. Some tentative plans are already being made for the orbiting observatories that will first go into service in 1967 and later years.

The successful flight of the first OSO is therefore highly portentous for all of astronomy. While it is still in orbit, and its successors are still abuilding, the time is ripe for an independent reassessment of the whole astronomy program. This has been one objective of the Space Science Summer Study. We have tried to steer a middle course between a study that is so broad that its conclusions find no applications in the present conduct of the NASA program, and one that is so detailed that it attempts to judge the scientific utility and/or engineering feasibility of every specific experimental proposal.

---

* See Appendix III for list of participants in the Working Group on Astronomy [appendix omitted].

FINDING: In broad outline we endorse the present NASA astronomy program.

[II-5]   III. Orbiting Solar Observatories

A. General Considerations

There is little disagreement among solar astronomers on the broad and important questions now awaiting solution. These questions can be listed briefly.
   (i) Evolution: What is the origin and course of evolution of the Sun and similar stars?
   (ii) Internal Structure: What are the details of the processes by which energy progresses outward from the center of the Sun? What is the composition and physical state of the Sun at all levels? What is the origin of the solar magnetic field and the solar cycle?

Answers to the foregoing questions lean heavily on theoretical studies and laboratory work on properties of atoms. The following questions are more directly related to present solar observations:
   (iii) Photosphere: What are the physical conditions in the photosphere? What is the spectrum of the turbulence observed there?
   (iv) Chromosphere: What is the structure of the chromosphere? In particular, how can one account for the increase of temperature with height which continues into the corona?

[2-6]
   (v) Solar Activity: What is the origin and energy supply of the many sporadic phenomena observed on the Sun — sunspots, flares, radio bursts, etc.?
   (vi) Corona: What is the form of the outer corona? How does it connect with the interplanetary medium?

Those scientists studying the Sun at the present time feel the need for improvements in observational capabilities in at least two respects that are uniquely attainable by space facilities. These are angular resolution and wavelength range.

   1. Angular Resolution. A number of problems of interpreting phenomena on the Sun depend at present on improved knowledge of the fine structure of the solar atmosphere. This is true for problems of photospheric turbulence, chromospheric models, the physics of active solar regions, and the heating of the chromosphere and corona. Most of the world's solar astronomers are now engaged in studying one or more of these problems. Solutions to these problems depend strongly on better angular resolution.
   There are some opportunities for improvement in the resolution in observations made from the ground. With full exploitation of such techniques as the utilization of good observing sites, the suppression of turbulent air currents in and around the telescope dome structure (sometimes done by discarding the dome altogether), shutter control by photoelectric seeing monitors, very short exposures, and possibly other techniques, the

number of photographs of solar features with resolution of 1 second of arc or better will be greatly increased.

In the special case of the height gradients in the solar chromosphere, a ground-based technique is available which allows even better effective resolution. During a typical solar eclipse, observations of the chromosphere closely spaced in time during totality record the emission from the narrow region of the chromosphere covered or uncovered by the Moon's motion during the short time interval between observations. The height resolution so obtained is equivalent to that which could be derived from direct observations with a resolution of 1/10 second of arc.

Experience to date with the balloon-borne telescope for photographing white-light features of the solar disk demonstrates that a resolution of 1/2 second of arc can be dependably obtained at 80,000 feet altitude. The limit arises in part from solar heating of the mirror and of the small amount of air still in and around the telescope.

Each improvement of angular resolution in the ground-based and balloon observations contributes to the understanding of solar phenomena. It is important to realize, however, that none of the improvements and techniques suggested in the preceding paragraph, with the special exception of the eclipse observations, can obtain a resolution smaller than the scale height of the phenomena observed. In the higher solar atmosphere the temperature and scale heights become very large and the resolution obtained with ground-based measurements may be smaller than the scale height; however, current disagreements in the temperature of this region as determined by different methods can be interpreted as being due to an as-yet-unobserved fine structure in the corona.

[2-7] The resolution needed to make major progress in an area is not always known. In the case of the spectrum of photospheric turbulence a resolution of 1/10 second of arc should be sufficient. Pictures of active features of the chromosphere and low corona made at the limb of the Sun with a resolution of 1 second of arc or slightly better show features at the limit of resolution of the picture. Similarly the quiet chromosphere has features which visual observers claim are finer than anything yet photographed.

> FINDINGS: In summary, it seems clear that the capability of photographing or otherwise recording solar features with an angular resolution of 1/10 second of arc is needed now and would produce great advances in solar physics. This requirement cannot be met from the ground, nor from any solar spacecraft now being planned. There seems to be no reason why the pointing and stabilization techniques being developed for the OAO cannot be adapted to a solar observatory. The development of such a spacecraft represents the next important task for solar astronomers to take up with the active support of NASA.

2. Wavelength Range. Almost all of the visible light from the Sun originates in a layer of the Sun's atmosphere, the photosphere, which is a few hundred kilometers thick. Higher layers are therefore difficult to observe using visible light, not only because of the obscuring effect of the bright photospheric light, but also because the higher layers, having conditions of temperature and density quite different from the photosphere, radiate mostly in wavelengths outside the visible band. Similarly, the active features of the

Sun, such as active prominences and flares, represent a wide range of physical conditions, and a complete description of these phenomena requires observations made over broad wavelength regions outside the visible.

For two reasons the extension of satellite observations to the ultraviolet and X-ray region seems more important at present than an extension to the infrared. First, the infrared radiation comes principally from the upper photosphere, and its observation may not tell us much that is not already deducible from observations in the visible. Infrared observations would actually favor regions of lower temperature than the photosphere, and most of the solar atmosphere above the photosphere is at a much higher temperature than the photosphere. Second, much of the infra-red emission from the Sun can be observed from the ground and, even better, from balloons through a series of atmospheric windows; this field is far from being fully exploited.

Much effort has already gone into ultraviolet and X-ray observations of the Sun from rockets and satellites. Rocket-borne instruments have sampled the ultraviolet and X-ray emissions, and have photographed and scanned the spectrum of the Sun from the visible down to about 100 Å. The first OSO has given repeated scans of the far-ultraviolet spectrum over several weeks and has thereby demonstrated that the line emission in this wavelength range is variable, some lines increasing in intensity at times of solar flares.

The needs for improved angular resolution and for extension of measurements to shorter wavelengths are not, of course, independent. We confidently predict that the need for angular resolution in the ultraviolet and X-ray wavelengths will be similar to those needs already developed in the visible wavelengths.

[2-8] B. <u>Present OSO Program</u>

The OSO (S-16) has opened up the ultraviolet and soft X-ray wavelengths for essentially continuous observation. The scheduled S-17 will add the capability of scanning the disk of the Sun at particular ultraviolet wavelengths with a scanning aperture of about 1-½ minutes of arc, and will provide routine monitoring at Lyman-∀, He I8584, and He II8304 with 1-minute resolution, and in two X-ray bands. The S-17 will also explore the possibilities for detecting the visible light from the outer corona, taking advantage of the lower intensity of interfering scattered light outside the atmosphere.

> FINDING: We foresee a continuing need for explorations of the solar spectrum and recommend that S-16/S-17 type flights be continued at the rate planned by NASA, about two launchings per year for several years.

The need for continued use of the OSO will not come exclusively from the requirements of solar physics. For example, the study of the ionosphere and the higher regions of the Earth's atmosphere is dependent on a quantitative description of ultraviolet radiation incident on the atmosphere. One sensitive technique for measuring the height gradients of terrestrial atmospheric constituents is the measurement of the solar ultraviolet spectrum as a function of height of the spectrometer or photometer carried in a sounding rocket. Success of this technique depends in part on a satellite monitor of the solar ultraviolet spectrum so that the height curves may be corrected for solar variability during the sounding rocket flight.

The OSO is especially needed in the next few years for observations to be used in studies of flare prediction. (See the section on ground-based solar observations for a discussion of this point. [omitted])

C. Advanced OSO Program

Possible design specifications for an advanced OSO, as explained by NASA representatives, assume a Thor-Agena B launch and a polar, slightly retrograde, orbit. They provide for payload space which would allow optics 10 feet long and up to 22 inches in diameter, pointing accuracy of 5 seconds of arc, pointing stability of 1 second of arc for 5 minutes, and various raster scans of selected regions of the Sun by the whole spacecraft. Data storage could be 40×10^6 bits/orbital revolution.

These specifications describe a spacecraft which is a major improvement over the S-17 in two respects: (i) in the pointing accuracy and stability and (ii) in the size of optics that can be carried. The great need for stable pointing stems from the requirement for improved angular resolution discussed above. The need for longer and heavier optics will certainly arise, both because of the long focal lengths required to make use of the high-resolution capabilities, and because of the need for high spectral resolution to analyze in detail the ultraviolet and possibly visible spectrum. The proposed spacecraft is therefore the next logical step needed to advance solar physics.

[2-9] RECOMMENDATION: We recommend that NASA develop, for solar observations, a spacecraft more advanced than the present OSO, to be ready for use in late 1965 or early 1966.

Certain auxiliary instrumental development must progress rapidly if full advantage is to be taken of the advanced OSO. Foremost among these is the consideration of the thermal problems of high-resolution optical systems pointed at the Sun. If the problem of thermal stability of the optics can be solved, the projected pointing stability of 1 second of arc for 5 minutes and the payload diameter of 22 inches hold the possibility of approaching, with short exposures, the desired 1/10-second resolution. Full utilization of such high resolution requires also the existence of electrical read-out image tubes of several-thousand-line resolution (see Section VII.A of this chapter [omitted]) and the commitment, at least occasionally, of the entire memory to this image device.

Document I-6

Document Title: Space Science Board, National Academy of Sciences—National Research Council, "Space Research: Directions for the Future," 1965.

Source: NASA Historical Reference Collection, History Office, NASA Headquarters, Washington, D.C.

The Space Science Board convened another wide-ranging space science summer study in 1965, this time at Woods Hole Massachusetts. Once again, a group interested in solar research was part of the summer study. This excerpt from the study report contains their assessment of the state of solar research and its future direction.

[cover sheet]

SPACE RESEARCH
DIRECTIONS FOR THE FUTURE

REPORT OF A STUDY
by the
SPACE SCIENCE BOARD

WOODS HOLE, MASSACHUSETTS
1965

Publication 1403
NATIONAL ACADEMY OF SCIENCES-NATIONAL RESEARCH COUNCIL
WASHINGTON, D. C. 1966

[177]

[chapter] III Solar Astronomy
1. INTRODUCTION

AIMS OF THE WORKING GROUP

As part of a summer study of space science by the National Academy of Sciences, the Solar Astronomy Working Group met at Woods Hole, Massachusetts, from June 21 to July 3, 1965. In many ways this study can be regarded as a follow-up to a similar conference held at the State University of Iowa in 1962, the proceedings of which have been published as "A Review of Space Research" (Publication No. 1079 of the National Academy of Sciences-National Research Council).

The direct objective of the Solar Astronomy Working Group was stated as follows: "To examine future needs of, and opportunities for research in solar astronomy, using all space techniques ranging from orbiting observatories more advanced than the Advanced Orbiting Solar Observatory (AOSO) and possible manned facilities, to small satellites and sounding rockets."

Scientists participating in the work of the Working Group on Solar Astronomy are listed in the Appendix. The recommendations of the Working Group are listed at the end of this Section; the discussion leading to these recommendations is found in Section 2.

SOLAR ASTRONOMY AND SOLAR SPACE ASTRONOMY

At the beginning of a report on solar space astronomy, it is appropriate to emphasize the unity of solar research, and to point out that observations from space and observations from the ground are simply two aspects of a [178] single venture. Clearly, if one wants to obtain all possible information about the Sun, one uses every means at his disposal; one must never forget that space and ground-based astronomy are partners–not competitors–and the knowledge to be gained from using the two together is very likely greater than the sum of the two used separately. If, for example, the time development of flares could be followed in the light of He I λ584 from a spacecraft and in the light of He I λ 10830 from the ground, it would be far more valuable to observe a particular flare simultaneously in both lines than simply to observe any flare in either line separately.

There are three major reasons why the Sun is important: first, because of its intrinsic interest as the only nearby star; second, because of its effect on the planets and, in particular, the Earth and the human race; third, because it is a valuable astrophysical laboratory.

As the only star on which detailed observations can be made, the Sun holds a unique position in stellar astronomy. Because the disk of the Sun can be resolved, single features – sunspots, prominences, granules, and flares – can be observed, and spectral scans across the disk from center to limb can be made, which in turn give directly the temperature and density in the photosphere as they vary with depth. The solar atmosphere can be used as a standard against which theories of stellar atmospheres and spectral-line formation can be tested. Stellar spots and cycles of activity can at present be studied in detail only on the Sun. Theories of convection can be tested on the Sun, where granulation and chromospheric microstructure can be seen. Chromospheres and coronas are seen nowhere else in as great detail as on the Sun. Finally, the Sun is a stable, main sequence star that serves as a photometric and spectroscopic standard for stellar work. Every new piece of information about the Sun contributes to our comprehension of the stars.

Knowledge of the Sun is indispensable to an understanding of the physics of the Earth and other planets. The Sun is the source of virtually all the heat and light a planet receives. Photon flux from the Sun causes dissociation, excitation, and ionization of atoms and molecules in planetary atmospheres, produces ionospheres, maintains planetary heat budgets, causes escape of atmospheres from the planets, is responsible for weather, and controls any life that exists on the planets. Particle flux from the Sun affects planetary ionospheres and magnetic fields, produces auroras, fills the interplanetary region, and changes the chemical abundances by producing nuclear reactions in planetary atmospheres. In addition, solar radiation (photon and particle) affects the Moon, comets, meteoroids, and asteroids, as well as grains in interplanetary space. In fact, the whole solar system can be viewed as being imbedded in the outer solar corona.

The importance of the Sun to life can scarcely be overemphasized: solar energy ultimately sustains practically every living organism – both plant and animal. Radiation from the Sun has also controlled the atmosphere and environment in which life has developed, and thus has been the principal determinant in shaping the course of evolution. Biology is [179] therefore interested not only in the present radiation (photon and particle) from the Sun but also in any changes that may have occurred during the

Earth's history. With the dawn of the space age, an upsurge of interest in exobiology – life outside the Earth – has taken place. This has increased the demand for an intimate knowledge of all solar radiation, especially in the ultraviolet and x-ray regions of the spectrum.

A subject deserving special consideration under the general topic of the effects of the Sun on its environment is the effect of the Sun on man. Living as he always has, on Earth, where the atmosphere shields him from the harmful radiations of the Sun, man has tended to look upon the Sun as completely benign. With the coming of the space age and man's emergence from within the protective atmospheric envelope, however, the problem of shielding astronauts from solar radiation must be faced. Solar protons (and possibly x rays) emitted at the time of solar flares pose the most serious danger. If man is to work in space, outside of a heavily armored module, the Sun must be constantly monitored so that the astronaut will have immediate warning of any solar event from which radiation might be sufficiently intense to force him to re-enter his spacecraft. Further knowledge about the nature of flare events, which would enable scientists to better predict their occurrence or better shield the astronaut, would be of substantial benefit.

Ever since Janssen's discovery of helium on the Sun, it has been apparent that the Sun can be used to complement the terrestrial scientific laboratory, for there are available in the Sun combinations of temperature, density, and path length quite beyond terrestrial capabilities. Although its potential value in this respect has not been fully utilized (and is perhaps not clearly appreciated by all scientists), the Sun has been exploited as a laboratory by workers in several branches of physics. The measurement of precise wavelengths of spectral lines from highly ionized elements in the corona was used by Edlen to determine a number of atomic parameters otherwise inaccessible. Recently, both in this country and abroad, scientists have used the Sun as a source for spectra of wavelengths that are not otherwise producible in the laboratory. In the same way, observation of many strong lines near 170 Å in the solar spectrum stimulated efforts in several laboratories to obtain spectra from highly ionized atoms in an attempt to identify these solar lines. As the ultraviolet spectrum of the Sun is examined more carefully, one may expect to gain important new knowledge of atomic physics and spectroscopy. The Sun has also been, and still is, important for studying nuclear reactions and testing theories of nucleogenesis. The solar atmosphere provides – in such features as spicules, the corona, and the convection zone – a large-scale laboratory in which phenomena of aerodynamics and hydrodynamics can be observed. In fact, observations of the outward particle flux from the Sun have stimulated a great deal of important work on the solar wind and interplanetary plasmas. Finally, the Sun provides unique opportunities for the study of plasma physics and magnetohydrodynamics. In addition to the solar atmosphere [180] itself, which can be regarded as a giant plasma, such features as sunspots, prominences, flares, and the corona are examples of the interaction between plasmas and magnetic fields. The Sun is a readily available source in which many waves and oscillatory phenomena of interest in magnetohydrodynamics–acoustic waves, magneto-acoustic waves, electromagnetic waves, and Alfvén waves–occur naturally. As solar research progresses, discoveries of interest to several disciplines will undoubtedly be made, and many new uses will be devised for our convenient astrophysical laboratory.

In summary, we emphasize once again the central position of solar astronomy and the contributions that it makes to such fields as stellar astronomy, radio astronomy, planetary atmospheres, interstellar matter, biology, atomic physics, and meteorology, as well as its influence upon such very practical matters as radio communication on the Earth and the safety of an astronaut.

Although the reasons for observing the Sun from above the Earth's atmosphere have already been listed in many places, we mention them once again in order to remove any existing doubt concerning the benefits to be gained by putting telescopes into space. The most obvious reason is the possibility of observing the ultraviolet and x-ray solar spectrum. Gases in the Earth's atmosphere completely block from our view all radiation with wavelengths shorter than about 2900 Å. Yet this obscured spectral band is vitally important in understanding the Sun for several reasons: the resonance lines of most elements lie within it; it is emitted by the interesting region of temperature inversion just above the photosphere; it contains many of the strongest coronal lines; and the most violent variations of radiation with solar activity are observed within it. The preliminary work that has already been done in the ultraviolet spectral band by means of sounding rockets and satellites has added greatly to our knowledge of the Sun. More complete observations (including x-ray and gamma-ray detection), offer the most fruitful means of understanding the processes taking place in the upper solar atmosphere. It is especially important for observations to be made during the years leading up to and through the coming solar maximum of 1967-1970.

A related advantage, and one that has not often been discussed, is the opportunity to observe the Sun in the far infrared – from 20 microns up to 1 millimeter. Our ignorance of this region is almost complete; we do not even know the energy distribution, much less any details of molecular bands or lines. Although the infrared is probably of less importance than the ultraviolet, and for this reason has often been neglected in space-science planning, there are several interesting observations to be made; for example, the detection of the radiation from the photosphere-chromosphere interface (somewhere between 20 and 200 microns) and observation of the molecular spectra of sunspots. Questions about the temperature structure of cool regions (such as above sunspots and near the temperature minimum) can perhaps be answered by observations at those [181] wavelengths. Again, since the atmosphere of the Earth entirely prevents us from detecting this radiation, such observations can only be obtained from space (perhaps, in this case, airplanes or balloons would suffice).

A crucial advantage that could be obtained from space observation is the increased resolution (clarity) of small features on the Sun. The developments of the past 20 years in the theoretical and observational study of the dynamical and magnetic properties of the quiet solar atmosphere and the many features of solar activity focus increasingly on the smallest observable structures. Every improvement we achieve in spatial resolution reveals new detail of the greatest significance. The size distribution of granulation, the fine structure of magnetic fields, the stranded structure of loop prominences, the local variations in line profiles, the turbulent velocity fields, and the minute structure in flares are examples. Ground-based observation has carried us to the verge of solutions, but in most instances falls short of the theoretically decisive resolution. While it is obvious that there will always be interesting details too small to be seen with any resolution we may attain, many features of the solar atmosphere should have a size scale in the neighborhood of the scale height.

Except for the very thin layer at the base of the photosphere, the scale height is of the order of 100 km or more. This corresponds to 0.14 sec of arc in angle. The practical resolution limits of ground-based solar observation are about 0.5 sec of arc for direct photographs and 0.8 sec of arc in the best spectra. Resolution of this grade is the kind one achieves only on very rare occasions in one lucky photograph or spectrogram. The limit is set by poor seeing in the Earth's atmosphere, and could be overcome by space observation.

Another limitation imposed by the Earth's atmosphere is the brightness of the sky. The bright sky near the Sun, at the best observing sites, is rarely less than 10 times the brightness of the inner corona. Nevertheless, ground-based observation has successfully detected the polarized component of the brightest streamers of the white corona out to a height of one solar radius above the limb; however, direct measurement of the electron-scatter brightness and detection of the corona beyond one radius are impossible except during a total solar eclipse. Any continuous watch from the ground for outward-moving plasma clouds is also out of the question. Since the sky is completely dark in space, we can expect to monitor the Sun's corona routinely from space vehicles, which will lead to major advances in our knowledge of coronal processes.

A final important benefit to be gained from observation in space is the continuity in time that can be obtained. Often, critical observations need to be extended for several hours, days, or even months, in order to study the time development of certain features, such as prominences and centers of activity. Such observations must ultimately be made from satellites where they will be independent of terrestrial meteorological and diurnal effects. Real-time monitoring of solar events will be accomplished by such satellites, making communication delays short or nonexistent as [182] telemetry reception becomes possible at the user's site. As space observations become routine, close cooperation between ground and space observations must be maintained for maximum benefit of such solar monitoring.

CURRENT SOLAR PROBLEMS

Solar physics has as an ultimate task of describing completely the structure, understanding thoroughly the dynamics, comprehending fully the origin and development, aid predicting exactly the future evolution of the Sun. These ultimate goals, which can never be fully attained, are perhaps best expressed in the form of more specific questions that might be asked about the Sun. The following are some of the major questions with which solar physics is now grappling. What are the details of the processes by which energy is transferred outward from the center of the Sun? What is the source of the sunspot cycle and solar activity? Why, and how, do solar flares occur? How are energetic particles and photons produced? What is the detailed structure of the chromosphere, and what is its connection with the corona and with magnetic fields on the Sun? What is the nature and cause of spicules and prominences? Why is there a solar corona, and how does it produce the solar wind and interplanetary medium? What is the origin and early history of the solar system? What produces the equatorial acceleration of solar rotation? How is the solar magnetic field produced and what is its effect on the solar activity cycle?

Such questions, which have been formulated and listed many times before, are the grand questions toward whose solution all solar physics is directed. But these are too comprehensive to be answered fully or even attacked intelligently. Instead, most solar physics is directed toward the solution of much more specific problems in the hope that the accumulated answers to many smaller problems may eventually provide answers to these larger questions.

The following are some of the more important specific questions with which present solar astronomy is engaged:

Photosphere. What are the size and velocity distribution of the solar granules? What are the source and structure of the weak magnetic fields produced at the surface of the Sun? What is the simplest nonhomogeneous model of the photosphere? What is the relationship between the granules and the super-granulation? Why is there a temperature inversion? What is the detailed variation of temperature with height through the temperature minimum? What is the vertical structure of features observed in this height range? Do these structures vary with the solar cycle?

[183] Chromosphere. What is the morphological specification of the vertical structure (including the size distribution)? What is the velocity distribution in this region and how does it vary both horizontally and vertically? What is the detailed structure of the cells of the chromospheric network? How does the magnetic field vary across the cell boundary? Which features inside the cells are periodic in time? Are any observed features rotating? Is the chromospheric oscillation vertical or horizontal? What determines its period? How does the temperature vary horizontally across the chromospheric cells? How are spicules produced? What is their microstructure? What are the magnetic fields in spicules? Why do spicules seem to favor the edges of the cells? What other aerodynamic phenomena take place in the chromosphere? What is the energy budget of the chromosphere? What types of waves are propagated in the chromosphere and what are their results? What is the temperature gradient as one passes from chromosphere to corona? What causes this temperature gradient? What is the nature of the transition from spicules to the corona? How should we interpret spectral lines formed in the absence of local thermodynamic equilibrium?

Corona. How is the corona heated? Is the corona localized over the edges of the chromospheric cells? Is the corona in equilibrium? What is an M region? What is a coronal streamer? What is the relation of streamers to the solar wind? What coronal phenomena give rise to radio bursts and gusts in the solar wind? What is the magnetic field of the corona and what is its effect on corpuscular streams? How do radio bursts escape from the corona? Is there a difference in chemical composition between corona and photosphere?

Flares. What is the primary flare phenomenon? How do flares occur? What is the source of energetic particles in a flare? Is the energy released primarily over a small hot kernel? What is the relation of the flares to the surrounding magnetic field? How are flares related to the coronal condensations? Are x-ray flares observable at extreme ultraviolet wavelengths? How do active regions with intense magnetic fields arise? What is the physical structure of active regions?

Sunspots. How is the observed brightness distribution across a sunspot maintained? What is the physical structure of sunspots? What is the relation of the magnetic field to the fine structure? How does the granulation behave inside the spots? What is the fine structure of the umbra below the resolving power of Stratoscope I? What system of gas motions exists around the sunspots?

Plages and Faculae. What is the microstructure of plages and of faculae? What is their relation to each other? What oscillation takes place in the faculae? How are the faculae related to the network cells? Is there a difference in coronal heating above faculae and above active regions?

[184] Prominences. Is there a basic lower limit to the size of filamentary structures in prominences? What is the magnetic field in prominences? What is the mass and energy balance between the corona, prominences, and the lower atmosphere? What causes eruptive prominences? Why are loop prominences so extremely hot? What are their special relation to solar flares?

ORGANIZATION OF THE PANEL

To meet its objective, the Working Group divided its assignment into three parts:
1) A critical review of the past and present NASA programs in solar physics and of the Prospectus 1965-1980;
2) An examination of major unsolved problems in solar physics and a specification of the instrumental requirements for their solution;
3) The recommendation of specific experiments that might be initiated in the three time periods 1965-1970, 1970-1975, and beyond 1975.

Certain other matters, such as the role of man in carrying out scientific observations in space and the relation of laboratory work to the NASA mission, were discussed as part of the solution of the problems of solar physics.

There are several alternative ways in which the Working Group could approach its assignments. The physically most meaningful is the problem-solving approach, in which the unsolved questions of solar astronomy (based on the list given above under Solar Astronomy and Solar Space Astronomy) are considered and then observations are sought that would aid in answering them. Experimenters generally prefer an instrumental approach, in which knowledge of the characteristics of an instrument is used to determine what observational data the instrument can obtain. A third possible approach might be termed vehicular, in that the capabilities (size and weight of payload, pointing accuracy, lifetime, orbit, power available, and data-storage and transmission capabilities) of the planned vehicles are first examined and then it is decided which observations could be made from them.

The Working Group favored the first approach, in general, as it felt that this approach would ultimately be most fruitful from a scientific standpoint. Although other viewpoints were adopted at times, an attempt was continually made to relate all discussions to the basic questions of solar physics.

To facilitate the work of the Working Group and ensure complete coverage of the field of solar astronomy, the Working Group was divided into four subcommittees. Since the detection and analysis of solar electromagnetic radiation forms the principal source of information about the [185] Sun, and since different regions of the spectrum come from different parts of the Sun, a division according to spectral region is to a large extent a division according to region or height on the Sun and hence according to basic questions about the Sun. This division of labor to some extent combines the problem-solving and instrumental points of view. The entire spectrum was therefore divided into sections and a subcommittee was assigned to study each section. Included in the charge given to each subcommittee was the request that it look ahead at least to 1975 and discuss problems, instruments (including the role of man), and vehicles. The assignments were as follows:

a) $\lambda < 500$ Å: Lindsay, Teske, Zirin;
b) $500 < \lambda < 1500$ Å: Athay, Firor, Orrall;
c) $1500 < \lambda < 3000$ Å: Johnson, Smith, Tousey;
d) $\lambda > 3000$ Å: Evans, Howard, Ney.

The reports of these groups were discussed by the Working Group as a whole and are contained in Section 3.

In addition, and partly as a result of the previous work, several informal subcommittees were designated to study and make recommendations about such other topics as the use of rockets, the role of man, the relation of the astronaut-observer to the scientist directing the experiment, the role of the ground-based laboratory in the space effort, and the question of a Moon-based observatory. Discussion and recommendations concerning the findings of all subcommittees are found in Section 2.

RECOMMENDATIONS

The full texts of the recommendations of the Working Group on Solar Astronomy are presented below.

Recommendation 1. (a) That the recommendation of the Iowa meeting (1962) concerning fine pointing be given immediate attention and that highest priority be given to the development of triaxially stabilized rocket attitude controls, leading, as soon as possible, to a fine-pointing system capable of an accuracy of 5 sec of arc and optimally designed for solar use. (This recommendation is essentially a reaffirmation of that made by the Iowa Summer Study, and is restated here to reflect the importance which the Working Group attaches to this matter: "We recommend that the sounding rocket program continue to receive full support; and that both the inertially guided Aerobee with fine pointing at selected stars, and the inertially guided Aerobee with fine pointing at the Sun controlled by an optical sensor be made available at the earliest possible time.")

[186](b) That other improvements (such as increased payloads and peak altitudes, increased reliability, and more dependable recovery techniques) be made in existing rocket systems;

(c) That the number of rockets available per year for research in solar astronomy be at least doubled;

(d) That funds for payload development be increased to an adequate level, especially when the triaxial pointing controls become available.

Recommendation 2. (a) That the presently approved Orbiting Solar Observatory (OSO) program be augmented by at least four additional launchings during the period 1970-1972 inclusive;

(b) That no decision be made to terminate the OSO program after 1972 without further review at an appropriate time;

(c) That NASA make every effort to implement such desirable improvements in the OSO spacecraft as increased power, offset pointing, localized raster scans, provision for slightly longer instruments, greater data capacity and more flexible data format, and improved pointing accuracy (15-30 sec of arc);

(d) That consideration be given to injection of one or more OSO spacecraft into a polar retrograde orbit in order to provide continuous surveillance of the Sun.

Recommendation 3. (a) That a satellite with Advanced Orbiting Solar Observatory (AOSO) specifications is an indispensable next step in NASA's solar program, and must be flown close to the coming solar maximum;

(b) That the AOSO program be accorded all the priority necessary to maintain the launch schedule shown in the Prospectus.

Recommendation 4. (a) That manned missions in the 1968-1972 time period, such as the Astronomical Telescope Orientation Mount (ATOM) in the Apollo Extension Systems, are desirable to supplement AOSO, but cannot replace it;

(b) That because it offers the prospect of providing answers to critical questions relating to the technology of manned space telescopes and data recovery, the ATOM concept merits vigorous support.

Recommendation 5. That solar space observation be included in the manned space science program of the Apollo Extension Systems in order to develop the technology of manned space astronomical operations. Such observations, which could attain resolving power of 1 sec of arc in the wavelength region 500-3000 Å, mark the next logical step beyond both AOSO and ATOM.

[187]Recommendation 6. That feasibility and design studies begin immediately on orbiting solar telescopes of at least 1-meter aperture designed to obtain a resolution of 0.1 sec of arc at visible wavelengths and 0.5 sec of arc at far ultraviolet wavelengths ($\lambda > 500$ Å). Very large and complex accessory instruments will be necessary to analyze the solar image. Erection, operation, and maintenance of this telescope will require full utilization of astronaut-engineers and scientists.

Recommendation 7. That provision be made for a continuing, uninterrupted experimental program while the more advanced manned flights are in preparation, with many flights of various spacecraft, so that a scientist will have frequent opportunities for observation.

Recommendation 8. That NASA find means to continue a strong program with relatively inexpensive rockets and small unmanned satellites at the same time the large manned projects are under way, since the former are indispensable to the latter.

Recommendation 9. That the relationship between scientists and astronaut-observers be studied and clarified. In particular, we recommend that when a single, large scientific instrument is carried, the scientific observation be designated the primary mission for the flight.

Recommendation 10. That NASA bring more scientists into the space flight program as astronauts or observers.

Recommendation 11. That NASA move to provide additional support for ground-based solar studies. As the flight program grows in sophistication and success during the next several years, the demands on ground-based work will also increase, and NASA should in turn anticipate an increased demand upon its resources for support of ground-based facilities and operations. In addition, in the next few years, NASA should expect, and respond favorably to, proposals for a few major ground-based solar installations.

Recommendation 12. That increased support be given to physical research in the laboratory, as required to develop improved space instrumentation for solar-physics research, to assist in the data reduction, and to make possible a full interpretation of the results.

Document I-7

Document Title: NASA Goddard Space Flight Center, *History of Orbiting Solar Observatory OSO-B,* **X-440-66-322, April 1966.**

Source: NASA Historical Reference Collection, History Office, NASA Headquarters, Washington, D.C.

On 14 April 1964, a disastrous accident killed three men when OSO-B's third stage ignited on the test stand at the NASA launch center at Cape Kennedy, Florida, deeply clouding the future of the program. Some of the parts of the craft were retrieved and refurbished, and a new satellite was built and flown successfully in February 1965. This excerpt from the history of the spacecraft details the causes of the accident and the steps taken to avoid a reoccurrence.
[2-1]

SECTION 2
OSO-B DISASTER

2.1 EVENTS LEADING UP TO THE DISASTER

The OSO-B spacecraft arrived at Cape Kennedy on 12 March 1964. Routine checkout and preparation of the spacecraft and experiments took place until 9 April 1964, at which time the payload was covered with a polyethelene bag, purged with dry nitrogen and placed in its shipping container. The payload was stored in hangar AE to await the arrival of the Delta third stage rocket motor.

Because of its heavier weight, the OSO spacecraft uses a rocket motor with a thicker wall casing. When the motor arrived, it was given a receiving inspection, and it was discovered that there was a defect in the rocket motor casing. This motor was rejected, and a new X-248 A-6 rocket motor was flown down from Wallops Island, Virginia on 9 April 1964. The igniter paddle was removed from the rejected rocket motor to be installed in the new motor. During the removal of this paddle it was damaged and a new Delta paddle was built by the Naval Propellant Plant. The new Delta igniter paddle was flown to Cape Kennedy on 11 April 1964 and installed in the third stage rocket motor. The rocket motor was transported to the Spin Test Facility on 12 April 1964.

On 13 April the payload was removed from the shipping container, placed on a truck and, at approximately 0400 hours, it was moved to the Spin Test Facility.

2.2 THE DISASTER

Between 0930 and 0939 hours EST on 14 April 1964, the third stage X-248 A-6 solid propellant rocket motor inadvertently ignited and burned in the Spin Test Facility at Cape Kennedy. The rocket motor with the spacecraft attached tore loose from the alignment fixture in which it was mounted and shot to the ceiling of the facility. When it hit the ceiling, the spacecraft was torn loose from the third stage motor and fell to the floor. The rocket motor continued on to the corner of the building and burned until its fuel was expended. Eleven men were burned - three fatally and eight others suffered injuries ranging from critical to minor. The three men who died were not killed immediately but died as a result of their burns within a couple days to a couple weeks after the accident occurred.

Eyewitness interviews after the accident indicated that the Douglas personnel had just completed their ordnance checks of the third stage/spacecraft [2-2] combination. One of the Ball Brothers Research employees stepped over to the spacecraft to adjust the polyethelene shroud which was placed over tile spacecraft and third stage as a dust protector and to purge them with nitrogen. As he touched the shroud a crackle was heard and the third stage ignited.

2.3 ACCIDENT INVESTIGATION

A Fact-Finding Committee was appointed by the NASA Goddard Space Flight Center Director. The committee was comprised of the following personnel:

D. G. Mazur, Chairman	NASA-GSFC
W. D. Baxter, Lt. Col	AFMTC
Dr. B. Bartocha	NPP
R. H. Gablehouse	Ball Brothers
E . E . Harton	NASA Headquarters
E. H. Helton	NASA-Wallops Island
L. T. Hogarth, Secretary	NASA-GSFC
R. J. Johnson	Douglas Aircraft Company

J. J. Nielon	NASA-GLOB
L. R. Piasechi	JPL
W. R. Schindler	NASA-GSFC
R. Steinberger	ABL
L. Swain	NASA-LRC

The committee investigated the following items at the Eastern Test Range (ETR) immediately after the accident to establish the circumstances surrounding the accident:

    a.   Hardware configuration at the time of the accident.

    b.   Eye witness testimony and reports.

    c.   Examination of the accident area including inspection and testing of significant items.

    d.   The time sequence of events leading up to the accident involving the rocket motor and the spacecraft.

[2-3]   e.   Review of procedures.

    f.   Determination of possible causes.

    g.   Plans for tests to verify the possible causes.

The committee divided the investigation into four areas: (1) heat, (2) RF signal, (3) electricity and (4) mechanical shock and/or vibration. The committee also undertook the investigation of a similar accident which took place a few months earlier at the Oklahoma Ordnance Depot at Pryor, Oklahoma, to determine if the two accidents were related. In the Oklahoma accident, the Douglas Aircraft Company was preparing a destruct test of the X-248 rocket motor in order to test a new Delta third stage destruct system. No one was killed in this accident, but test equipment and a crane used for moving the rocket motor were considerably damaged. One person received minor injuries.

The first of the investigative courses of action taken was to investigate electrostatic discharge. This investigation was to determine possible modes by which an electrostatic charge could have caused ignition of the X-248 A-6 motor. It was divided into five tasks which were directed toward developing a comprehensive picutre [sic] of the electrostatic characteristics of the spacecraft/motor configuration and the motor/igniter assembly. The first of these tasks was to determine the electrostatic sensitivity of the X-248 squib. The second task was to determine the electrostatic sensitivity of all bulk explosives in the X-248 motor. Another task was to determine total and inter-element electrical characteristics (resistance, capacitance, and charge storage) of the spacecraft/motor and the motor/igniter under the application of both static and transient electrostatic voltages. This task was also to establish the critical interelemental breakdown voltages and paths. Task number four was to determine the electrostatic potentials and energies that could have been present under the circumstances prevailing at the time. The final task was to try to duplicate the X-248 inadvertant [sic] ignition in both the Eastern Test Range and Oklahoma accident configurations.

Electrostatic sensitivity tests of the Delta X-248 squib were conducted by the Franklin Institute. These tests consisted of discharging, through the squib, incremental voltages of a 500 pico-farad capacitor through a 5000 ohm resistance to simulate the capacity and resistance properties of a human being.

Measurements of the electrostatic sensitivity of the X-248 bulk explosives were conducted by the Naval Propellant Plant. It was determined from these measurements that the X-248 bulk explosives could not have been a factor in the accident.

[2-4]   Determination of the electrical characteristics of the spacecraft/motor and motor/igniter were conducted by Cornell Aeronautical Laboratory. They found that relatively weak sources of electrical current, such as electrostatic charging phenomena, are capable of building up large potentials between the dome and the nozzle of the motor with the igniter assembly in place. The squib is polarized and has the ability to store electrical energy. At voltages sufficiently high (250 to 1000 volts) to break down the dielectric in the squib, the squib could supply sufficient energy to ignite the lead styphnate primer.

Determination of electrostatic potentials and energies that could have been present under the circumstances at the time of the accident were investigated. The polyethelene shroud used over the spacecraft was found to be an electrostatic generator and had charge-concentrating areas that could charge the spacecraft to 15,000 volts if a mechanism was available to transfer the charge from the cover to the spacecraft fast enough. Rolling the cover up and down in a manner identical to the operation performed at Cape Kennedy generated voltages on the spacecraft as high as 2500 volts and averaged about 1200 volts. Wiping the plexiglass plate over the nozzle exit of the X-248 redistributed the charge on the plate and induced transient peak potentials up to approximately 5000 volts on the igniter firing lead. It was also found that a man with or without a Clean Room Suit ("bunny suit") could easily generate, just by normal activity, a charge sufficiently large that, if placed on the X-248 motor/spacecraft assembly, it could increase the spacecraft potential about 3000 volts on contact with the man.

The tests were conducted with an inert motor and a live initiator. Six ETR configuration firings were all produced by friction or movement of the polyethelene film identical to that used for the OSO-B shroud around the body of the rocket motor.

Cornel [sic] Aeronautical Laboratory proposed several modifications to the igniter assembly. A modified igniter assembly was fabricated and tested for the committee. The assembly withstood 60 kilovolts discharged directly to the squib without ignition, and subsequent tests were successful up to 100 kilovolts.

From the RF tests conducted by the Franklin Institute and the Picatinny Arsenal, together with the RF data supplied by Eastern Test Range, it was concluded that the accident was not caused by RF energy. It was also improbable that the accident was caused by incompatibility or instability of the chemical characteristics of any of the igniter or motor components.

[2-5]
2.4 CONCLUSIONS AND RECOMMENDATIONS

The committee concluded that the cause of the ignition of the third stage rocket motor at Cape Kennedy was an electrostatic discharge through the igniter squib. Cornell Aeronautical Laboratory recommended the following changes to the X-248 rocket motor: (1) use of a squib insensitive to electrostatic energies up to 25 kilovolts and 500 pico-farad, (2) use of a resistive plug between the squib case and the bridgewire, (3) use of a Faraday cage covering all sensitive parts of the squib assembly and (4) use of a conductive spray on electrostatic sensitive portions of the paddle. Cornell Aeronautical Laboratory demonstrated that the last three of the forementioned [sic] changes are adequate to insure against accidental squib initiation due to electrostatic discharges up to 100 kilovolts. In all subsequent Delta launches, an X-258 rocket motor was used instead of an X-248 because it used an igniter assembly less sensitive to electrostatic discharges, and the squib can be inserted on the launch tower.

Precautionary measures were also suggested which would apply to any solid propellant rocket motor: (1) to avoid the use of non-conductive materials, especially plastics, (2) to use squib arrangements which would permit installation as late in the operation as possible, (3) to check the conductivity of each igniter–motor system planned for usage to verify a low resistive path between all conductive components, (4) to strictly adhere to proper grounding procedures whenever a solid motor is to be handled with or without an igniter installed. It was also recommended that procedures for grounding personnel, spacecraft, motor and associated systems and components should be carefully considered for future rocket motor handling.

2.5 SPLIT BALANCE FACILITY REWORK

The possibility of a new Spin Test Facility was investigated; however, a new facility could not be made ready until November 1965. It was decided to rework the damaged facility for use in the Delta program. During the rework of the Spin Test Facility, the following additional safety features were added to the building:

a. The pit in the southwest corner was floored over with portable decking which can be removed if the pit is required for future operations.

b. A new personnel door was placed in the center of the north wall.

c. Roll up doors on the east end of the building were replaced with two 6 by 10 feet swing-type doors. The remainder of the original open-ing was replaced with blast panels.

[2-6] d. Panic hardware was improved on all personnel egress doors. Blank latch facings were installed on the door frames.

e. The interior of the west wall of the building was covered with gypsum wallboard to provide sealing and to retard fires.

f. The protruding tracks on the exterior of the west end of the building were removed.

g. An emergency audible warning system was installed.

h. A "Cone of Protection Lightning System" was installed around the facility.

i. The existing communication system was removed and replaced by an explosion proof intercom system.

j. Conductive plastic mats were supplied for use in areas where ordnance is handled.

k. A sprinkler system was installed in the high pay area.

l. Placards denoting explosive materials, classes, personnel limits, etc. were installed.

m. Personnel safety showers were installed at all personnel egress doors.

n. An additional closed circuit TV system was installed with cameras in the high bay area and monitors in the office trailer and control room. Personnel can now witness operations without being physically present in the bay.

Procedures were changed so that all spacecraft testing necessary in the pit area is performed remotely.

o. The personnel trailer located at the west end of the building was removed to a more remote location.

p. The guard shack for the area was removed to approximately 350 feet from the facility.

q. The relative humidity inside the building was increased from 50% to 60%.

Document I-8

Document Title: Letter from Dr. Leo Goldberg, Harvard College Observatory, to Dr. Homer E. Newell, Associate Administrator, NASA, "Astronomy Missions Board Resolution," 25 January 1968.

Source: NASA Historical Reference Collection, History Office, NASA Headquarters, Washington, D.C.

NASA created an external Astronomy Missions Board in 1967 to advise the Agency with respect to its astronomy programs. The first chairman of the Board was Harvard astronomer Dr. Leo Goldberg. This letter from Goldberg to NASA Associate Administrator Dr. Homer Newell transmitted the Board's views on the proposed Apollo Telescope Mount program as part of NASA's post-Apollo activities to use human crews and the equipment developed for the Apollo program for other scientific purposes.

[no pagination]

HARVARD COLLEGE OBSERVATORY
60 GARDEN STREET
CAMBRIDGE, MASSACHUSETTS 02138
January 25, 1968

ASTRONOMY MISSIONS BOARD

Dr. Homer E. Newell
Code AA
National Aeronautics and Space Administration
Washington, D.C. 20546

Dear Homer:

At its third meeting on January 17-18, 1968, the Astronomy Missions Board adopted the attached resolution with respect to the ATM-A mission and requested that you also convey it to John E. Naugle (OSSA), George E. Mueller (OMSF), Werner von Braun (MSFC), Robert R. Gilruth (MSC), Charles W. Mathews (Director, AAP) and Dixon L. Forsythe (Manager, ATM).

I should also report the concern of the Board that the mission be accomplished by 1971, before the next solar minimum, without, however, sacrificing any of the scientific objectives of the experiments. The Board also strongly hopes that NASA will pursue with maximum vigor the two conditions concerning astronaut training and failure circumvention set forth at the end of the resolution.

Please be assured that the Board stands ready to render all possible assistance in facilitating the success of the ATM and other important astronomy missions.

Sincerely yours,

Leo Goldberg

****

RESOLUTION
adopted by the
ASTRONOMY MISSIONS BOARD
at its Third Meeting
January 17-18, 1968

The Astronomy Missions Board, during its past two meetings, has received briefings on the major aspects of the ATM-A project for solar research from the appropriate NASA officials. In addition several Board members have had access to further detailed information regarding this project.

On the basis of this combined information, the Board is now convinced that the ATM-A project is a logical and technically appropriate next step (a) for the purpose of gaining experience and insight in choosing the optimum role to be played by man in space in the operation of sophisticated technical and scientific equipment, and (b) for the purpose of achieving another important advance in solar research.

This present positive assessment of the ATM-A project by the Board is, however, conditioned on NASA's pursuing with maximum vigor activities in two specific areas. The first area refers to the selection and training of the Astronauts responsible for the operation of the scientific experiments carried on ATM-A. The Board has already recorded its recommendation regarding this area at its preceding meeting. The second area refers to efforts to increase the opportunities for the astronaut in flight to maximize the probabilities of experiment success by taking actions aimed at failure circumvention. The Board is much encouraged by current activities in this area, spearheaded by the Astronaut Office.

The Board believes that high priority has to be assigned to this type of activity if the ATM-A project is to fulfill its stated difficult purposes.

Document I-9

Document Title: Memorandum from Dr. Homer E. Newell, Associate Administrator, NASA, to Dr. John E. Naugle, NASA, response to letter dated 22 March 1968, from Dr. Leo Goldberg, Chairman, Astronomy Missions Board, 9 April 1968.

Source: NASA Historical Reference Collection, History Office, NASA Headquarters, Washington, D.C.

As NASA planned its program for the 1970s, it became clear that the aspirations of solar researchers related to the Apollo Telescope Mounts which were scheduled to be flown as part of the orbital workshop program (soon to be known as Skylab) were at variance with schedule realities. In this memorandum, NASA Associate Administrator Dr. Homer Newell asks Associate Administrator for Space Science Dr. John Naugle to investigate alternate paths for satisfying the scientific objectives established by the Astronomy Missions Board.

[no page number]

April 9, 1968

MEMORANDUM:
To: S/Dr. Naugle
From: AA/Associate Administrator
Subject: Response to letter dated March 22, 1968, from Dr. Leo Goldberg, Chairman,
             Astronomy Missions Board

I am forwarding the referenced letter for your consideration and for your preparation of a response for my signature.

In responding to this letter, I sense that we are faced with some difficult decisions and the need for a candid assessment of our space astronomy program, both in the unmanned and the manned spacecraft. As you are aware, present planning for the Saturn V workshops suggests that launches in 1973 and 1975 might be possible, but that these workshops are expected to be devoted primarily to the study of man himself. Thus, a requirement to fly an advanced solar ATM, a large UV stellar package and a so-called EMR package, is, at first appearance, in conflict with the intended direction of the '73 and '75 workshops. This is certainly so if one considers all three systems in the same workshop.

Thus, in order to carry out a meaningful and effective space astronomy programs as defined by the Astronomy Missions Board, we will have to provide other equally effective means.

In preparing a response, would you and your colleagues, soliciting the advice of the PSG/PCG working groups, please consider the following possibilities, individually and in concert:
[2]   a.   Extension of the OAO series beyond C
         b.   Modification of our policy on experimenters in astronomy, particularly those in the observatories (OAO and OSO), to require that all major instruments be operated on a "guest observatory" basis. This consideration should include the present OAO-B and C payloads.
              (You may wish to consult with AMB to their reaction and their individual interest in using such instruments. In any event, the proposition should be put to each of the Principal Investigators.)

> c. Expansion of the Astronomy Explorer program using Scout launched payloads, and/or
> d. Modification of the basic OSO to provide:
>   (1) for larger solar physics instruments
>   (2) for application to modest stellar astronomy instruments
> e. Solar OAO
> f. Use of the workshops to carry the x-ray and gamma ray (EMR) experiments in fully automated modes so that astronaut mainline activities are not impacted but the large weights and areas of these advanced instruments can be realized.
> g. A sharp expansion of the sounding rocket program using pointing controls with recovery capabilities. (The MSFC role in such an expansion should be considered.)

I would like to have an interim response to me indicating additional or other alternatives you consider worthwhile, how you intend to follow-up on these suggestions, and by what schedule you plan to reply to the Goldberg letter.

[3] We must find a number of means to make better use of our resources and to provide more astronomers more opportunities to carry out investigations in space.

Homer E. Newell

Cc:
M/Mueller
ML/Mathews
AAF/Frutkin
AAS/Stroud
SG/Smith
OV/Files

Document I-10

Document Title: NASA Headquarters telegram to Dr. Leo Goldberg, Harvard College Observatory and others, 10 May 1968.

Source: NASA Historical Reference Collection, History Office, NASA Headquarters, Washington, D.C.

Document I-11

Document Title: Letter from E. M. Reeves, Harvard College Observatory, to Mr. D. L. Forsythe, NASA, "Postponement of ATM-A launch into 1972, reference NASA telegram 1020362 May 1968," 20 May 1968.

Source: NASA Historical Reference Collection, History Office, NASA Headquarters, Washington, D.C.

Document I-12

Document Title: Letter from Dr. Leo Goldberg, Harvard College Observatory, to Dr. John E. Naugle, NASA, 21 May 1968.

Source: NASA Historical Reference Collection, History Office, NASA Headquarters, Washington, D.C.

The uncertainties surrounding the schedule for the launch of the first Apollo Telescope Mount (ATM) were a source of frustration to many solar researchers. This exchange of correspondence captures that frustration.

Document I-10

[no pagination]
[Western Union Telex Service]
WU TELTEX CAM

PASS TO DR RICCARDO GIACCONI AMERICAN SCIENCE & ENGINEERING INC
11 CARLETON ST CAMBRIDGE MASS
TLX072 VIA WESTOVER AFB MASS R3205 10/1958L
R 102056Z MAY 68

FM NASA HAS WASH D.C.
TO RUCIRNA/DR LEO GOLDBERG HARVARD COLLEGE OBSERVATORY
CAMBRIDGE MASS
RUEBJKA/NAVAL RESEARCH LABORATORY WASH D.C.
ATTN MR J D PURCELL
RUCIRNA/DR RICCARDO GIACCONI
AMERICAN SCIENCE & ENGINEERING INC
11 CARLETON ST CAMBRIDGE MASS
ZEN/DR GORDON NEWKIRK HIGH ALTITUDE 'OBSERVATORY BOULD& COLO
ZEN/NASA GODDARD SPACE FLIGHT CENTER GREENBELT MD

ATTN MR JAMES E MILLIGAN W Y
GRNC NASA
   BT
   UNCLAS SG-7170.

In letter of April 2, 1968 to Dr. Homer E. Newell, Dr. Leo Goldberg, chairman of the
Astronomy Missions Board stated that the board agrees that the Apollo Telescope Mount
(ATM-A) mission cannot be postponed until 1972 without greatly diminishing the
scientific importance of the payload. Further, this letter states that if the mission is
postponed the payload ought to be revised and optimized for the solar minimum. On the
other hand, Dr. E. M. Reeves at the request of Dr. Goldberg has contacted the ATM-A
Principal Investigators regarding their reaction to a possible postponement of the ATM-A
launch into the latter part of 1972 and would essentially go along with such a postponement. The one exception to this is the Harvard College Observatory (HCO), whose ATM-
A experiment is not appropriate for solar minimum. A postponement of the ATM-A
launch into 1972 would make desirable the substitution for the HCO-C Spectrometer of
an instrument proposed earlier – namely the HCO-A short-wavelength spectroheliometer.
   Current planning specifies an ATM-A launch in early 1971. Planning is also
proceeding to provide backup positions to maintain a capability for launching in 1971 or
1972 in the event of problems for the first workshop, whereby it cannot support the
planned ATM operations. Accordingly, we need more information than has been
provided by the ATM-A Principal Investigators to Dr. Reeves. Specifically, we need your
response to the following questions...
   1.   To what extent do the current scientific objectives require observations of
        large flares – importance 3 or 4, or the observations of many small flares
   2.   To what extent do the current scientific objectives require the observation of
        a wide variety of evolving solar active regions
   3.   If the present instruments fly at a time when the activity is low, what scientific
        questions can be answered with these instruments What would be the
        scientific objectives and the scientific validity and justification for
        these objectives
   4.   How would the astronauts role change in a changed scientific program as in
        item 3 above
   5.   If it is possible to adjust the launch time of ATM-A plus or minus two to three
        weeks to accommodate anticipated solar activity, how would you as Principal
        Investigator take advantage of this to optimize your research program

   The astronomy missions boards letter mentioned above also states that a 28-day
mission represents a hard-rock minimum. Could you please elaborate on this in terms of
the scientific objectives discussed in item 3 above, and in terms of the amount and value
of the data to be obtained.
   We request a reply to this by May 24, 1968.
   Response to be directed to Mr. D. L. Forsythe, Code MLA, NASA Headquarters,
Washington, D.C., 20546.
   SGD Harold T Luskin, Dir of Apollo Applications Progs OMSF
   SGD Jesse L Mitchell, Dir of Physics and Astronomy Progs OSSA

Document I-11

[no page number]

May 20, 1968

Mr. D. L. Forsythe
Code MLA
NASA Headquarters
Washington, D. C. 20546

Subject: Postponement of ATM-A launch into 1972, reference NASA telegram
1020362 May 1968

Dear Mr. Forsythe:

In your telegram of May 10, you requested our response to a specific list of questions regarding the reaction of the Principal Investigators to a possible postponement of the ATM-A launch into the latter part of 1972. When I surveyed the Principal Investigators and asked their reaction should such a postponement be inevitable, they agreed that although they would unanimously support an earlier launch as stated in the Principal Investigators' previous telegram and statement to the Astronomy Missions Board, they would go along with a postponement into the latter part of 1972 with their present experiments. The exception to this was the Harvard College Observatory.

We have stated previously and reiterated in our recent telegram to Dr. Naugle and Dr. Mueller of May 15, that we felt that since the ATM-A mission was currently envisaged in the first half of 1971 that a decision had to be made concerning the substitution of our A instrument for the C instrument completely on the basis of scientific yield. We are naturally prepared to discuss this decision at length and to justify the position which we have taken.

I will therefore answer your specific questions directed toward our HCO-A experiment. For a further and more detailed description of the aims and objectives of this experiment, I refer [2] you to our scientific proposal dated March, 1966, and our Application for Manned Space Flight Experiment (form 1138) dated August, 1966.

1. The extent to which the current scientific objectives require observations of large flares.

At the present time, there is a growing amount of information indicating that the importance number (1, 2 or 3) assigned to flares on the basis of the area observed in narrow band H-α may bear little correlation to the observations in the ultraviolet or X-ray regions. Some observations indicate that the intensity of H-α brightening may be a better parameter. Observations which we are making in association with the Lockheed Solar Observatory through 2Å H-α filter and video technique indicate that flares brighten over this wider wavelength range in relatively small areas of the order of a few arc seconds and that groups and sequences of these flares may be responsible for the larger area associated with center band H-α birefringent filter observations, upon which the flare classification is normally based.

The HCO-A experiment is not primarily a flare-observing experiment. However, observations of different types of flares with a spatial resolution of 5 arc seconds over a wide range of excitation energies would be most interesting, although not mandatory for the success of the experiment. Assuming a 100-hour period in which the HCO-A experiment controlled the pointing, then the expected number of flares of all types during this time would drop from 20 in late 1968, to 5 in late 1972 and 3 in late 1973. These very approximate calculations are based on the correlation described in Smith's book, Solar Flares, and assume that cycle 20 will continue to follow the mean of cycles 8 through 19 as is indicated to date. The probability of observing a major flare of class 2, 3 or 4 from so few occurrences is quite small. To increase the chance of catching a major flare through its course of development was described in some of the meetings where we presented our observing [3] requirements for the C experiment on ATM. In the latter part of 1969, many hundreds of hours would be required in order to raise the probability of observing such a flare to a significant level. Therefore, even for the C experiment, it was never an absolute requirement to obtain this type of observation. Nevertheless we felt it was interesting enough to merit a great deal of attention in order to bias the observations in favor of catching such an event, should it occur during the observing period of the astronauts.

2. The extent to which objectives require the observation of a wide variety of evolving solar active regions.

The HCO-A experiment was designed to observe with 5 arc seconds over a wide range of excitation energies of the Sun in order to probe the structure of various features of both the quiet and active solar atmosphere. The migration of active regions across the solar disc was one of the prime objectives of the HCO-A experiment. Observations of newly-forming active regions, the changes with time as these regions develop, and particularly the limb passage of active regions, would be heavily stressed in our observing program. (See question 5.)

3. The effect of low activity on the scientific value of the experiment.

Observations of the solar atmosphere at a resolution of 1 arc minute from OSO-IV are yielding a surprising amount of information on the structure of the solar atmosphere. When the experiment was proposed several years ago, it was questionable what degree of structure could be seen on the Sun with such coarse resolution. Nevertheless, we were most pleasantly gratified to see the great wealth of structure which could be discerned with a spatial resolution of 1 arc minute. The progression from 1 arc minute to 30 arc seconds is expected to yield even more interesting structure to be observed from OSO-G. However, we have continually pressed, in accordance with the Woods Hole Study Report, to achieve a resolution of the order of 5 arc seconds or better. On a perfectly quiet [4] solar atmosphere, center-to-limb investigations in a wide variety of lines can be used to assess the optical depth as well as the distribution of temperature with height in the solar atmosphere. The question of solar abundances can be investigated in more detail. The structure of the transition zone between the chromosphere and corona can be investigated and the interesting structure around the limb of the Sun, particularly in the polar versus equatorial directions, can be assessed. Spectroheliograms at this resolution can be used to examine the temperature changes across the boundary regions of supergranulation. Five arc second resolution should yield information about spicules and interspicular material, even if only the very largest spicules are resolved. Observations in the relative intensities of the Lyman α, ß, and continuum wavelengths will be particularly significant for a more detailed study of whether certain intensity changes result from changes in temperature or abundance. Therefore it is clear that a great deal of useful and otherwise unobtainable scientific data can be obtained from the ATM, even in perfectly quiet atmosphere. Since a certain amount of activity world be expected during the late part of the solar cycle, even in 1972 the study of active regions described in the previous section would be valuable even if the number of such active regions were limited.

4. Effect of the astronauts' role in the scientific program as a result of low activity.

The role of the astronaut would remain essentially unchanged. His duties, as currently described, would be to implement a prearranged observing profile to gather information on the quiet and active parts of the solar disc to the full capabilities of the scientific instrument. His role of active participation in the observing program would remain. Should active regions begin to develop, or should small flares begin to be seen, or in the particular chance occurrence of a large flare, he would change the observing program to take advantage of the rarer occurrences. His role as a scientific participant would not be diminished. Certainly the retrieval of data for pointing records from our H-α telescope would still be required. [5]

5. Adjustment of the launch window to accommodate solar activity.

It has been described in the previous sections how the balance between quiet and active solar regions forms the basis for the observing program in a given period of the solar cycle. If solar activity were extremely low, then an effort should be made to launch the ATM during a period when the Sun is in as active a state as possible. For example, the new arrival of an active region on the east limb could signal the launch of the ATM. The astronaut could then direct the experiment to this active region during the fourteen-day passage across the disc and the particularly interesting west limb passage in which the structure versus height can more easily be ascertained. During a 56-day mission, the recurrence of this active region on the east limb could be anticipated and observations directed to that area for several days prior to and following the second east limb passage. During extremely quiet portions of the 56-day mission, observations of the quiet solar structure could receive the greater emphasis.

6. 28-day minimum mission.

The presence of active regions on the solar disc can persist for several revolutions and the development of these regions would be most interesting. The 28-day mission represents an average solar rotation, during which two limb passages and a disc passage of a particular sunspot group could be followed. The quieter the solar disc, the more important it would be to achieve a longer mission, since there would be less activity to be observed.

We need not reiterate here our subscription to the usefulness of the ATM in the solar program, even if the ATM could not be launched until late 1972. There are a number of important questions concerning the structure of the solar atmosphere which cannot be answered without experiments of the size carried by the ATM. The ATM also provides the unique opportunity, [6] not present to nearly the same extent on the smaller OSO satellites, to observe solar structures over the widest possible range of energies and to observe simultaneously with a number of different techniques. There are a number of reasons, not the least of which is financial, which could delay the launch of tire ATM into this time period. Provided only that this is anticipated at an early time, and that the experiments are chosen to take advantage of this part of the solar cycle, then we will possess the only opportunity to take high resolution data on the Sun in the early 1970's. The orderly launch of several ATM's, distributed over part of the solar cycle, would be even more important in determining the dominant mechanisms and variations in the solar atmosphere. If only one ATM can be launched, and this occurs in the late part of 1972, then we reiterate our desire to participate in the mission.

Yours truly,

[signature]
E. M. Reeves
cc:
L. Goldberg
Dr. Tousey
Dr. Milligan
Dr. Newkirk
Dr. Reidy

Document I-12

[no page number]

May 21, 1968

Dr. John E. Naugle
Code S
NASA
Washington, D.C. 20546

Dear John:

I have the feeling that in the recent exchange of telegrams and telephone calls, I did not manage to convey fully the present feeling and attitude of my group towards our participation in the ATM-A mission. For example, you obviously understand that we do strongly recommend the reinstatement of our A instrument as the Harvard experiment. What is also implied by this recommendation, however, is that in our judgment the C instrument is not worth flying on the proposed new schedule and that the ATM-A mission would be better off without it even if the A experiment were not flown.

In all of the many discussions we have had about whether the C experiment would be of significant scientific value if it were flown in early 1971, late 1971, 1972, 1973, etc., I have simply been unable to get through with what to me is the most important consideration of all. This is that we agreed to fly the C experiment to help NASA out of what then seemed to be a very serious scheduling problem, but we did so only on two conditions, namely, that the mission would be launched in 1969 and that we would be permitted to continue the development of the A and B instruments for a second ATM mission. We were dismayed when the schedule was slipped to late 1970 or early 1971, but went along both because we could still expect significant solar activity in that time period and because the new schedule still did not make it possible to be ready with the A experiment.

It is now proposed that the mission be postponed until June 1971 and that work on the A experiment be terminated as soon as the present funds run out. Based upon our past experience, I think you will agree that we are justified in refusing to believe that there will not be a further slippage of at least two to three [2] months. This is all the extra time we feel we need to be sure of getting the A experiment ready. For all of these reasons, we feel that the best course of action both for NASA and for us would be to reinstate the A experiment.

I think it is time to face up to the realization that our participation in the ATM project has been guided more by circumstance and expediency than by the requirements of first-rate science. If we do not jointly take firm action now to reverse this trend we shall be doing astronomy and NASA both a great disservice.

As I have said many times before, the ATM project is an enormous consumer of manpower and takes up so much of the time of our key people that they have very little left in which to plan even new OSO experiments, which would be scientifically much more rewarding than the C experiment. It is patently unfair to ask my group to devote such a large fraction of their effort during the next three years to an experiment that has such a sizeable probability of being unproductive. What is more to the point, I fear that they will simply lose interest, and go elsewhere. Even the engineers can sense when the scientists they serve are unenthusiastic about their product.

Sincerely yours,

[signature]
Leo Goldberg
cc: J. Mitchell
E. Reeves/W. Parkinson
H. Smith
G. Mueller

Document I-13

Document Title: Astronomy Missions Board, NASA, "A Long-Range Program in Space Astronomy," July 1969.

Source: NASA Historical Reference Collection, History Office, NASA Headquarters, Washington, D.C.

This is one of a series of reports from NASA advisory groups that sets out both challenging research goals and mission requirements for the 1970s and beyond while also endorsing the current NASA program, and particularly the Apollo Telescope Mount.

[cover sheet]

NASA SP-213

A
Long-Range
Program
in
Space Astronomy

Position Paper
of the
Astronomy Missions Board

July 1969

NATIONAL AERONAUTICS AND SPACE ADMINISTRATION

* * *

A Solar Space Program, 1969

INTRODUCTION

[149]

Part 1, "Solar Space Astronomy" [not included] was an effort by the 1968 Solar Working Group (SWG) to explain what solar astronomy is about and the very prominent part observations from space will play in advancing our understanding of the Sun. The character of "Solar Space Program, 1969" is different. This is intended as an ongoing document to present the recommendations of the SWG for dealing with the current practical problems of obtaining solar space observations necessary to advance our understanding of the Sun in the most effective manner. Since current practical problems change continually, and sometimes abruptly, the program will doubtless require frequent revisions.

In "Solar Space Program, 1969," the objective is to assess the present situation from which the program must proceed; identify the specific observations and measurements needed, and designate their priorities; translate these into terms of spacecraft and instrumental requirements and priorities; and to make specific recommendations to NASA, including flight schedules. To this end the SWG divided into subcommittees corresponding to the major solar problem areas: photosphere, chromosphere, and corona. Each subcommittee then undertook to meet the objective in its own area.

The subcommittees then met jointly to combine their findings, and devise a program for space solar astronomy that makes the best use of the several classes of spacecraft to solve the problems of solar physics in the most orderly and efficient manner. The end product will consist of–

(1) A proper priority sequence of flights, timed with respect to larger and smaller spacecraft, rockets, ground-based observations, and the solar cycle.

(2) Minimal and maximal flight schedules (which have been requested by AMB) representing, respectively, the smallest effort in space observation that would enlist the efforts of good experimenters, and the most rapid rate of progress that can be sustained by the spacecraft designers and the experimenters.

(3) Recommendations to NASA for implementing the proposed program.

TERMINOLOGY-CLASSIFICATION OF EXPERIMENTS AND SPACECRAFT

Before launching into the discussion of the program, we digress for a moment to clarify some terms that we use repeatedly. We [150] have chosen to indicate the degree of sophistication of an observing instrument by stating its angular resolution ρ, and of a spacecraft by its peak-to-peak angular pointing stability π (surprisingly, ρ for "resolution" and π for "pointing"), or its peak-to-peak absolute pointing accuracy α (generally larger than π). We will use terms like a "5" (arcsec) experiment" or a "5" (arcsec) spacecraft." A 5" experiment is an experiment with $\rho=5$" designed with all the size and refinement of telescope and accessories necessary to achieve and fully utilize 5" resolution. Similarly, a 5" spacecraft is a spacecraft with $\pi=5$" and sufficient capacity in every respect to accommodate and support 5" experiments (including whatever α is required).

Generally, an observing instrument and spacecraft are compatible if $\pi<\rho$. There will be exceptions, when, for instance, a telescope has an internal guiding system more accurate than the spacecraft pointing stability.

Spectroscopic resolution is the customary $\lambda/\Delta\lambda$.

Duration of a series of observations is t.

Time resolution, the interval between successive observations in a sequence, is Δt.

We shall occasionally use the term "video systems." By it we mean a system consisting of–

(1) A photon sensing element that simultaneously and continuously senses the intensity of all discrete picture elements of a two-dimensional image (equivalent in this respect to a photographic film) and converts it into an electrical signal.

(2) A transmitting system that ends the signals, either in real time or from an onboard storage unit, to a ground station or an orbiting station.

(3) A receiving system to permanently record the data quantitatively with a minimum of degradation, and to display them for visual inspection.

In this report we define the wavelength regions of the spectrum as follows:

IR– 9000–Å to 1 mm (the region between the photographic limit and the "shortest radio waves")

V– 3000 to 9000 Å (limited by atmospheric absorption and the longwave limit of photographic material)

UV– 1500 to 3000 Å (from V to the shortwave limit of the solar continuum)

EUV– 300 to 1500 Å (from UV to extreme shortwave limit of normal reflection optics)

[151]   X-ray– 1 to 300 Å (requiring grazing incidence optics or other imaging devices)
XUV– 1 to 1500 Å (the region containing nearly all of the chromospheric and
coronal emission lines)
High-energy spectrum– <1 Å, or > 10 keV. This region includes three reasonably
distinct ranges:

> 10-300-keV bremsstrahlung continuum from active centers.
> 300-keV-10-MeV γ-ray lines from nuclear processes (predicted, but not yet
> observed in the Sun).
> ~100-MeV bremsstrahlung from extremely relativistic electrons and γ-rays
> from π° decays (predicted, but not yet observed).

For brevity, 1" or 1' means 1 arcsec or 1 arcmin, respectively (not 1 inch or 1 foot).

REVIEW OF ACTION TAKEN ON THE 1965 RECOMMENDATIONS
OF THE SOLAR PANEL AT THE WOODS HOLE SUMMER STUDY

The report of the Solar Panel at Woods Hole contained 12 recommendations
as follows:

Recommendation 1.– That the recommendation of the Iowa meeting (1962) concerning
fine pointing be given immediate attention and that highest priority be given to the
development of triaxially stabilized rocket attitude controls, leading, as soon as possible,
to a fine-pointing system capable of an accuracy of 5 sec of arc and optimally designed for
solar use. (This recommendation is essentially a reaffirmation of that made by the Iowa
Summer Study, and is restated here to reflect the importance which the Working Group
attaches to this matter: "We recommend that the sounding rocket program continue to
receive full support; and that both the inertially guided Aerobee with fine pointing at
selected stars, and the inertially guided Aerobee with fine pointing at the Sun controlled
by an optical sensor, be made available at the earliest possible time.")

Comment: The proposed improvement (1) in sounding rockets for solar research has
been largely accomplished, and the rocket experimenters are pleased with the technical
improvements. The recommended increase in the number of rockets available and
supporting funds for their payloads has not been realized. (See Comment on 1965
Recommendations 7 and 8).

Recommendation 2.– (a) That the presently approved Orbiting Solar Observatory
(OSO) program be augmented by at least four [152] additional launchings during the
period 1970-72, inclusive; (b) that no decision be made to terminate the OSO program
after 1972 without further review at an appropriate time; (c) that NASA make every effort
to implement such desirable improvements in the OSO spacecraft as increased power,
offset pointing, localized raster scans, provision for slightly longer instruments, greater
data capacity and more flexible data format, and improved pointing accuracy (15-30 sec
of arc); (d) that consideration be given to injection of one or more OSO spacecraft into
a polar retrograde orbit in order to provide continuous surveillance of the Sun.

Comment: NASA is planning for three additional OSO launches: OSO's I, J, and K.
Although the planning goes no farther at present, NASA has indicated a willingness to
consider further OSO's if there is further useful work for these relatively inexpensive
spacecraft. Some of the improvements recommended have been made. OSO now has the

capability for offset pointing and raster scanning. Further upgrading recommended by the Ad Hoc Committee of NASA's Solar Subcommittee is presently being considered for OSO's I, J, and K.

Recommendation 3.– (a) That a satellite with Advanced Orbiting Solar Observatory (AOSO) specifications is an indispensable step in NASA's solar program, and must be flown close to the coming solar maximum; (b) that the AOSO program be accorded all the priority necessary to maintain the launch schedule shown in the Prospectus.

Comment: AOSO was canceled, and no comparable ground-controlled solar spacecraft of the 5" guiding accuracy class has even been planned. Because of its relatively short life, during which continuous operation of several different instruments is hardly feasible, the ATM is not a substitute for AOSO.

Recommendation 4.– (a) That manned missions in the 1968-72 time period, such as the Astronomical Telescope Orientation Mount (ATOM) in the Apollo Extension Systems, are desirable to supplement AOSO, but cannot replace it; (b) that because it offers the prospect of providing answers to critical questions relating to the technology of manned space telescopes and data recovery, the ATOM concept merits vigorous support.

Comment: The ATOM spacecraft has been renamed "ATM (Apollo Telescope Mount)," and Apollo Extension Systems is now the Apollo Applications Program (AAP). The manned ATM has been approved and is scheduled for a 28- to 56-day flight in 1972.

Recommendation 5.– That solar space observation be included in the manned space science program of the Apollo Extension [153] Systems in order to develop the technology of manned space astronomical operations. Such observations, which could attain resolving power of 1 sec of arc in the wavelength region 500-3000Å, mark the next logical step beyond both AOSO and ATOM.

Comment: NASA has made definite and commendable progress in introducing the role of scientist-astronaut into the manned space astronomy program. However, utilization of man's unique capabilities in operating and maintaining a space observatory appears far from realization. For example, the ATM mission will provide an initial evaluation of men as onboard observers operating semiautomatic experiments. But man's unique ability to assemble, repair, and replace experiments or vital components has not been incorporated into the design of any existing or planned mission.

Recommendation 6.– That feasibility and design studies begin immediately on orbiting solar telescopes of at least 1-m aperture designed to obtain a resolution of 0.1 sec of arc at visible wave-lengths and 0.5 sec of arc at far-ultraviolet wavelength (Å>500Å). Very large and complex accessory instruments will be necessary to analyze the solar image. Erection, operation, and maintenance of this telescope will require full utilization of astronaut–engineers and scientists.

Comment: Meaningful feasibility and design studies of telescopes of the 1-m aperture and 0.1" guiding accuracy class have not begun.

Recommendation 7.– That provision be made for a continuing, uninterrupted experimental program while the more advanced manned flights are in preparation, with many flights of various spacecraft, so that a scientist will have frequent opportunities for observation.

Recommendation 8.– That NASA find means to continue a strong program with relatively inexpensive rockets and small unmanned satellites at the same time the large manned projects are underway, since the former are indispensable to the latter.
Comment (7 and 8):
NASA has provided rockets for most worthwhile experiments through fiscal year 1968, but has failed to meet the increased demand in fiscal years 1969 and 1970, a time period that is crucial to preparations for the more advanced solar spacecraft like ATM. The OSO series is to be extended by at least three additional spacecraft of improved performance.

Recommendation 9.– That the relationship between scientists and astronaut-observers be studied and clarified. In particular, [154] we recommend that when a single, large scientific instrument is carried, the scientific observation be designated the primary mission for the flight.
Comment: The designation of the acquisition of scientific observations as the primary mission objective has been generally followed, and scientific requirements have been met whenever the requirement can be achieved within certain general Apollo Applications Program constraints.

Recommendation 10.– That NASA bring more scientists into the space-flight program as astronauts or observers.
Comment: See comment on Recommendation 5.

Recommendation 11.– That NASA move to provide additional support for ground-based solar studies. As the flight program grows in sophistication and success during the next several years, the demands on ground-based work will also increase, and NASA should in turn anticipate an increased demand upon its resources for support of ground-based facilities and operations. In addition, in the next few years, NASA should expect, and respond favorably to, proposals for a few major ground-based solar installations.

Recommendation 12.– That increased support be given to physical research in the laboratory, as required to develop improved space instrumentation for solar-physics research, to assist in the data reduction, and to make possible a full interpretation of the results.
Comment (11 and 12): NASA has continued to support ground-based solar astronomy, astrophysical laboratory research, and theoretical investigations related to solar research. The dollar amounts have remained about constant, near 2 million/yr, with no provision for the increasing costs, although in the face of reduced solar program funds in fiscal years 1968 and 1969 have been reduced. Included in this support have been about 1.7 million for the construction and operation of two major ground-based solar installations.

PRESENT STATUS OF SOLAR SPACE RESEARCH

NASA's program in solar research has showed modest progress from the beginning. The opportunities for sounding rocket research have, at times, exceeded the demand, but the program is presently falling behind the demand rather seriously. The upgrading of sounding rocket capabilities appears to be well in hand, after a considerable delay. Aircraft have proved to be practical and very useful for observations of eclipses and the far-infrared solar spectrum. Although the OSO program is behind its originally planned schedule, it has gone ahead steadily. Five OSO spacecraft [155] have been successfully launched. There have been spacecraft and experiment failures but, on the whole, the OSO's are a tremendous success. They are the most advanced spacecraft so far flown for solar astronomy, and have amply justified the effort NASA and the many experimenters have put into them.

The first four OSO's were 60" spacecraft and those presently planned are upgraded to 30". The scientific results have been invaluable and the experience acquired has prepared us for the next step, observations from 5" spacecraft.

The current status of NASA's future program for orbiting solar spacecraft is as follows OSO's F, G, H are definitely approved and the construction of both experiments and spacecraft is underway. NASA plans to extend the series by three more spacecraft which will incorporate improvements recommended by the 1968 Ad Hoc Committee of the Solar Subcommittee. They are OSO's I, J, and K for the 1972-75 period.

ATM-A is definitely approved and scheduled for flight in 1972. This will be the first of the much-needed 5" spacecraft, and the first experiment in manned operation of solar observing instruments. It is unquestionably the most important solar spacecraft now planned.

Document I-14

Document Title: J. Allen Crocker, statement of Dr. Leo Goldberg, Director of the Harvard College Observatory, ATM/Skylab, 3 November 1970.

Source: National Archives and Record Administration, College Park, Maryland

As Congress questioned the scientific merit of launching the Apollo Telescope Mount (ATM) on the initial Skylab mission, NASA contacted a number of researchers to obtain statements of support for the ATM. This statement from Dr. Leo Goldberg is typical of the responses provided to NASA.

3 November 1970

MEMORANDUM FOR THE RECORD OF TELEPHONE CALL

SUBJECT:ATM/Skylab

Dr. Leo Goldberg, Director of the Harvard College Observatory, and Chairman of the NASA Astronomy Missions Board, called and dictated the following statement to me for Dr. Newell's use in preparation of budget backup material:

The solar observatory (Apollo Telescope Mount) to be flown in Skylab, under development and construction since 1965, is undoubtedly the most important solar spacecraft now planned. It will house by far the most powerful and sophisticated collection of solar instruments ever flown in a satellite and the combined payload is designed to investigate a number of very puzzling mysteries surrounding the Sun's behavior. For example, the Skylab seeks to discover the mechanism that creates the solar corona, an enormous expanding envelope of gas at a temperature of two million degrees which surrounds the earth and reaches out to the very boundaries of the solar system. The Skylab observations may also reveal how and at what circumstances the Sun manages so efficiently and quickly to transform vast quantities of stored magnetic energy into heat, as it does when a giant flare breaks out and bathes the earth and interplanetary space with x-rays and fast moving particles. The launching of Skylab will climax ten years of preparatory work which has been so successfully carried out in the series of small OSO satellites, largely by the same group of experimenters who are involved in Skylab. While solving many problems, the OSO experiments have also sharply defined a number of the most essential and critical measurements that can only be made with instruments as powerful as those projected for Skylab. The astronomical community is eagerly awaiting the results of these essential measurements and the expected breakthroughs in our knowledge of the Sun which the Skylab mission promises to bring about.

Finally I want to underscore the importance I attach to the contribution of astronomy that the astronauts will be making both by performing certain necessary and useful tasks in connection with the experiments, and in demonstrating man's capability as a scientific observer in space.

Document I-15

Document Title: National Academy of Sciences, "Scientific Uses of the Space Shuttle," 1974.

Source: National Academy of Sciences, Washington, D.C.

One implication of the 1972 approval of the development of the Space Shuttle was that NASA intended to use the Shuttle, once it became operational, to launch all of its missions, including space science missions. This report reflects the initial planning by the solar research portion of the space science community for such a situation.

[cover sheet]

Scientific Uses
of the
Space Shuttle

NATIONAL ACADEMY OF SCIENCES

[130]

[chapter] 7 Solar Physics
I. SOLAR-PHYSICS OBJECTIVES AND OVERALL PLAN

The outstanding scientific problems in solar physics derive their significance as much from their intrinsic interest as plasma phenomena of extreme complexity as they do from their importance for the study and elucidation of a range of basic questions arising in our efforts to understand the physical universe.*

In summary, these problems center around (a) the origin of solar activity and the mechanisms underlying its various manifestations (especially flares), (b) the nature and origin of the mass and mechanical energy flux from the Sun, and (c) physical problems of broad significance that can only be studied in the Sun. We discuss these broad areas below, giving particular emphasis to the progress to be anticipated from solar observations during the Shuttle era.

A. Solar Activity

The study of the formation, heating, and long-term development and decay of active regions requires spatial correlation of observations made over a broad spectral range and over consecutive periods of a few days. For example, in order to study the interaction of rising magnetic fields with the plasma of the solar photospheric layers, long-term time-lapse observations with high spatial resolution in the visible portion of the spectrum are needed of velocity fields, small-scale magnetic fields, and features reflecting different temperature and density conditions. These observations must be correlated with [131] the uv and x-ray observations of the same areas to yield parallel data on the higher levels in the Sun's atmosphere-the chromosphere, transition region, and inner corona. Data show the spatial structure of active regions to be extremely complex and to change completely in the higher layers, where the magnetic field dominates; however, the limited spatial resolution currently available severely restricts our ability to interpret such data fully. The evolution of activity and the details of magnetic-field development will almost certainly depend on magnetic-field measurements made with high spatial resolution and extending over periods of a week or more.

Little is known about the impulsive nonthermal phase of flare development during which energy is released and charged particles accelerated to very high energies. We would like to know the location of the primary acceleration, the magnetic- and electric-field configurations, and the time sequence of the energy release processes. X-ray and radio-wave observations provide essential information on the energetic electron population of a flare, while the white-light, gamma-ray, and neutron emission give clues to the acceleration of protons. Direct measurement in space of the isotopic content of energetic flare particles promises to add still another insight into the acceleration, containment, and release of charged particles. Because theory suggests that the energy release and subsequent thermalization must take place in an extremely small-volume, high-temporal and -spatial resolution is essential.

Another area of current interest is the state of an active region prior to the occurrence of a flare. There are periods of rapid magnetic change in an active region during several hours or days prior to a large flare, during which time the x-ray, xuv, and radio emission tend to increase in intensity. Accelerated particles of comparatively low energy are observed to escape from the buildup area into interplanetary space. This, with many other aspects of the buildup, is not understood, and further observations of particle densities, fluxes, temperatures, and magnetic fields–and the associated time variations–are needed.

B. Energy and Mass Flow in the Solar Atmosphere

The mechanisms that produce the large departures from radiative equilibrium that characterizes the chromosphere and corona are not understood. Compelling theoretical and observational evidence suggests that these levels are heated by mechanical disturbances such as acoustic, magnetoacoustic, and possibly gravity waves originating in [132] the subphotospheric convection zone. The principal mechanism has not been identified in spite of the fact that recent years have produced a wealth of data on the temperature structure of the chromosphere-corona transition region as well as microscopic motions in the lower atmosphere.

Future work must provide a complete specification of the temperature, density, velocity structure, and magnetic field over the entire atmosphere from the photosphere out into the lower corona. Because these layers contain an intricate fine-scale horizontal structure, closely associated with the concentration of the magnetic field into small columns, high spatial resolution at all wavelengths is essential. Without such resolution the critical effects of the channeling of the mechanical energy flux by the magnetic field cannot be determined.

The flow of mass and energy in the solar atmosphere continues into interplanetary space in the form of the corona and solar wind. The magnetic field plays a crucial, if incompletely understood, role in modulating the flow of the material and imprinting an intricate density, temperature, and velocity structure on the plasma as it rushes out from the Sun. Space probes have measured these at 1 AU; however, the connection between these observations and structures in the inner corona is just beginning to be established. Surprises, such as the recent realization that most of the solar wind originates in quite undistinguished regions of the corona, where the magnetic field is weak and open and the density is low, can be expected to be frequent and to lead to exciting revisions of our ideas on the structure of the outermost atmospheres of the Sun and stars.

Understanding these processes requires a complete specification of the density, temperature, and magnetic field in the corona and solar wind, with good temporal and spatial resolution, so that a full three-dimensional model can be established. Since the

---

* In considering the scientific motivation, we have drawn heavily on the reports of the NASA Payload Planning Working Group (Blue Book) on Solar Physics and of the ESRO-PASOL Group and particularly on the discussions of those problems that they believe should consume a major fraction of the best efforts in solar physics through the first decade of the Shuttle era.

medium is continually evolving, synoptic observations are necessary to describe the influence of activity in the lower atmosphere on the upper levels. Moreover, high time-resolution measures are required to investigate the response of the corona-solar wind plasma to solar flares. A variety of tools will be required. Spaceborne coronagraphs have demonstrated their power on the OSO and ATM; however, these data must be supplemented by x-ray, euv, ground-based radioheliograph and coronagraph, spaceborne radiospectrographs, and in situ solar-wind measures if a complete picture is to be obtained. We would particularly stress the need for coordination of ground and space observations for incisive attacks on particular scientific objectives.

[133] C. Physical Problems of Broader Significance

Solar activity originates below the visible levels of the solar atmosphere; our knowledge of the structure and dynamics of the interior is, at best, provisional. Models provide a basis for understanding the most obvious properties of the Sun—its mass, radius, and luminosity—and show that the presence of a chromosphere and corona depends on the existence of a convection zone, some of whose characteristics are reflected in the photosphere. Similar models applied to other stars provide insight into their evolution, variability, and the processes of element synthesis.

In all these investigations, comparison with the Sun furnishes a critical test; several tests lead to only a qualified confidence. For example, the currently accepted solar models predict a neutrino flux well in excess of the measured upper limit. Also, models incorporating convection in a rotating Sun are not yet sufficiently advanced to explain the observed differential rotation of the photosphere and the characteristics of the solar magnetic cycle. With these more obvious features of the Sun unexplained, it is small wonder that more subtle questions such as the nature of supergranulation cells, solar oblateness, and the role of the solar wind in the angular momentum history of the Sun remain subjects of speculation. Likewise, broader questions regarding the presence of similar phenomena on other stars remain uncertain. The constancy of the solar "constant"—a fundamental parameter in all studies involving terrestrial climate—appears to be an article of faith.

It is clear that little progress can be made until our ideas concerning the role of turbulent convection in determining the structure of the Sun and its interaction with solar rotation are clarified. Here, a fundamental advance in the theory of turbulent convection beyond the currently used mixing length models is essential. The application of modern computational tools to these problems will be essential but may be misleading without this fundamental knowledge.

A directly related problem is the operation of the solar dynamo and the production of the solar magnetic cycle. If the investigations mentioned earlier are successful, there should be no lack of fundamental mental knowledge that would impede progress in the study of the solar cycle. Advancing our knowledge of the stability of the Sun, and the consequent implications on the solar constant, and the neutrino deficit must proceed in concert with these studies. Although progress a can be made using the current models, the stability of the Sun is most [134] certainly dependent on the coupling between the energy generating core, the radiative envelope, and the convection zone. Since the characteristics of these zones are not fully known, the presence of a solar variability independent of the magnetic cycle remains uncertain.

D. Relation of Solar Physics to Other Disciplines

The outstanding problems discussed above have an importance far beyond solar physics. Thus, once the processes of mechanical energy production, transport, and dissipation are understood, observations of stellar chromospheres and coronae could be used for further studies of stellar structure and evolution, since the extent of the subphotospheric convection surely varies with spectral type and class. Since it seems clear that small-scale photospheric features are associated with production of the mechanical energy that heat the chromosphere and corona, such motions and fine structures should exist also in the atmospheres of stars exhibiting chromospheric features; the interpretation of the spectra of such stars must rest heavily on the solution of the mass and energy-flux problem of the solar atmosphere.

Continuing studies of the solar wind will find application in understanding stellar winds and mass-loss mechanisms. The process whereby the solar wind removes angular momentum from the Sun, thus slowing down solar rotation, is basic to an understanding of the origin and evolution of the solar system and of other stars and planetary systems. This mass loss is important in determining the composition of the interstellar medium and interplanetary plasma.

Solar flares exhibit a broad range of high-energy processes, including the generation of hard cosmic rays and associated radiation, extending over the spectrum from gamma rays to radio wavelengths. The Sun provides an opportunity for detailed study of the interaction of high-energy particles and magnetic fields, since both of these characteristics can be measured directly. Such studies have clear and direct relevance to the study of other energetic objects in the universe. Similarly, the study of solar-active regions and the long-term interaction of the solar plasma and magnetic fields should increase our understanding of the coupling between solar convection, differential rotation, and the loss of angular momentum, as well us cycles of stellar activity.

We can look to a continuing stimulation of many other areas in astrophysics coming from attempts to understand the complex questions posed by solar physics. As a single example. important [135] studies of atomic processes in low-density plasmas have followed efforts 'to account for the physical state of the solar atmosphere.

Finally, as man's technical achievements mount, the importance of a detailed understanding of solar-terrestrial effects will grow. The influences of solar activity on the upper terrestrial atmosphere are well documented, if insufficiently understood. The solar wind stands gut as the principal modulator of the magnetosphere. Significant progress has been made in our ability to predict the occurrence of major flares, and a capability for accurate prediction would have economic benefits and may determine the extent to which man can work in space above the atmosphere. Finally, a possible link between solar activity and large-scale terrestrial weather patterns suggests potential significance of solar space studies to all mankind.

I. PROFILE FOR A BALANCED PROGRAM IN SOLAR ASTRONOMY

With the above objectives as guidelines, we have developed a set of goals that we believe would provide a well-balanced program in solar astronomy through the 1980's. These are outlined briefly below; more detailed descriptions are set out in Section III.

A. Spaceflight Aspects

A solar maximum satellite for the 1978-1979 period would allow, in conjunction with ground-based studies, an incisive approach to the study of solar activity in its various manifestations. Furthermore, the basic spacecraft, through Shuttle recovery, relaunch, and revisit, could provide a free-flying payload for long-duration solar experiments in the 1980's.

Basic instrumentation for a Shuttle Sortie Solar Observatory (SSO) falls into two categories. First we envisage a set of major telescopes optimized for different wavelength regions and feeding interchangeable specialized instruments (spectrographs, direct cameras, magnetometers). Second would be a versatile, fine-pointed platform for mounting special-purpose instruments that may be incompatible with the larger feed telescopes or not require their power–examples [illegible] coronagraphs and polarimeters. The larger system, at least, should be started soon to provide the opportunity of studying problems of solar activity with more powerful instruments (even if narrower in scope) than those on the free-flying satellite.

[136] While the smaller fine-pointed platform should be developed on a single pallet as a module for the sortie solar observatory, we also see an attractive possibility in its use to carry payloads on a standby basis-an opportunity whereby an available payload could be carried on an otherwise unfilled sortie mission. This concept needs study to determine its feasibility.

Because the ultimate observational needs of solar astronomy may eventually require a free-flying Large Solar Observatory, we recommend that the National Academy of Sciences convene a panel of scientists to investigate all aspects of the need and specifications for, and use of, such a facility.

B. Other Necessary Components of a Balanced Program

1. OTHER SATELLITE OBSERVATIONS

A coordinated approach to a variety of solar-physics problems requires that numerous observations be made at the same time. Many of these must be made from spacecraft flying outside the magnetosphere. Particularly relevant are very-low-frequency radio measurements, *in situ* observations of solar-wind plasma and magnetic field, and high-energy particle measurements. Specific attention should be given to the scheduling of launches of such payloads to optimize the scientific returns coordinated with the solar Shuttle missions.

2. DATA ANALYSIS AND THEORETICAL STUDIES

Adequate and sustained support for the analysis of experimental data, as for parallel theoretical studies, is imperative if the data are to be used for increasing our understanding of the Sun. This support must be provided for at the earliest planning stages.

3. GROUND-BASED OBSERVATORIES

In the Shuttle era, solar astronomy will make increasingly heavy demands on the ground-based observatory capabilities at optical and radio wavelengths. The multiparameter observational detail required in order to develop an understanding of solar phenomena necessarily results in the integration of data from a broad variety of sources. Furthermore, as the understanding of basic solar processes unfolds, it is necessary to maintain the ground-based as well as the space-based solar capabilities at the forefront of technological sophistication.

[137] 4. ROCKETS AND BALLOONS

The return from the use of rockets and balloons for solar studies has far more than justified the cost. With the augmented payload capability and excellent pointing controls now available, these experiment platforms continue to provide an important part of a balanced solar-astronomy effort. As solar astronomy enters the Shuttle era, it is important that the rocket and balloon programs be continued, both for original solar studies and for the development of Shuttle-compatible instrumentation. Only after we are well into the operational Shuttle era will experience be available to permit a reassessment of the role of the rocket and balloon capability for solar studies.

5. SUPPORTING RESEARCH AND TECHNOLOGY

In the past, the SR&T program in NASA has been pivotal in developing and maintaining the solar-astronomy program and has underlain the excellent progress in understanding the Sun and its influences. Sadly, the decrease in this type of funding in recent years has not only had an impact on established research efforts but has curtailed the investigation and development of new ideas that represent investment in the future. We most urgently recommend that SR&T support be maintained and augmented as a balanced part of the total NASA program.

III. MISSION MODEL

Table 17 is the mission model that we recommend to achieve the goals outlined. It is designed to meet the anticipated needs of U.S., European, and other scientific groups. It envisages a launch of the Solar Maximum Mission (SMM) in 1977/78, a schedule of sortie missions starting in 1980 with a buildup to four missions a year from 1983, and, starting in 1980, an annual schedule of new flights, revisits, and refurbishments of the. free-flyer spacecraft originally designed for the SMM. A certain fraction of these would carry new payloads; some would simply replace consumables on the spacecraft. The initiation schedule (SMM in 1978, first sortie at the end of 1979) is set by the coming solar maximum and is more fully documented elsewhere.

TABLE 17 Mission Model

Item	Years 77	80	81	82	83	84	85	86	87
Missions on which SSO is prime payload[a]		2	2	2	4	4	4	4	4
Missions flying the SFPP only which are not included above[b]		1	1	1	3	3	3	3	3
Sortie flights of opportunity for the SFPP[c]		1	3	2	2	2	2	2	2
Solar Maximum Mission[d] (Large Solar Observatory)	1						1		

[a] These are dedicated missions for solar physics only. They might carry into space one of the four following packages:

Payload	Average Annual Rate (full level of activity)
STC + SFPP	2
LFPP + SFPP + FF	1
LFPP + SFPP + HESP	1

[b] These are missions for which solar-physics payloads carried by the SFPP will fly with payloads belonging to other disciplines.
[c] These numbers assume that the number of rocket payloads launched per year in the Shuttle area will be maintained at the present level of activity in the United States, Europe, and Japan.
[d] The Solar Maximum Mission satellite will be launched in 1978. It will be recovered by the Shuttle in 1980, refurbished, equipped with updated instruments, and launched by one of the dedicated missions once every year.

Over a 10-year period, the total number of dedicated missions will be 34 including the following: Solar Telescope Cluster (STC), 17 flights; Large Fine-Pointed Platform (LFPP), 17 flights; High-Energy Solar Package (HESP), 7 flights; Free-Flyer Satellite (FF), 10 flights (or revisits); Small Fine-Pointed Platform (SFPP), 34 flights. The total number of missions flying the SFPP only over 10 years is 24; the total number of flights of opportunity for the SFPP for the same period is 21.

The sounding-rocket program goal of 25 flights per year would continue through 1982 at least; its continuation beyond that must be a subject for study over the coming few years as the Shuttle sortie capability becomes more defined.

Also envisaged is a Large Solar Observatory program with annual revisits, although the need for closer definition of this program is reflected in our parenthetical entry of this item in Table 17.

A. The Pre-Shuttle Solar Maximum Mission

Because of the timing requirement imposed by the 11-year solar [139] cycle, we regard a free-flying satellite, with a carefully coordinated complement of instruments for the study of the next solar maximum, as the highest immediate priority item for solar physics.

Solar activity may be expected to return in 1977 and reach a peak approximately in 1979, with the likelihood of observing major flares in a seven-day mission decreasing rapidly after 1981. We believe that an immediate start on this project is required in order to use this opportunity, which will not be repeated until 1990.

The study of solar activity, especially flares, requires a wide range of instruments to cover the electromagnetic spectrum from visible wavelengths to several MeV, where solar nuclear gamma-ray lines have been observed. In particular, the study of the effects of nonthermal particles at high x-ray and gamma-ray energies requires specialized instrumentation that was not available during the last maximum in 1968 but that is now within the state of the art. Further, the high resolution that will become available simultaneously in spatial and spectral properties of the thermal flare plasma with the generation of x-ray and euv spectroheliographs, which we believe can be developed in ample time for the Solar Maximum Mission (SMM), will allow studies that can be achieved in no other way.

1. DESIGN OF THE SPACECRAFT

The SMM presents an opportunity to develop a standard solar free-flying observatory for the Shuttle era. The SMM satellite concept developed by the Goddard Space Flight Center seems to provide an excellent basic capability that can support the pre-Shuttle SMM and that has the growth capability to accommodate instruments of the class of the Shuttle Sortie Observatory in a free-flying mode. The SMM concept envisages a Delta-launched satellite with the capability of fine pointing of some 500 kg of instruments at the Sun. The SMM concept will, by 1977, provide a pointed payload four times greater than OSO-I, with over twice the power, more than 10 times the viewing area for pointed instruments, enhanced pointing accuracy, and comparable telemetry and command capability. Such capabilities, combined in a single spacecraft, will make it possible to achieve the scientific objects with a low-cost approach. We strongly endorse the SMM concept, not only for the pre-Shuttle SSM, which is of paramount importance, but as the basis for a flexible future series of the Solar Free-Flying Observatories.

[140] Alternative concepts for the spacecraft are not necessarily ruled out; however, it would be essential that the following requirements be met to provide a viable system:

1. The spacecraft should be available for use at the next solar maximum in 1977-1979.
2. The pointing stability should be better than 1 sec of arc over a period of 5 min.
3. The spacecraft should be designed as a revisitable and reusable free-flyer throughout the Shuttle era.

2. SELECTION OF PAYLOAD FOR SMM

While we recognize that there may be strong constraints on funding experiments for this mission, it is obvious that the best science, which must be the principal objective of the mission, will not be accomplished by simply reflying experiments that have already successfully returned data, simply in the name of economy. The design of the spacecraft and the ample size and weight provision should allow new experimental approaches. We, therefore, urge that NASA ensure that experiment selection follow the proven method of open competition and impartial review.

B. Use of the Space Shuttle for Solar Research

The following sections summarize our recommendations for use of the Shuttle as a base for solar experiments and as a transportation system for free-flying satellites.

I. SORTIE MODE

We have identified four basic solar-physics sortie payloads, which can provide the flexibility to accommodate the broad range of instrumentation required to implement the observational program outlined. Two payloads have been identified as basic multiuse facilities: the Solar Telescope Cluster, which provides a basic set of optical feeds for a variety of imaging, spectroscopic, and polarization studies between 8 Å and 10,000 Å; and a High-Energy Solar-Physics Package, which can carry out similar studies between 1 keV and 100 MeV and with the high time resolution necessary to study nonthermal events. We have also defined two different size fine-pointed platforms that can accommodate a variety of specialized instruments. These four basic experiment packages are described in this section, and representative instrumentation is presented in Appendix A.

[141]

TABLE 18   Characteristics of the Solar Telescope Cluster

Wavelength Range	Spatial Resolution (sec of arc)	Type	Aperture	Length[a]	Collecting Area
1200 Å	0.1 at 5000 Å	Gregorian	100-cm, $f/5$ primary, $f/5$ overall	6m	7500 cm^2
300-1600 Å	0.5	Normal-incidence mirror	40 cm, $f/10$	5m	1250 cm^2
140-600 Å	0.5 on axis	Wolter type II	80 cm	5m	1500 cm^2
8-300 Å	1 on axis	Wolter type I	80 cm	5m	450 cm^2 ($\lambda > 20$Å)
1-40 keV	4	Oda collimator	50 cm	6 m	1000 cm^2

[a]Includes anticipated focal-plane instrumentation.

(a) SOLAR TELESCOPE CLUSTER

Since radiation emitted over the entire wavelength range from below 1 Å into the millimetric range arises in different height and temperature regimes in the solar atmosphere, a battery of telescopes is required to carry out the needed research. Our recommendation for such a battery is summarized in Table 18, while a brief description of each component appears below. (Alignment of the entire battery on a given solar feature to within 1 sec of arc, as well as independent pointing of individual telescopes to any part of the solar disk, is required.) These specifications are presented as our desired goals; we welt recognize that funding or technical constraints may delay the deployment of some elements of the ultimate cluster.

(i) OPTICAL TELESCOPE A 1-m-diameter, $f/35$, diffraction-limited telescope yielding angular resolution of about 0.1 sec of arc at 5000 Å is desired. A design goal should be to extend the technology of surface finishing so that the system can operate with similar angular resolution down to Lyman-α. Such a system has the advantage that it builds upon the technology of intermediate-size systems of 65-cm aperture, which are planned for flight in the next several years in stratospheric balloons, while representing a reasonable advance of performance. As a general-use system, such a heliograph should be equipped with a variety of final image magnifications as well as auxiliary devices such as spectrographs, filters, polarimeters, cameras, and magnetographs for investigations in the wavelength range from 1200 Å to 1 mm. Such devices should be designed for modular installation of various combinations for [142] differing scientific objectives. The use of such a telescope for nonsolar observations should be considered in its design.

(ii) EUV TELESCOPE This instrument should be designed for maximum collecting area and greatest possible efficiency (i.e., minimum number of reflections), consistent with use of high-efficiency stigmatic spectrographs as subsidiary instrumentation. It should cover the range 300 to 1500 Å with normal-incidence optics, designed to produce image quality better than 0.5 sec of arc within 1 min of arc of the optic axis. Such a system could then feed, for example, a stigmatic spectrograph of ~1-m focal length, which also, by rocking the objective by ±15 sec, would produce high-resolution spectroheliograms in a variety of lines. Other possible instruments that could be placed at the focal plane include (1) narrow-band filters (e.g., for Ly-α); (2) special-purpose spectrometers for measuring velocities, particular line ratios, or line profiles; and (3) polarimeters.

Although the efficiency of normal-incidence optics drops seriously in the far uv, it is important that every attempt be made to extend the spectral ranges to include the strong He II 304 Å line. Consideration should also be given, however, to extending the long-wavelength limit of the grazing-incidence Wolter type II telescope described below to overlap the 300- 1600 Å range.

(iii) X-RAY TELESCOPES   Adequate coverage of the shorter wavelengths will require three individual telescopes. Two grazing-incidence imaging telescopes will operate longward of about 8 Å; the shortest wavelengths are probably best covered by a nonimaging mechanical Oda collimator, although this possibility needs further study.

The characteristics of the individual Wolter-type reflectors would be tailored to provide the maximum available effective aperture for each range. The short-wavelength limit of this system is strictly set by the brightness of the source; for solar flare studies, this telescope should be usable down to approximately 2 Å.

For imaging studies, the entire collecting area is available, and a spatial resolution of 1 sec of arc (or better) should be attainable over a 1-2 min of arc field. For spectroscopic studies, the different optics allow separate spectrometers to work in the wavelength ranges from 8 Å to ~50 Å and from 40 Å to ~300 Å. Limitations of collecting area may require spectroscopic or polarization observations to be carried out at lower resolution.

[143] The Oda collimator covers wavelengths too short for effective imaging, even at grazing incidence. It will require devices to raster or scan its field of view over the regions of interest; it normally will be used to feed spectrometers or polarimeters.

(b) COARSE-POINTED HIGH-ENERGY MEASUREMENTS

Comprehensive measurements of the characteristics of x-ray, gamma-ray, and neutron emission from the flaring and nonflaring Sun would give insight into the triggering mechanism and total energy content of a flare (in conjunction with other measurements) and into the acceleration, containment, and release of charged particles. The recent OSO-7 discovery of flare-excited nuclear gamma-ray lines is indicative of the expected new results from future high-energy studies. Use of the sortie mode for these high-energy experiments permits observations to be made simultaneously with longer-wavelength experiments and accommodation of high weight and data rates. These ends could also be accomplished by an appropriate scheduling of free-flyers. The cost-effectiveness of both modes should be investigated.

The intensity distribution and its variation with time and position should be measured for photons in the spectral range of 0.001 to above 10 MeV. The flux, spectrum, and time history of neutrons should be measured; a representative set of instruments is specified in Appendix A.

The measurements taken during flares will be of great significance when compared with simultaneous radio spectral and spatial measurements and with solar-particle measurements obtained by other spacecraft.

(c) GENERAL-PURPOSE, FINE-POINTED PLATFORMS

Several scientific disciplines will require oriented platforms for the Shuttle sortie mode. To carry the full range of possible solar experiments, we recommend that two pointed platforms be developed; the stability requirements for both platforms are 1 sec of arc, but they differ in size. The smaller fine-pointed platform should accommodate instruments up to 2 m long, and might, for example, be based on a half-pallet section. This unit could be flown on sortie launches with only a limited amount of unused space or load capacity. It represents an important component of the proposed facilities; it will be ideal for carrying the type of experiment now flown on rockets. It will accommodate larger and heavier payloads than do present rockets and will permit the evolution of current rocketborne experiments. Its [144] early deployment would allow smaller scientific groups to participate in early sortie flights.

The larger fine-pointed platform should accommodate instruments up to 2 m in diameter by 4 m long and weighing up to 3000 kg. As part of the Sortie Solar Observatory it would, for example, carry large special-purpose instruments not included in the Solar Telescope Cluster or a problem-oriented package of several experiments of a size intermediate between current rocket or OSO-type experiments.

The design of these platforms must be such as to allow payload development with minimal interaction with the Shuttle itself. A clean interface for power, thermal-control, data-transfer, and experiment control functions is important.

(d) DEPLOYABLE RECOVERABLE FREE-FLYER

A semiautomated free-flying pointing platform based on an evolution of the OSO series or the proposed SMM satellite is needed for some observational programs with duration well in excess of that of a single sortie flight. Additionally, some experiments require a higher freedom from contamination than is available on the Shuttle.

A deployment recovery mode is likely to be the most efficient way of serving this class of platform since (i) there is no need to build a new spacecraft for every mission; (ii) the instrumentation can be returned, updated, recalibrated (with a high degree of confidence never reached up to now), and flown again; and (iii) spacecraft consumables and components can be replenished, repaired, or replaced within a short lapse of time.

The SMM satellite should be designed with these needs closely in mind.

(e) OTHER ASPECTS

(i) SOLAR FLIGHTS OF OPPORTUNITY   The payload carrying capability of the Space Shuttle may be used to permit observations from space in a piggyback mode at modest cost and with great flexibility. In this mode, experiments of an exploratory or developmental nature may be carried out on a space-available basis.

For solar studies, this mode would make use of the general-purpose fine-pointed instrument platform described above. It is desirable that this platform be built as a modular independent facility to permit mounting into the Space Shuttle with minimum interference to the prime Shuttle mission. For effective and low-cost [145] utilization of this mode, it is essential that this facility have clean and standardized interfaces with the Shuttle orbiter.

We strongly *recommend* that the experiment accommodation management of this facility be as direct and informal as possible in order to promote maximum utilization at minimum cost and lead time.

(ii) CALIBRATION Accurate instrument calibration is especially critical for solar observations whose analysis demands high photometric accuracy. For example, a powerful method of determining density or temperature of the solar plasma makes use of the accurate measurement of ratios of spectral line intensities—often at widely separated wavelengths—and for this the absolute values of these intensities are essential.

The Shuttle sortie mode is well suited for achieving accurate calibration since, in principle, instrument calibration can be monitored during operation and a thorough recalibration made immediately after flight.

(iii) THE ROLE OF MAN IN SOLAR SORTIE OBSERVATIONS The recent successful operation of manned space solar-astronomy experiments of ATM during the Skylab SL/2 mission has given needed perspective on the role of man in future sortie solar observations. The man-instrument interaction on ATM takes three forms: as an observer, as an operator, and as a technician

As observers the SL/2 crew have shown themselves capable of educated and thoughtful choice of pointing coordinates within the solar features chosen for study and have made important real-time decisions such as when and how to observe transient phenomena such as flares. It is fair to say that the presence of educated observers at the telescopes has greatly enhanced the resulting data.

As operators the crew have skillfully initiated complicated observing sequences, many of which occurred out of reach of ground stations and therefore could not have been initiated from the ground. It is fair to note, however, that if the Skylab had been in continuous telemetry contact, these operations could have been accomplished from the ground.

        As a technician man has been essential in Skylab; the crew erected a thermal shield, deployed a faulty solar power panel, overhauled an inoperative stellar uv experiment, repaired faulty voltage regulators, cleared the optics of the ATM coronagraph, repaired faulty experiment doors, replaced two jammed film cameras, and returned exposed film to earth. It seems probable that the usefulness of man as a [146] technician will continue to be paramount in the Shuttle sortie mode. There seem to be very strong reasons, however, for carrying out observational and operational activities from the ground. These include the following:

1.   Ground support of scientific operations can continue 24 h per day by rotation of ground personnel, thus substantially increasing the total observing time.

2.   Consultations among a number of solar scientists on the ground before and during the observational sequences will improve the quality of the observations. Reasonably high data rates would be required to operate the experiments from the ground. However, this capability would also permit returning all or a sampling of the data to earth in real-time or near real-time. This leads to a third advantage of ground-based operation.

3.   Quick-look evaluation of the data within hours or at most a day of the observation will permit updating and improving observations planned for later in the same mission. Experience on OSO's and ATM have proven the worth of quick-look data evaluation for mission planning.

        Therefore, we believe that it is important to provide the capability for ground-based evaluation, through use of a Tracking and Data Relay Satellite or other continuous high-data-rate system. One crew member should be thoroughly competent to make technical adjustments to the solar instrumentation; if he is also a competent observer, he might carry out observations directly as time permits, in close collaboration with ground-based colleagues.

IV. REQUIREMENTS IMPOSED ON SHUTTLE AND SPACELAB BY THE SOLAR PROGRAM

A. Contamination of the Optical Environment

        The Panel is concerned that the Shuttle may contaminate the local environment and the optical surfaces of many of the experiments because of the extensive use of volatile materials and the uncontrolled dumping of wastes.

[147] The Panel *recommends* that NASA establish a Shuttle Contamination Control Board to examine all materials, engineering approaches, and inflight procedures that may have implications for the contamination problem. Such a group could recommend modifications to assure that tolerable limits of contaminating gases and particulates are maintained. A similar Board operated for Skylab, and a considerable body of observational data on this problem will be available from Skylab.

B. Scheduling of Solar Missions

Because solar studies typically make use of many coordinated observations, it will be advisable to schedule solar sortie missions to coincide with supporting ground-based observations. This will in general be during May–September, as most major solar facilities are in the northern hemisphere and are located at sites where the skies are clearer during the summer than the winter months.

C. Orbital Considerations

Solar sortie missions will in general make use of orbits requiring minimum fuel consumption in order to maximize available payload weight.

Solar free-flyers should be put into orbits that will maximize recovery and revisit opportunities. Sun-synchronus [sic] missions for studies requiring continuous coverage are also possible.

D. Tracking and Data-Relay Satellite System (TDRS)

We believe that a data-relay system permitting nearly continuous contact with the Shuttle sortie is essential for maximum scientific productivity. Further, we believe that the Orbiter/TDRS wideband data link, which in the present mission model is regarded as optional, is indispensable to the effective use of the sortie mode and should be a part of the Shuttle program from the beginning.

E. Payload Capacity

It appears that the solar sortie will be limited by return payload weight for most missions, which will limit the ability to conduct coordinated experiments. *We strongly urge* that the weight landing [148] capacity of the Shuttle be increased to as near the original goal as possible.

F. Mission Duration

Some solar missions will benefit greatly from longer missions, up to the full 30-day capability. The Shuttle should be designed to minimize the payload impact of such longer missions.

G. Use of the Payload Specialist Station

Flights of the Solar Telescope Cluster and other major solar payloads will utilize, on occasion, the payload specialist in an interactive role in the experiment. However, if the required console displays and controls are housed in a Spacelab pressurized module (as presently defined), it appears that the weight of that unit will seriously limit the size of the scientific payload and may, indeed, prevent flying the full Solar Telescope Cluster. Assuming that it is impractical to increase the permissible landing weight of the Shuttle, then the best solution seems to be to design the payload specialist console to allow adequate servicing of the scientific payload.

H. Data and Control Interfacing

It is recognized that the time available for payload integration with the Shuttle may be extremely limited. We suggest that these requirements may be met if the pallet itself includes a general-purpose computer of substantial capacity (e.g., 128 kbits of direct-access memory plus mass storage capability of at least 1010 bits) that is used for experiment control and data management and the payload specialist console serves primarily as a terminal for this computer. The console should also include video and CRT displays, the latter for display of information from the computer. Although most control functions would be derived from the pallet computer, it is advisable to have several analog servo-control circuits included in the console for instrument manipulation and limited analog readouts for critical experiment monitors.

With such a design, all experiment functions and computer software could be integrated and checked out using a Payload Specialist Console Simulator prior to mounting the pallet in the Shuttle. Also, this approach will minimize mission peculiar modifications of the Payload Specialist Station, requiring only that the terminal-computer interface be standardized.

[149] V. GENERAL CONSIDERATIONS

A. The Impact of Quality Assurance on Costs

The Space Shuttle could substantially reduce the cost of transporting payloads to orbit, as well as increasing the number of flight opportunities. To take advantage of these opportunities, the cost per pound of payload must be substantially reduced. Part of this saving may be achieved by substantially streamlining the documentation and verification requirements of present-day quality assurance procedures.

We recommend that a panel of experienced Principal Investigators and satellite and experiment program managers from the various NASA Centers and from NASA Headquarters be established to examine the problem of quality assurance in the Shuttle era and to make specific recommendations on procedures for sortie instruments and for instrumentation on free-flyers. We believe that the basic quality assurance approach recommended in the Shuttle sortie model presented by NASA is an excellent one and recommend that Principal Investigators work closely with NASA to implement this approach. We also recommend asking the proposed panel on quality assurance to consider the Solar Maximum Mission proposed for 1978, since this mission will be a prototype of the free-flyer of the Shuttle era.

B. Convening of a Shuttle Experimentation Planning Committee

Because the Shuttle and sortie laboratory are still in the planning stage, it is important to establish a continuing channel for exchange of information—e.g., payload accommodations, contamination control, and pointing requirements—between the scientific community and the Shuttle and sortie laboratory planners. Accordingly, we recommend that a committee of representative experimenters be set up for this purpose; this committee could be drawn, for example, from the existing U.S. and European working groups. We also recommend that these working groups be continued.

C. Selection and Responsibilities of Scientists

We consider that the successful construction and operation of individual instruments of any size is best accomplished under the supervision of a single responsible scientist. The process of selecting experiments must avoid conflict of interest, be open at all program [150] phases, and must reflect the requirement that observing time and data are to be made available to guest investigators. A promising start in defining management responsibilities in this area is set out in detail in the report of the NASA Payload Planning Working Group.

D. The Crucial Role of SR&T Support

Supporting research and technology provides, at modest cost, the basis from which flight programs grow. The Shuttle promises to provide a splendid opportunity for deployment of new and exciting instruments. To produce these in time for solar maximum, a start must be made now on instrument development. Because of the large payloads carried by the Shuttle, a substantial effort is needed, requiring a corresponding increase in SR&T funding or special allocation of funds for Shuttle instrument development.

VI. RECOMMENDATIONS

1. The occurrence of solar activity presents a unique opportunity to investigate a broad variety of energetic astrophysical processes and, in particular, to study the role played by magnetic fields in such phenomena. For this reason, it is crucial to exploit the forthcoming maximum in solar activity (anticipated for early 1979); an equivalent opportunity will not be repeated until at least 1990.

We, therefore, *recommend* that the highest priority be given to the implementation of a Solar Maximum Mission (SMM) satellite to be launched in late 1977 to observe the upsurge of solar activity and designed to permit uprating as a free-flyer in the Shuttle era.

2. The early data from the ATM have clearly demonstrated that instrumentation covering a wide range of the electromagnetic spectrum is essential for a broad attack on the fundamental problems of solar physics. This concept can be used to great advantage on the Shuttle because of its high-payload and data-return capabilities. Such a wide variety of problems can be approached in this wad that a series of missions is required, each having different specialized detectors at the focal planes of a cluster of generalized light collectors.

We, therefore, recommend that a flight program be initiated with the aim of development of a Shuttle Sortie Observatory consisting of (a) a solar telescope cluster of large collectors covering a wide range of the electromagnetic spectrum and designed to feed different focal-plane instruments on different flights; (b) a small, fine-pointed platform for experiments of the rocket class; (c) a coarse-pointed package for high-energy solar measurements.

[151] 3. Certain important needs of solar physics are not met by the Shuttle sortie. Among these are (a) long-term, synoptic observations of such long-lived phenomena as active regions and coronal structures where moderate data rates suffice; (b) rare events, such as major flares, which can be studied only by long-duration observations; (c) contamination-free observations; and (d) observations for correlative purposes with observations from other spacecraft or from the ground. All these needs can be met by a free-flying spacecraft. The concept of periodic recovery, refurbishment, and instrument

interchange on the SMM spacecraft offers an attractive, flexible, and inexpensive solution to this need.

We, therefore, *recommend* the creation of a solar free-flyer program based on Shuttle recovery and upgrading of the SMM spacecraft.

4. Certain important solar instruments such as coronagraphs and some special xuv devices require a large fine-pointed platform but are not adaptable to the general-purpose Solar Telescope Cluster.

We, therefore, *recommend* the development of a large fine-pointed platform to accommodate these larger instruments.

5. Considerable cost savings may be realized by developing instruments usable by different disciplines and programming observations so that some of the powerful hardware developed in one area of astronomy can be used for observations in others.

We, therefore, *recommend* that close attention be given at all planning stages to the possibility of development of modular instrument packages or interdisciplinary use.

6. The ultimate observational goals of solar studies make the eventual deployment of large instruments on a free-flying platform a most attractive possibility, particularly in view of recent spectacular a Skylab, OSO, and ground observations. With the availability of the Shuttle to carry such large loads, the time is ripe to begin planning for such a program.

We, therefore, *recommend* that a panel be convened under the auspices of the National Academy of Sciences to study all aspects of a Large Solar Observatory (LSO).

7. Because of the novelty and complexity of the Shuttle operation, we recommend the establishment of a representative Shuttle experimentation planning board drawn from the disciplines to work closely with the Shuttle and sortie laboratory planners in defining experiment accommodations to be required.

8. Because of the severe weight penalty presently imposed by the use of the sortie laboratory module, we *recommend* that, as a priority matter, adequate payload specialist console space be provided [152] along with sufficient and data storage in the orbiter/sortie pallet mode.

9. Because of the planned operational mode, detailed specifically in the above text, we *recommend* that the fundamental importance to solar-physics missions of a wideband Shuttle/TDRS relay satellite capability be kept closely in mind in all planning stages.

10. The extent to which the potential of the Shuttle is realized in advancing space science depends intimately on the degree of continued input of the scientific community, especially during the planning stages. The discipline working groups constituted by NASA have set a sound direction for such communication. We, therefore, *recommend* that a continued and close interaction between scientists and planners be recognized as an essential component in Shuttle development and that appropriate mechanisms (e.g., discipline working groups) be established to ensure this interaction.

Document I-16

Document Title: Andrea K. Dupree, Chairman, Ad Hoc Committee, "Report of the Ad Hoc Committee on Interaction Between Solar Physics and Astrophysics," 18 June 1976.

Source: American Astronomical Society, Solar Physics Division Papers, American Institute of Physics, Washington, D.C.

A persistent concern among the leaders of the U.S. solar research community was the possibility of the field's isolation from other areas of astronomy and astrophysics. To avoid such isolation, a Division of Solar Physics was established within the American Astronomical Society, and there were occasional ad hoc efforts to identify areas of common interest between solar physics and other areas of astronomical research. This report reflects one such effort.

[cover sheet]
DRAFT VERSION
June 18, 1976

Report of the Ad Hoc Committee on
Interaction between Solar Physics and Astrophysics

Committee Membership

J. Beckers
Sacramento Peak Observatory

A.K. Dupree, Chairman
Center for Astrophysics

L.W. Fredrick
U. of Virginia; A.A.S.

J.W. Harvey
Kitt Peak National Observatory

J.L. Linsky

Joint Institute for Laboratory Astrophysics

L.E. Peterson
U. of California, San Diego

A.B.C. Walker, Jr.
Stanford University

[1] I. Background and Introduction

Many solar astronomers are concerned about the existing separation between solar physicists and the rest of the astronomical community. In response to this concern, the Executive Committee of the Solar Physics Division suggested that an ad hoc committee be formed. This committee was asked to recommend appropriate action to the Solar Physics Division to enocurage [sic] increased interaction between solar phyiscs [sic] and astrophysics. This Report contains the recommendations of the committee.

It is useful to summarize here the existing problem. The study of solar physics, which once was an integral part of astrophysics and the astronomical community, appears now to be a distinctly separate and isolated field of astrophysical research. Communication and cross-fertilization among the subdisciplines of astrophysics has declined. The astronomical community is largely unaware of and maybe indifferent to current research in solar astronomy.

This present situation concerns solar physicists because of the adverse effects that could result from the loss of interest and hence support from the astronomical community. Lack of support for solar physics on a national and local level can endanger funding as well as encourage a further decrease in faculty positions in solar physics. Few students are then produced or even exposed to the problems and potential in [2] the study of the Sun. The implications and also the complexity, of the situation are clear.

In fact, that solar physics is a significant, active, and vital field is well known to Division members. Its strength, vitality, and broad interdisciplinary extension are also extensively documented in two recent reports: the Space Science Board Study on Solar Physics (the "Parker Committee" report) and the Report of the Solar Astronomy Task Force to the Ad Hoc Interagency Coordinating Committee on Astronomy (I.C.C.A.).

It is not the intent of this Committee to review or to update their conclusions except to note that the field of solar physics contains numerous substantial, exciting, and unsolved problems that relate to practically every aspect of the Sun and its environment. Additionally, the study of the Sun carries with it strong interdisciplinary relationships to stellar astrophysics, atomic and molecular physics, plasma physics, and magneto-hydrodynamics, to name but a few.

This Committee feels that positive steps can be taken by the Solar Physics Division as well as by individuals to improve the current situation. Our recommendations to the SPD follow. Of course, any Division strongly relies on actions of individual members, and in some cases individual actions would appear to be the more effective ones. Suggestions to individual members follow in III.

II. Recommendations to the SPD

1. Continue meeting jointly with the American Astronomical Society. Contributed solar papers should be mixed where appropriate with non-solar papers.

2. Invited review and topical sessions sponsored by the SPD and simultaneous with AAS meetings should be encouraged.

3. The SPD should be aggressive about its responsibility to provide invited speakers to the AAS. The SPD should emphasize specific topics of common interest to solar and non-solar astrophysicists.

4. The SPD should actively initiate, solicit, and support meetings, conferences, symposia, and. colloquia, among solar and non-solar astronomers on topics of mutual interest.

5. The SPD officers and members should be aware of other societies and meetings where solar physics can have an impact. The officers of the SPD should encourage and propose a selection of solar physicists for invited papers and perhaps attendance at appropriate meetings.

6. The officers of the SPD should capitalize on new areas of astrophysics where solar physics can strongly contribute. For instance, the launch of HEAO-B where X-ray spectroscopy will be carried out on non-solar sources: IUE and ST where the ultraviolet spectrum of non-solar objects will be generally available. Solar physicists are well-acquainted with the UV and X-ray regions as well as phenomena in high temperature plasmas. The SPD should encourage interdisciplinary lectures, [4] conferences, and meetings in these areas.

III. The members of the SPD can act individually in a number of ways. The following recommendations are offered.

1. Suggest, support, and encourage the appointment of solar physicists to faculty positions. This effort is one of the most important contributions that an individual can make. Astronomical research and the production of students centers on the university. Astronomers and students need exposure to active programs of solar research.

2. Publish solar papers of general interest in The Astrophysical Journal. This may require a new point of view to be developed among solar physicists. Specifically, we should always ask ourselves the question: What are the implications of a particular solar physics observation for the theory of stellar atmospheres? In research, we should try to emphasize fundamental rather than superficial results. Point out the parallels between phenomena of the physical principles underlying the phenomena in the Sun and in non-solar objects. Papers with broad interest should go to The Astrophysical Journal or, for instance, Astronomy and Astrophysics.

3. Apply theoretical and observational techniques developed in solar research to nonsolar problems.

4. Seek to publish semi-popular papers in Science, Sky and [5] Telescope, Scientific American, or other journals read by a large fraction of astronomers. Encourage editors to include papers relating to solar physics.

5. Volunteer for lectures and colloquia. In addition, encourage and publicize visits, discussions, lectures, etc. by solar physicists at your local (institutional) level.

6. Improve the public relations and publicity for solar physics. Think seriously about the idea of a general press release for a worthy and interesting result. Make an effort to point out, publicly and privately, the relevance of solar physics to the rest of astronomy.

7. To insure continued federal support for solar physics, it is obviously important to have strong representatives on advisory committees. Relations with Washington should be cultivated at all levels.

8. The journal Solar Physics has received some criticism. It is felt that only papers of strictly solar interest should be published there. Perhaps, to make the journal more inviting to non-solar astronomers, it would be useful to reorganize the sections which,

while they cannot parallel the divisions used by <u>Astronomy and Astrophysics</u>, could at least reflect the more general interest in physical and atmoic [sic] processes as studied on the Sun and in relation to other stars. It is also the responsibility of individual members to maintain high standards for refereeing papers submitted to <u>Solar Physics</u>.

Document I-17

Document Title: Space Science Board, National Research Council, "Report on Space Science," 1975.

Source: National Academy of Sciences, Washington, D.C.

This is another report from the scientific community in the mid-1970s, pointing out the interactions between solar research and other questions of scientific interest. In this case, the focus was on the contribution of solar physics to the understanding of solar-terrestrial relations.

[cover sheet]
Report on
Space Science
1975

Space Science Board
Assembly of Mathematical and Physical Sciences
National Research Council

NATIONAL ACADEMY OF SCIENCES
Washington, D.C. 1976

Participants
Solar Physics Study

Eugene Parker, Chairman
Jacques Beckers
Arthur Hundhausen
Mukul R. Kundu
Cecil E. Leith
Robert Lin
Jeffrey Linsky
Frank B. MacDonald
Robert Noyes
Frank Q. Orrall
Laurence E. Peterson

David M. Rust
Peter Sturrock
Arthur B. C. Walker, Jr.
Adrienne Timothy, NASA Participant
Kenneth A. Janes, Study Director

[147]
[Chapter] F
Solar Physics

I. SUMMARY

A. Introduction

Solar physics has become one of the most complex subjects in astrophysics. The Sun, our daytime star, is sufficiently near that its surface can be studied in detail, revealing the many phenomena that make up the complicated personality of an average stellar object. Similar effects and variations are part of the makeup of every star, and collectively of whole galaxies, but are blissfully suppressed in the unresolved radiation from the "night-time" stars. Observations of the Sun present a variety of phenomena that at first sight defy rational explanation in terms of the familiar concepts of physics but ultimately stimulate the theoretical understanding of new effects. Historically, the chromosphere and corona are outstanding examples, having stood in contradiction to the views of thermodynamics at the time their properties were determined.

The Sun presents such diverse phenomena as the cool prominence on phase immersed in the hot coronal gas phase, differential rotation, the 22-year magnetic cycle, the sunspot, the flare, and the predicted but still missing neutrino emission from the core (which provides the only independent test of the theory of stellar interiors). Each of these effects has proved to be an enigma, with only partial understanding available. Yet each limits the understanding of all stars. One of the [148] solar contributions of greatest importance to astrophysics has been the detailed study of nonequilibrium thermodynamic systems in the solar photosphere, chromosphere, and corona. The knowledge of such systems developed from the Sun is now applied to the less tractable problem of the analysis of the radiation from the unresolved disks of other stars.

The active ejecta of relativistic particles and hot plasma seen in solar flares provide the only examples of this violent phenomenon subject to detailed diagnostic examination. Many active stars and galaxies display ejecta of similar appearance. The solar wind is the only example of a stellar wind subject to comprehensive observation and theoretical modeling. The solar flare is the only example of the stellar-flare phenomenon (that appears in such colossal form in the dwarf-M emission stars) subject to close scrutiny and analysis. It is possible to diagnose the effects of many active phenomena only with the assistance of the closeup observations and theory of the solar analogue.

Recent studies of the Sun and its variable activity, together with the long record of variation of weather and climate of earth, show that there may be both a short- and a long-term relationship. Numerical modeling suggests a close connection between global circulation at low altitudes and the ozone-induced thermal gradients in the upper atmosphere. Modeling has also suggested that a minor change in solar luminosity or ultraviolet emission may produce a major change in climate, which indicates the importance of precision synoptic studies of the luminosity of the Sun. Indeed, the whole problem of variation of terrestrial climate and solar luminosity suggests that it would be important to monitor the luminosities of other stars of the same class as the Sun.

The convective zone beneath the photosphere is responsible for the chromosphere, the corona, and the solar wind. It is also responsible for the differential rotation and for the generation of the magnetic field of the Sun, whose emergence through the surface is the agent that converts the quiet Sun into an active Sun.

Many phenomena on the quiet Sun depart remarkably from thermodynamic equilibrium. The quiet Sun is active in many ways, but on so small a scale as to be inconspicuous. The convection beneath the photosphere causes superheating of the tenuous atmosphere above the photosphere, producing the chromosphere (10^4 °K) and the transition to the corona (10^6 °K) above. The quiet Sun exhibits magnetic fields over the entire surface, appearing either as the small ephemeral bipolar regions or as individual compressed flux tubes in the boundaries between the supergranule convective cells. The spicules evidently leap [149] up through the chromosphere along these flux tubes. The transition region is rendered extremely inhomogeneous by small-scale activity.

The conspicuous features of solar activity are large, and their forms are easily observed from the ground in visible light. Their general character was discerned decades ago. Their internal workings, however, are of small scale (100 km or less) and, in some cases, visible only in radio, ultraviolet, or x rays. Their general causes are conjectured from the general observations in visible light, but their particular effect can be probed in detail only from instruments carried out of the terrestrial atmosphere on spacecraft or by such sophisticated ground-based facilities as the Very Large Array (VLA). The high-resolution ground-based observations, together with the OSO series of spacecraft and the Skylab observations, have provided the exploratory observations that define the general nature of the complex atmosphere and activity of the Sun and have begun to probe the internal working of the various phenomena. The next stage is the detailed diagnostics, coordinating the necessary high-resolution observations to determine the precise physical nature of each phenomenon.

The variety of effects presented by the Sun has led to the development of a broad observational and theoretical program aimed at exploring and understanding this complex behavior. This is not the appropriate place to review the milestones that have already been passed. Suffice it to say that a variety of problems still lie ahead, baffling and challenging and beckoning to be solved. We can group these problems loosely into two categories, the quiet Sun and the active Sun, each of which presents many separate problems.

This report outlines a general, coordinated assault on the many questions and problems presented by the Sun. Sections II and III [no other sections included] explore the various problems, Section IV describes the observations and theoretical studies that are needed to probe the many phenomena, and Section V outlines the organized observational programs from the ground and from spacecraft that will provide the necessary information.

B. Recommended Program

1. SOLAR MAXIMUM MISSION

Of the several major space efforts required, the most timely is the Solar Maximum Mission (SMM), aimed at probing the active regions and particularly the solar flare. The instruments for the mission have been defined, designed, and in many cases already tested. The experimental groups are prepared, many of the needed diagnostic techniques are [150] tested, and the solar maximum of 1979 is approaching. We consider the SMM a pivotal step in space research. A new start is required in fiscal year 1977 in order to launch it by late 1979. The predicted behavior of the active regions and flares to be observed by the SMM is discussed in Section V.B. and in Appendix A [not included].

If it is not possible to begin funding in 1977, present experimental teams will dissolve and the cost of a later start will be increased, while the declining chances of obtaining complete coordinated observations of moderate-sized (importance 2) flares will seriously imperil the scientific objectives of the mission. In that case, we feel compelled to recommend abandoning the SMM and the serious study of flares until the next solar maximum in 1990. Since solar flares and active regions are central to the understanding of solar physics, the loss of the SMM would be a serious blow to the exploration of solar activity, leaving a conspicuous gap in our knowledge at a crucial point.

Whether or not the SMM is approved, we recommend proceeding with the study of the quiet Sun, and other forms of activity, with the combined Shuttle instruments and facilities in coordination with various free-flyer missions and ground-based observations.

2. EARTH-ORBITING FREE-FLYERS

The Solar Synoptic Satellite (SSS) series would be a series of free-flyers designed to achieve substantial improvements in the quality of observations and in the continuity of coverage of the evolution of quiet and active coronal structures such as coronal holes, active regions, and streamers. These satellites will provide (a) basic solar research information; (b) long-term history of features selected for high-resolution, short-term Shuttle sortie studies; (c) observations of coronal structure to complement the Solar Stereoscopic and out-of-the-ecliptic missions; and (d) monitor the long-term variability of the solar "constant" and ultraviolet emissions.

Possible vehicles for the SSS would be follow-on SMM-type spacecraft or Explorer-class spacecraft.

Other free-flying satellites to be considered for flight in the 1980's include the "Pinhole" satellite and the Large Solar Observatory (LSO).

3. SPACELAB INSTRUMENTS

NASA is currently studying five facility-class instruments for Spacelab flight in the early Shuttle era. These include a 1-m-class optical/uv [151] telescope facility, an xuv telescope facility, a soft x-ray telescope facility, a hard x-ray imaging facility, and an euv telescope facility. Their capabilities, presently under definition, will clearly make them the cornerstone of much of the observational research discussed in the body of this report. It is hoped that by mid-1980, one or more of the facility-class instruments will be available for flight during the period around the solar maximum and during the flight of the SMM. While such facility instruments could be an enormously valuable complement to the SMM, they could in no way be a substitute for it, because of the short flight time of

Spacelab missions and the fact that they lack the extremely broad range of instrumental capability contained on the SMM.

The special-purpose instruments on Spacelab are more specialized than facility instruments. Some, such as solar gamma-ray and neutron telescopes, should be flown as early as possible in Spacelab in order to overlap with flare studies on the SMM.

4. INTERPLANETARY MISSIONS

Interplanetary missions, such as the Interplanetary Monitoring Platforms (IMP), the International Sun-Earth Explorers (ISEE), the Solar Stereoscopic Mission, and the Out-of-the-Ecliptic Missions are of great importance for solar physics, both because of their direct measurements of particles and fields and (for the latter) their capability to view conditions over the poles of the Sun.

The missions perhaps most unique in their exploratory scope are the out-of-the-ecliptic missions (OOE). Their purpose is to probe the conditions in the space outside the thin layer to which satellite orbits are now confined. The OOE spacecraft would be instrumented to study the solar wind, cosmic rays, and fast solar particles at middle (solar) latitudes, looking directly into the most intense solar active regions and, over the poles of the Sun, looking directly into the large polar coronal holes. They would be instrumented to observe the azimuthal form of the coronal structures and to observe the convection, circulation, and magnetic fields in the polar regions in order to develop an understanding of the global structure of the circulation and convection in the Sun.

The possibility of sending a probe directly into the Sun (the solar "plunger"), in order to return information from distances as close as two or three radii above the photosphere, should also be studied carefully. Planning for the earth-orbital Solar Synoptic Satellites (discussed above) should ensure their operation concurrently with interplanetary missions to obtain coordinated observations.

5. LEVEL OF EFFORT PROGRAMS

The above recommendations stress the requirements of solar space research. Important research from ground-based observatories remains to be done even if all the space missions are implemented. The value of both space and ground observations should be measured by our ability to understand them, to interpret them in terms of basic physical mechanisms that are also applicable to other fields of science, and to relate them to other phenomena of interest to mankind. We recommend, therefore, a strong level of effort program in theoretical solar research, diagnostics, computer simulation of solar processes, and laboratory astrophysics to complement and enhance the space effort.

For ground-based observations, the following will continue to be of interest:

a. Small-scale dynamics in the lower solar atmosphere related to coronal heating, convection, waves, and mass loss. (If implemented, the large optical Shuttle telescope might reduce the significance of this type of ground-based effort.)
b. Large-scale global circulation on the Sun and the shape of the Sun itself. Both are related to solar structure, solar dynamo, and solar activity cycles.
c. Synoptic observations of solar variability.
d. Thermal radio emission of the outer solar atmosphere.

e. Nonthermal radio emission associated with solar transients.

f. Neutrino flux energy spectrum.

Many of these ground-based observational programs are an important part of the overall space effort, both as an essential and integral part of the SMM and as part of a level of effort program. They will also be an important complement to the other space missions discussed in this report. The threat to the continued existence of excellent ground-based facilities, and the actual closing of one facility, is a matter of extreme gravity for the space program. We urge, therefore, that adequate support be provided to ensure the continued pursuit of high-quality ground-based solar research.

Document I-18

Document Title: Memorandum from Noel W. Hinners, Associate Administrator for Space Science, to Dr. John F. Clark, Director, NASA Goddard Space Flight Center, "Solar Maximum Mission System Definition and Execution Plan," 15 January 1976.

Source: NASA Historical Reference Collection, History Office, NASA Headquarters, Washington, D.C.

A mission to study the Sun during its period of maximum activity in the late 1970s was the highest priority objective of the solar research community for the decade. NASA finally approved the development of such a mission, called the Solar Maximum Mission, in early 1976. The mission was intended to be the first to use a standard spacecraft bus, designed for Space Shuttle retrieval.

[no page number]
[Stamped JAN 15 1976]

MEMORANDUM

TO: Goddard Space Flight Center
           Attn: Dr. John F. Clark, Director
FROM: S/Associate Administrator for Space Science
SUBJECT: Solar Maximum Mission System Definition and Execution Plan
REF: Memo fr S/Hinners to Cooper dtd 2 Dec 75, same subject

In the process of formulating the FY 1977 NASA budget, several decisions have been made that affect NASA's planning for the Solar Maximum Mission.

The Execution Phase Project Plan for the Gamma Ray Explorer (GRE) has not been approved. The System Definition and Execution Plan for Solar Maximum Mission (SMM) as approved in the referenced memo is based on the GRE development preceding the SMM with the GRE initiating the development of the standard subsystems and a spacecraft bus. Since development of the GRE will not proceed in FY 1976, the SMM Project Plan should be revised to incorporate the following guidelines:

1. The SMM project will advance the concept of standardizing spacecraft subsystems. The SMM Project Manager should provide technical guidance which will yield, as an outgrowth of the SMM development, subsystems which can be considered as NASA standards and can be used by many of NASA's future missions. This activity should be carried out within the available resources and should not compromise the basic objectives of the SMM and its experiments.

2. The SMM should be designed to be launched on a Delta 2910 and with provisions that will permit retrieval from orbit by the Shuttle Transportation System (STS). Development of a [2] Flight Support System (FSS) will be deferred. In-orbit servicing of the SMM is not a requirement.

3. Subsystems and components originally identified for use on the GRE mission should be considered for use on the SMM.

4. The R&D resources budgeted to the SMM project for the execution phase of the SMM by fiscal year are supplied as an enclosure. Provision should be made in the scientific experiment accommodations to include a solar monitoring experiment of approximately 23 Kg, 18000 cm3 volume and a maximum power consumption of 10 watts. Resources should be allotted in the budget for the development and operation of this experiment.

It is requested that a revised SMM project plan, including schedules, procurement plans, resources, and manpower, be submitted to my office for review and approval by February 27, 1976.

Noel W. Hinners

Enclosure
cc: P. Burr

ST   SD
   HGlaser   ACalio

****

[enclosure]

SOLAR MAXIMUM MISSION
R&D RESOURCES GUIDELINES
NOA ($M) INCLUDING SAMSO INFLATION

ADVANCED TECHNICAL DEVELOPMENT PHASE

FY 76   TRANSITION

1.0   0.7

EXECUTION PHASE

FY 77	FY 78	FY 79	FY 80	FY 81	TOTAL
20.1	27.5	13.4	4.7	3.3	69.0

Document I-19

Document Title: Jet Propulsion Laboratory, "Project Plan for International Solar Polar 1983 Mission," November 1978.

Source: NASA Historical Reference Collection, History Office, NASA Headquarters, Washington, D.C.

The next planned solar research mission after the Solar Maximum Mission was to be a joint effort with the European Space Agency (ESA). Such a mission had been under study for several years. This project plan represents the results of those studies, and it was the basis for project approval. NASA and ESA were to launch identical spacecraft that would transit the Sun's polar regions and would provide stereoscopic images of solar phenomena. The United States in 1981 cancelled its spacecraft contribution. (See Volume II, Chapter 1.)

[cover sheet]

Project Plan
For
International Solar Polar
1983 Mission

Jet Propulsion Laboratory
California Institute of Technology
Pasadena, California 91103

November 1978

[1-1]

SECTION I
INTRODUCTION

A. IDENTIFICATION

International Solar Polar Mission (ISPM) is the Project title (UPN 836) designated by the NASA Office of Space Science in its request for a plan under the NASA Physics and Astronomy Program. The Program Project Approval Document (PAD) (Ref. 1-1) [no references included] notes that the Program comprises Astrophysics Programs, Solar

Terrestrial Programs, and Upper Atmospheric Research. The ISPM Project is an element of the NASA Solar Terrestrial Program.

B. SCIENTIFIC BACKGROUND

Studies of the Sun and heliosphere have a central role in the space program, as this area of science is one with vast practical benefits to man. The Sun provides the controlling influence on Earth's weather and climate. Since changes in solar conditions have the potential for causing variations in weather and climate, increased human knowledge of the Sun can allow increased understanding of these variations and their implications.

The Sun is a star, and is the only star close enough that we can resolve its surface structure. The heliosphere is the only large-scale astrophysical plasma that we can observe in situ. Then by analogy, observations of the Sun and the heliosphere serve as a basis for deciding what is possible in other astrophysical settings.

Numerous space missions have been flown to study the Sun and the heliosphere. Each of these missions has been limited in one major respect. To date no spacecraft has ventured off the solar equatorial plane by more than about 15° in heliographic latitude. This limitation is a serious one. As can be seen in the eclipse photograph shown in Figure 1-1 [figures not included] or in the X-ray photograph of the solar corona shown in Figure 1-2, the solar atmosphere exhibits [1-3] pronounced variations with latitude which should result in commensurate variations in the heliosphere. To date, then, we have studied only a non-representative sample of the solar wind. Similarly, our observations of the Sun are limited by projection effects because we have observed the Sun only from a narrow range of view angles.

Recent advances have now made it possible to explore the heliosphere and view the Sun over the full range of heliographic latitudes. The major advances of note are:

(1) Planetary payload injection capability, from the Titan/Centaur launch vehicle system of the 1970's to the Space Shuttle/Inertial Upper Stage combination of the 1980's.

(2) Precision space navigation and trajectory correction maneuver capabilities, such as made possible the multiple flybys of the planet Mercury by the MVM Spacecraft and the swingbys of the planet Jupiter by Pioneers 10 and 11.

A science rationale is provided in Section III-A [not included].

C. PROGRAMMATIC BACKGROUND

Out-of-ecliptic missions have been considered by NASA almost from its formation as an agency (Ref. 1-2, 1959). In its early years, NASA also sponsored scientific meetings, of relevance to such missions, that collected and disseminated the results of space research, such as the plasma space science symposium held at the Catholic University of America, Washington, D.C., on 11-14 June 1963 (Ref. 1-3), and the conference on the solar wind held at the California Institute of Technology, Pasadena, California, on 1-4 April 1964 (Ref. 1-4).

Various studies that lead to an out-of-ecliptic mission were published between 1963 and 1974, wherein the early studies are technology-oriented and the later studies focus on projects and system designs. Minovitch described in his now-classic report (Ref. 1-5, 1963) the use of gravity-assisted trajectories [1-4] to obtain multiple-body flybys. Biermann (Ref. 1-6, 1965) reviewed aspects of the physics of interplanetary space, such as the interplanetary plasma, magnetic fields and dust, and cosmic rays. Minovitch (Ref. 1-7, 1965) next provided details in his sequel to Ref. 1-5 of out-of-ecliptic trajectories which used Jupiter to perturb the spacecraft to achieve trajectories as described in Section IV-A [not included] herein. Hrach (Ref. 1-8, 1968) described an out-of-ecliptic probe mission which used electric propulsion. Simpson and others (Ref. 1-9, 1969) reviewed the potential of an out-of-ecliptic mission for fields and particles astronomy. Hrach and Strack (Ref. 1-10, 1970) described an early application of solar electric propulsion to a 1 AU out-of-ecliptic mission.

The NASA Ames Research Center (ARC) accomplished an in-house study during FY 1971 of a Pioneer Spacecraft out-of-ecliptic mission, with supporting tasks performed by the TRW Systems Group, and thus provided the first comprehensive report (Ref. 1-11, 1971) on the use of the Jupiter swingby mode. This mission was discussed in the context of an FY 1973 New Start with launch in May 1974.

NASA assigned lead Center responsibility for out-of-ecliptic missions to the ARC in the early 1970's.

JPL accomplished an in-house study during FY 1971 of a 3-axis stabilized, solar electric (propulsion) multimission spacecraft (SEMMS) capable of the following baseline missions:

(1) Out-of-ecliptic at 1 AU or less.
(2) Mercury orbiter.
(3) Outer planet orbiter.
(4) Comet and asteroid rendezvous.
(5) Close solar probe.
(6) Direct and swingby outer planet flyby.

A project plan (Ref. 1-12, 1971) was submitted for Phase B and for planning Phase C/D as an FY 1973 New Start with earliest launch in July 1975. The final report is in three volumes (see Ref. 1-13, 1971).

[1-5] European scientists had also realized the desirability of designing an out-of-ecliptic mission. During 1971 and 1972, a preliminary study of an independent out-of-ecliptic probe was carried out by a European Space Research Organization (ESRO) group composed of three scientists and five staff members (Ref. 1-14, 1972).

Wilcox (Ref. 1-15, 1973) considered specific aspects of space exploration with an out-of-ecliptic spacecraft and gave particular attention to the solar latitude interval from about 35° to about 65°, an area where projection effects begin to hamper earth-based observations, and to the polar regions above 65°. Wilcox used the ARC-prepared description of the trajectory of a single spacecraft. ARC continued discussions with NASA Headquarters on the use of Pioneer H (a refurbished Pioneer 10/11 prototype spacecraft with refurbished experiments) for an out-of-ecliptic mission and a project plan (Ref. 1-16, 1973) was submitted for an FY 1975 New Start with launch in July or August 1976.

The ESRO Solar System Working Group, having discussed the scientific priorities for the 1980's, recommended to the ESRO Launch Program Advisory Committee (LPAC) a solar-interplanetary mission aimed at investigating the heliographic latitude dependance [sic] of solar wind properties and at performing a stereoscopic study of solar activity (Ref. 1-17, 1973).

Shortly afterwards, when defining the guidelines for ESRO scientific mission studies (Ref. 1-18, 1974), the LPAC identified two candidate projects:
(1) A solar stereoscopic mission, requiring a space probe reaching an angular distance from the Sun-Earth line of at least 40°. This mission, according to the LPAC, could probably be carried out most cheaply in the case of a dedicated solar mission by a spacecraft in the ecliptic plane. However, the LPAC recognized that if an out-of-ecliptic interplanetary mission was planned, a stereoscopic view could be obtained using the separation in solar latitude thus achieved.

[1-6](2) An out-of-ecliptic mission, reaching at least 37° of heliographic latitude (direct injection), but preferably getting to higher latitudes.
   The two candidate projects above were considered during the NASA/ESRO Science Program Review held at the European Space Research and Technology Centre (ESTEC) on February 11, 1974, when the following conclusions were agreed:
(1) Solar Maximum Mission (SMM); NASA would continue independently to study the SMM, since it seemed quite possible to use the SMM as one half of a stereoscopic mission. ESRO would continue independently its studies on a spacecraft which would constitute the other half of a stereoscopic mission.
(2) Out-of-the-Ecliptic Probe; it was proposed that NASA and ESRO jointly study mission concepts which can achieve a higher heliographic latitude via the direct injection mode. The one possibility noted was to incorporate a solar electric propulsion (SEP) module in the spacecraft. NASA would use this mission as a test flight for SEP in preparation for a 1981 Encke rendezvous mission, therefore the out-of-ecliptic mission would be launched in 1979/1980.

   The ESRO Scientific Program Board, during its meeting on April 30, 1974, decided that a mission definition study should immediately be undertaken in connection with NASA, to study the scientific objectives and technical feasibility of a combined Out-of-Ecliptic/Solar Stereoscopic mission. NASA agreed, and a Joint Mission Definition Group was established in May 1974 with 3 European scientists and 4 U.S. scientists. Program guidance was provided by ESRO Headquarters and NASA Headquarters. Technical and programmatic support were supplied by ESRO, ARC and JPL.
   A symposium on "The Sun and Solar System in Three Dimensions", organized by ESRO, was held at the European Space Research Institute (ESRIN), Frascati, Italy, in July 1974. Written versions of the talks were made available to the Mission Definition Group. The Group met at JPL in August 1974 [1-7] when ARC first proposed the dual spacecraft concept. The report (Ref. 1-19) of the Mission Definition Group was published in December 1974, and it is noted that the dual spacecraft concept proposed by ARC is first described therein. The report described the use of both the SEP and Jupiter swing by options.
   The NASA/ESRO Science Program Review was next held 4-5 February 1975. NASA and ESRO agreed:
   (1) To accept the dual spacecraft mission as primary with a single spacecraft option.
   (2) To pursue preliminary spacecraft and mission definition studies.
(ESRO and the European Launcher Development Organization (ELDO) merged about this time to form the European Space Agency (ESA).)
   A workshop on mechanisms for solar Type III radio bursts was held at the University of California, Berkeley, California, on 8-9 May 1975. Baumbach and others (Ref. 1-20)

reported the use of plasma wave experiments on satellites to determine three-dimensional trajectories of such bursts.

NASA and ESA sponsored a symposium on the study of the Sun and interplanetary medium in three dimensions at the NASA Goddard Space Flight Center (GSFC) on 15-16 May 1975 (Ref. 1-21). Over 200 European and U.S. scientists attended to review the out-of-ecliptic mission and all aspects of related science.

To carry out the studies agreed upon, ESA contracted with the British Aircraft Corporation (BAC) for a detailed spacecraft definition study to start September 1975. ARC and JPL performed in-house studies, and ARC contracted with TRW and Martin Marietta for additional support. Joint ground-rules were:

    (1) Use Space Shuttle with 4-stage IUS.

    (2) Dual spacecraft concept.

[1-8]   (3) Backup option no. 1: a single spacecraft; and Space Shuttle with 2-stage IUS plus spinning injection stage (TEM-364-4).

    (4) Backup option no. 2: a single spacecraft; and Atlas/Centaur launch vehicle plus spinning injection stage (TEM-364-4).

The BAC 4-volume final report was published in April 1976 and provided a complete technical description of the ESA spacecraft and all of its interfaces, a system specification, system analysis and definition, subsystem analysis and definition, development plans and cost estimates (see Ref. 1-22). The ARC final report was also published in April 1976 (Ref. 1-23). ESA published its own final report in May 1976 (Ref. 1-24).

NASA assigned overall project management responsibility for continuing the out-of-ecliptic mission studies to the JPL on 1 July 1976. NASA and JPL formed an Out-of-Ecliptic Science Working Group (SWG) in August 1976 to provide science guidance to JPL in-house studies during FY 1977. (The SWG report is found in Ref. 1-25, and the JPL final reports are found in Refs. 1-26 and 1-27.) ESA contracted with BAC for additional studies. The JPL and ESA/BAC study groundrules were:

    (1) Use Space Shuttle with 4-stage IUS (no change).

    (2) Dual spacecraft concept (no change).

    (3) No backup options.

    (4) Two modes of NASA/ESA cooperation:

        (a) Each to provide one spacecraft.

        (b) NASA to provide selected subsystems and ESA to provide both spacecraft. (This mode was eliminated from further consideration in January 1977.)

The NASA Cost Review was held at JPL in May and July 1977. Subsequently the Project was submitted to the Congress as a (proposed) New Start for FY 1979.

ESA plans to contract competitive studies for the period January through July 1979, followed by study by a single contractor from September through [1-9] December 1979. The ESA contract for development of the ESA Spacecraft is planned to be signed in January 1980. The details of the planned JPL contract-ing effort for the NASA Spacecraft are provided in Table 6-2 [not included] herein.

D. SCOPE OF PROJECT PLAN

This Plan defines the Joint NASA/ESA International Solar Polar 1983 Mission. Two spacecraft will be launched from Cape Canaveral, Florida, by a single NASA Space

Transportation System in February 1983, to a Jupiter gravity–assist swingby over the poles of the Sun. The spacecraft provided by the United States is designated the NASA Spacecraft. The spacecraft provided by the European Space Agency is designated the ESA Spacecraft.

Technical and implementation information related to the NASA spacecraft contained in Sections IV and V, such as the descriptions, mission design, trajectory characteristics, mission events and maneuvers, is the result of studies at JPL and elsewhere and must be considered as representative or typical data which is not final in any sense. Technical information on the ESA Spacecraft is provided where appropriate, again with recognition that this information is representative, but not final, data.

Technical information on the NASA and ESA science payloads is based upon the proposals submitted by those scientists who were conditionally selected as Principal Investigators, and is more subject to revision before science confirmation than afterwards.

This Plan presents neither requirements for the system design concept study contractors nor conclusions of the system design concept studies.
[2-1]

SECTION II
PROJECT PLAN SUMMARY

The primary mission objectives of the ISPM Project are to extend scientific knowledge and understanding through exploration of the Sun and its environment, and to investigate possible mechanisms coupling solar variability to terrestrial weather and climate by studying the Sun's structure and emission as a function of latitude from the solar equator to the solar poles. The secondary mission objectives are to perform investigations of interplanetary physics during the initial Earth-Jupiter phase and the Jovian magnetosphere during the Jupiter flyby phase.

The ISPM Project uses a NASA Spacecraft and an ESA Spacecraft. The two spacecraft are launched from Cape Canaveral, Florida by a single NASA Space Transportation System (STS) with a 3-stage Inertial Upper Stage (IUS) combination into an interplanetary orbit toward Jupiter. Shortly after injection the two spacecraft are separated from the third stage of the IUS. The flight paths take both spacecraft nearly in the ecliptic plane to Jupiter with a small separation in Jupiter arrival time achieved by post-launch maneuvers. By choosing the proper Jupiter encounter strategy, the gravitational field of Jupiter is used to deflect both spacecraft into out-of-ecliptic trajectories, one north and the 'other, south. After Jupiter flyby, both spacecraft travel in heliocentric out-of ecliptic orbits with high heliographic inclination and passages over the rotational poles of the Sun at 1.3 AU to 2.0 AU from the Sun.

Launch opportunities to Jupiter occur every 13 months. The more favorable geometries, however, occur on approximately six year centers, when Jupiter crosses the ecliptic plane. The most favorable geometry occurs next in December 1981/January 1982. The following opportunity (February 1983) will be used by the ISPM Project with a 10-day launch period. End-of-Mission occurs on or before 30 September 1987.

[2-2] The Announcement of Opportunity for the ISPM mission was issued jointly by NASA and ESA on 15 April 1977. NASA followed the guidelines and procedures for acquisition of investigations defined in Refs. 2-1 and 2-2, with minor modifications occasioned by the

joint process. A conditional selection of the science payloads for each spacecraft was announced jointly on 15 February 1978. In addition to the hardware investigations, NASA and ESA also conditionally selected scientists for a joint Radio Science Team and for participation in theoretical and interdisciplinary science investigations. NASA and ESA then formed a joint Science Working Team from among the conditionally selected scientists.

NASA Spacecraft conceptual designs are currently being developed by two aerospace contractors. Each configuration under study incorporates significant elements of that firm's previous spacecraft designs. A final competition will be conducted furing [sic] FY 1979 between these two firms and will lead to the selection of a single contractor to accomplish the design and development of the NASA Spacecraft.

The ISPM is a joint international project. The United States will supply:
(1) The NASA Spacecraft.
(2) One partial and five complete science experiments for the ESA Spacecraft.
(3) Department of Energy radioisotope thermoelectric generators for both spacecraft, and one electrical thermoelectric generator (a simulator for test purposes) to be shared by NASA and ESA.
(4) Launch operations services and launching by the NASA Space Transportation System.
(5) Tracking and data acquisition (TDA) for Earth-orbital checkout of both spacecraft and TDA from deep space.
(6) A mission control and computing facility with hardware, software and personnel to conduct flight operations.
(7) Appropriate processed data records to scientific and engineering personnel.
(8) Technical advice and consultation.

[2-3] The ESA Member States (Belgium, Denmark, France, Germany, Ireland, Italy, Netherlands, Spain, Sweden, Switzerland, and United Kingdom) will supply:
(1) The ESA Spacecraft.
(2) Software and personnel to manage and support ESA flight operations and data processing at the U.S. facility.
(3) Technical advice and consultation.

The Federal Republic of Germany (FRG), through its Federal Minister for Research and Technology (BMFT), will supply three complete science experiments for the NASA Spacecraft.

The Project plans to use existing technology, equipment, NASA standard subsystems, NASA data system standards, and existing NASA and contractor facilities to the maximum extent possible. Development is limited mainly to the science instruments and to their data processing.

The Jet Propulsion Laboratory has been designated to manage the Project for the NASA Office of Space Science. JPL will manage the NASA Spacecraft System, which will be procured in a system procurement mode. The NASA Space Transportation System will be managed by the NASA Johnson Space Center. The STS/ISPM Project management interface is being defined. The JPL has been designated by OSTDS the lead Center responsible for all Tracking and Data Acquisition support for the ISPM Project. The data collecting and processing activity will use the NASA Spaceflight Tracking and Data Network (STDN), the TDRS, the JSC Mission Control Center, the Deep Space Network, the NASA Communications Network, and the JPL Mission Control and Computing Center facilities.

Joint management relationships between NASA and ESA, formal systems reviews and design reviews are discussed herein.

[2-4] No environmental issues have developed to date.

At the July 1977 NASA Cost Review the planning estimate for development of the total baseline project through launch plus thirty days, excluding STS costs, was established at $178.1 M, inflated. The planning estimate for operations of the total baseline project was established at $37.0 M, inflated.

Document I-20

Document Title: Memorandum from Goetz Oertel, NASA Deputy Chief, Solar Physics, to Director, Physics and Astronomy Programs, "Priorities in Instrumentation Development for Solar Astronomy," 7 October 1979.

Source: NASA Historical Reference Collection, History Office, NASA Headquarters, Washington, D.C.

High-resolution observation of solar phenomena was a top-priority objective of solar researchers from at least the late 1960s. (See Document II-13.) At the end of the 1970s, NASA began to plan for instruments that would facilitate such observations. The mission to incorporate such capabilities became known as the Solar Observation Telescope.

[no page number]
[Stamped OCT 7 1979]

TO: SG/Director, Physics and Astronomy Programs
FROM: SGS/Deputy Chief, Solar Physics
SUBJECT: Priorities in Instrumentation Development for Solar Astronomy

1. Goals in Solar Astronomy

Research in Solar Astronomy falls into two categories, depending upon the angular and time scales of the phenomena under study. Large scale aspects of the Sun and phenomena can be studied with presently available angular resolution from space and from the ground depending upon the absorption properties of the earth [sic] atmosphere. Thus, the structure and physics of the solar atmosphere near the temperature minimum can be studied with suitable problem-oriented payloads such as on the pointing section of OSO-I and the general solar magnetic field can be studied by ground-based techniques and by measurements in situ in the interplanetary minimum. On the other hand, there are phenomena on the Sun which can be understood only if angular and time resolution are significantly improved while maintaining a high spectral resolution. Examples include flares, spicules, intergranular space, certain filaments, and other phenomena or structures which we have identified but which – by all indications in hand – have a significantly smaller fine structure than is presently resolvable. The indications include glimpses of smaller detail at rare moments of superb seeing, magnetic field gradients of tens of kilogauss per arc second, and rapid time variations of flare related emissions.

While problem oriented payloads can be chosen for sufficiently large scale phenomena, solar physics is still in the exploratory phase when it comes to small scale phenomena which require angular resolution of better than one arc second. Physical [2] understanding of these phenomena requires as a first step the exploration of their detailed structure. Spectral studies of these structures will then identify the physical mechanisms at work and will lead to the understanding we seek.

2. Recommendation 1

It is therefore my position that the development of a capability to study structures with angular sizes of .1 to 1 arc seconds has highest priority in the exploratory portion of the solar physics program. The long range goal is a 1.5 meter diffraction limited solar telescope in space. A diffraction limited solar telescope of half this aperture is the logical intermediate step, not because it may happen to be possible on some particular mission, but because an intermediate step is in my opinion necessary for the attainment of the goal.

The "photoheliograph" has been under development for a few years now and has had the benefit of continued support by Hal Zirin and his group at the California Institute of Technology. It represents the logical next step to take and should be implemented. With a universal filter it will be capable of resolving structures about 3 times smaller in diameter, or 9 times smaller in area, than have been resolved so far. It may in fact be sufficient to complete the exploratory phase of the study of many solar phenomena. It will be sufficient, in all likelihood, to discover new ones and perhaps give surprises, and certainly a great deal of excitement.

It happens that the photoheliograph will fit the present ATM canister. While the flight of a second solar ATM with this instrument would be desirable and an excellent way to take this important step on a timely basis, the photoheliograph as an advanced solar space telescope is not dependent upon a second solar ATM, but upon the availability of any mission with the necessary capabilities.

The development of a diffraction limited photoheliograph will take a considerable effort in time. It is impossible to come up with such a system on short notice. It is necessary to continue the development at Cal Tech vigorously so that a flight unit can be produced on relatively short notice when an opportunity arises.

[3] 3. Recommendation 2

Next highest priority is given to an X-ray telescope system with angular resolution higher than 1 arc second, with large effective aperture to assure high time resolution, and with associated spectroscopic equipment to provide spectral information.

[signature]
Goetz K. Oertel

Cc: SGT/Aucremanne
     SGT/Chase
     MLA/Forsythe

Document I-21

Document Title: Letter from Harold Glaser, Director, Solar Terrestrial Division, NASA, to Dr. Thomas A. Mutch, Associate Administrator for Space Science, 30 May 1980.
Source: NASA Historical Reference Collection, History Office, NASA Headquarters, Washington, D.C.

Document I-22

Document Title: Letter from James M. Beggs, NASA Administrator, to the Honorable Larry Winn, Jr., House of Representatives Committee on Science and Technology, about the Solar Maximum Mission, 16 April 1982.

Source: NASA Historical Reference Collection, History Office, NASA Headquarters, Washington, D.C.

The Solar Maximum Mission was launched in February 1980, and it began to return a constant stream of high-quality data. NASA soon considered whether it was desirable and feasible to retrieve and refurbish the satellite for a second launch for additional research missions. A plan to do that was developed in May 1980. When the satellite lost its fine-pointing capability in December 1980, its retrieval and repair became of even greater interest, since its other instruments were still operating well. NASA had to fight to justify the retrieval mission, which finally took place in April 1984. The satellite was repaired on orbit and returned to useful service.

Document I-21

[no page number]
[stamped 30 MAY 1980]

TO: S/Associate Administrator for Space Science
FROM: ST-5/Director, Solar Terrestrial Division
SUBJECT: Solar Maximum Mission Retrieval

In response to your request, I have developed a Division recommendation on the possible retrieval of the Solar Maximum Mission spacecraft.

The Solar Maximum Mission development program, authorized in 1976, has been funded for orbital operations through February 1981. NASA has submitted a funding request for two additional years of operation. Based on the February 14, 1980 launch, the three years of operations will extend through February 1983. Due to orbital decay the SMM will have a finite active lifetime. Actual orbital data yield an earliest possible reentry time of early 1983. The dispersion on this projection is quite high, but if retrieval is incorporated as a firm requirement, we must assume the earliest possible reentry date. At first look the likely end of useful lifetime and the planned operational time (through Feb 83) are compatible.

Requirements were originally placed on the SMM that the spacecraft be designed to be compatible with both Shuttle and expendable launch vehicles and also that the SMM be capable of retrieval by the Shuttle. The SMM has complied with these requirements in that it has been designed to be mechanically and electrically compatible for retrieval by Shuttle. The OSS is managing the development of a Flight Support System (FSS) to facilitate such retrieval. The FSS was designed with SMM funds and development is funded by STS, to provide a retrieval capability for the Shuttle.

[2] To retrieve SMM there are several elements of the Space Transportation System (STS) that must be provided. First, a Shuttle launch opportunity during the period must be identified. Second, the STS must certify that the orbiter has the capability to rendezvous with the spacecraft at approximately 250 n mi. Finally, the Flight Support System development must be completed, tested and interfaced with the orbiter. Prior to a commitment to retrieve, there should be a study to define the specific requirements for retrieval and the associated costs and to determine the disposition of the retrieved spacecraft. It should be noted, that there are no authorized or funded retrieval plans for the SMM by either OSS or STS. In fact, the most recent STS manifest does not include the SMM retrieval.

One possible utilization for the retrieved SMM would be for the Solar Corona Explorer/Mission that is currently undergoing scientific and technical study. This project will study origin of the solar wind using several newly developed diagnostic techniques, and the three dimensional evolution and structure of the Sun's corona. The strawman payload includes a coronagraph. Thus, a recovered SMM would provide a spacecraft bus, solar pointing and guidance systems, instrument mounting systems and at least one of the four or five experiments required for an SCE. The tentative launch period would be 1987 or 88, in time to phase with the solar polar passages of the ISPM.

There appears to be center support for a retrieval of the SMM. Staff of the Goddard Space Flight Center have briefed the Office of Space Transportation Operations on a proposed plan to demonstrate in-orbit servicing and retrieval utilizing the SMM Spacecraft.

Conclusion

The SMM is currently returning high quality data and should be operated as long as meaningful scientific data can be obtained. The retrieval of the SMM will require an intensive effort over the next three years. Firm program commitments will have to be made for manpower and resources in FY 81.

[3] Retrieval would be a new and unique undertaking involving the Space Transportation System in a substantial way. As a first step in determining an Agency policy on SMM

retrieval I suggest that you meet with Glynn Lunney to discuss potential benefits and management structure of such a retrieval. Funds will have to be budgeted in FY 80 for definition studies and in FY 81 and subsequent years for the actual retrieval. If the Agency, for policy reasons, determines that a retrieval program is desirable to recover a reusable spacecraft and to demonstrate STS performance we should support the effort but with the provision that the SMM be operated through its useful lifetime. If there is a decision to retrieve SMM, I recommend consideration of reuse of the spacecraft by the SCE.

[signature]
H. Glaser

Document I-22

[no page number]
[Stamped APR 16 1982]

Honorable Larry Winn, Jr.
Committee on Science
  And Technology
House of Representatives
Washington, D.C. 20515

Dear Mr. Winn:

I understand that the Committee on Science and Technology will soon consider the recommendations of its Subcommittees on the FY 1983 budget for programs of the National Aeronautics and Space Administration. The Space Science and Applications Subcommittee has recommended a significant change in the program proposed by the President by deleting FY 1983 funding for the Solar Maximum Repair Mission and recommending that NASA's FY 1982 reprogramming request for this demonstration be rejected. I strongly urge you to reconsider these proposed actions when the full Committee meets on this subject.

As described in our reprogramming request dated February 19, 1982, the proposal to rendezvous with the Solar Maximum Mission spacecraft and perform on orbit repair has a very high priority in the total NASA program. We recognize and appreciate the Committee's concern about the availability of resources. We have attempted to address this concern and, indeed, as described further below, have managed to significantly reduce the total cost of the mission such that the previously identified FY 1982 requirement of $35 million (the amount deleted by the Space Science and Applications Subcommittee) is now approximately $18 million.

I am also concerned that we have perhaps failed to adequately convey the scientific potential of this mission. During the period of full operation of the Solar Maximum

Mission, the scientific data were unsurpassed; they provided an important link in the continuous study of the Sun and its interrelationship with our environment. However, many of the phenomena discovered during the first phase of the mission require deeper investigation. For example, the mission has shown that the amount of energy emitted by the Sun changes with time; these changes have strong consequences which could effect the Earth's climate. Continuation of these measurements through the solar cycle would be of great scientific importance. For instance, a repaired Solar Maximum Mission would be able to continue the study of solar flares through the observation of the very strong flares which usually occur a few years following the sunspot maximum. A repaired SMM could also carefully test the various flare models and theories which have been derived as a result of the original mission.

Although the Solar Maximum Mission was an unqualified success for seven months, the observatory was expected to remain fully functional for three or more years. Reacquisition of high quality solar data through repair of the SMM is much more cost effective and timely than attempting to replace the spacecraft which cost more than $300 million in terms of equivalent FY 1983 dollars.

[2] Additional benefits could also be realized by restoration of the SMM spacecraft. An operational SMM in 1986 would provide the capability to observe Halley's Comet in some detail while the comet is in its closest proximity to the Sun. The repair mission, planned for early 1984, would occur at a time when the Tracking and Data Relay Satellite System (TDRSS) will have become operational and, because the Solar Maximum Mission spacecraft has provisions for use of the TDRSS, deployment of its antenna will significantly improve the science data acquisition from the observatory. Operating techniques developed for the SMM-TDRSS combination would provide an important foundation for a similar system planned for Space Telescope operations.

Space Shuttle operational experience gained on the first three orbital test flights has made a very positive contribution to the assurance that necessary Shuttle systems required to successfully perform the Solar Maximum Repair Mission are available. In addition, the opportunity to demonstrate and validate the Space Shuttle as something more than just a payload launcher could have international implications. At a time when international competition for launching space systems is becoming formidable, demonstration of the versatility of the Space Shuttle as a servicing and repair "work bench" could be instrumental in convincing future users that the Shuttle system provides added assurance that their payloads can be made to operate successfully after enduring the rigors of launch and space exposure.

As mentioned earlier, the resource requirements for the Solar Maximum Repair Mission have been significantly reduced through the adoption of an ascent trajectory which obviates the need for the use of the Orbiter Maneuvering System Payload Bay Kit. This revised plan has reduced the necessary FY 1982 funding to approximately $10 million as opposed to the earlier estimate of $35 million. The total mission costs for the SMM repair are estimated at $45-55 million, which is comprised of specific repair costs as well as planned generic STS hardware. While this estimate excludes Shuttle launch costs, the launch costs assignable to the SMM repair are minimal because the SMM support hardware would occupy only approximately twelve feet of the sixty-foot payload bay of the Shuttle. The proposed SMM repair mission is manifested on a shared basis with the deployment of the approved Long Duration Exposure Facility (LDEF), and the LDEF will occupy most of the payload bay of the Shuttle Orbiter.

In closing, let me say that I have been very impressed by the response from the scientific community in support of this mission. At a meeting in San Diego during the week of March 22, a distinguished group of solar physicists prepared a unsolicited white paper on the benefits of restoring the Solar Maximum Mission. I am enclosing a copy of the Summary Report of this group, as well as a summary of the scientific opportunities which would be lost if the repair mission is not undertaken. Additional endorsements of the science benefits of the mission have been made by the Committee on Solar and Space Physics of the Space Science Board of the National Academy of Sciences.

I urge you to reconsider the Subcommittee's recommendation, and to support the Solar Maximum Repair Mission as submitted in the President's budget request. I am available for further consultation concerning this matter at any time.

Sincerely,

[signature]
James M. Beggs
Administrator

****

[no page number]
ENCLOSURE 1

Scientific Importance
of the
REPAIR OF THE SOLAR MAXIMUM MISSION

Summary of the Report of the Ad Hoc Committee

The Solar Maximum Mission was launched February 14, 1980, following a decade of planning by NASA and the solar physics community. Intended for an operational life of 2-3 year, its scientific usefulness was seriously impaired after nine months because of failures in the spacecraft attitude control system. There is strong scientific justification for its repair and return to full operation. Reactivation of the spacecraft is necessary to complete the original objectives of the mission, and offers a unique opportunity to address a new set of objectives. These have arisen in part from discoveries made by the Solar Maximum Mission, in part from the existence of new theoretical and observational tools. Renewal of the mission will provide valuable data concerning solar plasma phenomena, solar-terrestrial relations, solar internal structure and dynamics, and variability in the Sun's energy output.

Loren W. Acton   Lockheed Palo Alto Research Laboratories
R. Grant Athay   National Center for Atmospheric Research
Jacques M. Beckers   University of Arizona
Richard C. Canfield   University of California, San Diego
John A. Eddy   National Center for Atmospheric Research
David J. Forrest   University of New Hampshire
John W. Harvey   Kitt Peak Observatory
Thomas E. Holzer   National Center for Atmospheric Research
Lewis L. House   National Center for Atmospheric Research
Hugh S. Hudson   University of California, San Diego
Gordon J. Hurford   California Institute of Technology
Richard E. Lingenfelter   University of California, San Diego
Robert M. MacQueen   National Center for Atmospheric Research
Peter A. Sturrock   Stanford University
Roger K. Ulrich   University of California, Los Angeles
Arthur B. C. Walker   Stanford University

March 25, 1982
La Jolla, California

****

[no page number]
ENCLOSURE 2

SOLAR MAXIMUM MISSION

Impact of Failure to Repair

The SMM was launched in 1980 as a two to three year mission focused on the study of solar flares. Because of failure of the coronagraph/polarimeter after seven months and the loss of fine pointing after nine months, many scientific opportunities were lost.

Following are examples of scientific opportunities which will be lost if the SMM spacecraft and coronograph are not repaired:

Comprehensive observations of great, terrestrially important flares.
Observations, by the SMM imaging and spectroscopic instruments, of X-ray producing flares.
Observations relating flares to transients affecting the outer corona and interplanetary space.
Precision data on variability in the total radiative energy of the Sun over a long time base.
High resolution X-ray spectra of diagnostic importance to understanding solar plasma and plasma processes.
Observations tracing magnetic fields in the outer solar atmosphere.

Observations on the sunspot cycle dependence of flare, solar wind, solar constant and
  solar oscillation properties.
Complete many guest investigator programs on both flare and non-flare topics.

Document I-23

**Document Title: Letter from Peter A. Gilman, Senior Scientist, Head, Solar Variability
Section, High Altitude Observatory, National Center for Atmospheric Research, to Dr.
David Morrison, Acting Deputy Associate Administrator for Space Science, NASA,
8 April 1981.**

Source: Federal Records Center, Suitland, Maryland

Document I-24

**Document Title: Letter from Dr. David Morrison, Acting Deputy Associate Administrator
for Space Science, NASA, to Dr. Peter Gilman, High Altitude Observatory, National
Center for Atmospheric Research, 30 April 1981.**

**Source: NASA Historical Reference Collection, History Office, NASA Headquarters,
Washington, D.C.**

*NASA in February 1981 was forced by the incoming Reagan Administration to cancel one of its
approved scientific missions as a budget-cutting measure, and it chose to cancel the U.S. spacecraft
that was to be part of the NASA-ESA International Solar Polar Mission. This action produced anger
and dismay in the U.S. solar research community. This exchange of letters captures the intensity of
the feelings involved.*

Document I-23

[no pagination]

April 8, 1981

Dr. David Morrison
Acting Deputy Associate Administrator
for Space Science
NASA
Washington, D.C. 20546

Dear Dr. Morrison,
  I just received your "dear colleague" letter of 16 March concerning the amended FY
81 and 82 NASA budgets. I must tell you that in light of recent actions taken by NASA

management to attempt to scuttle the International Solar Polar Mission, including particularly the way the various relevant scientific advisory groups were treated, I take an extremely cynical view of the pious statements of your opening paragraph, to wit "Communication between those of us in Washington and the real world of science in the universities and NASA centers is particularly important in the current environment of budget cuts and reexamination of program priorities at NASA. We should all work together to use the limited resources available to the best advantage and to continue to make the case for a strong space science program in the future."

In the ISPM matter, I believe NASA acted most unwisely in either failing to seek or flagrantly disregarding the advice of every science advisory body that had any interest. As a member of one of these bodies, the Committee on Solar and Space Physics of the National Academy, I felt our committee was unable to deal with the most important budgetary issues because we were largely kept ignorant of NASA management's overall strategic thinking. Thus members were forced to respond after the ISPM cuts were announced and attempt to light a back fire. I would hardly call this "good communication" or "working together."

This experience raises serious questions in my mind as to whether NASA is really interested in the advice of scientists in the community on important matters, or whether we are just ornaments added to lend credibility. Your testimony before Congress concerning ISPM cuts implies you believe this program has not been seriously harmed - and yet you know it has, and you know the space science community both in the U. S. and Europe believes it has. A congressman reading your testimony, knowing NASA is supposed to listen to outside scientific advice, might think you had credible scientific backing for the judgment that ISPM remains a strong scientific program – when nothing could be further from the truth.

[2] I could go further into the arguments as to why crippling ISPM was so unwise even in a time when we are forced to face the budget setbacks of the current Administration, but I won't, because you have heard them all already. By this letter, I am instead expressing my deep concern over what I see as a serious erosion of the credibility of the process by which NASA seeks scientific advice and acts upon it. If such erosion continues, it will greatly impair the ability of NASA and the scientific community to achieve its common goal of a strong, productive space science program.

Sincerely yours,

[signature]
Peter A. Gilman
Senior Scientist
Head, Solar Variability Section

cc:
Dr. Louis Lanzerotti
Dr. Richard Hart
Dr. Alastair Cameron
Dr. Alan Lovelace
Mr. Andrew J. Stofan

Dr. Franklin D. Martin
Dr. J. David Bohlin
Mr. Hugh Loweth

Document I-24

[no page number]
National Aeronautics and Space Administration
Washington, D.C. 20546

April 30, 1981

Dr. Peter Gilman
High Altitude Observatory
National Center for Atmospheric Research
P. O. Box 3000
Boulder, CO 80307

Dear Dr. Gilman:

Thank you for your letter of April 8 expressing concern about the International Solar Polar Mission (ISPM) and registering your unhappiness at the way NASA has dealt with this issue. As you may be aware, the circumstances surrounding the FY 1982 Budget Amendment have been extremely difficult for us in the Office of Space Science (OSS). The President's policy of reducing Federal expenditures, as applied to us by the Office of Management and Budget, mandated a 23% cut in OSS funds for Fiscal Year 1982. After considering a great many options, we concluded that a cut of this magnitude could not be absorbed without major impact on one of our three large ongoing flight missions – Space Telescope, Galileo or ISPM. I do not believe this conclusion is disputed by the science community. We, in OSS, felt the elimination of the U.S. spacecraft in the ISPM was the least harmful option available to us, and we have acted accordingly.

All of these decisions involving the FY 1982 Budget Amendment had to be made under circumstances where it was impossible for the Agency to consult with any scientific advisory groups or with the European Space Agency, our partners in the ISPM. This restriction was placed upon us by OMB and was not NASA's choice; however, contrary to the assertion in your letter, our decisions have been generally supported by the scientific community, including the Space Science Board and the Space Science Advisory Committee. The fact that NASA management must occasionally make major decisions of this sort without the direct participation of scientific advisory groups makes it all the more important that we routinely maintain a close working relationship with those groups. Only if there is "good communication" and if we are regularly "working together" do we have much chance of success.

I assure you that we at NASA are well aware of the severe restriction in the science capabilities of ISPM inherent in the elimination of the U.S. spacecraft, and we are continuing to work with ESA to try to find a solution that will permit ISPM to continue as a dual spacecraft mission, although in a descoped mode. However, it is also true that this mission is a truly exploratory one, and it is our belief that even just the ESA spacecraft with its current payload [2] will be able to achieve a significant scientific return. Concerning the tone of the written congressional testimony, I can only remind you that as a part of the Executive Branch, we at NASA support the President's Economic Recovery Plan.

I regret very much that you perceive a serious erosion of the credibility of the process by which NASA seeks scientific advice and acts upon it. I personally would not be at NASA Headquarters if I did not believe that the Office of Space Science was committed to carrying out a sound and innovative program of space science and to working closely with the science community through a variety of advisory groups. I stand by my sentiment expressed in my 16 March Dear Colleague letter, that "we should all work together to use the limited resources available to the best advantage and to continue to make the case for a strong space science program in the future."
Sincerely,

[signature]
David Morrison
Acting Deputy Associate
Administrator for Space Science

ccs:
Dr. A.G.W. Cameron
Dr. Richard Hart
Dr. Louis Lanzerotti
Dr. A. M. Lovelace
Mr. Andrew J. Stofan
Dr. Franklin D. Martin
Dr. J. David Bohlin

Document I-25

Document Title: Memorandum from Edmond M. Reeves, Acting Deputy Director, Shuttle Payload Engineering Division, NASA, to Director, NASA Goddard Space Flight Center, "High Resolution Solar Observatory (HRSO) as a Space Station Program," 4 November 1986.

Source: NASA Historical Reference Collection, History Office, NASA Headquarters, Washington, D.C.

Document I-26

Document Title: Letter from Dr. Eugene N. Parker, The Enrico Fermi Institute, University of Chicago, to Dr. Lennard A. Fisk, Associate Administrator for Space Science and Applications, NASA, 4 June 1987.

Source: NASA Historical Reference Collection, History Office, NASA Headquarters, Washington, D.C.

Document I-27

Document Title: Letter from Dr. Lennard A. Fisk, Associate Administrator for Space Science and Applications, NASA, to Dr. Eugene N. Parker, The Enrico Fermi Institute, University of Chicago, 8 July 1987.

Source: NASA Historical Reference Collection, History Office, NASA Headquarters, Washington, D.C.

NASA struggled in the mid-1980s to find a way to get a significant new solar research mission initiated. What had been known as SOT, the Space Optical Telescope, was renamed High Resolution Solar Observatory, and briefly was planned as an attached Space Station payload. This concept had only a brief lifetime, and then the mission was re-redefined as a separate robotic spacecraft. These shifts were disheartening to the solar research community, which doubted whether NASA would ever proceed with the kind of mission its members believed was needed. Despite its advocacy for a next solar research mission, NASA in 1988 was unable to gain White House approval for such an undertaking, and it subsequently disappeared from NASA plans.

Document I-25

[no pagination]
National Aeronautics and
Space Administration
Washington, D.C. 20546

[Stamped: NOV 4 1986]

TO: Goddard Space Flight Center
          Attn: 100/Director

FROM: EM/Acting Deputy Director, Shuttle Payload Engineering
           Division

SUBJECT: High Resolution Solar Observatory (HRSO) as a Space Station Program

Recent changes in planning here at Headquarters now require that HRSO be planned as a Space Station payload. There continues to be strong scientific and programmatic support for the new HRSO program, and sufficient funding is being provided in FY 1987 to make the needed progress towards development phase approval for FY 1988. GSFC has done an outstanding job of restructuring the Solar Optical Telescope (SOT) program, and we look forward to implementing this mission at long last.

In addition to the continuation of advanced definition phase activities for the HRSO program, GSFC is requested specifically to update three study areas which were previously conducted or initiated for SOT. These separate but related studies need to be in the areas of: 1) accommodation to Space Station; 2) contamination (with and without a potential ultraviolet strap-on instrument); and 3) servicing on-orbit (including, but not limited to, telerobotics). These studies should be based on the assumption that HRSO will fly on the earliest planned configuration of Space Station, but should give consideration to the subsequent Space Station growth. Close coordination with all appropriate elements of the Space Station office will be essential.

Since interface specifications are not available for Space Station at this time, you are requested to proceed with HRSO using currently defined Spacelab/Shuttle interfaces. At such future time as one or more interfaces cannot meet this assumption, then some intermediate device or change to HRSO will have to be considered. Nothing in this directive shall preclude the option of a precursor flight of HRSO on the Shuttle.

Preliminary results from the studies are requested to be part of the reconfirmation review of HRSO required by the Non-Advocate Review Board and which will be conducted in June or July 1987. Please keep this office informed on progress towards these objectives.

[signature]
Edmond M. Reeves

Document I-26

[no pagination]   June 4, 1987
Dr. Lennard A. Fisk
Associate Administrator for
Space Science & Applications
Mail Code E
National Aeronautics and Space
Administration Headquarters
Washington, D. C. 20546

Dear Len,

I appreciated very much the opportunity to discuss the High Resolution Solar Observatory with you last week. Since that time, I have been thinking about the situation and I have a number of comments. Or should I say lamentations? First of all, there is a growing feeling in the solar physics community that the essential solar observations that need to be done from space (e.g. high angular resolution, UV, etc.) will not be done by NASA, if, indeed, they are done at all. This conviction is based on the continuing delays in the high resolution telescope, which was initiated in 1974, as I recall, and has subsequently been redesigned, descoped, and greatly reduced in cost in response to various requests from NASA. Recently, a collaborative program with the UK has been developed, further cutting the cost of the mission to NASA. What is more, HRSO presently enjoys a certain congressional sympathy. And in spite of all this, HRSO is again put off and continues to be the orphan in the NASA management structure. Charlie Pellerin, reflecting the majority opinion of the AMOWG, never showed any real interest to SOT or HRSO. He generally forgot to mention HRSO, unless one of us solar types reminded him. Indeed, over the last year, it has become clear in my attendance at AMOWG meetings that, in spite of occasional protestation to the contrary, the solar telescope is considered a nuisance. High Energy Astrophysics is the real name of the game in NASA astrophysics. Big, expensive, sexy and phenomenological.

I do not understand the scientific basis, if indeed there is one, for the communal disinterest in the hard science at which HRSO is aimed. X-ray astronomy is limited to phenomenology because it cannot be shown why an ordinary star is obliged to produce a corona hot enough to emit X-rays. One would think, therefore, that X-ray astronomers would be intensely interested in putting their field on a firmer footing. The HRSO provides the opportunity to go after the crucial observational facts for a star like the Sun. It is the dynamics of the fibrils that is responsible, apparently, for both the X-ray corona and the coronal holes, i.e. responsible for the X-ray emission and the mass loss.

It is clear that the AMOWG is not intending to further HRSO under Pellerin's guidance, and there is nothing to be achieved by the continuing participation of solar physics types in Pellerin's AMOWG. I do not intend to waste more time attending future AMOWG meetings. Your proposal to establish a Solar-Terrestrial Physics Division is a step that is desperately needed. Equally important, in the short term, is the task to get HRSO going again. So I have talked to Dan Spicer, with regard to the important question of launching HRSO as a free flyer. He informs me that there are three serious options, centered on a Delta launch, using the space bus design employed for the SMM. These options have, apparently, been choked off before being communicated to your office. Noel Hinners has the dope on the launch possibilities and would be happy to describe and explore them with you.

The fundamental importance of the HRSO observations to the physics of stars cannot be over-emphasized. Nor can we overlook the fact that, as of the moment of this writing, there is nothing on which the solar physics community can base a belief that NASA will carry out the HRSO observations in the foreseeable future. Things have deteriorated over the years to the point where the young blood is avoiding the field because nothing is happening. I think, however, that the means to remedy this unfortunate situation may be available.

Sincerely yours,

[signature]
Eugene N. Parker

Document I-27

[no page number]
National Aeronautics and Space Administration   [stamped "Jul 8 1987"]
Washington, D.C. 20546

Dr. Eugene N. Parker
The Enrico Fermi Institute
The University of Chicago
933 East 56th Street
Chicago, IL 60637

Dear Dr. Parker:

I would like to thank you for your long and thoughtful letter of June 4, 1987, in which you discussed a number of aspects of the High Resolution Solar Observatory (HRSO) program. I assure you that the difficulties in achieving a start of this most important program have caused as much frustration here at NASA as in the solar community. Your comments about the importance of understanding the exact physical processes in stars is, of course, exactly why HRSO is of such high intrinsic value for astrophysics. There are few things that would give me, and many others here, as much pleasure as seeing this program finally approved for development, including a firm plan for its operations phase. Let me now address briefly the other specific points that you raised.

I am aware of the idea of the use of a free-flying satellite bus for HRSO. Dr. Charles Pellerin, Director, Astrophysics Division, initiated such a study this spring in an effort to provide a backup plan. However, at that time, HRSO was actively being considered by Congress with regard to the Fiscal Year (FY) 1988 budget submission as a Spacelab/Space Station payload, consistent with the program plan for the last 5 years. Therefore, it was judged unwise to surface an alternative plan which might be perceived to be at variance to the one which Congress had specifically directed in the course of last year's budget appropriation. Enclosure 1 [not included] is a program plan/information package developed this spring which concludes with an outline of the two options (Spacelab vs. free-flyer) and the position adopted this spring by the HRSO Program Office while the FY 1988 budget was under active discussion. If HRSO fails to gain definitive support in the final version of the FY 1988 budget, alternative approaches will be examined.

While I cannot speak directly to your perceptions regarding the Astrophysics Management Operations Working Group (AMOWG), I would like to point out two facts concerning the support by Dr. Pellerin for the HRSO program. Enclosure 2 [not included] is a copy of a letter recently sent by Dr. Pellerin to all members of the Space and Earth Science Advisory Committee (SESAC). In it, HRSO is given the highest priority ranking in the Astrophysics [2] program, along with the Advanced X-ray Astrophysics Facility (AXAF). The second factor that should be kept in mind is that, owing to the current organizational structure within the Office of Space Science and Applications, the HRSO program appears in the budget of the Shuttle Payload Engineering Division, not that of the Astrophysics Division. I assure you that HRSO is prominently displayed therein for our consideration in preparation of the FY 1989 budget.

Finally, I appreciate hearing your views on the possibility of the creation of a new division for the solar and space plasma physics disciplines. This concept has been under very active consideration here, and I hope to be able to make a decision on it in the near future. In closing, I would like to thank you for the very active and steady support that you have given to the space science program over the duration of your career. We share the mutual regret that not all of the worthwhile programs that should be flown as part of this nation's space program can be accommodated within the budgets made available to us. I can only ask your patience and continued support as we try to initiate the ones we can within the limitations of financial resources and flight opportunities.
Sincerely,

[signature]
L. A. Fisk
Associate Administrator for
Space Science and Applications

Document I-28

Document Title: Space Studies Board, National Research Council, "Space Science in the Twenty-First Century: Imperatives for the Decades 1995 to 2015," 1988.

Source: NASA Historical Reference Collection, History Office, NASA Headquarters, Washington, D.C.

Beginning in the mid-1980s and extending over the period of rethinking the NASA space science program in the aftermath of the Challenger *accident, the renamed Space Studies Board (formerly the Space Science Board) made a comprehensive assessment of the desirable scientific objectives for the future space program. This excerpt summarizes the Board's thinking with respect to solar research.*

SPACE SCIENCE IN THE
TWENTY-FIRST CENTURY:
IMPERATIVES FOR THE
DECADES 1995 TO 2015

SOLAR AND SPACE PHYSICS

Task Group on Solar and Space Physics
Space Science Board
Commission on Physical Sciences, Mathematics, and Resources
National Research Council
NATIONAL ACADEMY PRESS
Washington, D.C. 1988

[33]4
New Initiatives: 1995 to 2015

Many of the most novel and exciting initiatives in solar and space plasma physics for the period 1995 to 2015 will involve combining remote sensing or imaging techniques in situ. Observations for the study of the Earth's magnetosphere and the analysis of the solar corona, as well as other solar phenomena. These new initiatives fall into four groups, corresponding to the following subdisciplines of solar and space physics: heliospheric physics, terrestrial magnetospheric physics, terrestrial atmospheric physics, and planetary science.

SOLAR AND HELIOSPHERIC PHYSICS

Local Measurements in the Solar Atmosphere

At present, our understanding of the origin of the solar wind is based entirely on theory and remote sensing. Direct measurements of the solar wind plasma, the interplanetary magnetic field, the energetic particle population, and associated wave-particle interactions are available, but only at distances greater than the 0.3 AU perihelion distances of Helios 1 and 2. The task group recommends a Solar Probe mission whose primary objective is to carryout the first in situ observations of the solar wind plasma and fields (electric and magnetic) near the source of the wind in the [34] solar atmosphere.

FIGURE 4.1 A trajectory for the Solar Probe.

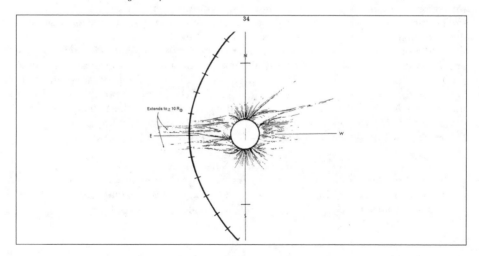

Included will be a detailed study of energetic particles, which will yield important diagnostic data on particle acceleration processes and coronal structure.

The spacecraft must be placed in an orbit that will bring it as close to the Sun as possible and still survive to provide useful data near closest approach. A perihelion distance of 4 solar radii is anticipated, with a local wind speed of about 50 km/s, electron and ion plasma temperatures of about 106K, and plasma density and magnetic field strength of less than 10' electrons/cm 3 and 10' gamma, respectively. A drawing of the Solar Probe trajectory is shown in Figure 4.1.

Theories of solar wind origin place the transition region from subsonic plasma flow to supersonic flow somewhere between 1 and [35]10 solar radii. Radio scattering experiments on Viking during superior conjunction suggest a critical point closer to 10 solar radii. In situ measurements should clarify this issue.

The location of the critical point and the plasma properties (speed and temperature) of the supersonic wind will depend greatly on the physical processes that heat the corona. Theoretical studies suggest that the proton temperature profile is very sensitive to these heating processes. It is not clear whether the corona contains an extended region of heating (out to as far as 20 solar radii) or undergoes adiabatic expansion beyond the solar surface. Plasma temperature data and observations of the wave types and amplitudes should lead to the identification of the important heating and acceleration mechanisms.

Many other important problems can be studied with Solar Probe, including a detailed characterization of coronal streamers, the place of origin and the boundaries of high-speed and low-speed flows close to the Sun, the extent of heavy element fractionation and elemental abundance variations, and the scale sizes of inhomogeneities and the development of the magnetohydrodynamic turbulence that characterizes the solar wind near 1 AU and beyond. The Solar Probe mission can also study the solar spin down rate through measurements of solar wind angular momentum flux.

Further study needs to be carried out to determine the best method of designing detectors that are required to look in the direction of the Sun.

In the original study, it was assumed that the spacecraft would go to Jupiter, where a gravity assist would send it on course to the inner corona. Our task group learned of a possible alternate trajectory involving a hypersonic flyby in the upper atmosphere of Venus; the two possibilities are sketched in Figure 4.2. It should be possible to add low-thrust propulsion in order to attain an it ecliptic orbit around the Sun with a 1-year periodicity so that the probe enters the vicinity of the Sun several times. As shown in Figure 4.2 , the Venus flyby technique also yields a very short orbital period.

High-Latitude Solar Studies

The heliosphere is known to have a complicated three-dimensional structure. The magnetic field is a tight spiral near the solar end equatorial plane, but is expected to be essentially radial over the [36]

FIGURE 4.2 Two concepts for the trajectory of a Solar Probe mission.

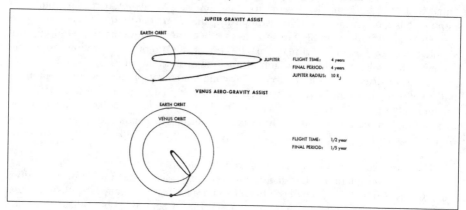

solar poles. Coronal holes, one of the sources of high-speed solar wind, are expected to produce quasi-steady high-speed flows over the solar poles during much of the solar cycle, whereas at low latitudes interacting high- and low-speed flows predominate.

To understand heliospheric conditions at low solar latitudes has required numerous missions, e.g., Explorers, Pioneers, Mariners, and Voyagers. To understand heliospheric conditions at high latitudes will similarly require repeated missions. NASA and ESA will fly the first exploratory mission over the solar poles (Ulysses). However, as with most exploratory missions, Ulysses will probably uncover more questions than it will answer, and follow-on missions will be required.

The objective of the Solar Polar Orbiter (SPO) would be to provide a detailed, repeated study of conditions at all heliographic latitudes. In circular orbit, SPO will observe the heliosphere at constant radius and thus will distinguish latitude from radial effects. With

FIGURE 4.3 The orbit of the Solar Polar Orbiter shown together with locations of other solar measurement platforms (Starprobe, the 1-AU observing network, and the Heliosynchronous Orbiter).

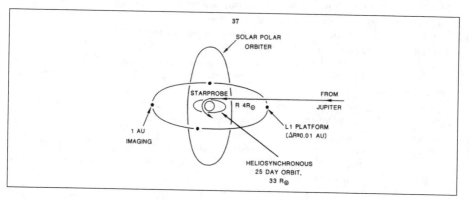

a circular orbit at less than or equal to 1 AU, and thus [37] an orbital period less than or equal to 1 year, SPO should be able to make several passes over the solar poles in a nominal mission lifetime, and thus distinguish spatial from temporal effects.

No detailed study of an SPO mission has yet been done. However, the required orbit should be achievable through the use of a low-thrust, continuous acceleration propulsion system such as solar-electric propulsion with a final orbit as shown in Figure 4.3.

The SPO spacecraft should carry a full complement of plasma, energetic particle, magnetic field, and radio wave instruments, similar to what is to be flown on ISPM. In addition, SPO should have pointing capability, through the use of a despun platform on a spinning spacecraft, or as a three-axis stabilized spacecraft, for detailed solar observations using a coronagraph, x-ray telescope, and similar photon observing instruments.

The principal technical development required for SPO is a solar-electric propulsion system, or its equivalent, for low-thrust, continuous acceleration. In the cost projections for SPO it is assumed that such development will not be charged against the mission costs, because the need is common to several proposed [38] programs. Also, studies need to be conducted on the impact of a continuous propulsion system on particle, field, and photon instrumentation, and on the measurements these instruments make.

Outer Coronal Physics

The Heliosynchronous Orbiter (HSO), as described in the ESA document Horizon 2000, is an instrumented probe orbiting the Sun at about 30 solar radii with a 25-day period, synchronous with the rotation of the Sun (see Figure 4.3).

This mission will be able to address a very broad range of scientific objectives from solar physics, physics of the interplanetary medium, and high-energy astrophysics to relativity:

• Investigation of the morphology and dynamical development of all solar structures from the photosphere to the outer corona from a vantage point close to the Sun (0.15 AU) over a large range of solar latitudes, with frequent access to the solar polar regions. Stereoscopic viewing of structures through motion of the spacecraft. The understanding of the relationship between the thermal structure and heating of the solar corona will ultimately permit the identification of the physical nature of the solar wind acceleration. Imaging of the coronal structures could be achieved by observations at 1 AU.

• Investigation of the three-dimensional structure of the inner heliosphere near or even outside the region where the wind is accelerated.

• Measurements of solar wind particle fields and waves; studies of the heating and acceleration of the solar wind (thermally or wave-driven wind?) with the advantage of a wide latitude coverage.

• Studies of the propagation, acceleration, and modulation of solar energetic particles including the significant reduction of propagation effects with respect to 1 AU. Study of shock wave acceleration.

• Radio sounding of the solar corona as the spacecraft passes behind the Sun.

• Correlative studies of expanding and traveling solar structures and their manifestation in interplanetary space.

• Investigation of the three-dimensional distribution of mass and velocity of interplanetary dust in the inner heliosphere.

[39]• Investigations of the Hermean magnetosphere and remote sensing of Mercury during flybys early in the mission (the last in situ measurements date from Mariner 10 in 1974-1975.)

• Establishment of a reference observatory for other missions in the heliosphere, in particular for solar optical remote sensing missions near 1 AU.

• Baseline observations of galactic gamma-ray bursts.

• Performance of relativity experiments (if possible, e.g.), determination of J2 (the second gravitational moment) in the case of a highly elliptic orbit; frame dragging experiments.

The technological problems (propulsion, thermal design, data transmission pose a considerable challenge. The mission concept is certainly not only attractive to a large scientific community, but it would also be appealing to the general public and from the technological point of view.

1-AU Observing Network

The Sun is the only star that we can observe from different directions, i.e., from any position within the heliosphere. This provides a stereoscopic view of structures whose geometry and energy content cannot be determined because they are either optically thin or because parts of them are not entirely visible from one single viewing condition. In addition, simultaneous observations at different positions inside the heliosphere provide three-dimensional snapshots of the magnetic field and the solar wind, important observations that will give new insight into the mechanisms that govern the wind generation, acceleration, and propagation. Similarly, simultaneous measurements of the irradiance with a set of several spacecraft would allow us to infer what mechanisms induce variations in the solar constant, whether they are due to sunspot luminosity deficiencies compensated by equivalent increases on the hidden solar hemisphere or whether they are in phase over the whole surface and due to global variation of the solar volume. It should also be noted that a 360° network for ecliptic monitoring of flare events might become an indispensable element in any manned mission to another planet.

A set of 4 1-AU spacecraft positioned at 90° in the ecliptic plane and augmented by another one in a solar polar orbit should (see Figure 4.3) provide the necessary means to conduct [40] these measurements. They should be equipped with coronagraphs, XUV and x-ray telescopes, particle detectors, magnetometers, and radiometers. The 1-AU spacecraft near Earth could be at L1; the Space Station could service an L1 platform essentially as well as a geosynchronous one.

Additional Solar and Heliospheric Studies

There is now a considerable body of evidence to suggest that all scales of structure on the Sun, as well as other astrophysically interesting objects, are ultimately governed by small-scale processes associated with intermittent magnetic fields and turbulent stresses. The understanding of the physics of the creation and decay of these dynamical structures is essential to a proper description of large-scale structures (such as coronal active regions, flares, and the solar wind) and their effects on interplanetary space and the near-Earth environment.

The interplay between processes occurring on vastly different spatial scales is ubiquitous in astrophysics. Whether in accretion disks feeding black holes at the center of active galaxies or quasars, in the magnetospheres of neutron stars, or in the x-ray coronae now known to surround a wide range of stars, small-scale magnetohydrodynamic processes are thought to influence and sometimes control the behavior of the object.

In these astrophysical situations, observations using even the most advanced technology currently conceivable will not allow us to directly observe the controlling small-scale processes. Using the Sun, however, we can indeed imagine direct observations. The Sun is therefore a unique tool for advancing our understanding of a broad class of astrophysical phenomena, if we can penetrate to the domain of underlying processes that often operate on spatial scales of 1 to 100 km.

An orderly progression of goals that could realize much of this 21 promise would include the following:

1. Development of the successor to the Solar Optical Telescope and its integration into the Advanced Solar Observatory on the Space Station, along with the development of 0.1-arcsec ultraviolet and x-ray solar instruments on the Space Station.

2. Interferometric experiments in the ultraviolet and extreme ultraviolet, aimed at a preliminary reconnaissance of solar features at angular sizes much less than 0.1 arcsec. [41]3. Development of new 1-m class facilities, utilizing the emerging multilayer coating technologies, designed to obtain resolution in the 0.01-arcsec regime at extreme ultraviolet or soft x-ray wavelengths.

4. Improvement of the angular resolution of 1-m class telescopes by the use of multiaperture arrays to achieve baselines of order 10 m. Further details are contained in Appendix D.

Chapter Two

Space Physics

by James Green and Brian Dewhurst*

1. Origins of Space Physics

Space physics, the study of the particles, magnetic fields, and electric fields that surround Earth and extend to the Sun, has been a central portion of NASA's science program since the Agency's inception; indeed, carrying out space physics experiments (James Van Allen announced the discovery of what came to be known as the Van Allen belts in May 1958) preceded the 1 October 1958 opening of NASA. (See Documents II-1 and Volume V, I-5.) The study of the Sun itself, the interactions of its layers, the mysteries of the corona, the origin of sunspots, and other topics make up the field of solar physics, and were discussed in the previous chapter of this volume.

Although physicists had conducted upper atmosphere research using balloons and sounding rockets for some years before 1957, the scientific field of space physics began with the space race. The Soviet Sputnik 1 and 2 missions were launched into orbit around Earth in October and November 1957.

Rather than wait for the troubled Vanguard program to produce a working launch vehicle and satellite, the Department of Defense decided in November 1958 to use the Army's Jupiter-C rocket to place an American satellite in orbit. The Technical Panel for the Earth Satellite Program of the U.S. National Committee for the International Geophysical Year was given the task of selecting the scientific investigations that would go on the Explorer 1 spacecraft. The Secretary of Defense had promised the President that the satellite would launch in ninety days, so there was very little time in which to make this selection.

The panel decided to use experiments that had already begun development for launch during the Vanguard program, selecting a meteoritic dust detector, temperature sensors, and a cosmic ray measurement experiment designed and built by Van Allen, who in preceding years had been in touch with the von Braun rocket team in Huntsville with respect to the requirements for flying his instrument in a satellite developed by the Army's Jet Propulsion Laboratory, launched on a booster developed by von Braun. (See Volume V, Documents I-6, I-7, and I-10.) Van Allen had mapped cosmic ray intensities throughout the upper atmosphere using a campaign of sounding rocket experiments. He had discovered that the levels of cosmic rays evened out at altitudes above 55 km, leading him to believe that he had discovered what their intensity was in space. The Explorer satellite gave him the opportunity to carry his experimentation into space and validate his conclusion.

* We would like to gratefully acknowledge our discussions with Joseph Alexander, David Bohlin, Michael (Mike) Calabrese, David Cauffman, the late Burton Edelson, Robert Farquhar, Kent Hills, Noel Hinners, John Hrastar, Frank McDonald, Norm Ness, Charles Pellerin, Guenter Reigler, Tyco VonRosenvinge, and William Worrall. All contributed to the development of this essay.

In January 1958, the first U.S. space mission, Explorer 1, carried Geiger Muller tubes; again in March 1958, Explorer 3 (after Explorer 2 failed) carried them. Both missions detected an inner radiation belt in which the fluxes increased with altitude, unlike the cosmic-ray flux.[1] In May 1958, Van Allen announced the discovery of this belt. (Sputnik 2 was equipped with Geiger Muller tubes and actually made the first observation of the inner radiation belt; but due to the U.S.S.R.'s policy of secrecy, Soviet scientists were not able to have access to the data until much later.)

While U.S. scientists were analyzing their data and speaking about their initial results (See Volume V, Document I-12), Congress passed the Space Act of 29 July 1958, establishing NASA. The Space Act provided NASA with wide latitude for the definition and establishment of its missions. NASA continued to launch missions to the inner radiation belt (the term "belt" was actually coined by a news reporter during a news conference on 1 May 1958), discovering more of its nature and structure. (See Document II-2.) Scientifically, the space race was moving quickly, and once again the Soviet mission Sputnik 3, launched in May 1958, observed the outer radiation belt before Explorer 4, launched in late July 1958. But based on the time of published articles about these discoveries, James Van Allen will always be remembered as the discoverer of both radiation belts.

The discovery of the two Van Allen radiation belts brought the new field of space physics a great deal of attention. The belts had implications for everything from proposed Department of Defense (DOD) satellites to the survivability of humans in orbit. The discovery also had a substantial impact on the direction of space science research. In early 1958, the newly established DOD Advanced Research Projects Agency (ARPA) had approved plans for five Pioneer satellites designed to fly by the Moon and to transmit television pictures of the Moon's surface. When the inclusion of a television camera proved infeasible, ARPA managers decided to use the Pioneer spacecraft to expand on Van Allen's discovery. Pioneers 1, 2, and 3 failed to achieve escape velocity and fell back to Earth, though they produced some useful data during their brief flights. Pioneer 4, however, escaped Earth's gravity well and flew by the Moon, eventually settling into an orbit around the Sun. Pioneer 4 also returned excellent data on the inner and outer Van Allen belts.

As the ARPA-managed Explorers and Pioneers were proceeding, NASA began to organize itself. Homer E. Newell, a veteran upper atmosphere physicist who moved to the new space agency from the Naval Research Laboratory and was put in charge of NASA's space science activities, laid out NASA's new program in space research. In a speech before the Royal Society in London on 12 November 1958 (See Document II-3), Newell presented a relatively comprehensive plan for space research in Earth's upper atmosphere, ionosphere (including the radiation belts), and astronomy.[2] The responsibility for carrying out much of this program was given to the new Goddard Space

---

1. James A. Van Allen, "Energetic Particles in the Earth's External Magnetic Field," in *Discovery of the Magnetosphere*, C. S. Gillmor and J. R. Spreiter, eds. (Washington, DC: American Geophysical Union, 1997), pp. 235–264.

2. Homer E. Newell, *Beyond the Atmosphere: Early Years of Space Science*, NASA SP-4211 (Washington, DC: Government Printing Office,1980). This book is an excellent summary of the early years of space science at NASA, written by the individual most responsible for the space science program in the first fifteen years of NASA.

Flight Center (GSFC) in Greenbelt, Maryland.[3] Newell brought to NASA a number of Naval Research Laboratory scientists who became the core of the Agency's space science capabilities at GSFC. The Jet Propulsion Laboratory of the California Institute of Technology was transferred from Army to NASA management and focused its efforts on lunar and planetary science missions.

In the fall of 1958, Leslie H. Meredith (the newly appointed head of the Space Sciences Division at GSFC and Van Allen's first graduate student) invited John Naugle and Frank McDonald to join the new GSFC science team; these two individuals were to be central to the NASA space science effort for many years. The two scientists began their tenure at GSFC on 1 July 1959. McDonald's first assignment was to develop a satellite that would later be designated Explorer 12. One of McDonald's first decisions was to split the management of the project into two positions—the project manager and the project scientist. The project manager would oversee the implementation of the program, while the project scientist made the necessary scientific decisions. McDonald recognized that the day-to-day management of a scientific project did not have to be done by a scientist, and that good scientists were not necessarily good project managers. It made much more sense to have a strong project manager responsible for the overall success of a mission and a strong project scientist who had the power, or could invoke the power, to overrule the project manager if the scientific objectives of a mission were in jeopardy. This simple management structure became the standard for all subsequent NASA science programs.

Explorer 12 was launched on 16 August 1961. For three months the satellite provided cosmic-ray data, the most important of which was the first observation of high-speed particle streams coming from what was later believed to be a solar Coronal Mass Ejection (CME) which occurred on September 28. The dispersion of the arrival times of various solar particles and the observation of a large geomagnetic storm associated with the CME were groundbreaking in recognizing the relationships of Earth's magnetic storms with events originating at the Sun. Explorer 12 later failed due to a short circuit in the transmitter, but a spare satellite was launched on 2 October 1961 as Explorer 14 and continued the study.

A year later, on 10 July 1962, the United States launched Telstar 1, the first communications satellite. Unfortunately, the launch was only 24 hours after the Starfish test in which the American military detonated a nuclear device in the upper atmosphere. The resulting explosion inflated the Van Allen belts, causing large amounts of penetrating radiation that led to substantial damage to the Telstar satellite. This event brought about much interest in the effects of such radiation in the upper atmosphere and low-Earth orbit. The Defense Department asked NASA to provide a satellite to observe these effects during the next nuclear test, which was scheduled to occur in approximately two months. GSFC decided to use the spare Explorer 14 satellite and pack it with off-the-shelf radiation experiments. Known as Explorer 15, this satellite was launched on 27 October 1962. The satellite was a success, mapping the extent and composition of the artificial radiation belts created by the test, as well as the decay rates of these belts. In addition, Explorer 15 produced the first detailed maps of the ion and electron populations near Earth. While

---

3. The original location of GSFC was a vacant building at the Naval Research Laboratory until the facility at Greenbelt, Maryland, was constructed.

the Explorer nomenclature continued to be used to designate U.S. satellites for years, Explorer 15 was the final first-generation American space physics satellite.

Another important line of missions that was established early in the space program was the Injun series of satellites meant to study Earth's magnetosphere. Injun was not an acronym, but rather the name assigned by James Van Allen and his students to the series. All of these spacecraft were designed and built at the University of Iowa, where Van Allen was the head of the Department of Physics. During most of the time period of the Injun series, the University of Iowa was the only university able to produce satellites in-house.

This series contained six missions and extended from June 1961 to the launch of the Injun-6 (also known as Hawkeye) in June of 1974. The objective of this series of missions was to study natural and artificial trapped radiation belts, auroras, airglow, and any other geophysical phenomena they might discover. These satellites were jointly funded by the Naval Research Laboratory and NASA.

2. Origins of the Interplanetary Monitoring Program (IMP)

As NASA planned a response to the launch by the Soviet Union of the first man in space, Yuri Gagarin, on 12 April 1961, the Agency convened a meeting in early May in which the various science disciplines discussed their potential contributions to the human spaceflight program, and particularly to a potential lunar-landing program. Frank McDonald was asked to provide the "particles and fields" perspective. McDonald realized that the radiation environment between Earth and the Moon would need to be fully understood in order to determine what protection the astronauts would need. To achieve this understanding, he proposed the Interplanetary Monitoring Platform (IMP) program, quickly approved at NASA Headquarters.

The IMP program[4] had several scientific objectives. The primary objective was to study and monitor the radiation environment in Earth's magnetosphere and near lunar space. Because continual monitoring was part of Apollo program requirements, a whole series of IMPs was planned. As a part of this effort, further goals were to develop the capability to predict solar activity and flares, to study the relationship between the Sun and Earth, and to determine the quiescent properties of interplanetary magnetic fields and how they are related to solar particle fluxes.

McDonald was responsible for the design of the scientific payload for the first three IMP missions. He chose a range of experiments that would investigate magnetic fields, cosmic rays, and the solar wind, the flow of particles away from the Sun's corona.[5] This suite of instruments was the last set of experiments to be chosen unilaterally by a project scientist.[6] In mid-1960, Newell stated that instrument selection on future NASA missions would be made by NASA Headquarters based on peer review by the scientific community. By this time, NASA Headquarters had established a series of advisory committees. At the

---

4. Paul Butler, *Interplanetary Monitor Platform: Engineering History and Achievements*, NASA TM-80758, May 1980.

5. Frank McDonald, "IMPs, EGOs, and Skyhooks," *Journal of Geophysical Research*, 101, 10521-10530 (1996).

6. John Naugle, *First Among Equals: The Selection of Space Science Experiments*, NASA SP-4215 (Washington, DC: Government Printing Office, 1991).

fall 1961 Particles and Fields Subcommittee meeting, the committee accepted McDonald's recommendations for the initial IMP missions, but made it clear that it supported the concept that future selections would be made at the Headquarters level.

After a substantial delay (due to the problems with the Thor-Delta launch vehicle), IMP 1 was launched on 27 November 1963. In order to sample as much of interplanetary space as possible, the IMP satellites were placed into a highly elliptical orbit (with apogees at ~ 15 to 30 Earth radii). The spacecraft were able to provide substantial insight into the magnetosphere and magnetosheath, and data from IMP-1 was used to discover the existence of the geomagnetic tail. Over the next four years, the IMP series provided a substantial amount of data.

As the IMP program was underway, Norman Ness of GSFC, the Principal Investigator (PI) of the IMP-1 Magnetic Field Experiment, and discoverer of the geomagnetic tail, came up with the concept of having an IMP placed in orbit around the Moon.[7] This was the beginning of the Anchored IMP (AIMP) effort. The first AIMP overshot the Moon, due to more powerful performance from the Thor Delta booster than was anticipated. AIMP-1 was placed into a highly elliptical Earth orbit, however, and provided much useful data on Earth's magnetic bow shock and geomagnetic tail. The next AIMP attempt achieved lunar orbit on 21 July 1967 and provided a wealth of data on the particle and field environment in lunar orbit.

After the success of AIMP-2, the IMP line returned to using highly elliptical Earth orbits. On IMP-7 and -8, Ness suggested using a secondary motor to put the spacecraft into a more circular orbit. This innovation removed the limit on lifetime caused by atmospheric drag and enabled the IMPs to remain in orbit indefinitely. IMP-8 provided useful data throughout the rest of the twentieth century and was still in operation in 2002.

As NASA began to send probes to Mars and Venus and to plan for exploratory missions to the outer planets and Mercury, space physicists were eager to measure the "particles and fields" associated with planets other than Earth. They were concerned that NASA was not giving adequate attention to the space physics aspects of solar system exploration, a concern that NASA claimed was not warranted. (See Documents II-7 and II-8.)

Data about the structure of the magnetosphere and the radiation belts were rapidly accumulating from the numerous NASA missions, but access to the data for the whole science community was very difficult. On 15 August 1960, Newell told Abe Silverstein, head of the NASA Office of Space Flight Programs, that NASA would release the reduced data from its missions to the broader scientific community and would archive the data for future users. (See Document II-4.) Newell subsequently developed a policy for handling data from NASA missions, giving initial access to investigators who had conceived space experiments or who had developed instruments to gather data in space. In 1966, Newell established the National Space Science Data Center (NSSDC) at GSFC. The establishment of the NSSDC enabled U.S. investigators to archive their data at a central location and provided a mechanism to distribute the data to any U.S. scientist upon request.

---

7. Norman F. Ness, "Pioneering the Swinging 1960s into the 1970s and 1980s," *Journal of Geophysical Research*, 101, 10497-10509 (1996).

But the international science community still could not obtain access to this archival data. In September 1968, Dr. Merrill A. Tuve, Chairman of the Geophysics Research Board of the National Academy of Sciences (NAS), asked NASA Administrator James Webb to approve the transfer of the responsibilities of the U.S. Rockets and Satellites Center from NAS to NASA. (See Document II-6.) The Rockets and Satellites Center was part of the World Data Center system that maintained important international agreements for the exchange of archival data with the international science community. The task of responding to this request fell to Newell who, on 12 November 1968, assigned the management responsibility to the NSSDC. (See Document II-7.) As part of the World Data Center system, the NSSDC could make its archival data available to foreign scientists upon request. This arrangement greatly facilitated the distribution of space physics data and stimulated important national and international scientific collaborations in space physics.

3. The Shift to Multiple Observations

By the early 1970s, NASA had an advisory committee structure at every level. Individuals like Arthur C. Clarke were advising the Administrator on broad space issues, another committee of scientists was advising the Associate Administrator for Space Science on overall research strategy, ad hoc subcommittees were evaluating concepts for specific missions, and working groups were interacting with project directors and major office heads. In addition, of course, the Space Science Board of the National Academy of Sciences and its various committees provided external advice to NASA. Through these multiple channels, any scientist could propose an idea for a space experiment or mission. Many of these ideas found their way into a NASA Announcement of Opportunity (AO), the formal mechanism by which scientists competed for access to space. To get an idea into an AO, it was necessary to have the advocacy of an established scientific institution or a NASA Field Center, support from a NASA program office, and support from the appropriate advisory committees. (See Document II-9.)

By the 1970s, space physicists had developed a basic understanding of the environment in the neighborhood of Earth from the analysis and publication of data from their individual instruments. However, only a small percentage of the research that was accomplished was done by relating data from multiple instruments on various spacecraft or with ground-based measurements. A major change in direction in space physics research occurred when the Scientific Committee on Solar-Terrestrial Physics (SCOSTEP), which operated under the auspices of the International Council of Scientific Unions, established an international program designed to understand magnetospheric processes by coordinated observations with space and ground-based instruments. The International Magnetospheric Study (IMS) period ran from 1976 to 1979. NASA scrambled to participate in this effort. (See Document II-10.) The principal NASA contribution consisted of correlative data from operating spacecraft, data from the International Sun-Earth Explorer (ISEE) program (which had three spacecraft), and support from the NSSDC. During and shortly after the IMS period, the NSSDC held a series of intensive data comparison workshops enabling scientists to use data from many sources. These workshops were called Coordinated Data Analysis Workshops[8] and were designed to understand the response of the magnetosphere to changes in the solar wind. The IMS period and the workshops had a major effect on the thinking of the science community, turning its attention from exclusive analysis of local, in situ measurements to more global science questions that needed to be answered using multiple data sources. Attempts to respond continued into the 1980s and 1990s.

4. The Tortuous Path to a Comet's Tail

In 1977 and 1978, three ISEE spacecraft were launched to study the solar wind and its impacts on Earth's geospace environment during the maximum period of the sunspot cycle. The ISEE-1 and -2 spacecraft orbited Earth in a highly elliptical orbit and remained mostly in the magnetosphere, but ISEE-3 was placed into a halo orbit (as observed from Earth) at the Sun-Earth libration point, L1 (the location between the Sun and Earth where the pull of the gravity of the two bodies effectively cancel one another out, allowing an object in that position to remain there without expending station-keeping energy). The value of ISEE-3 in measuring solar wind input into Earth's magnetosphere was immediately appreciated by the entire space physics community. Within a few years of operations, the three spacecraft had completed an impressive list of space physics firsts. By August 1981, in the declining phase of the solar cycle, they had completed their primary mission. ISEE-1 and -2 were the last two missions in the IMP series and provided data for over eight years after they were launched. The long-term fate of ISEE-3, however, would be determined in a very different, highly political way.

Comets were created at the birth of the solar system and as such have been of considerable interest to planetary scientists. The most famous comet has probably been Comet Halley, which is visible once every 76 years. In the late 1970s, the Japanese space science agency ISAS, the Soviet bloc grouping Intercosmos, the European Space Agency, and NASA were all planning missions to Comet Halley at the time of its March 1986 passage by the Sun. Though the rationale offered in support of NASA's planned mission to explore Halley was a combination of scientific curiosity and national prestige, obtaining the scientific priority and, ultimately, the funding proved to be impossible.[9] After more than three years of planning, Halley rendezvous mission was never approved. NASA's dropping out of the race to a comet did not affect the efforts of the other space agencies; in fact, it stimulated ESA to plan, on its own, what became the Giotto mission. Previously, ESA had planned to collaborate with NASA in a single Halley mission.

Even though NASA did not gain approval for a dedicated comet mission, the goal of being the first nation to visit a comet was quite important to some U.S. scientific working groups. In addition, many American scientists were surprised and even embarrassed to learn that NASA would not be first to a comet. It was in this climate that in March 1981,

---

8. J. I. Vette, D. M. Sawyer, M. J. Teague, and D. J. Hei, "The Origin and Evolution of the Coordinated Data Analysis Workshop Process," *The IMS Source Book* (Washington, DC: American Geophysical Union, 1982), pp. 235–241.

9. John M. Logsdon, "Missing Halley's Comet: The Politics of Big Science," *Isis*, June 1989.

Fred Scarf (Principal Investigator for ISEE-3) and Robert Farquhar (ISEE-3 orbital engineer) devised a plan for ISEE-3 to move from the L1 point to deep into Earth's magnetospheric tail, and then on to Comet Halley using lunar swingby trajectories. ISEE-3 still had onboard fuel remaining after it had been maneuvered to L1, and it had maintained its halo orbit for three years.[10] Even before the launch of ISEE-3, plans for a magnetotail excursion had been discussed, but using the remaining fuel for a Comet Halley encounter was a new twist brought about by the circumstances of not having a NASA mission to Halley.

This intriguing proposal was investigated by NASA's space science leaders. On 27 April 1981, Andrew Stofan, NASA Associate Administrator for Space Science and Applications, authorized Thomas Young (Director, GSFC) to conduct a feasibility study for the Comet Halley encounter. (See Document II-11.) During this initial study, it became clear that although reaching Halley in March 1986 was possible, a better option was an encounter with Comet Giacobini-Zinner (GZ) in September 1985.

On 8 February 1982, the plan for ISEE-3 to encounter Comet GZ after its deep-tail mission in Earth's magnetosphere was presented to the ISEE-3 science working team. The science working team rejected the Comet GZ mission option and voted to send ISEE back to L1 after the deep-tail encounter. But Scarf and Farquhar remained undaunted; they sought and received the endorsement of their magnetotail-GZ plan by the Comet and Asteroid Working Group of the Solar System Exploration Committee (SSEC), which reported to Burton Edelson (who in 1982 had replaced Stofan as Associate Administrator for Space Science and Applications). David Morrison, SSEC Chair, stated, "This use of the ISEE-3 spacecraft [for encountering Comet GZ] in its extended mission mode has our warm endorsement, and it is hard to imagine that NASA would decline this opportunity to obtain new and unique data with such a small expenditure of resources." An endorsement for the magnetotail-GZ encounter plan from the powerful Space Science Board of the National Academy of Sciences followed. The Board also recommended that NASA should take steps to replace the monitoring functions of ISEE-3 of the upstream solar wind at L1, but the majority of the ISEE-3 investigators still remained against the Comet GZ option. In the face of this conflict, NASA Headquarters remained silent, and no decision was made.

A major decision point finally appeared that would either doom the Comet GZ encounter option or maintain it. For ISEE-3 to even make Comet GZ, it had to be repositioned in its halo orbit about L1. The Director of Space Sciences at GSFC, George Pieper, although initially against the Comet GZ option, approved the repositioning of ISEE-3. (See Document II-12.) In a memo to Frank Martin, Office of Space Science and Applications at NASA Headquarters, Pieper stated that "unless you advise me to the contrary, the 10 June 1982 orbit maneuver will be carried out." This gave NASA Headquarters ten days to decide; again it remained silent. With the repositioning of ISEE-3 completed, Pieper bought time for the study team to complete its work and Headquarters to make a final decision.

On 6 August 1982, the decision meeting was held at NASA Headquarters. Charlie Pellerin, Deputy Director of the Astrophysics Division of NASA Headquarters, made the

---

10. Robert W. Farquhar, "The Flight of ISEE-3/ICE: Origins, Mission History, and a Legacy," American Institute for Aeronautics and Astronautics paper 98-4464, August 1998.

presentation to Sam Keller, Deputy Associate Administrator of the Office of Space Science and Applications (OSSA), to obtain approval for in situ exploration of Comet GZ. (See Document II-13.) The pros and cons were discussed extensively, with the focus of the discussion centering on the risk of losing ISEE-3. Pellerin argued that "the uncertainties/risks cited . . . appear to me to be less than those that generally exist when we start a project" and that two major risks already had been overcome; there had been no launch vehicle failure and no "infant mortality" (failure during the early days of spacecraft operation). ISEE-3 was operating nominally and was in space ready to go. In summary, Pellerin argued that the "return justifies the risk—NASA should do it."

From the viewpoint of the ISEE-3 Principal Investigators, Fred Scarf, Ed Smith, and Sam Bame all supported the GZ encounter; Ed Stone, Dieter Hovestadt, and Harry Heckman supported it, but with conditions; and Keith Oglivie, Robert Hynds, Jean-Louis Steinberg, Kinsey Anderson, Tyco Von Rosenvinge, Peter Meyer, Bonnard Teegarden, and John Wilcox all opposed it. Weighing all the facts in a controversial decision, Burt Edelson approved the Comet GZ mission on 30 August 1982. (See Document II–14.) On 22 December 1982, ISEE- 3 was renamed the International Cometary Explorer (ICE).

To ensure rapid delivery of the data to the ICE European investigators, a transatlantic link was established connecting GSFC, a main SPAN (Space Physics Analysis Network) routing center, to the European Space Operations Center (ESOC) in Darmstadt, Germany.[11] ESOC then directly connected all the ICE investigators in Europe together, thereby creating ESA's first Internet; the network was called European-SPAN or E-SPAN.[12] On 11 September 1985, ICE made history as the first spacecraft to fly through a comet tail. For the first time in NASA's history, real-time spacecraft data flowed directly from the tracking stations to the central SPAN network hub, across the United States to ESOC, and then to all the ICE investigators for rapid analysis.[13] The scientific return from ICE was impressive, but the space physics community would have to wait nine years before it would be able to replace ISEE-3 at the L1 location, a loss that some still feel was an enormous blow to making rapid progress in the understanding of solar wind interactions with Earth's magnetosphere.

5. Saving Space Physics: The Selling of ISTP

NASA organized a study group in 1977 to define a major program in space plasma physics for the 1980s. The result of this effort led to the Origins of Plasmas in the Earth's Neighborhood, or OPEN, program. OPEN had plans to launch four spacecraft, one each in solar wind (Wind), polar (Polar), and equatorial (Equator) orbits in the inner magnetosphere, and one in the deep geomagnetic tail (Geotail) region. Similar studies were being conducted by ESA and ISAS at the time. The concept behind OPEN was to observe the solar wind energy, from the Wind spacecraft, as it flowed into the deep tail of

---

11. T. Sanderson, S. Ho, N. Van der Heijden, E. Jabs, and J. L. Green, "Near-real-time data transmission during the ICE-Comet Giacobini-Zinner Encounter," *ESA Bulletin*, 45, p. 21, February 1986.

12. T. Sanderson, M. Albrecht, W. Baumjohann, P. Benvenuti, J. Franks, G. Green, J. L. Green, M. Hapgood, C. Harvey, N. Van der Heijden, E. Jabs, P. A. Lindqvist, D. de Pablo, F. Pasian, and G. Veldman, "The European Space Physics Analysis Network," *ESA Bulletin*, 53, February 1988, p.45.

13. J. L. Green and J. H. King, "Behind the Scenes During a Comet Encounter," *EOS*, March 1986, pp.67, 105.

the magnetosphere where magnetic reconnection, to be observed by Geotail, was believed to start the flow of energized plasma from the tail into the inner magnetosphere, where Equator and Polar would observe the results. This overall explosive flow of energy into the inner magnetosphere, producing an intense current around Earth (called the ring current) and an intense aurora, was called a geomagnetic storm.

NASA, ESA, and ISAS hoped that there could be convergence among their separate plans for solar-terrestrial work, but getting approval for a new space science mission, much less a coordinated set of missions among international partners, was a complex process. In the first stage, the separate scientific communities involved established their highest-priority scientific objectives. They agreed on the scientific questions that needed answers and the logical steps to get those answers. A certain amount of logic should have prevailed in this stage. In the second stage, the scientific community, NASA, and, in the case of the joint project, the space agencies of the other countries involved decided on the basic instruments, the kinds of spacecraft, the funding, and the missions to be flown. This process had some logic, but generally was chaotic. The project team had to consider national priorities, several bureaucracies, and the state-of-the-art technology of the instruments and the spacecraft, just to name a few of the issues. "Selling the mission," or obtaining the necessary funds and authority from multiple governments to conduct the mission, constituted the third and most chaotic and unpredictable stage. All of these patterns were manifested in creating the International Solar-Terrestrial Program (ISTP).

In 1979, NASA released an Announcement of Opportunity for OPEN. Investigators who provided instruments and participants in the science working groups on these missions were selected in 1981. Even with all this effort on the part of NASA and the space physics community, OPEN did not receive a new start in the President's budget for 1983; its future was quite uncertain.

The Solar-Terrestrial and Astrophysics Division within OSSA, home of the OPEN program, was reorganized and became the Astrophysics Division in late 1981. This reorganization eliminated the Plasma Physics Branch and the Solar and Heliospheric Physics Branch. Key space physics personnel were dispersed into other divisions. Not only was the OPEN program in jeopardy, but the discipline of space physics also no longer had any advocacy within the NASA management. Without such advocacy, the entire space physics community would not endure long.

Against this backdrop, Burt Edelson, an engineer with long experience in communications satellites, became the Associate Administrator for Space Science and Applications in February 1982. During Edelson's first month at NASA, a number of scientists put pressure on him to get some form of space physics program restarted. In response, Edelson initiated another reorganization and brought back the Solar and Heliospheric Physics Branch, but now within the Astrophysics Division, and created an Upper Atmospheric and Magnetospheric Branch within a new organization named the Earth Science and Applications Division. In addition, Edelson decided to bring to NASA someone from the community who was dynamic and could take the OPEN program to the next step. He contacted James Van Allen from the University of Iowa for suggestions. Van Allen recommended Stanley Shawhan, an outstanding professor and Principal Investigator on the Dynamics Explorer 1 mission at the University of Iowa. Shawhan accepted an IPA (Intergovernmental Personnel Act) position at NASA Headquarters and, in September 1983, joined the Upper Atmospheric and Magnetospheric Branch.

Shawhan's first assignment from Edelson was to get the OPEN program started through some sort of international cooperation; the new Associate Administrator was a strong advocate of such collaboration. Shawhan initiated a series of bilateral meetings between NASA and ISAS, and between NASA and ESA. In Shawhan's first year at Headquarters, he merged the NASA OPEN program with NASA's participation in ISAS and ESA solar-terrestrial missions into a single endeavor and established a project office at GSFC. This new program was called the International Solar-Terrestrial Physics (ISTP) Program. As the program developed, NASA would be responsible for the Wind, Polar, and Equator missions, along with fast-paced, ground-based, and theory activities. ESA was responsible for the SOHO and Cluster missions, with ISAS responsible for the deep-tail Geotail mission.

In addition to these ISTP project activities, it was clear that a strong archive and communications capability had to be a key element of an infrastructure necessary to facilitate data exchange between the agencies. Neither ESA nor ISAS maintained a space physics archive, and the NASA archive, the NSSDC, was, as discussed in a 1984 report[14] from the NAS Joint Data Panel of the Committee on Solar and Space Physics, in desperate need of overhauling. In response to this NAS report, Edelson stated that NASA would undertake a number of important activities that would establish improved access to archives and data. (See Document II-15.) In particular, Edelson redefined the NSSDC and provided baseline funding for the SCAN network, which was later renamed the Space Physics Analysis Network (SPAN). SPAN became NASA's first Internet-based communications network.

With these activities initiated, all seemed ready for implementation. However, as part of the budget process, NASA's upper management determined that ISTP was too costly in 1984. In late summer of that year, the NASA Administrator challenged OSSA to restructure the NASA contribution to ISTP for about half its original proposed cost.

By 24 August, Shawhan and Mike Calabrese (a space physics program manager) developed and documented[15] possible approaches for reducing the costs of the NASA part of the ISTP program. The key element of reducing the costs was the dropping of the Equator spacecraft. At that time, NASA was negotiating with the Air Force on a new mission called the Combined Release and Radiation Effects Satellite (CRRES); NASA decided that this mission would serve as the replacement mission for Equator. Shawhan broke the news to the selected investigators on Equator that the available budget could not support three NASA missions, and Equator would be deleted with the equatorial measurements to be provided by CRRES. He indicated that this action was needed to keep the rest of the program intact. Shawhan was persuasive. Even though the investigators were tremendously disappointed, they supported his overall goal of a coordinated set of spacecraft.

The scientific goals of the revamped ISTP program were to develop a comprehensive, global understanding of the generation and flow of energy from the Sun and through

---

14. National Research Council, *Solar-Terrestrial Data Access, Distribution, and Archiving* (Washington, DC: National Academy Press, 1984).

15. Stanley Shawhan and Michael Calabrese, "Overview of Possible Approaches to a Reduction-in-Scope of the International Solar-Terrestrial Physics Program," 24 August 1984.

Earth's space environment (called geospace). Shawhan redefined NASA's portion of the ISTP program into two distinct parts. The Global Geospace Science (GGS) program would contain Wind, Polar, ground-based analysis, theory, and NASA's portion of the CRRES program. The Collaborative Solar-Terrestrial Research (COSTR) program would contain NASA's portion of Geotail, SOHO, and Cluster.

NASA's budget for fiscal year 1985 contained only funds for further design and mission analysis of ISTP; there were no funds for a new start. But Shawhan was undaunted; he told the Director General of ISAS, Professor Minoru Oda, that NASA would fully support its role in the Geotail mission. (See Document II-16.) NASA's commitment enabled Oda in April of 1986 to receive approval for ISAS to build Geotail, the most expensive ISAS mission to date. ISAS was the first agency to gain budget approval for an ISTP mission, which showed its trust in NASA and its reliance on Shawhan to keep the U.S.-Japan bargain.

Prior to Shawhan's joining NASA, Edelson had become involved in the international Inter-Agency Consultative Group (IACG).[16] The science heads of the four major space agencies, NASA, ESA, ISAS, and Intercosmos, were the members of the IACG. The IACG had been formed in 1981 specifically to better coordinate the set of international missions that were to encounter Comet Halley in order to enhance their scientific return.

After the successful Halley encounter, Edelson wanted to keep the IACG going. He could see that international cooperation, at the right level, had a role to play in selling certain programs, and that solar-terrestrial physics was the next logical program for the IACG. Edelson noticed that each of the agencies had space physics projects, and they all had a number of political problems, especially NASA's ISTP program. At the 1985 meeting of the IACG, Edelson suggested that if the IACG were to continue to work as a multi-agency coordinating body, a permanent "Terms of Reference" would need to be developed. The first post-Halley mission meeting of the IACG was held in Padua, Italy, in November 1986. Using the principles that the agencies had already worked out for coordinating the Comet Halley encounters, a set of general IACG principles were drafted. These new Terms of Reference were quickly signed by all agency space science leaders. The IACG decided that solar-terrestrial physics would be its next area of focus and created the following three working groups to support this effort: Science, Data Exchange, and Mission Design and Planning. Members of these three working groups were from the science communities and space agencies of the participating countries.

At IACG meetings, the chairman from each working group would make recommendations designed to better coordinate the efforts in each of the agencies to further facilitate ISTP science. Each recommendation was either approved, not approved, or modified before approval by unanimous agreement of each agency space science chief. Under the IACG auspices, international science communication networks between NASA, ESA, ISAS, and Intercosmos were put into place or maintained. With NASA's SPAN well established and an international link to Europe already operational, one of the first IACG recommendations was to maintain this transatlantic link. By 1987, the NASA SPAN network was connected to ISAS. The establishment of Internet-based communications

---

16. Joan Johnson-Freese, "A New Model of International Cooperation: The Inter-Agency Consultative Group," *Changing Patterns of International Cooperation in Space* (Melbourne, FL: Krieger Publishing Company, 1990), pp. 101–111.

between the agencies provided a revolution in the way scientists and mission planners worked together. International cooperation was at its highest, but NASA still had no firm commitment for its own ISTP missions.

The space physics community in the U.S. was vigorously pushing for a new start for ISTP, which scientists believed was vital for the survival of the space physics discipline. Although Shawhan supported a new start for the entire ISTP program, he was unable to obtain administration support for such an initiative. Consequently, Shawhan developed a strategy that would present to Congress the COSTR program one year before the GGS program. By starting the NASA portion of the cooperative ISAS and ESA programs ahead of NASA's own programs, Shawhan would be able to make good on his commitment of NASA support to those efforts. Although the science community had hoped that the entire ISTP program would be approved at one time, there was confidence that Shawhan's leadership could deliver the program in a two-phase approach. The initial COSTR arrangements were for NASA to provide 40 percent of the Geotail payload, 25 percent of Cluster and SOHO payloads, data support, and a NASA launch of Geotail. The selling of the GGS program to Congress involved NASA retaining a leadership role. In the COSTR program, NASA was playing only a supporting role. This strategy worked; the White House and Congress approved a new start for COSTR in 1986 and for GGS in 1987. With these two programs approved, the entire NASA ISTP program was now in place, with the IACG providing a capability to facilitate the international collaboration on both operational and scientific levels. Once COSTR was approved, international support became solid. In March 1987, the joint ESA/NASA Solar-Terrestrial Science Programme (STSP), Cluster, and SOHO Announcement of Opportunity was released. Within a year, Geotail investigators were selected and confirmed for flight.

Meanwhile, at NASA Headquarters, Edelson left, and Lennard Fisk became Associate Administrator for Space Science and Applications on 6 April 1987. A new Space Physics Division was created in 1988 and, within a year, included the Cosmic and Heliospheric Physics Branch, the Solar Physics Branch, the Magnetospheric Physics Branch, and the Ionospheric Physics Branch. In addition, NASA received a budget augmentation for the COSTR program which increased NASA's contribution from 25 percent to 33 percent of the SOHO and Cluster missions. This augmentation was necessary in order to preserve collaboration with ESA in SOHO and Cluster, and ensure the development of both missions. The GGS investigators were selected based on the original OPEN AO and confirmed for flight. Instrument investigations were jointly selected for definition for SOHO and Cluster. By the end of 1989, ESA confirmed SOHO and Cluster investigations for flight. With that confirmation, the complex national and international components for ISTP were in place.

By the late 1980s, space physics was a discipline brought back from near extinction by the sheer energy, devotion, and leadership of Shawhan; the support of the NASA ISTP team of Mario Acuna, Joseph Alexander, Dave Bohlin, Mike Calabrese, Art Poland, and Ken Sizemore; the overall guidance and management insight from Edelson and Fisk; and the support of the science community.

For NASA, in general, this was a period of undertaking large missions such as the Great Observatories. (See Volume V, Chapter 3, for a discussion of these missions.) The ISTP program became the "Great Observatory" set of missions for space physics. In this era, with adequate budgets available, spacecraft and instrument performance was the highest priority. ISTP was now established and under construction, but one more major challenge lay ahead for NASA's GGS mission before it would finally become a reality.

6. The Cost Conscious 1990s—Trading Performance and Cost

Lennard Fisk had not been at NASA Headquarters a month in 1987 when he was faced with a major decision. One of the instruments on the Magellan mission to Venus was significantly over budget. At an earlier time in NASA's history, money would just be "found" to cover instrument overruns, since performance was the highest priority. In contrast, Fisk decided to cancel the instrument on Magellan based on the projection of its high cost to completion. Fisk told his staff that he did not want to cancel another instrument and that they had to figure out a way to keep costs under control. NASA budgets were getting tighter, with much less flexibility to accommodate cost growths. Fisk challenged his program managers in OSSA to come up with some sort of science investigation cost-control procedures. The result was a Cost Control Plan (CCP) for each proposed mission which contained a series of de-scoping options. The CCP became a requirement for mission approval and was to be created at the onset of the development phase for every new mission and implemented when the investigation was confirmed. The de-scoping options in the plan were identified on an instrument-by-instrument basis, with the key options delineated in the confirmation letters to the investigators for each instrument. If unforeseen costs occurred during the development phase of an instrument, CCP procedures would be implemented. Decreasing instrument performance or reducing redundancy were the typical compromises outlined in the CCP. By January 1990, the CCP for NASA's ISTP program was completed. (See Document II-19.) This document was also NASA's first CCP.

A number of changes in key personnel at NASA Headquarters occurred in the early 1990s. The space physics field experienced a major setback with the sudden death of Shawhan in 1990 from a heart attack. Shawhan was well-respected, and his death was a major loss to the community, both as a person generally well-liked because of his fairness and as a person who made significant contributions in saving the field and in creating a number of new missions. Fisk moved quickly to replace him; George Withbroe came on board within a year as the new head of the Space Physics Division. Shawhan was a tough act to follow, but in time Withbroe would prove to be another major strength for the community. In 1992, Daniel Goldin became the new NASA Administrator. Goldin was known for his ability to make broad and sweeping changes in an organization and to force it to reinvent itself throughout. In October 1992, Goldin split Fisk's jurisdiction into three separate offices, one for Earth Science, one for Life and Microgravity Science, and one for Space Science. Fisk retained control over only a third of these new units. By 1993, Fisk had departed and was replaced by Wes Huntress as the new Associate Administrator for Space Science.

Goldin was determined to usher in a new era of more cost-control measures and approaches. He believed that such steps were necessary for the Agency to survive past the end of the Cold War. Goldin told both the White House and Congress that he was going to implement a system to better establish and control NASA's ongoing programs. Goldin's "faster, better, cheaper" approach to new missions was first delineated in the 1992 Discovery Program Handbook (see Volume V, Document II-39) and continued into the Explorer program (see below), but he did not stop with these approaches.[17]

Goldin established Program Commitment Agreements (PCAs) between him and his Associate Administrators to do a particular mission with a delineated set of requirements in science and cost. In a 15 June 1993 memo to his staff, all Center Directors, and the Director of JPL, Goldin established a Program Management Council (PMC), headed by his Deputy, General Jack Dailey. (See Document II-22.) The Council would provide an additional program review at the Agency level for "addressing strategic planning, implementation, and management of all major Agency programs." A review would be initiated by the Deputy Administrator when the cost to completion, as estimated by the Comptroller, was projected to exceed the established baseline cost of a mission by more than 15 percent, or when any other PCA threshold requirement, as determined by the PMC chair, was being violated. The PMC was directed specifically to recommend "cancellation or continuation of programs and projects." This approach was designed to prevent going back to Congress for more money to solve the problem of escalating costs in the mission development phase. It was commonly believed among the NASA staff that Goldin intended to cancel a project or mission to demonstrate his serious intent.

The first project to be seriously reviewed for cancellation was GGS (Wind and Polar). By this time, the workload of the ISTP development had brought about the establishment of a GGS (Wind and Polar) and a COSTR (Geotail, SOHO, and Cluster) program office at GSFC to manage the ISTP program. Bill Huddleston served as Program Manager for GGS at NASA Headquarters, while John Hrastar was the GGS Project Manager at GSFC. The Wind and Polar spacecraft were being developed by Martin Marietta and were both experiencing a variety of development problems and cost overruns. The GGS review occurred on 5 April 1994. In an effort to save the program, John Hrastar presented to the PMC a series of options. Hrastar's preferred option was for a program re-baseline in which Wind would move forward while Polar would be put in storage; this was seen as a way to move the GGS project forward and get costs under control. Wind was still a very viable scientific mission without Polar. Hrastar argued that if Wind was launched and operating successfully and there were enough funds remaining, the project office would then consider restarting Polar. This approach was accepted by the PMC. (See Document II-23.) Both Wind and Polar were finally completed. They did not exceed 15 percent above the baseline cost ceiling. Wind was launched on 11 November 1994 and Polar on 24 February 1996. The cancellation of GGS would have been devastating to the space physics community and might very well have ended serious research in the field in the United States.

---

17. For an assessment of this approach, see Howard McCurdy, *Faster, Better, Cheaper: Low-Cost Innovation in the U.S. Space Program* (Baltimore, MD: Johns Hopkins Press, 2001).

7. The New Explorers

With the 1980s emphasis on performance versus cost and schedule, the development of NASA's flagship missions were costing almost a billion dollars each. However, overall funding for NASA space science missions was being reduced by the end of the decade. The impact of the high costs of large space science missions even influenced the progress of the long-running Explorer program of small satellites, which had been providing relatively low-cost access to space for physicists since NASA's early days.

Due in part to the *Challenger* accident in 1986 and the complexity of the existing approved missions, the stretching out of all programs undergoing development in the Explorer line was increasing their costs significantly and thereby delaying the new Announcements of Opportunity that were necessary for initiating new missions. In essence, the OSSA budget in the post-*Challenger* period was overcommitted. At that point in time, Charles Pellerin headed the Astrophysics Division where the Explorer program resided. He recognized the need for major changes in this program. His goal was to shorten the time from mission selection to flight. He also recognized that given the number of missions already in development, any new announcement at that time would lead to "five missions being selected," and that such an action could possibly close the Explorer opportunity to the end of the 1990s. (See Document II-17.) Pellerin made a decision to hold off on any new large Explorer missions, to have a competition for Small Explorer missions (SMEX), to concentrate on launching the existing backlog of approved missions, and to use the time gained to decide on how to restructure the Explorer program. (Dan Goldin's 1992 push for faster, better, cheaper was in line with what Pellerin had decided he needed to do six years earlier.) After intense discussions with the science community, the new Explorer program began to emerge. (See Document II-18.)

In September 1994, the Office of Space Science announced that the restructured Explorer Program would emphasize faster, better, and cheaper missions, and would consist of three classes. The new Explorer classes were the medium-sized or MIDEX, cost capped at about $70 million; the small Explorers or SMEX, capped at about $35 million; and the university class or UNEX, capped at about $25 million (all excluding the launch vehicle costs). The restructuring also caused the Far Ultraviolet Spectroscopic Explorer (FUSE) to be re-baselined from its then current estimate of nearly $300 million to $125 million to avoid cancellation. The last large Explorer mission would be the X-ray Timing Explorer (XTE), which was launched in 1995. In addition to the cost caps, the restructured Explorer program had a plan for launching two Explorer missions (nominally one space physics and one astrophysics mission) per year and included a Principal Investigator mode in which the PI could take full responsibility for all aspects of the mission. PIs could be from a government or industry center, laboratory, or university.

With the restructuring of the Explorer program, space physics benefited enormously through the successes of a number of small focused science missions that complemented NASA's flagship GGS program and thereby provided a more comprehensive ISTP program to better understand the solar-terrestrial system. These missions included Solar Anomalous and Magnetospheric Particle Explorer (SAMPEX, launched 1992); Fast Auroral Snapshot Explorer (FAST, launched 1996); Advanced Composition Explorer (ACE, launched 1997); Transition Region and Coronal Explorer (TRACE, launched

1998); and the Imager for Magnetopause-to-Aurora Global Exploration (IMAGE, launched 2000).

8. Mission Operations and Data Analysis

By the early 1990s, NASA was entering a time of the highest launch rate for science missions in the history of the space program. Cost controls and procedures for missions in the development phase were being initiated, but a significant fraction of the OSSA budget was going into operating its ongoing missions. In 1992, Lennard Fisk asked Guenter Riegler (Chief, Space Science Operations Branch, NASA Headquarters) to put together a Mission Operations and Data Analysis (MO&DA) "Blue Team." The charter of the team, issued on 17 June 1992, was to seek efficiencies and reduce the cost of operating missions. (See Document II-20.) Fisk's rationale for this activity was to create a funding wedge for new initiatives.

This effort highlighted the large infrastructure that NASA had developed to operate its missions. There were a number of multimission facilities, designed to handle a large number of missions, that were not cost effective due to the small number of missions that actually used the facilities. It became nearly impossible to accurately determine all the costs for a mission because of the overhead related to existing infrastructure; few potential savings were identified. It was realized that the greatest cost saving would be obtained from ensuring that the developing missions were also developing a cost-effective mission operation system. Based on the results of the Blue Team, Riegler issued an implementation plan on 6 August 1992. (See Document II-21.)

In the FY93 budget, Congress mandated a cut of 10 percent in NASA's MO&DA costs. In addition, in such a limited budget environment with many operating missions, funding was needed to operate new missions during their prime phase. The situation finally came to the point at which NASA had to find a way to stop operating missions whose prime objectives were accomplished. It was no longer acceptable to continue a mission beyond its prime phase just because it was operating nominally.

In December of 1992, Riegler was again asked to reconstitute the Blue Team and create a Red Team (to critique the Blue Team's work) to make other recommendations for MO&DA reductions. A requirement that resulted from Blue/Red team reports in 1993 became known as the Senior MO&DA Program Review. The first Senior Review was implemented in 1994 for astrophysics missions; in June 1997, a review was implemented for space physics missions. The Senior Review process was designed to evaluate scientific return from each operating mission and to cut costs accordingly. All missions had to be put through the Senior Review process in order to be approved for an extension. Beginning in 2000, space science archival data centers like the NSSDC, funded by NASA, were also added to the list of projects to be reviewed.

The 1997 Senior Review approved an extended mission for ISTP but with much reduced funding. Over several years, major efforts were undertaken to reduce the cost of the GGS mission operations by nearly 70 percent. This was accomplished by dismantling the multimission infrastructure, paying for only the services the Wind and Polar missions needed, and by taking greater risks in operating ongoing missions.

9. Solar-Terrestrial Probes and Living With a Star

In the mid-1990s, Administrator Goldin continued to make a number of dramatic changes in the structure of NASA by moving several key functions, including program management, out of NASA Headquarters to the Field Centers. Within the Office of Space Science (OSS), the discipline divisions and their associated branches were abolished. Headquarters space science staffing was reduced from more than two hundred civil servants to fewer than seventy persons. The Office of Space Science was restructured. Associate Administrator for Space Science Wes Huntress replaced the discipline divisions of Astrophysics, Space Physics, and Planetary Exploration by a functional organization, with all program scientists in the Research Division and all engineers and program managers in the Flight Division.[18]

By the summer of 1996, the heads of the science disciplines became members of a Science Board of Directors, each representing a specific science theme. Withbroe went from the Chief of the Space Physics Division with a staff of twenty-six to an advisor to Huntress for a new Sun-Earth Connections theme. Withbroe had no budget authority in his new role. Undaunted, he continued to hold regular meetings with his former staff, sharing goals and information. It was in this atmosphere of organizational chaos that the former members of the Space Physics Division had to work with the science community to develop follow-on programs to ISTP. This matrix approach to organization continued until 2001, when the original line organization was reconstituted by new Associate Administrator for Space Science Ed Weiler; the Sun-Earth Connection theme became a separate division with OSS. By mid-1996, the last mission in the ISTP queue, Polar, was successfully launched.

In 1998, Wes Huntress left NASA. The new Associate Administrator, Edward Weiler, encouraged Withbroe to continue to push a new concept he had developed, in which space physics missions would perform the scientific research necessary to support a variety of practical applications with respect to space weather and its effect on human society and life. (Space weather is the term describing the conditions in space, including the behavior of the Sun and its interactions with Earth's magnetic field and atmosphere, that influence Earth and its technological systems.[19])

Withbroe saw a strong linkage between NASA's research into the causes and effects of solar variability and societal needs in the areas of space weather, human exploration, and climate change. He employed the Pasteur Research Model[20] in promoting a new initiative that expanded the traditional rationale for space science from one that stressed increased understanding to one that also included applying that understanding in useful ways. Withbroe worked with the interagency National Space Weather Program Council members that included the Departments of Commerce, Defense, Energy, Interior, and Transportation, along with NASA, the National Science Foundation, and the Federal Coordinator of Meteorology, to develop interagency linkages to the space weather initiative and to ensure it was responsive to national needs. Planning for the initiative,

---

18. G. Riegler, "Reorganization of NASA's Office of Space Science," *AAS Newsletter*, Vol. 109, March 2002, p. 16.

19. See *www.nas.edu/ssb/swwhat.html* for more information on space weather.

20. See Donald Stokes, *Pasteur's Quadrant* (Washington, DC: Brookings Institution Press, 1997), for an explanation of this model.

which became known as "Living With a Star" (LWS), focused on human radiation exposure related to spaceflight and high-altitude flight, the impact on space assets, satellite operations, communication systems, terrestrial power grids, and the effects of solar variability on terrestrial climate change.

Weiler wanted to pitch this concept to Goldin; Withbroe prepared the presentation. The first opportunity Withbroe had to present the LWS program to Goldin was at a 3 August 1999 Science Council Meeting. (See Document II-24.) Goldin enthusiastically supported the concept. By August 1999, the main elements of NASA's FY01 budget request to the White House had been finalized. LWS had to be proposed as an "add-on" option. Fortunately, the Clinton administration liked the concept and folded it into its proposed NASA budget to Congress. In this way, LWS became a new initiative for NASA starting in FY01.

The major elements of LWS included coordinated science missions, targeted research and technology (TR&T), and space environment test beds. The initial LWS missions included a Solar Dynamics Observatory aimed at providing the solar observations to understand the origins and flow of energy from the Sun and the solar drivers of variability at Earth. Additionally, Ionosphere Thermosphere Storm Probes and the Radiation Belt Storm Probes would determine the solar variability effects on Earth's space environment. Finally, Sentinels would connect eruptions and flares on the Sun with Geospace disturbances. An International LWS Working Group was formed in 2002 to pursue international cooperative missions within the LWS program.

The combination of Withbroe's leadership, the match-up of NASA's scientific research capability to the societal needs, and the importance of the relevancy of NASA programs all contributed to the approval of the LWS initiative. The solar maximum in the 1999–2000 period also cooperated in demonstrating the relevance of the LWS initiative, with solar storms and Earth effects that were monitored by the ISTP missions, along with other spacecraft.

During this period, Withbroe also was working hard to provide a continuing space physics research capability. He proposed a line of missions which would provide in situ and remote-sensing observations from multiple platforms for sustained study of the Sun-Earth system. A new start for the Solar-Terrestrial Probes (STP) program was included in the FY 1998 budget. Six STP satellites were planned; they continued research on how and why the Sun varies, and the effects of that variation on Earth and other planets. The first STP mission, TIMED (Thermosphere, Ionosphere, Mesosphere Energetics and Dynamics), was launched in December 2001. TIMED was scheduled to be followed by SOLAR-B, Solar-Terrestrial Relations Observatory (STEREO), Magnetospheric Multiscale (MMS), Global Electrodynamic Connections (GEC), and Magnetospheric Constellation (MC). The initiation of the STP program represented a major programmatic success in space physics. Planning for LWS and STP missions anticipated a number of constellations providing multi-spacecraft-coordinated observations extending to 2025.

10. Conclusion

NASA space physics activity clearly has evolved over the years. In the late 1950s and 1960s, a series of mission lines (e.g., Explorer, IMP, Injun) were created to explore unknowns about space physics, beat the Russians in doing so, and improve the nation's space capability. Cost was constrained to the rate authorized by Congress, so the schedule was fairly flexible. Each series was built in an overlapping way. In a line of missions with similar instruments, a mistake in an instrument design on mission A would be found when it reached orbit, but mission B was already in testing, so a complete fix could only be made with mission C. The fact that launch vehicles were not reliable was probably behind the acceptance of this mode of operation.

Up until the mid-1970s, investigators were publishing results from single instruments that reflected the exploring phase of space physics research. But in the late 1970s, a set of smaller, focused science Explorer missions, such as ISEE (three satellites), were designed to work together as a coordinated multispacecraft effort taking simultaneous data. A shift in research emphasis began to take place, away from initial discovery and more toward understanding the physics of the magnetosphere. Theorists began to take a greater interest in space plasma physics. As this transition was occurring, the field of space physics almost died in NASA.

By the late 1980s and into the early 1990s, a series of connected missions were implemented in the ISTP program. These missions had a well-defined purpose and produced coordinated observations that were designed to monitor energy flow into the magnetosphere. The faster, better, cheaper era began in the early 1990s as a reaction to the end of the Cold War, declining budgets, and the excesses of the previous era. Daniel Goldin may have "saved the Agency" by insisting that smaller, cheaper missions get done more rapidly.

In the late 1990s, the STP and LWS mission lines were developed as part of a more strategic approach to planning NASA's missions. This required looking at the Sun-Earth system as it really was, a connected system of actions and responses. These missions made coordinated sets of observations from a constellation of spacecraft designed to perform fundamental research in space physics and to develop the tools and techniques to understand and predict space weather. Today it is increasingly recognized that space physics is a fundamental science discipline that also has the potential of utility for addressing issues on Earth.

Document II-1

Document Title: **"Proposal for Satellite-Borne Cosmic-Ray Experiments," The University of Chicago, 1 April 1958.**

Source: **John A. Simpson Papers, University of Chicago Library, Chicago, Illinois**

John Simpson, a physicist at the University of Chicago, was one of the scientists who took an early lead in developing proposals for conducting science in space. He remained active in space science for almost three decades. This excerpt from a 1 April 1958 proposal which was widely distributed is one of the first comprehensive plans for a series of space science missions. It led to the participation of Simpson and his associates in the Thor-Able 1, or Pioneer space probe, built by TRW and sponsored by the Air Force.

CML-PR-E-127

Copy No. _ of _ copies

Consisting of _ sheets
Proposal for
SATELLITE-BORNE COSMIC-RAY EXPERIMENTS

Submitted by
THE UNIVERSITY OF CHICAGO
Chicago 37, Illinois

1 April 1958

[iii] PREFACE

A significant portion of the physical sciences at the University of Chicago is devoted to fundamental questions concerning the physics of the earth and astrophysics–especially the problems of interplanetary and interstellar space, the origin of the elements, cosmic rays, stars and the galaxies. Contributions of the University of Chicago in these fields have been unique. To its present facilities for the pursuit of these researches the University considers essential the addition of artificial satellite instrumentation. With this new tool at the disposal of its physicists, astrophysicists and meteorologists, it is anticipated that significant strides in science will be made which would be otherwise unattainable within the next few years.

The incentive to take this step immediately has come from physicists in the Enrico Fermi Institute for Nuclear Studies who have devised experiments to extend their research in cosmic rays and in interplanetary magnetic fields which require satellites and rockets.

It is proposed to begin by undertaking these specific cosmic ray experiments and this request for financial support is directed toward that end. However, at a later time it is anticipated that other scientists in the Departments of Physics, Astrophysics and Meteorology will wish to participate in additional experimental observations not now included in the present budget.

The technical staff, laboratories, and instrumentation shops of the Chicago Midway Laboratories of the University of Chicago will be available for this enterprise. No new laboratories or large additions to the staff are required to accomplish the objectives outlined in this proposal.

The cosmic ray experiments are the logical outgrowth of studies at the Enrico Fermi Institute for Nuclear Studies of the University of Chicago. Over the past years the University's scientists have been concerned with:

1. The origin of cosmic radiation.
2. The production of cosmic radiation in solar flares.
3. The influence of the sun upon cosmic ray intensity.
4. The magnetic fields which exist in interplanetary space.
5. The electrical and magnetic properties of the interplanetary medium surrounding the earth and the changes in these conditions with time, using techniques derived from cosmic ray observations.

6.   Hydromagnetic phenomena within the solar system.

[iv] The new experiments are directed toward answering crucial questions which have arisen as a consequence of these experimental and theoretical research programs. The new experiments can only be carried out successfully in satellites and rockets. The proper selection of orbits is an essential requirement in the successful prosecution of the experiment. One of the required orbits will enable the satellite to co-rotate with the earth.

It is proposed that the Chicago Midway Laboratories of the University of Chicago undertake the necessary instrumentation development, design, construction, and preparation for satellites and rockets to be launched by an assigned agency of the U. S. Government. The Chicago Midway Laboratories will be responsible for the data-handling systems and for the reduction of the data after they have been recorded at earth stations. The Cosmic Ray Group of the Enrico Fermi Institute for Nuclear Studies will plan the experiments and undertake the analysis and scientific interpretation of the experiments when they are completed.

Details of the experiments, their required instrumentation development, construction problems, orbit requirement, data handling, etc. , are considered in some detail in the body of this proposal. The appendixes treat some of the major engineering problems in further detail.

It is emphasized throughout the document that these experiments are to be performed at different phases of the solar activity cycle. Consequently, to obtain the full fruits of the observation, it is essential that the work begin as soon as possible to take advantage of the current high solar activity. It is also pointed out that the observations are necessary over a long period of time and it is planned to telemeter to the earth over periods of approximately 400 days.

[v] ABSTRACT

Proposed herein is a program to undertake special cosmic ray and astrophysical experiments as an extension of the cosmic-ray research program of The University of Chicago. The proposed experiments require the use of rockets and earth satellites.

A detailed outline of the desired data, required instrumentation and communications link and the time and cost estimate for the program are included with the recommendation that the measurements be obtained while solar activity is still at a high level.

[1] 1. INTRODUCTION

This is a proposal for fundamental physical and astrophysical experiments to be carried out with earth satellites and high-altitude rockets, with the purpose of increasing basic understanding of the origin of cosmic rays, the electrodynamics of the solar system, geomagnetism, and the properties of interplanetary space, in the environments of both the earth and the sun.

Over the past decade various experimental and theoretical studies, mainly concerning the origin of cosmic rays and their intensity variations, have been carried out in the Enrico Fermi Institute for Nuclear Studies at The University of Chicago (see the attached bibliography of published research). These researches have led to many new concepts, which bear on the nature of magnetic fields in the vicinity of the sun, the earth, and throughout interplanetary space; they stress the importance of solar electromagnetic phenomena in our understanding of cosmic ray effects and the properties of the interplanetary medium.

We know that there are rare but important occasions when the sun, through a sudden tremendous release of energy, can produce cosmic ray particles with energies up to or exceeding 20 Bev. The associated electromagnetic phenomena have a profound influence upon the earth's magnetic field and its interplanetary environment. Through The University of Chicago's continuing cosmic-ray research program, we have also learned that those cosmic rays which come to us from beyond the solar system may undergo radical changes in their energy spectrum as a result of electromagnetic processes within the interplanetary space of the solar system. Current theoretical ideas indicate that high-temperature gaseous plasma emissions from our sun are the cause of these changing electromagnetic conditions in interplanetary space, resulting in such terrestrial effects as magnetic storms and auroral displays. Clearly, these phenomena are most important in explaining many of the observations in experimental cosmic-ray physics and in astrophysics. So far, cosmic-ray experiments have been the most successful approach to the investigation of the interplanetary magnetic conditions.

Thus the problems which are under current attack fall into two broad categories. First, there are the questions associated with the nature and origin of cosmic rays. They are of the following nature: Does the sun only rarely produce cosmic ray particles in solar flares, or does the sun produce low-energy cosmic rays also through other phenomena? Is the entire range of the energy spectrum due to cosmic rays coming from our galaxy or are they produced locally? The second category of questions is [2]concerned with the electrodynamics of the solar system. How are the magnetic fields distributed in interplanetary space? How do these fields change over the years between maximum and minimum solar activity? These questions are open to investigation because the charged particles of the cosmic ray act as probes to detect the presence of magnetic fields in the interplanetary medium and in the region of the earth. As investigations in both areas of research are so intimately related, most of the proposed experiments will yield answers in both.

Cosmic-ray research also has a significant impact upon some of the earth sciences–particularly our understanding of the shape of the earth's magnetic field above the surface of the earth. For example, experiments performed by The University of Chicago with aircraft and surface vessels from 1954 to 1956 proved that the dipole field analysis used during the past 100 years to determine the distribution of the earth's magnetic field is incorrect for describing the motion of cosmic ray particles approaching the earth. We expect that the observations of the current satellites will soon extend those results.

Through techniques evolved in the Fermi Institute, The University of Chicago established by 1951 a series of continuously operating neutron monitor piles extending from as far south as Peru to as far north as Chicago to analyze the energy spectrum of the incoming cosmic rays as a function of time by utilizing the earth's magnetic field. Aircraft and balloon observations supplemented this world-wide network. Always the problem has been the fundamental limitation in overcoming the effects of our own atmosphere and the earth's magnetic field. Now we wish to press with alacrity to the final step: experiments with satellites and rockets.

With such experiments, which may avoid the limitations of our atmosphere and reduce the influence of the earth's magnetic field, we wish to study the properties of the low-energy cosmic radiations, including their charge distribution, energy spectrum, and changes with time. We also wish to search for radiations the existence of which is not yet established. For example, we plan to search for high-energy gamma rays of galactic and solar origins and for neutrons of solar origin; i. e. , neutral radiations. In addition, we plan to explore the possibilities of detecting electrons and positrons of nonterrestrial origin. These experiments have a bearing on the nature of our galaxy and its genesis.

The proposed experiments will include the launching of three satellites and two very-high-altitude rockets. Each of the satellites is to transmit information for approximately 400 days. The experiments should

[3]begin at the earliest possible moment (not later than 1959) so as to cover as much as possible of the present maximum of the sun's activity, and extend through the forthcoming minimum (about 1964) . Only the first two-year program is covered by this proposal.

These experiments depend upon the proper selection of satellite orbits; this matter is discussed in Sec. 3.

Development of new types of detectors and information storage systems will be necessary, although preliminary investigations have shown that an appreciable amount of the instrumentation already developed by the Fermi Institute, or available elsewhere within the International Geophysical Year satellite program, can be directly applied to the proposed experiments. Development work will be necessary to achieve instrument packages which can provide continuous information for approximately 400 days and at the same time meet the weight requirements for launching.

Obviously a program of this scope and character must be a joint effort between scientists and engineers. The University of Chicago recommends that its Chicago Midway Laboratories undertake the development program and that scientists in its Fermi Institute plan the experiments, maintain close association with Chicago Midway Laboratories, and interpret the data obtained from the satellites and rockets. It is also likely that the Yerkes Observatory, the Department of Meteorology and other groups within the Physical Sciences Division of the University will contribute ideas for experiments and instrumentation requiring future satellite observations. Furthermore, it is possible that experiments involving mass spectroscopy may be considered later.

This is a program designed to give our scientists the opportunity for full-scale exploration of the fundamental problems. Although this is not a "crash" effort, these problems are of such concern than the experiments will receive the most careful attention and highest priority.

The proposed experiments are the logical outgrowth of a decade of research which has led to many new results in cosmic rays and astrophysics. This is not a data-collection program but rather a series of carefully designed experiments which we anticipate will yield decisive results in both cosmic-ray physics and astrophysics.

Document II-2

Document Title: Memorandum from Presidential Science Advisor James R. Killian, Jr., to the President, 23 August 1958.

Source: Dwight D. Eisenhower Presidential Library, Abilene, Kansas

This concise summary of the results of the first four U.S. satellites, all of which carried out space physics investigations, was prepared for President Dwight D. Eisenhower by his assistant for science and technology and former President of MIT, James Killian. Killian did not hesitate to provide a great deal of technical detail on the satellites and their investigations to the President.

[no page number]

THE WHITE HOUSE
WASHINGTON

August 23, 1958

MEMORANDUM FOR THE PRESIDENT

I inclose [sic] a brief summary of information which has been obtained by the four earth satellites we have placed in orbit. One of the most significant findings is the unexpectedly high radiation intensities at altitudes greater than 620 miles.

The "air" temperature has not yet been directly measured, but there are reports on satellite surface temperatures. These have varied from –13°F to 167°F.

It is interesting to note that the air density at sea level is 92 billion times greater than at 230 miles altitude.

[Original Signed by J. R. Killian, Jr.]

J. R. Killian, Jr.

1 Incl: Notes on U.S. Satellite Measurements

[no page number]

NOTES ON U. S. SATELLITE MEASUREMENTS

1.   The U.S. has successfully placed in orbit four earth satellites (Explorers I, III and IV, and Vanguard I). All except Explorer III are still in orbit. Explorer III, because of its low altitude perigee (118 miles) and the resulting relatively high air resistance, remained in orbit for only three months. Some pertinent information about these satellites is contained in the following table:

Satellite	Launching Date	Expected Life- Time (Years)	Perigee (Miles)	Apogee (Miles)	Payload (Pounds)
Explorer I	Jan. 31, 1958	3-5	224	1573	30.8
Vanguard I	Mar. 17, 1958	200	404	2465	3.3
Explorer III	Mar. 26, 1958	.25	118	1740	31.0
Explorer IV	July 26, 1958	.7-1	157	1380	38.4

2. The most extensive scientific measurements that have been conducted with the satellites have been in the area of cosmic radiation. In addition, measurements have been made of micrometeorite impact and the temperatures reached by the satellites. The measurement of the path of the satellite, which has been done by both electronic and optical means, has also allowed a determination of the air density at various altitudes.

3. Cosmic radiation measurements have been made with Explorers I, III and IV. New and unexpected results have been obtained from these measurements. Explorers I and III each carried a single Geiger-Mueller counter. These counters indicated radiation intensities as expected up to about 620 miles. This information correlates with data obtained in previous rocket tests. [2] At altitudes greater than 620 miles, however, unexpectedly high radiation intensities were measured which exceeded the measuring capacity of the counters.

    Explorer IV carried more extensive instrumentation which in addition to measuring the radiation intensities, also allows insight as to the type and characteristics of the particles causing the high measurements. Preliminary results from Explorer IV indicate substantiation of the estimates made from the Explorers I and III data. The radiation intensity increases by a factor of several thousand between 180 and 1000 miles. The intensity continues to rise to 1500 miles, the maximum altitude for which data is currently available. At 1000 miles the exposure level would be 2 roentgens per hour, as compared to the permissible human rate of approximately .002 roentgens per hour. This does not necessarily preclude manned space travel, but does make imperative proper shielding. In addition, the high radiation levels need to be considered in relation to the heating of the high atmosphere, the amount of visible light, radio noise and ionization produced.

4. Temperature measurements have been made on Explorers I and III, and in addition have been derived from transmitter frequency changes on the Vanguard satellite. The measurements have been made on the surface or within the satellites themselves, using "thermometers" whose electrical resistance changes with temperature. No attempt has been made as of yet [3] to measure the "air" temperature in space. The temperature measurements on the satellites have been made during the early state of satellite operations for engineering rather than scientific purposes. The purpose of the measurements was to ascertain the reliability of the method used to control the temperature of the satellite and therefore the internal electronic components.

In order to operate properly, these electronic components must be maintained within certain temperature limits, hence the concern over the temperature of the satellites. The temperature of the satellite will be determined by the quantity of heat transferred to it by means of conduction, convection or radiation. Since there is no body in contact with the satellite, there is no heat transferred by conduction. Convective or frictional heating resulting from the satellite passing through the "air" is negligible. It is not meant to imply that the "air" temperature is negligible, but that the heat transferred from the "air" to the satellite is negligible. This is because the heat transfer rate is dependent upon the density of the air, and the density at these altitudes is extremely small. Consequently, the predominating heat transfer process will be radiation. This radiation will be from the sun, either directly or reflected from the earth, and infrared radiations from the earth. A relatively small amount of heat is also generated within the satellite due to dissipation of power, but the temperature effects of this source of heat are small compared to the radiative heat transfer. The radiative heat transfer rate is directly [4] affected by the emissivity of the surface of the satellite. Consequently, by proper choice of surface material and finish the heat transfer rate, and thereby the temperature, can be controlled. For the Explorers, the stainless steel surface was partially covered by eight equally spaced, aluminum oxide strips running longitudinally along the cone-cylinder body. Calculations indicated that the resulting average emissivity would result in maintaining the electronic components between the desired 23°F and 113°F. The success of the method used is indicated by the measured temperatures in the transmitter which varied from 32°F to 97°F. The satellite surface temperature varied from -13°F to 167°F. The temperature variation results from the fact that part of the time the satellite is in the shadow and therefore not subject to direct radiation from the sun, resulting in a cyclic temperature history. The smaller variations in temperature in the components, as compared to the satellite surface, result from the fact that the rate of heat transferred from the satellite surface to the internal electronic components is retarded by the use of insulators between the surface and the internal components.

5. Explorers I and III carried wire grids for the detection of micrometeorite impacts. Explorer I also carried a microphone for micrometeorite detection. In one thirty-two day period the wire grid registered only a single impact, whereas the microphone detected seven hits. Taking into account the sensitivity of the instrument, the average influx of particles 4 ten-thousandths of an inch in diameter or greater was 8 per square foot [5] per twenty-four hour day, whereas the average influx of particles 1.6 ten-thousandths of an inch in diameter or greater was 80 per square foot per day. These values are in fair agreement with previous predictions, but are too limited for statistical analysis. While the probability of micrometeorite impacts of a size large enough to do structural damage is extremely small, the sandblasting effect of small particles over long periods of time could affect the satellite's surface condition. Surface conditions are important for temperature reasons as indicated previously, and in addition solar battery "windows" may be sensitive to surface conditions.

6. The density of the air determines the air resistance to the satellite, and in turn affects its path and orbital characteristics. Conversely, a knowledge of the change in orbital characteristics will allow the density of the air to be derived. All the satellites are

tracked electronically and optically, and from these measurements various investigators have derived values of air density in the altitude range of 110 to 230 miles. Previous measurements with vertical sounding rockets extended up to 135 miles. Above this altitude estimates of density were obtained from the observation of meteors. The satellite density data falls roughly between the measurements obtained with these two methods, and varied from 4170 X 10-14 pounds per cubic foot at 110 miles altitude to 88 X 10-14 pounds per cubic foot at 230 miles altitude. The air density at sea level, therefore, is 92 billion times greater than at 230 miles altitude. [6] The data is quite sketchy at this time, but as tracking techniques improve and data is accumulated, the satellites will provide a powerful tool for the accurate determination of density over a wide range of latitude, altitude and time.

Document II-3

Document Title: Homer E. Newell, Deputy Director, Space Flight Programs, NASA, "The United States Program in Space Research," 27 October 1958.

Source: NASA Historical Reference Collection, History Office, NASA Headquarters, Washington, D.C.

Dr. Homer Newell was one of the pioneers of the U.S. space science program. When NASA opened for operation on 1 October 1958, he was one of the people who transferred from the Naval Research Laboratory to the new space agency. Newell was given primary responsibility for shaping NASA's initial space science efforts. This talk, dated less than a month after NASA was created, reflects his thinking on those efforts.

27 October 1958

THE UNITED STATES PROGRAM IN SPACE RESEARCH*

by
HOMER E. NEWELL, JR.
National Aeronautics and Space Administration

*Prepared for presentation at the Royal Society, London, on 12-13 November 1958.

[no page number]

On July 29, 1958, the President of the United States signed into law a bill passed by the Eighty-fifth Congress and entitled the "National Aeronautics and Space Act of 1958". The Act calls for an integrated national program under civilian direction and control, of research and development in the fields of aeronautics and space, and specifically declares "that it is the policy of the United States that activities in space should be devoted to peaceful purposes for the benefit of all mankind."

The form and conduct of any program depends on the objectives of that program. The U.S. National Aeronautics and Space Act of 1958 specifically states that objectives of aeronautical and space activities of the United States shall be:

  (a) The expansion of human knowledge of phenomena in the atmosphere and space;
  (b) The improvement of the usefulness, performance, speed, safety, and efficiency of
      aeronautical and space vehicles;
[2] (c) The development and operation of vehicles capable of carrying instruments,
      equipment, supplies, and living organisms through space;
  (d) The establishment of long-range studies of the potential benefits to be gained
      from, the opportunities for, and the problems involved in the utilization of
      aeronautical and space activities for peaceful and scientific purposes; and
  (e) Cooperation by the United States with other nations and groups of nations in
      aeronautics and space activities and in the peaceful applications of the results
      thereof.

These objectives call for an extremely broad program that will take considerable time
to get fully underway. To conduct such a program, the Space Act created a National
Aeronautics and Space Administration (NASA), which it vested with the necessary power
and authority. Dr. T. Keith Glennan was appointed the Administrator of NASA.

Since the NASA program is still in its planning and organizational states, it will not be
possible to present you with a detailed description of the program. Instead, it will be
necessary to speak in general terms of areas of interest and the immediate and long-range
plan of attack. I shall further confine myself to remarks on the space science portion of
the [3] program, and in particular to those aspects of space science involving the use of
rockets, satellites, and space probes.

The present U.S. activity in space science has grown out of the rocket upper
atmospheric research of the past twelve years. Prior to the start of the International
Geophysical Year several hundred high altitude sounding rockets had been fired to study
the upper atmosphere and the sun. Although some of the soundings reached the F2-
region of the ionosphere, for the most part the studies were confined to the E-region and
below. For the IGY, now drawing to its close, a couple hundred additional rockets have
been launched, including a special effort to probe the F-region. The IGY has also seen the
launching of the Explorer and Vanguard satellites, and probes far out toward the distance
of the moon.

NASA plans to continue the use of sounding rockets and satellites in exploring the
earth's atmosphere and the space beyond the atmosphere, and to add deep space probes
for the study of cislunar and interplanetary space. Without attempting to give a complete
review, we list here some of the areas of major scientific interest.

[4] (a) <u>Atmospheres</u>. The earth's upper atmosphere continues to present many unsolved
problems. Although the work of the IGY has moved us a long way toward a full
understanding of the atmosphere up through the E-region of the ionosphere, much
remains to be done. This is all the more true of the F-region and beyond. In particular a
considerable amount of work remains on the dynamical aspects of the high atmosphere,
and on the geographic, daily, and seasonal variations of atmospheric parameters and
phenomena. There is also the question of where and in what manner the earth's
atmosphere merges with that of interplanetary space.

In addition to the earth's atmosphere, the atmospheres of the moon and planets
present fascinating and challenging problems. Those of Venus, Mars, and Jupiter are of
special scientific interest and challenge. There is, of course, the ever present question of
whether or not the moon has any atmosphere; if it does have any, how far does it extend

and of what is it composed?

Finally, the sun's atmosphere is of major interest. How far does it extend? Is the earth enveloped in it outermost reaches? If so, what is the temperature of the solar atmosphere at the distance of the earth?

[5] (b) Ionospheres. Recent work in the investigation of the earth's ionosphere, particularly during the IGY, both from the ground and in rockets, has brought about a considerable advance in our understanding of the ionosphere. But here too, as in the case of the more general aspects of the atmosphere, there is need for further measurements in the F-region and beyond, and to determine geographical, daily, seasonal, and solar cycle effects. How does the ionosphere merge with the medium of outer space? What is its relationship to the Van Allen radiation belt? To the aurora? To the solar atmosphere?

What sort of an ionosphere, if any, does the moon have? What sort of ionospheres do the planets have, particularly Venus, Mars, and Jupiter?

(c) High energy particles. A strong attack is underway with satellites and space probes on the study of energetic particles from cosmic rays to those of kilo-electron-volt energies. The discovery of the Van Allen radiation belt is one result of these studies. What are the total spatial distribution of such particles, their fluctuations with time, their nature and spectral distribution, and their ultimate origin? The answers to these questions are not only of great [6] scientific interest, but also may reveal whether or not it will be possible for man to fly out into space.

(d) Fields. What are the details of the gravitational, magnetic, and electric fields that will be encountered in interplanetary space? Do the moon and planets have magnetic fields? What is the detailed structure of the moon's gravitational field? How is the earth's magnetic field modified and distorted by clouds and streams of charged particles from the sun? These are questions whose answers can be sought by means of satellites and vertical probes.

(e) Photons. We can study the sun, stars, and galaxies because of the radiations from them. With satellites and space probes it will be of interest to observe and measure throughout the spectrum from gamma ray wavelengths all the way to the radio wave region. Also of interest are the airglow and auroral radiations from the earth's atmosphere in both the visible and ultraviolet wave lengths; infrared emissions from our atmosphere; atmospheric radiations from other planets; and radiations from the material in interplanetary space.

(f) Astronomy. There is especial interest in ultraviolet astronomy, which can be pursued only from above the [7] atmosphere. The small amount of work done to date in sounding rockets has shown the existence of many strong ultraviolet sources in the sky, and this work should be continued. An obvious extension of this work will be to observe in the x-ray and possibly gamma ray regions.

(g) Controlled experiments. The high atmosphere and outer space provide a huge laboratory in which it will be possible, by means of satellites and space probes, to expose materials and biological specimens to conditions unattainable in laboratories on the ground. One can conduct tests in extreme vacuums and over long periods of weightlessness. The possibility also exists of using for special studies the highest energy cosmic rays, which are many orders of magnitude more energetic than those obtainable in man made accelerators. Likewise it may be possible, with highly accurate clocks installed in orbiting satellites, to provide additional checks on the general theory of relativity.

The developing NASA program will include experiments and observations in the areas listed above.

Prior to the establishment of NASA, rocket and satellite research had been carried out on a rather [8] broad base, with government agencies, universities, and industry participating. The major strength of the effort lay in the programs of the Department of Defense. Now NASA provides a natural center for such activity, and it is the plan to build up a strong space research operation within NASA itself. This internal operation will, however, be only a portion of the total national space science program, for it is planned to maintain and if possible to extend the current base of the program. The best talents will be utilized wherever they exist, and NASA will provide financial support where necessary.

The National Academy of Sciences has established a Space Science Board, with Lloyd Berkner as Chairman, and drawing upon scientific leaders in a wide variety of fields for its membership. The Board has had set before it the task of reviewing and assessing the relative scientific value of proposals for space research that have been submitted to the Board, NASA, the National Science Foundation, and the Advanced Research Projects Agency of the Department of Defense. Through a number of committees in such areas as meteorology, the ionosphere, astronomy, the physics of field and particles in space, and geodesy, further [9] scientific talent is drawn into the deliberations of the Space Science Board. The conclusions and thinking of the Space Science Board will be available to NASA and should be of invaluable assistance in planning the total program and in deciding what proposals to support.

On the international side, there is now the Committee on Space Research (COSPAR) of the International Council of Scientific Unions. It is the hope of NASA that through bodies like COSPAR scientific thinking of an international scope will be available to NASA. Such an input will help the NASA to guide the U.S. space research program so that it will contribute effectively to the benefit of all mankind.

Document II-4

Document Title: "Memorandum from Homer Newell to Dr. Silverstein," (Abe Silverstein, Director, Space Flight Operations, NASA) Transmitting "Policy on Release of Reduced Data Acquired from Experiments Carried on NASA Sounding Rockets, Satellites, and Space Probes," 15 August 1960.

Source: NASA Historical Reference Collection, History Office, NASA Headquarters, Washington, D.C.

An important issue from the start of the NASA space science program was access to the data acquired in the course of NASA's scientific efforts. NASA wanted to make sure that these data were well archived and eventually made available for analysis to all in the scientific community with an interest in using them. NASA's policy on data release was developed by its Space Science Steering Committee.

[cover letter]
MEMORANDUM To Dr. Silverstein

SUBJECT: Policy on Release of Reduced Data Acquired from Experiments Carried on
           NASA Sounding Rockets, Satellites and Space Probes

1. The Space Sciences Steering Committee has considered the subject of the release of reduced data acquired in experiments flown on NASA sounding rockets, satellites, and space probes. The policy which it approved on 9 August and recommended for approval as a NASA policy is attached for your consideration.

2. The Space Sciences Steering Committee in its consideration of a policy kept two points in mind. The first was to give the individual scientific experimenter maximum freedom in the use of data that he obtained from his own experiments. The second was to avoid loss of data by an experimenter's inability to release them in the technical literature, and to specify the conditions under which NASA itself would release the data in advance of any actual case.

3. The policy recommended is practical, and it is simple administratively. It agrees with traditional scientific practice in giving the experimenter the right to his own data, but it does set up provisions for NASA to recover data that might otherwise be lost and for NASA to release the data for general use after the experimenter has finished with them himself.

Homer E. Newell
Deputy Director
Space Flight Programs

Enclosure
Dr. Stauss/jn/15 Aug 60

[no page number]
POLICY FOR RELEASE OF DATA ACQUIRED ON NASA SOUNDING ROCKETS, SATELLITES AND SPACE PROBES

Background
An important goal of the NASA is the encouragement of the widest possible participation in the space sciences program. Within the scientific community the individual scientist is of prime importance to the space science program. He is responsible for the conception of experiment, for carrying out the experiment, and for interpreting its results. To support the space sciences program the NASA makes available vehicles and funds for research in space and assists in the dissemination of the resulting experimental information.

In assigning the use of its facilities for space research the NASA endeavors to choose the best experiments and experimenters available. In order that new knowledge in the space sciences may be disseminated as widely and as rapidly as possible it is essential that the experimenters publish their findings as soon as they can.

This paper defines the procedures which NASA will follow in its relations with the individual scientists or groups participating in the NASA Program with regard to the publication of experimental data obtained in NASA sounding rockets, satellites, space probes or other spacecraft operations.

Basis of Policy

Scientists as a group feel a responsibility for making the results of their efforts known to the worldwide scientific community. Scientists have been the most vigorous proponents of free dissemination of the knowledge gained from their investigations and, in general, can be expected to release experimental scientific information rapidly.

In the usual case, an agreement that will assign the responsibility for the analysis and publication of the data to the scientist will exist between the NASA and the scientist. In the period before the publication of the results of the experiment it may happen that another scientist may wish to have access to the experimental data in its preliminary form. This is generally arranged informally by a request made directly to the experimenter.

[2] Occasions may arise when it will be necessary for NASA to release the results of experimental investigations made under its sponsorship, without waiting for publication by the cognizant scientist. It is expected that this will happen only when the responsible scientist is unable to meet his commitments within a reasonable time.

In many instances the data acquired from space vehicles, like the observations made in astronomy, will have cumulative value. Whenever this is the case, it is desirable that such data be made available for the general use of the scientific community.

The original data received from a satellite or space probe is meaningless until the various channels are separated or demultiplexed, the calibration of the instrumentation is introduced, and the data put in chronological order. Provisions for these operations must be included in the arrangements made for supporting the applicable experiments.

Policy

The original data recordings received from a satellite or probe will be stored at the cognizant NASA center.

Secondary records containing reduced data will be prepared under the cognizance of the responsible NASA center. On occasion this operation may take place at contractor's or experimenter's establishments. In the secondary records all extraneous material and all confusion of data will be removed to the fullest extent possible. The records will be prepared by the NASA activity or contractor with the cooperation of the experimenter so as to satisfy the experimenter's and NASA's needs for reduced data.

The experimenter will be given whatever original or reduced data from his experiment that his agreement with NASA calls for as soon as possible after they are received. The cognizant NASA center will retain copies of the reduced data furnished the experimenter. The experimenter will have sole use of the reduced data for study for a period previously decided upon between him and the cognizant research center, with the approval of Headquarters.

[3] No fixed time is set for the period of exclusive use of reduced data by the experimenter. Each case will be considered individually and no general rule is established.

If, during the period in which the experimenter has not reported the results of his experiment, information on the progress of the experiment or portions of the reduced data are requested by another scientist for designing space science experiments, or for the preparation of a technical paper, informal arrangements for the release of the needed data should be made by the experimenter concerned. Both the cognizant NASA center and the NASA Headquarters should be informed of these arrangements.

At the end of the agreed upon period, during which the experimenter will have the opportunity to analyze and report on the reduced data, the reduced data may be released for general use if such a course seems best for developing the space sciences. An extension of the period of sole use requested by the experimenter will require the approval of Headquarters.

All releases of the reduced data by the NASA centers to other than the original experimenter must be approved by Headquarters.

[no page number]
Definition of Terms
Used in the Policy Statement

Original data records
Those records made by the various telemetering and/or tracking stations as part of the basic field operations. Those recordings will generally require editing for the removal of useless segments, playback to introduce calibration factors, correlation with related recordings and the application of specialized processing techniques to recover weak signals from noisy recordings.

Reduced data records
Those records prepared from the original data records by editing, introduction of calibration factors and inter-record correlations. These will contain a minimum of extraneous information and/or noise and will generally present the value of the physical quantity measured as a function of time and position. It is from these records or the tabulations or graphs prepared directly therefrom that the responsible scientist will form his analysis and conclusions.

Analyzed data
Those data which have been reviewed and correlated by either the original experimenter or another scientist so as to form the experimental basis for a technical paper submitted for publication to either NASA or a scientific journal.

Document II-5

Document Title: Letter from M. (Merrill) A. Tuve, Chairman, Geophysics Research Board, National Academy of Sciences to NASA Administrator James E. Webb, 13 September 1968.

Source: Archive of National Space Science Data Center, NASA Goddard Space Flight Center, Greenbelt, Maryland

Document II-6

Document Title: Letter from Homer E. Newell, Associate Administrator, NASA, to Merrill A. Tuve, Chairman, Geophysics Research Board, National Academy of Sciences, 12 November 1968.

By the mid-1960s, data about the structure of the magnetosphere and the radiation belts were rapidly accumulating from the numerous NASA missions, but access to the data for the whole science community was very difficult since the data was largely held by a group of instrument teams distributed across the United States. Finally in 1966, NASA established the National Space Science Data Center (NSSDC) at Goddard Space Flight Center. The establishment of the NSSDC enabled U.S. investigators to archive NASA data at a central location and provided a mechanism to distribute the data to any U.S. scientist upon request. But the international science community could still not obtain access to this archival data. In September 1968, Dr. Merrill A. Tuve, Chairman of the Geophysics Research Board of the National Academy of Sciences (NAS), asked NASA Administrator James Webb to approve the transfer of the responsibilities of the U.S. Rockets and Satellites Center from NAS to NASA. The Rockets and Satellites Center was part of the World Data Center system that maintained important international agreements for the exchange of archival data with the international science community. The task of responding to this request fell to Newell, who assigned the management responsibility to the NSSDC. As part of the World Data Center system, the NSSDC could then make its archival data available to foreign scientists upon request. This arrangement greatly facilitated the distribution of space physics data and stimulated important national and international scientific collaborations in space physics.

[no page number]

NATIONAL ACADEMY OF SCIENCES
NATIONAL RESEARCH COUNCIL
OF THE UNITED STATES OF AMERICA

GEOPHYSICS RESEARCH BOARD
13 September 1968

Mr. James E. Webb
Administrator
National Aeronautics and
    Space Administration
Washington D. C. 20546

Dear Mr. Webb:

    The Committee on Data Interchange and Data Centers has approved the proposal to transfer the Subcenter for Rockets and Satellites of World Data Center A from the National Academy of Sciences to the National Aeronautics and Space Administration. As you know, discussions on this have been held between members of the GRB staff and NASA. The following comments seem appropriate.

WDC-A and its subcenters are committed to international arrangement that must be honored. The arrangements in some countries are crucial to their over-all ability to cooperate in international programs and evens, in some countries, for carrying on what we consider normal programs of observations and measurements in geophysics. Thus continuity as well as adequacy of support of the subcenter is required. Moreover, the data in question are important to our domestic interests. The WDC system now provides for the flow of certain data, compilations, and reports to the world data centers on a voluntary basis; disruption of the system could undermine this approach, and we could be faced with the spectacle of public and private groups engaged in unilateral purchasing efforts.

Within the Academy, the GRB is responsible for the general guidance of the subcenters of WDC-A. This is accomplished through an advisory Committee on Data Interchange and Data Centers (Data Committee) and through the GRB staff (in particular, the Director of WDC-A). In addition to its international responsibility in the conduct of WDC-A, the Data Committee is concerned with the national aspects of interchange of geophysical data. An assessment of national aspects of collection of space data was made by an ad hoc Committee chaired by A. H. Waynick (copy of report enclosed [report not included]). The Data Committee is continuing to look into ways to enable the WDC's and the relevant national data centers to best serve the scientific community.

We ought soon to formalize the transfer of WDC-A for Rockets and Satellites to NASA, taking into account the attendant responsibilities. The essential considerations in the transfer of the subcenter from the Academy to NASA are the following:

1. International Provisions. The general provisions of such understandings are published in the ICSU Guide to International Data Exchange. They include: (a) collection and suitable archiving of specified material, (b) interchange of such material with the other World Data Centers, (c) issuance [2] of catalogues of data, (d) provision of data to scientists, and (e) free access to the data collection of the subcenters by visiting scientists from nations participating in appropriate ICSU endeavors. In contrast to most discipline areas, you will note the simplicity of provisions in the rocket and satellite areas, where raw data and even processed data are not presently involved; we would hope that more useful arrangements might evolve over the coming years.

The Academy's responsibility for supervision of the WDC subcenters includes an obligation to ensure that the subcenters meet these international commitments. The Academy is particularly concerned about free access to the data in the WDC subcenters; this means that visitors to WDC-A subcenters should have free access not only to all data in the subcenters, but also to the subcenter itself. Therefore, the subcenter should be free of complex restrictions that might apply to other activities in the organization in which the center is located. In this connection, physical arrangements will be required that will clearly distinguish the WDC subcenter from other activities in the building in which it is housed in order to ensure that free access to the subcenter is not complicated by access requirements of other facilities.

2. Head of the subcenter. The head should be a technically qualified individual with adequate authority to conduct the subcenter as indicated in (1) above, and to represent

the Academy in domestic and international matters as called upon. He should understand the distinction between his role as head of the subcenter, when he functions as an international official on behalf of the Academy, and other duties he may have as a NASA employee.

Formal concurrence in these provisions by NASA would permit the Academy to transfer the activities of the WDC-A subcenter in the near future.

Sincerely yours,

M. A. Tuve
Chairman

Enclosure
cc   Donald R. Morris
     James I. Vette

Document II-6

[no page number]
[NASA logo]
[stamped NOV 12 1968]

NATIONAL AERONAUTICS AND SPACE ADMINISTRATION
WASHINGTON, D.C. 20546

OFFICE OF THE ADMINISTRATOR

Dr. M. A. Tuve, Chairman
Geophysics Research Board
National Academy of Science
2101 Constitution Avenue
Washington, D. C. 20418

Dear Dr. Tuve:

In reply to your letter of 13 September 1968, to Mr. Webb, we agree with the transfer of the Subcenter for Rockets and Satellites of World Data Center A (WDC-A) to the National Aeronautics and Space Administration, and are willing to assign management of the subcenter to the National Space Science Data Center (NSSDC) at the Goddard Space Flight Center (GSFC). Such a move has distinct advantages for both the subcenter and our NSSDC. In operating the subcenter it is definitely our intention to continue to honor the existing international arrangements, and to be responsive to the general policy guidance as it pertains to international affairs from the National Academy of Sciences in the future. In addition, we will provide adequate support to the subcenter so that the identity of the subcenter will be preserved.

We agree with the essential considerations cited in your letter. Although I believe we understand these provisions, I think it is advisable to detail the responsibilities we expect to discharge as the operators of the WDC-A for Rockets and Satellites to insure agreement on both sides. According to the Committee on Space Research (COSPAR) Guide to Rocket and Satellite Information and Data Exchange adopted by the Xth COSPAR Plenary Meeting, London, July 1967, the WDC for Rockets and Satellites endeavors to collect and maintain rocket and satellite data in the form of reports and reprints from publications that receive limited distribution. Samples of analyzed data appear in these reports, but the subcenter does not collect all of the analyzed data from a particular experiment. On the other hand, NSSDC is mainly concerned with collecting all the reduced and analyzed data available regardless of the recording medium. It is our desire that [2] data in the national archive - the NSSDC - be made available to foreign scientists through the World Data Center network for data exchange. However, it should be clearly understood that this data collection in the NSSDC will [sic] not be considered a part of WDC-A for Rockets and Satellites. It would be impractical and extremely expensive to reproduce and furnish copies of all the data to the other WDCs. The new Guide recognizes this fact and indicates that the WDCs will assist scientists in obtaining copies of original or calibrated data held by experimenters in national archives. It is clear that the NSSDC and WDC-A for Rockets and Satellites would remain separate entities with complementary functions. In addition to providing copies of reports and reprints to the other WDCs and to individual requestors, we expect to carry out the following principal duties of the subcenter for producing catalogues and sounding rocket listings as given in the Guide:

1. Prepare and distribute every six months, catalogues of materials received during the previous six months. These catalogues will be accumulated every two years.

2. Prepare and distribute to COSPAR a monthly summary of sounding rocket launchings.

3. Prepare and submit to WDC-A a summary of rocket launchings to be included in the U. S. report to COSPAR.

I certainly share your concern over providing access to the subcenter free of complex restrictions. We anticipate no problems in providing free access to the subcenter for recognized scientists and administrators from nations participating in the International Council of Scientific Unions' endeavors. The subcenter will be housed in a single large room within the NSSDC (Building 26) at GSFC. The room and building will be marked with appropriate signs. For your information, the GSFC has established a new procedure for foreign visitors to the subcenter which will be free of restrictions as long as they identify themselves properly as WDC-A visitors. Although we intend to announce in the various WDC-A for Rockets and Satellites catalogues that visitors should provide the subcenter with advance notification of a visit, it will not be absolutely necessary to do so.

The mailing address of the subcenter will be:

[3]
World Data Center A for Rockets and Satellites
Code 601
Goddard Space Flight Center
Greenbelt, Maryland 20771
U.S.A.

The Director of the NSSDC (currently Dr. James I. Vette) will be appointed as the Director of WDC-A for Rockets and Satellites. It is felt that the technical qualifications necessary for the Director, NSSDC, are commensurate with those you listed. In addition, he would be sufficiently versed in both domestic and international matters so that he can function as an official on behalf of the Academy. Consequently, you can be assured that the appropriate leadership will be provided regardless of the actual incumbent. The Office of International Affairs will continue to monitor NASA's responsibilities to the Academy in the international aspects of data exchange.

If you agree with our understanding of the responsibilities for the operation of the subcenter, I suggest that the transfer of the subcenter be made effective January 1, 1969, or as soon as possible thereafter.

Sincerely yours,

[stamp, "Original Signed by Homer E. Newell]

Homer E. Newell
Associate Administrator

cc: GSFC/Clark
    GSFC/Vette
[stamp of arrow pointing to Vette]

Document II-7

Document Title: Dr. Leo Goldberg, Harvard College Observatory, to Homer E. Newell, Associate Administrator, Office of Space Science, 23 April 1969.

Source: NASA Historical Reference Collection, History Office, NASA Headquarters, Washington, D.C.

Document II-8

Document Title: Donald P. Hearth, Director, Planetary Programs, Office of Space Science and Applications, to Dr. Leo Goldberg, Harvard College Observatory, 3 June 1969.

Source: NASA Historical Reference Collection, History Office, NASA Headquarters, Washington, D.C.

Scientists working on space physics questions did not limit their interest to Earth's magnetosphere. As NASA began to send probes to Venus and Mars, they were eager to have their instruments aboard and to take measurements in the interplanetary medium and in the vicinity of planets other than Earth. They also perceived that their interests were not being given high enough priority as NASA selected scientific payloads for solar system missions. Leo Goldberg, chair of NASA's Astronomy Missions Board, expressed his Board's concern in the following letter to Space Science Chief Homer Newell. Responding for Newell, Solar System Exploration Director Donald Hearth suggested that interplanetary particle and field measurements may have gotten too high a priority in early missions to Venus, and the interests of this area of space science were adequately accommodated in the planned NASA missions to Mars.

Document II-7

[no page number]
[stamped "Action Copy to AA Info Copy to S" unreadable]
[stamped "RECEIVED MAY 9 04 AM '69 NASA CODE S]

HARVARD COLLEGE OBSERVATORY
CAMBRIDGE, MASSACHUSETTS 02138

April 23, 1969

Dr. Homer E. Newell
Code SG
NASA
Washington, D. C. 20546

Dear Homer:
    At the last AMB meeting, the Board expressed some concern about the methods used by NASA to assign various fractions of the budget of joint planetary-interplanetary missions. Some members suggested budget quotas, fixing the assignment of funds on a joint mission at the outset, so that specific areas, such as fields and particles of the planets, are not squeezed out as mission costs increase. However, the Board as a whole was reluctant to attempt to advise NASA on such specific management techniques, and voted instead to endorse the recommendation below. The Board also requested that you consider placing this subject on the agenda of the joint meeting of some members of the AMB, the L&PMB, and NASA administration in May.

The Astronomy Missions Board identifies the following problem, that during the development of planetary missions interplanetary experiments seem to be preferentially dropped to meet budgetary requirements. The Board asks that NASA examine this situation, and if they agree that it is a problem, report back to the Board their proposed solution.

Sincerely,

[original signed by Leo Goldberg]

Leo Goldberg

LG:lf

Document II-8

[no page number]
[stamped 3 JUN 1969]

SL (DPH: mag)

Dr. Leo Goldberg
Harvard College Observatory
Cambridge, Massachusetts 02138

Dear Dr. Goldberg:

This letter is relative [sic] to your letter of April 23, 1969, to Dr. Newell. I understand your concern regarding the apparent tendency of NASA to delete particles and fields experiments on planetary missions when mission costs increase. However, when one examines the record in the context of the objectives of the missions I believe a different picture emerges.

First, let's look at the Venus program. Mariner II had a plasma probe, magnetometer and two energetic particle experiments. Mariner V's payload included a plasma probe, magnetometer, energetic particle experiment, and a dual frequency receiver. Neither of these spacecraft included a TV, infrared spectrometer, or an ultraviolet spectrometer, despite the fact that the Venus atmosphere is very important and these three are prime tools for atmospheric investigations. In retrospect, I believe that too much emphasis may have been given to particles and fields, and not enough to atmospheric experiments. I might note in passing that the mission objectives of the Venus Planetary Explorer in 1972 gave first priority to the interaction of the solar wind with the ionosphere and second priority to the ionosphere and atmosphere.

The Mars exploration program has developed somewhat differently due to different scientific objectives. NASA has stressed Mars exploration because of the possibility of making discoveries important to biology. This, in turn, has influenced the decisions on the mission payloads.

Our first Mars mission, Mariner IV, carried a plasma probe, magnetometer, and two energetic particle experiments. Thus, both interplanetary and planetary particles and fields experiments were conducted. The next spacecraft, Mariners VI and VII, due to fly by Mars later this year, are [2] atmosphere and surface oriented, with no particles and fields experiments included. The decision to take this course was based on the prime objective of exobiology and on the funds available. The same philosophy was originally adopted in planning the Mariner 1971 orbiters. However, in putting the payload together NASA decided to include, as a secondary priority, two experiments to measure the cosmic ray gradient. Subsequently, it was recognized that the Mariner 71 project would cost significantly more than the funds budgeted. After efforts to completely resolve the problem without affecting the scientific payload failed, we had no recourse but to delete the experiments in the second priority category. We sincerely regretted this action, but the alternative of removing experiments related to the prime objective of exobiology was unacceptable. This decision does not mean that we have no desire to make particles and fields measurements in the vicinity of Mars. The Planetary Explorer program, which we hope to initiate in Fiscal Year 1971, includes missions to Mars which will be largely in this scientific area.

One final comment should be made on the interplanetary measurements. The Pioneer series, A-E, is exclusively this type of mission. Furthermore, on Pioneers F and G these experiments are a major objective and comprise a major portion of the payload.

I trust that when you examine this history, you will agree with my initial remarks. We all must recognize that scientific priorities must be established and that some decisions will inevitably be made that will be displeasing to a particular scientific group. Otherwise, prime objectives will be compromised and the total program will suffer.

Sincerely yours,

[stamped, "Original signed by Donald P. Hearth"]

Donald P. Hearth
Director of Planetary Programs
Office of Space Science and Applications

bcc :   S / Naugle
SL /Chron
    SL /Hearth:Official File
[handwritten SS/Smith SL/Rea]

SL/DPHearth:mag 6/2/69

A23370

Document II-9

Document Title: H. (Harold) Glaser, Director, Solar-Terrestrial Program, to Dr. Lewis (sic—correct spelling is Louis) Lanzerotti, 20 November 1975.

Source: NASA Historical Reference Collection, History Office, NASA Headquarters, Washington, D.C.

Harold Glaser was brought to NASA Headquarters in the mid-1970s to revitalize what was then known as the Solar-Terrestrial Physics Program. He began his efforts by reaching out to leading non-NASA scientists interested in the field.

[stamped] Nov. 20, 1975

Dr. Lewis J. Lanzerotti
Bell Laboratories
600 Mountain Avenue
Murray Hill, NJ 07974

Dear Lew:

I am very interested in having an opportunity to discuss with you and a few other members of the Physical Sciences Committee (PSC) a number of issues regarding the new Solar Terrestrial Programs. I have discussed the desirability of doing this with George Field and Ichtiaque Rasool and both are very warm towards the idea of such discussions.

The items I wish to discuss are:

1. Should the new division have a standing committee which would advise me, as Director, on the balance between various program elements; and provide an overview for a more detailed look at the scientific merit of the various program elements? Would such a committee, with the proper membership, be useful in pointing out scientific gaps in solar-terrestrial physics that NASA should fill? If there is such a committee, who would you suggest be on it? This committee would not be part of the PSC.

2. The field of magnetosphere [sic] physics seems to be on the defensive. I think it is extremely desirable to take a careful look at the field; to understand its present status and to evaluate its future, and to attempt to define an appropriate role for NASA in this field. If such an examination is warranted, how can it best be performed? Should NASA ask the Space Science Board to do this? Should it be done by an ad hoc group?

3. Should an examination similar to that of 2 above be done for the atmospheric sciences? As was mentioned to you at the last Physical Sciences Committee meeting by Noel Hinners, this division will absorb the Upper Atmospheric Research Office. It may be that such an examination of the atmospheric sciences is not warranted.

I would like to have an opportunity to discuss these items with you and others on the occasion of the next PSC meeting. Would it be possible for us to have dinner on either December 10 or December 11 for this purpose?

Sincerely,

[signature]
H. Glaser
Director
Solar Terrestrial Programs

Cc:
George Field/Harvard
SL/Thomas E. Burke
SS/Ichtiaque Rasool

Identical letter sent to: Drs. Donahue, Sturrock, Vogt

ST(HG/aser:svab) – 11/19/75 x51790

Document II-10

Document Title: Memorandum from Chief of Magnetospheric Physics (E. R. Schmerling) to the Director of Solar-Terrestrial Programs, "Status of International Magnetospheric Study," 8 December 1975.

Source: NASA Historical Reference Collection, History Office, NASA Headquarters, Washington, D.C.

A major change in direction in space physics research occurred when the Scientific Committee on Solar-Terrestrial Physics (SCOSTEP), which operated under the auspices of the International Council of Scientific Unions, established an international program designed to understand magnetospheric processes by coordinated observations with space- and ground-based instruments. The International Magnetospheric Study (IMS) period ran from 1976 to 1979. The principal NASA contribution consisted of correlative data from operating spacecraft, data from the International Sun-Earth Explorer (ISEE) program (which had three spacecraft), and support from the National Space Science Data Center.

[no page number]
8 December 1975
MEMORANDUM

TO: ST/Director of Solar Terrestrial Programs
FROM :   ST/Chief of Magnetospheric Physics and NASA Representative/IMS
SUBJECT: Status of International Magnetospheric Study (IMS)

The IMS Steering Committee met at Goddard Space Flight Center 4-6 December 1975 under the chairmanship of Dr. J. G. Roederer, University of Denver. The IMS period will start 1 January 1976. Owing to the delay in key spacecraft launches, especially the ESA-GEOS (now scheduled for launch April 1977), it is expected that the peak IMS activity will be in the 1978-1979 period, and that the 1976 activities will center on the ground-based observations coordinated with data from existing spacecraft.

The principal NASA contributions consist of the Satellite Situation Center (SSC) at the Data Center, GSFC, and correlative data to be contributed by operating spacecraft and spacecraft which are part of our on-going approved program.

The SSC has already started to operate, and has provided predicted positions of spacecraft making magnetospheric measurements in 1976. From these, key times have been recommended for special attention, when these spacecraft are in especially favorable positions for determining events at the main magnetospheric boundaries. These total 493 hours, or about 6% of the available time. We should be able to obtain summary data for these periods for release through the Data Center without using resources beyond those already anticipated.

The International Sun-Earth Explorers (ISEE/A,B,C) are expected to be close to schedule, with launches of ISEE/A,B in mid-1977 and ISEE/C in mid-1978. NASA project costs are in the $42-46M range. The science working groups are planning to release summary data tapes to the IMS community within two to three months.

[2] U. S. Government agencies will be asked to re-affirm their support at the meeting on 9 December of the Federal Council for Science and Technology. The NASA contributions, as summarized here, are well in hand, and we do not anticipate problems in meeting them. The proposed press release is quite consistent with this position.

[signature]
E. R. Schmerling

cc: A/Fletcher
     AA/Naugle
     I/Frutkin
     S/Hinners
     SS/Rasool

Document II-11

Document Title: Letter from Andrew Stofan, Acting Associate Administrator for Space Science, to A. Thomas Young, Director, Goddard Space Flight Center, 27 April 1981.

Source: NASA Historical Reference Collection, History Office, NASA Headquarters, Washington, D.C.

Document II-12

Document Title: Memorandum from George Pieper, Director of Sciences, GSFC, to Dr. Franklin Martin, Office of Space Science and Applications, NASA Headquarters, "International Sun-Earth Explorer-3," June 1982.

Source: Robert Farquhar

Document II-13

Document Title: Presentation by Charles J. Pellerin, Jr., Director, Astrophysics Division, to Sam Keller, Deputy Associate Administrator for Space Science and Applications, "ISEE-3: Obtain Approval for In-Situ Exploration of Comet Giacobini-Zinner," 6 August 1982.

Source: NASA Historical Reference Collection, History Office, NASA Headquarters, Washington, D.C.

Document II-14

Document Title: Memorandum from Associate Administrator for Space Science and Applications (Burton Edelson) to Director, Goddard Space Flight Center (Noel Hinners), "ISEE-3," 30 August 1982.

These documents trace the 1981–1982 debates within NASA over the possibility of sending an existing spacecraft, the International Sun-Earth Explorer-3, to intercept a comet as it made its way through the inner solar system in the mid-1980s. The initial impetus for this suggestion was the lack of a U.S. mission to Halley's Comet during its 1986 encounter, but additional study showed that Comet Giacobini-Zinner was a better target for a comet encounter.

Document II-11

National Aeronautics and
Space Administration
Washington, D.C.
20546

[stamped] APR 27 1981

Mr. A. Thomas Young
Director
Goddard Space Flight Center
Greenbelt, MD 20771

Dear Tom:

I have been made aware of the possibility of deflecting the ISEE spacecraft to intercept comet Halley. Since we are continually assessing what our total involvement should be with respect to the apparition of Halley's comet, it behooves us to understand all of the options available to us. While I am, in no way, endorsing or supporting the potential of an ISEE intercept with comet Halley, I would like to have the information as to whether the mission is even physically realizable. On this basis, I would support a small study at GSFC, to be led by Dr. Farquhar, to evaluate the feasibility of the mission he proposed. Dr. Farquhar should understand that our desire is solely to gain information, and he should not represent the authority to conduct the study as an endorsement of this mission by the Office of Space Science.

This study will be funded and managed by the Advanced Programs area of the Solar System Exploration Division, and would, therefore, be under the direct cognizance of Mr. Diaz.

Sincerely,

[signature]
Andrew J. Stofan
Acting Associate Administrator
for Space Science

Document II-12

[no pagination]
National Aeronautics and Space Administration
Goddard Space Flight Center
Greenbelt, Maryland
20771

TO: NASA Headquarters
    Attention: Office of Space Science Applications
    Dr. Franklin D. Martin/EZ-7

FROM: Director of Sciences

SUBJECT: International Sun-Earth Explorer-3 (ISEE-3)

A small but significant orbit maneuver for ISEE-3 is planned for June 10, 1982. The ISEE-3 is, as you know, in a halo-type orbit in the vicinity of the sunward libration point, L1, of the Sun-Earth system. The spacecraft has periodically required station keeping maneuvers to keep its orbit properly aligned with the Sun and Earth to allow the gathering of scientific data. For this particular maneuver I have authorized the maneuvering of the spacecraft in such a way as to keep open several extended mission options which are presently being actively investigated.

The forthcoming maneuver which consists of adding a velocity increment propulsion of 4.6 meter per second will, if not rescinded, cause the spacecraft to start a trajectory which will carry it toward and subsequently into the geomagnetic tail. The enclosed sketch (Figure 1) [no figures included] shows the orbital path.

It is important that this maneuver be done at this time because it keeps open the opportunities for the following extended mission options:

a. geomagnetic tail excursion followed by return to halo orbit,

b. geomagnetic tail excursion and remain in the tail,

c. geomagnetic tail excursion followed by an encounter with the comet Giacobini-Zinner, and

d. remain in halo orbit.

The key features of the timing of the maneuver are subsequent necessary lunar flybys and perigees when the spacecraft can use the Earth and Moon for trajectory modification for either a return to the halo orbit at L1 or the start of a trajectory which would lead to the cometary encounter (options a. and c. above).

The cost, in onboard propellant, of rescinding the planned mid-course maneuver increases with time to the beginning of August after which it would be too costly to stop the spacecraft from going on a trajectory to the geomagnetic tail. Figure 2 shows the velocity increment required to rescind the planned maneuver increases from 4.6 meters per second to about 130 meters per second by mid-August. The present total maneuver capability of ISEE-3 is about 330 meters per second. The cost per year of remaining in the geomagnetic tail orbit is between 10 and 100 meters per second depending on the kinds of orbit selected. The risk associated with going to the geomagnetic tail is about the same as leaving the spacecraft in its present orbit. In either case, the propulsion system will have to be used. There are no significant ground system requirements for the geomagnetic excursion. Everything would be about the same except that the spacecraft would be in an interesting and unexplored region of space.

From a physical point of view, the decision concerning the comet encounter does not need to be made until April 1983. Budgetary planning will force it to be made this summer, I believe.

Unless you advise me to the contrary, the June 10, 1982 orbit maneuver will be carried out.

[signature]
George F. Pieper

Document II-13

[handwritten note, "1"]

ISEE-3
AUGUST 6, 1982

PURPOSE

OBTAIN APPROVAL FOR IN-SITU
EXPLORATION OF COMET GIACOBINI-ZINNER

[handwritten note, "2"]

MEETING AGENDA

(RM. 226 A, 4:00-5:30 AUGUST 6, 1982)

- OVERVIEW   C. PELLERIN

  - PURPOSE
  - BACKGROUND
  - MISSION PROFILE
  - SCIENCE
  - DSN
  - RISK ASSESSMENT
  - PROGRAMMATIC SUMMARY
  - SUMMARY

- ADDITIONAL DETAILS AS REQUIRED

  - COMET SCIENCE   E. SMITH
  - HALO SCIENCE   T. VON ROSENVINGE
  - SPACECRAFT TECHNICAL   S. PADDACK
  - DSN TECHNICAL   W. MARTIN
  - FLIGHT ASSURANCE   A. JONES
  - TRAJECTORIES   B. FARQUHAR

[handwritten note, "3"]

ISEE-3 BACKGROUND

- EXPLORER-CLASS SPACECRAFT
- LAUNCH VEHICLE: DELTA-2914
- LAUNCH DATE: AUGUST 12, 1978
- LOCATION: HALO ORBIT AROUND SUNWARD LIBRATION POINT
- NETWORK: STDN
- MISSION OBJECTIVES
  - SUPPORT ISEE-1 AND –2 SPACECRAFT
  - MEASURE ISOTOPIC COMPOSITION OF SOLAR AND GALACTIC COSMIC RAYS
  - STUDY DIVERSE INTERPLANETARY AND SOLAR PHENOMENA
  - PROVIDE BASELINE DATA FOR DEEP-SPACE PROBES (E.G., PIONEER, VOYAGER)

- PRIMARY MISSION COMPLETED: AUGUST 1981

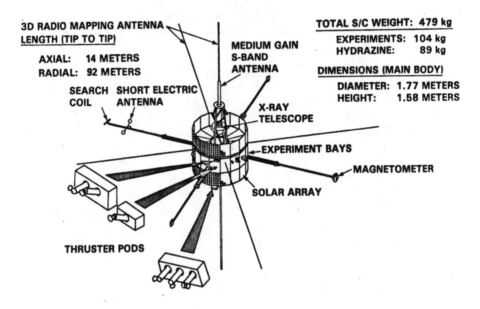

[handwritten note, "4"]

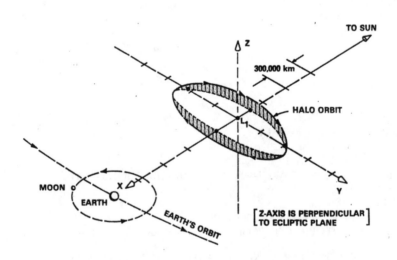

[handwritten note, "5"]
[handwritten note, "6"]
BACKGROUND

- AUGUST 1981 – BOB FARQUHAR PROPOSED COMET RENDEZVOUS AS PART OF ISEE-3 EXTENDED MISSION

- DECEMBER 21, 1981: LETTER FROM DIRECTOR, ASTROPHYSICS DIVISION, NASA HQ, TO GSFC, REQUESTS FEASIBILITY STUDIES FOR EXTENDED MISSION OPTIONS FOR ISEE-3
  - GEOMAGNETIC TAIL AND BACK TO HALO ORBIT
  - GEOMAGNETIC TAIL FOLLOWED BY INTERCEPTION OF COMET GIACOBINI-ZINNER

- MARCH 24, 1982: FEASIBILITY STUDY RESULTS REPORTED TO NASA HEADQUARTERS

- JUNE 10, 1982: SMALL BUT SIGNIFICANT ORBIT MANEUVER TO CAUSE TRAJECTORY TO GEOMAGNETIC TAIL

- JUNE 14, 1982: OSSA, OSTDS, AND USAF REPRESENTATIVES REVIEWED ISEE-3 MISSION OPTIONS

- JUNE 15, 1982: PRESENTATION TO NATIONAL ACADEMY OF SCIENCE

- JUNE 29, 1982: AA/OSSA BRIEFED ON ISEE-3, GEOTAIL EXCURSION APPROVED

- AUGUST 3, 1982: FLIGHT ASSURANCE OFFICE REVIEWED COMET MISSION PROPOSAL

- AUGUST 6, 1982: OSSA DECISION MEETING AT NASA HQ

[handwritten note, "7"]

ISEE-3 EXTENDED MISSION STUDY TEAM

CHARLES PELLERIN	HQ, ASTROPHYSICS DIVISION
ERWIN SCHMERLING	HQ, PROGRAM SCIENTIST
STEVE PADDACK	GSFC, STUDY MANAGER
BOB FARQUHAR	GSFC, PRIME MOTIVATOR
TYCHO VON ROSENVIGNE	GSFC, PROJECT SCIENTIST
JIM TRAINOR	GSFC, SCIENCES DIRECTORATE
ED SMITH	JPL, COMET SCIENCE SUBCOMMITTEE

FRED SCARF   TRW, GEOTAIL SCIENCE SUBCOMMITTEE
KEITH OGILVIE   GSFC, HALO SCIENCE

BILL WOOD   HQ, NETWORKS DIVISION
CHARLES FORCE   HQ, MANAGER, DSN
ROMO CORTEZ   JPL, MANAGER, REAL TIME SUPPORT
WARREN MARTIN   JPL, DSN PLANNING

VERN MALAHY   USAF
GEOFF BRIGGS   HQ, SOLAR SYSTEM EXPLORATION DIVISION
SHELBY TILFORD   HQ, ENVIRONMENTAL OBSERVATION DIVISION

AL JONES   GSFC, FLIGHT ASSURANCE

[handwritten note, "8"]
MISSION PROFILE

- SPACECRAFT TRAJECTORIES
- MILESTONES

[handwritten note, "9"]

ISEE-3 EXTENDED MISSION (TOP VIEW)
TRANSFER FROM L1 HALO ORBIT TO GEOMAGNETIC TAIL ORBIT. LEG 1

PERIGEE-0   64 RE
APOGEE-0   188 RE
PERIGEE-1   12 RE DV= 0.283 M/SEC
APOGEE-1   221 RE DV=32.664 M/SEC

DATES D1: MAY 23, 1982   DATES A1: FEBRUARY 8, 1983
      D2: SEPTEMBER 30, 1982   S1: MARCH 30, 1983
      PO: OCTOBER 17, 1982   P2: APRIL 3, 1983
      AO: NOVEMBER 23, 1982   PERILUNE RADIUS S1: 21,782 KM
      P1: DECEMBER 22, 1982   AT SWINGBY

ISEE-3 EXTENDED MISSION   (TOP VIEW)
TRANSFER FROM GEOMAGNETIC TAIL TO LUNAR SWINGBY ON
DECEMBER 30, 1983, TO IMPART SUFFICIENT EXCESS HYPERBOLIC VELOCITY
TO ENCOUNTER COMET GIACOBINI-ZINNER ON SEPTEMBER 11, 1985

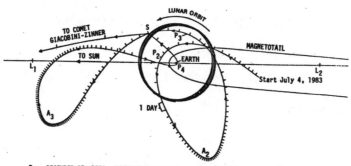

P_4: DECEMBER 27, 1983   PERIGEE RADIUS 3 R_E ΔV = 12 M/SEC

S: DECEMBER 30, 1983   PERILUNE RADIUS 2580 KM, HEIGHT 840 KM

[handwritten note, "10"]

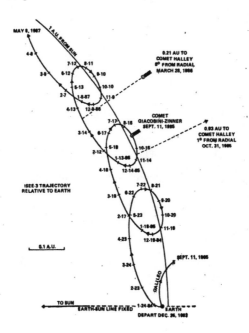

[handwritten note, "11"]
[handwritten note, "12"]

ORBIT PROFILE

- DEPART HALO JUNE 10, 1982

- GEOTAIL EXPLORATION DURING 1983

- DEPART FOR COMET (BEGIN "CRUISE PHASE") DECEMBER 1983

- DSN SUPPORT BEGINS MARCH 1984

- COMET GZ ENCOUNTER SEPTEMBER 1985

- MEASURE SOLAR WIND INPUT TO HALLEY MARCH 1986

[handwritten note, "13"]

ISEE-3 SCIENCE

- COMET
  - SCIENTIFIC OBJECTIVES
  - ANTICIPATED RESULTS
  - NAS REVIEW RESULTS

- HALO
  - SCIENCE

- INVESTIGATOR'S VIEWPOINT
- USAF VIEWPOINT

[handwritten note, "14"]

SCIENTIFIC OBJECTIVES OF COMET EXPLORATION

1. TO DETERMINE THE COMPOSITION AND PHYSICAL STATE OF THE NUCLEUS

2. TO DETERMINE THE PROCESSES THAT GOVERN THE COMPOSITION AND DISTRIBUTION OF NEUTRAL AND IONIZED SPECIES IN THE COMETARY ATMOSPHERE; AND

3. TO INVESTIGATE THE INTERACTION BETWEEN THE SOLAR WIND AND THE COMETARY ATMOSPHERE.

[handwritten note, "15"]

ANTICIPATED SCIENTIFIC RESULTS

- EXISTENCE, LOCATION, PROPERTIES OF REGIONS AND SURFACES:
  BOW SHOCK
  SUPERSONIC FLOW
  CONTACT SURFACE
  PLASMA TAIL

- IDENTIFICATION OF PHYSICAL PROCESSES:
  IONIZATION OF COMETARY IONS BY COLLISIONS
  PICK-UP OF HEAVY IONS BY SOLAR WIND ("MASS LOADING")
  PARTICLE ACCELERATION
  PLASMA INSTABILITIES AND THE GENERATION OF WAVES

- QUANTITATIVE INFORMATION:
  MAGNETIC FIELD STRENGTH AND DIRECTION
  PLASMA FLOW SPEEDS
  PLASMA DENSITIES, COMPOSITION
  ELECTRON, HEAVY ION TEMPERATURES
  BASIC PLASMA PARAMETERS:
    BETA
    WAVE SPEED
    MACH NUMBER

[handwritten note, "16"]

AD HOC SPACE SCIENCE BOARD ISEE PANEL

A.G.W. CAMERON, <u>CHAIRMAN</u>, HARVARD COLLEGE OBSERVATORY

W. BOYNTON, UNIVERSITY OF ARIZONA

G. CARIGNAN, UNIVERSITY OF MICHIGAN

C.F. KENNEL, UNIVERSITY OF CALIFORNIA, LOS ANGELES

L. J. LANZEROTTI, BELL LABORATORIES

M. NEUGEBAUER, JET PROPULSION LABORATORY

A.B.C. WALKER, JR., STANFORD UNIVERSITY

S.E. WOOSLEY, UNIVERSITY OF CALIFORNIA, SANTA CRUZ

[handwritten note, "17"]

SPACE SCIENCE BOARD

"THUS, IN SUMMARY, WE CONCLUDE:

1. USE OF ISEE-3 TO EXPLORE THE EARTH'S MAGNETOTAIL IS OF HIGH SCIENTIFIC IMPORTANCE.

2. BOTH OF THE SUBSEQUENT OPTIONS, RETURN TO THE HALO ORBIT AND A MISSION TO COMET GIACOBINI-ZINNER, ADDRESS SCIENTIFIC GOALS OF THE HIGHEST IMPORTANCE IN DIFFERENT DISCIPLINES.

3. IN ADDITION TO ADDRESSING SPECIFIC QUESTIONS OF SCIENTIFIC IMPORTANCE, SPACE MISSIONS OFTEN SERVE BROADER GOALS. ONE SUCH GOAL, IN THIS CASE, IS EXPLORATION–TO INVESTIGATE NEW REGIMES OR NEW PHENOMENA FOR THE FIRST TIME. THE EXPLORATION OF A COMETARY ENVIRONMENT FOR THE FIRST TIME IS A SCIENTIFIC GOAL OF THE HIGHEST PRIORITY."

[handwritten note, "18"]

SPACE SCIENCE BOARD

"FINALLY, WE <u>RECOMMEND</u>:

1. ISEE-3 SHOULD CONTINUE ON ITS PRESENT TRAJECTORY TO EXPLORE THE MAGNETOTAIL.

2. SUBSEQUENTLY, ISEE-3 SHOULD BE SENT TO EXPLORE THE TAIL REGION OF THE COMET GIACOBINI-ZINNER.

3. NASA SHOULD TAKE STEPS TO REPLACE THE MONITORING FUNCTIONS OF ISEE-3 IN A HALO ORBIT."

[handwritten note, "19"]

<u>HALO EXTENDED MISSION SCIENCE</u>

- SIGNIFICANT SCIENCE
  - SOLAR WIND
  - SOLAR TERRESTRIAL
  - GAMMA RAY BURSTS

- HIGHLY DESIRABLE SUPPORT FOR OTHER MISSIONS
    - ISEE 1 AND 2 (TO MID 1987)
    - DE-1
    - AMPTE (JUNE 1984 - JUNE 1985)
    - SIS/ISPM (JUNE 1987-1990)
    - VOYAGER/PIONEER 10
    - GRO

- USAF WEATHER SERVICE
    - 30 MINUTE LEAD TIME ON SUBSTORMS
    - RADIATION HAZARDS (STS)

(MISSION FROM DEC. 1983 TO ?)

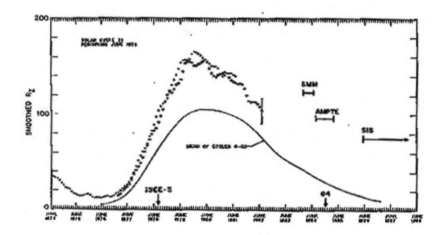

[handwritten note, "20"]

[handwritten note, "21"]

ISEE-3 INVESTIGATORS RE COMET

ISSUES

- GAIN/LOSS FOR EACH EXPERIMENT

- PROBABILITY OF CONTINUED MO&DA

- COVERAGE (CRUISE PHASE VS. HALO)

- PROBABILITY OF REPLACEMENT

- POLITICS (GIOTTO, SIS, OPEN)

- EXPLORATORY SCIENCE VS. ANALYTICAL SCIENCE

(CURRENT?) VIEWPOINT

- SCARF, SMITH, BAME - VERY STRONG SUPPORT

- STONE, HOVESTADT, HECKMAN - CONDITIONAL

- OGLIVIE, HYNDS, STEINBERG, ANDERSON, VON ROSENVINGE, MEYER, TEEGARDEN, WILCOX - OPPOSE

[handwritten note, "22"]

ISEE-3 INVESTIGATIONS

TITLE	PRINCIPAL INVESTIGATOR	EXPERIMENT AFFILIATION	STATUS
SOLAR WIND PLASMA	BAME	LOS ALAMOS NATIONAL LAB	ELECTRONS ONLY (ION PORTION FAILED)
PLASMA COMPOSITION	OGILVIE	GSFC	OPERATIONAL
MAGNETOMETER	SMITH	JPL	OPERATIONAL
PLASMA WAVES	SCARF	TRW SYSTEMS	OPERATIONAL
ENERGETIC PROTONS	HYNDS	IMPERIAL COLLEGE, LONDON	OPERATIONAL
RADIO WAVES	STEINBERG	PARIS OBSERVATORY, MEUDON	OPERATIONAL
X-RAYS, LOW	ANDERSON	UCB	X-RAYS AND EE > 200 KEV ENERGY ELECTRONS (LOW ENERGY ELECTRON PORTION FAILED)
LOW ENERGY COSMIC RAYS	HOVESTADT	MPI	PARTIAL FAILURE (ULEZEQ)
MED IUM ENERGY COSMIC RAYS	VON ROSENVINGE	GSFC	OPERATIONAL
HIGH ENERGY COSMIC RAYS	STONE	CIT	PARTIAL FAILURE (ISOTOPE PORTION)
HIGH ENERGY COSMIC RAYS	HECKMAN	UCB/LBL	PARTIAL FAILURE (DRIFT CHAMBER)
COSMIC RAY ELECTRONS	MEYER	UNIVERSITY OF CHICAGO	OPERATIONAL
GAMMA RAY BURSTS	TEEGARDEN	GSFC	PARTIAL FAILURE (PHA MEMORY)

[handwritten note, "23"]

USAF VIEWPOINT

- USAF <u>PREFERS</u> ISEE-3 REMAIN IN HALO ORBIT

- AVAILABILITY OF USEFUL DATA WILL CEASE IN SEPT. 1982

  - WILL RESUME IF RETURNED TO HALO
  - WILL CEASE TOTALLY IF SENT TO COMET

- CONTACT HAS BEEN COL. VERN MALAHY; CHAIRMAN, JOINT ENVIRONMENTAL SATELLITE COORDINATING GROUP, OFFICE OF THE UNDERSECRETARY OF DEFENSE

- USAF APPROACH

  - USAF DESIRES SHOULD BE HEARD AND CONSIDERED
  - NO PLANS TO INTERPOSE

[handwritten note, "24"]

DSN IMPLICATIONS

- REQUIREMENTS

- IMPACTS/CAPABILITIES

[handwritten note, "25"]

ISEE-3 REQUIREMENTS
(SIRD)

- FROM NOW UNTIL DEC, 1983 - 2KBPS - GOAL 90% COVERAGE

- CRUISE PHASE - 2KBPS AS LONG AS POSSIBLE - GOAL 60% COVERAGE

- GOLDSTONE AND MADRID SUPPORT FOR ENTIRE VIEWING PERIOD ON DAY OF ENCOUNTER (512 BPS EXCEPT 1KBPS FOR 2 1/2 HOURS)

ADDITIONAL DESIRES

- ADDITION OF OTHER STATIONS (E.G. JAPANESE, GERMAN) TO ALLOW CONTINUOUS COVERAGE NEAR ENCOUNTER

- ADDITION OF ARECIBO TO ALLOW 2KBPS OPERATIONS

[handwritten note, "26"]

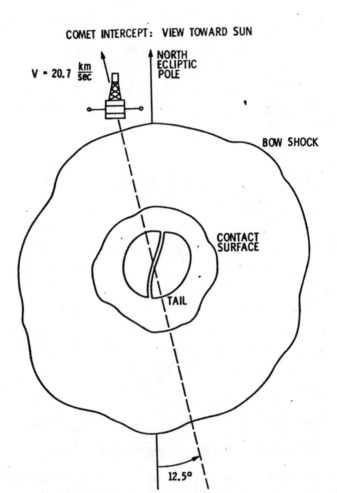

COMET INTERCEPT: VIEW TOWARD SUN

[handwritten note, "27"]

ISEE-3 GIACOBINI-ZINNER DSN SUPPORT
REQUIRED GROUND STATION IMPLEMENTATION

- DSN 64-MEER STATIONS .
  - 512 B/S, 2270.4 MHz
    - ADDITION OF AN LS 4815 SEQUENTIAL DECODER

- 1024 B/S, 2270.4 AND 2217.5 MHz
- ADDITION OF AN LS 4815 SEQUENTIAL DECODER
- CONSTRUCTION AND INSTALLATION OF A 2217 MHz MASER
- RELOCATION OF SECOND MASER INTO THE CONE
- 2217 MHz RECEIVER .
- BASEBAND COMBINER (MARK IVA DSN BBA OR A RESISTOR)

- ARECIBO
  - 2048 B/S, 2270.4 MHz LISTEN ONLY
    - 2270 MHz LINE FEED AND POLARIZER
    - 2270 MHz MASER OR COOLED FET AMPLIFIER
    - DSN RECEIVER, DEMODULATOR, LS 4815 DECODER
    - DATA HANDLING EQUIPMENT
    - LINK TO PROJECT OPERATIONS AND CONTROL CENTER

[6483-2242

WLM-6 3-24-82]

[handwritten note, "28"]

ISEE-3 GIACOBINI-ZINNER DSN SUPPORT
GROUND STATIONS TO BE CONSIDERED

- DSN GOLDSTONE COMPLEX
  - 64-METER (DSS 14)
    - 2217 MHz AND 2270 MHz DOWNLINKS ARRAYED
  - 34-METER (DSS 12) - POSSIBLE
    - 2270 MHz DOWNLINK ONLY

- DSN SPANISH COMPLEX
  - 64-METER (DSS 63)
    - 2217 MHz AND 2270 MHz DOWNLINKS ARRAYED
  - 34-METER (DSS 61) - POSSIBLE
    - 2270 MHz DOWNLINK ONLY

- ARECIBO - POSSIBLE
  - NEEDED FOR 2048 B/S DATA

- EFFELSBERG - POSSIBLE

  - 100-METER
    - POSSIBLE ALTERNATIVE TO ARECIBO

- JAPAN - POSSIBLE
  - 64-METER
    - ENABLES CONTINUOUS COVERAGE

[6483-2242   WLM-6 3-24-82]

[handwritten note, "29"]

RISK ASSESSMENT

- EXTENDED MISSION REVIEW TEAM CONVENED (GSFC)

ALTON E. JONES, CHAIRMAN	FLIGHT ASSURANCE
JAMES A. FINDLAY, DEPUTY CHAIRMAN	FLIGHT ASSURANCE
ROBERT L. SEGAL	FLIGHT ASSURANCE
JAMES TRAINOR	SCIENCES
EDWARD POWERS	ENGINEERING
HENRY PRICE	ENGINEERING
ALBERT YETMAN	ENGINEERING

- REVIEW AREAS

    - MISSION PROFILE
    - SPACECRAFT SYSTEMS
    - TRACKING AND DATA ACQUISITION
    - MISSION AND DATA REQUIREMENTS

[handwritten note, "30"]

FINDINGS

- NO CURRENT PROBLEMS WITH S/C SYSTEMS IDENTIFIED

- SLIGHT INCREASE IN RISK TO S/C DUE TO 30-MINUTE ECLIPSE

- DATA ACQUISITION AREA IS AREA OF BIGGEST QUESTION

    - 512 BPS AND 1024 BPS NOT USED SINCE LAUNCH
    - SIGNAL MARGINS ARE CLOSE (GSFC PERSPECTIVE)

-   WITHOUT ADDITIONAL STATIONS (ARECIBO, JAPAN, GERMANY), COVERAGE GAP MAY RESULT IN LOSS OF BOW SHOCK DATA.

•   ADDITIONAL CONCERNS TO NOTE

-   CRITICAL MANEUVERS DURING 1983 MAY CONFLICT WITH STS (COORDINATION SHOULD RESOLVE)

-   RESULTS OF THESE STUDIES SHOULD BE CAREFULLY DOCUMENTED

[handwritten note, "31"]

RISK SUMMARY

"ALL-IN-ALL, THE UNCERTAINTIES/RISKS CITED ABOVE
APPEAR TO ME TO BE LESS THAN THOSE THAT GENERALLY
EXIST WHEN WE START A PROJECT. THE DIFFERENCE IS,
IN THIS CASE, WE CAN ONLY FIX GROUND SYSTEM."

NOTE: TWO MAJOR RISKS HAVE BEEN OVERCOME:

-   LAUNCH VEHICLE FAILURE
-   INFANT MORTALITY

[handwritten note, "32"]

COMET PROGRAMMATIC SUMMARY

(THOUSANDS OF FY 83 DOLLARS)

	83	84	85	86	87	88	89
OSSA-DELTAS							
SCIENCE/OPERATIONS	+135	+135	+290	+525	-1000	-1700	-2300
OSTDS-DELTAS							
DSN	+1700	+1000					
OPERATIONS	+195	+95	+70	-225	-450	-450	-450

OSSA LEVEL FOR HALO = 2300/YEAR
OSTS LEVEL FOR HALO = 450/YEAR

[handwritten note, "33"]

SUMMARY

- RECOMMEND APPROVAL FOR ISEE-3 MISSION TO EXPLORE TAIL OF COMET GIACOBINI-ZINNER

  - FIRST IN-SITU MEASUREMENT OF A COMET

  - UNIQUE OPPORTUNITY FOR IN-SITU MEASUREMENT OF COMET/SOLAR-WIND INTERACTION

  - EXCELLENT TIMING FOR. CORRELATIVE REMOTE OBSERVATIONS (HALLEY WATCH, IUE, ST, OSS-3)

  - SCIENCE COMPLEMENTS THE HALLEY PROBES (ESA, JAPAN, USSR)

  - RETURN JUSTIFIES THE RISK–NASA SHOULD DO IT

  - A MOTIVATED, ENTHUSIASTIC TEAM SUGGESTS HIGH SUCCESS PROBABILITY

[FIGURE 2, HANDWRITTEN PAGE 5]

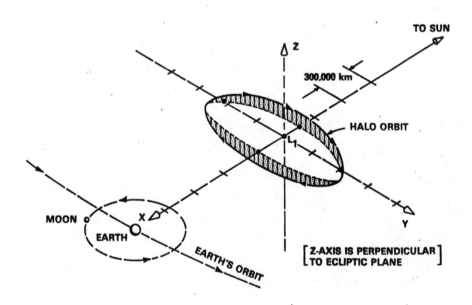

Document II-14

[no page number]
[stamped AUG 30 1982]
[handwritten note – "Boss – Advance copy of approval letter. This has not been sent as of 9/8/82." signed "JW"]

National Aeronautics and
Space Administration

Washington, D.C.
20546

Reply to the Attn. of: EZ-7

TO: Goddard Space Flight Center
    Attn: 100/Director

FROM: E/Associate Administrator for Space Science and Applications

SUBJECT: ISEE-3

The mission of ISEE-3 to explore the comet Giacobini-Zinner is approved. This approval is based on the presentation and discussion of August 6, 1982, and the draft Project Plan provided by GSFC. If there are major changes in the technical risk, or in the funding requirements for either OSSA or OSTDS, please assure that I am promptly informed.

[Original signed by Samuel W. Keller for B.I. Edelson]
B. I. Edelson

Document II-15

Document Title: Letter from Burton I. Edelson, Associate Administrator for Space Science and Applications, to Dr. Herbert Friedman, National Research Council, 4 May 1984.

Source: James Green

By the mid-1980s, NASA realized that it needed a fresh look at how it archived and made available data from space science missions, particularly those involving solar-terrestrial physics. NASA asked the National Academy of Sciences to review NASA data activity. This letter to space science pioneer Herbert Friedman of the Naval Research Laboratory communicated NASA's appreciation of the review and indicated how its findings and recommendations would help NASA.

[no page number]
[stamped] May 4, 1984

Dr. Herbert Friedman
Chairman
Commission on Physical Sciences,
    Mathematics and Resources
National Research Council
2101 Constitution Avenue, NW
Washington, DC 20418

Dear Dr. Friedman:

Congratulations to the Joint Data Panel of the Committee on Solar and Space Physics and the Committee on Solar-Terrestrial Research for producing such a thorough and useful report—Solar-Terrestrial Data Access, Distribution, and Archiving. I thank you for transmitting this report to us.

Receipt of this report at NASA is very timely since we are undertaking several activities which could benefit from its recommendations:

- Redefining the role and the scope of the National Space Science Data Center (NSSDC).
- Recognizing the benefits of the Space Plasma Computer Analysis Network (SCAN) and providing baseline support.
- Evaluating the methods for handling data processing and dissemination of data from Spacelab and other Shuttle science missions.
- Establishing a Central Data Handling Facility for the Upper Atmosphere Research Satellite (UARS) and planning a similar facility for the proposed International Solar-Terrestrial Physics Program (ISTPP).
- Discussing means for highly coordinated data access, distribution, and archiving with the Japanese ISAS and the European Space Agency as part of the ISTPP plan.

Obviously, with these current activities and those planned for the future, the whole question of data access, distribution, and archiving is important to our science community. We intend to pay particular attention to the issues cited in your report over the next few years.

[2] I would be pleased to discuss these topics with you, with the Committee on Solar and Space Physics, and with the Committee on Solar-Terrestrial Research.

Sincerely,

[signature]
B. I. Edelson
Associate Administrator for
Space Science and Applications

Document II-16

Document Title: Stanley Shawhan, Director, Space Physics, Office of Space Science, NASA, to Professor Minoru Oda, Director General, Institute of Space and Astronautical Science, 1 November 1984.

Source: Personal Files of Stanley Shawhan

Even though NASA's own planning for solar-terrestrial physics missions was somewhat in disarray in 1984, maintaining the international character of the program was very important to NASA's Director of Space Physics, Stanley Shawhan, and his boss, Associate Administrator for Space Science and Applications Burton Edelson. This letter to the Director General of Japan's space science agency was intended to assure that organization that NASA was committed to supporting its portion of the joint Japanese-U.S. Geotail mission.

[November 1, 1984]

Prof. Minoru Oda
Director-General
Institute of Space and
    Astronautical Science
Tokyo, Japan

Dear Prof. Oda,

Thank you very much for the meeting we had on Monday, October 29, 1984 to discuss a number of topics related to collaboration between ISAS and NASA in the Solar-Terrestrial Physics discipline.

This note is to clarify the position and the action being taken by my office toward the joint ISAS/NASA Geotail mission. In looking at options for the NASA contributions to the International Solar-Terrestrial Physics Program, we have carefully considered the scientific merits and the programmatic implications of each mission. The NASA ISTP Program and Project Office conclude that the joint ISAS/NASA Geotail mission would provide a very significant enhancement to our knowledge of the configuration and dynamics of the geomagnetic tail region and that NASA should proceed as speedily as possible to implement its portion of the joint Geotail mission plans.

During the NASA fiscal years 1985 and 1986, Advanced Technology and Development (ATD) funds have been set aside for further design and mission analysis of the NASA ISTP Program. Because of the proposed ISAS Geotail Program schedule, the NASA ISTP Program and Project Offices carry as their highest priority the further definition, design, and analysis of the planned NASA contributions to the Geotail Program. The work that we plan to carry out through ISAS/NASA joint planning meetings, through advanced instrument design and through analysis of launch, tracking and data analysis requirements will parallel much of the work you propose to carry out next year.

Sincerely,

[signature]
S. D. Shawhan

Document II-17

Document Title: Charles J. Pellerin, Jr., Director, Astrophysics Division, to Dr. James L. Matteson, Center for Astrophysics and Space Sciences, University of California, San Diego, 30 September 1986.

Source: NASA Historical Reference Collection, History Office, NASA Headquarters, Washington, D.C.

Document II-18

Document Title: Charles J. Pellerin, Jr., Director, Astrophysics Division, to Dr. Laurence E. Peterson, Assistant Director for Science, "Explorer Program Fact Sheet," 23 October 1986.

Source: NASA Historical Reference Collection, History Office, NASA Headquarters, Washington, D.C.

The Space Shuttle Challenger *accident in January 1986, and the delays that it would cause in NASA's space science program, aggravated an already difficult situation with respect to the continuing series of smaller Explorer-class missions. That situation was the slow pace in launching approved missions and thus the long time between approval and launch, often eight to ten years. This led the head of NASA's Astrophysics Division, Charles Pellerin, in 1986 to decide to hold off on any new large Explorer missions, to have a competition for Small Explorer missions (SMEX), to launch the backlog, and to use the time gained to decide on how to restructure the Explorer program. This letter to James Matteson is an early indication of the thinking behind Pellerin's decision. The October fact sheet sets out the reasoning in more detail.*

Document II-17

[no page number]
[stamped SEP 30 1986]

National Aeronautics and
Space Administration

Washington, D.C.
20546

Reply to Attn of: EZ (LE P)

Dr. James L. Matteson
Research Physicist
Center for Astrophysics
and Space Sciences (CASS)
C-011
University of California, San Diego
La Jolla, CA 92093

Dear Jim:

This is in response to your letter of September 13, 1986, regarding evaluation of the proposals submitted for the Explorer Concept Study Program. As you indicated, we have discussed not evaluating the proposals received last month. We have had considerable internal discussion on this matter and have consulted several of our advisory groups. No firm decision has been made, and we are also considering various alternatives between a complete selection and a complete return.

While I am sympathetic to many of the points you raised and have a feel for the effort the community has expended on these proposals, you must be cognizant of the impact of the STS 51-L disaster on the NASA science programs. If we were fully aware of this impact at the time of the "Dear Colleague Letter" in March, we would undoubtedly have delayed its issuance. In the Explorer Project alone, there is an impact of some $100M caused by having to convert the COBE to a Delta launch, the likelihood the SMM will not be retrieved, and the stretchout in the EUVE and XTE. This is a 2-year delay for any new major mission, at the present funding level of approximately $50M per year. I believe you are aware that the Explorer Program has been under considerable criticism from the scientific community because of the long time between selection and flight. While I appreciate your noting that 8-10 years is, in fact, the recent norm, and the scientific community may be adjusted to this, it is not healthy for the vitality of science. Commitment to future Explorer missions now would only continue the present situation. As you indicated, there is much enthusiam [sic] for the Explorer Program; however, with a 6-year backlog, I do not perceive it having much momentum. It is the restoration of the momentum that is our goal.

[2] I agree that many of the investigations proposed are likely to be "timeless," in that the objectives remain first rate and are unlikely to be accomplished by other means. However, selecting a mission for conceptual study does imply a priority and a commitment to an implementation mode. Such commitments and priorities are usually made in the context of today's fiscal, political, and managerial environment; this may not be the same environment some 6 years in the future when the first significant funding for a major new Explorer start becomes available.

I have here indicated some of my thoughts on the Explorer selection dilemma; as I have said, no firm decision has been made, and we are looking into various methods of keeping a viable scientific establishment intact until the real opportunity for a flight series reopens.

Best regards,

[originally signed by Charles J. Pellerin, Jr.]

Charles J. Pellerin, Jr.
Director
Astrophysics Division

cc.:
Harvard Observatory Center
for Astrophysics/Dr. Grindlay

Jet Propulsion Laboratory/Dr. Jacobson

Document II-18

[no page number]
[originally stamped OCT 23 unreadable]

EXPLORER PROGRAM FACT SHEET

- When were Explorer missions last competitively selected?

  The last time Explorer missions were selected was in the 1974-1975 time frame. These were from a pair of Announcements of Opportunity (AO's) ("AO 6 & 7") which solicited Scout and Delta-class missions.

- Which missions were these?

  The missions were a mix across several disciplines. They include: the Dynamics Explorer (DE), the Infrared Astronomical Satellite (IRAS), the Dynamics Explorer (DE) [sic], the Active Magnetospheric Particle Tracer Experiment (AMPTE), the Extreme Ultraviolet Explorer (EUVE), and (although delayed a bit) the X-Ray Timing Experiment (XTE). [handwritten note "+ SME"]

- What has been the approach since then?

  We have assumed that these proposals were of higher priority than new, noncompetitively selected missions, and tried to complete them first.

- What new things have been added since AO 6 and 7, and why?

  Five efforts have been added since that time. They are:

- Roentgensatellit (ROSAT) - A focal plane instrument for the German ROSAT mission was added to the Explorer Program in 1982.

- Cosmic Ray Isotope Experiment (CRIE) - The Cosmic Ray Isotope Experiment of John Simpson's was added in 1983 and intends to fly on the USAF CRRES mission. This experiment was removed from the ISEE program for nonscientific reasons.

- San Marco - A joint Italian/U.S. activity in which U.S. instruments are flown on an Italian satellite launched by Italy to conduct atmospheric research.

- Combined Radiation Release and Effects Satellite (CRRES) - This effort is not managed as an Explorer, but because of the unique opportunity with the USAF, the Explorer Program provided funds to the Earth Sciences Division.

- [2] High Energy Solar Platform (HESP) - A "Soft X-ray" Telescope is being added to the Japanese High Energy Solar Platform (now renamed SOLAR-A mission).

• Were these additions reviewed and supported by the NASA Advisory apparatus?

Yes.

• What advice has been given for future Explorer strategies by the Space Science Board (SSB)?

Both the SSB Committee on Space Astronomy and Astrophysics (CSAA) and the Committee on Solar and Space Physics (CSSP) have reviewed and made specific reports on the role of the Explorer Program. Besides listing accomplishments, reaffirming the validity of the present program, and indicating important scientific efforts for the future, these committees:

a. Recommended a two step competitive selection process.

b. Recommended an augmentation to achieve, as a goal, one new Explorer/year in each area.

c. Recommended a management approach be found to ensure timely implementation of the selected missions.

d. Recognized an augmentation was needed to achieve the above objectives.

• What was the plan for new Explorers?

Over the past year, we began to implement a plan to select new Explorers. The first step, a "Dear Colleague Letter" which solicited short mission concept proposals for "Phase A" studies, was released last spring.

- Was this consistent with the advice of the scientific community?

Yes.

- [3] What has been the response?

About 43 proposals have been received to date. The greatest number have been in Astrophysics, with significant, but lesser numbers in Space Plasma Physics, Solar Physics, and thermospheric and mesospheric physics.

- What is the situation with regard to starting any new Explorer missions?

At the time that we began the "Dear Colleague Letter" process, we expected that new missions could be started in the near future. However, several events have conspired to impact the Explorer Program in a negative way as follows:

- Gramm-Rudman - $7M

- Challenger - Loss of opportunity to recover SMM and refurbishment for EUVE - $60M

- USAF - Vandenberg activation delays - impact to COBE - $40M

- Improved cost estimate of U.S. telescope system on the Japanese HESP mission - $8M

- Schedule delays in remainder of Explorer ongoing, efforts - $10M

- Overall, the Explorer Program has suffered about $140M of cost impacts in the last year.

Our current budgetary estimates indicate that there will be no significant new funds for missions following EUVE and XTE until approximately 1993.

- Have you reviewed the scientific merit and timeliness of EUVE and XTE in the context of the passage of time and in the context of the new proposals?

In addition to the original scientific peer reviews both EUVE and XTE have been recommended to NASA in the Report of the Astronomy Survey Committee (Field Committee) of the National Academy of Sciences. Internal scientific reviews have been conducted which examined the scientific importance of EUVE and XTE in the current context. The reviews found that the scientific problems addressed by these missions have only grown in importance over time, [4] and none of the proposals received offered missions that could meet the objectives of EUVE or XTE. Accordingly, our recent efforts have stressed completing these missions in a timely way and at lowest cost.

• What implementation options have been examined?

We have examined three spacecraft alternatives–the use of "throw-away" spacecraft; the use of spacecraft reuse and on-orbit servicing; and the capability for dual (i.e., Delta and STS) compatibililty [sic].

• What has been selected for the program baseline?

We have decided to procure a "clone" of the Solar Maximum spacecraft. The plan is to launch EUVE using either a Delta or the STS and to conduct the EUVE mission for 2 years. Then, we plan to rendezvous with the spacecraft (just like the SMM repair mission) and replace the EUVE instruments with the XTE instruments.

• Would it be less costly in the near term to launch EUVE on a "throw-away" spacecraft and worry about XTE later?

No! We have surveyed all known applicable spacecraft and estimated the cost of a new build. The SMM/MMS option is cheapest even in the near term because we are able to buy the spacecraft modules as part of a USAF block buy and get advantageous prices.

• What is the cost of this spacecraft?

The cost is $64M in FY-86 dollars for a dual launch capability spacecraft to perform either the EUVE or XTE mission. If we were to limit the spacecraft capability for EUVE only, we could reduce the cost by about $5M.

This latter action would necessitate ground refurbishment for XTE use or another spacecraft development. There are several additional costs to be added to the spacecraft development cost of $64M, including inflation, center IMS, and reserves, which bring the total cost of the dual-compatible, dual-use spacecraft cost to $100M. This cost also includes about $10M of program stretchout cost which is caused by funding limitations. We are working to reduce the costs of other ongoing Explorer elements to allow an optimum funding profile and reduce or eliminate this extra cost.

• [5] Under this approach, what is the total cost of EUVE and XTE?

The cost to complete EUVE and XTE using the "Explorer Platform" approach is as follows:

EUVE instrument/mission $ 60M

XTE instrument/mission $ 75M

Explorer Platform $ 100M

In addition, there is the capability of implementing two additional missions through the Explorer Platform after XTE.

- How does this compare with the "rule of thumb" that Explorer missions should not often exceed 2 years of the Explorer budget?

The Explorer budget in this time frame is $65-$70M/year.

The EUVE program alone slightly exceeds the 2 year rule of thumb; the EUVE and XTE programs are significantly below this level if we amortize the spacecraft development cost between the two missions.

- How does this program concept depend on launch capability?

Both missions utilize the easiest orbit inclinations (28.50) to obtain with U.S. launch facilities, and neither mission requires upper stages or additional propulsion systems. The spacecraft is compatible with Delta or STS launch with the EUVE payload. The XTE payload geometry is incompatible with a Delta shroud and is planned for STS launch.

- What are the technical risks and how have they been analyzed?

Risk 1 - Loss of payload during launch phase. This is an approach independent risk for the most part. If we were to lose the EUVE mission in a launch failure, the only additional cost incurred in order to ensure mission flexibility is the approximately $5M expended to make the spacecraft capable of supporting XTE also.

Risk 2 - Loss of instrument or spacecraft systems on orbit. This risk is mitigated because of the large experience base with MMS spacecraft and the fully redundant and safe hold system. Should mission [6] threatening failures occur, there are three alternatives to continue the mission: (1) on-orbit MMS servicing; (2) on-orbit instrument exchange; or (3) recovery and return for ground refurbishment.

- The Explorer Program seems preoccupied with astronomy missions. Why?

As indicated earlier, a mix of missions was selected from AO 6 and 7. Because of community pressure, lower technical complexity, and international considerations, the nonastronomy missions were flown early (i.e., DE, SME, and AMPTE). Also, the three recent additions (i.e., CRRES, San Marco, and HESP) are both in the areas of Solar and Space Plasma Physics.

- What are your plans for the new proposals?

As mentioned earlier, the Explorer Program has suffered many setbacks in the past year. We have halted the evaluation process until we can develop a new program baseline which accommodates the already approved and ongoing efforts. We hope to complete this effort this fall and will return our attention to the new proposals as soon as we can estimate when new resources will become available.

• Wouldn't it be a good signal to the scientific community to start some new Explorer projects?

Yes, it would if such projects could be initiated without deferring the launch dates or increasing the costs of the ongoing efforts. Our current analysis shows that we are now overcommitted to the point that runout costs are being driven upwards because of schedule delays caused. by inadequate resources. We are not in favor of further cost increases or delays to EUVE and XTE for the sake of new additional Explorers at this time. These missions are already about 10 years delayed in their initiation from the period of AO 6 and 7. We note that these missions remain scientifically compelling; also, the PI's from both missions have offered one of their instruments to the general community via "General Investigator" programs similar to IUE in recognition of the sparsity of new flight data to the space science community.

• [7] What does the scientific community think about additional delays in selecting new Explorer missions from the new proposals now at NASA?

-   This is a complicated question with many facets.

-   The scientists who have been involved in proposal submissions are urging a selection at this time and initiation of study activities.

-   We are inclined to be more cautious in view of the concerns voiced by the community over the past decade. These include:

"Quick is beautiful" - The time span from selection until flight needs to be shortened. Also, missions are executed more efficiently when schedules are not perturbed by delays injected by considerations external to the program. A selection now would not allow the first of the new missions to begin development until 1993 with launch 4-5 years later. Further, there is every possibility that political forces may result in as many as five missions being selected and possibly closing the Explorer opportunity until the next century.

We hope for and are pressing for an Explorer augmentation which would alleviate our financial problems and allow an expansion of near term Explorer opportunities. However, a realistic assessment suggests that there is probably an equal chance that resources may be diminished. The launch schedule for ongoing efforts like COBE and ROSAT may be extended beyond our current assessments, and the recent history of Congressional budget cuts may recur.

- Aren't you concerned about sending a discouraging signal to the scientific community?

Yes, but we feel that the community deserves an honest and truthful signal. We are also concerned about the Agency's recent history of raising false expectations and are trying to more closely align the expectations we raise with our ability to carry the programs to completion. We think that NASA should place more stress on credibility in the wake of such events as the cancellation of the U.S. spacecraft on ISPM, the [8] selection of 40 Spacelab instruments in 1978 of which only a few have entered development, the recent tragedy in the Spartan program, as well as the extended delays in launching the Hubble Space Telescope and the starts of the OPEN, AXAF, and SIRTF programs.

Finally, we are in the midst of a rapidly evolving environment with regard to launch opportunities, launch vehicles, and Space Station opportunities. Selections made now might look very different in the context of the NASA program in the mid to late 90's.

Also, the invitation of a new round of mission studies would represent an additional drain of the very limited Explorer resources.

- But, what about an augmentation?

Since the NASA budget for FY-88 has already been submitted to OMB, the earliest Explorer augmentation in the context of the usual budget process would be in FY-89.

- Ok, what are the arguments for proceeding with the evaluation at this time?

First, it would please those who have worked hard to submit their proposals. Thus, those involved in future extensive predevelopment work would have more than ample time to plan.

- What plans do you have for consultation with the scientific community before a decision is taken?

We plan to discuss these issues in detail with the CSAA and CSSP within the [sic] several weeks. Both committees have recently published reports on the Explorer Program, and the program interest focusses [sic] well there.

Document II-19

Document Title: Michael A. Calabrese, ISTP Flight Program Manager, "An Introduction to Science Investigation Cost Control," 22 January 1990.

Source: NASA Historical Reference Collection, History Office, NASA Headquarters, Washington, D.C.

In 1987, NASA Associate Administrator for Space Science and Applications Lennard Fisk challenged his program managers to develop an approach to containing cost growth in space science missions. The response was a Cost Control Plan (CCP) for each proposed mission which contained a series of de-scoping options. The CCP became a requirement for mission approval and was to be created at the onset of the development phase for every new mission and implemented when the investigation was confirmed. The de-scoping options in the plan were identified on an instrument-by-instrument basis, with the key options delineated in the confirmation letters to the investigators for each instrument. If unforeseen costs occurred during the development phase of an instrument, CCP procedures would be implemented. Decreasing instrument performance or reducing redundancy were the typical compromises outlined in the CCP. By January 1990, the CCP for NASA's ISTP program was completed. This document was also NASA's first CCP.

[no page number]
[stamped JAN 24 1990]

National Aeronautics and
Space Administration

Washington, D.C.
20546

Reply to Attn of: E S

To: Distribution

From: ES/ISTP Flight Program Manager

Subject: Science Investigation Cost Control Procedures

In response to Dr. Fisk's request to institute science investigation cost control procedures within the Space Flight Programs of NASA's Office of Space Science and Applications (OSSA), the Space Physics Division has prepared a [sic] investigation cost control methodology for the International Solar Terrestrial Physics (ISTP) Program. The attached report, which is intended to be a management aid, describes this methodology and provides directions for implementing a cost control plan at the onset of the investigation development phase. The actual plans developed for ISTP, which includes [sic] the Global Geospace Science Program and the Solar Terrestrial Science Program, are provided in Enclosures 1 and 2 of the report. A typical Investigator confirmation describing these plans is also provided in Enclosure 3.

[originally signed by M. A. Calabrese]
M. A. Calabrese

Distribution:
E/A. Diaz
   J. Alexander
EB/A. Nicogossian
EC/R. Arnold
EE/S. Tilford
EL/G. Briggs
EM/R. Benson
EN/R. Schmitz
EP/K. Schmoll
EZ/C. Pellerin

[no page number]

An Introduction to
Science Investigation Cost Control

As Practiced by

The National Aeronautics
and Space Administration (NASA)
Office of Space
Science and Applications (OSSA)
Space Physics Division

January 22, 1990

[no page number]

Table of Contents

I. [1] PURPOSE

The purpose of this paper is to describe the method used by the Space Physics Division (SPD) to control the science investigation development costs of the International Solar Terrestrial Physics (ISTP) Program.

II. BACKGROUND

The SPD developed a Global Geospace Science (GGS) cost control plan for the investigations being flown on the Wind and Polar spacecraft. A similar plan was also developed for the U.S.-supplied investigations to the Solar Terrestrial Science Program (STSP). The STSP investigations will be on the Cluster and Solar and Heliospheric Observatory (SOHO) missions. These cost control plans were prepared in response to Dr. Fisk's request to institute cost control procedures for investigations at the onset of the development phase. The goal of these procedures is to avoid the need for budget augmentations to resolve cost problems during the development of an investigation. This assumes the establishment of adequate reserves in the program baseline.

Cost control procedures are used to identify cost increases at an early stage so that corrective action can be taken while minimizing serious impact to the investigation or program. The strategy begins with monthly reporting of cost performance, as well as regular cost reviews conducted during the usual design reviews. At any sign of significant cost growth, appropriate actions (e.g. use of available reserves and/or predetermined descoping options) are taken to eliminate rising costs or to mediate cost impacts.

III. IMPLEMENTATION

Cost control plans are usually developed during the investigation definition phase, and are implemented when the investigation is confirmed. During the definition phase, each Principal Investigator (PI) is informed in writing of their funding cap, which is based on their selected proposal, and agreed to by NASA and the PI. Each PI is asked to prepare a scientific and engineering descoping plan detailing how internal cost changes will be accommodated during the development phase while remaining within the funding cap. This plan, which establishes and documents the priorities and impacts for descoping science investigations, is submitted as part of the [2] Instrument Development Plan (IDP) for NASA approval. The IDP is a definition phase deliverable to NASA.

Investigations are confirmed for flight only after a NASA cost-risk assessment has demonstrated that sufficient reserves exist to accommodate the risks and that the development cost is within the funding cap. The approved descope plan then becomes part of NASA's contract with the PI and is included in the PI's letter of flight confirmation. The mission investigation cost control plan also becomes part of the Project Implementation Plan, the Field Center's commitment to Headquarters for the development phase.

During the development phase, the progress of each investigation is assessed monthly by the usual program/project management tools at the Field Center and Headquarters, taking into account monthly reports on cost, labor, and schedule progress. If unforeseen events occur which cause an investigation's internal costs to rise (i.e., design changes, parts procurement delays, or a change in subcontractor), the PI is expected to make every attempt to resolve them. The fixes, which may include performance, redundancy, or reliability trades, should be implemented within available resources, using the predetermined internal descoping plan. Some internal problems which are beyond the control of the PI may be fixed using project office contingency reserves, as appropriate and available. Changes in program scope imposed by external circumstances (i.e., changes in launch date or launch vehicle) are corrected with the Headquarters program office Allowance for Program Adjustment (APA), as appropriate and available. When project reserves or Headquarters APA are not available, the PI should be prepared to descope the investigation, as required in accordance with the descope plan.

Periodically, at key decision points such as the Preliminary Design Review (PDR) and the Critical Design Review (CDR), the project office reassesses the run-out cost for the entire project. If problems are identified which result in unresolvable cost increases in the current fiscal year or out-years, the project office can then either descope the element responsible for the increase (i.e., the science investigation, spacecraft, or ground data system), or reapportion the resources between the project elements with appropriate descopes. For cases where descoping is insufficient to resolve the cost problem, the option of dropping an investigation must be considered. If a cost problem can be isolated to a particular investigation, its removal may be recommended. If a problem develops at the mission level (e.g., spacecraft [3] subsystems), then individual investigations can be recommended for removal according to a predetermined priority list established by NASA Headquarters. Usually, the investigations are ranked in order of the priority of the science to be addressed in meeting prime mission science objectives.

IV. SUGGESTIONS FOR IMPROVING COST CONTROL

A significant part of science investigation cost growth can be attributed to the use of an excessive number of Co-Is whose teams must be supported. A technique for control is to group the Co-Is into three categories:

- Category A: Hardware suppliers

- Category B: Non-hardware design support

- Category C: Post-launch data analysis only

This technique forces the PI to justify the unique contribution of each supporting Co-I and allows NASA better control of C0-I [sic] Funding by category, particularly in the pre-launch phase.

There are other important points that are critical to a cost control plan's success. First, the PIs must be convinced as early as possible (preferably in the AO) and reminded throughout the development phase that cost control will be enforced. Timely and accurate reporting of investigation progress along with good communications are also crucial. Finally, timing is very important in any cost control scenario as it is much more difficult to reduce costs using descoping after the CDR. As the ISTP program progresses through the development phase and more cost control experience is gained the Program Office will report on and recommend areas where improvement can be made.

[4]

• V. ENCLOSURES

1. GGS Cost Control Plan

2. STSP (Cluster/SOHO) Cost Control Plan

3. Typical Confirmation Letter (SOHO/Brueckner/LASCO)

[5]
1. GGS Cost Control Plan

[6]
[stamped FEB 24 1989]

National Aeronautics and
Space Administration

Washington, D.C.
20546

Reply to Attn of: ES

TO: E/Deputy Associate Administrator for Space
Science and Applications

FROM: ES/Program Manager

SUBJECT: Global Geospace Science (GGS) Cost Control Plan (Final)

We have developed the enclosed GGS Cost Control Plan in response to Dr. Fisk's request to provide an agreed and understood methodology for cost control of investigations at the onset of the GGS program. Our objective is to identify problems at an early stage such that corrective action can be taken without serious impact to the investigation or the program. We are prepared, however, to make the hard decisions described in the plan if required after suitable discussion with your office.

The Cost Control Plan will be incorporated in the GGS Project Plan. Also, we will describe the plan to the GGS Investigators at the next Science Working Team meeting on March 8-10, 1989 at GSFC.

[originally signed by M. A. Calabrese]
M.A. Calabrese

cc:
E/Mr. Alexander
ES/Dr. Shawhan
GSFC/407/Mr. Sizemore(w/o Encl. 2)

[7] GLOBAL GEOSPACE SCIENCE (GGS) COST CONTROL PLAN

The following steps and priorities are to be implemented to keep the Global Geospace Science (GGS) Program, consisting of the Wind and Polar spacecraft [in coordination with Geotail and the Combined Release and Radiation Effects Satellite (CRRES)] within the runout budget envelope.

HQ Program Office Allowance for Program Adjustment (APA) is to be utilized for changes in program scope imposed by external circumstances, such as change in launch date or launch vehicle

GSFC Project Office Contingency Reserves is to be utilized to fix usual internal problems for the PI's or spacecraft contractor, such as design changes, parts procurement, alternate approaches, and change of subcontractors.

Progress of the GGS Program is to be assessed monthly by the usual program/project management tools at GSFC and at HQ taking into account monthly reports on cost, manpower and progress.

In addition to the usual project controls, the following plan is to be instituted:

STEP 1. Implement Plan

* Develop plan concept with the HQ Office of Space Science and Applications and with the GSFC International Solar Terrestrial Physics (ISTP) Project Office

* Establish and document the priorities for descoping of science investigations and other elements of the program

* Inform Principal Investigators of cost control plan via the development phase confirmation letters (signed 12/27/88)

[8] STEP 2. Review Science Investigations and Institute PI Fixes
* PI's provide monthly Reports on controlled items, such as mass, power, manpower, schedule milestones and cost, for Project review to forecast problems, above and beyond the monthly and quarterly financial reports

* If problem develops that would lead to unacceptable cost increase

(1) Each PI attempts to fix problem within available resources with internal descoping that may mean tradeoffs in performance, redundancy or reliability, etc. If the particular troublesome investigation had an improved enhancement (see Enclosure 1 [not included]), consider descoping the enhancement

(2) Project uses Contingency Reserves or GSFC manpower, as appropriate

STEP 3. Review Project Status and Institute Descope Measures

• Periodically and at key review times (Requirements Review, Preliminary Design Review, Critical Design Review, Flight Readiness Review, etc, see enclosed Milestone Schedule) the Project reassesses the runout cost for the entire project

• If problem develops that may lead to unresolvable fiscal year or total cost increase then identify the element (science investigations, spacecraft, ground data systems) of the Project that is causing the problem and descope or reapportion the resources between the project elements with appropriate descopes

• If problem develops with science investigations or science element must otherwise be descoped

(1) First, descope the approved performance enhancements listed in Enclosure 1 [not included], if feasible (#1 is first to be considered for descoping). If a particular investigation is causing the problem, it is descoped. If the problem develops on the Wind or [9] Polar mission, then use the entries, Wind or Polar, on the priority list, in priority order, to bring cost into limits. If problem is generic, then proceed down the priority list

(2) Second, for the case that descoping of enhancements is insufficient to resolve the cost problem, consider dropping entire investigations from the mission according to the priority lists in Enclosure 2 [not included]. If a particular investigation is causing the problem, it is dropped. If the problem develops on the Wind or Polar mission, then use the priority list for that mission (#1 is first to be considered for descoping) to bring cost into limits. If problem is generic, then proceed down both the Wind and Polar priority lists in priority order

[10] WIND and POLAR Investigation Drop Orders, not included]

[11] 2.   STSP (Cluster/SOHO) Cost Control Plan

[12]
[stamped JAN 19 1990]

National Aeronautics and
Space Administration

Washington, D.C.
20546

Reply to Attn of: ES(MAC)

TO: E/Deputy Associate Administrator for Space
         Science and Applications

FROM: ES/ISTP Program Manager

SUBJECT: Solar Terrestrial Science Program (STSP) Cost Control Plan

We have developed the enclosed STSP Cost Control Plan for the U.S. supplied payload elements on the Cluster and SOHO missions. This plan was prepared in response to Dr. Fisk's request to provide an agreed and understood methodology for cost control of investigations at the onset of development. Our objective is to identify problems at an early stage such that corrective action can be taken without serious impact to the investigation or the program. We are prepared, however, to make the hard decisions described in the plan if required after suitable discussion with your office.

The Cost Control Plan has been incorporated in the ISTP Project Plan and a description of the plan has been enclosed in the confirmation letter for the U.S. investigators.

[originally signed by M. A. Calabrese]
M.A. Calabrese

Enclosure .

[13]
CC:
E/Mr. Alexander
ES/Dr. Shawhan
ES/Dr. Bohlin
GSFC/407/Mr. Sizemore(w/o Priority List)

[14] SOLAR TERRESTRIAL SCIENCE PROGRAMME (STSP) COST CONTROL PLAN

The following steps and priorities are to be implemented to keep the STSP investigations on the Cluster and on the SOHO missions within the runout budget envelope.

HQ Program Office Allowance for Program Adjustment (APA) is to be utilized for changes in program scope imposed by external circumstances, such as change in launch date or launch vehicle

GSFC Project Office Contingency Reserves is to be utilized to fix usual internal problems for the PI's or spacecraft contractor, such as design changes, parts procurement, alternate approaches, and change of subcontractors

Progress of the STSP Program is to be assessed monthly by the usual program/project management tools at GSFC and at HQ taking into account monthly reports on costs, manpower and progress

In addition to the usual project controls, the following plan is to be instituted:

STEP 1 Implement Plan

- Develop plan concept with the HQ Office of Space Science and Applications and with the GSFC International Solar Terrestrial Physics (ISTP) Project Office

- Establish and document the priorities for descoping of science investigations and other elements of the program

- Inform Principal Investigators of cost control plan via the development phase confirmation letters

[15] STEP 2   Review Science Investigations and Institute PI Fixes

- PI's provide monthly Reports on controlled items, such as mass, power, manpower, schedule milestones and cost, for Project review to forecast problems, directly to the project office above and beyond the monthly and quarterly financial reports submitted through the contracts office

- If problem develops that would lead to unacceptable cost increase-

(1) Each PI attempts to fix problem within available resources with internal descoping that may mean tradeoffs in performance, redundancy or reliability, etc.

(2) Project uses Contingency Reserves and/or GSFC manpower, as appropriate

STEP 3 Review Project Status and Institute Descope Measures

- Periodically and at key review times (Requirements Review, Preliminary Design Review, etc, see enclosed Milestone Schedule) the Project reassesses the runout cost for the entire project

- If problem develops that may lead to unresolvable fiscal year or total cost increase then identify the element (science investigations, spacecraft, ground data systems) of the Project that is causing the problem and descope or reapportion the resources between the project elements with appropriate descopes

- If problem develops with science investigations or science element must otherwise be descoped

(1) Descope Phase—If a particular investigation is causing the problem, it is descoped. If the problem develops on the Cluster or SOHO mission, then use the entries, on Enclosure 1 [not included] priority list, in priority order, to bring cost into limits.

(2) Removal Phase—For the case that descoping of investigations is insufficient to resolve the cost problem, consider dropping entire investigations from the mission according to the priority lists in Enclosure 1. If a particular investigation is causing the problem, it is dropped. If the problem develops on the Cluster or SOHO mission, then use the priority list for that mission (#1 is first to be considered for descoping) to bring cost into limits. If problem is generic, then proceed down both the Cluster and SOHO priority lists in priority order

[16] SOHO and CLUSTER Investigation Drop Orders, not included]

[17] 3. Typical Confirmation Letter (SOHO/Brueckner/LASCO)

[18]
National Aeronautics and
Space Administration

Washington, D.C.
20546

Reply to Attn of: ES

Dr. Guenter E. Brueckner
Code 4173
Naval Research Laboratory
Washington, DC 20375

Dear Dr. Brueckner,

NASA has completed the definition phase activities in support of the Solar Terrestrial Science Program (STSP) for the Cluster and Solar and Heliospheric Observatory (SOHO) Missions, jointly sponsored with the European Space Agency (ESA). It is my pleasure to inform you that your investigation "Wide Field White Light Spectrometric Coronagraph, for SOHO (LASCO)" originally proposed in response to the Announcement of Opportunity OSSA-1-87 and modified by your definition phase study, has been confirmed for the development, mission operations, and data analysis phases of the STSP SOHO mission. The confirmed Co-Investigators (Co-I's) on your investigation team, along with confirmation assumptions and descope elements, are identified in Enclosure 1. This confirmation is contingent on our ability to conclude a satisfactory agreement with the European funding agencies for their support of your European Co-I's.
    The confirmation of your investigation is being made as a result of the careful assessment of the scientific, technical, and financial requirements developed during the

definition phase. We feel that your investigation can be performed within the resources agreed upon with the NASA Project Office at the Goddard Space Flight Center (GSFC) during the definition phase, the engineering and conceptual design review, and the cost confirmation meeting held in the summer of 1989. Your investigation request for additional spacecraft resources will be negotiated during the ESA spacecraft prime contractor definition phase.

The NASA Project Office will work closely with you to ensure that your investigation is implemented within the agreed resources. However, because the NASA budget is fixed, problems may arise which may cause NASA to request descope of your investigation, beginning with the descope elements contained in Enclosure 1 and detailed in your Development Phase Plan submitted to the NASA Project Office. If these measures are unsuccessful, we may be forced to remove your investigation from this program. Therefore, I urge you to work closely with the Project Office to control your costs and schedules during the implementation of your investigation. Enclosure 2 describes the Cost Control Plan that I have approved for this mission to be implemented through the NASA Project Office.

I must stress that no PI is empowered to either enlarge the scope of his accepted investigation or commit to any increase in NASA funds beyond that in the revised proposal, as amended by the definition phase and the confirmation assumptions contained in Enclosure 1, without prior consultation with and approval by the NASA Program Office at Headquarters and the Project Office at GSFC.

[19] Enclosure 1 defines, by three broad categories of participation, the roles of each of your U.S. Co-I's, as based upon our understanding of your revised proposal:

(A) U.S. Co-I's with responsibility for prelaunch development of flight hardware and postlaunch data analysis;

(B) U.S. Co-I's with responsibility for prelaunch preparation for mission operations and/or data management critical to the execution of the investigation during the operational phase of the mission, and postlaunch data analysis; and

(C) U.S. Co-I's with responsibility for prelaunch science definition and postlaunch data analysis only

Owing to the restrictions on NASA's budgets, we wish to inform you and your U.S. Co-I's, by way of copy of this letter, that our policy regarding support for the U.S. Co-I's, as based on these three categories, will be as follows.

Prelaunch: With regard to categories (A) and (B), NASA is prepared to provide you with funds for U.S. Co-I costs that are directly related to the production of flight hardware and/or activities that are necessary to prepare for the mission operations and/or data management, not to exceed that defined in the revised version of your proposal as amended by the definition phase. With regard to category (C), we are prepared to provide you with the funds for U.S. Co-I's, as agreed to, only for participation in your science team meetings.

Postlaunch: With regard to all three categories, NASA will negotiate with you regarding the support of your U.S. Co-I's for their participation in mission operations, data management and/or reduction, and data analysis activities consistent with their defined role in your revised proposal as amended during the definition phase and, most importantly, the extent allowed by NASA's budget for this phase of the mission.

Consistent with our policies for non-U.S. Co-I's on U.S. investigations, a letter of agreement will eventually be executed between their national funding agency and NASA's International Relations Division confirming our mutual commitments.

You are reminded that NASA Management Instruction (NMI) 8030.3A, "Guidelines for Acquisition of Investigations," which was enclosed in my letter of April 25, 1988, continues to govern data use and data rights for your investigation.

The overall program management for the STSP/SOHO mission has been assigned to the Space Physics Division of the Office of Space Science and Applications at NASA Headquarters. Mr. M. Calabrese is the International Solar Terrestrial Physics (ISTP) Program Manager and Dr. J. D. Bohlin is the Program Scientist for the SOHO mission. Project management responsibilities have been assigned to GSFC, where Mr. K. Sizemore is the STSP Project Manager, Dr. M. Acuna is the Project Scientist, and Dr. A. Poland is the Deputy Project Scientist for the SOHO mission. The GSFC/STSP Project Office will provide you with more details about the development phase as part of the contract negotiation process.

[20] NASA joins ESA in congratulating you and your team for submitting an outstanding investigation for the STSP. We look forward to working closely with you for its successful completion.

Sincerely,

L. A. Fisk
Associate Administrator for
Space Science and Applications

Enclosures

[21] cc:
J. Bohlin/HQ/ES
M. Calabrese/HQ/ES
M. Gaskins/HQ/EPS
L. Cline/HQ/XI
K. Sizemore/GSFC/407
M. Acuna/GSFC/695
M. Goldstein/GSFC/692 (Cluster) or A. Poland/GSFC/682(SOHO)
G. Cavallo/ESA HQ
P. LoGalbo/ESTEC

[22]
[p. 1 of 2]

ENCLOSURE 1

- INVESTIGATION: LASCO

- PRINCIPAL INVESTIGATOR: G. E. Brueckner/NRL

- U. S. CO-INVESTIGATORS:

  (A) Design, Development, Hardware Provision and Data Analysis

  J. D. Bartoe/NRL
  D. J. Michels/NRL
  G. A. Doschek/NRL
  R. A. Howard/NRL
  M. J. Koomen/Sachs-Freeman
  D.G.Socker/NRL

  (B) Design, Support without Hardware Provision, Pre-Launch Preparation for Mission Operations and/or Data Management, and Data Analysis

  K. P. Dere/NRL
  F. Giovane/UFL, Gainesville
  W. J. Wagner/NOAA SEL

  (C) Data Analysis Only

  S. K. Antiochos/NRL
  C. C. Cheng/NRL
  S. W. Kahler/Emmanuel College
  J. T. Mariska/NRL
  E. N. Parker/U. Chicago
  N. R. Sheeley/NRL
  R N. Smartt/NSO

- EUROPEAN CO-INVESTIGATORS:

J. L. Bougeret/Observatoire de Paris, Meudon
P. Daly/Max-Planck-Institut
R Giese/Ruhr-Universitat, Bochum
B. Inhester/Max-Planck-Institut
H. Keller/Max-Planck-Institut
S. Koutchmy/Rastitut d'Astrophysique
J. Kramm/Max-Planck-Institut
P. Lamy/Lab d'Astronomie Spatiale
A. Llebaria/Lab d'Astronomie Spatiale

A. Maucherat/Lab d'Astronomie Spatiale
J.-C. Noens/Observatoire du Pic du Midi

[23]
[Brueckner p. 2 of 2]

- EUROPEAN CO-INVESTIGATORS - CONTINUED:

M. Pick/Observatoire de Paris, Meudon
H. Rosenbauer/Max-Planck-Institut
R. Schwenn/MPAE

- CONFIRMATION ASSUMPTIONS:

  - Allowance for critical additional spacecraft resources by ESTEC
  - Reduced data analysis support
  - UK provision of structure & adjustable mounting legs
  - German provision of C1 and aperture door mechanism for C1, C2 & C3
  - French provision of C2 and filter/polarizer wheel mechanism for C1, C2 & C3
  - NRL supplied electronics box common with EIT
  - NRL provision of C3 instrument System, I & T, F-P for C1, all CCD cameras & shutters

- DESCOPE ELEMENTS:

  - Selected performance reductions
  - Reduced electronic effort/redundancy
  - Reduced number of Fabry-Perot models
  - Relaxed criteria for CCD selection
  - Reduced instrument calibration/testing

[24]   ENCLOSURE 2

SOLAR TERRESTRIAL SCIENCE PROGRAMME (STSP)
U. S. INVESTIGATION DEVELOPMENT PHASE
COST CONTROL PLAN

OBJECTIVE
    EACH U.S. PI AND U.S. CO-I TO A EUROPEAN PI COMMUNICATES AND WORKS
WITH THE NASA PROJECT OFFICE TO CONTROL COSTS AND MAINTAIN
SCHEDULE THROUGHOUT THE DEVELOPMENT PHASE OF THE INVESTIGATION
IN RESPONSE TO INTERNAL AND EXTERNAL CHANGES
PROCEDURE
    U.S. PI/CO-I PROVIDES MONTHLY REPORTS ON CONTROLLED ITEMS (MASS,
POWER, MANPOWER, COST, SCHEDULE, ETC.) AND FORECASTS PROBLEM AREAS

IN ADDITION TO REASSESSMENT OF ALL ASPECTS INCLUDING COST/
COMPLEXITY TO COMPLETE AT MAJOR MILESTONES (PDR, CDR, ETC.)

U.S. PI/CO-I WORKS WITH NASA PROJECT OFFICE TO ACCOMMODATE
INTERNAL CHANGES WITHIN FUNDING CAP BY INVOKING DESCOPE IDENTIFIED
IN INSTRUMENT DEVELOPMENT PLAN

NASA WILL AUGMENT FUNDING WITH OVERALL PROJECT RESERVES TO
ACCOMMODATE EXTERNAL CHANGES, AS APPROPRIATE AND AVAILABLE,
GIVING FULL CONSIDERATION TO NATURE, SOURCE, EXTENT, AND
CRITICALITY OF PROBLEM WITH THE UNDERSTANDING THAT INDIVIDUAL
EXPERIMENTS WILL NOT BE ALLOWED TO DOMINATE USE OF RESERVES

IF COST CONTROL IS UNSUCCESSFUL, U.S. PI INVESTIGATION OR U.S. CO-I
SUPPORT OF EUROPEAN INVESTIGATION MAY BE RECOMMENDED, AFTER
CAREFUL AND THOROUGH CONSULTATION WITH ESA, FOR REMOVAL FROM
PAYLOAD PENDING REVIEW AND APPROVAL BY AA/OSSA

Document II-20

Document Title: Guenter Riegler, Chief, Space Science Operations Branch, NASA
Headquarters, "Charter: OSSA Operations and Data Analysis (MO&DA) Blue Team," 17
June 1992.

Source: Guenter Riegler

Document II-21

Document Title: Guenter Riegler, "OSSA MO&DA Efficiency Improvement Program:
Implementation Plan," 6 August 1992.

Source: Guenter Riegler

*By the early 1990s, NASA was entering a time of the highest launch rate for science missions in the
history of the space program. Cost controls and procedures for missions in the development phase were
being initiated, but a significant fraction of the space science budget was going into operating its
ongoing missions. In 1992, Associate Administrator for Space Science and Applications Lennard
Fisk asked Guenter Riegler (Chief, Space Science Operations Branch, NASA Headquarters) to put
together a Mission Operations and Data Analysis (MO&DA) Blue Team. The charter of the team,
issued on 17 June 1992, was to seek efficiencies and reduce the cost of operating missions. Fisk's
rationale for this activity was to create a funding wedge for new initiatives. Based on the work of the
Blue Team, Riegler issued a plan for improving the efficiency of NASA's MO&DA activities in
August 1992.*

Document II-20

OSSA Mission Operations and Data Analysis (MO&DA)
Blue Team

Charter

Charter: The charter for the OSSA MO&DA Blue Team is to seek efficiencies and reduce cost, while maintaining our science productivity. To this end we must be willing to take more risks, accept more data outages, take advantage of multi-mission operations and multi-mission science centers, and simplify the flow of information from missions and centers to the science community. This Blue Team is concerned with MO&DA for Solar System Exploration, Earth Sciences, Space Physics, and Astrophysics.

The objective of the cost reduction is to create a funding wedge for new initiatives. The Associate Administrator for Space Science and Application will hold and apportion the resulting savings for new OSSA projects.

Cost Reduction Baseline: The baseline for MO&DA cost reductions is the President's Funding Plan (PFP-93), as described in the POP 92-2 guidelines to the NASA centers. Program augmentations (which have been considered by the AA/OSSA after issuance of the POP 92-2 guidelines for submission in the FY94 budget to the Administrator) will not be included in the baseline, but will be considered later independent of the Blue Team effort. Lists of PFP-93 and POP 92-2 Guidelines are included as Attachment 1.

Cost Reduction Guidelines: Reductions below the above-mentioned baseline are guidelined at 3% in FY93, 6% in FY94, 9% in FY95, 12% in FY96, and 15% in FY97 and thereafter. As a starting point, the science grants programs are to be kept intact, and significant reductions in the science content are not acceptable. As a further starting point, the reduction guidelines apply to the total MO&DA program at each center, i. e. at GSFC and separately at JPL.

Organization: The MO&DA Blue Team is chartered by the Associate Administrator for Space Science and Applications. This Blue Team works with the Red Team for "Robotic Exploration of the Solar System and the Universe," chaired by Peter Burr (GSFC).

Team Membership: Team membership will consist of NASA civil servants and JPL employees. The core of the MO&DA Blue Team consists of three persons from NASA Headquarters and six persons each from GSFC and JPL. A comparable number of persons is responsible for critical support and backup functions. Membership consists of a mix of personnel from Programs, Projects, Science, and institutional support directorates. The full list of MO&DA Blue Team participants is shown as Attachment 2.

Process: In order to meet the stringent schedule guidelines, the MO&DA Blue Team began on June 1 to work initially as two subteams at GSFC and JPL, under HQ guidance, with frequent telephone coordination between HQ and the leaders of the GSFC and JPL subteams. The separate studies concentrated on the MO&DA programs at GSFC and at JPL, respectively. Beginning on June 10, the full team met via videoconference to review status and plans for a joint presentation to the Red Team on June 15, and a first status presentation to the Associate Administrator for Space Science and Applications on June 16. The Final Report for the first phase of this effort is due on August 1, 1992.

Program Elements: MO&DA programs consist of three major elements:
- "operations" including the use of space and ground networks, mission control, flight operations, and flight project support,
- "science centers", also referred to as "science analysis centers", including project or mission data processing, data analysis, archiving, and the use of infrastructure for computing, science communications and archives, and
- science "grants programs", including support for instrument investigators, guaranteed-time investigators, guest observers, and guest investigators.

Implementation: The MO&DA Blue Team is chartered to
- review current program plans and implementation practices,
- determine how the programs can be executed leaner, more efficient, cheaper,
- assess the balance between the "operations", "science center" and "grants programs" elements and between different programs,
- develop a longer-range plan for continuous improvement in order to identify further efficiencies, and to accomplish a 15% cost reduction in FY97 and thereafter.

Attachments:
Att. 1: PFP-93 and POP 92-2 Guidelines
Att. 2: Blue Team Participants

Document II-21

OSSMA MO&DA
EFFICIENCY IMPROVEMENT PROGRAM

IMPLEMENTATION PLAN

G. Riegler
August 6, 1992

Summary

Objective:
    The objective of this program is to answer the question "Is there a better (more efficient) way for NASA to conduct mission operations and data analysis (MO&DA)?"
Approach:
    The approach for this program is to
    1. Identify innovations in programmatic, organizational, technical, and motivational aspects of MO&DA projects, which are likely to generate efficiencies that cannot be obtained by traditional cost-cutting measures, and assess their anticipated cost benefits or science benefits;

2. Implement all necessary programmatic, organizational, technical, and motivational changes for a select subset of MO&DA programs, and assess their actual impact, as compared to the anticipated benefits.

About this Implementation Plan:
The work of the "OSSA MO&DA Blue Team" during June and July 1992, as well as previous studies, resulted in a large number of ideas for improved efficiency of MO&DA efforts. This Implementation Plan describes a transition from idea gathering to a controlled test.

The plan contains background information, three steps towards establishing a continuing program and identification of innovative efficiency measures and expected benefits, and finally the implementation of those measures.

Background

Content of MO&DA Projects
The MO&DA budget line is now responsible for roughly one-quarter of the total OSSA budget. Most MO&DA projects can be separated into three major elements: mission operations, science operations and data processing, and science (data analysis and interpretation). On average, the first two elements – "operations" comprise roughly one-half of the non-development Mission Operations and Data Analysis (MO&DA) budget. Support for "science" (funding for instrument PIs and their teams, guest observers, archival researchers, other members of the science community) amounts to toughly one-half of the non-development MO&DA budget.

Results from Phase 1:
- Science Loss: Budget reductions in the 3%-6% range are achievable by traditional methods with acceptable impacts to science. Reduction off 9% or more will result in a serious reduction in science.
- No "Fat": No glaring inefficiencies in the current conduct of mission operations were discovered. Some minor efficiencies will be achieved by accepting higher risk levels and obtaining "learning curve" efficiencies.
- Pre-Launch Actions: Only modest improvements seem achievable by missions currently in their operational phase. The areas offering the greatest potential for improvement require pro-launch actions.

Details of the Phase 1 results, including efficiency concepts and mission-by-mission impacts, can be found in the OSSA MO&DA BLUE TEAM Phase 1 Final Report.
The OSSA MO&DA Efficiency Improvement Program

Blue Team Recommendation for the Next Phase:
The Blue Team believes that it is imperative to proceed with
- A more detailed assessment of the recommended innovative long-term changes and efficiencies,

- Estimates of benefits to be obtained, and
- Implementation of selected non-traditional changes for selected MO&DA projects, coupled with monitoring of actual impacts versus expected improvements.

We suggest that, in order to benefit from the potential for continuous improvement in MO&DA performance, the ad hoc Blue Team needs to evolve into a continuing, formal MO&DA Efficiency Improvement Program.

Primary and Secondary Objectives:

For "operations", the primary goal of the formal MO&DA Efficiency Improvement Program is to pursue efficiencies and cost reductions to (1) create new budget flexibility, and (2) lessen the science impact of budget reductions already mandated. As a secondary objective, the MO&DA Efficiency Improvement Program will also assess the productivity and efficiency of the "science" part of MO&DA, and improve the connection between science operations and "science".

Step 1: Establish the NO&DA Efficiency Improvement Program

Program Implementation Options: How should this Program be Structured?

Members of the NO&DA Blue Team at NASA HQ, GSFC, and JPL will extract from the results of Phase 1 the most promising set of efficiency improvement concepts and use it as a "strawman program plan". With this strawman plan, they will evaluate options for implementing a continuing MO&DA Efficiency Improvement Program, assess pros and cons, and recommend a specific implementation option to AA/OSSA. Program options to be considered will include

- Continuation of the present distributed approach (MO&DA associated with discipline programs), with
- via the present "Blue Team" approach, or
- Coordination via an OSSA Staff function; and
- Formation of a centralized MO&DA program

Independent Review:

For the selected program structure, a detailed implementation plan will be generated. The implementation plan will specify objectives and responsibilities, define products and schedules, and provide a periodic independent oversight function. The Review Board will include developers (Project and Program Managers), operators (flight operations and science operations teams), and users (scientists). It will include personnel from NASA and other government facilities, universities, and industry.

Schedule and Outcome:

This step should take roughly two months from initiation; the target date for completion should be the beginning of FY93. The outcome of this step will be the establishment of the MO&DA Efficiency Improvement Program, and appointment of key persons.

Step 2: Characterize the Elements of the MO&DA Efficiency Improvement Program

Objectives:
This task builds on the work accomplished during Phase 1 of the MO&DA Blue Team effort. Its objective is to identify the basic elements which comprise MO&DA (mission operations, data processing and science operations, and science) and determine the characteristics of those elements which most affect cost and science return. This will provide the context for the Efficiency Improvement Program.

An important product of this step is the development of metrics and benchmarks for operations and program efficiency. Recommendations and metrics will be collected from past studies, and ideas will be solicited from a wide spectrum of sources, including working level operations personnel, experienced operations managers, the science community, and persons with no direct involvement in MO&DA.

Schedule:
Most of the work in this step is expected to require approximately two months from initiation. A small part of this work will probbly [sic] be executed concurrent with the next step.

Step 3: Identify Candidate Efficiency Improvements and Assess their Impacts

Objectives:
During Phase 1 of the MO&DA Blue Team effort, a large number of ideas and concepts for efficiency improvement were identified. They can be grouped according to programmatic, organizational, technical, and motivational aspects.

The objectives of this step are to identify those concepts with the largest potential for realizing efficiency improvements, and specific scenarios for test implementations. Some of the evaluations will be in the form of side-by-side assessments, e. g. for life-cycle cost caps vs. development cost caps. Analysis will include determination of potential cost savings, cost-benefit assessments, the probability of successful implementation, the impact on science output, and the required changes for their implementation (for example: organizational realignments, up-front funding needs, construction of facility requirements).

Four Categories of Efficiency Improvement Approaches:
We group the efficiency improvement concepts from Phase 1 into four categories. All four appear inter-related – none can improve efficiency without the others.
- Programmatic: Remove development programs from MO&DA program budget line; life cycle cost caps; changes in NASA risk policies and risk acceptance; requirement for detailed review of operations concept at the time of the non-advocacy review (NAR); charge back and full cost accounting (inter-office "purchases", purchases from multimission facilities, or purchases from outside sources); continuity (rather than transfer) of Project management responsibility for the development and operations phase.
- Organizational: Modifications of NASA and/or institutional organizations which reduce cost. Issues include: centralized or distributed mission operations centers; centralized or distributed science operations centers; multimission or mission-specific

operations support; centralized (OSSA) or distributed (Divisions) NASA management of MO&DA; efficiency of multi-layer management structures for science facilities (e.g. Space Telescope Science Institute); and contracting philosophy (e.g. for mission-unique vs. multi-mission flight operations team contracts).

* Technical: Operability; adoption of standards; use of expert systems and other emerging technologies, partnering with industry and DoD, and sharing techniques and expertise across NASA and the Flight Centers.
* Motivational: Policies which provide natural incentives for projects to improve cost efficiency; incentives for the use of multimission standards and capabilities; post-launch multi-year budget cap (science incentive to stretch observing phase by lowered annual operation cost); incentives in individual and project performance plans.

Results of this Step: Selection of Projects and Efficiency Improvement Approaches

The result of this part of the program should be a set of recommendations to AA/OSSA for specific efficiency improvement approaches, to be applied to specific missions/projects. At the present time we expect to focus on a small subset of OSSA's programs, selected according to:
* Mission size: small (SMEX- or UNEX-type), medium (Explorer, Discoverer) and large (observatory-class) ;
* Type of mission (earth-orbiting, deep-space, surveying, observatory-style mission); and
* Mission phases: in early definition phase; during development; in primary mission phase or extended mission phase.

The rationale for the recommendations, pros and cons, and expected results, as well as success criteria, will be presented to AA/OSSA for selection.

Schedule:

This step is expected to require approximately six months from initiation. Some of the recommendations might emerge earlier, and could be implemented earlier.

Implementation Phase

Implementation of the MO&DA Efficiency Improvement Program will be compatible with the principles of TQM. Specifically, within the context of the Plan-Do-Check-Act cycle of Continuous Improvement, our Management Approach stresses planning and doing, while the Reporting and Reviewing function stresses checking and acting.

Management Approach:

The form of management, and the roles of Headquarters and the Centers, depend on the scope of the approved selections. We envision that – in order to avoid creating another level of management – the most promising ideas will be assigned to specific individuals for detailed "planning" and "doing". These individuals may be located at NASA HQ, within a flight project, or elsewhere at a Flight Center. These individuals must already have, or must be specifically given, all necessary authority to turn into realities the efficiency improvement ideas they have been assigned.

Reporting and Reviews

Monthly reports will be presented at the Flight Program and OSSA Monthly reviews. All programs will report on the efficiency and performance metrics developed in Step 2.

These metrics will be widely reported to stimulate discussion and encourage the adoption of beneficial results. Periodic reviews of MO&DA Efficiency Improvement Program results will be held quarterly. These reviews will compare metrics from all programs and evaluate the success of specific ideas. Success must be reinforced and failure acted upon.

Participation:

Ideally, individuals "from the trenches" of MO&DA should participate in the process. Given the size of the OSSA MO&DA program, this participation must necessarily be limited to keep the size of the MO&DA Efficiency Improvement Team at a reasonable level. Two actions are planned to offset the relative lack of participation by individuals "from the trenches": 1) The MO&DA Efficiency Improvement Team should obtain at least 10% of its members from the "hands-on" operators of flight and science operations centers, 2) local arrangements should be made to provide a forum for all who want to contribute ideas to the MO&DA Efficiency Improvement Program. These could be intensive "quality circles" or more ad-hoc "open mike" events. Thousands of civil service and contractor personnel support OSSA MO&DA – it is important that all of their ideas be brought forward.

Outcome

The outcome of this MO&DA efficiency improvement program should be conclusions regarding those MO&DA approaches (programmatic, organizational, technical, motivational) which did produce improvements versus those which did not. In either case, the investments and results must be clearly understood and documented, and disseminated for potential adoption in other projects.

Document II-22

Document Title: Memorandum from Daniel S. Goldin, NASA Administrator, to John Dailey, Acting Deputy Administrator, "Program Management," 15 June 1993.

Source: NASA Historical Reference Collection, History Office, NASA Headquarters, Washington, D.C.

Daniel S. Goldin became the NASA Administrator on 1 April 1992 with a mandate to reform the way in which the Agency carried out its activities. Goldin mandated a "faster, better, cheaper" approach to space science missions. In addition, he wanted to ensure that once a mission was approved, it would be managed according to an agreed schedule and budget. His instrument for achieving that objective was a top-level Program Management Council, which was established by this directive.

[no pagination]
National Aeronautics and
Space Administration
Office of the Administrator
Washington, DC 20546-0001

June 15, 1993
TO: AD/Acting Deputy Administrator
FROM: A/Administrator
SUBJECT: Program Management

As the Agency's Acquisition Executive, you are hereby directed to institute a NASA Program Management Council. This council should function in accordance with the enclosed charter and will be the chief governing body responsible for ensuring excellence in program management throughout NASA. I will entertain proposed revisions to this charter following a pilot implementation period.

In addition, you are assigned responsibility to direct the development of specific implementing instructions for program management within the Agency. These instructions should draw upon the efforts of the Program Excellence Team led by Dr. Howard Robins. You are delegated authority to approve these instructions.

[signature]
Daniel S. Goldin

Enclosure

cc: Officials-in-Charge of Headquarters Offices
Director, NASA Field Installations
Director, Jet Propulsion Laboratory
Officials-in-Charge of Headquarters Offices:
AC/Gen. Dailey
AF/Gen. Dailey
AK/Dr. Fisk
AG/Dr. Griffin
B/Mr. Peterson
C/Mr. Rack
D/Mr. Aldrich
E/Dr. Freeman
F/Gen. Armstrong
G/Mr. Frankle
H/Ms. Lee
I/Ms. Finarelli
J/Ms. Cooper
K/Mr. Thomas
L/Mr. Lawrence
M/Gen. Pearson
O/Mr. Force
P/Mr. Vincent
Q/Mr. Hertz
R/Dr. Harris
S/Dr. Huntress

T/Dr. Broedling
U/Dr. Holloway
W/Mr. Colvin
Y/Dr. Tilford

Directors NASA Field Installations:
ARC/Dr. Compton
GSFC/Dr. Klineberg
JSC/Mr. Cohen
KSC/Mr. Crippen
LaRC/Mr. Holloway
LeRC/Mr. Ross
MSFC/Mr. Lee
SSC/Mr. Estes

Director, Jet Propulsion Laboratory:
Dr. Stone

[enclosure]
PROGRAM MANAGEMENT COUNCIL CHARTER

Background
   NASA requires a well-defined system of integrated planning, approval and
implementation for Agency programs to assure that the Agency initiates programs
Consistent with its strategic plain and available resources and conducts them in
accordance with the commitments made for each program it initiates. The Agency also
requires a forum to involve the highest level of program officials in efforts to address
issues pertaining to program/project management policy and implementation.

Objective
   The objective of the PMC is to provide an Agency-level forum for addressing strategic
planning, implementation, and management of all major Agency programs. It shall
support the Deputy Administrator in:
   (1) Assuring that the Agency functions as an integrated system in planning,
       approving, and implementing its mission to meet its commitments within
       available resources; and
   (2) Meeting his/her functional management responsibilities for program/project
       management policy development, maintenance and oversight.

Functions
a. Major System Program/Project Formulation and Implementation - the PMC will
    accomplish integrated:
    (1) Planning, approval, and prioritization recommendation to the Administrator for
        all candidate major system programs and projects: and
    (2) Review and assessment of those subsequently approved programs and projects (at

key decision points requiring Administrator approval) from their initiation through completion of mission operations.

In accomplishing these functions, the Council will address the following Program/Project Formulation and Implementation related matters:

(1) Compatibility of Phase 8 candidates with the NASA Strategic Plan and with projected resources availability (funding and institutional).

(2) Adequacy of proposed:

    (a) Program and project organizational structure (including assignments and interfaces); and

    (b) Program control policies and systems.

(3) Priority and readiness for initiation of Phase B (mission need established, initial specification of performance requirements).

(4) Readiness of programs and projects to submit new start request for Phase C/D (maturity of requirements and corresponding adequacy of technical and management definition; validation of technology readiness, cost estimates, and resources availability).

(5) Conformance of programs/projects to their PCA's [Program Commitment Agreements].

(6) Recommended cancellation or continuation of programs and projects, as required (including whenever the PAA or CFO/Comptroller projects an EAC that exceeds the PCC or DCC component of the PCC by more than 15%).

(7) Special issues arising in the planning and execution of the agency major system program.

b. <u>Technology and Advanced Development Program Review and Assessment</u> - the Council will review, assess, and make recommendations regarding major technology and advanced development programs after considering future mission need, the potential for multi-mission or multi-program use, the potential return on investment (in terms of both cost and performance) for both agency and commercial use, readiness to begin the proposed phase of RAT development, and the realism of the funding profile and identified Agency benefits.

c. <u>Program/Project Functional Management Support</u> – the Council will serve as the agency's highest level forum for addressing issues related to program/project management (including acquisition) policies, systems and processes. For this purpose, the Council's programmatic functions will include providing oversight to insure conformance with these agency policies, systems and procedures. The Council will identify needed revisions resulting from this oversight and will review and assess proposed revisions from all sources.

<u>Membership & Operation</u>

The Program Management council will consist of the following:

- Deputy Administrator, Chairperson
- Associate Administrator for Advanced Concepts and Technology
- Associate Administrator for Space Systems Development
- Associate Administrator for Space Flight
- Associate Administrator for Space Communications
- Associate Administrator for Aeronautics
- Associate Administrator for Space Science

• Associate Administrator for Life and Microgravity Sciences and Applications
• Associate Administrator for Mission to Planet Earth
• Associate Administrator for Safety and Mission Quality
• Comptroller
• Associate Administrator for Procurement
• General Counsel

As Chairperson of the Council, the Deputy Administrator is authorized to convene the Council, as necessary, to discharge the responsibilities and perform the functions of the Council. Attendance and participation by others will be as determined by the Chairperson. The results of the Council's assessments will be presented to the Administrator in the form of findings and recommendations.

Document II-23

Document Title: J. R. Dailey, Acting Deputy Administrator and Chairman, Program Management Council, to Daniel Goldin, NASA Administrator, "Special Review of Global Geospace Science (GGS) Program," 20 April 1994.

Source: NASA Historical Reference Collection, History Office, NASA Headquarters, Washington, D.C.

The first program to be reviewed for possible cancellation by the Program Management Council established in 1993 by NASA Administrator Daniel Goldin was the Global Geospace Science Program. Rather than cancel the program, the Management Council recommended to Goldin a phased approach to carrying out the program.

[no page number]
[stamped] APR 20 1994

TO: A/Administrator

FROM: AD/Chairman, Program Management Council

SUBJECT: Special Review of Global Geospace Science (GGS) Program
On April 5, 1994, the Program Management Council (PMC) conducted a special review of GGS program options. An attendance list is enclosed.
Problems associated with the GGS program were first noted by the PMC in the first Quarterly Program Status Review of March 10, 1994. Concerns were reaffirmed in the Independent Annual Review (IAR) report presented to the PMC on March 24, 1994. It is noted, that during the FY 1995 budget formulation process, Code S, in coordination with Code B, has been reporting concerns which might lead to serious GGS program problems during the first and second quarters of FY 1994. Congressional notification was included in the first FY 1994 Operating Plan letter.

The program office presented a comprehensive briefing of the program's history and science objectives. Particular attention was given to the essential role of the Wind and Polar spacecraft in the International Solar Terrestrial Program (ISTP). It was pointed out that there are 37 international agreements for GGS alone with an aggregate partner investment of approximately $140 million. This does not include the substantial contribution of ESA in Solar and Heliospheric Observatory (SOHO) and Cluster, in which the United States participates through the Collaborative Solar Terrestrial Research (COSTR) Program (another part of ISTP).

The GGS Project Manager presented an overview and status of the Wind and Polar spacecraft, including their instrument complement. He highlighted the problems associated with the Martin Marietta East Windsor facility where the two spacecraft are being built. A Launch-1-year independent review, conducted in the summer of 1993, identified a number of concerns. The project accepted and addressed the review recommendations while acknowledging a threat to the launch readiness dates. Then, in August and September 1993, three spacecraft, built at the East Windsor facility (Mars Observer, NOAA-13, and Landsat 6), failed following launch. These failures have resulted in a series of focused efforts to assure mission success on the Wind and Polar spacecraft.

[2] A Wind Laboratory (spacecraft) Independent Readiness Review (IRR), chaired by Code Q, was chartered by the Associate Administrator for the Office of Space Science. This technical review was conducted concurrent with the IAR. These reviews were limited to the Wind spacecraft. The IRR team assessed both the thoroughness of the development, integration, and verification process and the ability of laboratory hardware, software, and ground systems. The IRR team reported seven significant findings and provided associated recommendations. The Wind Laboratory IRR team's conclusion, "assuming no further major technical issues are uncovered either in current on-going reviews or in the proposed testing program: risk associated with the Wind mission meeting the mission success criteria is acceptable."

Examples of confidence were cited in four areas:
(1) The problem-plagued power supply of NOAA-13 shares little direct heritage with Wind.
(2) The orbit of Wind is relatively benign compared to the NOAA-13 orbit.
(3) The stable-spinner spacecraft design has proven reliability.
(4) The flight software is modest in scope and criticality (updatable from the ground).

In addition to the Wind Laboratory IRR conducted by Code Q, a mission-success review team from Martin Marietta and Goddard Space Flight Center (GSFC) is examining the spacecraft design.

The project office presented three options which were considered for continuing the program:
Option A - Continuation of the full program (unmodified) – launch both spacecraft as soon as possible.

Option B - Launch Wind as soon as possible; store Polar.

Option C - Continuation of full program (modified) – launch Wind as soon as possible; delay Polar, pending results from Wind.

Option C was the recommendation selected by the GGS Program Associate Administrator for the following reasons:
(1) It permits single-string processing, Wind then Polar, removing pressure of simultaneous development;

[3] (2) It allows Martin Marietta more time to implement organizational changes to benefit Polar;

(3) It allows a full year of flight experience on Wind before Polar is launched;

(4) It retains a full commitment to the science community; and

(5) It reduces FY 1994 need for additional funds.

The additional forecasted funding does not violate the program's 15-percent development cost threshold. The baselined launch schedule, however, is broken. The new proposed launch readiness dates are November 4, 1994, for Wind, and November 21, 1995, for Polar.

The PMC members proffered a number of challenging and penetrating questions and comments. Responses often elicited further questions. The remarks were directed to the reasons for the current situation in an effort to avoid repetition.

In response to one question, the PMC was told that NASA's portion of the East Windsor Facility's business was relatively small. NASA, therefore, had been unable to affect needed infrastructure changes. It was only after the three failures that management recognized the threat to the total business complement.

It was commented that, although we have fixed or are fixing problems and will verify all known problem resolutions, there is no guarantee we have identified all problems. There is still risk.

In executive session, the PMC members agreed to endorse a plan to proceed with Option C. The following actions and assumptions are added to the endorsement:

(1) The Comptroller will review and verify the cost to complete funding requirements by April 15, 1994.

(2) OSS will absorb the cost differential within the OSS budget. The specific recommended actions to absorb the cost differentials will be identified by July 1, 1994.

(3) OSS will alert the science community that continuing GGS will reduce the funds available for new starts by June 1, 1994.

(4) GSFC will rework the fee structure to focus on on-orbit performance by August 1, 1994.

(5) GSFC/Code B/Code G/Code H/Code S will develop a contract strategy for Polar, e.g., capping cost, by August 1, 1994.

(6) Code G/Code H to review NASA procurement policy to place additional emphasis on the evaluation of cost realism (underbidding by 50 percent in this case) in the mission suitability evaluation of contract proposals by August 1, 1994.

(7) The Wind Laboratory IRR team will be available to the GGS program for special tasks such as a Polar review.

In addition to the above actions, the Option C plan is modified to specifically include:

(a) The Wind Laboratory IRR team will conduct a delta IRR as part of the System Acceptance review.

(b) The Wind Laboratory IRR team will conduct an IRR of Polar.

(c) After the successful launch and initial operation of Wind, the Polar budget and schedule will be verified by Code B.

(d) The verified Polar plan will be reviewed with the science community specifically delineating the impact of continuing Polar on other existing or planned programs.

(e) The coordinated Polar plan will be presented to the PMC for recommendation to the administrator.

In conclusion, the PMC recognized the Agency's risk in continuing GGS. Option C offers NASA the best opportunity for success. The PMC recommends that the GGS Program be permitted to proceed with Option C, modified by the aforementioned applicable actions and assumptions.

The PMC will continue to monitor the GGS progress through the Quarterly Program Status Reviews.

[signature]
J. R. Dailey

Enclosure

PROGRAM MANAGEMENT COUNCIL
Global Geospace Science (GGS) Special Review
April 5, 1994

ATTENDEES
AD/Gen. Dailey
AT/Mr. Mott
AS/Dr. Cordova
ABA/Mr. Lee
B/Mr. Peterson
CF/Mr. Levine
D/Mr. Aldrich
G/Mr. Frankle
HS/Mr. Fournier
J/Ms. Cooper
M/Mr. Mann
O/Mr. Force
Q/Mr. Gregory
S/Dr. Huntress
U/Dr. Nicogossian
Y/Mr. Townsend

Document II-24

Document Title: George Withbroe, NASA, "Living With a Star: The Sun-Earth Connection," 3 August 1999.

Source: NASA Historical Reference Collection, History Office, NASA Headquarters, Washington, D.C.

Associate Administrator for Space Science Edward Weiler in 1999 encouraged George Withbroe, who was in charge of NASA's Sun-Earth Connection space science theme, to continue to push a new concept he had developed, in which space physics missions would perform the scientific research necessary to support a variety of practical applications with respect to space weather and its effect on our society and life. (Space weather is the term describing the conditions in space, including the behavior of the Sun and its interactions with Earth's magnetic field and atmosphere, that influence Earth and its technological systems.) Weiler wanted to pitch this concept to NASA Administrator Daniel Goldin; Withbroe prepared a presentation for this new concept, named "Living With a Star" (LWS). The first opportunity Withbroe had to present the LWS program to Goldin was at a 3 August 1999 Science Council Meeting. Goldin enthusiastically supported the concept, and it became part of NASA's space science program.

[slide]

Living With a Star
The Sun-Earth Connection

August 3, 1999

George Withbroe
NASA Headquarters

[slide]

The Sun-Earth Connected System

Variable Star   Planet

Varying
• Radiation
• Solar Wind
The Sun   • Energetic Particles   Earth's Magnetosphere

Interacting
• Solar Wind
• Energetic Particles

Interacting   Interacting
• Magnetic Fields   • Magnetic Fields
• Plasmas   • Atmosphere
• Energetic Particles   • Plasmas
                                          • Energetic Particles

Questions:
• How and why does the Sun vary?

• How do the Earth and planets respond?
• What are the impacts on humanity?

[slide]
Sun-Earth System – Driven by 11 Year Solar Cycle

Solar Maximum:
• Increased flares, solar mass ejections, radiation belt enhancements.
• 100 Times Brighter X-ray Emissions 0.1% Brighter in Visible
• Increased heating of Earth's upper atmosphere; solar event induced ionospheric effects.

Declining Phase, Solar Mnimum [sic]:
• High speed solar wind streams, solar mass ejections cause geomagnetic storms.

[slide]
Why Do We Care?

• **Solar Variability Affects Human Technology, Humans in Space, and Terrestrial Climate.**
• **The Sphere of the Human Environment Continues to Expand Above and Beyond Our Planet.**
    - Increasing dependence on space-based systems
    - Permanent presence of humans in Earth orbit and beyond

[slide]
Solar Variability Can Affect Human Space Flight

• Radiation Protection operations for future human missions both to the ISS and to Mars, as would be provided by SOHO and STEREO; exposure applies to high latitude airline flight.

• Space Station Orbit is Exposed to High Energy Solar Particles
Issue: Requires focused research effort to improve knowledge about risk levels and possible risk mitigation techniques.

[slide]
What can we do about it? Goals of the Living With a Star Initiative

1. Quantify physics and dynamics of the Sun-Earth connected system due to the 11 year solar cycle.
    - Obtain improved and continuous measurements of system disturbances.
    - Understand the solar cycle. For long-range forecasting & assessing solar role in climate change.

2. Develop predictive models for the system that:
    -   Demonstrate understanding of physics.
    -   Have utility for prediction of space weather.

3. Minimize impact of space weather on technology and human space and aero flight.
    -   Determine space environmental conditions.
        Needed for design of systems to minimize sensitivity to space weather.
    -   Develop improved techniques for predicting space weather events and their access to ISS and to human explorers in deep space.

Apply a systems approach.

[slide]
I. Accelerate: Solar Terrestrial Probes

Why?
-   Studying Sun-Earth connected system requires simultaneous observation of interacting regions
-   FY00 budget: Missions have 2 yr design life and are launched at 2.5 yr intervals.
    -   Limits synergism between missions studying different regions of Sun-Earth system.
-   Goal: 1.5 year interval between missions.

[slide]
What is Still Missing?

1. Missing: Detailed information on dynamics of solar interior and the dynamo that generate and control solar variability, especially long term cycle irradiance.

    Solution: A Solar Dynamics Observatory (Next Generation SOHO) providing high time & spatial resolution data to probe:
    -   Solar interior & the subsurface structures underlying regions generating solar disturbances.
    -   Dynamics of magnetic structures in solar atmosphere where these disturbances occur.

2. Missing: Continuous observations of solar regions generating solar disturbances ("solar weather patterns"); measurements of solar interior from other side of Sun.

    Solution: Solar Sentinels:
    -   To observe the entire solar surface, including far side from Earth.

- To observe globally & in stereo solar wind disturbances from Sun into interplanetary space.
- To obtain "missing" seismology data from solar far side.

3. <u>Missing</u>: Detailed information on the dynamics of the terrestrial space environment during geospace disturbances from short space weather to climactic time scales.

<u>Solution</u>: A Geospace Dynamics Network:
- Network of spacecraft to provide data with sufficient spatial and temporal coverage to specify the dynamics of disturbances affecting geospace and the neutral atmosphere.

Chapter Three

Life Sciences in Space

by Joan Vernikos

Introduction

Pursuing research in the life sciences in an engineering agency like NASA is unlike pursuing any other science in both its breadth and culture. Life sciences encompass every aspect of research and technology involving living organisms that has an impact on, is affected by, or requires the ability to leave Earth and go into space. Space provides the only milieu to study how environmental variables characteristic of Earth, such as gravity, have shaped life on Earth. Such knowledge could be applied to developing treatments for diseases on Earth or to preparing astronauts for a healthy return to Earth after flight. The eventual ability to study how life may evolve in the absence of gravity by following multiple generations of life forms in Earth orbit would direct biologists more precisely to what evidence of life forms they may seek on other planets. Going into space is ultimately needed to answer the question of whether we are alone in the universe.

Operating in the alien environment of space requires systems developed to support life forms in this environment, whether they are cells, plants, animals, or humans. Technologies and research methods must be miniaturized and improved, and, where possible, automated, because of weight and power constraints. Although NASA is the primary customer for these technologies, instrumentation and life-support systems developed for space research purposes find significant technological, scientific, and engineering applications in many non-NASA domains. The development of reliable, closed ecological life-support systems has provided a small test bed of that larger life support system, Earth.

To understand life sciences in space, one must appreciate their significant cultural differences with other sciences carried out by NASA missions. The diversity of life sciences ranges from the search of how life forms to understanding how the human body changes away from Earth, and how best to help astronauts readjust to Earth after spaceflight. Biologists see these apparently diverse questions as a logical continuum in the development of a knowledge base, requiring similar expertise and approaches. Nonbiologists, organized around their facilities, whether telescope or spacecraft missions, tend to separate their research questions according to the facility needed to answer the question.

The dispersion through repeated reorganizations over the past forty years of life sciences among NASA Headquarters functions and offices has its roots in the conflict of these cultures. On the one hand, space missions and other sciences require support from life sciences. On the other hand, the science community, through its many reviews and consistent recommendations, sees frequent access to space as what life sciences most need to advance understanding and knowledge of life on Earth and the universe. A persistent issue also has been the relationship between NASA Headquarters and various NASA Field Centers, and the rivalry between these Centers. Centers with lead roles in life sciences have usually reported institutionally to other Headquarters programs. Life sciences have never had complete authority over the space missions they needed.

This essay chronicles the fluctuating fortunes of life sciences caught between these two views. Medical care of the crew and other operational issues relating to astronaut health will be dealt with in later volumes in this series. After discussing the origins of space life sciences research and of the NASA life sciences program, the essay traces the long-running debate over the best way to organize for life sciences within NASA. It then reviews various life sciences research efforts, first those conducted on robotic spacecraft, and then those involving human presence in orbit. The essay concludes with a discussion of various other elements of the space life sciences effort and brief reflections on the future of this area of space research.

Before Sputnik

Almost as soon as humans learned how to propel bodies upward into the atmosphere, a fly, worm, or other living thing was attached to whatever payload was being launched "to see what happens." At the end of World War II, balloons rising to 97,000 feet were used to study the effects of radiation on living organisms. Air Force scientists at the Holloman and Wright-Patterson Air Force Bases used such balloons to provide up to twenty-eight-day exposures in "space," flying fruit flies, fungus spores, mice, and even monkeys.[1] In addition to the effects of radiation, they studied changes in respiration and blood pressure. Later they used rockets equipped with a camera to observe animals in microgravity freefall at altitudes up to 236,000 feet. At the Army Ballistic Missile Agency in Huntsville, Alabama, Wernher von Braun and the other Germans who came to the United States after the war were working with recoverable nose cones. Von Braun's deputy for research was Ernst Stuhlinger. Although a physicist, he had an undergraduate degree in biology and was interested in biological studies in space. The following poem he wrote for von Braun in 1959 exemplifies Stuhlinger's enthusiasm and vision of space biology:

> I doubt that you remember me,
> I am an urchin from the sea
> Where once my unlaid eggs did swell
> Is nothing but an empty shell.
> These eggs though, and God bless your heart,
> In Jupiter they had a part,
> Where in their own peculiar ways
> They registered some cosmic rays

---

1. A complete survey of this research at Holloman during this period is in David Bushnell's, *History of Research in Space Biology and Biodynamics, 1948–1958* (Holloman AFB, NM, Air Force Missile Development Center, 1958) and J. H. Hanrahan and David Bushnell, *Space Biology: The Human Factor in Space Flight*, pp.139–149 (New York: Basic Books, 1960).

And had, while on a weightless ride,
Their tiny little cells divide.
They cut, though modest still and frail,
For living man an open trail
To Space.
Their trail, we hope, will grow
And help man's daring goal: to go
Perhaps in ten years or eleven
Directly to the Gates of Heaven![2]

In December 1957, soon after the Soviet Union launched Sputniks 1 and 2, the Rocket and Satellite Research Panel, a group of scientists who were among the first to plan experiments in space, called a meeting in Washington to discuss how to best organize for an expanded U.S. space program. (See Volume V, Document I-11.) The panel, chaired by James Van Allen, included scientists from universities, industry, government agencies, and the Huntsville group. There Stuhlinger met Richard S. Young, who was working with sea urchins at Woods Hole. He asked him to come to Huntsville to develop biological flight experiments. The Army's Redstone rockets gave about 30 seconds of microgravity, useless for biological experiments. Nevertheless, they provided a learning opportunity to build hardware and environmental control systems for temperature, atmosphere, pressure, and fluid handling necessary to maintain living systems in the hostile environment of space. Recovery ships would sail from San Juan, Puerto Rico, and Young would set up a rudimentary laboratory on the ship. This was truly a new frontier. No one had ever had the ability to study the effects of less than one gravity on living systems before.

The advent of the orbiting spacecraft made such research possible, but getting the gravitational field in the spacecraft down closer to zero was (and remains) a serious challenge. There were always things going on in a spacecraft that impart gravity-producing accelerations to the vehicle, such as motors starting and stopping, firing of positioning rockets to keep the vehicle stable and prevent it from tumbling, valves opening and closing, not to mention the far greater disturbances when a human is also in the spacecraft.

Questions asked in those early experiments were fundamental and important to biology. (See Document III-1.) What was the threshold of sensitivity of living plant and animal cells to gravity? While scientists knew that both plants and animals had gravity sensors and respond to gravity, what were those sensors, how sensitive were they to the amount and duration of gravity input, and how important were those inputs to the functioning and evolution of organisms? Suppose an egg was fertilized in 0.001 or 0.0001 of Earth's gravity and allowed to develop, what would happen? Some predicted the answer would be nothing. Others thought otherwise. More than forty years later, there is still not a complete answer to these questions, in large part because of limited experimental flight opportunities.

Meanwhile, the race for putting humans in space was on. Beginning in 1951, the Soviets were flying a variety of living specimens and animals such as mice, rats, and dogs to determine what was needed for survival. Most experiments were not designed to allow recovery of the experimental animal. The dog Laika was launched on Sputnik 2

---

2. Autobiography of Richard S. Young, Cape Canaveral, May 1996, unpublished manuscript, p. 38.

on 3 November 1957. No recovery was attempted, but the animal survived one week in space. The U.S. Air Force also successfully flew rhesus monkeys on sounding rockets, but did recover them after flight. First to fly was Albert in 1948, later Ham and Able, and then Baker, who returned to live a long life at the Primate Center at Holloman Air Force Base.[3] Much was learned about space not being life threatening as long as a reliable life-support system was provided. Prediction that eyes would pop out or of an inability to swallow were refuted by these first biological flights and paved the way for the first humans venturing into space.

After Sputnik—The Creation of NASA

The successful launch of the first two Sputniks and the disappointing failure of the first Vanguard provided the impetus in the United States to create an organized space program. As with other events that are seen as hurting American pride and its standing in the world, congressional hearings were held and expert advice sought as to how best to rally the country and organize for space.[4] President Eisenhower proposed that the National Advisory Committee for Aeronautics (NACA) serve as the core of a new space agency. He signed the National Aeronautics and Space Act on 29 July 1958, creating NASA, the National Aeronautics and Space Administration.

Even before the President signed the Space Act, the National Academy of Science (NAS) moved quickly to take a leading role in shaping the new Agency's space science program. On 4 June 1958, Detlev Bronk, President of the Academy, established the Space Science Board (SSB, renamed the Space Studies Board in 1986), chaired by Lloyd V. Berkner; on 27 June, the Board held its first meeting. It was quick to begin identifying scientific areas and experiments to be carried out in space and formed twelve ad hoc committees. The ad hoc Committee on Psychological and Biological Research was chaired by H. K. Hartline, with S. S. Stevens as Vice Chair. This was the first attempt at a space life sciences planning and advisory function of the NAS. The Board was charged by Bronk to address "payload composition, relative importance of experiments, expectancy, timing, environmental effects, orbital requirements, and so forth, in relation to the effort and cost involved." Harold C. Urey, who had a strong interest in the origins of life, emphasized the need for the committee to develop expertise in the area through symposia, publications, and appropriate membership. Short-, intermediate-, and long-term plans were developed, and experiments in the growth of living tissue and psychology were put forth as areas to pursue. Hartline and Stevens identified Orr Reynolds at the Office of Naval Research (ONR) and W. R. Lovelace as being excellent sources of information. Lovelace was highly respected for his work in aerospace physiology and medicine.

The SSB went beyond Bronk's charge, and, by 3 July 1958, he sent out a telegram solicitation asking for experiment ideas intended to broaden the base of interested

---

3. Hanrahan and Bushnell, *Space Biology*, pp. 139–149.
4. The essays and documents in John M. Logsdon, et al., *Exploring the Unknown: Selected Documents in the History of the U.S. Civil Space Program, Vol. I*, "Organizing for Exploration," NASA SP-4407, 1995, provide background information on the formation of NASA.

scientists. Furthermore, it was prepared to review and select proposals—the first attempt at setting up a peer-review system for space science. However, this was not to be. The new NASA leadership quickly took charge, making sure the SSB understood its role was to recommend program "guiding principles" both in content and in policy, whereas NASA's responsibility was the formulation and management of the program. (See Volume V, Chapter 1, and Documents I-13 through I-17 for a discussion of NASA-SSB interactions at that time.) The SSB and its Subcommittee on Biology and Medicine had served as the independent voice of the scientific community and made continuing recommendations as to what life sciences should emphasize and what its role within NASA and its relationship to other NASA functions should be. NASA, unfortunately, only partly heeded its advice.

Birth of NASA's Space Life Sciences Program

When T. Keith Glennan became the first NASA Administrator in October 1958, there were no life sciences programs at any of the research centers that formed the core of the new Agency. He had the option of turning this research area over to other agencies such as the National Science Foundation (NSF) or the National Institutes of Health (NIH), or to the military, in particular the U.S. Air Force (USAF), which had an active biomedical program in anticipation of sending humans into space.

Glennan did not wish to enlarge his staff, but he saw the need to ensure coordination with the military when presenting life sciences requirements and activities to Congress and the President, and to have a single point of contact in NASA for scientists from the external community. He formed a Special Advisory Committee for Life Sciences (SACLS) using external consultants. Lovelace was asked to chair it. He and his committee foresaw the continuing need for life sciences input into the space program and recommended a high-level life sciences office and a centralized life sciences research facility. A review by W. L. Hjornevik, staff assistant to Glennan, reached the same conclusion—NASA was underestimating the importance of biomedicine to the effective management of spacecraft projects. In November 1958, Glennan hired Douglas L. Worf as Chief of Biology and Life Support Systems. He reported to Abe Silverstein, the Director of the Office of Space Flight Development. Worf was a biophysicist from the Atomic Energy Commission and had previously been at the ONR, where he worked with Orr Reynolds. However, his efforts to develop a biology program compatible with NASA engineering goals were soon thwarted.[5]

By early 1959, it was apparent to Glennan that NASA was in danger of becoming totally dependent on the military for life sciences research and development (R&D). A tug-of-war began with the Air Force, which was maneuvering to gain control of any post-Mercury human spaceflight program. In March 1959, Glennan appointed to his staff Special Assistant in Life Sciences Clark T. Randt, a respected neurophysiologist and clinical researcher from Case Western Reserve University Medical School. He gave him

---

5. For an extensive early history of life sciences in NASA, see John Pitts, *The Human Factor*, NASA SP-4213, 1985.

the charge of developing NASA's long-range requirements in life sciences. Glennan also formed the Bioscience Advisory Committee (BAC), composed of consultants with basic research backgrounds. Its purpose was to recommend to NASA appropriate roles and responsibilities for life sciences, and to suggest organizational changes that would improve the disciplines.

Randt claimed NASA needed to give high priority to basic biomedical research and to integrate all life sciences research and development. Noting that the life sciences comprised a continuum from basic research in biology to clinical practice, he suggested that the organization and management of NASA's programs should reflect this. In practical terms, this meant that the four primary activities—space biology, human research, biotechnology, and space medicine—should be administered within a single life sciences program office. He also urged the creation of a life sciences research program at a NASA Field Center. An active program of grants and contracts to life scientists would fall under the jurisdiction of the Director of the Life Sciences Program Office. (See Document III-2.)

In February 1960, Randt produced what could be considered as the first Strategic Plan for life sciences. He proposed a progressive and integrated approach of ground and flight research, including flight missions ranging from short (two to seven days), to intermediate (ten to thirty days), to long duration (beyond six months). These missions would include both subhuman and human organisms. He further identified a program of biological research focused on the origin and evolution of life, and the search for extraterrestrial life forms. The wisdom and logic of this plan stood the test of time. It has been followed, perhaps unknowingly, yet at least in principle, by the successive life sciences management organizations. However, it could not be followed in toto because access to flight for life sciences research had not provided the appropriate mission duration nor the needed research capabilities.

The BAC was chaired by psychiatrist and basic researcher Seymour Kety. Kety's committee produced the first external report on how best to organize life sciences within NASA. (See Document III-4.) Of all the external reviews of NASA's life sciences efforts, this is perhaps the most significant report historically, not only because it is comprehensive, but also because subsequent committees consistently produced similar recommendations. However, only Glennan came close to implementing them. In March 1960, Glennan established the Office of Life Sciences Programs (OLSP), with Randt as its Director, placing the OSLP on the same level as the other program offices at NASA Headquarters. This changed the initial NASA Headquarters organization from three program offices to four, with OLSP on par with the Office of Launch Vehicle Programs, the Office of Space Flight Programs, and the Office of Advanced Research Programs.

Randt moved ahead on the recommendations of the Kety report, dividing the Office of Life Sciences Programs into the following three divisions: Space Biology, which included ground and flight research investigations; Space Medical and Behavioral Science, a program in ground and flight research in human physiology and behavior; and Flight Medicine, the operational research in support of human flight. He then developed a ten-year plan, strongly justifying a NASA life sciences research program. Randt also lobbied for jurisdiction over the biomedical aspects of the Mercury Project, but this initiative was rejected by Abe Silverstein, the Associate Administrator for Space Flight

Programs. Thus began a constant argument between engineers and physical scientists on the one hand and life scientists on the other that has caused tensions ever since. Randt feared that Silverstein's haste to put a human in orbit would result in losing a man in space and set back human spaceflight indefinitely. Silverstein and his engineers feared that Randt would interpose research flights with animals to verify a human's ability to live in microgravity, thereby delaying Mercury and giving the Soviets an opportunity to orbit the first human. (See Document III-3.) Physical scientists joined with the engineers in this attack, since animal research flights would spend funds which they believed they could use to better advantage. Glennan began to waver in his commitment to the life sciences program and, by the end of 1960, slowed down Randt's proposed schedule.

Nevertheless, Glennan encouraged Randt to move ahead on the development of a life sciences program at a NASA Center. By late June 1960, Randt had identified the Ames Research Center in California and the newly constructed Goddard Space Flight Center in Maryland as candidate locations. Glennan indicated he would select Ames as the site, but he made no firm commitment.[6] Both Centers were surrounded by strong academic and research institutions, but Ames had research expertise in human factors and biotechnology from its NACA days. To encourage regular interaction with the human spaceflight program, Glennan had first considered moving all aspects of life sciences research to be near the main human spaceflight activity, then based in the Space Task Group (STG) at the Langley Research Center in Hampton, Virginia. When he suggested moving it all to California, he met political resistance.

Opposition to NASA's life sciences plans also came from congressional concern about possible duplication of effort between the Air Force and NASA.[7] The Air Force was continuing to question the scope of NASA's life sciences program, and the military recruited congressional support in the persons of Rep. Emilio Daddario (D-Connecticut) and Sen. Hubert Humphrey (D-Minnesota). The scope and quality of NASA's life sciences research also came under attack by critics in the scientific community, who saw little hope of getting funded by the Agency. The Space Science Board reviewed NASA's life sciences planning and gave notice of what would be a long-standing concern with both the risks associated with contaminating other bodies in the solar system with microbes from Earth and bringing microbes from other celestial bodies back to Earth. (See Document III-5.) The commitment to the gigantic effort to land man on the Moon forced competition between traditional space science missions and the human spaceflight budgets. Life sciences programs, needing higher budgets to assure credibility, were caught between these many conflicting forces.

---

6. Memorandum from T. Keith Glennan to Smith J. DeFrance, Director, Ames Research Center, 15 March 1960. (ARC Director Files)

7. The most extensive study of U.S. capabilities in space biology and medicine at this time is included in the U.S. Senate Committee on Aeronautical and Space Science, "Space Research in the Life Sciences," 86th Congress, 15 July 1960.

What To Do with the Life Sciences?
A Recurrent Theme

Throughout the existence of NASA, how to organize the Agency's life science efforts has been problematic. While life scientists, both internal and external to NASA, have preferred to concentrate the efforts in a single organization under high-level leadership, most NASA managers have seen elements of life sciences activities primarily as support for other NASA functions, and thus have spread those activities in various offices throughout NASA.

Change in Fortunes

With the end of the Eisenhower administration, Administrator Glennan resigned in January 1961. President John F. Kennedy appointed James Webb to be the second NASA Administrator. Hugh L. Dryden continued as Deputy Administrator and Robert C. Seamans, Jr. as Associate Administrator. Randt submitted a plan to Seamans, proposing once again to consolidate the total NASA life sciences program. When nothing happened, Randt resigned. Charles Roadman was selected by Seamans as the next Director of the OLSP. He was an Air Force general, skilled bureaucrat, and experienced aerospace physician, but he was not the research-oriented person Randt had been. Roadman was mission-oriented and saw little need for research on animals prior to human spaceflight. He thus recommended the separation of life sciences into a human-oriented program under the Office of Space Flight Programs and a nonhuman program, which he called space biology, to join other nonhuman programs in space science. The major Soviet success in the space race with Yuri Gagarin's orbital flight on 12 April 1961, and the two successful suborbital U.S. launches on 5 May and 21 July 1961 added fuel to Roadman's argument that animals need not be flown before humans. What Roadman may not have known or acknowledged was the very extensive series of animal flights the Soviet program had carried out before Gagarin's successful mission.

Thus began the long-running saga of separating various components of the life sciences in different organizations, demonstrating the lack of understanding between engineers and medical practitioners on the one hand and of space scientists on the other, that has haunted life sciences in NASA to this day. Not least of the problems was the resulting inability to consolidate life sciences efforts long enough to grow and maintain an adequate life sciences program budget. It was not in the interest of other factions within NASA, who had been trying to get a foothold for their own efforts, to see this new entity get established, grow, and prosper, competing for program dollars. Space life sciences also received no help from the broader life sciences community. For the space science community, NASA was their primary source of funding, and going into space provided the opportunity to leap into a new era of discovery. However, the broader life sciences community was skeptical of the value of going into space to advance knowledge or develop new health applications. Funding, predominantly from the NIH and NSF, was plentiful or, at least, adequate, whereas the paucity and instability of its budget resulted in the deserved image that NASA was unreliable as a source of life sciences research funding.

In August 1961, Webb disbanded the Office of Life Sciences Programs, in accordance with Roadman's views. By the end of 1961, he reorganized NASA, creating five program offices: Space Science (OSS) to direct basic investigations in space; Advanced Research and Technology (OART) to support advanced aeronautical and human spaceflight programs; Manned Space Flight (OMSF) to manage approved manned space programs; Tracking and Data Acquisition (OTDA); and Applications (OA). This last office was combined in 1963 with OSS to form the Office of Space Science and Applications (OSSA). Webb's intention was to improve communications between NASA Headquarters and Field Centers, and to promote coordination between Program Directors at Headquarters and Center Directors, where the projects were carried out. He intended to reduce Center autonomy with respect to research programs, especially those inherited by the NACA R&D centers, and relate research efforts to future flight programs.[8]

The War of the Centers

Glennan's decision to establish a space life sciences research program at Ames in California, separate from the STG, the principal human spaceflight activity, was reinforced by the splitting off of Space Medicine from the other life sciences activities at NASA Headquarters. Life sciences no longer reported directly to the Administrator and lost its status as one program, leaving the external community confused. Smith "Smitty" DeFrance, Director of Ames, wanted a Headquarters commitment to support his fledgling life sciences program.

Since 1958, Robert Gilruth had been the Director of the STG, which included military biomedical personnel on assignment supporting Project Mercury. In 1959, they became a permanent operational component of the STG, absorbing a life-support capability from Langley Research Center. When the human spaceflight operations were moved to Houston in 1962, the STG evolved into the Manned Spacecraft Center (MSC), and Gilruth became its first Director. Gilruth, echoing Roadman's position at NASA Headquarters, insisted he should retain authority over biomedical research for human spaceflight programs. Seamans convened a Life Sciences Working Group (LSWG), under the chairmanship of Bernard Maggin, to resolve this inter-Center dispute. Its report identified three life sciences program categories: 1) space biology, including effects of space on biological systems and the search for extraterrestrial life; 2) human research and life-support needs of man in space; and 3) operational needs to support man in space, including life-support and medical operations. The LSWG report emphasized the need for a high level of coordination among these three elements. In 1962, Seamans distributed the components of the space life sciences program across three offices: bioscience in OSS, human research and support in OART, and operational medicine in support of approved human spaceflight programs in OMSF. With minor changes, this split prevailed through 1970.

---

8. This reorganization further separated the human from the robotic programs, causing much frustration over the years. See R. C. Hall, "NASA: Thirty Years of Space Flight," *Aerospace America*, December 1988, pp. 6–9.9.

OART was chosen to provide coordination across these components, but the scope and authority recommended by the LSWG were never given by Seamans to the new Biotechnology and Research Directorate in OART. Orr Reynolds, who had come from the ONR to be Bioscience Director in OSS, was not convinced that OART, an applied research and engineering development office, could coordinate basic biological research. Roadman, in OMSF, was concerned that OART would retard the biomedical aspects of the approved human spaceflight programs. The belief by NASA management that through collocation this reorganization would boost support for life sciences in each of the major programs while at the same time advancing Agency-wide objectives did not materialize.

In November 1961, Ames Director DeFrance brought in Webb E. Haymaker, a highly respected neuropathologist from the Army Institute of Pathology, to direct the new life sciences organization at Ames. Haymaker hired young and enthusiastic scientists who brought with them their research culture, independence, academic customs, unusual work habits, need for a library, and, worst of all from the perspective of conservative NASA engineers, biological specimens. The culture shock for traditional engineers at Ames was immense.

Whereas Ames in its NACA days had always organized itself around facilities, it had begun, by 1963, to organize around scientific disciplines. In December 1965, Ames inaugurated its life sciences research laboratory. A range of intricate centrifuges to carry everything from small organisms to humans was built, as were state-of-the-art laboratories. A well-constructed animal shelter housed a variety of animals, including a pigtail macaque primate colony, for use in ground-based control experiments prior to the planned Biosatellite missions.

While MSC near Houston screened and trained individual astronauts, Ames developed the fundamental science underlying Houston's work. Funded by OART, the Ames Biotechnology Division studied the psychological and physiological performance of pilots and human volunteers in conditions simulating the effects of spaceflight. OART also supported research with plants, rats, bacteria, and other organisms in different gravitational loads, as well as potential radiation damage from exposure in space to high-energy solar rays, usually filtered by Earth's atmosphere. The Ames instrumentation group applied its expertise in building sensors for aircraft to building sophisticated biological sensors and clever telemetry devices. These were later used in flight experiments to record physical and physiological data in unrestrained animals. The Bioscience Office in OSS supported exobiology, the research on life in outer space and other planets. Harold P. Klein arrived from Brandeis University in 1963 to head the Ames Exobiology Branch and was instrumental in guiding the construction of specialized laboratories and quarantine facilities for analyzing lunar or other samples. In 1964, Klein replaced Haymaker as Director of Life Sciences at Ames.

Compared to other space science programs, life sciences were always underfunded. From 1962 to 1964, they received about 1.6 percent of the Agency's R&D budget. The situation was particularly bad in 1964. While NASA management attempted to obtain larger sums for life sciences, Congress cut the total life sciences request by 31 percent. No other R&D area in NASA had a budget request reduced by more than 18 percent. Congress justified these cuts on the recurring theme of comparable capabilities in space medicine and human factors existing in other government agencies.

Whenever budgets decreased, bickering among internal factions increased. Conflict between OSS and OART developed over the management of Biosatellite, the first life sciences flight project. Attempts at cooperation failed, and OSS, the originator of the concept, prevailed even though Ames, an OART Center, was responsible for the project. OART, on the other hand, was pressuring Ames to devote more of its facilities to biotechnology and human factors research. Tensions developed between Ames and MSC as the latter increased its human research capabilities over the objections of OART. NASA management had not specified roles and missions for biomedical research. No one in authority had precluded MSC from following this independent course. Those who did collaborate across Centers were viewed with suspicion. Memoranda of Agreement had to be countersigned by the respective Center Directors to allow such collaborations to proceed.

No Agreement on the Best Organization

Culminating with the drastic life sciences budget cuts of 1964 and increasing criticism from Congress and the scientific community, NASA's top management came to realize that all was not well. From 1962 to 1964, three external reviews of NASA's life sciences efforts were conducted.[9] All reviews yielded the same conclusions and recommendations. The life sciences programs were disorganized due to a lack of decisive leadership at top management levels. A common recommendation was that NASA should appoint a prominent life scientist to a high-level position (preferably as a Deputy Associate Administrator) with authority to plan and direct overall life sciences programs and act as the single point of contact with the external science community.

Webb and his associates could not see how they could live with this recommendation. Seamans' assessment of the situation was that it was unlikely an individual competent in fields ranging from fundamental biology to life-support technology could be found. However, the main issue appeared to be the discrepancy between the consistent request from the outside community for a high-level administrative position and the management view of life sciences as a support function to all NASA activities. Top NASA managers continued to believe the latter would be better served by distributing life sciences throughout the Agency. Webb, Seamans, and Dryden felt that a concentrated life sciences focus was inconsistent with NASA's broader programs and that the distributed program had "not proven an obstacle and made life sciences more responsive to the needs of program directors." (See Documents III-7 through III-10.)

Seamans compromised by forming a Life Sciences Directors Group (LSDG), chaired by W. R. Lovelace, the Director of Space Medicine, thus deftly finessing the issue of a new high-level scientist appointment. It was only because of Lovelace's good sense and charisma that the LSDG succeeded where other similar attempts had failed. Lovelace also instituted a Manned Flight Experiments Board as a subelement of the LSDG to select and coordinate the planning of proposed experiments. The value of this Board concept across program offices has withstood the test of time.

---

9. In addition to a Space Science Board review, the life sciences program was examined by the President's Science Advisory Committee and by Nello Pace, a consultant hired by Administrator James Webb.

NASA's leaders felt the life sciences management problem had been resolved, but after Lovelace's death in 1965, Orr Reynolds succeeded him, and the bickering resumed. Col. Jack Bollerud, acting Director of Space Medicine on assignment from the Air Force, yet again repeated to Webb the recommendations of previous panels that NASA establish a Life Sciences Associate Administrator.[10] Once more, no action was taken.

Organizing Life Sciences in the 1970s

Public interest in space waned after the Apollo successes. Budgetary support for NASA was greatly reduced, especially for the human flight programs. President Nixon and his administration favored a science-oriented program over a new major human flight goal. NASA had to reconsider its priorities. Plans for the construction of permanent lunar bases after 1975 and a permanent space station to serve as a research laboratory and transfer platform to the Moon and Mars had to be set aside. Humans in space were limited to the completion of the three Skylab Earth-orbiting research missions in 1973, and the space station was deferred in favor of a recoverable reusable Space Transportation System (STS), the Space Shuttle.

As the Apollo program neared its initial lunar-landing missions, NASA began to prepare a comprehensive plan for its post-Apollo activities; life sciences activities were part of this planning effort. (See Document III-11.) After the Apollo successes in 1969, the Space Sciences Board once again examined the NASA life sciences effort. H. Bentley Glass, a biologist at the State University of New York, Stony Brook, chaired the review committee. The Glass Committee report was released at the end of 1970. (See Document III-13.) Yet again the advice was for the creation of an Office of Space Biology and Medicine that would consolidate all the life sciences programs under the authority of a director at an Associate Administrator level or as a Deputy in the office of the overall NASA Associate Administrator. The report also recommended the creation of a permanent life sciences advisory board composed of members of the scientific community "with no ties to NASA."

In response to the Glass report, NASA top management consolidated most, but not all, of its life sciences activities in OMSF. (See Documents III-14 and III-15.) Although space biology was moved into the new Life Sciences Division (LSD), exobiology remained in OSSA, and OART retained human factors, life support, and biotechnology responsibilities. The Director of the LSD reported to the Associate Administrator for Manned Space Flight. He also was subordinate to other program Associate Administrators who had components of life sciences activities under their control. The charge to the new Director of the LSD to coordinate life sciences activities throughout NASA was unworkable. NASA did, however, create a Life Sciences Advisory Committee (LSAC), reporting to the NASA Associate Administrator. NASA management continued the practice of not giving the Director of its new life sciences organization the authority to execute his responsibilities. General J. W. Humphreys served as the first LSD Director from 1970 to 1972, followed from 1972 to 1974 by Charles A. Berry, who came from MSC.

---

10. Col. Jack Bollerud, "Staff Study of the Structuring of Life Sciences Activities within NASA," 1966.

Neither had any experience or interest in research. The LSAC complained to Associate Administrator Newell that the coordination of life sciences programs was unworkable because other program offices were taking actions in their life sciences program elements without the LSD Director's knowledge.

New NASA Administrator James C. Fletcher became concerned in 1973 with this impasse and asked William Barry to conduct yet another review of the program. In February 1974, Barry presented to Fletcher his report outlining the same fundamental flaws raised by all previous reviews—responsibility without authority and fragmentation of programs. He also identified as a problem "bureaucratic inertia" from those seeking to maintain 1960s relationships while the program was trying to change to new priorities of the 1970s. (See Document III-19.)

Barry's report proposed again the formation of a directorship independent from OMSF, with independent budget authority and reporting to top management. He insisted that all NASA Headquarters life sciences components, not just programs, including occupational medicine and environmental health, be placed under the control of one director. He went further by proposing that NASA should reorganize the life sciences activities at the Field Centers, thereby supporting earlier findings of the LSAC. In mid-1974, Fletcher began realigning Headquarters responsibilities to meet NASA's needs for the Space Shuttle operations era. He replaced the old program offices with two Associate Administrators, one for Headquarters who would oversee the five program offices, and the other for Center management. This radical move abolished the alignment of the old Headquarters program offices with specific Field Centers. The eight-year absence of human flights between Skylab and the Shuttle saw a large increase in ground research activities at MSC, renamed the Johnson Space Center (JSC). Life sciences activities also sprouted at many other Centers.

Following another of Barry's suggestions, Fletcher appointed David L. Winter as the Director of the new Life Sciences Division in September 1974. Winter had come from Walter Reed Army Medical Center, where he performed basic and applied research in neurophysiology relevant to military operations. Before coming to NASA, he was Deputy Director of Space Life Sciences at Ames. In September 1975, in an attempt to further integrate all activities of life sciences into one office, the LSD was moved from OMSF to OSS. (A separate Office of Applications had been created.) Among the new responsibilities of the division were planning research activities to be conducted on the Shuttle and developing Shuttle life-support systems and EVA suits. Even though some bits of life sciences, like the test beds for advanced life-support systems and EVA glove research, still remained in OMSF, Fletcher had essentially pulled the management of biological and medical research and life sciences experiment flight payload definition into one place at Headquarters.

Occupational and aeronautical medicine were also added to the division, since the Space Shuttle was expected to carry passengers in addition to flight crews. But by late 1976, Winter ran head on into the different scientific culture of OSS in the form of John Naugle, then NASA Chief Scientist, and Noel Hinners, the Associate Administrator for OSS. Winter saw his function as that of designing a balanced program for high-caliber ground research in life sciences that would enjoy the respect and support of the external community and that would lead to flight experiments. He understood the need for fundamental research in human and nonhuman organisms as the foundation for the operational needs of medical support on future human spaceflights.

In contrast, Naugle and Hinners considered that Apollo and Skylab had demonstrated human ability to operate for six months in orbit and that no further research was needed. They saw the development of the ground research program as distinct and unrelated to flight or applied research questions. They did not see the need for a continuum of research and the integrating significance of ground and flight research required for life sciences. They essentially ignored recommendations of all earlier expert reviews. Furthermore, the cultural difference between an "event" science, such as astronomy and astrophysics, and an experimental science, like most life sciences that require continuing research, became an obstacle to program design. The reliance of life sciences on confirming results of research that depends on probabilities and statistics, and therefore requires relatively large numbers of experimental subjects, was (and is to this day) alien to the "event science" way of thinking. In a 1977 memorandum to the Administrator, Winter outlined his frustration with this situation. (See Documents III-20 and III-21.)

On 17 November 1978, Winter was moved to Hinners' staff; he resigned three months later. After a lengthy search by a committee chaired by Gerald Soffen, Hinners appointed Soffen on 31 March 1979 as the next Director of the LSD. Soffen was a biophysicist and had been a project scientist for Viking. Soffen oriented the program toward the impact of the space environment on basic biological processes, an area somewhat neglected in the past.

Organizing Life Sciences: 1981—1992

In 1981, Administrator Beggs appointed Andrew J. Stofan as Associate Administrator for OSSA (once again, applications work had been combined with space science). Soffen continued as Life Sciences Division Director. Stofan identified LSD responsibilities as the "direction and management of a national life sciences research program which includes the medical aspects of manned spaceflight operations, biomedical and bioscience research, advanced bioinstrumentation, and space biology. The responsibility also includes the development of an integrated program for all of NASA's medical efforts other than occupational medicine." Soffen transferred physico/chemical life-support and spacesuit research programs to the Office of Advanced Science and Technology (OAST), arguing that these programs were more technology-oriented than research-oriented, and so once more split off components of the life sciences program.

In mid-1982, Stofan was replaced by an engineer, Burton Edelson. In November 1983, Soffen was replaced by Arnauld E. Nicogossian as Director of Life Sciences. Nicogossian was a flight surgeon from JSC and had been Chief of Operational Medicine under Soffen. The pendulum had swung again. True to OSSA goals, Edelson favored robotic programs, but Nicogossian shifted emphasis back to medical support of Space Shuttle operations. So Edelson constrained the life sciences program budgets. In 1987, a physical scientist named Lennard A. Fisk replaced Edelson. Overall, he was supportive of life sciences and effective in increasing its budget during his tenure.

The years 1976–1993 provided relative stability and some growth for the life sciences. In particular, the long-proven policy and procedural practices OSS/OSSA had established and used for managing its science programs became part of the way the life sciences did business. The LSD was part of strategic planning. It regularized the process for research solicitation, peer review, and selection; adopted OSSA rules for planning, developing, advocating, and prioritizing new missions; and generally improved NASA's relationship with the external scientific community.

A New Regime: Dan Goldin and "Reinventing" NASA

In April 1992, Daniel S. Goldin was appointed by President George H. W. Bush as the new NASA Administrator. In October 1992, Goldin began a series of radical steps in his desire to "reinvent" NASA. He split up OSSA into three components—one focusing on traditional space science, one on Earth science, and one on life and microgravity sciences. The new Office of Life and Microgravity Sciences and Applications (OLMSA) was headed by psychiatrist Harry C. Holloway from the Uniformed Services University of Health Science. With the creation of OLMSA at the program office level, an individual representing life sciences was reporting directly to the Administrator for the first time since the early 1960s. (The author of this essay came to Headquarters from Ames on 10 April 1993 as Director of the Life Sciences Division (LSD), reporting to Holloway.)

Things were looking promising for life sciences. Here was the opportunity to consolidate life sciences activities. Since the previous consolidation of 1975, elements had been transferred to other parts of NASA; by 1992, when Goldin took office, they were spread across OSSA, the new Office of Exploration, OAST, and the International Space Station program. Advanced life support, with partial funding, and human factors research, with no funding, were transferred from OAST to OLMSA. However, the exobiology program that had been part of the LSD since 1972 was transferred out of the LSD into OSS in January 1993. Thus, life sciences remained fragmented. Despite its program office status, OLMSA had no institutional authority over a Field Center and no coordinating authority over the life sciences programs elsewhere in NASA.

In 1993, Goldin initiated a strategic planning activity that divided NASA's programs into various Enterprises. Life and microgravity sciences were not given separate Enterprise status; rather, they were subsumed into the Human Exploration and Development of Space Enterprise, which was led by the Office of Space Flight. Finally, OLMSA in 2000 was renamed the Office of Biological and Physical Research (OBPR) and given separate Enterprise status. Unlike other Enterprises, however, it still had no jurisdiction over flight platforms or over any Center.

Life Sciences Research on Robotic Platforms

Even in the face of continuing organizational turmoil, the NASA life sciences program has carried out a continuing program of research on both robotic and crewed missions. While flight opportunities have been too scarce, there have been some impressive research achievements.

Biosatellite

Despite the fragmentation and disorganization of the overall life sciences program, the bioscience program was sheltered within OSSA. It developed a reputable research program and generally enjoyed a period of relative stability. Orr Reynolds took advantage of flight opportunities for simple biological experiments on whatever platform was available, including the two-astronaut Gemini missions and the frog egg development and sea urchin experiments that were flown on Gemini IV, VI, and XII. Technical complications were ever present, but learning how to fly biological payloads was a big part of these missions. Although several biology flight experiments were planned for the Apollo program, most were canceled in restructuring the program following the January 1967 Apollo fire. A pocket mouse requiring little water flew successfully on Apollo 17. It was implanted with dosimeters to study radiation exposure outside Earth's atmosphere. The postflight analysis showed radiation damage only to the olfactory system, with little damage to the brain. Scout free-flyers launched from Wallops carried more sophisticated experiments with frogs implanted with electrodes in their inner ears.

By far the most ambitious set of experiments, however, were those proposed for the Biosatellite project. Biosatellite started when Headquarters asked Ames what science might come from launching leftover Mercury capsules. Conceived by OSS in 1962, Biosatellite was the first purely life sciences flight research project and was strongly supported by academic life scientists.[11] It was planned as a long-term project of three to six flights to begin in 1966, with experiments on single-cell organisms progressing to primate flights of fifteen to thirty days. In October 1962, Ames was assigned management responsibility for the Biosatellite project. OART, which had failed to get approval for a flight program, argued it should be involved in managing Biosatellite, especially the primate flights. To resolve the issue, a joint OSSA/OART working group began meeting in January 1963.[12] It divided responsibilities so that OSS would manage the development of the basic spacecraft and nonprimate experiments, and OART would develop life-support systems and later flights with primates. But this effort at cooperation failed due to a lack of communication. OSS retained overall responsibility. Biosatellite 1, launched on 15 February 1967, orbited for three days, but its retrorockets did not fire, and thus the capsule containing biological specimens eventually crashed to Earth in Australia. Biosatellite 2, launched on 7 September 1967, carried a variety of biological specimens, including frog eggs, amoeba, bacteria, plants, and mice; the mission lasted three days. Each experiment was self-contained, some with photographic capability. Biosatellite 3 included a highly instrumented pigtail monkey (Bonny) that was due to fly for thirty days. The mission was terminated on 6 July 1969, just nine days after launch, after telemetered data indicated significant physical deterioration of the animal. By that time, 80 percent of the anticipated data from the mission had been gathered. Bonny died approximately eight hours after recovery, apparently from dehydration.

---

11. Space Science Board, "Report on the Organization of the Life Sciences in NASA," 7 August 1962.

12. Memorandum from Benny B. Hall to Robert Seamans, "Minutes of the First Meeting of the Joint OSS/OART Biosatellite Working Group," 24 January 1963.

A variety of programmatic and physiological reasons led to the failure of this mission. Cost overruns, funding problems, and mission failures plagued the Biosatellite project from the outset. Brain and behavior experiments designed for one mission were combined with cardiovascular and metabolic experiments planned for another primate on a later mission. The experiment ended up being overloaded and over-instrumented. This was an overwhelming physiological stress burden on the animal, which was smaller than required in order to fill the capsule and thus prevent unwanted motion. The extent of space motion sickness, reduced blood volume, and dehydration experienced by this animal were not appreciated at the time and presaged what was later documented in humans in longer-duration flight. The loss of Bonny was a great disappointment to the life sciences community. NASA's decision to terminate Biosatellite flights after this mission was seen as further evidence of the lack of recognition of the scientific nature of the program and of NASA's desire to subordinate biological science issues to engineering and operational considerations. Opposition to the use of animals in NASA research by animal activists also was beginning to rear its head.

In 1969, Hans Mark, a physicist from the University of California, took over as Center Director at Ames. A major internal review of the Biosatellite program produced a thorough and excellent report.[13] Congress also held hearings. Rep. Joseph Karth's (D-Minnesota) Subcommittee on Space and Applications recommended the Biosatellite Project be reinstated, that the "role of science be uprated as a mission objective," that NASA "conduct a new and higher level of biomedical experiments on the astronauts," and that it implement "to fullest practicable extent" recommendations of the President's Science Advisory Committee report.[14]

Although the Biosatellite project was not reinstated, practical issues of managing an animal flight experiment identified during the project resulted in a new way of developing and managing space biology flight programs. The Space Life Sciences Payloads Office developed at Ames in the mid-1970s was in many ways an implementation of this new way of developing biological payloads and supporting flight research.

The Cosmos/Bion Program

Biosatellite 1, 2, and 3, and the advanced preparations for the Mars lander Viking (discussed below) were valuable learning experiences. They taught life scientists that improved telescience, robotic, recoverable space platforms provided many advantages for space biology research using nonhuman organisms. No value was added by having humans with no life sciences expertise present, since there was either no interaction between the astronaut and the experiment, or, when there was, it was of limited or sometimes negative value. U.S. life scientists looked with envy at their Soviet and Eastern European colleagues, who in 1973 were beginning a steady program of regular Cosmos Biosatellite missions. Every flight used a Vostok spacecraft—8 feet in diameter, with a volume of 140 cubic feet, and with excellent environmental control able to accommodate

13. J. W. Dyer, Manager, Project Biosatellite, "Historical Summary Report," 1969 (Biosatellite Project Files, NASA Historical Reference Collection, History Office, NASA Headquarters).

14. U.S. Congress, House of Representatives, Subcommittee on Space Science and Applications, "The Future of the Bioscience Program," Hearings, 12–18 November 1969, p. 39.

2,000 pounds of payload. The Eastern European scientists' involvement was arranged by Intercosmos, an umbrella organization of the Soviet Academy of Science. The program offered a reliable fleet of spacecraft and regularly scheduled spaceflights solely for research, with the ability to repeat experiments.

The Soviets had already flown two successful biosatellite missions, Cosmos 605 (Bion 1) in 1973 and Cosmos 690 (Bion 2) in 1974, when Intercosmos, surprisingly, invited U.S. and French scientists to propose experiments for Cosmos 782 (Bion 3). The U.S./Soviet space relationship in the 1960s had been characterized by fierce competition and high levels of secrecy, although there also had been ongoing discussions of potential cooperation. (See Volume II, Documents I-38 through I-43.) But by 1971, space research representatives hammered out significant civil space agreements. One of these between NASA and the U.S.S.R. Academy of Science covered Space Biology and Medicine; it paved the way for closer cooperation in this field.[15]

The first biological payload on the Shuttle was not due to fly until 1983. The Soviet invitation was too good to refuse. On 4 November 1974, David Winter, NASA's Director of the Life Sciences Division, and Dr. N. N. Gurovsky, Head of the Space Biology and Medicine Office of the U.S.S.R. Ministry of Health, signed the agreement to participate in this joint research.[16] There was no budget for such a commitment nor, because of the poor communications between Washington and Tashkent in central Asia, where the meeting was held, was Winter able to communicate with NASA Headquarters to obtain its approval prior to signing. But the Soviets were paying all spacecraft and launch costs, and thus Winter judged the risks involved acceptable. The cooperation extended from sharing biospecimens on experiments conducted by Soviet scientists to the United States eventually designing and building experiment payloads. Participation of the seven Ames scientists on Cosmos 936, which landed in Central Asia 19.5 days after launch, cost nothing except for travel. During the rest of the 1970s, the cost to NASA of this cooperation never exceeded $1 million per mission.

Although, in the beginning, U.S. scientists did not know what experiments the Soviet scientists were conducting, this partnership evolved into a true and seamless collaboration. The results of the Cosmos/Bion series allowed U.S. scientists to transition from earlier observations of what happened to humans in space to an understanding of why these changes occurred. With the ability to study a range of

---

15. The 1971 U.S.-U.S.S.R. and Applications Agreement led to the creation of five Joint Working Groups (JWGs), including Space Biology and Medicine, Meteorological Satellites, Meteorological Soundings, the Natural Environment, and the Exploration of Near-Earth Space, the Moon, and the Planets. Most of these JWGs ceased to exist in 1979 in the aftermath of the Carter administration's disapproval of the Soviet invasion of Afghanistan. The umbrella U.S.-Soviet Space Cooperation agreement, signed in 1972 and renewed in 1977, was allowed to expire in 1982; a new agreement was reached only in 1987. During this period of cool U.S.-Soviet relations, only the JWG on Space Biology and Medicine continued to meet on an annual basis.

16. Protocol, Fifth Meeting of the Joint U.S.-U.S.S.R. Working Group on Space Biology and Medicine, Tashkent, USSR, 26 October–4 November 1974. (Joint Working Group Meeting Protocols folder, Cosmos Biosatellite files, ARC History Project archive, ARC.) Paragraph seven of the Protocol reads "The Joint Working Group considered possible areas of joint space biological investigations on available flights and agreed that such investigations are very important for the solution of fundamental biological problems and practical problems of space biology and medicine. The U.S. side accepted with satisfaction the proposal to perform joint experiments onboard the Soviet biological satellite."

organisms in space, life scientists began building the knowledge base for the universal role gravity plays across the phylogenetic scale. The collaboration also paved the way for the much more sophisticated experimentation that would later be done in the human-tended Shuttle-Spacelab missions.

The regularity of the Cosmos/Bion flights taught life scientists to constantly improve payload design protocols, ground controls, and data analysis. New collaborators from the science community were added, using new types of organisms or tissue cultures. During the 1980s, as experiments became more sophisticated and more scientists participated, the cost to NASA for participating in these missions averaged $2 million per launch, all of which was spent in the United States.

Beginning with the Cosmos 1514 flight in 1983, the Soviets began to fly two primates in addition to rats and other biological payloads. They had never flown primates before, so this first mission in 1983 was only five days long. By 1989, Cosmos 2044 flew two primates for fourteen days, together with rats and other specimens, studying microgravity and radiation effects. Twenty-nine U.S. experiments were accepted. Together with the biospecimen sharing program, over eighty investigators throughout the United States participated. The sophistication of the instrumentation and the experiments were continuously improving. As none of these missions overlapped with the first life sciences payloads on the Shuttle, hardware and instrumentation concepts destined for later Shuttle missions were first tested on the Cosmos/Bion missions.

The political upheaval that led to the breakup of the Soviet Union in 1991 did not seem to affect the course of these missions. However, by the launch of Cosmos 2299 in December 1992, it was becoming apparent that the Russians were experiencing budgetary constraints. This meant that spacecraft and flight costs would have to be shared by the United States in the future. After almost twenty years and eight Cosmos/Bion missions, the best bargain with free access to flight for life sciences had come to an end. (This essay is not the place to summarize the immense scientific and technological value to life sciences spaceflight research that was made possible by this collaboration.)[17] Suffice it to mention that from the very first U.S. participation in 1975, results in rats showed that less new bone was being formed in space. This finding, later confirmed in ground and flight studies in humans, pointed to the complexity of the action of gravity loading in maintaining healthy bone on Earth, and provided clues to potential countermeasures for astronauts on long missions.

During the early 1990s, the Life Sciences Division was also working on another set of primate experiments due to fly on a Shuttle Space Life Sciences mission. This was a collaboration with French scientists; the French Space Agency CNES was building the containers and life-support systems that were to fly in the pressurized Spacelab, carried in the Shuttle payload bay for seven days. In January 1994, this mission was canceled together

---

17. For such a summary, see J. P. Connolly, M. G. S. Kidmore, D. A. Helwig, eds., *Final Reports of the U.S. Experiments Flown on the Russian Biosatellite Cosmos 2229*, NASA TM-110439, April 1997; R. C. Mains and E. W. Gomersall, eds., *Final Reports of Monkey and Rat Experiments Flown on the Soviet Satellite Cosmos 1514*, NASA TM-88223, May 1986; J. P. Connolly, R. E. Grindeland, and R. W. Ballard, eds., *Final Reports of the U.S. Experiment Flown on the Soviet Biosatellite Cosmos 1887*, NASA TM-102254, February 1990; and J. P. Connolly, R. E. Grindeland, and R. W. Ballard, eds., *Final Reports of the U.S. Experiments Flown on the Soviet Biosatellite Cosmos 2044*, NASA TM-108802, September 1994, Volumes 1 and 2.

with other projected life sciences missions in a sweeping deficit-reducing budget-cutting activity. Its cancellation terminated the Rhesus project, which also was running into technical problems with the animal life-support system and rising costs for both NASA and CNES. NASA suggested that CNES might transfer the project's experiments to the next Cosmos/Bion missions, Bion 11 and Bion 12. For these two Bion missions, a contract was negotiated for the first time between NASA and the Russian Space Agency (RSA). These missions were planned for fourteen days with a total of four animals and would therefore yield far greater scientific return than earlier, shorter missions.

Bion 11 was launched on 24 December 1996 for a fourteen-day mission with its payload of two primates, seeds, beetles, newts, and moss. It was recovered successfully on 7 January 1997. The payload was transferred back to Moscow for postflight evaluation. On the day after landing, anesthesia was administered to the primates for routine biopsy. One of the two animals died as it recovered from anesthesia, despite revival efforts by experienced Russian veterinarians and NASA's Chief Veterinarian.

The death of this one monkey essentially signaled the end of the Cosmos/Bion program. Stirred by the relentless campaign of the People for the Ethical Treatment of Animals (PETA), U.S. Congressmen questioned why NASA was spending money to help the Russians send monkeys into space.[18] NASA Administrator Goldin convened another independent review panel to investigate the monkey's death. The panel concluded that the risk of anesthesia for surgical procedures after fourteen days of flight raised the mortality risk of animals, but it did not recommend cancellation of the next mission, Bion 12. In life sciences research, it is not unusual to lose an experimental animal despite all precautions, and the report pointed out the important implications this finding had to possible postflight emergencies in humans.[19] NASA became increasingly nervous at the attention it was receiving from animal rights activists, and, on 22 April 1997, a press release was issued "discontinuing NASA's participation in the Bion 12 mission." Russia was unable to continue on its own and canceled the mission.

LifeSat

The NASA experience with the Cosmos/Bion program was so rewarding that the Life Sciences Division began considering a program using similar U.S. robotic spacecraft. Hundreds of U.S. scientists and engineers had been initiated to the world of flight experiments and eagerly sought new opportunities. Small payloads on the Shuttle became the best substitute for dedicated life sciences missions. These were self-contained experiments designed to fit in Shuttle middeck lockers. They mostly required no intervention except occasionally making sure cameras were functioning or water bottles were full. On the other hand, having a series of successive flights like the Russian Cosmos would enable the program to provide steady access to space opportunities and increase

---

18. PETA is an organization "that operates under the simple principle that animals are not ours to eat, wear, experiment on, or use for entertainment." See *www.peta.online.org.*

19. Letter from Ronald Merrell, M. D., Yale University, to Arnauld Nicogossian, NASA, with attached report, 21 April 1997.

both the size of the involved community and the quality of the science. Planning workshops were organized by Ames, which had the firsthand experience with both Biosatellite and the Cosmos/Bion programs. Out of these discussions, LifeSat was born.

The *Challenger* accident and grounding of the Shuttle brought about another hiatus to spaceflight experiments. The Office of Space Science and Applications was receptive to such robotic missions, and an Earth-orbiting, multipurpose reuseable satellite fit in with its philosophy. It would be launched using expendable launch vehicles. The appeal was even greater because it was proposed that half of the LifeSat flights would be devoted to life sciences and the other half to material and Earth science research. Two or three pieces of standard hardware would be designed to accommodate either plants or small mammals, such as rats or mice, in interchangeable modules. The spacecraft would be capable of long-duration flights of up to forty days and fly directly into trapped radiation belts and in circular or eccentric polar orbits. It could provide artificial gravity, impose fewer restrictions than the Space Shuttle on the use of hazardous materials, such as chemical fixatives and radioisotopes, and could be used to test hardware concepts and prototypes. Documentation requirements would therefore be less cumbersome than for the Shuttle, and shorter proposal-to-launch intervals would be possible for researchers, making the mission cheaper, faster, and more appealing to the science community. LifeSat was warmly received by NASA's Life Sciences Advisory Committee. (See Document III-22.) The committee worked closely with the Committee on Space Biology and Medicine (CSBM) of the Space Studies Board throughout the process of developing the LifeSat concept. The CSBM stated in its 1987 annual report that "one of the major reasons for the slow progress in space biology and medicine over the last decade has been the lack of flight opportunities, especially in the U.S. space program. The lack of flight opportunities is all the more unfortunate because biology is largely an empirical science. Replication is needed to verify results. If significant progress is to be made by the end of the century, then the space biology and medicine scientists will need continuous access to spaceflight opportunities."[20] LifeSat was given the necessary priority and was included in the 1988 OSSA Strategic Plan as a future mission.

LifeSat also received congressional support. The House Committee on Space Science and Applications in the NASA authorization bill for fiscal year 1989 stated "a major and essential component of a balanced science program must be a provision for frequent and sustained flight opportunities for each major discipline. In addition to the Explorer and Observer mission, the Committee has directed the establishment of an Earth Probes program for Earth Science and a LifeSat program for Life Sciences."

Thus everything looked promising, but the project was short-lived. Construction responsibility for LifeSat was assigned to JSC. More and more requirements were added. The price soon escalated to over $100 million per mission, an amount judged too high by NASA management, and LifeSat was terminated in 1991. There have been no subsequent proposals for robotic spacecraft dedicated to life sciences research; that research was planned as a primary activity aboard the International Space Station.

---

20. Committee on Space Biology and Medicine, Space Studies Board, National Research Council, *A Strategy for Space Biology and Medical Science for the 1980s and 1990s* (Washington, DC: National Academy Press, 1987), p. 196.

Exobiology: Early Activities

Another aspect of the space life sciences, the bioscience program in OSS, dealt with the origins of life and the search for life elsewhere in the universe. In 1962, the Space Science Board had identified the search for extraterrestrial life, particularly on other planets in the solar system, as the highest scientific priority in the space biology program. (See Document III-6.) These were and are some of the most compelling scientific, moral, and political questions that can be posed, and only can be answered by exploring other life-supporting planetary candidates. Orr Reynolds argued in a presentation to NASA upper management in 1966 that "if life is unique to this planet and does not exist anywhere else in our universe, then the improbability of its occurrence is something like 10^{23} 'chances' for life to have developed on the Earth and nowhere else. If, on the other hand, life is encountered on some other planet in our solar system, it would appear to be a very probable event, and must have occurred many, many times in the history of our universe. As . . . Pittendrigh said, 'this question is what makes the search for life outside this Earth so fascinating. It amounts to no less a question than the fundamental one of man's place in nature."[21]

Interest in the exobiology field was remarkable and international. Programs sprang up in France, Germany, Great Britain, Holland, Japan, Spain, Sweden, and the U.S.S.R. The Committee on Space Research (COSPAR) served as a major forum for discussing this work. Perhaps the most important meetings of the U.S. academic community related to exobiology were convened by the SSB at Woods Hole, Massachusetts, during the summers of 1964 and 1965 to discuss the scientific aspects of Mars exploration. One product of the Woods Hole meetings was the formulation of a strong case for sending spacecraft to Mars for biological investigations on the planet. (See Volume V, Document 1-23.)

This recommendation was warmly received by NASA, as many planetary scientists both within and outside NASA were already developing plans and instruments for a Mars landing mission. By early 1966, the feasibility of a mission of this type, named Voyager, was being discussed. But the ambitious mission concept ran into trouble in the U.S. Congress and died in August 1967. Nevertheless, interest in a Mars landing mission continued. By 1969, NASA went back to Congress with fresh plans for a Mars mission with a new objective—the search for life—and a new name, Viking. This mission won congressional approval.

Viking: An Exobiology Mission

Preparations for executing the Viking missions resulted in the 1969 release by NASA of an "Announcement of Flight Opportunity for Viking 1973." This was the first Announcement of Opportunity (AO) for flight experimentation in life sciences and set the precedent for future ways of doing business. In contrast, flight experiments in human

---

21. Orr Reynolds, personal notes of presentation to the NASA Administrator, 13–14 December 1966.

research were selected either from unsolicited proposals or those submitted or sponsored by scientists at NASA Centers, or from scientists at other government agencies. By the end of 1969, NASA selected four scientists to develop the instruments for the search for life on Mars.

The Viking program was the largest robotic program undertaken by NASA up to that time. It ultimately cost $1 billion over a seven-year span. The missions, designated Viking 1 and 2, each included an orbiting probe and a lander. The Viking project was managed by Langley Research Center. Richard Young, Chief of the Exobiology Program in OSSA, had personally been very active in the scientific movement to go to Mars to search for evidence of life. Young was appointed Program Scientist for the Viking program. Gerry Soffen of the Jet Propulsion Laboratory (JPL) became Project Scientist. They co-chaired the Viking Steering Committee responsible for the development of an interdisciplinary team of about seventy scientists working on thirteen major experiments from the United States and abroad. One of these investigators was Chuck Klein, the Director of Space Life Sciences at Ames. Since he was a skeptic about the existence of life on Mars, he was selected to head the Viking Biology Team. JPL built the life-detection instruments and the imaging systems.

After a two-year delay because of funding problems, Viking 1 was finally launched on 20 August 1975 and Viking 2 on 9 September. The hope was that Viking 1 would land on Mars on 4 July 1976, the 200th anniversary of the Declaration of Independence. However, this was not to be. When Viking 1 reached Mars and went into orbit around the planet, it began taking pictures of the surface that showed the preselected landing site was rougher than had been predicted. Jim Martin, the Langley Mission Director, would not take the risk of crashing into a rocky landing site, and Viking 1 continued to orbit and take pictures to find a better landing site. Finally, a decision was made to land at a site known as Chryse. Viking 1 touched down at 5:12 A.M. on 20 July 1976, exactly seven years to the day after the Apollo astronauts landed on the Moon. Viking 2 landed in September.

After Earth, Mars is the most likely planet in the solar system able to support life. Viking carried a small gas exchange laboratory built around a gas chromatograph to measure respiration in the soil when treated with biological nutrients. An arm extended to collect a sample, place it in a jar, mix it with chemicals, and identify the resulting gas. The Viking program was ambitious and flawlessly executed. From the development of the experiments, assembling the spacecraft, launch from Cape Kennedy, almost a year in transit to Mars, the incredibly successful landing on Mars, the first pictures from Mars, the first sample of the Mars surface, and then the three-year operation of the experiments studying the chemistry, biology, weather, temperature, pressure, seismic activity, and photographic documentation by the two cameras, the Viking project delivered an incredible bounty.[22] (See Volume V, Document II-26.) The Viking experiments found no clear evidence of life, though questions about what it did find have excited planetary scientists for years.

Planetary Protection

One of the major concerns of biologists planning planetary missions, whether fly-by or landers, was the issue of biological contamination. The community of biologists interested in studying the possibilities of the existence of life on the Moon, Mars, and other planets was concerned from the start of the space program that Earth organisms carried by a spacecraft would contaminate these destinations, destroying any chance of discovering life forms that may have existed there. Biologist Joshua Lederburg was the first to voice these concerns. In 1959 and 1960, the Space Science Board (see Document III-5) and the International Council of Scientific Unions addressed this problem and urged the United States and the Soviet Union to adopt policies and take measures to prevent contamination of the solar system with organisms from Earth. As a result, both countries developed procedures to ensure compliance with the recommendations of national and international advisory bodies. The SSB revisited in a 1963 report the importance of sterilization of space probes specifically with regard to lunar and Mars probes.[23] NASA responded to these recommendations by developing its own policy and appointing Lawrence B. Hall as the Planetary Quarantine Officer in the Bioscience Programs Division of OSS. This policy was updated in 1963 and at regular intervals thereafter. (See Document III-12.) It was critical in planning the Apollo, Mariner, and Viking missions.

More recently, resurgent interest in human exploration and a series of robotic missions to the Martian surface have raised to a higher level of awareness the issues of spacecraft sterilization and planetary protection. The possibility of human planetary exploration is increasingly considered as the next logical step for humans in space. Such consideration creates an urgency on the part of NASA biologists studying the origins of life on other planets to press forward with Mars missions before humans step on the planet with the accompanying inherent greater difficulty of sterilization and therefore greater risk of contamination. Currently, all solar system exploration spacecraft are sterilized before launch. (See Document III-27.)

Exobiology and SETI

After the Viking missions, the Exobiology Program did pathbreaking work on organic material and water in meteorites, studied the intricate lives of some of Earth's most primitive micro-organisms, and studied fossil markers for extinct microbial life. Scientists turned to extreme environments on Earth, such as frozen lakes, hot springs, and the Antarctic for developing the methods needed to search for life away from Earth. It also took advantage of advances in computational capability at Ames and in the Silicon Valley to start planning a concerted effort to search for extraterrestrial intelligence (SETI). The logic was: if you could not go there, maybe you could use technology from Earth to scan for signals coming toward Earth.

John Billingham at Ames was a moving force behind the birth of the SETI project. In 1971, he teamed with Bernard Oliver, an electrical engineer and former Vice President for Research at Hewlett Packard and a technical expert in microwave signal processing. Together with Frank Drake, the astronomer who had first developed a

---

22. For summaries of the scientific results of the Viking missions, see *Science* (193), 27 August 1976; *Science* (194), 10 October 1976; and *Science* (194), 17 December 1976.
23. Space Science Board, "Space Probe Sterilization," 5 August 1963.

mathematical equation containing the relevant factors involved in determining the likelihood of communicative civilizations in our galaxy, they organized a summer study to plan a project for accomplishing their goals. The result of their efforts, titled Project Cyclops, "envisioned a detector consisting in its final stages of an 'orchard' of perhaps 100-meter antennas covering a total area some 10 km in diameter."[24] When Project Cyclops was presented to NASA's leadership as a possible large-scale initiative (the cost projection for the total project was in the range of $8 billion), the concept received a sympathetic reception on intellectual grounds, but was thought to be too politically risky and too uncertain of success to justify significant NASA funding. (See Documents III-16 through III-18.)

In 1976, Hans Mark, the Director of Ames, set up a SETI Program Office headed by John Billingham within the Extraterrestrial Division (previously Exobiology). The SETI program also had the support of Bruce Murray, who in 1976 became Director of JPL. Developing this Ames-JPL partnership became a major feature of NASA's formal SETI program.[25] SETI was starting to appear on NASA's screen. Billingham was developing his concept of space life sciences encompassing all aspects of the program. He called it "Life in the Universe." In June 1979, he organized a major conference by that name at ARC in which SETI's program was laid out in detail.[26] In 1980, NASA appointed a thirteen-member SETI Science Working Group (SSWG) of scientists and engineers from NASA, JPL, ARC, and other federal agencies, universities, and private industry. The SSWG assessed the feasibility and scope of a formal SETI program within NASA.[27] NASA began to fund SETI more seriously in 1981 at an average of $1.9 million per year projected over the next decade.

The scientific value of SETI was constantly challenged. Senator William Proxmire had bestowed a Golden Fleece Award on the SETI Program in 1978 and, in 1981, successfully proposed an amendment to the NASA appropriations bill deleting SETI's FY82 funding. Many scientists, including Carl Sagan, met with Proxmire to argue the merits of the science; ultimately, Proxmire agreed to no longer oppose the program.

This lesson in politics stimulated SETI proponents to become more politically active. In 1984, they formed the SETI Institute near ARC, a research nonprofit organization that could raise funds from sources in addition to NASA. Jill Tarter from the University of California, Berkeley, joined the SETI Institute as its Chief Scientist and became active on Capitol Hill. The SETI advocates encouraged university researchers to join in their work

---

24. Barney Oliver and John Billingham, *Project Cyclops: A Design Study of a System for Detecting Extraterrestrial Intelligent Life*, NASA CR114445, 1972. The report has been recently reprinted by the SETI Institute. See also Steven J. Dick, "The Search for Extraterrestrial Intelligence and the NASA High-Resolution Microwave Survey (HRMS): Historical Perspectives," *Space Science Reviews*, 64: 93–139 (1993). Among the conclusions of the Cyclops Report was that "signaling was vastly more efficient than interstellar travel."

25. NASA, *Outlook for Space*, report to the NASA Administrator, 1976, pp. 38, 145–149. This report was prepared by members from all NASA Centers to guide NASA's thinking for the next 25 years. It considered research into the origin and existence of life, microbial or intelligent, as an important part of NASA's space objectives through the end of the century.

26. John Billingham, *Life in the Universe*, NASA CP-2156, 1981.

27. Frank Drake, et al., "SETI Science Working Group Report," NASA Technical Paper 2244, 1983. The report identified scientific objectives, organizational structure, system requirements, costs, schedules, and applications.

and collaborated with Soviet scientists to obtain data from their studies. They found applications for the technologies they had developed for the program; the National Security Agency saw applicability in code breaking, the FAA was interested in using frequency analyzers developed for SETI, and an algorithm developed to pick up weak signals against a dense background was incorporated in mammography technology to improve its sensitivity.

The most significant time in getting SETI accepted as an ongoing project was in 1989, when NASA got approval from the Office of Management and Budget to use existing funds for a new start to support SETI. In January 1993, the Bush Administration requested $12 million for FY91, up from $4.2 million in FY90, to support a fully fledged Microwave Observing Project search. Renamed the High-Resolution Microwave Survey (HRMS), the program in exobiology was finally approved for funding to the tune of $11.5 million in FY92. It looked, at last, as if the search for intelligent life in the universe was becoming a reality.

Yet less than a year later, Congress terminated the SETI/HRMS program. Reasons for termination ranged from concerns over the federal deficit to the unfounded perception that associated the program with UFO encounters and "little green men." In January 1993, OSSA decided to transfer the program (together with the rest of Exobiology) from the Life Sciences Division to the Solar System Exploration Division and made it part of its TOPS (Toward Other Planetary Systems) program. The House Authorization and Appropriations Committees and Senate Authorization Committee tried to cancel it, but it was saved by Senator Jake Garn (R-Utah) and was included in the NASA appropriations bill that was sent to the Senate floor for debate. However, on 22 September 1993, Senator Richard Bryan (R-Nevada) offered a last-minute amendment to kill SETI; the full Senate concurred. A House-Senate conference approved only $1 million for program termination costs. Bryan issued a press release saying "This hopefully will be the end of Martian hunting season at the taxpayer's expense." Due to the foresight of its creators, SETI has survived and prospered with private funding in the form of the SETI Institute, while the cancellation of NASA's SETI program significantly curtailed the scope of the search.[28]

From Exobiology to Astrobiology

In 1993, an intense strategic planning activity was initiated at Ames in parallel with a Headquarters-organized Zero-Base Review to assess the activities, resources, and roles of the NASA Centers. Since the reduction in force activities of 1972, the closure of one NASA

---

28. Stephen J. Garber, "Searching for Good Science: The Cancellation of NASA's SETI Programs," *Journal of the British Interplanetary Society*, 52: 3–12 (1999) discusses the political battles that led to the cancellation of the program. Garber refers to a conversation with Kevin Kelly on 2 July 1997. Kelly was then the majority clerk and staff director for the Senate appropriations subcommittee dealing with NASA. "Kelly criticized the claim by SETI supporters that 'if this doesn't get funded by Congress, it won't get done' as being false since the SETI Institute was able to continue Project Phoenix with private funds. Project Phoenix, however, only continued the targeted search portion of NASA's SETI program; the all-sky survey was dropped for lack of money. Additionally, Kelly stated that doing ground-based astronomy is not part of NASA's prime mission, but even most casual observers would probably concede that looking for ETI fit into NASA's overall mission more appropriately than that of any other agency."

Center, the Electronics Research Lab in Massachusetts, and the attrition of the aeronautics activities, there had been a resurgence of human flights, and the Stennis and Dryden organizations, which previously had reported to one of the existing Field Centers, gained separate Center status.

Under the intense pressure of the Zero-Base Review, Ames perceived itself as very vulnerable to closure. The future of space science there looked bleak, as Ames was unlikely to receive new missions to manage. NASA's managers decided that the Agency could only support two Centers pursuing space science missions; Goddard was strong in space and Earth science missions and JPL in planetary missions. Ames had leadership in information technology and its work in human factors, and air traffic control was highly regarded. But its life sciences program had been allowed to erode. There was no longer a Space Life Sciences Directorate at the Center, and the number of life scientists had dwindled. Life sciences research had been amalgamated in the early 1980s into a Space Research Directorate consisting of space, Earth, and life sciences components. The Zero-Base Review concluded that Ames should get out of Earth science, major elements of space science, and major elements of life sciences and technology work.

Under threat of Center closure, Ames scientists repackaged their range of interests in all three science areas into a more cohesive vision very similar to Billingham's "Life in the Universe" concept. (See Document III-27.) Associate Administrator for Space Science Wes Huntress liked the idea. He suggested "Astrobiology" as an encompassing name for the broader concept. Intense planning activities and workshops with the science community followed. The Astrobiology program was born, and Ames was given the lead role. The astrobiology concept incorporated research earlier considered exobiology—the origin of life and how life evolved within Earth's harshest environments. It also encompassed the distribution, evolution of life, and future of life in the universe, and how to search for other biospheres in the solar system and beyond. The range of topics covered under the theme of "astrobiology" thus went well beyond traditional "exobiology" and included Earth science components as well as other life sciences areas such as gravitational and evolutionary biology.

Human Spaceflight and Life Sciences Research

After the Skylab program ended in 1973, there were no more flight research opportunities involving human presence in orbit in sight for life sciences. This meant that for about ten years, until research could be done on Space Shuttle flights, life sciences research, to the degree it involved humans, was essentially grounded. This was not all bad news, because there was much ground research to be done in preparation for future flight opportunities.

In addition, Europe decided to contribute a major element to the post-Apollo human spaceflight program in 1973. Europe chose to build Spacelab, which was composed of a pressurized module that fit into the Shuttle bay and additional elements that would allow other science that did not require a pressurized environment to be done there. Thus Spacelab would make it possible for the Shuttle to act as a usable orbital laboratory. It provided life scientists with a very sophisticated facility that could be used to conduct flight experiments. By being outfitted for power, water, plumbing, and other utilities as an integral part of the laboratory, Spacelab could accommodate interchangeable payload hardware and more sophisticated experimentation than had ever been carried out in space.

Although a great deal has and can be accomplished on robotic space platforms, there were some experimental procedures that needed expert human operators. The Spacelab series of Shuttle flights dedicated to life sciences research carried payload specialists rather than career astronaut mission specialists. The use of academic scientists as payload specialists, who were expected only to fly once, and who were selected to perform a set of experiments in flight related to their academic training, provided a crucial advantage over telescience.

The success of the Spacelab series of Shuttle flights dedicated to space life sciences research using payload specialists enabled a significant increase in experimental sophistication. Dedicated life sciences missions, like SLS-1 in 1991, SLS-2 in 1993, and Neurolab in 1998, proved to be immensely productive because the advantages of the prior training of the payload specialist on the mission was maximized. Unfortunately, the astronaut category of payload specialist came to an end with Neurolab and was not planned to be employed on the International Space Station (ISS).

The other advantage of human missions was when astronauts served as test subjects for the purpose of human biology research. From the very first Mercury flight, astronauts have been tested for medical evaluation purposes. The 1973 Skylab series of three missions was the first to use experimental protocols to test hypotheses derived from observations on Gemini VII, ground research, and nonhuman flight experiments regarding how the absence of gravity would affect calcium balance and the neurovestibular system. The building of the life sciences community over the nine years between Skylab and the Shuttle, through support of focused ground research and the use of the Cosmos/Bion space platform, was an important and serendipitous step in the development of hypotheses and methodologies to get the most out of the Spacelab missions dedicated to life sciences research.

The use of astronauts as research volunteers (in addition to their other duties) was not without its problems. Ethical practices governing the use of all human research subjects had to apply. Even though the pool of potential subjects is limited by the number of astronauts on a mission, precautions had to be taken to ensure they were truly volunteers and did not feel coerced in any way.[29] The limited number of Spacelab life sciences research missions and the paucity of astronaut volunteers who could be studied concurrently constrained what could be accomplished. Despite limited opportunities to study large numbers and repeat and confirm results, the remarkable consistency of the data was attributed to the fundamental importance of gravity to biology.

SLS-1, the first dedicated Life Sciences Spacelab mission, flew in June 1991. The Announcement of Opportunity (AO) for this mission had been released in 1978, but the flight was delayed after the 1986 *Challenger* accident. SLS-2 was launched on 18 October

---

29. For a discussion of this issue, see Laurie Zoloth, "The Ethics of Human Spaceflight," in Stephen J. Garber, ed., *Looking Backward, Looking Forward: Forty Years of U.S. Human Spaceflight*, NASA SP-2002-4107.

1993; among its experiments were several that required the killing in orbit of laboratory rats, an issue of high political sensitivity. (See Document III- 26.) The AO for Neurolab was released in 1993; the mission flew in May 1998. (See Document III-24.) Life sciences experiments also flew on another eight multidisciplinary Spacelab missions. Delays between selection of experiments and flight were very costly, but more important was a major source of frustration among better scientists, who, in response, could have turned away from the program. It also was tragic for graduate students who hoped to get degrees on research results from flight experiments.

Yet the history of Spacelab was impressive. NASA and ESA signed the Memorandum of Understanding for Spacelab development in 1973. Eleven European nations came together to develop a pressurized laboratory at the same time as the Shuttle was moving forward in its development. The space station was decades away, and there was a pressing need to introduce laboratory capabilities on the Shuttle. The first Spacelab mission, STS-9, flew in November 1983, and the last, Neurolab, in April 1998. Elements of the Spacelab system, including unpressurized pallets as well as the pressurized laboratory, flew thirty-six times in the seventeen years from 1981 to 1998, with a variety of international hardware spanning all space science disciplines and involving investigators and crew from around the world. The Spacelab program paved the way for international space research collaboration aboard the International Space Station (ISS). When the Shuttle-Mir program started in 1995, Mir had no life sciences research capabilities. A Shuttle with Spacelab aboard docked to Mir, carrying instruments and allowing some experiments to be conducted. The scientific accomplishments of Spacelab research were recognized in 1999 in a special forum held at the National Academy of Sciences.[30]

Mir was later outfitted by NASA with limited life sciences research capability. Because it provided U.S. experimenters extended duration in space for the first time, some research was carried out. For instance, it was demonstrated that a plant could go from seed to seed, while circadian rhythms away from Earth's gravity were studied in beetles.

Together with a complementary ground research program, Spacelab and Mir findings opened up a Pandora's box of expectations in the life sciences community. The International Space Station was to be the laboratory every space life sciences researcher could have hoped for, with its proposed suite of state-of-the-art facilities, including an onboard centrifuge providing an Earth-like one-gravity control with all it could reveal about gravity thresholds and the evolution of gravity sensors.

But it was not to be. These hopes for the space station have been dampened by mismanagement, budgetary woes, hardware cutbacks and delays, reduced crew size and crew time available for research, missing skills, and onboard capabilities. Interim Shuttle flights providing the occasional small payload opportunity also receded—victim to related budget setbacks.

Yet another hiatus in flight opportunities loomed. The International Space Station could not be justified on space life sciences research alone. Just as life sciences research was not the only or even the primary beneficiary of Spacelab, equally it could not rely on ISS alone to accomplish a flight research program that responded to the

---

30. J. Edmond, ed., *The Spacelab Accomplishments Forum*, NASA CP-2000-210332, 2000.

enthusiasm generated by the Neurolab model in the broader, sophisticated science community developed and nurtured during the past ten years. With the completion of ISS receding into the future, plants and animals such as mice and rats could not be accommodated until 2008 at the earliest. It seemed that a complementary robotic and human-tended research capability was the only way to build on the exciting potential of the space life sciences.

Other Developments

The relationship of the life sciences program to the rest of NASA has always been uneasy. Attempts to establish space life sciences as a legitimate research area with the broader life sciences community have met with the expected resistance of investigators having to deal with any new principle, in this instance, gravity. There have been fundamental cultural differences between its experimental orientation and the space scientists' event-science approaches. Settling comfortably within an engineering agency has been impossible. Unlike other science where the "what" and the "why" drives how science was done and what space platforms it needed, life sciences has usually been subordinate to engineering agency priorities. In addition, it has constantly had to justify to engineers the need for an integrated ground and flight research program whose sole index of success was survival of the specimen, or to the medical operations community who could not see how research-derived knowledge helped them to do their job. Notwithstanding this environment, or maybe because of its difficulties, life sciences leaders have creatively sought out methods and partnerships to promote their goals.

International Collaboration

Following on David Winter's active support in the 1970s of the U.S.-Soviet collaboration in the Cosmos/Bion program, Arnauld Nicogossian expanded international collaboration by forming the International Life Sciences Strategic Working Group (ILSSWG) in 1989 to bring in non-Soviet space agencies who were collaborating on Spacelab missions. In the 1990s, the group developed an approach to selecting jointly solicited peer-reviewed flight experiments and established hardware-sharing mechanisms for all available space platforms. It became crucial to the international scientific collaboration on the ISS and ensured cost savings to the United States through facilities provided by international partners. Most bilateral life sciences Joint Working Groups (JWGs) were abolished to force multilateral collaboration rather than bilateral agreements with individual space agencies. The JWG with Russia continued on a separate basis, since the Russians stated their preference for bilateral collaboration, which in many instances involved exchange of funds.

Interagency Collaboration

In the 1990s, partnerships with other U.S. agencies were fostered. An agreement was signed in 1992 between Administrator Goldin and Bernadine Healy, Director of the National Institutes of Health (NIH), the result of many years of discussion between Life Sciences Division staff and NIH directors. This agreement led to joint funding of

NASA Specialized Centers of Research and Training (NSCORT) for research in the Vestibular System. An agreement in 1993 with the National Science Foundation (NSF) led to the joint funding of an NSCORT devoted to plant research. An agreement with the Department of Energy (DOE) in 1997 supported the construction of a heavy ion beam facility at Brookhaven National Laboratory which made possible much needed research on the effects of space radiation in living organisms, as well as the testing of shielding materials.

NASA's Life Sciences Division and NIH Institutes, as well as NSF, the Office of Naval Research, and international agencies, joined forces to fly Neurolab as NASA's contribution to the congressionally mandated "Decade of the Brain" in the 1990s. The 1998 Neurolab Spacelab mission was devoted to research on the nervous system. Experiments were solicited and selected jointly. NIH managed the international scientific review. International partners provided important equipment that made the research possible. They also supported investigators and research from their countries. The U.S. investigators were supported by NIH, NSF, or ONR for ground research and NASA for flight experiment components. Experience gained on previous dedicated Spacelab missions enabled NASA to fly a highly sophisticated set of life sciences experiments on Neurolab. For instance, on SLS-1, the rats flown were only studied before and after the seven-day flight. On SLS-2, rats flown for fourteen days were studied before, during, and after the flight because of the surgical expertise of the payload and mission specialists. The ability to evaluate the effects of microgravity during spaceflight was of fundamental significance to biology because postflight data were confounded by exposure of animals, humans, and other life forms to hypergravity forces during reentry. One significant finding was that the nervous system, especially the inner ear vestibular gravity-sensing system, was much more plastic than previously known. It showed adaptive structural changes to both microgravity and hypergravity. For the first time on a (sixteen-day) Neurolab mission, the electrical activity of sympathetic nerves in the legs of crewmembers were recorded directly before, during, and after flight. This was done to determine whether circulation and muscle changes were caused by reduced electrical activity in the absence of gravity loading. Nerve electrical activity also was recorded directly from the brains of rats and the vestibular nerves of fish. The number of brain cells of mouse embryos was greater in space than their counterparts on the ground. Muscles of developing rats were smaller, weaker, and less able to support body weight after spaceflight.

This productive collaboration with NIH paved the way on 28 October 1998 for the successful return to space of 1962 Mercury astronaut Senator John Glenn (D-Ohio) on STS-95. Glenn was 77 at the time of his launch. NASA and the National Institute on Aging jointly selected the peer-reviewed flight experiments that would best address the effects of spaceflight on older humans; these included sleep, muscle, cardiovascular, and balance studies. Much criticism was levied by some in the science community that no significance could be placed on just one test subject if he differed from his younger crewmates. In fact, Glenn's response to spaceflight was no different from those of younger astronauts, suggesting that state of health and activity, rather than age, were more relevant to one's adaptability to space. His flight also highlighted the value of space research to understanding age-related body changes.

After years of skepticism of the need or value of space life sciences research, NIH put out its own Program Announcement in 1999 for space-related ground research.[31] From NASA's perspective, this announcement introduced to the 20,000 or so NIH-supported investigators the potential research advantages that the space environment had to offer and helped develop the research community for the ISS.

Strategies for Research—Science Policy

Over the years, excellent strategic plans and research strategies have been developed for life sciences. They have not been fully implemented because requests for additional research funding were always hard to defend, and the organizational instability of program elements made it even harder to sustain a stable strategy. When Arnauld Nicogossian was Director of the Life Sciences Division under OSSA, he asked Frederick C. Robbins to convene a distinguished panel and review the life sciences yet again. This was probably the most comprehensive review yet, including science, management, and Center issues. The NASA Life Sciences Strategic Planning Committee Report in June 1988 established the basis for a strategic plan for life sciences. It recommended a significant increase in ground-based and flight research efforts in order for Life Sciences Division to meet its obligations to ongoing and future programs. (See Document III-23.) Nicogossian used this report as the basis of his successful advocacy of an increase in the life sciences budget. He saw the success OSSA had by organizing research around missions or facilities. He proposed a series of dedicated Spacelab missions, LifeSat, and a suite of sophisticated hardware for the then-planned Space Station Freedom, including an onboard centrifuge. Nicogossian was successful within NASA, where so many others had failed; his interactions with the external science community were not as productive. Then, the budget cuts of FY94 eliminated SLS-3, 5, 6, and 7, in essence gutting Nicogossian's plans.

No sooner was OLMSA formed in 1993 than a memo from Senator Barbara Mikulski (D-Maryland), then Chair of the Appropriations Subcommittee that oversaw NASA's budget, arrived at NASA. It reflected concerns of the life sciences research community and criticized the quality of the scientific peer-review process and management of grants by the Life Sciences Division. So significant was the concern of the scientific community that grants were managed at NASA Field Centers and not at NASA Headquarters. The memo requested reports from NASA every three months on changes and progress.

In response, a thorough review and revision of policy and process was put into effect. Grants management was returned to NASA Headquarters. An annual NASA Research Announcement (NRA) reflecting the advice of the CSBM research strategy reports was put into place. Policies on conflict of interest of peer reviewers were enforced, and the segregation of intramural and extramural budgets were abolished so that scientists from both communities could compete openly and fairly in the same arena. (See Document III-25.)

---

31. This announcement stated that "The purpose of this PA is to stimulate ground-based research on basic, applied, and clinical biomedical and behavioral problems that are relevant to human space flight or that could use the space environment as a laboratory. Although none of the research supported under this initiative would be conducted in space, it is anticipated that it would form a basis for future competitively reviewed studies which could be conducted on the International Space Station, or other space flight opportunities, by skilled on-board specialists." More information is available at: *http://grants.nih.gov/grants/guide/pa-files/PA-00-088.html*

Science Institutes

In 1995, following on the Zero-Base Review, one suggested organizational change was the transfer of scientists out of NASA by the formation of "privatized" institutes for focused science activities. An Agencywide committee, led by Al Diaz of OSS, conducted a benchmarking review and comprehensive analysis of a variety of existing science institutes around the country to develop approaches to the institute form that would best suit NASA. Recommendations and criteria were presented to France Cordova, Chief Scientist, and Administrator Goldin. The recommendations included authorities and responsibilities, methods of solicitation and selection, and emphasized the importance of the quality of the Director to the success of the institute.

However, not all institutes that were actually established followed the same process. The Lewis Research Center Institute, the first to be formed in March 1997, was the result of an unsolicited proposal. The second institute, established in May 1997, was the National Space Biomedical Research Institute (NSBRI), a multiuniversity "virtual" institute with its main base at Baylor University. It was solicited, competed, and peer reviewed before selection. Neither the NSBRI nor the Lewis Institute required the transfer of civil servants. Others, like one for the biological science at Ames, hit a roadblock over the problem of moving civil servants into a private institute. Congress balked at passing legislation to ease post-employment restrictions for NASA employees or allow them to transfer their pensions. Without these provisions, universities were reluctant to take on such a big task. The SSB recommended caution to the space science community, and the momentum to form institutes subsided.[32]

The 1996 discovery of hints of fossils in a meteorite from Mars and photos from the Galileo orbiter showing that Jupiter's moon Europa could have ice or liquid water led NASA Administrator Goldin to name Ames as the Lead Center in Astrobiology. He also asked Ames to consider forming an institute to promote collaboration with the broader scientific community. The result was the creation in 1998 of the NASA Astrobiology Institute. This "virtual" institute was very different from the others because it had its institutional center within NASA, and the director was a NASA employee. It supported collaborative science, providing an information link between scientists, mission planners, and technologists while explaining field-related questions to the educational community and the public. This was an ambitious experiment of managing science and seemed particularly relevant to interdisciplinary astrobiology interests. The institute has leveraged the participation of 700 members, including fifteen Principal Investigators and ninety graduate students. Nobel Laureate Baruch Blumberg was its first Director. The National Astrobiology Institute showed evidence of tremendous value added and promised to be an interesting model for managing science within NASA. Of the three institutes, it has been the most successful in recruiting talent and resources internationally.

Unlike the SETI Institute, none of these three new institutes was truly private, since they were almost wholly funded by NASA. Neither cost savings nor the transfer of NASA personnel to the private sector have materialized as a result of the science institute initiative.

---

32. Space Studies Board, *Managing the Space Sciences*, 1995.

The Contributions of Dan Goldin

The new emphasis on biology in NASA would not have happened were it not for Daniel Goldin's active and unprecedented advocacy. Goldin, a visionary, stood up for his intellectual investment, for his constant desire to be in the forefront of knowledge, and for his enthusiasm. He was quick to recognize the impact the revolution in biological research worldwide was having and the leaps NASA life sciences could make in health applications and biology-based technologies. He wanted NASA to be in the vanguard of that revolution.

In 1999, he brought NSF biologist Kathie Olsen to NASA as Chief Scientist and forged partnerships with the National Cancer Institute. In June of 2000, he asked Olsen to become Acting Associate Administrator of OLMSA when Nicogossian left to assume full-time Chief Medical Officer responsibilities. A new NASA Strategic Plan was in the works, and Olsen persuaded Goldin and NASA upper management to recognize OLMSA as a separate Enterprise. The name was changed to Office of Biological and Physical Research (OBPR). On 30 August 2000, when the author of this essay retired, the Life Sciences Division was divided into a Fundamental Space Biology Division and a Bioastronautics Research Division. Despite this significant upgrading in status, life sciences programs were still to be found in many other offices in NASA.

The Future

From the recommendations in 1958 that space life sciences should be part of NASA rather than that of other agencies, recommendations for its organizational status and its management have been remarkably consistent. Essentially these included consolidation of life sciences efforts in one program office with appropriate management and budget authority reporting directly to the Administrator. Apart from Administrator Glennan's actions in 1960, at no time in NASA's history has the sum of these recommendations been followed. Each reorganization has responded only in part, leaving the new life sciences organizational structure incomplete or handicapped. Unlike other NASA science organizations, it had neither authority nor responsibility for planning, developing, and managing its assets, whether flight platforms or Centers. As a result, its flight program has been opportunistic rather than systematic. Yet the life sciences program has led the way in interagency and international partnerships and leveraging. Recently, with ever-improving scientific standards, excellent external advice, and effective outreach through education, perceptions of the life sciences community among Congress and the public have been turned around and have become increasingly positive. (See Document III-29.)

This is a great time in the history of the world to be a biologist. It is also the best opportunity NASA has had to capitalize on its own life sciences research potential.

Document III-1

Document Title: **J. W. Joyce, Head, Office for the International Geophysical Year, National Science Foundation, ONR Plans for a Biological Experiment in Early U.S. Satellite Flights, 4 November 1957.**

Source: **National Archives and Record Administration, Washington, D.C.**

[no page number]
[all pages stamped "ADMINISTRATIVE-CONFIDENTIAL"]

This is a record of a meeting in the weeks following the launch of Sputnik 1 to discuss the possibility of flying a life sciences experiment on one of the initial Vanguard satellites that at that point were the only approved U.S. scientific satellite efforts. The National Science Foundation was overseeing the experiments to be flown on Vanguard satellites, which were being developed at the Office of Naval Research. Notable here is the concern that there be no possibility of linking a simple life sciences experiment in space with preparations for biological warfare.

DIARY NOTE

November 4, 1957

Subject: ONR Plans for a Biological Experiment in Early U.S. Satellite Flights

On Friday, November 1, I received a call from Captain Phoebus of the Office of Naval Research advising me of a meeting to be held the following morning, Saturday, November 2, at 10 o'clock in Room 1515 T-3 Building. The announced purpose of this meeting was to consider possible release of information on planned biological experiments that might be undertaken with the six-inch test satellites and later on with the 20-inch spheres.

In response to my queries, Captain Phoebus indicated that contact had already been made with Reid and Odishaw at the USNC-IGY. I suggested also that he advise Hirsch of the meeting, and we agreed that I would mention it to Hirsch myself.

The meeting was held as scheduled, and Sprugel, Krats, and I attended from the National Science Foundation.

There were about 30 people present with Captain Metzger acting as the master of ceremonies, since Admiral Bennett had been unexpectedly detained on some other work. Metzger introduced Captain Phoebus, who proceeded to read a proposed press release announcing the initiation by the Office of Naval Research of a step-by-step program which would eventually lead to space flight by human beings. Stripped of the fancy language, the proposal revealed that as a first step in this admittedly difficult undertaking small samples of simple cellular organisms would be sent up under reasonably controlled conditions and environment to determine their behavior.

Briefly, Captain Phoebus, and later Dr. Sidney Galler, indicated that a group of top biophysicists and biologists had met on an ad hoc basis the previous week to explore the problem. The numbers of substances that might be utilized had been narrowed down to a relatively few, including certain yeasts and certain bacteria. The idea was to use a small cylinder of about two inches in length, one inch in diameter, which would contain a sample of the organisms and which would provide as nearly as possible an environment in which they might be able to survive and actually reproduce. This, of course, implied reasonable temperature excursions, pressure control, and protection against ultra-violet light and some components of X-rays. There would undoubtedly be some X-ray exposure, to say nothing of cosmic ray exposure which cannot be controlled. On the ground at the same that the sample was in flight a control batch of the organisms would be maintained under carefully regulated conditions, and comparisons would then be made between the earth-bound sample and the material in the satellite.

Variations, if they occurred, would then be traced down to see what their causes might have been. Properties that might have been telemetered from the satellite would include (a) rate of CO_2 production, (b) resistance- [2] measured accumulation of metabolic products, (c) changes in optical properties of the nutrient medium. This latter measurement would tie to (b) to some extent.

The reported advantages of using simple cells, in addition to a significant reduction in the number of variables that would apply for a more complex organism, are that (a) several generations can be studied during a single flight, and (b) reasonably good statistical data can be obtained.

The total weight of the biological sample with a simple electronics modulator to convert sensing data to electronic signals for telemetering purposes is estimated at three and a half ounces. One of the ground rules which was clearly stated was that this experiment would in no way interfere with or compromise the geophysical experiments already planned for both the six-inch sphere and the 20-inch satellite. The only penalty might be a slightly reduced battery life (perhaps a day in some three-weeks life for the six-inch ball was a quantitative figure mentioned.)

The VANGUARD contingent (Hagen, Newell, et al) have apparently reviewed the proposal carefully and are convinced that this is a fundamental and worthwhile, and that they are therefore quite willing to see it put into the spheres if possible.

Odishaw, Reid, and later on Porter were present from the USNC-IGY and there was no basic disagreement as far as their reactions were concerned provided, of course, that the geophysical work was not compromised. Odishaw asked whether there might not be some other geophysical experiment that could be flown which would match the three-and-a-half ounce requirement of the biological experiment. Newell stated that there was no such experiment available at this time.

Several points were raised in the ensuing discussion, including the danger propaganda-wise of being tied in with biological warfare if it was announced that bacteria were being flown in the sphere. While everyone recognized the untenability of such a claim on the basis of scientific and technological considerations, it was equally clear that propaganda-wise it can be used in spite of this to influence certain uninformed populations. For this reason it was urged that yeast be used as the biological material since this, of course, could be associated with such everyday concepts as food and drink (bread, beer, etc.).

Although the meeting was supposed to be primarily for the purpose of informing the various interested groups about the planned activities, it was decided that the proposed press release was far too lengthy and grandiose. It was further decided that nothing should be said until after November 7 when more is known of Soviet activities and intentions. (The November 3rd release on Sputnik II answers this very effectively.) When any statement is made there will be a simple unaffected announcement in a low key and it may include all of the experiments that will be undertaken by the particular satellite involved. Every effort will be made to use yeast instead of bacteria and an active liaison with the USNC-IGY Satellite Panel will be started immediately. This latter step will consist of a briefing of the [3] internal instrumentation group on Tuesday, November 5, and a further briefing of the entire Earth Satellite Panel at its Wednesday meeting, November 6.

In an effort to circumvent adverse propaganda which may be attempted even if yeast is the material utilized, it was suggested that arrangements be made to bring in suitable world-renowned biologists from countries such as India who might take an active part in the planning and certainly would be used in the inspection of the proposed experiment in order to confirm that it is, in fact, a basic scientific effort, in no way tied to warfare, bacteriological or otherwise.

Another item briefly considered was the cost of this experiment, and the indications were that ONR would pick up the tab.

Following the meeting on Saturday, I gave a brief oral report to Dr. Waterman.

J. W. Joyce

cc: Director
     Odishaw
     Carothers [sic]
     Wilson
     Sprugel

Document III-2

Document Title: Objectives, Problems, Program, and Budget for the NASA Biosciences Program, 3 March 1959.

Source: NASA Historical Reference Collection, History Office, NASA Headquarters, Washington, D.C.

Within six months of its opening for business, NASA had developed initial plans for a life sciences research program. It is outlined in this document.

[no page number]

NATIONAL AERONAUTICS AND SPACE ADMINISTRATION OFFICE OF SPACE
SCIENCES BIOSCIENCES PROGRAM

30 March 1959

OBJECTIVES

To determine the effects on living terrestrial organisms of conditions in the earth's upper atmosphere, in space, and in other planetary atmospheres. To determine the effects of flight through space on living organisms. To investigate the existence of life throughout the solar system, and to study extraterrestrial life forms in detail.

PROBLEMS

The opportunity now exists to conduct fundamental life sciences research in satellites and space probes. In such research conditions in space and conditions of flight through space enter as new experimental variables. Principal interest lies in the components of the space and space flight environment that are irrevocably different from the terrestrial, such as radiation and the altered gravity state. The former is qualitatively and quantitatively different from terrestrial radiation and looms as a far more important problem than originally thought. The latter offers the possibility of elucidating biological processes known to be affected by weight or apposition of one part against another, such as [2] cell division and organogenesis. These and other areas of study could be carried out in satellites or probes from which specimens under study could be recovered.

More specifically, fundamental experiments on living organisms (from simple cell systems to the monkey) in the satellite environment or in other types of space vehicles and stations should be conducted along the following general lines:

Animal navigation and orientation
Mitosis and embryology
Plant morphogenesis
Geotropic response of plants to altered gravitational fields
Biological rhythms and cycles
Altered gravitational effects on blood circulation in animals

Vestibular physiology
Neuropharmacology–effects of tranquilizers, motion sickness drugs, etc.
Effects of cosmic radiation
Tolerance limits for combined stresses
Physiological and psychological deterioration
Effects of acceleration and deceleration

There has been considerable speculation about the possibility of the existence of extraterrestrial life forms. The [3] debate ranges from whether or not life exists on Mars and Venus, and if so in what form, to whether or not spores and similar life forms can survive the rigors of interplanetary space. Investigations of these topics must for the most part await observations which can be made on direct samples and studied in space laboratories of the future. It is conceivable, however, that with proper and specialized instrumentation and remote sampling techniques it may be possible to obtain evidence of primitive or complex organic material without the presence of a scientist and a laboratory in the sample area. Elaborate instrumentation and the recovery of records, or advanced telemetering techniques could provide information about the nature of biochemicals on the planets.

PROGRAM

The long range program in biosciences will involve the use of two satellites or space probes per year devoted primarily to biosciences research. In addition experiments will be conducted in other vehicles in which space may be made available for biosciences instrumentation. In general, until manned orbiting laboratories become available, biosciences experiments will. require the recovery of exposed specimens for further study in the laboratory.

Immediate steps in the biosciences program will be: [4]
(a) A series of group discussions with leading bioscientists to firm up the important initial lines of attack.
(b) The establishment of one or more Space Sciences Working Groups in the biosciences area, to work on the preparation of satellite and space probe payloads for biosciences experiments.
(c) The establishment of a Biosciences Branch of modest size in the Beltsville Space Center.

Throughout the country there is a great potential of scientific capability for biosciences space research. At the present time, however, the attention being given to the area is very limited. The discussion groups proposed under (a) will stimulate interest in the field, will acquaint bioscientists with the possibilities that exist for fundamental biosciences research in space, will make known to them the existence of NASA support for such research, and will bring their collective thinking to bear upon the many problems in the area.

The Working Groups to be established as stated in (b) will undertake the experimentation determined upon as a result of the discussion group meetings.

Most of the space research in biosciences will be done by industrial, university, private, and other government teams already in existence. Nevertheless, it is essential for NASA [5] to have a comprehensive, even though modest, in-house program in biosciences. Such an activity is necessary to attract to NASA the competent scientists whose leadership is needed to ensure that the NASA biosciences program is sound and effective.

The Headquarters staff needed in support of the bio-sciences program will consist of 3 or 4 scientists in the Office of the Assistant Director for Space Sciences.

BUDGET

The annual budget for the biosciences program would be roughly as follows:

	M$
Beltsville group (incl. sals)	1
Two satellites	
(a) vehicles	5
(b) payloads	2
(c) launch, track, telemeter, reduce data	1
Supporting research and development, partici-pation in other satellites, etc.	2
TOTAL	11

Document III-3

Document Title: G. (George) M. Low, "Biological Payloads and Manned Space Flight," 16 November 1959.

Source: NASA Historical Reference Collection, History Office, NASA Headquarters, Washington, D.C.

George Low in 1959 was in charge of planning the U.S. human spaceflight effort at NASA Headquarters. This memorandum reviews both early U.S. and Soviet plans for launching animals and humans into space and concludes that achieving human spaceflight was primarily an engineering challenge; there seemed to be no major life sciences obstacles to be overcome.

[no page number]

[marked "Confidential"]

G. M. Low
16 November 1959

BIOLOGICAL PAYLOADS AND MANNED SPACE FLIGHT

TABLE OF CONTENTS

[no page number]

BIOLOGICAL PAYLOADS AND MANNED SPACE FLIGHT
by
G. M. Low

I. UNITED STATES PROGRAM

A. PAST ACHIEVEMENTS

Biologic research in this country dates back to the high altitude rocket experiments performed from 1948 to 1952. In these experiments 5 V-2's and 3 Aerobee rockets were used to send animal (monkeys and mice) payloads to altitudes approaching 100 miles. Although recovery was attempted in all cases, only the monkeys and mice in the last two Aerobee flights were recovered successfully. An average of 2 1/2 minutes of weightlessness was achieved, during which arterial and venous pressures, EKG, pulse, and respiration values were recorded.

No further biological flights were conducted, in this country, until 1958. During the spring and summer of that year, three separate Thor-Able nosecones were launched, containing mice as a payload. Peak altitudes in excess of 1000 miles were reached during a flight profile that included 38 minutes of weightlessness. Recovery was not accomplished.

In December of 1958, a South American squirrel monkey was launched along a ballistic trajectory in the nose cone of a Jupiter missile. During the flight, heart sounds, EKG, respiratory rate, and skin temperature were recorded. An altitude of 300 miles was reached, with a weightless period of 13 minutes. The attempted recovery failed.

In May 1959, two monkeys, Able and Baker, were flown in another Jupiter nose cone. Able was a rhesus monkey, while Baker was a squirrel monkey. Both were recovered safely after having reached an altitude of 300 miles, and flown in the weightless state for about [2] 4 1/2 minutes. (Able died later, but not as a result of the space flight.) No adverse reactions to space flight were observed.

In June 1959, an attempt was made to launch four mice into an earth orbit, using the Discoverer vehicle. The orbital attempt failed, but 13 minutes of data were received, including 8 minutes of weightlessness. Heart rate, EKG, and respiration, all appeared to be normal. No recovery was made.

The following table summarizes the results of U. S. biological space flights to date.

DATE	BOOSTER	ANIMAL	ALTITUDE	WEIGHTLESS PERIOD	RECOVERY
1948-50	V-2 (5)	Monkeys and mice	100 mi.	4-5 min.	No
1950-52	Aerobee (3)	Monkeys and mice	40 mi.	2-3 min.	Yes
1958	Thor-Able (3)	Mice	1000 mi.	38 min.	No
1958	Jupiter	Monkey	300 mi.	13 min.	No
1959	Jupiter	Monkeys	300 mi.	4 1/2 min.	Yes
1959	Discoverer	Mice	140 mi.	8 min.	No

B. PRESENT PROGRAMS:

1. X-15 PROGRAM - The X-15 airplane is designed to study the problems of piloted flight at hypersonic speeds. Although the X-15 is an airplane, and not a spaceship, its flight trajectory will leave the earth's atmosphere for short periods of time. It will have the capability of reaching altitudes approaching 60 miles, or speeds of 6600 feet per

second. The initial phases of flight testing are now underway, and maximum performance flights are scheduled for the fall of 1960.[3]

2. PROJECT DISCOVERER - The bioastronautics phase of the Discoverer project has the following objectives:

(a) Recovery of living specimen from satellite flight;
(b) Study of the psychophysiologic response of animals to launch, orbit and recovery conditions;
(c) Gain confidence and build experience in biomedical ground support and recovery techniques.

To date, the Discoverer project has been plagued with engineering difficulties, and none of the instrumented payloads have yet been recovered. Present plans call for an animal flight immediately after the first successful recovery of an instrumented payload. Assuming a successful recovery late in November 1959 (the next scheduled flight), an animal flight could take place in December. The second, and only other animal flight, is scheduled for one or two months after the first.

The flights will have the following characteristics:

Orbit altitude	140 mi.
Orbit period	93 min.
Time in orbit	27 hrs.
Launch azimuth	173°

The animals used in these flights will be rhesus monkeys, trained to perform tasks during the flights; the usual biomedical data will be taken.

The following summary table includes the earliest possible flights scheduled for Discoverer:

DATE	BOOSTER	ANIMAL	ALTITUDE	WEIGHTLESS PERIOD	RECOVERY
Dec 59	Thor-Agena	Rhesus	140 mi.	27 hrs.	Yes
Jan 60	Thor-Agena	Rhesus	140 mi.	27 hrs.	Yes

[4]

3. PROJECT MERCURY - Project Mercury has the objectives of achieving manned orbital flight and successful recovery at the earliest practicable date; and of studying man's capabilities in a space environment. In order to accomplish these objectives, a step-wise buildup of capability is employed. Thus, for example, full-scale models of the Mercury capsule are first launched along ballistic trajectories, in order to determine heating and stability characteristics. Later on, production capsules will be launched along ballistic paths, before orbital flight is attempted.

Some of the preliminary Mercury flights will carry live payloads - monkeys, chimpanzees and men - in order to qualify the systems, obtain physiological data, and, in the case of manned ballistic flights, provide pilot training.

When the Mercury program was established, it was recognized that the engineering problems, and not the biomedical problems, would be the difficult ones. The flight schedule, therefore, reflects a desire to obtain engineering data at an early date, but no attempt was made to perform either animal or manned ballistic flights at the earliest possible date. For example, an important milestone in Project Mercury was the "Big Joe" Atlas-launched ballistic flight of a full-scale Mercury capsule model in September 1959. This payload, which was recovered, could easily have contained a monkey or chimpanzee; yet no animal was included because it was believed that an animal in this flight would not have contributed to the ultimate objectives of Project Mercury.

The manned ballistic flights that are apart of Mercury were also programmed only as a step toward orbital flight, and not as an end in itself. Manned ballistic flights using the Redstone booster, would, we believe, have been accomplished even before now if these flights had been programmed as an end result, and were not to be performed with hardware capable of orbital flight and post-orbital recovery.

[5]
The present target dates for animal and manned flights of Project Mercury are as follows:

DATE	BOOSTER	PAYLOAD	ALTITUDE	RANGE	WEIGHTLESS PERIOD
Dec 59	Little Joe	Monkey	75 mi.	200 mi.	4 min.
Mar 60	Little Joe	Chimpanzee	9 mi.	8 mi.	1 min.
Jul 60	Redstone	Chimpanzee	125 mi.	200 mi.	5 1/2 min.
Aug 60	Redstone	MAN	125 mi.	200 mi.	5 1/2 min.
Sep 60	Atlas	Chimpanzee	120 mi.	4500 mi.	15 min.
Jan 61	Atlas	Chimpanzee	120 mi.	3-orbit	4 1/2 hours
Mar 61	Atlas	MAN	120 mi.	3-orbit	4 1/2 hours

All payloads will be recovered.

4. DYNA-SOAR PROGRAM - Although the Dyna-Soar vehicle is not an orbital vehicle, its flight profile will encompass all the problems of a launching into orbit, and recovery from orbit. Whereas in Mercury the pilot's functions are primarily of a backup nature, the Dyna-Soar concept relies entirely on the pilot's ability to fly the vehicle. Maximum performance flights are scheduled for the mid-1960's.

C. FUTURE PLANS:

It is difficult to formulate precise plans for the future at this time, when the first manned space vehicle has not yet flown. But certain general plans regarding future manned space missions can be specified.

In the immediate future, certain extensions of the Mercury program should be pursued. In particular, a controllable Mercury vehicle, capable of landing within a small preselected area, should be constructed.

The next major milestone, after Mercury, is manned circumlunar flight. Such a mission should again follow a step-wise program, including unmanned recoverable flights in highly elliptic orbits to develop both [6] aerodynamic and guidance concepts; and long-term flights to develop appropriate life-support systems.

The schedule for the final accomplishment of manned circumlunar flight will be dictated not by the payload, but by the availability of the booster vehicle. The Saturn booster, with appropriate upper stages, should be capable of performing this mission. An immediate and vigorous development of upper stages for Saturn is required to perform manned flight around the moon within a reasonable period of tune.

II. SOVIET PROGRAM

A. PAST ACHIEVEMENTS:

Between 1949 and 1957 Russian scientists launched twenty high altitude rockets, each containing two dogs. Initial flights (1949-1952) were to altitudes of 60 miles, with correspondingly short periods of weightlessness. Later flights were launched to increasingly higher altitudes, with three flights in 1957 reaching an altitude of 130 miles, and sustaining a weightless period of six minutes.

Apparently recovery attempts were made in all flights, but success was only achieved in the latter flights of each series. Two different recovery methods were used. In 11 flights, attempts were made to recover the payload intact, with the dogs descending to earth in a sealed cabin. In the nine remaining flights (1953-1956), one of the two dogs was ejected from the capsule at 50-55 miles altitude and descended from there with a double parachute system; the second dog was ejected at 24 miles altitude, fell freely to 20,000 feet, and then descended by parachute. In these experiments the dogs wore pressure suits.

The usual biomedical measurements, including EKG, pulse, blood pressure, respiration and body temperature were made. The general conclusion was reached that there were no seriously adverse effects that could be attributed to the environment.

On November 3, 1957, the dog Laika was launched into orbit aboard Sputnik 2. The life-support capsule contained, in addition to the 13.2 pound Laika, an air regeneration system, temperature regulating system, food troughs, [7] sanitary arrangements and medical instrumentation. Measurements taken included EKG respiration rate, arterial pressure, and motor behavior. The basic results of the flight were that, although there were some effects on the heart and respiration during launch, these returned to normal during weightless flight. There has been some speculation that the oxygen generating system was of a chemical type, not yet fully developed in this country.

During the summer of 1959, the Russians performed two more ballistic flights with animal payloads. In one of these, two dogs were used, and in the second, two dogs and a rabbit. The announced weight of the payload was of the order of 4000 pounds. Peak altitudes have been estimated to be between 130 and 500 miles. It was announced that the animals were recovered alive.

The Soviet achievements are summarized in the following table:

DATE	NO. of FLIGHTS	ANIMALS	ALTITUDE	WEIGHTLESS PERIOD	RECOVERY
1949	1	Dogs	60	2-3 min.	Sometimes
1950	1	Dogs	60	2-3 min.	Sometimes
1951	1	Dogs	60	2-3 min.	Sometimes
1952	2	Dogs	60	2-3 min.	Sometimes
1953	2	Dogs	70	3-4 min.	Sometimes
1954	3	Dogs	70	3-4 min.	Sometimes
1955	3	Dogs	90	3-4 min.	Sometimes
1956	4	Dogs	120	3-5 min.	Sometimes
1957	3	Dogs	130	6 min.	Sometimes
1959	2	Dogs, Rabbit	130-500?	?	Yes
11/57	1	Dog	Orbital	Days	No

[8]
B. PRESENT PROGRAM:

There is, of course, no doubt that the Russians plan to place a manned satellite into orbit at an early date. But, aside from the fact that they, too, have selected their astronauts, nothing has been published regarding their program.

III. COMPARISON OE UNITED STATES AND SOVIET PROGRAMS

A simple comparison of numbers indicates that the Russians have announced 22 ballistic biomedical flights, while we have had 14. They have used dogs in all their experiments, while we employed monkeys and mice. Both countries had their share of successful and unsuccessful recoveries, and both started the experiments at approximately the same time.

In addition, the Russians have observed one animal, Laika, in. an earth orbit about two years ago. As yet, we have not attempted to fly an animal payload in orbit.

All of these flights have supplied physiological data that can be applied to a man-in-space program. But in general, the information has served only to indicate that one need not be concerned about the feasibility of manned space flight; and this information has been equally available to both the United States and the Soviet Union.

One cannot, therefore, reach a conclusion concerning manned space flight, based on an analysis of past accomplishments in rocket flights with biomedical specimens. The accomplishment of manned space flight will be largely an engineering accomplishment, depending upon the size and reliability of the booster system, a knowledge of reentry parameters, and perhaps most important, the development time available from the formal initiation of the project.

However, one might speculate on where the Soviets stand today, based upon an extrapolation of their biomedical and other achievements.

[9]

An important point of departure is last summer's Soviet ballistic flights with a 4000 pound payload, containing animals. If this payload is the same one that they plan to use in their manned satellite flight, then the flight would correspond to the Redstone flights of the Mercury program. We will not be in a position to duplicate that achievement until July 1960, or one year after the Russians did it. One cannot help but wonder why the Russians have not yet made a manned ballistic flight; perhaps they do not consider such a flight as a prerequisite to manned orbital flight, or perhaps last summer's animal flights were not as successful as their announcements led us to believe.

In our program, the first manned orbital flight is scheduled nine months after the first Redstone ballistic flight with an animal. Again, if we assume that last summer's Soviet flights were the equivalent of our Redstone flights, and if their program is similar to ours, their manned orbital flights could take place in the Spring of 1960, or about a year before ours.

One of the most difficult conditions imposed upon our Mercury program is that of remaining within the severe weight limitations of the Atlas booster. The Soviets have demonstrated many times that they have no such weight limitations. On the other side of the ledger, we have demonstrated reentry and recovery of a Mercury type capsule with our "Big Joe" shot; the Russians have publicized no such achievement.

One might also speculate that when the Soviets accomplish manned orbital flight, they will put two men into orbit, as opposed to our one. Their entire history of biological flights in space is based on having two animals in each payload. They most certainly have the weight-lifting capacity for a two-man capsule. And the propaganda value of flying two men, with the knowledge that we cannot do this for many years to come, would be a great one.

Document III-4

Document Title: NASA Bioscience Advisory Committee Report, 25 January 1960.

Source: NASA Historical Reference Collection, History Office, NASA Headquarters, Washington, D.C.

This is the first in what over the subsequent decades would be a series of reports on how best to organize the NASA life sciences effort. It is notable because its recommendations were echoed by most subsequent reviews, but only the first NASA Administrator, T. Keith Glennan, implemented them in their totality.

[no page number]

National Aeronautics and Space Administration
1520 H Street Northwest
Washington 25 [sic], D.C.

REPORT OF
NATIONAL AERONAUTICS AND SPACE ADMINISTRATION
BIOSCIENCE ADVISORY COMMITTEE
JANUARY 25, 1960

Wallace O. Fenn, Ph.D.
David R. Goddard, Ph.D.
Seymour S. Kety, M.D., Chairman
Donald G. Marquis, Ph.D
Robert S. Morison, M.D.
Clark T. Randt, M.D., Executive Secretary
Cornelius A. Tobias, Ph.D.

[Table of Contents not included]
[no page number]

Summary and Recommendations

The role of the life sciences in the National Aeronautics and Space Administration Program was evaluated by the Bioscience Advisory Committee at the request of the Administrator.

The objectives of space research in the life sciences are twofold: (1) investigation of the effects of extraterrestrial environments on living organisms including the search for extraterrestrial life; (2) scientific and technologic advances related to manned space flight and exploration.

The same reasons which prompted the establishment of NASA and gave it responsibility for all space research and development devoted to peaceful purposes require that NASA assume responsibility for leadership, coordination and operation of the biomedical aspects of the national space program.

Present and future needs were considered in three broad categories:

1. Basic biologic effects of extraterrestrial environments, with particular emphasis on those phenomena associated with weightlessness, ionizing radiation, and alterations in life rhythms or periodicity as well as the identification of complex organic or other molecules in planetary atmospheres and surfaces which might be precursors or evidence of extraterrestrial life.

2. Applied or technologic aspects of medicine and biology as they relate to manned space flight including the effects of weightlessness on human performance, radiation hazards, tolerance of force stresses, and maintenance of life-sustaining artificial environments.

3. [no page number] Medical and behavioral scientific problems concerned with more fundamental investigation of metabolism, nutrition, blood circulation, respiration, and the nervous system control of bodily functions and performance in space equivalent situations.

The Bioscience Advisory Committee makes the following recommendations:

1. That NASA establish an Office of Life Sciences having the responsibility and authority for planning, organizing, and operating a life sciences program including intramural and extramural research, development, and training.

2. That a Director of Life Sciences be appointed who is directly responsible to the Administrator of NASA in the same manner and at the same directional level as the other program directors.

3. That the internal organization of the Office of Life Sciences include Assistant Directors for Basic Biology, Applied Medicine and Biology, Medical and Behavioral Sciences, and the Life Sciences Extramural Program.

4.That an intramural life sciences program and facility be established with three sections:

    a. Basic Biology
    b. Applied Medicine and Biology
    c. Medical and Behavioral Sciences

5. That the Director of Life Sciences recommend advisory committees made up of consultants outside of NASA to be appointed by the Administrator.

6. That maximum integration of the personnel and facilities applicable to the space-oriented life sciences in the Military Services and other Government agencies be arranged in the most appropriate manner indicated by the nature and extent of the specific problem at hand.

7. [no page number] That the Office of Life Sciences assume proper responsibility for education and training in the space-oriented life sciences through post-graduate fellowships, training grants to institutions, and short-term visiting scientist appointments to be integrated with other NASA efforts in this area.

8. That the NASA Life Sciences Program place special emphasis on the free exchange of scientific findings, information, and criticism among all scientists.

9. That security regulations be exercised with great caution and limited to matters in which national security is clearly involved.

10. That the NASA life sciences facilities be considered a public trust in implementing national anal international cooperative efforts.

[1] I. The Role of the Life Sciences in the National Space Effort.

The Congress of the United States has given to the National Aeronautics and Space Administration the responsibility for all space research and development devoted to peaceful purposes. NASA has begun the fulfillment of this responsibility with an emphasis on the physical and engineering sciences which occupy a fundamental position by virtue of their pertinence to the design, launching and control of all vehicles, whatever their ultimate scientific purpose. With this aspect of the total program well under way, attention is properly being directed to other disciplines which, though dependent on the engineering sciences, will in turn give scientific meaning to the national effort. The biological, medical and behavioral sciences are among these disciplines. The Bioscience Advisory Committee has been appointed to aid in representing them adequately within the NASA program.

The reasons which prompted the Congress to create NASA as a civilian space agency and to give it responsibility for achieving the peaceful purpose of the national effort in space argue equally strongly for the creation in NASA of a strong division of life sciences. As set forth below, two major areas represent the role of the biological sciences in the national space effort and should form the core of the proposed program in the life sciences of NASA. These are the fundamental biological questions relative to extraterrestrial environments and the scientific and technologic aspects of manned space flight.

It is altogether fitting that these matters, both of which involve man's curiosity about himself and his environment in their broadest and most fundamental sense, should be placed in the hands of an agency broadly representative of society as a whole. The military agencies which have so soundly laid the groundwork for much of existing space technology must properly give primary attention to the development of weapons systems and the national defense. Although the military effort in astronautics should not be arbitrarily restricted by narrow definitions of military relevance, the broader implications of extraterrestrial exploration demand the attention of an organization unhampered by such predetermined objectives.

[2] Space exploration has captured the imagination of men the world over to an extent which was not, perhaps, anticipated. These activities have became representative of technological superiority among nations. The United States must maintain its international role as a strong and self-confident but basically peaceful and benevolent power. This requires that the first of her citizens who enter space do so as representatives of the scientific aspirations of all men and not as a symbol of military strength.

The basic study of extraterrestrial environments is ultimately likely to be most productive in furthering an understanding of the fundamental laws of nature. Among the most perplexing questions which have challenged men's minds are the nature and origin of life and the possibility of its presence elsewhere in the universe than on the earth alone. For the first time in history, partial answers to these questions are within reach. Limited knowledge acquired over the past century concerning atmospheric and climatic conditions on other planets, the topographical and seasonal variety in color of the surface of Mars, the spectroscopic similarities between scattered sunlight from portions of that planet and those demonstrable from algae and lichens on earth have suggested the presence of extraterrestrial environments suitable for life and permitted the formulation of hypotheses for the existence there of some forms of life at present or in the past. These hypotheses may, within the foreseeable future, be tested, at first indirectly by astronomical observations made beyond the interference of the earth's atmosphere and by samplings taken mechanically from various celestial bodies, and finally, by direct human exploration. The discovery of extraterrestrial life and a description of its various forms, knowledge of the presence and types of complex molecules based on carbon or other elements, or conversely, the absence of living organisms or of their traces in environments conducive to life will have important implications toward an ultimate understanding of biological phenomena.

[3] These studies will not be complete until the scientist himself is able to make meticulous investigations on the spot. This is true, not only for the biological, but, also, for many other physical, chemical and geological problems which are involved. Although significant engineering achievements in automation, sensing, recording, programming and telemetering have been realized and considerable future development is in prospect,

the indispensability of the human observer in much of space exploration is well established. Man's versatility and selectivity, his ability to perceive the significance of unexpected and unprogrammed findings or to react intelligently to unanticipated situations have not been simulated by any combination of physical devices, however complex, which have been' developed or are even contemplated. Human intelligence and manual skill in servicing the complicated mechanisms of space vehicles or repairing breakdowns in flight are not readily dispensed with or replaced. When along with these attributes are considered his weight of 70 kg, his total resting power requirements of 100 watts, his ability to function for years without maintenance or breakdown, then even the most elaborate provisions for his sustenance, welfare and safety are amply justified simply in terms of engineering efficiency. A national program in space science which does not recognize the essentiality of the human observer and does not plan to utilize him most effectively may wait indefinitely for the automatic devices to replace him or be limited to incomplete and opportunistic observations.

Putting a man into space, especially if he is to stay for long periods, is a task which involves considerable attention and effort from a wide variety of biological, psychological and medical specialties. It will require careful planning and extensive basic and developmental research. Together with the effort in astrobiology it should constitute a substantial part of the total space research and development enterprise.

It comes as no surprise to find that the early stages of space research have been primarily concerned with engineering matters. To many responsible people it seems premature if not actually presumptuous to think about what man will do in space until we are sure that we can actually put him there. But the validity of even the earliest of engineering decisions must be continually appraised in terms of their capacity to maintain man comfortably and effectively in space and increase his knowledge of its properties. Failure to meet the numerous and often subtle physiological and psychological needs of the [4] human organism, or premature decisions to send man off into an unknown universe can have disastrous effects not only on the individuals concerned, but on the nation's political and moral position in the eyes of the world. The scientific objectives of the program and especially the determination of the nature of extraterrestrial life may be forever rendered impossible if vehicles containing complex organic molecules are carelessly allowed to contaminate celestial bodies before science has had a chance to study the original conditions. Nor can we simply ignore the perhaps remote possibility that infective organisms brought back from space to earth may cause human disease or destroy food crops essential to human life. How is the necessary biological wisdom to be brought to bear in planning the space effort?

As pointed out later in this report, the nation's best scientific brains are already organized in the form of advisory committees to study and consult on every detail of the space problem. However, such outside bodies, no matter how soundly constituted, cannot have effective impact on day-to-day decisions within the space agency unless the agency itself is provided with a sensitive and powerful administrative mechanism for receiving the advice and translating it into the energy of decision.

To implement this program in the life sciences, appropriate in size and importance to its responsibilities, it is essential to have in NASA a Director of Life Sciences reporting directly to the Administrator of NASA so that the biomedical interests and skills will have adequate representation in important decisions. The director of the life sciences program, therefore, must have broad biological training and interests. He must be able to understand the physicists and engineers as well as have the ability to present biomedical aspects of combined problem areas effectively to his colleagues so that he can have appropriate influence on comprehensive policies and decisions.

These reasons compel the committee to emphasize that the NASA Life Sciences Program requires and deserves strong financial support and adequate administrative representation.

[sections on "Present Status of Life Science Activities" and "Present Needs" not included]

[14]   IV. Recommendations for a NASA Program in Life Sciences.

A. Organization of the Office of Life Sciences

This office should have the responsibility and authority for planning, organizing and operating the Life Sciences Program of NASA, including intramural and extramural research, development and training. This office would also advise and consult with the other divisions of NASA and with the Administrator in matters involving biology, medicine, and psychology. It should have the responsibility for safeguarding the welfare of human subjects and the public health as well as definitive participation in those projects which might jeopardize satisfactory investigation of possible extraterrestrial life.

1. The Director of Life Sciences would be vested with the responsibility and authority of the office of Life Sciences and should be responsible directly to the Administrator of NASA in the same meaner and at the same directional level as the other program directors. The calibre [sic] of the incumbent is obviously of fundamental importance. He should be a man of high scientific stature, an able administrator with demonstrated capability in the selection and direction of staff. It is probable that the Director will be found among physicians who have had considerable experience in the basic medical sciences, although there are others who are not physicians who might have the requisite background.

2. The Internal Organization of the Office of Life Sciences

The Committee proposes that the Office be organized in four Sections, each with an Assistant Director responsible to the Director of Life Sciences.

a. Section on Basic Biology
b. Section on Medical and Behavioral Sciences
c. Section on Applied Medicine and Biology

The substantive nature of the program of each of these three sections is indicated under the respective heading in Section III of this report [not included], although considerable latitude in planning should be given to each assistant director.

[15]   d. Section on Extramural Program

This section should be responsible for the administration and in collaboration with the other assistant directors, and the Director of Life Sciences, the planning of the extramural program.

3. Advisory Committees

The Director of Life Sciences may desire an advisory committee made up of consultants outside the NASA, recommended by him and appointed by the Administrator. Such a committee would normally report to the Director of Life Sciences or on occasion directly to the Administrator of the NASA.

The assistant directors of the four sections may well need advisory committees for their activities. These could be made up of NASA personnel plus outside consultants.

B. Intramural Program of the NASA Office of Life Sciences

For a number of cogent reasons, an intramural program in the life sciences of significant size, diversity, and excellence should be established by the NASA. It is urgent that this program be initiated without delay.

1. Objectives

The present research effort in this field within NASA appears to be concentrated upon a single specific goal, exemplified by Project Mercury, at the possible expense of broader, more remote, but fundamental aims. It is important that the biomedical aspects of the Project Mercury be placed squarely under the jurisdiction of the Office of Life Sciences and that it be coordinated with other aspects of the Life Sciences Program. The remainder of the national space biomedical effort, as found in military, industrial and academic laboratories, is sporadic and incidental to other primacy interests or responsibilities. These efforts are, on the whole, of excellent quality and

[16]

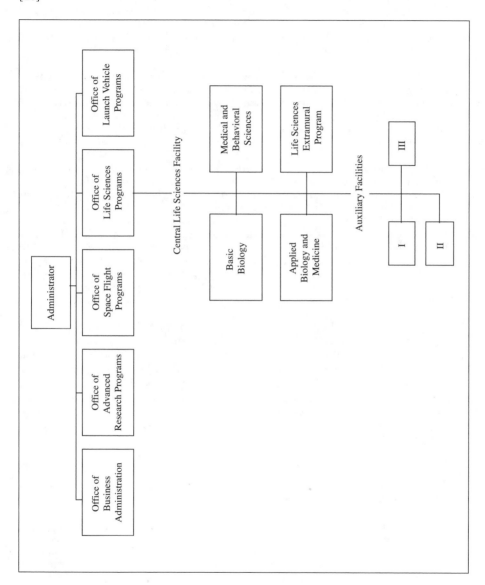

[17] should be maintained and supported; there is need, however, in addition to these and, coordinated with them, for a broad and thoughtfully planned biomedical program of research extending from the most fundamental aspects to their most practical applications. The nucleus of such a national undertaking should be the NASA intramural program in biology, medicine, and psychology.

The number of competent biological, medical and psychological scientists motivated toward space research and skilled in its special problems and techniques is, at present, seriously limited. It is necessary to create a number of career opportunities in these fields on a long-term, full-time basis and to increase the number of laboratories and facilities in which post-graduate training for such careers may be accomplished.

An important ingredient of a productive and creative research effort is the opportunity for interaction among scientists in all the relevant disciplines; between those whose interests are in the fundamental areas and those working in the applied aspects of the problem. The need for interaction has become essential in recent years as the result of the high degree of specialization which modern science and technology demands and the accelerating rate at which new knowledge is accumulating. The older formula for creativity which depended upon the accumulation by a single mind of all the information necessary to a new concept is becoming increasingly difficult to achieve; it may, partially at least, be replaced by the daily contact and collaboration among scientists within a single institute.

An active and distinguished research program in the biological, behavioral and medical sciences within NASA should provide an atmosphere of knowledge and responsibility is which the national effort in these fields can best be planned, administered and coordinated. It should be represented at the highest administrative levels within NASA and should participate is the planning and direction of the entire space program. Its members should be available for consultation and should be given appropriate responsibility and authority in all of NASA activities which involve biology, medicine or psychology.

[18]   2. Scope of Intramural Life Sciences Facilities

To fulfill these objectives the Committee recommends an intramural research program whose ultimate dimensions may be envisioned as follows:

a. A broad central facility with laboratories ranging from the most basic biological, behavioral and medical disciplines through their highly applied aspects. A site at Goddard Space Flight Center at Greenbelt, Maryland or adjacent to the National Institutes of Health recommend themselves, each for somewhat different reasons. The latter would offer the advantages of the unsurpassed facilities of the National Library of Medicine and of interaction with basic and clinical medical, behavioral and biological scientists at the National Institutes of Health, the adjacent Naval Medical Center, and the Walter Reed Army Medical Center and Armed Forces Institute of Pathology which are only a few miles away. All of these opportunities would make the NIH site especially advantageous and attractive to scientists in the medical, biomedical, and behavioral fields. The Greenbelt site, on the other hand, would offer active interfaces both with the space sciences and space technologies and also with the basic biological sciences represented in the laboratories of the Department of Agriculture. This site has a further advantage in the potential for expansion as a national space center which the greater congestion and the different orientation of the NIH could not make possible. Further plans regarding the facility should be the responsibility of the Director of Life Sciences and his staff who will make specific recommendations to the Administrator.

b. A limited number of additional facilities situated at some of the present or future NASA installations and possibly an institute at one or two universities. Each of these accessory groups would be somewhat differently oriented depending on the special functions and the variety of competences represented in their environs. Thus, an institute located at a university with an important biological tradition should be more heavily weighted toward basic astrobiology, while one situated where astronomy and physics were [19] emphasized should reflect an orientation toward astrophysics. The groups to be incorporated into NASA installations, on the other hand, should be primarily representative of the technological and engineering aspects of biology and medicine. They would thus be in a position to utilize the unique facilities of these installations in the furtherance of astronautical research and, conversely, this would ensure that engineering development of space vehicles would be carried out with due regard for the requirements of future occupants.

The Committee is reluctant to stipulate the dimensions which these facilities should attain or to indicate more precisely their scientific complexion. It would suggest, however, that the directors would give prior consideration to high quality rather than quantity, realizing that excellence is not necessarily proportionate to size.

It would emphasize, however, that at least some of the peripheral units, as well as each of the three units of the central facility, be planned in terms of a minimum critical mass, defined as an adequate variety of disciplines and number of professional personnel and their necessary supporting staff and physical facilities to constitute a self-sufficient, mutually interacting and sustaining unit. It is of interest that the varied experience of the Committee members converged on an estimate of 20 scientists and 30 to 35 supporting personnel as constituting such a minimal staff. An annual budget of $800,000, exclusive of permanent equipment but including overhead or reimbursement, would probably be required to support such a minimum unit, and a facility of 30,000 square feet, over all, to house it, based upon acceptable standards of biomedical research in other fields. This would indicate therefore an annual budget for the central facility of the order of 2.4 million dollars and a total of 90,000 square feet.

Even where the program of a unit were oriented to one or another aspect of the field as would be the case in the accessory laboratories, the scientific staff should be representative of numerous disciplines, basic as well as applied.

[20] 3. Development of the Intramural Program

The rate of growth toward an intramural program of the scope outlined above will perforce be limited by the total budget and the competitive needs of the parent agency.

More important, perhaps, may be the limitation, self-imposed by the program's directors in recognition of the paucity of adequately trained personnel and the other national needs for such individuals, including the needs by the military departments for national defense and security and the needs of academic and other institutions for teaching and research.

The Committee recomends [sic], therefore, that the development of the intramural program be deliberate and gradual with cooperative utilization of presently available manpower and facilities which are outside of NASA and their judicious duplication or replacement by the intramural NASA program as those facilities become obsolete or overutilized and as the total resource of competent and motivated scientists is augmented by a training program which NASA itself will substantially support.

The immediate and most pressing need of the program is the appointment of a Director of Life Sciences and, on his recommendation, the Assistant Directors. (See Section IV - A. Organization [not included].) It should be the responsibility of the Director and his Assistant Directors, in consultation with an Advisory Council, should one be appointed, to plan a national program for NASA in the Life Sciences, to determine its complexion, establish its philosophy, recruit its senior personnel, and guide its development.

The Committee recommends that the Director of Life Sciences and his staff in their initial planning select those segments of the national program which are currently being carried out by existing facilities in the military services, in universities and research institutes, and by industry, or which certain of these facilities are capable of carrying out in the immediate future. By appropriate contracts, transfer of funds, construction grants or other mechanisms of support, participation of these existing facilities in a coordinated national program should be invited and made feasible.

[21] At a very early date, the Director of Life Sciences and his staff should begin the planning, construction and organization of the central and certain of the auxiliary facilities, concentrating on those areas of basic and applied science not adequately provided for in existing programs.

As major physical facilities utilized by the NASA biomedical program on a cooperative basis and of primary concern to that program (i.e. centrifuge and controlled environmental chambers) become obsolete or overutilized, or as completely new designs become necessary and feasible, these may be constructed by NASA within its intramural program and maintained as national and international facilities. This should not prevent the construction of similar facilities by other agencies where necessary to the execution of their respective responsibilities. Present cooperative arrangements are a fitting precedent for the continuance of the concept that these expensive facilities should be shared wherever possible both in cost and usage, but that the initiative and responsibility for the construction of any one of them should lie with the agency which has the greatest need.

C. Extramural Program of the NASA Office of Life Sciences

Investigations in extraterrestrial biology and resolution of problems related to manned space flight provide an area for research and development necessitating many diversified contributions. An optimum rate of achievement will require further cooperation with other government agencies. Important contributions are expected to come from scientists working in universities, research institutes, and industry. Thus a strong extramural program is an essential aspect of the activities of NASA in the Life Sciences: (1) to mobilize the relevant research talent; (2) to obtain ideas, information and participation essential to the activities of NASA from the best qualified available sources; (3) to generate among the scientific and industrial communities an awareness of the activities of NASA and to secure support of its programs.

1. Grants

NASA should set up a system of research grants for individual scientists or groups of scientists working in [22] universities or nonprofit research institutes based on original research proposals and with appropriate means for their review and approval. Such grants should be for the support of basic or applied research in areas of interest to NASA. These areas of its interest should be broadly interpreted. Proposals from well qualified interdisciplinary groups should be encouraged.

2. Contracts

We believe that the NASA should enter into contracts with industrial corporations and governmental agencies for specific research needs, particularly in the field of technology but also in fundamental research. Such research contracts are particularly favorable for the solution of short-term problems which might be inefficiently studied in an intramural program and which might require the hiring of specialized scientists or the building of particular equipment that would have no long-term value to the agency.

3. Timing

The Committee strongly recommends that the grant program and perhaps research contracts should be initiated so that money is available to the recipients at the earliest practicable date. The initiation of this program need not await the setting up of the permanent organization of the Life Sciences. The NASA may be able to borrow an experienced official from the United States Public Health Service, the Office of Naval Research, or the National Science Foundation to get this program under way. Alternately, the NASA could delegate the approval of such grants to the National Research Council - National Academy of Sciences. The sort of study section mechanism used by the United States Public Health Service could serve as an excellent model.

[subsections on NASA/military relationship, training, and communication and information not included]

[30]   G. NASA Life Sciences Facilities as a Public Trust

Although much basic research related, to problems of space can be conducted in appropriate facilities on earth, it is apparent that many observations must be made in space vehicles. The study of the effect of weightlessness is an obvious example; spectrographic analysis of the surface of the planets from platforms high above the disturbing influences of the earth's atmosphere is another. For some time to come, the space available for scientific instruments in space vehicles is likely to be strictly limited. At the present time almost all such space and bandwidths available for telemetry are being absorbed by the equipment necessary to monitor the function of the vehicle itself or to make limited physical observations of its immediate environment. Prospective improvement of propulsion systems will soon provide more commodious vehicles, but for years to come the supply of facilities is likely to be far less than the demand. Proper allocation of such space facilities will be very difficult to arrange and certainly cannot much longer be left solely to the good will of those responsible for the design and operation of launching equipment, or to random excitement as to who can inject the largest mammal into orbit.

Attention may be drawn to the fact that at present two great powers between them enjoy a monopoly on operations in space. Although this list may be expanded somewhat in years to come, the extensive resources needed to support such missions make it likely that they can be carried out only by the very largest nations. It is a tradition of long standing in the United States that a monopoly position carries with it the obligation to conduct affairs with due regard to the public interest. In the present instance the monopoly is essentially worldwide, since it includes the control not only of the vehicles themselves but of the most suitable launching sites throughout the world. It follows that these facilities should be administered so far as it is possible in the public interest of the world at large. The Committee is heartened by the provisions which the NASA has made towards greater international cooperation. As man stands before the moment when at last he may break the bonds which have chained him to a single planet, it seems fitting and proper to ensure that all mankind, and not two nations alone, should have the opportunity to meet this momentous challenge.

Document III-5

Document Title: "Interim Report to Space Science Board of Biology Committees," 1960.

Source: NASA Historical Reference Collection, History Office, NASA Headquarters, Washington, D.C.

The possibility of both contaminating the Moon or planets with organisms brought from Earth aboard a landing spacecraft (forward contamination) and of bringing an alien organism back to Earth as a spacecraft returned from the Moon or another planet (back contamination) has been, from the start of space exploration, a concern of the scientific community. This 1960 report to the overall Space Science Board from its Committee on Psychological and Biological Research, and that committee's Exobiology Subcommittee, and the attached resolution, were discussed at the Space Science Board's seventh meeting in the spring of 1960 and forwarded to NASA in May 1960.

National Academy of Sciences   Attachment A
National Research Council
2101 Constitution Avenue, N. W.
Washington 25, D. C.

Interim Report
to
SPACE SCIENCE BOARD

of

Biology Committees (11 and 11-A)

The Space Science Board Committee on Psychological and Biological Research, and its Exobiology Subcommittee has given careful consideration to the tasks outlined in SSB-139 and its attachments. The Committee is now reorganizing to improve the biology coverage in the space science programs of the 1965-70 and 1970-80 periods. Completion of this reorganization will make possible the broad assessment necessary to provide a properly balanced biology input to program formulation.

1970-80 Period and the late 1960's

Within the foreseeable future, the cost of sending an experimental device through space and receiving information from it will be many times that of using comparable analytical instruments in the laboratory. For many other reasons, the retrieval of samples of the planets would ultimately be the most informative means for the advancement of planetary science. This self-evident design has been and should continue to be foremost in the long-range planning for the scientific utilization of spacecraft with the requisite capabilities. However, such missions also introduce the risk of back-contamination, a risk that cannot [2] be decisively evaluated within the framework of our present knowledge of planetary biology. The same missions, as well as the one-way probes that will precede them, also entail the possibility of contaminating the targets; however, these missions can be programmed so as to minimize the carriage of "samples" of the earth, and to disinfect the spacecraft by methods of known efficacy for terrestrial organisms. Furthermore, as a matter of policy, acceptable risk figures for contaminating a planetary target must be substantially higher than for bringing trouble home.

From this standpoint, it may be fortunate that the vehicles for one-way missions will (as is obvious) become available first. We must make every effort to develop experiments that can be flown on such missions and give telemetered information on planetary life and life habitats (with particular reference to exobiology).

1960-65 Period

During this time, Centaur will be available, but Saturn will not. The most significant experiments in exobiology are not possible with these capabilities: these require planetary soft landings, however, considerable preparatory work must be done now. This consists of two parts–both, for this period, independent of sequence.

1) Space Experiments

    a) Lunar soft landing
    b) Planetary fly-bys
    c) Satellite-borne telescope
d) Recoverable nose-cone or satellite capsule

2) Back-up research, for space flight experiments and for general scientific information

    a) instrumentation
    b) Earth-based experimentation

Many aspects of the planetary chemistry program are of particular interest to the biology groups; in fact, there is no chemical information not useful to planning biological experiments.

A. Exobiology

    1) Space Experiments

        a) Lunar Landing - Biologists await anxiously the chemical analysis of the lunar surface and of deeper layers. These results are essential for planning any biological approach [3] [U]ntil this information is available, however, we continue to stress the necessity for minimizing contamination of the moon by terrestrial organisms.

        b) Planetary Fly-bys

            i) Determination of the temperature profile of Venus

            ii) Examination of Mars in the infra-red as proposed by Davis and Gumpel in their COSPAR Symposium paper. This experiment was conceived by Jet Propulsion Laboratory in collaboration with Messrs. Calvin and Weaver of the University of California. It is understood that the instrumentation is essentially ready.

        c) Satellite Telescope - This experiment can, under certain conditions of technological development, provide essentially the same information as the fly-bys, either would be helpful depending on the solution of their respective technology problems. It is very important to have planetary study capability included in the satellite telescope.

d) Recoverable Nose-cone or Satellite Capsule

i) Temperature relations of small particles in vacuum under incident radiation. Particles whose diameter is less than IR wavelengths may reach very high equilibrium temperatures owing to their difficulty of reradiation. It may be possible to simulate the reaction of, say, bacterial spores (d = 1 μ) in vacuum in the terrestrial laboratory, but it may be difficult to prevent interaction between particles and container [sic] wall. Space experiments, e.g. recoverable satellite capsule and nose-cone flights (giving as much as 30 minutes exposure) may be the most convenient way to determine the effect.

ii) Study of the effects of the low pressures of free space on the survival of terrestrial organisms. To a considerable extent it may be possible to simulate on the ground.
(Halverson of Wisconsin is interested in both i and ii.)

As with radio biological experiments final assessments should be made empirically in space with test systems that have been thoroughly worked out on the ground.

[4]
2) Back-up Research

Throughout this period, continuing research on every aspect of the control of contaminating organisms must continue in anticipation of soft landings on the moon, and more particularly the planets.

a) Instrumentation for Space Flight Experiments

i) Gathering of samples of planetary surface materials for analysis. This problem needs fundamental review.

Electrostatic collection of falling dust Scoops or augers

ii) Chemistry of planetary surfaces

Application of mass spectrometer techniques to solid as well as gaseous samples
Automatic spectrometry in infrared and ultraviolet
Handling of materials for chemical analytical reactions
Detection of micron-sized particles giving volatile products

iii) Protection of samples

Encapsulation of samples in hermetically-sealed containers for (a) local protection for later analysis and (b) ultimately for safe return to earth.

iv) Biological detection

Culture and micro chemistry of micro-organisms (Vishniac and Lederberg both interested)

v) Space Craft (and components)

Means for subsequent location and identification of space craft used in early planetary landings

vi) Surveys for macroscopic exobiota

Means for vidicon survey at various magnifications and focal lengths for macroscopic exobiota (and for other planetary data)

Detectors for enzymatic catalysis
Interaction of planetary organisms, with tissue cells
Response of planetary organisms to disinfection [5]

b) Ground-based Experimentation for General Scientific Information

i) Telescopy - particularly continuing planetary studies

ii) Analysis of meteorites

iii) Profile of earth's atmosphere with respect to microbe-bearing dust

Sample collection by high altitude balloons, and sounding rockets; also micrometeorites thus collectec [sic] can be analyzed for abundances of carbonaceous materials.
The transport of micron particles can be determined through studies of high altitude atmospheric turbulence.
Consideration of the mechanisms necessary for organisms to escape the earth's gravitational field.

iv) Interpretation of the appearance of earth from high altitude for detection of biological phenomena-both high resolution photography and in IR.

v) Simulation of planetary environments as habitats for terrestrial organisms.

vi) Conditions for the origin of life: simulation of planetary atmospheres and characterization of products of biochemical interest resulting from their irradiation.

B.   Radiation Biology

1) Space Experiments - More detailed physical measurements of the radiations in space. When known reliably, these measurements can be translated easily into biological terms by (a) laboratory simulation and (b) calculations from fundamental parameters.

Cosmic Ray Effects. These effects at a cellular level are pretty well understood, but their heterogeneous distribution over the animal (needle-like lesions 25 μ diameter, 1 mm long) gives effects on performance of the whole animal not yet easily calculated.

2) Back-up Research - Experiments with available accelerators and microbeam equipment will enable selection of the best biological indicators to be used later in flight tests.

[6]

C.   International Co-operation

1) Development of acceptable method of decontamination

2) Exchange of ideas concerning the objectives of exobiological research and discussion of the instrumentation

3) Depending on success of these enterprises negotiation of tacit agreements through COSPAR.

D.   Environmental Biology

No serious proposals for rhythms or gravity-free state experiments were at hand.

[no page number]

Attachment B

POLICY RESOLUTION CONCERNING QUARANTINE OF MATERIALS OF PLANETARY ORIGIN

WHEREAS

A.   The technological capability for roundtrip space missions may be achieved by this and several other nations within the next ten to twenty years.

B.   Such return missions from other planets may allow for the implantation on the earth of organisms whose biological capacities are at present unknown.

C. Such organisms may include many types unable to survive on earth, or having a neutral ecological value, or even of important benefit to industry, agriculture or medicine; however, they may also include types that might be seriously deleterious to the health of [sic] economy of man. (Footnote)

D. The protection of public health and the maintenance of agricultural quarantines are presently within the jurisdiction of several national agencies which have had considerable experience and expert knowledge in dealing with related problems; however, they have not evidently actively concerned themselves with the impending problems of planetary quarantine.

E. Policy on space missions, as it concerns the activities of other sovereignties and the welfare of all people, must involve our international relations. In due course, the World Health Organization will inevitably take an interest in interplanetary quarantine, and other organs of the United Nations may also be expected to do so. The United States should take statesmanlike leadership in such an area, rather than be made to seem to comply reluctantly with the pressure of external opinion.

F. The development of sound policy on interplanetary quarantine has important implications for the planning of future space missions. On the other hand, present scientific knowledge limits the reliability of any conclusions, and must be exploited to the fullest possible extent.

G. The National Aeronautics and Space Administration has primary operational responsibility for space science and exploration; however, it cannot be expected to take responsibility for policy questions primarily concerned with public health.

[2]
H. Although one or more decades may elapse before return vehicles actually can function, considerable time is needed for the assessment of policy, for planning space missions, and for intercurrent research.

THEREFORE BE IT RESOLVED by the Space Science Board of the National Academy of Sciences that the following recommendations be transmitted to the Administrator of the National Aeronautics and Space Administration.

1. That he join with the Surgeon-General of the United States Public Health Service in establishing an inter-agency committee on interplanetary quarantine, with representation from such agencies as National Aeronautics and Space Administration, the Public Health Service, Department of Agriculture, Department of Defense (Biological Warfare Defense), Department of Sate [sic], and others; that this committee be charged with the formulation and timely review of a national policy on interplanetary quarantine; and the committee be advised by experts in the various relevant sciences from within the agencies and from civil life.

2. That the requisite organization be established within the National Aeronautics and Space Administration to represent it in the formulation and administration of policy in space biology, and to develop the research programs that are therefore urgently needed.

Footnote

While many scientists have already concluded that back-contamination is a serious and tangible threat, or should be regarded as such until we can be sure otherwise, others feel that the risks are very small and should be disregarded. Except to report that the magnitude of the risk is at least controversial, we need not anticipate the further findings of the committee whose establishment is being recommended herewith. The main arguments take the following form:

1. Are other planets in fact inhabited by micro-organisms?

    The evidence at least for Mars is sufficiently encouraging to warrant substantial effort in constructing experiments to detect life there despite great technical difficulties, however, at least a landing will be required to be sure.

2. Could such organisms grow on earth?

    This cannot be predicted in advance. However, many species of terrestrial bacteria would grow on Mars, as far as we can judge by our knowledge of their requirements and of the environment.

[3]
3. Could such organisms be harmful to man?

    This question elicits the sharpest division of opinion.

    Pathogenicity for man on the part of most organisms seems to require the evolution of very elaborate adaptations to allow for transmissions from one infected individual to a new host, and to invade the host tissues despite natural defenses. Microbial pathogenicity would have evolved only in company with quasi-human hosts, and even so these pathogens would be poorly adapted to attack terrestrial organisms which would be biologically novel for them.

    The counterarguments would be that our natural defenses against infection represent our own evolved adaptations against terrestrial bacteria and viruses; they may require the presence of familiar proteins and carbohydrates to recognize the invading organisms as foreign. Planetary organisms with a distinctive chemistry might not be recognized as foreign, and therefore not elicit an adequate response. Furthermore, new organisms might cause serious economic harm even if they are not pathogens for man.

Document III-6

Document Title: A Review of Space Research, Report of the Summer Study conducted under the auspices of the Space Science Board of the National Academy of Sciences, 17 June–10 August 1962.

Source: NASA Historical Reference Collection, History Office, NASA Headquarters, Washington, D.C.

The Space Science Board in 1962 organized a comprehensive examination of NASA's research program. The study took place at the State University of Iowa under the general chairmanship of Professor James Van Allen. The Board's assessment of NASA's life sciences research is contained in this excerpt from the study.

A REVIEW
OF
SPACE RESEARCH

The Report of the Summer Study conducted under the auspices of the
Space Science Board
of the
National Academy of Sciences

at the

State University of Iowa
Iowa City, Iowa
June 17 - August 10, 1962

Publication 1079
National Academy of Sciences-National Research Council
Washington, D. C.
1962

Chapter Nine [9-1]

BIOLOGY*

I. Introduction

The biologist's task in space science may be divided arbitrarily into three categories. First and - for reasons developed later - foremost, is the search for extraterrestrial life. Second is the immense task for the biological engineer of putting man into space adequately protected from the peculiar hazards of space and adequately sustained by a good simulacrum of his terrestrial environment. Third is an exploitation of special features of the space environment as unique situations for the general analysis of the organism-environment relationships including, especially, the role environmental inputs play in the establishment and maintenance of normal organization in the living system.

Superficially these three aspects of space biology are very different in character, in importance, and in justification. They are, furthermore, so disparate in their emotional and intellectual appeal that we have a heavy responsibility to appraise their relative importance as carefully as possible without at the same time - by the very exercise of sober care - stifling that element of challenge and high adventure that has characterized the scientific enterprise since it was put on its modern course by the brave minds of the Renaissance.

Later sections of this report address themselves to details pertinent to each of the three aspects of space biology. The purpose of this general preamble is to explicate, as we see it, the general philosophy that should guide the over-all enterprise of space biology.

If one looks at the third category - the utilization of space environments as tools for analyzing the organism-environment-relationship - he is faced with a prospect of certain dividends but at huge cost, and dividends, moreover, that insofar as they are certain are neither large nor revolutionary. If we limit our outlook to such dividends (those that are certain) and center our attention on the nonsense that a program like this must inevitably invite in its early stages, it is hard to escape cynicism about the vast sums that will be spent and a revulsion from the mixture of mediocrity, nonsense, and huge cost in which the NASA program could become involved if extreme care is not taken. A similar cynicism about putting man in space derives on the one hand from the staggering cost involved, and on the other from the project's early association with cereal boxes and comic strips. Most of us at one time or another have been beset by such cynical doubts. However, we have undertaken a careful scrutiny of the over-all potential for space biology in the NASA program. It is our considered judgement that the exploration of space will prove to be one of man's truly great adventures.

*See Appendix I for list of participants in Working Group on Biology [not included].

[9-2] To delimit the scope of opportunities for environmental biology in the NASA program seems deceptively simple. In the space environment we can identify several factors of obvious biological importance which differ quantitatively and even qualita-tively from those which obtain [sic] on Earth. Space can provide us with unique environmental variables (or with otherwise unattainable ranges of a familiar variable) to accomplish critical tests of existing biological theory. We cite briefly as examples the attainment of near weightlessness and the imposition of an environment unequivocally disconnected from the Earth's rotation. In both instances there exist theoretical bases for predicting more than one pattern of growth and behavior to be expected of organisms exposed to these features of the space environment. Patterns of orientation, growth, and other

responses to the Earth's gravitational field are identifiable in all phyla of both animal and plant kingdoms. Circadian rhythms (those having a period of about one day) in divers [sic] biological parameters also have been observed in a large variety of plants and animals. The mechanism of the biological sensing apparatus, whether for gravity or for time, is unknown or incompletely known. In the case of gravitational responses there are a number of sensing mechanisms which differ in principle. For the apparently ubiquitous circadian rhythms displayed by different species in many different ways we have no sound basis for deciding whether in principle nature employs one or more than one fundamental mechanism to achieve the responses we can observe. Regretfully we recognize that theoretical analysis fails to provide neat predictions of the behavior of biological systems in the space environment on which most biologists are agreed. There is still so much ignorance of biological organization and so little theory in the proper sense of that term that we are much more likely to encounter unexpected results, to detect relationships we do not know how to anticipate, and to be surprised by the behavior of biological systems in respect to the unique aspects of the space environment, than is the case for non-biological systems. Therefore, it seems much more than idle speculation to suggest that, insofar as space provides opportunity for truly new types of environmental exposure of known biological systems, it not only offers means of testing - sometimes critically - those relationships we now think are understood, but it also invites the discovery of unforeseen relationships. Space experiments are too expensive and too difficult to be designed in cavalier fashion by trying out anything and everything in hope of a new discovery of fundamental importance; nevertheless, programmed serendipity is not really a contradiction and it should play a role in our long-range thinking about environmental biology in space.

We are, however, in general agreement that this category of effort in the space program, environmental biology, is subsidiary in importance to the search for extra-terrestrial life. The peculiar point is that in this search we can be even less certain of the dividends which the cost in money and effort will return. It may be that Mars, the only serious candidate for another home of life in our solar system, will prove barren when we get there. But we have not the slightest hesitation, while facing fully our responsibility as members of society, in urging strongly that the venture be undertaken. More than that - unless we wholly mistake the nature of man and science - we are not really free to eschew the challenge and the opportunity to pursue this goal that rocketry has placed within our grasp.

We are fully aware that what follows - the justification for setting the search for extraterrestrial life as the prime goal of space biology - has a far greater philosophic appeal than has been fashionable in scientific discussion for some time. But it is not since Darwin - and before him Copernicus - that science has had the opportunity for so great an impact on man's understanding of man. The scientific question at stake in exobiology is, in the opinion of many, the most exciting, challenging, [9-3] and profound issue, not only of this century but of the whole naturalistic movement that has characterized the history of western thought for three hundred years. What is at stake is the chance to gain a new perspective on man's place in nature, a new level of discussion on the meaning and nature of life.

The Darwinian revolution in biology had as one of its implicit propositions the notion that the development of life itself was only one chapter in the natural history of the planet as a whole. Oparin later made this notion explicit in his view that the origin of life was a fully natural - and in some sense inevitable - step in the ontogeny of the Earth. Those chemical systems capable of replication and controlled energy transfer which we call living systems had their origin in the sequence of chemical changes that were part of the planet's early history. This view has inevitably led to questioning the uniqueness and centrality of man in the universe in an even profounder way than the Copernican and Darwinian insights forced upon us. If planetary systems like our own are at all common in the vast reaches of the universe, life may also be almost as common in nature as a whole. It is surely unnecessary to belabor the immensity of this prospect for any man's philosophic position; or to belabor the pusillanimous and provincial viewpoint that would shrink from pursuing it.

It is useful to recall here a few of the more specific points that the biologist himself finds so exciting. The exploration of Mars, which the new technology makes an almost certain prospect within a decade or two, will allow us to pursue the question of how common life is. Given the estimate that there are something like 1020 stars in the universe, we need to know first, what fraction of these have planetary systems comparable to ours;* and second how "inevitable" - or how probable - is the origin of life in the natural history of a planet. It is the second question which in principle we can now begin to answer by an empirical exobiological research [sic]. If there is life on Mars, and if we can demonstrate its independent origin (from its chemical basis, etc.), then we shall have an enlightening answer to our question of improbability and uniqueness in the origin of life. Arising twice in a single planetary system it must surely occur abundantly elsewhere in the staggering number of comparable planetary systems.

If there is life on Mars, the biologists' interest will not stop with its discovery. In one sense the quest will have only begun. The analysis of Martian life - or even Martian paleontology, if the planet has passed through that phase of its development during which it could sustain life - will be a grand extension of the comparative method that characterizes most of biological analysis. Shall we find on Mars some of the unexplained peculiarities of terrestrial biochemistry? If amino acids are involved, are they D or L forms, or both? Does adenine, for instance, again play so many roles? Is the energy transfer system again based on phosphorus? What is the genetic mechanism like? What information storage and replication systems are involved? Has sex again been evolved or have some other devices arisen to engender genetic variability? At higher levels of organization, how have these systems adapted to peculiar features of the Martian environment including its ranges of temperatures and moisture, its thin atmosphere, and its low gravitational field? What has been the course and mechanism of Martian evolution?

*The estimate of Huang puts the number at 1018. Huang, Su-Shu, "Occurrence of Life in the Universe," American Scientist, 47, 3, pp. 397-402 (Autumn, 1959). If the upper limit of star systems that can support life is 5% and if there are 1020 stars in the observable universe, then 1018 systems capable of supporting life is a reasonable guess.

[9-4] This search for life elsewhere will inevitably demand that man must get into space himself. The search for life on Mars, for example, ultimately will require that Martian samples be studied in manned - preferably terrestrial - laboratories. Many of us believe that the retrieval of Martian samples should be recognized explicitly as the ultimate objective of exobiological missions. It is doubtful if such retrieval can be accomplished satisfactorily by an unmanned expedition. Of course we shall get on with the job using remotely controlled life-detection systems even before this venture is fully possible. But we shall never be satisfied with negative results from our instrumental life-detectors because they are intrinsically hampered in their scope by our current ignorance of the nature of whatever extraterrestrial life there may be. Should these preliminary sallies give positive results, the urgency for man to get there will only increase. Therefore, while the intellectual appeal and status of the extraterrestrial life question stands out above all else in space biology, we are also confronted with the important task of providing the physiological basis for prolonged manned space flight.

Finally, it becomes clear that what the physiologist has to do to put man in space for any reasonable length of time is to learn far more fundamental environmental physiology. The three aspects of space biology enumerated at the onset are not, after all, so very different in aim nor disparate in importance. In a real sense they are three aspects of one and the same undertaking - a general extension of the scope of nature that the biologist can bring within his grasp.

In addition to cosmobiology, man into space, and environmental biology, we feel that a fourth point warrants emphasis, not because it is logically coordinate in importance but because of its practical, economic, and scientific impact on experimental biology of all kinds. We predict as confidently for space biology as for other space sciences that the economic costs will be amply repaid in the long run by applications of space-oriented biotechnology to other fields of biology and medicine. We are not unmindful of these inevitably substantial though indirect contributions of NASA's continuing efforts in space biology.

Document III-7

Document Title: Dr. Robert C. Seamans, Jr., Associate Administrator memo to Albert F. Siepert, Director of Administration, "Bioastronautics," 6 July 1962.

Source: NASA Historical Reference Collection, History Office, NASA Headquarters, Washington, D.C.

Document III-8

Document Title: Memorandum from Albert F. Siepert, Director of Administration, to Dr. Robert C. Seamans, Jr., Associate Administrator, "Coordination of NASA's Interests in Bioastronautics," 19 July 1962.

Source: NASA Historical Reference Collection, History Office, NASA Headquarters, Washington, D.C.

Document III-9

Document Title: Letter from James E. Webb, NASA Administrator, to Dr. Norton Nelson, Institute of Industrial Medicine, New York University Medical Center, on NASA's Life Sciences programs and organization.

Source: NASA Historical Reference Collection, History Office, NASA Headquarters, Washington, D.C.

Document III-10

Document Title: Memorandum from Nello Pace, Consultant for Life Sciences to the Administrator, to Associate Administrator, 4 October 1963.

Source: NASA Historical Reference Collection, History Office, NASA Headquarters, Washington, D.C.

These documents reflect the deliberations, at the top levels of NASA, regarding how best to organize for life sciences. While external advisory committees and consultants consistently recommended centralizing all NASA life sciences activities in a single management structure, NASA management resisted these recommendations.

Document III-7

July 6, 1962

MEMORANDUM: to Mr. Albert F. Siepert
Director, Office of Administration

From: Associate Administrator

Subject: Bioastronautics

Reference: Memo June 29, 1962, to Seamans from Dr. E. C. Welsh, Executive Secretary, National Aeronautics and Space Council, subject: Bioastronautics

There is attached a copy of the above referenced memorandum [not included]. In this memorandum, Dr. Welsh refers to a recommendation made by the Bioastronautics Panel of the President's Scientific Advisory Committee that NASA appoint a national leader in biomedical sciences as Deputy Associate Administrator.

As stated in the memorandum, we have directors of biotechnology, bioscience, and aerospace medicine, each in separate offices, and it is felt by the Bioastronautics Panel that policy coordinating work necessitates a special Deputy Associate Administrator. We established three directors for bioastronautics because of the breadth of the bioastronautics discipline and the difficulty experienced in obtaining a single individual with competence in fields ranging from fundamental biology to life support technology. In addition, if this rationale were followed to the limit, we would have deputy associate administrators for propulsion, electronics, and a variety of other discipline that cut across our program lines.

However, I certainly recognize that we have had difficulty planning and implementing our effort in bioastronautics and properly relating our effort to that of the Department of Defense, National Institutes of Health, and non-government agencies. Undoubtedly, one of the medical scientists listed in Dr. Welsh's memorandum would strengthen our organization. With these factors in mind, I would like your recommendations as they relate to Dr. Welsh's memorandum.

[stamped "Original Signed by Robert C. Seamans, Jr."]
Robert C. Seamans, Jr.
Associate Administrator

Attachment a/s
RCS:flp
cc: Mr. Webb w/ inc
[handwritten note "cc Dr. Dryden w/inc"]

Document III-8

July 19, 1962

MEMORANDUM: to Dr. Robert C. Seamans, Jr.
Associate Administrator

Subject: Coordination of NASA's Interests in Bioastronautics

I am in agreement with the reasoning set forth in the second paragraph of your memorandum to me of July 6, 1962. Our experience with a Life Science Program Office indicates that the concept of the life sciences as a coherent field around which to set up an effective program office was not sound. Instead of placing the various elements of the life sciences in an organizational posture where they could have an effective voice in the various programs and projects to which they were integrally related, just the opposite occurred. Most of the time the Office of Life Sciences was on the outside looking in. Our reorganized structure attempts to fit into the direct operating structure the various aspects of Life Sciences. This arrangement still looks promising, and the problems of program coordination envisioned in Dr. Welsh's letter are not likely to become a serious internal operating problem. The real problem is that this arrangement gives NASA a fragmented external image which is not reassuring to the outside bioastronautics community. Unfortunately, the medical, dental, and some scientific disciplines still place considerable importance on the device of an across-the-board official to look out for their interests at a high level in the organization even if he has no direct operating responsibility.

In several situations where other organizations found it obligatory to create this type of high level staff symbol, the results were either disruptive or ineffectual. Similar arrangements for chief professional officers in the Public Health Service, in the medical corps of the several Armed Services, and in the legislatively created post in HEW of an Assistant Secretary for Medical Officers, all have suffered from the inability to keep a good man in such a position from asserting line responsibility. The purely advisory function seldom is sufficiently appealing to attract an adequately distinguished outsider to serve as NASA's image.

In view of these considerations, I would suggest that in replying to Dr. Welsh you make the following points:

1. NASA's experience in trying to establish a single focal point in Headquarters to coordinate all NASA's interests in the life sciences.

2. Desire to test the effectiveness of the present organizational arrangement over a longer period of time before making additional changes. Only with the recent availability of Dr. Konecci have all three key positions in the Life Sciences area been staffed.

[2]

3. That you will keep an open mind on the suggestion that a Deputy Associate Administrator (under you) or an Assistant Administrator (under Mr. Webb) be appointed to give over-all leadership and coordination to NASA's interests in the Life Sciences.

If the actual need for over-all coordination at the Associate Administrator's level, both internally and externally, becomes a real burden, you may wish to appoint a life sciences specialist under Mr. Fleming, or the external pressure for a larger symbol may dictate that this individual be designated an Assistant Director of the Office of Programs in this area. This would have advantages over adding another staff person you will have to supervise and communicate with directly on a day-to-day basis. One caveat needs to be recognized. It may still be necessary for developing some assistants working directly with you to handle material and special assignments falling within certain functional areas, e.g., Field Center Relations, DOD Relationships, Industrial Relationships. While I would not consider Bioastronautics a logical cross-sectioning under such an arrangement, you might need to accommodate it in view of the external concerns. I personally believe there would be more external satisfaction, and less internal disruption, if such a position were created as a staff assignment directly under the Administrator rather than attached to the critical focus of operations in our agency.

[signed]
Albert F. Siepert
Director of Administration

Document III-9

[no page number]

IN REPLY REFER TO:
Office of the Administrator   November 14, 1962

Dr. Norton Nelson
Institute of Industrial Medicine
New York University Medical Center
550 First Avenue
New York 16, New York

Dear Dr. Nelson:

Dr. Dryden, Dr. Seamans and I have had several discussions recently on your recommendations concerning NASA's life sciences programs and organization. I appreciated your discussing your ideas with Dr. Seamans who has taken a deep and personal interest in this subject.

In considering these questions over the past year, we have been striving to incorporate our life sciences, activities within meaningful program contexts. For example, our space medical activities which are under the direction of General Roadman are an integral part of the manned flight program. The formulation and execution of space medicine efforts are interwoven into the broader mission planning and operations involved in manned flight. Thus we are assured that our energies are focused upon space medicine technology directly associated with manned flight and that these tasks receive the resource support and decisive management direction required. Similarly, our bioscience efforts under Dr. Reynolds, and our biotechnology and human factors research under Dr. Konecci, have also been fully integrated within the broader space science and advanced technology programs. We are convinced that the benefits accruing from such close association and identification with on-going development activity are substantial.

With respect to the question on the coordination of these three related activities, I would like to point out that the absence of a single office to oversee these efforts has not proven to be an obstacle. General Roadman, [2] Dr. Reynolds, and Dr. Konecci meet at fixed and frequent intervals to review and consider their program accomplishments and plans. In addition, they receive important support from our outstanding technical staffs at Houston, Ames and other NASA and DOD centers. This type of responsible interchange of ideas and data promotes a healthy overall life science program which is responsive to the needs of our program directors.

We recognize that NASA does not present a single "Life science" face to the nation's scientific community. In our judgment, however, the auxiliary benefits of a "single face" must not be overestimated since in the final analysis, such benefits are direct functions of the program activity planned and underway. Organizational patterns in and of themselves will contribute little in terms of meaningful contributions to our national capability in the life sciences without the right arrangements for programmatic support. We have been steadfast in our determination to avoid the establishment of a figurehead position in this area in order not to disrupt or dilute the very real authority delegated to General Roadman, Dr. Reynolds, and Dr. Konecci. Each of these individuals has a major programmatic responsibility and has the authority and resources needed to execute his programs in an expeditious and comprehensive manner.

Although we are convinced that the existing arrangement is suited to our internal requirements in terms of program planning and im,lementation [sic], Dr. Dryden, Dr. Seamans, and I are agreed that we can and ought to do more to develop and crystallize the identification of NASA's life sciences programs with the outside scientific and medical community. We are currently considering the assignment of one or more consultants to general management to determine how this relationship can be strengthened, particularly in terms of involving in a very active manner a greater portion of our truly outstanding medical and scientific talent. We have asked several of our program managers to consider this question and to give us the benefit of their thinking. In this regard your letter and listing of eminently qualified leaders in the life sciences has been particularly helpful. As a consensus develops on this question, we will be happy to discuss our judgments with you and other participants in the Summer Study.

[3]May I again express my appreciation to you and to each member of the Working Group on the NASA Biological Management for a thorough and most helpful series of recommendations.

Sincerely yours,

James E. Webb
Administrator

Document III-10

[no page number]

Report No. 6
4 October 1963

Memorandum to   AA/Associate Administrator

From   A-6/ Consultant for Life Sciences to the Administrator

Subject   Reorganization of NASA Life Sciences Activities

Careful consideration of the objectives of the National Aeronautics and Space Act of 1958 indicates that the broad responsibilities of NASA lie in three major areas:

1. The pursuit of research on phenomena of the atmosphere and space.

2. The design, construction and operation of vehicles for carrying out such research, and for transportation generally through the atmosphere and space.

3. The dissemination of results of these research and technological developments.

That the research an development to be conducted under the provisions of the ACT includes the life sciences (taken herein to comprise the various branches of biology and medicine, including psychology) is implicit in its language defining the objectives of NASA: Sec. 102(c)(1), "The expansion of human knowledge of phenomena in the atmosphere and space"; and Sec. 102(c)(3), "The development and operation of vehicles capable of carrying instruments, equipment supplies, and living organisms through space". The inclusion of life sciences research and development activities in the NASA program is also made explicit by the annual Congressional Authorization and Appropriation acts which include funds for this purpose.

[2] Although life sciences activities were being carried out by NACA as studies of "human factors" as early as 1943, and although Project Mercury, initiated in October 1958, had a strong life sciences component, the development of a substantial and well-rounded life sciences program has been slow. This may be compared with the spectacularly successful and far larger physical and engineering sciences program which has grown so rapidly under the initial powerful impetus of the International Geophysical Year. As detailed in my Report No. 4 to you, dated 17 September 1963, in FY 1963 life sciences total research in NASA was only 2.1% of the entire NASA total research obligation, whereas physical and engineering sciences research accounted for 97.8% of the NASA research dollar.

The reason for the unduly protracted growth of the NASA life sciences program are, but for the consideration at hand suffice it to say that it has involved such factors as the newness of the concepts involved, lack of experience in a new field, greater complexity of the field generally, misunderstanding of the distinction between headquarters staff versus line management functions, and to a certain extent unfortunate personality differences. Regardless of past history, however, it is clear that the life sciences must be regarded as full partners with, rather than adjuncts to, the physical and engineering sciences in the total NASA program.

It should be pointed out that the original NASA Bioscience Advisory Committee under the chairmanship of Dr. Seymour S. Kety produced a brilliant justification for the establishment of a strong, centralized authority in the form of a Headquarters Office of Life Sciences in a report to the Administrator dated 25 January 1960. Largely on this basis the Office of Life Sciences Programs was established on 1 March 1960 with Dr. Clark T. Randt as Director. After a brief and somewhat stormy tenure, however, Dr. Randt resigned on 1 April 1961, and was succeeded by Brig. Gen. Charles H. Roadman, USAF.

Brig. Gen. Roadman strongly favored dismemberment of the Office, with complete dissociation of operations activities from research activities in the life sciences area. Accordingly, on 1 November 1961 the Office of Life Sciences Programs was abolished, leading to the present structuring of life sciences in NASA Headquarter which comprises the following independent components:

[3] Office of Manned Space Flight:
      Aerospace Medicine (MM), Dr. George M. Knauf
      Acting Director
      Systems Studies (MG), Dr. William A. Leo, Director
      Spacecraft and Flight Missions, Human Factors (MSH),
      Maj. Leroy Paige, USAF
      Air Force Systems Command, Space Medicine (MAF),
      Col. John M. Talbot, USAF

    Office of Advance Research and Technology:
      Biotechnology and Human Research (RB), Dr. Eugene B. Konecci, Director

    Office of Space Sciences:
      Bioscience Programs (SB), Dr. Orr E. Reynolds, Director
      Grants and Research Contracts (SC), Dr. Thomas L. K. Smull, Director
      Manned Space Sciences Division (SM), Dr. Verne C. Fryklund, Acting Director

    Office of Administration:
      Safety and Health Office (BY), Mr. George D. McCauley
      Occupational Medicine (being established)

    Office of Technology Utilization:
      Life Sciences Evaluation (being established)

Another result of the dissolution of the Office of Life Sciences Programs has been the development of independent research grants and contract negotiating activity in the life sciences area in some of the field centers, particularly Ames and Houston, as well as by most of the above listed Headquarters components.

Because of the wide dispersal of life sciences responsibility throughout NASA, both for funding and for operations, it is not surprising that the development of a sound, broad program in this area has been slow, with many instances of needless duplication and serious gaps.

After careful and intensive consideration of the situation for the past three months, I should like to propose reconsolidation of the management of life sciences activities. The consolidation should not only substantially improve the effectiveness of current operational practices, it should materially [4] improve NASA's image in the academic-industrial scientific community and in Congress, and perhaps most importantly it should expedite the development of forthcoming flight programs which have major or entirely life sciences objectives. In this category I would include the Gemini, Apollo, Biosatellite, Mariner and Voyager series as well as more advanced concepts such as Manned Orbital Research Laboratory, Manned Lunar Base and Manned Mars Mission.

Before elaborating on the proposed reorganization, I should mention that I have talked with a number of people both from within and from outside NASA about it, as appropriate. Among the former are Drs. Bisplinghoff, Mueller and Newell, as well as Drs. Knauf, Konecci and Reynolds, because of their obvious primary interest and role in the matter. Among the latter were representatives of universities, industry and other Federal agencies. Thus, the proposed plan takes into account information and opinions from a wide variety of sources.

One of the major weaknesses of the current position of the life sciences in NASA is the arbitrary split of primary funding authority into three autonomous programs: Aerospace Medicine (MM), Biotechnology and Human Research (RB), and Bioscience Programs (SB).

Living organisms, whether man, amoeba or bread-mold, represent enormously complex, highly integrated systems whose nature is understood only fragmentarily and dimly by scientists. There are few generalizations or laws in biology which have predictive value; hence, specific answers to new problems can usually be obtained only empirically.

This is quite the opposite situation from engineering where known materials are combined according to exact laws to achieve a desired end effect. Even better, many times engineers can specify the exact nature of previously non-existent materials required to do a job, and confidently expect the development of such materials.

Biology is not this way. In reality, no distinction can be made between "basic" and "applied" research on a human being, for instance, because all scientific data obtained on man are useful in on medical application or another. By the same token, there is no "advanced research" in biology. We know [5] far too little to put any research area in a special category. Therefore, as with the living organisms themselves, attempts at overcompartmentalization or isolation of parts in life sciences administration can lead to loss of viability of the whole.

Accordingly, in order to achieve the requisite coordination of life sciences activities within NASA, it is proposed that a Deputy Associate Administrator for Biospace Missions be appointed for overall management coordination of life sciences activities at the Headquarters level.

It is further proposed that all present life science program functions in the Office of Space Sciences and the Office of Advanced Research and Technology be combined into an Office of Biospace Missions. In effect, this would be accomplished by coalescing the current activities and personnel of Bioscience Programs (SB) and Biotechnology and Human Research (RB) into a single Office.

Because of his broad experience, his excellent reputation in the National bioscience community, and his demonstrated high level of administrative and scientific competence as exemplified in the past year by his extremely effective development and management of Bioscience Programs, I strongly urge the dual appointment of Dr. Orr E. Reynolds as Deputy Associate Administrator for Biospace Missions and Director of the Office of Biospace Missions, to be effective at the earliest possible time.

Parenthetically and by way of contrast, I should like to mention that there has recently occurred an international display of poor scientific judgement on the part of Dr. Eugene N. Konecci, which serves to illustrate the potential hazards of lack of adequate coordination and competent overall review in the life sciences area. The incident concerns the press publicity resulting from Dr. Konecci's unfortunate remarks on the subject of mind reading while serving as Chairman of the Bioastronautics Committee of the International Astronautical Federation in Paris on 26 September. It is my understanding that formal censure of Dr. Konecci's action is being considered by the Space Science Board, although on being asked my opinion by the Chairman of their Life Sciences Committee I urged a less drastic treatment of the incident. In any case, it seems quite evident that Dr. [6] Konecci has not yet developed the level of scientific maturity required for independent program decision making in an important segment of the national scientific economy.

One of the major management problems which arose early in the history of the old Office of Life Sciences Programs was the conflict over ultimate decision authority concerning selection and flight activities of the astronauts in the manned space flight program. Clearly, the enormously formidable operational engineering problems to be solved required total control over all aspects of a manned flight by the responsible engineering authority. Yet equally clearly the man is a living organism with enormously complex characteristics, and poorly defined tolerance limits, the assurance of whose effective performance, well-being and safety require total control over the man-environment interface by the responsible medical authority.

I propose that the dilemma be resolved by retention of the present Directorate of Aerospace Medicine (MM) in direct line responsibility to the Deputy Associate Administrator for Manned Space Flight for astronaut selection and flight operations matters, but at the same time the Directorate also be place under direct line responsibility to the Deputy Associate Administrator for Biospace Missions for biomedical research and development matters. Thus, a smooth coordination can be effected between practical engineering end-item production and physiological research input so essential in this unique man-machine assemblage.

Implicit in the changes suggested is the very important feature that all life sciences Supporting Research and Technology funds would be administered by the Office of Biospace Missions. This is absolutely essential to a coordinated effort in the total life sciences area so that wasteful duplication and hurtful gaps will be avoided.

Besides the major collaborative role with the Office of Manned Space Flight, the Office of Biospace Missions would have as its other primary role the "expansion of human knowledge of phenomena in the atmosphere and space" in the life sciences area. These activities fall quite naturally into two broad sub-units: the search for extraterrestrial life and the elucidation of the mechanisms of the origin of life generally, which has been termed conveniently "Exobiology", and investigation of the [7] effects of the space environment in all of its characteristics on the life functions of terrestrial organisms, plant and animal alike, which may be described conveniently by the broad term "Environmental Physiology". It is to be emphasized and re-emphasized, however, that a perfect continuum exists between the theorizing concerning biochemical evolution and the origin of life at one extreme, and the medical monitoring of an astronaut during a flight mission at the other extreme. Any fragmentation for administrative convenience of this broad consideration of the various facets of the life process serves only to delay our ultimate understanding of the phenomena involved and the solution of many practical medical and engineering problems.

Another change which I propose is that the intramural Life Science Research Laboratory, currently scheduled for construction at Ames, be given the status of a Field Center and that the Director report directly to the Deputy Associate Administrator for Biospace Missions. This would accomplish two important results, and it would permit the acquisition of a higher level bioscientist to direct NASA's intramural research program than is now possible. This proposed change would apply whether the Laboratory remains physically at Ames as a Life Science Research Center, or not. It would also apply if a Human Factors Research Center were instituted separate from an Exobiology Research Center, except that in such a case there should be two Directors rather than one.

Another suggestion I make is that an effective, coordinated procedure be adopted at the Headquarters level, i.e., in the Office of Biospace Missions, to provide expert outside review of all research and facilities proposals in the life sciences area, intramural and extramural alike, for scientific quality and to determine acceptability from a scientific point of view. It would then remain the responsibility of the NASA cognizant scientific administrators in the Office of Biospace Missions to determine priorities and mission appropriateness among the scientifically acceptable proposals. Additionally, the review machinery established for this purpose should easily be able to handle any review requirements in the life sciences area generated by the Office of Technology Utilization or by the Office of Research Grants and Contracts.

[8] It is further proposed that the Headquarters Occupational Medicine and Safety and Health activities be transferred from the Office of Administration to the Office of Biospace Missions. This would again make possible the establishment of a smooth liaison between research in radiobiology and toxicology on the one hand, for example, and industrial hygiene practices on the other.

Another factor of importance in achieving the most effective program in the life sciences is the matter of primary responsibility for flight projects. In order to assure optimal experiment control, it is essential that the cognizant program activity have primary flight project control and responsibility. Therefore, I propose that flight projects relating entirely or primarily to biospace missions shall be assigned to the Office of Biospace Missions for management control. Thus, at present, the Biosatellite Project would fall in this category as would some of the Martian exploration proposals. Flights involving biospace missions only in part, such as the Gemini and Apollo Projects, would continue to be the joint concern of the program offices involved.

Finally, because of the scope and importance of NASA's manifold responsibilities and activities in the life science area, which cannot help but increase in the future, I urge most strongly that an Office of the Life Sciences Advisor be established at the Administrator's staff level. In order to maximize the utility of this Office in advising the Administrator on policy in the life sciences area, I recommend that prestigious bioscientists be invited to spend not less than one year and not more than two years as the Life Sciences Advisor to the Administrator. Furthermore, the Office should have a small permanent staff which would include one medical man and one engineer, each experienced in space activities, who could provide the practical, applied information background and perspective for the incumbent.

Such an arrangement should make available to the Administrator a continuous flow of new ideas from the best bioscientists of the country, yet would not interfere with the managerial need for continuity in the operating program in life sciences, which would be provided by the Deputy Associate Administrator for Life Sciences at the level of the Associate Administrator.

[9] As epilogue I should like to make apology for the rough-hewn nature of this communication. However, as a major reorganization of the Headquarters management structure is rumored to be imminent, I feel it is urgent and essential that the NASA life

sciences activities receive appropriate consideration in any new arrangement. They form many important threads of the NASA fabric, and are long overdue for top management and policy level representation. I shall of course be pleased to discuss the proposal further as you may wish.

Nello Pace

cc: A/Webb
     AD/Dryden

Document III-11

Document Title: Letter from Homer Newell, Associate Administrator, NASA, to Dr. Harry Hess, Chairman, Space Science Board, 16 April 1969.

Source: NASA Historical Reference Collection, History Office, NASA Headquarters, Washington, D.C.

Beginning in 1968 and picking up momentum as the first lunar landing approached, NASA began to develop a comprehensive plan for its post-Apollo activities. Homer Newell was in charge of this planning effort, which involved participants from all parts of the Agency. As those who were planning the space life sciences program developed a statement of NASA's future objectives in that area, Newell asked the Space Science Board for its assessment of NASA's planning.

[stamped APR 16 1969]

Dr. Harry H. Hess
Chairman, Space Science Board
2101 Constitution Avenue, N.W.
Washington, D.C. 20418

Dear Dr. Hess:

Although currently we are having some difficulty funding the kind of space bioscience program that we feel the nation should have, we are, nevertheless, attempting to develop a long range plan that can be used to put our space bioscience program on a sound basis and will lead to getting the best returns we can from our investment in the program. To this end, the Office of Space Science and Applications has assembled a set of goals and objectives for space biology, a copy of which is attached. I would very much appreciate the comments of the members of Dr. Allan Brown's Committee on Life Sciences of the Space Science Board, on the appropriateness and adequacy of these goals and objectives as a starting point for NASA's long range planning.

In particular, is the stated position of value appropriate? Is it achievable within the next one or two decades, assuming a reasonable application of resources? Is it too optimistic? Or is it just plain not achievable? Is there perhaps some other position of value that is more appropriate to state as the motivation for our space biology program?

Are the stated objectives the right ones on which to base a national space biology program? Are the objectives oriented toward the right questions and are they sufficiently aggressive to take advantage of our capability to advance in this area? Are there any gaps that should be filled in? With regard to the last question, it should be emphasized that exobiology is included elsewhere in our planning. Under space biology we include the use of space for fundamental research into the fundamental nature and processes of life, using the space environment and techniques for whatever advantage they can give us.

[2] Any assistance the Committee on Life Sciences can give us in fixing upon the best set of goals and objectives for planning our national space biology program will be greatly appreciated.

Sincerely yours,

[stamped "Original signed by Homer E. Newell"]
Homer E. Newell
Associate Administrator

Attachment:
Space Biology Report
Dated 1 April 1969

CC: Dr. Allan H. Brown

Prepared by: SB/Jenkins/Arney
ext. 24621   4/11/69

[3]
[stamped APR 1 1969]
                        SPACE BIOLOGY
GOALS

* To contribute substantially to the development of a fundamental unifying theory of biology, by increasing our understanding of the influence of gravity and time on life processes and structures.

* To utilize this biological theory in combination with the unique capabilities of space flight for the advancement of such fields as medicine, public health, and agriculture.

FUTURE POSITION OF VALUE

The significant position of achievement, to which our space biology program will contribute substantially during the next fifteen years, is the development of a coherent and unifying theory of biology.

The possession of such a theory will make it possible to increase by several-fold the contribution of biological research to the solution of contemporary social problems.

Although biological knowledge to date has resulted in some of the most useful applications in the whole of human existence, such as a phenomenal increase in longevity and in food production, the knowledge has come from empirical research approaches which are very slow and which are low in predictive value. In other fields of science, such as physics and chemistry, unifying theories have greatly increased the ability to predict events. Therefore, we can logically expect a new theory in biology to result in a much improved ability to produce useful benefits for man. Since many of our major problems are biological in origin (population, food, health, pollution), it follows that many of the solutions will be biological in nature.

Two missing links in our development of biological theory are the operation of gravity and time (Earth-induced periodic influences) on living processes. Space flight frees experiments from the regulating influences found in laboratories on Earth, allowing us to accurately study the mechanisms involved. Thus, space flight can provide a unique environment for [4] obtaining these missing links in our knowledge, speeding our theory-building and vastly improving our ability to place biological research at the service of pressing human needs. It is in support of this area of research that the Space Biology Program is dedicated.

Exobiology, a related area of research, is expected to provide the third key ingredient from the space program to the building of a general theory of biology. This activity is supported under the Lunar and Planetary Exploration progams [sic]. The discovery of life elsewhere in the Universe would allow us to compare its chemistry, structure, and function, with that of life on Earth. It would allow us to derive universal concepts of biology and broaden our understanding of the laws governing the science of life. It would provide us with a much better understanding of the relationship of life and life processes to the laws governing the related sciences of physics and chemistry. Much can be learned from an independent set of molecules and molecular systems, whether they are, or have been, alive. In particular, extraterrestrial genetic codes or morphologies could provide, through comparisons, opportunities to understand the origin and evolution of life on Earth.

BROAD AND SPECIFIC OBJECTIVES

Broad Objective

1. Understand the biological roles of the universal force of gravity on life and the capability of life to adapt to gravitational changes.

Supporting Specific Objectives

a. To determine the effects of gravity on maintaining normal organization in living cells.

b. To determine the gravity sensors and the effects of gravity on plant growth, develop-ment, physiology, and morphogenesis.

c. To determine gravity sensing thresholds and gravity selection preference for animals.

d. To determine the effects of gravity on the development, growth, metabolism, physiology, and behavior of animals. [3]

e. To determine the effects of weightlessness in producing and correcting abnormalities and to modify or alter mechanisms in organisms.

Broad Objective

2. Understand the roles of time expressed in oscillations, rhythms, and life spans on living organisms.

Supporting Specific Objectives

a. To determine the influence of Earth and lunar environmental periodicities in inducing and maintaining oscillations and biorhythms in living organisms using Earth orbital, lunar, and heliocentric missions.
b. To determine the role of oscillations and biorhythms in maintaining normal organization and life, and determine whether they are innate or environmentally induced.
c. To determine the interaction of time phenomena with gravity and organism volume (space).

Broad Objective

3. Determine the potential applications and develop techniques to modify or alter living systems to produce changes or correct abnormalities to advance medicine, public health, agriculture, and space exploration, including modification of planetary atmospheres or surfaces.

Supporting Specific Objectives

a. To determine the capabilities of life to change and adapt to environmental factors.

b. To develop techniques to use time and weightlessness in producing and correcting abnormalities and to modify or alter mechanisms in organisms.

c. To utilize the experimentation techniques and capabilities developed in the program, for short-leadtime problem-solving in biology and medicine.

d. To explore the potential use of Earth organisms to modify or alter environments on planetary or lunar surfaces. [4]

PROGRAM THRUST

Pursue a program to make a broad survey, and carry out selected intensive studies on the biological effects of space environmental factors on a variety of types of Earth life. Studies will be carried out initially in Earth orbit and then in interplanetary space, and on lunar and planetary surfaces. The program will be balanced, including: studies on Earth; automated recoverable biological satellites; non-recoverable Earth orbital and heliocentric missions; and manned flight experiment programs, both manual and automated.

Document III-12

Document Title: NASA Policy Directive 8020.14, 16 July 1969.

Source: NASA Historical Reference Collection, History Office, NASA Headquarters, Washington, D.C.

NASA, since its early years, has had a policy regarding measures to avoid both forward and back contamination, resulting from robotic and human visits to other celestial bodies. With the 16 July 1969 launch of the Apollo 11 mission to land the first humans on the Moon, an updated policy was issued.

[1]

July 16, 1969
Effective date

Policy Directive

SUBJECT: EXTRATERRESTRIAL EXPOSURE

1. PURPOSE

This Directive establishes:

a. NASA policy, responsibility and authority to guard the Earth against any harmful contamination or adverse changes in its environment resulting from personnel, spacecraft and other property returning to the Earth after landing on or coming within the atmospheric envelope of a celestial body.

b. Security requirements, restrictions and safeguards that are necessary in the interest of the national security.

2. SCOPE

The provisions of this Directive apply to all NASA manned and unmanned space missions which land on or come within the atmospheric envelope of a celestial body and return to the Earth.

3. DEFINITIONS

a. "NASA" and the "Administrator" mean, respectively, the National Aeronautics and Space Administration and the Administrator of the National Aeronautics and Space Administration or his authorized representative (see NMD/A 8020.15).

b. "Extraterrestrially exposed" means the state or condition of any person, property, animal or other form of life or matter whatever, who or which has:

(1) Touched directly or come within the atmospheric envelope of any other celestial body; or

[2]
July 16, 1969
(2) Touched directly or been in close proximity to (or been exposed indirectly to) any person, property, animal or other form of life or matter who or which has been extraterrestrially exposed by virtue of subparagraph (1).

For example, if person or thing "A" touches the surface of the Moon, and on "A's" return to the Earth, "B" touches "A" and, subsequently, "C" touches "B," all of these – "A" through "C" inclusive – would be extraterrestrially exposed ("A" and "B" directly; "C" indirectly).

c. "Quarantine" means the detention, examination and decontamination of any person, property, animal or other form of life or matter whatever that is extraterrestrially exposed, and includes the apprehension or seizure of such person, property, animal or other form of life or matter whatever.

d. "Quarantine period" means a period of consecutive calendar days as may be established in accordance with paragraph 5a.

e. "United States" means the 50 States, the District of Columbia, the Commonwealth of Puerto Rico, the Virgin Islands, Guam, American Samoa and any other territory or possession of the United States, and in a territorial sense all places and waters subject to the jurisdiction of the United States.

4. AUTHORITY

a. Sections 203 and 304 of the National Aeronautics and Space Act of 1958, as amended (42 U.S.C. 2473, 2455 and 2456).
b. 18 U.S.C. 799.
c. Article IX, Outer Space Treaty, TIAS 6347 (18 UST 2416).
d. NMIs 1052.90 and 8020.13.

5. POLICY

a. Administrative Actions. The Administrator or his designee as authorize by NMD/A 8020.15 shall in his discretion:

(1) Determine the beginning and duration of a quarantine period with respect to any space mission; the quarantine period as it applies to various life forms will be announced.

[3]
July 16, 1969   NPD 8020.14

(2) Designate in writing quarantine officers to exercise quarantine authority.

(3) Determine that a particular person, property, animal or other form of life or matter whatever is extraterrestrially exposed and quarantine such person, property, animal or other form of life or matter whatever. The quarantine may be based only on a determination, with or without the benefit of a hearing, that there is probable cause to believe that such person, property, animal or other form of life or matter whatever is extraterrestrially exposed.

(4) Determine within the United States or within vessels or vehicles of the United States the place, boundaries, and rules of operation of necessary quarantine stations.

(5) Provide for guard services, by contract or otherwise, as may be necessary to maintain security and inviolability of quarantine stations, quarantined persons, property, animals or other form of life or matter whatever.

(6) Provide for the subsistence, health and welfare of persons quarantined under the provisions of this Directive.

(7) Hold such hearings at such times, in such manner and for such purposes as may be desirable or necessary under this Directive, including hearings for the purpose of

creating a record for use in making any determination under this Directive or for the purpose of reviewing any such determination.

(8) Cooperate with the Department of Health, Education and Welfare and the Department of Agriculture in accordance with the provisions of paragraph 6.

(9) Take such other actions as may be prudent or necessary and which are consistent with this Directive.

b. Quarantine

(1) During any period of announced quarantine, the property within the posted perimeter of the Lunar Receiving Laboratory at the Manned Spacecraft Center, Houston, Texas, is designated as the NASA Lunar Receiving Laboratory Quarantine Station.

[4]
July 16, 1969

(2) Other quarantine stations may be established if determined necessary as provided in paragraph 5a(4) .

(3) During any period of announced quarantine, no person shall enter or depart from the limits of any quarantine station without permission of the cognizant NASA quarantine officer. During such period, the posted perimeter of a quarantine station shall be secured by armed guard.

(4) Any person who enters the limits of any quarantine station during the quarantine period shall be deemed to have consented to the quarantine of his person if it is determined that he is or has become extraterrestrially exposed.

(5) At the earliest practicable time, each person who is quarantined by NASA shall be given a reasonable opportunity to communicate by telephone with legal counsel, or other persons of his choice.

6. RELATIONSHIP WITH DEPARTMENTS OF H.E.W. AND AGRICULTURE

a. If either the Department of Health, Education, and Welfare or the Department of Agriculture exercises its authority to quarantine an extraterrestrially exposed person, property, animal or other form of life or matter whatever, NASA will, except as provided in subparagraph c, not exercise the authority to quarantine that same person, property, animal or other form, of life or matter whatever. In such cases, NASA will offer to these departments the use of the Lunar Receiving Laboratory Quarantine Station and such other service, equipment, personnel and facilities as may be necessary to ensure an effective quarantine.

b. If neither the Department or Health, Education and Welfare or the Department of Agriculture exercises its quarantine authority, NASA shall exercise the authority to quarantine an extraterrestrially exposed person, property, animal or other form of life or matter whatever. In such cases, NASA will inform these departments of such quarantine action and, in addition, may request the use of such service, equipment, personnel and facilities of other Federal departments and agencies as may be necessary to ensure an effective quarantine.

c.   NASA shall quarantine NASA astronauts and other NASA personnel as determined necessary and all NASA property involved in any space mission.

[5]
July 16, 1969   NPD 8020.14

7. COOPERATION WITH STATES, TERRITORIES AND POSSESSIONS

Actions taken in accordance with the provisions of this Directive shall be exercised in cooperation with the applicable authority of any State, territory, possession or any political subdivision thereof.

8. COURT OR OTHER PROCESS

a.   NASA officers and employees are prohibited from discharging from the limits of a quarantine station any quarantined person, property, animal or other form of life or matter whatever during an announced quarantine period in compliance with a subpoena, show cause order or other request, order or demand of any court or other authority without the prior approval of the General Counsel and the Administrator.

b.   Where approval to discharge a quarantined person, property, animal or other form of life or matter whatever in compliance with such a request, order or demand of any court or other authority is not given, the person to whom it is directed shall, if possible, appear in court or before the other authority and respectfully state his inability to comply, relying for his action upon this paragraph 8.

9. VIOLATIONS

Whoever willfully violates, attempts to violate, or conspires to violate any provision of this Directive or any regulation or order issued under this Directive or who enters or departs from the limits of any quarantine station in disregard of the quarantine rules or regulations or without permission of the NASA quarantine officer shall be fined not more than $5000 or imprisoned not more than one year, or both (18 U.S.C. 799).

[signed by "Thomas O. Paine"]
Administrator

DISTRIBUTION:
SDL 1
Published in the Federal Register under Title 14 , Chapter V , Part 1211
(34 F.R. 11975 - 11976, July 16, 1969) .

Document III-13

Document Title: "Life Sciences in Space," Report of the Study to Review NASA Life Sciences Programs, Space Sciences Board, National Academy of Sciences, 1970.

Source: Space Studies Board, National Academy of Sciences, Washington, D.C.

Document III-14

Document Title: Homer E. Newell, Associate Administrator, NASA, Memorandum on Organizational Alternatives for NASA's Life Sciences Activities, 9 November 1970.

Source: NASA Historical Reference Collection, History Office, NASA Headquarters, Washington, D.C.

Document III-15

Document Title: Memorandum from George M. Low, Acting Administrator to Associate Administrator for Manned Space Flight, 3 December 1970.

Source: NASA Historical Reference Collection, History Office, NASA Headquarters, Washington, D.C.

As it began to plan and execute its programs in the post-Apollo period, NASA in the spring of 1970 asked the Space Science Board to undertake a comprehensive review of the NASA life sciences efforts. The review committee was chaired by H. Bentley Glass, a biologist at the State University of New York, Stony Brook. Once again, as had most prior reviews of space life sciences, the Glass Committee recommended the centralized management of most NASA life science activities, and once again NASA managers decided not to accept the totality of that recommendation.

Document III-13

Life Sciences in Space

Report of the Study
to Review NASA Life Sciences Programs
Convened by the
Space Science Board
National Academy of Sciences
National Research Council

H. BENTLEY GLASS
EDITOR

NATIONAL ACADEMY OF SCIENCES
Washington, D.C. 1970

[1] 1. Summary and Major Recommendations

The successful Apollo landings on the moon mark the attainment of the primary goal of the National Aeronautics and Space Administration for the 1960's and necessitate a reappraisal and reorientation of the objectives for the next decade. The objectives indicated by President Nixon in his statement of March 7, 1970, reduce to three: exploration, scientific knowledge, and practical application. Because the enlargement of human understanding and the increase of scientific knowledge occupy a position of growing importance in the official justification of the support of NASA's programs (see pages 9-10) [not included], the need to reevaluate the NASA programs in the life sciences is clear.

The present Study was convened under the auspices of the Space Science Board in the spring of 1970 to undertake this task. The Committee has visited major laboratories of NASA where life sciences work is being conducted and has conferred with numerous consultants now or previously engaged in the various aspects of life sciences work supported by NASA. It also examined the more recent surveys and critiques of NASA life sciences programs. Among these, Space Biology, the report of a study convened by the Space Science Board in Santa Cruz, California, in the summer of 1969, and the report (1969) of the Space-Science and Technology Panel of the President's [2] Science Advisory Committee, entitled The Biomedical Foundations of Manned Space Flight, received particular attention. The present Committee, composed for the most part of biologists not previously engaged in NASA programs and largely unfamiliar with the scope and the details of the life sciences programs supported by NASA, must disclaim any readiness in so short a time to provide a definitive evaluation of the total program in all details.

Nevertheless, in the three months of our preliminary planning and site visits and the more intensive two-weeks' study conducted at Woods Hole, Massachusetts, July 12-25, 1970, we have been able to arrive at a consensus about broad priorities in the life sciences as well as about the effectiveness with which NASA is currently able to pursue its life sciences objectives. The present report will be limited to these matters.

First, our Committee agrees that exobiology – the inquiry, into the existence of life elsewhere in the universe and the scientific explanation of the origin of life – represents the most important basic question in the life sciences disciplines associated with the space program as well as one of the great scientific questions of our time. In this we concur with the views expressed in the NASA Report for the Space Task Group, 1969 (see page 14 of the present report) [not included]. We are not optimistic, however, about finding living things elsewhere in the solar system: the moon is quite lifeless, and Mars, which presents the greatest probability for life of any of the planets, has an environment hostile to life as we know it. We believe, nonetheless, that the search for extraterrestrial life is a prime scientific goal, one that must be a part of any program of planetary exploration. Exobiology embraces considerably more than the simple question, "Does life exist elsewhere than on our earth?" The further questions that must be asked, and that must determine experimental strategy in this area, include the following: Has life ever existed, in former ages, upon the celestial body being examined? Are there indications of prebiological chemical evolution that would support or clarify present ideas about the origin of life? How does life develop? What are the environmental conditions that would prevent terrestrial organisms from populating the planet? Can a lifeless moon or planet, by controlled modification of the environment, become a laboratory for the adaptation and evolution of terrestrial life? In pursuing answers to these question, NASA can also make very important contributions by support of fundamental studies here on earth. Nevertheless, earth-based studies must remain preliminary. [3] They can only provide useful guides to the definitive explorations to be made in space itself.

Second, we agree that if manned space flight is to continue and be further developed, then there must be a much stronger and more broadly based program of research in the physiology and psychology of man in space, over and above the aspects of biomedicine concerned with the safety and efficiency of the astronaut or space passenger. We regard this consensus of our Committee to be in agreement with and parallel to the President's Science Advisory Committee Space Science and Technology Panel's emphasis upon the need to "qualify man for space." Unless this can be done, all other missions depending upon man in space must fail or be severely handicapped. We agree with that report's view that NASA has not had an adequate scientific program directed toward this goal.

Third, if the space station and space shuttle represent the technological goals of the coming decade, then such facilities should certainly be adapted to include an appropriate program in space biology. As citizens and scientists, we cannot avoid uneasiness over the large costs involved relative to the prospective gains in scientific knowledge. We have asked ourselves whether a better understanding of biological rhythms, radiation effects upon man and other organisms, and the biological effects of gravity and weightlessness justifies so great an expenditure of public funds in comparison with other fundamental biological problems and critical needs for federal support of the life sciences. Yet we also realize that Skylab, the space station, and the space shuttle will be programmed or abandoned for reasons other than the expectation of making important biological findings. We therefore reiterate our conviction that if the new space facilities are to be developed, they should provide for well-chosen and well-designed biological experiments.

Fourth, we emphasize the excellent opportunity for NASA to promote international cooperation within the life sciences. Much has already been done to encourage international participation in the planning and execution of certain kinds of experiments conducted in space. This is commendable. Nevertheless, more can be done in biological and biomedical experimentation, and, in view of the extensive Soviet manned and unmanned flights, cooperation with Soviet scientists is of significant mutual interest and should be earnestly sought.

Fifth, even a preliminary survey of the organization of the life sciences programs within NASA has disclosed grave defects, resulting from overlapping authority, insufficient [4] internal communication, a multiplicity of advisory groups, each with a very limited purview, inadequate programmatic involvement on the part of the life sciences community, and lack of any strong representation of the interests of the life sciences at high administrative levels. We are convinced that before any reordering of other priorities within the life sciences, there must be a thoroughgoing reorganization of NASA's administration of its life sciences programs. The present system exists for historical reasons which were presumably compelling at the time but which no longer obtain. Repeated recommendations for a reform of the administrative structure have remained unheeded. Without reorganization along such lines as our Committee and others have recommended, it is folly to expect any major improvement in implementation of goals and in the development of the life sciences within NASA. If the programmatic objectives stated above are worthwhile, then a better organization with a strong central voice for the life sciences must be sought to achieve them. The matter cannot be postponed without the gravest future damage.

MAJOR RECOMMENDATIONS

1. We recommend the appointment of a Life Sciences Advisory Board at a high administrative level within NASA, to review programs on a continuing basis and to recommend policies and priorities. This Board should have rotating-term appointments of individuals who are not currently staff members, grantees, or contractors of NASA.

2. We recommend the creation of a new Office of Space Biology and Medicine (OSBM) headed by an Associate Administrator for the Life Sciences or, alternatively, of a Deputy Associate Administrator for the Life Sciences in the Office of the Associate Administrator, in that order of preference.

3. We recommend, in either case of Recommendation 2 above, a functional reorganization of the life sciences programs, along disciplinary lines, into four Units: Exobiology and Planetary Ecology, Space Biology, Human Biology and Aerospace Medicine, and Personnel Health and Evnironmental [sic] Medicine. A biotechnology research group to coordinate the life sciences programs with physical science technology should be made an integral part of the life sciences organization.

[5] 4. We recommend a simplification of the advisory struc-ture and a reduction in the number of advisory committees and panels. Each of the four Units proposed in Recommendation 3 should have a single advisory committee to review plans and projects and to evaluate applications in support of in-house programs. Advisory groups or panels, organized along disciplinary lines, would be named to evaluate proposals for support of outside investigators.

5. We recommend the establishment of better criteria for the selection of flight experiments based on a careful review of existing criteria. The proposed Life Sciences Advisory Board should have a major role in the final decisions with respect to these criteria. As a continuing body of experts, it should be able to advise and recommend priorities in programmatic strategy and planning which an ad hoc committee such as the present one is unable to do in a few months of study.

6. We recommend the establishment of a NASA Life Scientist Program, in which up to 40 appropriately selected life scientists would receive six-year appointments (renewable once), with salary and minimal supporting research funds. Appointees would agree to spend one third of their six-year term (suitably arranged) at one or more of the NASA Centers or field stations. We are unanimous in the belief that no similar expenditure of funds by NASA will do more to generate the increased involvement of the scientific community, and that no other program will so quickly redound [sic] to the benefit and improvement of the life sciences programs of NASA.

7. We recommend that NASA inform the life sciences community, especially university staffs and students, more generally of its future plans and of opportunities to contribute to the planning of life science objectives. When definite plans and programs are formulated, the life sciences community should be informed well in advance of target dates, and specific research projects should be solicited with greater lead-time than in the past.

8. We recommend that an additional vigorous effort be made by NASA to encourage international participation in the planning and conduct of experiments in its life sciences programs and especially to seek active participation by the Soviet Union.

9. We recommend that the search for life on other worlds and for deeper understanding of the origin of life (exobiology) remain the prime scientific priority of NASA life sciences, one commensurate in importance (though not necessarily in expenditure or immediacy) with other primary scientific objectives [6] of NASA. This recommendation does not imply endorsement of any specific presently planned experiments (e.g., Viking): our Committee does not feel that it should endorse or criticize such experiments without a much more exhaustive study.

10. We recommend that, if manned spaceflight is to continue, then, in order to ensure the safety and efficiency of man in space and to "qualify man for space," a far broader program of space biomedicine and human biology should be undertaken. In particular, programs in space physiology and psychology must be selectively strengthened. The criteria for selection of astronauts and space passengers must be reevaluated periodically and should envisage a progressive transfer of the tasks of collecting data and conducting experiments from astronauts to persons trained as scientists.

11. We recommend that fundamental studies in space biology, and the supporting ground-based experiments in particular, continue to receive support, and that, because funding for and space within spacecraft must inevitably remain limited, only definitive experiments should be given preference for flight. Accordingly, 90 percent or more of the preliminary work necessary for experiments selected for flight, and probably most of the controls, should be ground-based. It is therefore appropriate for NASA, even within the strictest mission orientation, to support extensive ground-based biological research.

Document III-14

[no page number]

NATIONAL AERONAUTICS AND SPACE ADMINISTRATION
WASHINGTON, D.C.
20546

OFFICE OF THE ADMINISTRATOR   [stamped] NOV 9 1970

MEMORANDUM

TO: A/Acting Administrator

SUBJECT: Organizational Alternatives for NASA's Life Sciences Activities

The Glass Committee, as one of the strongest recommendations in its report on the recent summer study on the overall NASA life sciences program, urges NASA to draw together into a single organizational entity, under the direction of a competent life scientist, all of the research and research-related program in NASA's life sciences efforts. Alternatively, if NASA finds that this cannot be done, the Committee recommends that NASA reduce the undesirable effects of the current fractionation of the life sciences effort by appointing a Deputy Associate Administrator for Life Sciences responsible for the total integration of the NASA life science program.

The Glass Committee is firmly convinced that an effective integration of NASA's life sciences programs is essential to generate the interdisciplinary interaction between the various elements needed to produce the high quality results that should be obtained from the substantial investments made in the space program. Moreover, the Glass Committee is strongly of the opinion that such an improvement in program quality is going to be an essential factor in NASA's conduct of long-duration manned space flight missions. The Glass Committee arguments are quite persuasive, and moreover the NASA Directors of our Medicine and Biotechnology Programs agree with the Glass Committee that a substantial integration of the total NASA program is essential to achieving success and the required quality.

[2]On the other hand, NASA is not a life sciences agency, and its organization is determined primarily by other factors. Under these circumstances, there are persuasive reasons why different portions of the life sciences activity should be in different elements of the NASA activity. These would be cases where the value of interaction between the life science activity and another part of the NASA program overrides the importance of having all life sciences together.

In my view, we have several such cases in NASA. The medical flight operations, because of their criticality to manned space flight operations, must be located within the Office of Manned Space Flight; there appears to be universal agreement on this. Likewise, the life sciences work in the aeronautics program, because of its criticality to aeronautical flight and research operations, must be within the Office of Advanced Research and Technology, and must be staffed with enough medical research capability to be able to carry out an effective program. The reasoning behind this point of view is illustrated in the attached letter [not included] from Dr. Hans Mark, Director of the Ames Research Center. I feel, also, that having exobiology and planetary quarantine in the Office of Space Science and Applications is beneficial because of the central role that exobiology plays in planetary exploration at the present time, a role that is best carried out with the very closest of association between the biology, engineering, and physical sciences people involved in the effort.

The rest of the life sciences research could well benefit by being pulled together into a single office. Moreover, if we placed all of this research under the direction of a Deputy Associate Administrator (of the office) for Life Sciences, and assigned to that individual both the responsibility and the authority to oversee the total NASA life sciences research program, we will have gone a long way toward meeting the recommendations of the Glass Committee. By "oversee" I mean, working with the program offices, to develop an integrated life [3] sciences research program for all of NASA, to present to the Administrator each year a proposed program plan and budget for the integrated program, to review the conduct of the entire NASA effort and make periodic reports to the Administrator, to recommend on they filling of key life sciences positions, and to serve as NASA's principal contact with the outside life sciences community in life sciences matters.

After discussions with a number of people, including Dale Myers, Roy Jackson, John Naugle, General Humphreys, Walt Jones, and Hans Mark, I have come to the following conclusions. First, it does not seem to me that we want to load OSSA with responsibility for biotechnology and bioengineering, man-machine integration, biomedical research, or even with advanced bio-instrumentation Thus, I would eliminate placing the nucleus of life sciences activity within OSSA.

Secondly, if this point of view is accepted, then there appear to be two principal alternatives to consider. These are set forth in Attachment 2, entitled: CURRENT AND PROPOSED LIFE SCIENCES PROGRAM LOCATIONS [not included]. In both alternatives, medical flight operations would be in OMSF, life science research associated with aeronautics would remain in OART, and exobiology and planetary quarantine plus the very small activity in earth resources would remain in OSSA. In alternative 1 all of the remainder would go to OART, while in alternative 2 the remainder would be placed in OMSF. In either of the two alternatives I would recommend establishing the position of Deputy Associate Administrator (of the office) for Life Sciences, with the total life sciences oversight responsibilities discussed above. (Some other name such as "Aerospace Medicine and Biology" might be preferred over "Life Sciences".)

Advantages of alternative 1 are:
– A large part of the work involved is in the nature of advanced technology work, or applied research leading to the advancement of technology, and this is precisely the sort of research OART was set up to accomplish.

[4] – This arrangement would maintain the closest of association with the aeronautics aspects of our life sciences work, in a period when we are increasing our aeronautics effort, moving into more sophisticated man-machine systems, and increasing our work with flight prototypes.

– OART has one of the two major centers engaged in life sciences research, namely Ames Research Center, under its cognizance.

– OART has had extensive experience with the university community in the support and conduct of academic research programs, which form a very important segment of NASA's advanced life sciences research effort.

The disadvantages of alternative 1 are:

– Biomedical research, which the Glass Committee has emphasized will in the long run be fundamental to developing capabilities for long-duration manned space flight, will be in a separate office from medical flight operations, with which it must continually interact in order to realize the full returns desired.

– The Manned Spacecraft Center, which is engaged in a major portion of NASA's life sciences activities, is not directly under OART cognizance.

– Procedures for moving from the definition of flight experiments for manned missions to the engineering development and conduct of those experiments will be more complicated than if the total life sciences activity were under OMSF.

[5]
The advantages of alternative 2 are:

– Biomedical research will be closer organizationally to medical flight operations with which it must interact intimately.

– The process of moving from definition of flight experiments to their engineering, development, and conduct should be more straightforward than in alternative 1.

– The Manned Spacecraft Center, which has a large segment of NASA's life sciences activity, is directly under the cognizance of OMSF.

– The life sciences research program activities within OMSF would give the office an opportunity to wrestle internally with the conduct of a scientific program in the shuttle and space station; this experience should prove of value, in working on scientific programs in other areas with OSSA and OART.

The disadvantages of alternative 2 are:

– The major nucleus of life sciences research is separated from the aeronautics program, with which it should interact intimately.

– Responsibility for a research effort is placed in a development and operations office, where special care will have to be taken to ensure that the ever-present urgent demands of development and operations do not act so as to short-change the research.

– The Ames Research Center, which has a large segment of NASA's life sciences activity, does not come under the cognizance of OMSF.

[6] Both of the foregoing alternatives are based on the premise that we wish to pull together as much as possible of the life sciences activity, excluding only those activities that must, for reasons stated earlier, be located in specific program offices. Based on past history and the various discussions we have had on the subject, the choice of either alternative would produce strains in NASA, leaving some key people strongly dissatisfied. On the other hand, to fall back from these two alternatives would certainly be viewed by the Glass Committee as leaving the NASA life sciences program substantially as fractionated as before, and would be viewed as a rejection of the Committee's very strong recommendation.

With respect to medicine and medical research, the major and most visible activity within NASA will be associated with manned space flight. The Glass Committee has said that it is undesirable to separate clinical medicine from medical research. Rather, the two should be brought together so that both can benefit from the interaction between the practical urgencies of the clinician and the exploratory viewpoint of the research scientist. Also, it is important that OMSF direct its thinking and planning toward the research that will be necessary if the future needs of manned space flight are to be fulfilled. Thus it would appear, on balance, that alternative 2 (combining all life sciences, other than certain unique programs, in OMSF) should be chosen. I therefore recommend that NASA take appropriate steps to implement alternative 2.

[Original Signed by Homer E. Newell]
Homer E. Newell
Associate Administrator

Document III-15

[no page number]
[stamped "DEC 3 1970"]

NATIONAL AERONAUTICS AND SPACE ADMINISTRATION
WASHINGTON, D.C. 20540

OFFICE OF THE ADMINSTRATOR

TO: M/Associate Administrator for Manned Space Flight

FROM: A/Acting Administrator

SUBJECT: Establishment of a NASA Director of Life Sciences in the Office of Manned Space Flight

The position of NASA Director of Life Sciences is established, effective this date, in the Office of Manned Space Flight with responsibility and authority to oversee the total NASA life sciences program.

I   FUNCTIONS AND RESPONSIBILITIES

The NASA Director of Life Sciences, working with the Program Offices will develop an integrated life sciences research program for all of NASA, present to the Administrator each year a proposed program plan and budget for the integrated program, review the conduct of the entire NASA effort and make periodic reports to the Administrator, recommend on the filling of key life sciences positions and serve as NASA's principal contact with the outside life sciences community in life sciences matters.

The NASA Director of Life Sciences will be responsible for providing Biomedical Flight Operations support to project offices and for the direction of all life sciences research and technology programs including Biomedical Research, Bioscience Research, Life Support and Protective Systems, Man Machine Integration, Advanced Bioinstrumentation and related flight experiment definition.

The following Life Sciences activities will remain assigned as follows:

a. Exobiology and Planetary Quarantine will remain within OSSA.

b. Research relative to man's operating environments in aircraft and the techniques and equipment required to improve and measure his performance, noise research and simulation technology, will remain within OART.

[2]
c. Occupational Medicine and Environmental Health will remain within the Office of Administration.

II   BUDGET

FY 1971 program in areas impacted by this reorganization will continue as planned except that funds approved for these programs will be transferred to the OMSF consistent with the revised functions and responsibilities.

The proposed FY 1972 budget will be reconstructed to reflect the realignment of responsibilities and a line will be established within the Space Flight Operations budget authorization to cover OMSF life sciences responsibilities.

III ORGANIZATION

Details including the restructuring and the identification of key positions will be developed and submitted for approval as soon as possible.

IV MANPOWER RESOURCES

Negotiations to identify and transfer manpower resources from OSSA and DART will be completed as soon as possible.

V APPOINTMENT

Dr. J. W. Humphreys, Jr., is appointed as NASA Director of Life Sciences.

[signed]
George M. Low

Document III-16

Document Title: Meeting Record, "Proposal to Undertake Search for Extraterrestrial Intelligence," 11 September 1973.

Source: NASA Historical Reference Collection, History Office, NASA Headquarters, Washington, D.C.

Document III-17

Document Title: Letter from Bernard M. Oliver, Hewlett Packard Company, to James C. Fletcher, NASA Administrator, 20 September 1973.

Source: NASA Historical Reference Collection, History Office, NASA Headquarters, Washington, D.C.

Document III-18

Document Title: Memorandum from James C. Fletcher, NASA Administrator, to Dr. Homer Newell, Associate Administrator, "Cyclops Proposal," 4 October 1973.

Source: NASA Historical Reference Collection, History Office, NASA Headquarters, Washington, D.C.

John Billingham at NASA's Ames Research Center was the moving force behind the birth of the SETI project. In 1971 he teamed with Bernard Oliver, an electrical engineer and former Vice President for Research at Hewlett Packard and a technical expert in microwave signal processing. Together with Frank Drake, the astronomer who had first developed a mathematical equation containing the relevant factors involved in determining the likelihood of communicative civilizations in our galaxy, they organized a summer study to plan a project for accomplishing their goals. The result of their efforts, titled Project Cyclops, "envisioned a detector consisting in its final stages of an 'orchard' of perhaps 100-meter antennas covering a total area some 10 km in diameter." As these documents show, when Project Cyclops was presented to NASA's leadership as a possible large-scale initiative—the cost projection for the total project was in the range of $8 billion—the concept received a sympathetic *reception on intellectual grounds, but was thought to be too politically risky and too uncertain of success to justify significant NASA funding.*

Document III-16

[no page number]

MEETING RECORD

DATE: September 11, 1973

TIME: 10:00 am. - 12:15 p.m.

ATTENDEES: Dr. Fletcher, Dr. Newell, Dr. Barry - HQ Dr. Mark, Dr. Billingham, Dr. Wolfe, Dr. Johnson - ARC Dr. Oliver - Hewlett Packard

SUBJECT: Proposal to Undertake Search for Extraterrestrial Intelligence

COMMENTS:

The ARC personnel, supported by Dr. Oliver, outlined a proposal that NASA undertake a study to establish the feasibility of initiating a serious search for extraterrestrial intelligence. Described in general terms, the feasibility study would involve the selection of a search mode, an examination of engineering alternatives, the selection of a preferred system, and the preparation of a work plan for detailed engineering design. The study would cost $300K in FY 74, and $1100K in each of FY's 75 and 76 for a total of $2.5M.

Specific discussion centered on the evolution of an earthbound searching system which might eventually consist of microwave receiving and recording equipment in conjunction with a clustered array of 1000 to 2500 antennae having an effective area ranging from 7 to 20 km2. This facility would conceivably detect microwave transmissions from selected solar systems on a target-by-target basis out to a distance of 1000 light-years (the thickness of our disc-shaped galaxy), assuming the existence of a 1000 MW omnibeacon at 1000 light-years broadcasting in the frequency band 1420-1660 MHz. Such a facility would cost approximately $8B to plan and construct.

Several potentially advantageous features of the proposed study were mentioned in the presentation. The overall program would span a considerable period of time, and would require a limited initial investment. It would be a new application of space technology which would ostensibly generate widespread public support and interest, and, at the same time, it would [2] serve as a possible avenue for international participation. If extraterrestrial contact should be made, the event would have great scientific, philosophical, and social significance. New and useful knowledge would be forthcoming to help provide answers to questions associated with galactic history, longevity of a civilization, and the biochemical nature of alien life. If extraterrestrial contact should not be made, the most significant benefit of the overall program would be an enhanced capability in the field of radioastronomy.

In discussion following the formal presentation, Dr. Fletcher based his remarks on anticipated public opinion. He posed the crucial question of whether the proposed program would actually generate widespread public support and interest. Dr. Fletcher stated that it would be quite undesirable to fund such a program over a 10 year period with no significant results having been realized. He said that the general public were [sic] not greatly interested in radioastronomy, and that if the chance of achieving contact were something on the order of 1 in 10, he would not want to support the effort.

Drs. Mark and Oliver could not offer positive assurance that extraterrestrial contact would be made. Instead, they pointed out the relationship between the longevity of a civilization and the probability of a successful contact. For a longevity of 10^3 years, the contact probability was considered to be small. For a longevity of 10^6 years, the contact probability was assessed as high. If the entire $8B program were funded, 30 years would be required to search the entire galactic thick.

Dr. Fletcher stated that, in his opinion, the public would not be willing to take a long shot on a civilization lasting 10^6 years. He felt that the proposal would be more palatable if other desirable spin-offs and side effects in addition to an enhancement of radioastronomical capability could be identified and emphasized.

[signed]
Acting Executive Secretary: Harvey W. Herring

Document III-17

HEWLETT-PACKARD COMPANY
1501 PAGE MILL ROAD
PALO ALTO, CALIFORNIA 94304

BERNARD M. OLIVER
VICE-PRESIDENT
RESEARCH AND DEVELOPMENT

20 September 1973

Dear Jim:

I want to thank you and Homer for taking the time to discuss the future of interstellar communication with us last week.

Your concerns over possible adverse political reaction to the proposed study are understandable. We are in complete accord that if a final program is to succeed, we must proceed openly and with full congressional and popular understanding of the magnitude of the effort, of the time scales involved, of the consequences of success and the possibility of failure. It occurs to me that, not being a NASA employee, I may in some instances be in a better position to help win that support that [sic] if I were. If you have any suggestions I would like to hear them.

Jim, I am very worried, as I know you must be, over the long term future of NASA. The shuttle program is a means to an end – the end of getting sizable payloads into orbit more cheaply. Whether the missions needing these payloads will be approved remains to be seen, and the shuttle itself may encounter technical problems. If it gets shot down, NASA will lose a great deal of talent and momentum that took years to assemble. Small probe explorations can continue, of course, but after we've explored Mars and the outer planets, what then?

At the brink of space – real space – our missions must stop, but our curiosity will not. The only way I can see of greatly extending NASA's lifespan is to extend the radius of its sphere of influence a million fold; from a milli-light year to a kilo-light year, with a program like Cyclops. Surely this possibility must be explored further.

I am convinced that the real hidden motive that lay behind NASA's early popularity would be restored if we could once more entertain the expectation of finding other life. You asked what we really thought the chances of success were and I hesitated because I couldn't give a precise figure. Let me say that I believe the probability of success is on the [2] order of unity, that is, not 10^{-1} or 10^{-2} or less. Knowing that great discoveries in astronomy and cosmology will be realized in any event, this is enough for me.

I strongly hope Ames can proceed with this program. In my view, the organizational risks of not having a strong backup program ready outweigh the political risks of going ahead. In addition to giving us a better fix on the task of the search, the study represents an excellent way to test the political waters with our little toe, so to speak, before diving in. As I said, if there is any way I can help to warm these waters for you, let me know.

With best personal regards,
[signed]
B. M. Oliver

Dr. James C. Fletcher
National Aeronautics and Space
Administration
400 Maryland Avenue, SW, Room 7137
Washington, D.C., 20546

Document III-18

[no page number]
[stamped OCT 4 1973]

MEMORANDUM

TO :   AA/Dr. Homer Newell

FROM: A/Administrator

SUBJECT: Cyclops Proposal

After screening the Cyclops proposal and participating in a discussion with Hans, Barney and others, I have come to the following tentative conclusions:

1) The ideas is a very intriguing one, particularly if there is a real possibility that in a reasonable length of time we would indeed be able to pick out signals transmitted by a society similar to our own somewhere within 100 million light-years of earth.

2) The calculated statistical probability of being able to so gives me great concern. As I understand it, we need to assume that the life expectancy of a high technological society, such as our own, must be of the order of several thousand years to get somewhere near unity probability of detecting a similar society with the proposed system.

3) The assumption of several thousand years may appear to some to a reasonable one, but it would certainly be unprecedented compared with seemingly more stable societies (e.g., Byzantine Rome and Egypt), and would certainly be considered unduly optimistic by most scientists.

4) Even if this assumption is made, we must still assume that the other society is willing to transmit with a multi-directional antenna the required megawatts for the full several thousand years of its existence and in the particular band that we happen to have chosen as being optimum.

[2]
5) There is even some question in my own mind as to whether indeed we have picked the optimum frequency, although this amendable to further study.

Putting all of the above together, I have serious doubts regarding whether Cyclops makes sense at NASA's present budget levels. However, if Ames feels strongly that a further study of this should be made, I would offer no objection. If you, George Low or Hans would like to discuss this further, I would be happy to do so.

[stamped "Original signed by James C. Fletcher"]
James C. Fletcher

cc: AD/Dr. Low

Document III-19

Document Title: Report, **"The Life Sciences Program of the National Aeronautics and Space Administration,"** 1 February 1974.

Source: NASA Historical Reference Collection, History Office, NASA Headquarters, Washington, D.C.

NASA Administrator James C. Fletcher became concerned in 1973 with conflicts within the life sciences program and asked William Barry of Ames Research Center to conduct yet another review of the program. In February 1974, Barry presented to Fletcher his report outlining the same fundamental flaws that were raised by all previous reviews—responsibility without authority and the fragmentation of programs. He also identified the problem of "bureaucratic inertia" from those who were seeking to maintain 1960s relationships while the program was trying to change to the new priorities of the 1970s.

[no page number]

THE LIFE SCIENCES PROGRAM

OF THE

NATIONAL AERONAUTICS AND SPACE ADMINISTRATION

Headquarters, NASA
Washington, DC
1 February 1974

[1]

I. Historical Perspective

The problems inherent in conducting an agency-wide Life Sciences Program have long been recognized. Since the establishment of the National Aeronautics and Space Administration in 1958, at least 13 major studies have been undertaken to give top NASA management guidance and direction in the conduct of its Life Science Program (Attachment A).[not included] After reading these reports in their entirety, some recurring recommendations become readily apparent. The first of these is the repeated counsel to consolidate the management of all elements of the life sciences in a single office located at NASA Headquarters. The administrative level of this office, and its location within the agency's organizational structure have been widely debated. The most common suggestions have been to establish the position of Director at either the Associate Administrator or Assistant Administrator level, and the proposed location within the organization has varied with the chronological age of the agency, the perspectives of the respective advisory committees, and apparently with the perceived actions and reactions NASA had taken to previous studies.

An early abortive, attempt at consolidation was the Office of Life Science Programs established on March 1, 1960. It was headed by Dr. Clark T. Randt and reported to the Administrator of NASA "in the same manner and at the same directional level as the other program directors." A significant distinction in this program office however, was the mechanism of funding. Although detailed evaluation is difficult, it appears that at the time of abolishment on November 1, 1961, this office still had only minimal budgetary responsibility. It existed largely as an advisor organ available as counsel to the "real" program managers, and affecting coordination only when input was provided from the line organizations.

The parochial requirements of the major program elements of that time (OMSF, OART, and OSSA) continued to support independent efforts. For several years after 1962, there was an agressive [sic] build-up of NASA funds in the SR&T programs of both OART and OSSA. Much impetus was apparently provided by the Bioastronautics Panel Report of the President's Science Advisory Committee, in February of 1962. During this period Drs. Lovelace and Pace served as consultants to the NASA Administrator, in repeated [2] attempts to consolidate the agency-wide life science effort.

As the agency effort in manned space flight gained momentum with the termination of Project Mercury and the beginning of Project Gemini, the role of man-in-space took the attention of most of the NASA Life Sciences establishment regardless of where in the organizational chart they were located. Flight programs which were evisioned [sic] to have life science objectives included Gemini, Biosatellite, Mariner, Voyager, Apollo and more advanced concepts such as the Manned Orbital Research Laboratory, the Manned Lunar Base, and the Manned Mars Missions.

An anomaly in the organization had developed by 1963, the Headquarters Occupational Medicine function. This had been separated and was located in the Office of Administration. Despite a knowledge of the personalities involved in the establishment of this office, and the temporal expediency once served by its separation, it has persisted. Previous reports are practically unanimous in their recommendation that it be aligned with the rest of NASA's life science organization.

As the Apollo Program progressed to a successful conclusion, and as the Skylab Program with its significant life science experiment role approached, NASA once again sought the assistance of the external scientific community regarding appropriate directions for its life sciences effort. The Space Science Board of the National Academy of Sciences-National Research Council convened an ad hoc panel in response to a formal request by NASA. Its mandate was to review all previous NASA Life Sciences Programs, and to make recommendations aimed at shaping agency goals and priorities for the 1970's. Despite the fact that none of the numerous appraisals of these activities since 1963 had really addressed itself to the entire range of NASA's life sciences work, considerable reluctance to participate was evident on the part of some previous NAS consultants. The reasons cited centered on the widespread perception that NASA had shown no obligation or inclination to implement recommendations of previous studies. The findings and recommendations of this ad hoc panel chaired by Dr. H. Bentley Glass, were published in a final report entitled Life Sciences in Space, in December of 1970.

[3]

Shortly after completion of that study, agency top management established the position of NASA Director for Life Sciences. Management Instruction 1138.14, dated May 28, 1971, described the roles and responsibilities of this position. Specifically, to develop an integrated life sciences research program for all of NASA, by working closely with the program offices involved. A requirement to present a proposed plan and budget to the Administrator each year was also established. The responsibilities for this position included a line function in the Office of Manned Space Flight to plan, budget, justify, and manage a space life research, technology, and experiment definition program. This program mandate included biomedical, behavioral and biological sciences; life support systems and crew equipment; atmosphere, food, water and waste management subsystems; man-machine integration, human augmentation devices; advanced bioinstrumentation and habitability. In addition, an overview function with responsibilities for "guidance, review, and recommendations" was outlined for the following agency activities:

    a. Office of Space Science and Applications (1971) (exobiology, planetary quarantine, and ecology).

    b. Office of Advanced Research and Technology (1971) (aeronautical life sciences, human performance augmentation, stress measurement, bioengineering, noise research and simulation technology).

    c. Office of Administration (occupational medicine and environmental health).

It is significant to note however, that this latter category of overview responsibility included no budgetary mechanism by which the Director could effect the degree of integration and prioritization apparently envisioned. The Director was further cautioned in paragraph 4a. of the NMI, that in exercising his responsibilities related to the life sciences activities in OART, OSSA, and the Office of Administration he should "be mindful of the line responsibilities of these offices over the people working in life sciences activities assigned thereto." Final direction stated that in the performance of his duties, the NASA Director for Life Sciences would function as the equivalent of a program office deputy associate administrator. The position of Director since the time of its establishment [4] (May 1971) has been held by two physicians. The first was USAF Major General James Humphreys; the current Director is Dr. Charles Berry.

II. The Current NASA Life Sciences Program

The recent budgetary restraints on NASA have been reflected in its life sciences program. The level of effort showed a decline from FY '73's $66 million total to a level of approximately $57.3 million in FY '74. (The R&T base declined in the same period from $39.2 million to $35.5 million.) This represents roughly 2% of the annual NASA budget. Of the SR&T resource total, less than one quarter is spread between Lewis Research Center, Langley Research Center, Kennedy Space Center, Dryden Flight Research Center, Goddard Space Flight Center, Jet Propulsion Laboratory, Marshall Space Flight Center and Wallops Station. Essentially all of the agency life science direction comes from Headquarters, Johnson Space Center, and Ames Research Center.

Life sciences manpower is fairly evenly divided between civil service (441) and support contractor (392) personnel. This represents slightly less than 2% of NASA's current manning. These manpower proportions are similarly reflected at ARC and JSC, with well over half of the total NASA effort being located at these two centers.

The current Occupational Medicine and Environmental Health program is conducted on an individual basis at each Center. It is managed by a small Headquarters staff, with line responsibilities in the Office of Administration. The need for standardization is reflected by the widely varying patient costs and benefits among the Centers. Each office's integration into the overall activity of its respective Center in areas such as environmental health and safety, job site surveys, hazardous area monitoring, protective equipment for known occupational exposures and worker outpatient health programs varies widely. The demands of the Occupational Health and Safety Act of 1970 (OSHA) alone, should prompt some closer, coordination in the preparation of each Centers' required Environmental Impact Statements. These are currently being [5] written by staff engineers without appropriate medical review. Although attempts at closer coordination between individual center efforts have been undertaken by the use of such techniques as annual occupational medicine meetings, the wide variety of each office's structure and function (civil service vs. support contractor, etc.) markedly resists standardization. In some cases, the medical personnel are closely related to the center mission with its respective hazards, requirements and problems; while in others they merely function as private practitioners, essentially unaware of the clinical importance of the hazards inherent in the highly complex NASA industrial work environment. Standardization of this $3 million per year effort should be a significant element of any proposed reorganization process.

A. The Headquarters Life Science Organization

At the time of its establishment in 1971 after the NAS-NRC study, the office of the NASA Director for Life Sciences appeared to have been structured around individual strengths of the personnel transferring in from other program offices involved in the consolidation (Attachment B [not included]). Because of the attendant parochial loyalties (programmatic as well as personal) involved in the consolidation process, apparently no zero-base exercise in organizational structuring was attempted. As a result of the requirement to accommodate the ingress of key personnel from what had been, in effect, three independent life science programs, all of the current division heads are very senior. Most appear to have very little grass roots program experience, and there seems to be a distinct absence of feeling for problems at the Primary Investigator and Center level.

From the time of consolidation under Dr. Humphreys, through Dr. Berry's tenure, the lack of dialog between the life science communities at each Center has been striking. The deficit of cross-fertilization can be cited in numerous examples. ARC personnel have only recently been invited to participate in Skylab debriefings at JSC, despite an extensive basic science research effort aimed at assisting the development of solutions for current and future operational problems. Similarly, JSC has had considerable concern regarding the non-invasive techniques currently being used to measure cardiac output and other cardiovascular parameters; because of this they have gone to [6] the NIH with a request to conduct a study regarding the use of echo equipment. During this same period, ARC in concert with researchers at Stanford, had become one of the nation's recognized centers for state-of-the-art expertise in doppler echo techniques. The lack of rapport is obvious. Much of what is transmitted between centers (leaving Headquarters out of the loop) occurs at the P.I. level. Although this coordination is hit-and-miss, some excellent examples of cooperation can be found. The efficiency of utilizing another center's well-developed laboratory competence in processing sophisticated clinical samples is particularly noteworthy; it underlines the need to extend such cooperation.

An aspect of routine operational technique on the part of the entire Headquarters staff, which is readily obvious to even the most cursory observer, is the attitude of OMSF program preeminence. The appropriateness of this pervasive approach is beyond the scope of this review; its effect on the total NASA Life Sciences Program however, demands that the question of organizational structure be raised. The perception that the MM Directorate (even though charged with the additional responsibility of being total program coordinator) is primarily OMSF line organization, is ubiquitous. This attitude is noted in the Headquarters review of program/project PAD's which have a life sciences component. These PAD's are required to have the concurrence of the NASA Director for Life Sciences. Direction from the NASA Comptroller on March 19, 1973, outlined the seven unique project numbers (UPN's) in which Life Sciences were to be conducted. In the areas of Space Life Sciences, Planetary Biology, and Planetary Quarantine the effort had already been defined at the UPN level; and no major restructuring of the PAD elements was required. However in the case of Aeronautical Life Sciences, Medical Engineering, Ecological Studies, and Technology utilization, specific action was required

to break out the life sciences component at the unique project level and to incorporate those projects in the applicable PAD. In keeping with the objective of a total overview of all NASA Life Sciences, the Manned Space Flight PAD was extended to cover not only its own area of responsibility (i.e. Space Life Sciences) but also to summarize and integrate the total NASA Life Sciences effort. The improvements noted since these implementing directives were published, are modest at best.
[7]
A system which had been utilized with success by OAST, involving review by task discipline (RTOP), was adopted by the other Associate Administrators for their life science projects. A widely held opinion however, is that the OMSF projects (Space Life Sciences) for which the MM Directorate retains direct line responsibility, receive a predominance of management attention. A corollary to that observation is that the other major program offices' life science projects receive only secondary emphasis. The present Life Science Director's concern in this regard is considerable. But in light of his dual mandate of line responsibility on one hand (with associated budgetary authority), and his "guidance, review, and recommendation" function on the other (with a caution to "be mindful of the line responsibilities OA, OSS and OAST"), the direction of his attention and emphasis are predictable.

An additional overview function which the Director has been involved in since the NAS-NRC study, has been with the Office of University Affairs. The Life Scientist grant program suggested by the study was begun on a limited basis with the selection of three university-based life scientists in 1972. NAS-NRC recommended 40 such Life Scientists, who would spend one-half to one-third of their time at NASA Centers. This demonstration program however, has not been effective. Although continuation for an additional year at current funding levels is planned, resolution of the program's future remains unclear.

B. Ames Research Center

The roles and missions of each individual NASA Center are understandably modified by the interest and support of its respective Center Director. Nowhere in NASA is this more apparent than at ARC. Dr. Hans Mark's knowledge, participation in, and personal support of the Life Science Program which has developed under his aegis is exceptional (Attachment C [not included]). ARC has developed strong working relationships with both nearby colleges and universities, and the external scientific community. These relationships in the life sciences are practically ideal. ARC, an OAST Center, is currently engaged in extensive life science research for OSS, OA, OAST, and OMSF. The extensive degree of of [sic] OMSF research which has developed in the past few years is striking. This growth was prompted to some degree by an interest [8] of the Headquarters Life Sciences Office aimed at improving the overall agency competence in operational space-related research. A more forceful etiology however, was JSC's preoccupation with current on-going mission requirements, and a predisposition to engage in only the most hardware-related applied research. This predisposition by all reports was pervasive, and extended even to exclude important basic projects which were demanded by the very flight program to which their support was dedicated. In light of such circumstance, ARC acted to fill the void, and develop the basic life science research capability which the agency required. This was culminated in FY '73, with OMSF research representing some $8 million, roughly 73% of the total ARC life sciences budget. While some elements of this research were no doubt appropriate for ARC direction, a significant percentage could probably have been more effectively located at JSC. As the Skylab Program draws to a close, alternative roles and missions for the sizeable proportion of the JSC life sciences staff currently devoted full-time to operational aerospace medicine, will have to be established. An organization with obvious overlap, in the spectrum extending from purely basic to totally applied research, is ARC's Biomedical Research Division.

A significant part of the research currently being conducted by the Biomedical Research Division is closely akin to the applied aspects of space life science. Organizationally, this has clearly been described as a prime JSC mission function. While much of this type of preliminary research (endocrinology, metabolism, environmental physiology, etc.) could be termed basic–it takes an agency program which is closely integrated in fact as well as in name, to avoid costly overlap. The Headquarters Life Sciences Office must maintain cognizance over such research, to shift the thrust of activity from ARC to JSC as it becomes more closely related to the flight program phase. The recent female bedrest studies utilizing Air Force active and reserve flight nurses provides an excellent example of this jurisdictional problem. Much of the animosity which resulted between the Centers could easily have been avoided by Headquarters Life Science Office leadership and coordination. Resolution of this problem continues to be a matter of primary concern, as planning for extensive follow–on protocols is proceeding apace. In an attempt at objectivity however, it should be added that similar problems will no doubt [9] recur as long as JSC remains indifferent to all but unequivocal operational support, and ARC continues to agressively [sic] fill the vacuum.

A recently approved addition to the ARC Life Sciences directorate has been a flight experiments office (Attachment D [not included]). This office should allow a close overview by the Director and Deputy Director of Life Sciences of the entire pipeline of Experiment development: from inception, through proof-phasing, to the delivery of a piece of flight ready hardware. If the Headquarters Life Sciences Office develops the degree of coordination outlined above, it would have the ability to shift the control of these developing experiments to the applicable project/program office as required. Whether this final integration process occurred at JSC, or at MSFC (with its developing responsibilities in this area for the Shuttle), the NASA Director of Life Sciences could support it appropriately.

A significant parameter in ARC's studied development of a capable in-house organization has been their interaction with the academic community. The utilization of grants, with only occasional use of contracts, has worked well. The augmentation of in-house personnel by research associates, post-doctoral researchers, grant-supported students, and other technical personnel has been highly effective. The development of the ARC in-house staff of civil service professionals has been exceptional. This staff is now over twice the size of JSC's (175 to 70); with manpower tables similarly reflecting JSC's disparate dependence on support contractor personnel (100 to ARC's 60). Leadership in all three divisions (Planetary Biology, Biomedical Research, and Biotechnology) is good. The Division Directors besides being professionally capable, have had sufficient management experience and tenure to develop considerable talent at the Branch level. Two of the divisions manage their own extramural grants, while the other does not. The Exobiology Division does not because its responsible Headquarters Program Office (OSS) has retained central control. This variance in policy is repeated in the disparity noted between major program office reporting requirements. These differences in administrative procedures are most apparent at the Center Life Sciences Office, where time and effort are lost accommodating non-standardized systems.

[10]
Some aspects of current ARC research deserve comment. The first is in the obvious lack of any biomedical research programs in the aeronautical life sciences area. Concentration on the hardware aspects of the man-machine interface is virtually complete. While this has led to some promising work in such areas as new cockpit displays, simulation, crew-tasking, avionics integration, flight-path prediction, and ADI/map/HSI presentations, the lack of biomedical programs is striking. ARC personnel attribute this lack to policy developed at Headquarters OAST, by a single life science manager (Gene Lyman), who is not responsible to Dr. Berry's office. When one considers the outstanding simulation techniques being developed at Ames, and the wide variety of testbed aircraft available on station, failure to capitalize on the obvious opportunities is particularly regrettable. The correlation of "flight truth" for highly advanced aircraft simulation systems to the "ground truth" of ERTS data is virtually one-to-one. Life Science personnel at ARC describe the flight testing phase as the ultimate dream; scientists involved with analogous research at other locations take it for granted as a necessity in each phase of development. A cursory sampling of ARC flight test personnel seems to indicate that sufficient flexibility exists in the operational flight schedule to accommodate a considerable number of these missions, especially if the piggybacking of experiments on already-scheduled missions is acceptable. When ARC test pilots participated in previous life science research (biological sampling on 4 hour U-2 flights while in partial pressure suits, etc.), pilot acceptability was excellent, and rapport regarding research participation was instant. Another personnel resource for aeronautical life science projects is the pool of experienced USN aircrews at NAS Moffett. Cooperative participation on such research projects could serve as a positive example for other USN-ARC joint programs.

Another locus of expertise which has developed at ARC is the group working in the area of vestibular physiology. In light of the current Russian work in preconditioning, and their extensive human and animal research program, their early reports of widespread motion sickness seem plausible. Considering our Skylab experience, crew selection criteria must be improved; and the portent for shuttle's short seven day mission profiles seems obvious. As with other problems, JSC has been prone to [11] look outside of NASA for assistance in this perplexing dilemma. Unfortunately the consultants most often selected (Naval Aeromedical Institute, et al) have a uniformity of experience in angular disturbance (slow rotation rooms, etc.); while ARC's Neurosciences Branch has been aggressive in looking into more widely variable stresses. Once again the problem of NASA-wide coordination seems at issue.

Other continuing problems which ARC might profitably pursue are: the venous compliance changes seen on SL-4, to include the development of appropriate countermeasure techniques; hematologic studies to address the open question of bone marrow regulation in weightlessness (erythropoietin studies, etc.); and the search for sensor etiologies for the significant red cell mass reductions noted during all of the Skylab missions.

C. Johnson Space Center

The life sciences activity at JSC has been focused on operational support requirements for ongoing flight programs since the beginning of the Center in the early 1960's. The rapid growth of life support technology required for Projects Mercury, Gemini, and Apollo appropriately dictated the development and capabilities of the JSC life sciences staff. Because of the systems approach to spacecraft hardware development, and the lack of scientific and technical expertise on the part of what was then, in effect, the "Flight Surgeon's Office;" decentralization of effort was complete. With some insight regarding that legacy, and the attendant expertise which then developed in a wide variety of Center organizations that had both systems responsibility and developmental funding, JSC's current organizational dilemma is understandable.

There are currently a minimum of twelve major JSC divisions or offices which have some ongoing life sciences effort, outside the control of the Director of Life Sciences (Attachment E [not included]). Examples include the widespread activities noted in Bioenvironmental Technology and Bioengineering. Divisions working in these areas include three of those in the Life Sciences Directorate (Bioengineering, Biomedical Research, and Health Services), as well as three in the Directorate of Engineering and Development (Crew Systems, Spacecraft Design, and Information Systems)–each under a separate Assistant Director, and another [12] in the Directorate of Flight Crew Operations (Crew Training and Simulation). This excessive degree of decentralization has long been recognized. An attempt to improve the abysmal level of coordination between all of the diverse elements at JSC operating in life science areas was outlined in a memorandum dated March 29, 1973, establishing a JSC Life Sciences Steering Committee (Attachment F [not included]). The functions of this committee were ostensibly to improve the management and review of all life science activities at the Center. It was planned to meet twice a year to carry out an in-depth program review and make recommendations on the content of the RTOP program. An additional inclusion in the memo was the note that JSC should attempt to organize a management team which more closely aligned with the Headquarters Office of Life Sciences. Accordingly, it appointed JSC technical managers for the three major areas of Bioresearch, Bioenvironmental Technology, and Bioengineering. Although some improvement may have been derived from such structuring, widespread dissipation of effort and uncoordinated, redundant activities are at once apparent to the objective observer.

The reasons for JSC's difficulties in the area of Life Sciences are manifold and diverse. The morale effect of fewer and fewer planned manned space flights, on an operational support organization must be recognized. In the development of alternate Center roles and missions for the post-Skylab era of the Space Shuttle, an integral part of the change envisioned for JSC must include the Directorate of Life Sciences. The progressive elimination of research talent during the operationally-oriented decade of Apollo, must be reversed.

The life science staff at JSC is significantly limited in scope. This stems from two major factors. The first has been outlined above; it is the pervasive operational orientation of the Johnson Space Center, from the Center Director on down. While the reasons for this have similarly been alluded to, the success of any conscious effort to reorient JSC's role in the future demands its change. The overwhelming successes of Project Apollo were based largely on management decisions which were sensitive to the constraints which existed at that time (performance, safety, time, etc.). NASA successes in the period of the Space Shuttle will similarly be based on [13] a sensitivity to changing constraints (demand for increased productivity, decreasing budgets, etc.). The JSC life sciences organization will have to improve its research capabilities to remain viable. The second major factor in the organization's current debility is the obvious lack of leadership (Attachment G [not included]). Despite any bias induced by discipline, speciality, or parochial loyalty, the professional qualifications of the Director of Life Sciences must be addressed. The minimum qualifications for this position should be: (1) a professional Doctoral degree in medicine or an allied research discipline, and (2) some demonstrated scientific/technical management ability. Without such leadership, discussion of further organizational improvements becomes moot. Since appointment of the current Director, the already diminished morale of the few remaining professionals has ebbed even lower. It is anarticle of faith that in scientific, professional or organizations of competence, the tone must be set at the top. This lack of both technical and leadership ability is similarly reflected in every division; it is perhaps most acutely noted in the Biomedical Research Division, also headed by an unqualified non-professional. Though this situation is apparently universally deplored throughout the agency, even JSC operational leadership seems unable to present a cogent scenario regarding how it was allowed to develop.

Any discussion of developing closer research ties with the external academic and scientific communities would serve no useful purpose, as the minimum requirements for preceptorship and residency training affiliations would preclude any participation on the part of the excellent institutions in the Houston area. Conversely, however, should reorganization proceed in a manner which placed qualified life scientists in the positions of Director and Division Chiefs–the development of ARC-style academic affiliations should be aggressively pursued.

Because of the requirement to integrate the operational and research efforts of such a wide variety of scientific disciplines, it would be ideal to find a Director with degrees in both engineering and medicine. At the very least however, the Director of Life Sciences should have the M.D. or Ph.D. degree.

It is against the background outlined above, that a review of the expenditure of funds in the life sciences should be approached. In the NASA total of $57.3 million expended last

[14] year, $35.5 was spent on R&T, and $21.8 was spent on projects. The significant majority of both categories was administered by JSC. A striking aspect of the funding mechanism at JSC is the ability to pay for, what is loosely called "supporting research," from either of these two major funding categories (project or R&T). A brief survey appeared to reveal that itemizations for virtually similar expenditures could be from either category. Another subject which this review found perplexing was the wide variance. in perceptions throughout JSC of what constituted life sciences. The Office of Safety, Reliability and Quality Assurance; the Earth Resources Program Office; the Urban Systems Project Office; the Future Programs Division; and the Telemetry and Communications Systems Division are all devoting effort to what could be described as Life Sciences Applications. Roughly 10% of JSC's life sciences personnel are said to be involved in these Ecology Projects. Some better measures of control in these areas of ecology, technology utilization and applications would seem to be in order. This is of particular importance when one considers the variety of disciplines that have congregated at JSC, which have only the most peripheral of mission support or related research functions.

One division alone has some ten veterinarians, making it one of the largest complements of these specialists in the Federal Government outside of the USDA. Such abuses of staffing are not only questionable by any standard; they also serve to defeat the medical research environment being sought. In addition they are a forceful lobby for the expensive extension of more esoteric research, which if funded could compromise the fundamental human research which NASA requires. A cursory review of the vast number of research proposals engendered by these veterinarians, would be cause for serious concern on the part of any functional program manager. Another aspect of JSC staffing which deserves review is the policy of continued utilization of life sciences personnel from other branches of government, for prolonged periods. Although the continuous infusion of new ideas is highly desirable, the policy is defeated when DOD or USPHS detailees are allowed to extend and re-extend. Some of these "temporary" personnel have been at JSC since it was opened; this ossification process must be reversed.

[15]
The persistent lack of center-to-center communication among life sciences personnel was again noted at JSC. The individual professional, primary investigator or scientist knew little of the life sciences activity at other NASA centers. Because of their proximity to the Houston medical complex, JSC personnel were conversant with the surgical materials which had been developed as spin-offs from previous space research. Yet they were totally unaware of the materials research being done at Lewis Research Center (total hip replacement prosthetic devices, cardiac valve seats and races, etc.), and similar research being done elsewhere in NASA.

The intellectual climate which has existed for years in the JSC Life Sciences Directorate, which has suppressed the publishing of professional articles, required the researcher to include the Director of Life Sciences as senior author on key publications, and withheld data for which no clear explanation was available, has been stifling. The current Director has little insight into either the subtle interfaces each scientist must maintain with his parent specialty, or the need to get preliminary data into the open literature quickly–a fact which was stressed by the NAS-NRC study of 1970. An additional suggestion of the Bentley Glass Committee was to convene an annual meeting of all key NASA Life Sciences personnel, to exchange information and ideas. Investigations reveal that this has never been done; but that key personnel would welcome such an opportunity to improve communications.

The three medical scientist-astronauts merit some discussion in this review, because of their recent incorporation into the JSC life sciences organization. Because of their individual concerns regarding participation on shuttle missions as the payload specialist only (not as pilot, or even as the mission specialist), and the fact that they represent a minority when compared to the physical scientist-astronauts, their acceptance of the recent reorganization has been guarded. This has led to an interesting phenomenon regarding their physical location at JSC; culminating in a strong stance to keep their offices immediately adjacent to the Astronaut office in the Directorate of Flight Operations. Given the state of leadership and the Center image of the Life Sciences Organization, their reluctance is understandable. The assimilation of their respective talents into the organization however, should be viewed as highly desirable. Their lack of clinical expertise [16] and their absence from the field of medicine for a protracted period of time notwithstanding, they represent a natural strength in the areas of flight crew rapport and operational mission support. Their failure to complete residency training or to become Board Certified in a medical specialty should in no way detract from their utilization in either applied research or flight hardware development. Their full incorporation into the Center life sciences organization should be aggressively pursued.

III. Options for the Space Shuttle Era

The NASA Life Sciences Program for the period post-Apollo and Skylab, must be approached as a single entity. The integration of efforts at the various NASA centers must be closely coordinated to avoid costly reduplication and overlap. Any efforts to restructure the headquarters organization must take into consideration aspects of the main operational units of the present Life Sciences' triad: Ames Research Center and Johnson Space Center. Replacement of the current Director at JSC can only succeed if Headquarters support is provided by an effective, capable new NASA Director of Life Sciences. Conversely the most competent new NASA Director cannot expect success without a strong operational arm at JSC, adequately supported by the JSC Center Director. These two positions are unquestionably the key slots upon which the future success of the NASA Life Sciences Program depends.

The Headquarters Life Sciences Office (Attachment B) [not included] could be consolidated. The Director of Bioenvironmental Systems recently retired, without subsequent replacement. This function could easily be absorbed by the Directorate of Bioengineering, at some considerable savings. In addition, the functions of the Bioresearch Division could be included in a Directorate of Aerospace Medicine and Biology. Assuming that the Occupational Medicine function were [sic] absorbed into the Headquarters Office, a relatively streamlined three division organization could result:

- Bioenvironmental Engineering

- Aerospace Medicine and Biology

- Occupational Medicine and Environmental Health

[17]

The NASA Director, with a Deputy and a small administrative staff to manage program planning and control, could interface most effectively with the center organizations via this organization.

The location of the Headquarters Office remains an enigma. Unless a major NASA reorganization were contemplated, to somehow modify the line missions of the Associate Administrators as they presently exist, there does not appear to be a logical slot for the Life Sciences at that level. Placing it under the control of any of the five major program offices (OMSF, OSS, OA, OTDA, or OAST), is equally bad; because what is required above all else is independence. An option which has not been discussed is establishing this office, like that of the Comptroller or General Counsel, as a separate operating office. It could report directly to top management and yet not compromise the efficiency of span-of-control considerations, by interfacing with the Deputy Administrator in the same manner as an Assistant Administrator. This would have a number of salutory [sic] effects. The first would be to affirm its independence, while operating to integrate the total NASA Life Sciences Program. The second could be the complete standardization of the life sciences budget, utilizing the PAD's, UPN's, and RTOP's. Examples of such dual line and staff responsibility are found throughout the federal government. Individual military hospital, research laboratory, and medical center commanders in the field have primary line responsibility to the unit commander they support; yet they receive professional program guidance, coordination, and funding through the office of an independent Surgeon General. Such a system acts to preserve and protect the medical and research functions which might otherwise be sacrificed for line mission objectives (e.g. airplanes and weapons instead of X-ray machines and pharamaceuticals [sic]). Even in the academic environment, basic research must be protected. To assume that NASA major program or Center Directors would not sacrifice life sciences efforts first during periods of restricted funding, would be contrary to past experience. The autonomy envisioned would allow the Director of Life Sciences to develop long-term prospective planning to a much greater degree, including those C of F items necessary to support major NASA projects and programs. This in turn could avoid experiences such as the one at Langley, where the Center [18] Director's decision to disband his life sciences effort resulted in the inefficient conversion, of a costly new life science facility to other purposes. Not only are such abuses difficult to explain to Congress, they also result in the loss of an invaluable facility to the NASA Life Sciences organization as a whole. An attendant benefit in such consolidation should also be the elimination of the current "Life Sciences Principals" in OAST, OSS, and OA; or their incorporation into the Headquarters Life Sciences Offices. The number and responsibility of these characters is not insignificant.

The qualifications for the position of NASA Director of Life Sciences have been widely discussed. There is much to be said for the recommendation of bringing a capable scientist from outside of NASA. His stature in the national scientific community could significantly add to the success of NASA's relationships with the USSR, as well as with international colleagues in the European Space Research Organization (ESRO). These associations will assume even greater importance with the approach of the Apollo-Soyuz Test Project (ASTP), and the use of the Spacelab on the Space Shuttle transportation system. A desirable but not necessarily required characteristic would be some experience on previous NASA programs.

The reappraisal and reorientation of the main objectives of the National Aeronautics and Space Administration, after the successful Apollo moon landings, were indicated by President Nixon in his statement of March 7, 1970. These were three in number: scientific knowledge, exploration, and applications. In a similar fashion, the new Director of Life Sciences should be charged with addressing major issues of important future significance:

1. Technology development to support the search for and identification of extraterrestrial life.

2. Defining the physiological parameters for space shuttle scientist-passengers, and developing appropriate selection techniques.

3. Definition, preparation, and coordination of Life Sciences payloads for the Space Shuttle, including Spacelab.

[19]

4. Reduction of the inflight data obtained on Projects Apollo and Skylab, and its prompt publication in the international scientific literature.

5. Conducting research aimed at better understanding the physiological problems associated with weightlessness (vertigo, motion sickness, bone marrow regulation, reduction of red blood cell mass, fluid and electrolyte imbalances, etc.); and developing effective countermeasures for long duration space missions.

6. Developing operational procedures and research protocols in concert with Soviet Physicians, for the US-USSR Apollo-Soyuz Test Project.

The success of such a program is heavily dependent upon the authority of the new Director to assemble the appropriate Headquarters staff, and to have some measure of participation in selecting key life sciences personnel in the field. The recent emphasis on personnel mobility should assist such efforts; and the wealth of talent at ARC should be drawn upon heavily in any scheme to revitalize JSC. Above all else however, the new Director should be given independent budgetary authority. Without it no significant measure of program integration can be expected, as has been demonstrated by previous abortive attempts at consolidation. The dual issues of independent fiscal and personnel authority, and separation from any singular program office (OMSF, etc.), are the two parameters which will most directly determine the caliber of the man ultimately selected.

Finally the present consultation mechanism deserves comment. The NASA Life Sciences Committee is perhaps the least effective organ of the Space Program Advisory Council (SPAC); the reasons for this are manifold. The external life sciences community is probably the most heterogeneous assembly of diverse scientific disciplines that can be described as comprising a single category. Therefore by definition, all specialities [sic] cannot be proportionately represented. The makeup of the committee should be shaped in large measure by the NASA Director of Life Sciences. The broadest representation of all disciplines is highly desirable, but the needs of the Director for counsel in specific areas should also be a strong consideration. Given the close working relationship between the Director of Life Sciences and the Chairman of the LSAC, top NASA management [20] might wish to consider delaying appointment of a successor to Dr. Shields Warren, until the inputs of the new Director are obtained. Should that not be feasible an alternative solution might be to hold the currently-vacant seats on the committee in abeyance, until the new Director's preferences are known. The predominant representation on the committee should continue to be those medical specialties most directly related with manned space flight (aerospace medicine, physiology, cardiology, hematology, etc.), but increased effort should be undertaken to include young scientists in the prime of their research career. All members should have a clear understanding that a lack of interest, as evidences by repeated absences at committee meetings, cannot be condoned and will result in replacement. This committee should have representation by some members with significant operational experience, as well as some with research management success. Both of these parameters will be important in overseeing a successful NASA program in the 1970's and 1980's.

There also presently exists no formal mechanism to include life sciences participation on the Donlan ad hoc shuttle payloads applications team. Although a significant proportion of the shuttle payloads will involve life sciences, including a proportion of roughly 50% of those missions in excess of 30 days, no effort has been made to include life science input on a continuing basis. This oversight is indicative of the lack of visibility of the current Headquarters Office, and reflects the repeated frustration of career NASA life scientists to provide effective counsel in their professional area.

IV. Summary of Recommendations

1. Reorganize the Headquarters Life Sciences Office into three Directorates (replacing the current four and absorbing the Office of Administration's Occupational Medicine function):

- Aerospace Medicine and Biology

- Bioenvironmental Engineering

- Occupational Medicine and Environmental Health

[21]

2. Establish the office of the NASA Director of Life Sciences as an independent office, reporting directly to top management.

- Establish its location separate from OMSF, pre-ferably in FOB 6.

- Provide an independent budgetary mechanism to allow it to develop a NASA-wide integrated Life Sciences Program, including support for the development of a prospective three to five year research plan.

- Rewrite NMI 1138.14 to clearly reflect these new mandates.

3. Appoint a capable medical or biological scientist of national stature as the NASA Director of Life Sciences.

- Provide the personnel mechanisms to allow him to establish a capable staff at the Headquarters level (the appointment of three strong Directors for the organizational structure of 1 above, as a minimum).

- Clearly identify the significant issues to be addressed by the agency, and provide his mandate in the most specific terms possible (see section III).

- Allow him to actively participate in restructuring the JSC life sciences organization, to include the appointment of all key personnel (including the intra-agency transfer of ARC personnel, if required).

- Solicit his inputs prior to filling the position of Chairman, and the vacant member seats of the a Life Sciences Advisory Committee.

- Establish the requirement for an annual meeting of all key NASA life sciences personnel; with a record of the formal proceedings to be forwarded to the Administrator.

[22]

4. Consider the appointment of Dr. David L. Winter as the Director of Life Sciences at Johnson Space Center. His knowledge of the research capabilities at ARC, his extensive personal experience with the problems of JSC, and his knowledge of NASA life sciences personnel, should provide a unique ability to blend the strengths of both Centers. His experience in the basic research environment should contribute heavily to the effective restructuring of JSC, to reflect the new roles and missions required for the shuttle era. An exposure to the operational aspects of major flight programs would also serve to broaden his personal experience, and qualify him for future leadership positions.

- Obtain a clear mandate from the Center Director at JSC, regarding such matters as: the appointment of key life sciences professionals, the elimination of unqualified incumbent non-professionals in leadership positions, an equitable agreement on center housekeeping matters for the Directorate, and support for expanded affiliations with the scientific and academic communities.

• Centralize the coordination of all life sciences activities, at JSC in the Directorate of Life Sciences. Such action would eliminate the need for a cumbersome ad hoc steering committee with the same purpose, and would allow coordination on a continuing basis.

• Tighten the fiscal reporting procedures between JSC and Headquarters, to require the funding of supporting research from either SR&T or Project sources, with clear accountability for each.

• Allow the prompt termination of all exchange personnel from other governmental organizations, who have served over three years with NASA (Department of Defense, U.S. Public Health Service, etc.).

• Selectively eliminate all ancillary personnel not directly contributing to primary mission functions (e.g. veterinarians, dentists, foreign exchange officers, etc.).

[23]

• Actively incorporate the three medical scientist astronauts into the formal JSC Life Sciences organization by eliminating the isolated separate office stigma; and integrate them into the medical staff in consonance with their individual talents.

5. Consider expanding the activities of the Ames Research Center to include the following:

• Increased biomedical research in the areas of aeronautical life sciences.

• Increased use of flight testing for applicable phases of life sciences research.

• Development of joint ARC-USN research programs, to take advantage of P-3 testbeds and a large pool of experienced aircrew personnel.

6. Consider modifying the membership of the Space Program Advisory Council's Life Sciences Advisory Committee to include:

• Rotating appointments with limited tenure and renewal provisos, to insure the continued infusion of new ideas and perspectives; with mechanisms for revocation of appointment if participation is ineffective.

• Representation by young scientists and physicians in the prime of their research and clinical careers.

• Attempts to include participation by a few life scientists with aeronautical or space flight program experience, to insure the balanced representation of the operational point of view.

7. Establish a formal mechanism to insure life sciences participation on the Donlan ad hoc shuttle payloads applications team. [24]

- Insure that this representation is delegated from the office of the NASA Director for Life Sciences.

- Insure participation on team visits to the Centers, to allow cogent, realistic answers to the significant number of life sciences issues in question throughout the agency.

Document III-20

Document Title: Memorandum from Dr. David L. Winter, Director for Life Sciences, to James C. Fletcher, Administrator, "Report on the Status of the Life Sciences of NASA," 16 February 1977.

Source: NASA Historical Reference Collection, History Office, NASA Headquarters, Washington, D.C.

Document III-21

Document Title: Memorandum from John E. Naugle, Associate Administrator, to the Deputy Administrator, 24 March 1977.

Source: NASA Historical Reference Collection, History Office, NASA Headquarters, Washington, D.C.

David Winter was named as the NASA Headquarters Life Sciences Division Director in September 1974. His memorandum to NASA Administrator James Fletcher summarizes his reflections, and frustrations, in executing the responsibilities of his position. NASA's number-three official, Associate Administrator John Naugle, provided a differing perspective on the situation. Within a year after having written the memorandum, Winter had left NASA.

Document III-20

National Aeronautics and
Space Administration

Washington, D.C.
20546

16 February 1977

MEMORANDUM

TO: A/Administrator

FROM: SB/NASA Director for Life Sciences

SUBJECT: Report on the Status of the Life Sciences of NASA

REF: NMI 1138.14, Roles and Responsibilities, NASA
               Director for Life Sciences

In my role as NASA Director for Life Sciences (NDLS), I have the responsibility to report periodically on the status of Life Sciences within the Agency. I also have the responsibility to represent the NASA focal point for external interactions on Life Sciences issues (Attachment 1 [not included]).

In the past year and a half, we have seen many changes take place in the Agency, our scientific constituency, and the field of Life Sciences. It is my feeling that Life Sciences now constitutes a viable and highly relevant activity of the Agency. This has been evidenced by the public reaction to the biology experiments on Viking, the very large response to our Announcements of Flight Opportunities, the joint biological missions with the U.S.S.R., and cooperative efforts with the National Institutes of Health and other parts of the Department of Health, Education, and Welfare.

Concomitant with this highly desirable public interest and acceptance of the Agency's activities in Life Sciences has come increasing contact with the external community. However, the multiplicity of contact points within the Agency is making it increasingly difficult to assure that a coherent policy posture is maintained. Accordingly, there have developed a number of potentially serious problems.

In response to various demands, the Agency has established offices within Headquarters to function in broad generic areas encompassing many disciplines; e.g., Applications, [2] Energy, Technology Utilization (TU), and the Low Cost Systems Office (LCSO). The structure of most of the offices has required matrix-type technical/scientific support from discipline offices. In this typical matrix-management concept, top management always has the option of staffing the generic offices by colocating specialists or by maintaining the specialists in their peer groups and providing support to the generic offices on an "as required" basis.

In the first case, the specialists tend to drift away from peer contact with concomitant reduction in proficiency and increased affiliation with the generic office's problems and parochialisms. In the second case, without colocation, the generic office tends to build a psuedo [sic] capability within their own staff, thereby ignoring proper specialist and technical support.

It appears to me these basic organization ambiguities may be contributing to the Life Sciences' problems. Functional authority has been delegated to these several organizations within the Agency, and all of these functional entities have a requirement to interface with the outside world in a broad capacity. Fine print in their Roles and Responsibilities statements require "recognition of responsibility and authority of other officials... and proper coordination with other groups...." In the day-to-day conduct and handling of large, diverse activities, it is understandably easy to lose sight of a need to coordinate with other parts of a complex interface matrix such as exists in NASA. Case two seems to apply in regards to Life Sciences.

Clearly those Supporting Research and Technology (SR&T) programs and projects currently funded in the Office of Space Science (OSS) are accepted Life Sciences activities. They include Planetary Biology, Planetary Quarantine, Space Medicine Research and Operations, Advanced Teleoperator Technology, Space Biology, Bioinstrumentation, and Life Sciences Payloads. These items cover a wide range of disciplines, but are all part of the Life Sciences organization and are reviewed externally by Life Sciences Review Panels and Advisory Committees. They are closely coordinated within my OSS Division. However, there are additional Agency efforts, funded and managed by other offices, which clearly fall within the Life Sciences disciplines. These efforts are not reviewed nor coordinated effectively with the NASA Director for Life Sciences.

It is these areas that are causing a concern at this time. I do not have sufficient staff at my disposal to colocate [3] personnel in each of these offices and am dependent upon the various organizations to recognize when their plans, and particularly their activities, require a Life Sciences perspective.

A clear example of the problem is in Space Bioprocessing. In spite of repeated attempts over some three years, Life Sciences' input into this program has been discouraged. Only in the last few months has any effort been made to consult with our staff on the biological aspects of processing. The program planning, the overselling of commercial benefits, and the partial (I hope) alienation of the scientific community, done without our participation, has not enhanced the NASA image and has certainly tarnished the NASA Life Sciences image (Attachment 2, for example)[not included].

The area of animal tracking is also one of concern. Repeated requests from outside groups such as the Marine Mammal Commission, the Council on Environmental Quality, and the Fish and Wildlife Service of the Department of the Interior for NASA help in technology and, of course, funding have been received. Additional requests from university scientists, either in the form of unsolicited proposals or responses to AO's, were and still are being received. NASA's involvement in this area has all but ceased. At a time when NASA technology could impact an area of great economic potential, and certainly from my viewpoint public and scientific impact, NASA has "dropped out" of this area. Admitting that NASA cannot do everything for all people, I am concerned that the decisions made, concerning an essentially biologically related program, were made without any Life Sciences input.

A closely related area, again involving remote sensing techniques, is one Life Sciences calls Public Health Ecology. This involves indirect assessments of geographic locales to empirically relate habitat conditions to disease carrying vectors of public health concern. Some initial enthusiasm in this program has been rapidly cooled by mixed outside reviews of one NASA-supported task. Without debating the details of this task, the basic concern I have is that, again, programmatic decisions have been made without consulting the Life Sciences Directorate in Headquarters. As a result, I personally believe that some very great potential benefits have been missed, not necessarily in the previously funded task, but in the development of an approach to deal with a large series of generically similar tasks. I can think of no more important benefit to mankind than to aid in the eradication of disease vectors. This is not to say that NASA definitely can do this now, but the technology presently [4] developed may do this. I believe that the potential cost/benefit ratio is certainly worthy of some effort on NASA's part–again, a Life Sciences' judgment which has not been heard, much less assessed.

A recent experience with the Energy Office illustrates the complexities of external interfaces with the Life Sciences community. The concept of a Space Power Station is a very important initiative. At present, there are three potential "show stoppers" to this concept. Two of them are Life Sciences' problems: particulate radiation at geosynchronous orbit as it may effect a crew or work team, and microwave radiation as it may effect biology and the environment on Earth. This latter problem is highly complex and frought [sic] with controversy (see the recent New Yorker articles for a lengthy appraisal of the issues).

A presentation was made to Electromagnetic Radiation Management Advisory Council (ERMAC) about a NASA program to study the biological effects of microwave exposure; this presentation was less than adequate. The main issue here is that the Life Sciences Office in NASA knew nothing about this presentation to ERMAC and had no input on the style, format, or content. The NASA Office of Life Sciences suffers the consequences of activities outside its operational purview, yet clearly within its NASA Management Instruction charter. In this specific case, NASA itself suffered. Recent contacts with the Energy Office offer potential solutions to this problem.

On the other side of the coin, there are some good examples of coordination, and I would like to discuss these briefly. Good coordination exists between Life Sciences and the Office of Space Flight (OSF). Our people regularly attend OSF staff meetings and, I, OSF Management Council Meetings. The Life Sciences role in the Shuttle program is well understood, and a large segment of our activities is directed towards Shuttle support. The irony here is that we are able to work effectively with the four Centers concerned with Shuttle operations even though we are not a formal part of OSF. Obviously this could not happen if we did not have the support and backing of OSF management and an understanding of our role in coordinating the individual Center responsibilities in this large program.

The fact that Life Sciences was formally a part of OMSF has no doubt been a major factor in this coordination. However, I am certain that the mature management philosophy in

OSF has been equally as important and demonstrates that it is possible for Life Sciences to work across organizational lines to carry out its Agency responsibilities. [5]

Another example of a good coordination can be found in the TU-Life Sciences interface. There is continuing close contact between these offices and, in fact, one Life Sciences person was, until recently, spending part time colocated in the TU office. Although there are some fundamental differences in philosophy between the Life Sciences and the TU approaches, these are well understood by both offices and appear not to affect our interaction. Such differences that do exist dwell primarily about the point at which TU efforts should begin and the manner in which commercialization should proceed. I personally believe that TU should be more involved in the developmental work of medical devices; that is, to step in a little earlier and to support efforts a little longer in order to facilitate the tortuous task of insuring medical community acceptance of such devices. Acceptance by the medical community is a complicated procedure which requires tact and interaction with specific elements; notably, governmental scientific bodies such as the National Institutes of Health and the other parts of the Department of Health, Education, and Welfare. This approach is necessary as opposed to interaction with commercial interests. In the biomedical field, physician acceptance drives the product as opposed to other areas of commercialization where the product paces the user. These comments are made only to improve what I consider a very good relationship between two offices to allow NASA to take more advantage of biomedical spin-off.

This office continues to support and interact with the Occupational Medicine Program, another example of effective coordination. In my opinion, the Occupational Medicine Program is a model government program which is cost effective and sensitive to the legal and ethical considerations of the day. As you know, this office was instrumental in designing the new Occupational Medicine NMI and initiating the forward looking program that NASA now expresses.

In summary, I have tried to point out a number of failures and successes which the Office of Life Sciences has been party to, either directly or indirectly. I believe that some fundamental questions should be asked regarding the future role, if any, Life Sciences should play within NASA. The NMI which outlines our activities and responsibilities is rather clear (Attachment 1) [not included]. Either the NMI should be rewritten to be consistent with NASA management's view of the Life Sciences role, or methods should be found to enable this office to carry out its responsibilities. I offer several options (A through D) that could be considered.

[signed]
David L. Winter, M.D.

Enclosures

[no page number]
Option A

"REDUCED LIFE SCIENCES ROLE"

In this option, I propose that the Office of Life Sciences be eliminated. Those SR&T activities associated with Planetary Biology and Planetary Quarantine could be retained in OSS in another discipline office. A flight surgeon could be assigned to OSF to coordinate operational requirements. All other Life Sciences activities could be eliminated and NASA refrain from presenting itself to the external community as a Life Sciences-related organization.

This option would provide the "service" functions which have been required over NASA's two decades of existence. It presumes no requirement exists for SR&T and Advanced Studies to prepare for longer duration flights. It also presumes that Life Sciences payloads can be managed by non-Life Sciences organizations to the satisfaction of the national and international Life Sciences community.

[no page number]
Option B

"MODIFIED STATUS QUO"

In this option, I propose the responsibilities of the Life Sciences Office be reconsidered in light of its current hierarchical location. The functions of the NASA Director for Life Sciences cannot be performed satisfactorily given the limitations on personnel, travel, and autonomy that go with what is rapidly becoming OSS "Division" status. The offices that I am required to overview do not appreciate an OSS input into areas they do not consider to be OSS related. This places my organization in an untenable situation.

If the status quo is maintained, then I recommend my non–"Division" responsibilities be reduced to a consulting, vis-a-vis overview responsibility.

[no page number]
Option C

"INCREASED VISIBILITY"

In this option, I propose that the prescribed functions of Life Sciences remain, but that they be made known to NASA management. As indicated above, the position of NASA Director for Life Sciences (NDLS) has been downgraded in OSS to that of a Division Director. Communication across organizational lines has been discouraged and both manpower and financial resources have been reduced. It is unworkable to perform one role as a "Deputy Associate Administrator" and another as a Division Director. This option proposes that some arrangement be made, either within the existing or some other structure, so that the NDLS has visibility and, more importantly, access to NASA top management so that there is truly a single unfiltered voice representing the Agency on Life Sciences issues. At the present time, far too many managers consider themselves as adequate spokesmen on Life Sciences problems, while admiting [sic] a substantial parochialism to causes.

[no page number]
Option D

"INCREASED LIFE SCIENCES ROLE"

In this option, the various Life Sciences activities within the Agency are put under single management. This option has been recommended by at least five outside committees who have reviewed the Life Sciences operations in NASA over the past ten years. It is time to reconsider this possibility. This option presumes that NASA considers Life Sciences as an important part of its developing programs and is willing to give it visibility and resources to do its job properly. This point is of particular concern in regards to present activities (Shuttle Biomedical Payloads) and future activities (space station and other concepts). In my opinion, as long as the country pursues manned flight, Life Sciences is an absolutely essential element in the NASA program and requires status comparable to OE, OA, TU, and LCSO to function effectively.

Document III-21

[no page number]

NATIONAL AERONAUTICS AND SPACE ADMINISTRATION
WASHINGTON. D.C. 20546

OFFICE OF THE ADMINISTRATOR   March   24, 1977

MEMORANDUM

TO: AD/Deputy Administrator

FROM: AA/Associate Administrator

SUBJECT: NASA Life Sciences

Enclosed is background on Life Sciences in NASA and my comments and recommendations on the issues raised by Dave Winter. Since the memo comes out longer than the one page you and Jim want, you may choose to skip the background and go directly to the issues and recommendations. I wanted you to have this information prior to our next weekly meeting.

[signed]
John E. Naugle
Enclosure

[1]
3/24/77

BACKGROUND ON LIFE SCIENCES IN NASA

There is a long and controversial history to the establishment of the position "NASA Director for Life Sciences" reporting to the Deputy Administrator.

The life scientists involved in the Space Program have always felt that they should be represented by a program office reporting directly to the Administrator's office as the physical scientists have in OSS. There was even such an office for a short time under Glennan. They feel that their research objectives are usually not understood and often looked down upon by the physical scientists and that the only way their objectives will be met is by having their "own" program. There is some justification in their view. There is also a need to closely examine the caliber of people involved in life sciences and the value of their experiments. We have generally not had the same caliber of research in their area. There has also been a continuing controversy between the Office of Space Flight and both the "flight surgeon" portion of the life sciences community and the bioscientists using animals for research in space. The external medical scientists have been unhappy because not enough data was collected on the reaction of the astronauts to O-G. The astronauts resented being considered guinea pigs and subjected to a variety of sensors and painful tests during hazardous and [2] hectic missions. The bioscientists felt that NASA should precede manned flight by elaborately instrumented animal flights and were jeopardizing the health of the astronauts by not doing so. NASA was concerned that if we waited until these adequate animal experiments we would never get off the ground. Gagarin's flight eliminated that disagreement.

There was also a continuing internal hastle [sic] between the bioscientists in OSS, the aeronautical life scientists in OAST, and the "flight surgeons" in OMSF as to who should be doing what.

This all came to a head shortly after Low came to Headquarters. We had demonstrated man could survive in space and on the moon. We had just completed a rather disastrous primate flight in which we had chilled the primate during flight as well as losing several controls due to surgical procedures. Skylab was coming up, the future of bioscience was at stake and everybody in NASA was fed up with the internal haggling. We has the SSB convene an external study group to review what we should do and how we should organize to do it. That group chaired by Harry Eagle,[actually H. Bentley Glass] recommended that we combine all the life science activity into one organization reporting to the Administrator's office. I do not remember their program recommendations but shortly thereafter Low decided that the only [3] life science work NASA would do would be to determine the effect of the space environment on humans. Since the only place where such work could be done was in OMSF on the remaining Apollo flights and Skylab, I recommended and Low agreed that as much of the life science activity as possible be centered in OMSF. OSS kept only the exobiology associated with Viking and OAST the life science work associated directly with aeronatuics [sic]. To satify [sic] the reporting requirements, Low gave the responsible individual in OMSF the title NASA Director of Life Sciences. This was a director-level job in OMSF reporting to the Associate Administrator of OMSF. There was a dotted line to Low and an understanding that the person was to oversee all of the life science activity in NASA and pull it all together annually in the form of a Life Sciences PAD.

The above situation prevailed until the completion of Skylab, the establishment of Petrone as the person responsible for all of NASA's R&D, and the retirement of Chuck Berry who had been the NASA Director of Life Sciences up to that time.

We made a decision at that time to move life sciences from OMSF to OSS and to reexamine the objectives of life sciences research. We hired Dave Winter just before the transfer. He was aware of the impending transfer and that he had a dual [4] responsibility but reporting to Petrone rather than Low. We have not rewritten the NMI. There was also an explicit agreement between Yardley, Hinners, Winter, Petrone and me regarding certain activities that Winter would do for Yardley. There is an MOU in the system which details that responsibility.

The decision to move OSF out from under AA's responsibility once more justified Winter's reporting to you in his role as the NASA Director of Life Sciences since there are life sciences activities in OSF. So much for background.

ISSUES

Winter states that there is no problem in OSS life sciences work; that there are good relations with OSF and TU; and problems with OA and OEP. I generally agree with his observations. He offers four options:

    a. Reduced Life Science Role
    b. Modified Status Quo
    c. Increased visibility
    d. Increased Life Science Role

Option A is unacceptable to me since he proposes the elimination of the Life Sciences organization. That is not necessary, desirable or appropriate. We are in the process of developing a Life Sciences Program for the 1980s. We do not know whether we will succeed but we have every reason to believe we will. [5]

Option D is also unacceptable to me. If we were to separate Life Sciences from OSS then we would have to create an organization to handle their Spacelab missions and I would have one more organization to personally worry about staffing and interfacing with STS. Also, there is very limited experience in the life science community in developing experimental space flight equipment. It would be an organization which would need close supervision for the next several years. I am looking to Hinners and Calio to provide that oversight.

There does need to be a better interaction between Applications and Life Science. I do not intend to have Johnston create a separate life science capability in his organization. I intend to have him use the OSS/Life Science capability. Aside from the one instance cited by Winters, I believe OEP is looking directly to OSS to handle their environmental problems.

I would propose the following course of action.

a. We agree to keep Life Sciences as a division of OSS and you assign to me the action to work with Hinners and Winter to resolve the problems.
b. I will work the OSS/OA problem to involve Winter's organization in space processing.
c. I will work the problem, if any exists, between OEP and Winter's organization. [6]
d. I will use my regular biweekly meeting with Yardley to resolve any issues with OSF. (None have arisen since we have been meeting.)
e. In the event there is an issue that Yardley, Hinners, Winter and I cannot resolve we will promptly involve you.
f. We will plan to keep Life Sciences as a division of OSS for the next five years with the understanding that we will review the need for exercising option D at the end of five years or when the Life Science Budget exceeds $100M annually, whichever occurs first.
g. Hinners, Winter and I recognize that the Administrator's office (A, AD, AA) may from time to time consult Winter on the over-all NASA Life Science Program.
h. Winter prepare an annual report on the NASA Life Sciences Program for the Administrator summarizing the major achievements of the past year and the main activities that will take place in the coming year.
i. We abolish the requirement on Winter to prepare an integrated Life Sciences PAD. It serves no useful purpose that cannot be served as well or better by the annual report.
j. Winter be responsible for intergrating [sic] the total Life Sciences Program into the five-ten year planning process.

Document III-22

Document Title: Letter from Dr. Robert H. Moser, Chairman, Life Sciences Advisory Committee, to Dr. Lennard A. Fisk, Associate Administrator for Space Science and Applications, 30 July 1987.

Source: NASA Historical Reference Collection, History Office, NASA Headquarters, Washington, D.C.

The NASA experience with U.S.-Soviet cooperation in the Cosmos/Bion program was so rewarding that NASA's Life Sciences Division began considering a program using similar U.S. robotic spacecraft. Planning workshops were organized by Ames, which had the firsthand experience with both Biosatellite and the Cosmos/Bion programs. Out of these discussions, the LifeSat program was born. LifeSat was warmly received by NASA's Life Sciences Advisory Committee.

The NutraSweet Company
Box 1111, 4711 Golf Road, Skokie, Illinois 60076
Telephone: 312/982-8383

July 30, 1987

Dr. Lennard A. Fisk
Associate Administrator
Office of Space Science and Applications
National Aeronautics and Space Administration
Washington, D.C. 20546

Dear Dr. Fisk:

The Life Sciences Advisory Committee (LSAC) has met several times to discuss the current concept of a recoverable, reusable biosatellite for application to life science research issues. The Committee is in unaminous [sic] agreement that this capability is required by Life Sciences to fill both a science and time void in our program.

We are all painfully aware of the science hiatus brought by 51L. When Dr. Culbertson met with LSAC in March of this year he stressed the importance of off-loading payloads to expendable launch vehicles (ELV) and preserving the Shuttle for missions which require its unique capability. In life sciences we not only have many payloads which can be off-loaded but much science which can, in fact, be done better utilizing ELVs to orbit a LifeSat. Additionally, the planned Spacelab missions are too few to accomplish everything we must do and science should no longer he forced to rely on a single system which continues to be plagued by uncertainties of launch date, crew time, mid-deck space, etc. Life Sciences requires and would appreciate alternative means of pursuing its flight objectives–means which OSSA has provided to its other Divisions.

I believe you understand the salient science features of the LifeSat. There are many biological processes for which the Shuttle offers only the first few data points during its 7 to 10 day flight, while the planned biosatellite provides up to at least 60 days. Although the Space Station offers us indefinite low gravity conditions, it will not be operational until the mid-1990's. Data from the LifeSat would scope our science and hardware for the [no page number] Station. If we are to consider seriously long term human flights and science bases on the moon during the early 21st century we need to explore the CELSS option early–much of its effort requires more than 10 days aloft. The capability of LifeSat to access unique orbits would enable radiation biological studies in space. Finally, there are a class of investigations requiring studies of toxic substances or use of radio-tracer labels which are difficult to carry out on manned vehicles. If you so desire I would be most willing to discuss these science benefits with you.

From your candid discussion with LSAC on July 17, 1987 regarding a LifeSat approval I understand and appreciate your two major concerns, ELV availability and satellite cost. We were subsequently informed in a review of the status of LifeSat that a fully satisfactory biosatellite can be launched as a partial payload (30%) on a Delta II at an apportioned cost equivalent to that of a current Scout, which cannot provide adequate science capability. Further, I understand that the planned Phase II Mixed Fleet acquisition would be able to provide a Delta II in early 1992, corresponding to the first planned LifeSat flight and that ELVs will probably be costed in the same manner as Shuttle opportunities. For the satellite itself, I agree that its cost should be weighed carefully relative to other life science and Space Station initiatives. We saw a range of costs from $20 to $50 millions of dollars for a LifeSat, but even at the top end the scientific benefits far exceed the cost.

Speaking for all of the LSAC members, I urge you to commit at least to a design effort for LifeSat which can provide the cost numbers required for an informed decision. Please let me know your decision on this matter.

Sincerely yours,

[signed]

Robert H. Moser, M.D.
Chairman, Life Sciences Advisory Committee

Document III-23

Document Title: NASA Life Sciences Strategic Planning Study Committee Report, "Exploring the Living Universe: A Strategy for Space Life Sciences," June 1988.

Source: NASA Historical Reference Collection, History Office, NASA Headquarters, Washington, D.C.

When he became NASA's Associate Administrator for Space Science and Application in 1987, Lennard Fisk decided to develop an overall strategic plan for all of NASA's space science programs. Life Sciences Division Director Arnauld Nicogossian created a Strategic Planning Study Committee, chaired by Frederic Robbins, to help his division in developing such a plan for the life sciences.

[25]

2. Findings and Recommendations

This section is the central part of the Life Sciences Strategic Planning Study Committee (LSSPSC) report, for it highlights the Committee's overarching recommendations, strategic milestones for achieving those recommendations and findings and recommendations related to particular subject areas. The material emerged from the LSSPSC Study Group reports given in section 3 [not included], which present corresponding and more detailed findings and recommendations. It is organized in the categories itemized below, incorporating the subjects explored by the Study Groups:

- Human Space Flight focuses on the physiological and psychological challenges to humans in space and on the research and facilities necessary to overcome factors that may limit the success of manned missions, especially of extended duration.
- Gravitational Biology is concerned with the influence of gravity on the structure, development, and function of plants and animals.
- Planetary Biosciences Research concentrates on scientific issues pertinent to the origin, evolution, and distribution of life in the universe and the relationship of a planet's biota to its biosphere.
- Flight Programs emphasizes the need for flight opportunities for life sciences research, including dedication of Space Station laboratories for clinical and basic biological research.
- Program Administration itemizes administrative and organizational issues important to strengthening the work of NASA in the life sciences.

Overarching Recommendations and Strategies

In developing their summary papers, the LSSPSC Study Groups came to a number of parallel conclusions about life sciences at NASA and devised several similar recommendations. The LSSPSC determined that these recommendations were basic to the success of the Life Sciences program and, by extension, to the achievement of NASA's overall goals and long-range strategies, particularly as they affect human exploration of the solar system. The Committee presents these recommendations in the box on the next page.

The LSSPSC devised strategic milestones for fulfilling the requirements that are part of the overarching recommendations. These milestones, itemized according to 3-, 5-, and 15-year periods, emphasize the need to initiate work immediately, in the 1989 fiscal year.
[26]

Overarching Recommendations

To resolve life sciences issues critical to the success of the civilian space program, NASA should:

- Maintain and expand the Nation's life sciences research facilities located at the Agency's field centers, universities, and industrial centers by:

    - Establishing a mechanism for attracting promising young scientists to work on NASA projects and developing additional training programs at major universities and appropriate NASA installations

    - Establishing a program of NASA supported professorships in space life sciences at selected universities

    - Encouraging industries to develop capabilities in space life sciences through technology research and development.

- Assure timely and sustained access to space flight, thereby facilitating the conduct of critical life sciences experiments. This should be accomplished through:

    - Accumulating state-of-the-art instrumentation

    - Flying an augmented series of Spacelab missions

    - Using a series of autonomous bioplatforms to study radiation and variable-gravity effects

    - Dedicating suitable facilities on the Phase 1 Space Station complex for life sciences research

    - Conducting a major augmentation of life sciences capabilities during the early Post-Phase 1 period.

- Synergize the presently independent research activities of national and international organizations through the development of cooperative programs in the life sciences at NASA and university laboratories.

- Complete and consolidate the unique national data base consisting of basic life sciences information and the results of biomedical studies of astronauts conducted on a longitudinal basis. This data base should be expanded to incorporate information obtained by other spacefaring nations and be available to all participating partners.

Strategic Milestones for 1989-1991

Life sciences research requires replication to verify experimental results, a process that involves considerable time in planning and conducting the investigations, as well as in developing advanced technology. Working from this understanding, the Committee recommends that NASA should do the following in the next 3 years:

- Strengthen the planning process of the Life Sciences Division by assuring its timely integration into the Agency's overall strategic planning process.
- Augment life sciences research programs to establish the base of scientific knowledge required by planners and engineers to conduct missions relevant to agency goals.
- Provide adequate funding to develop new state-of-the-art flight hardware for upcoming manned and unmanned life sciences missions in space. Such an investment will have a significant impact on the field of biomedicine not unlike the impact of the Apollo Program on medicine and space science.
- Initiate advanced technology development in the areas of minimally invasive biomedical instrumentation, biological remote sensing, exobiological flight instrumentation, and microwave signal processing.
- Increase the frequency of life sciences data acquisition on the Space Shuttle and international missions.
- Conduct a study to determine the requirements for extravehicular activity (EVA) for the next 20 years, to delineate innovative options, and to identify needed technologies.

Strategic Milestones for 1989-1994

The next milestones are gauged for completion of life sciences preparations for the Space Station, scheduled to begin operations in the mid-1990's, and to implement a project requiring immediate action. For 1989-1994, the LSSPSC urges the Agency to:

- Operate reusable biosatellites to obtain environmental, radiation, and artificial variable-gravity data on plants and animals.
- Achieve ground-based validation of major physiological and psychobiological countermeasures for long-duration missions.
- Conduct ground-based research on bioregenerative life support systems to achieve 90-percent closure.
- Initiate the Microwave Observing Project of the Search for Extraterrestrial Intelligence (SETI) Program.

Strategic Milestones for 1989-2004

The 15-year plan looks to missions beyond the Space Station and asks the Agency to:

- Establish a combined national and international life sciences research facility on the Space Station. This facility must support basic research on plants, [28] animals, and humans necessary to develop an understanding of the fundamental biological processes affected by gravitational forces.
- Develop an advanced biomedical research facility in space to investigate and verify technologies and medical support necessary to enable the planning and implementation of human exploration of the solar system.
- Develop and test in space a fully operational bioregenerative life support system(s) for future use in solar system exploration.

- Conduct cooperative missions with other national and international organizations to study the behavior of the biosphere and the origin, evolution, and distribution of life on Earth and in space.

[pp. 28-37 not included]

[37] Program Administration

The coordination of life sciences activities at NASA is a challenging task. The research is multidisciplinary in approach and involves many other organizations – both within and external to the Agency – that are pursuing similar interests. The findings and recommendations given below identify the administrative challenges, acknowledge recent progress, and specify resources requirements.

Findings

- During the course of this study, the life sciences have received increased attention within NASA.

  - Concern about the effects of long-duration space flight has given life sciences a higher priority in the Agency and has provided the program with an opportunity to articulate its own goals more clearly.

  - At the same time, however, senior managers have not always appreciated that life sciences concerns are unique in the study and maintenance of life in space and that this uniqueness creates special administrative challenges for the program.

- The Life Sciences Division does not have sufficient resources in funds, staff, and facilities to realize its own objectives or the objectives set for the program by senior managers.

- The dispersion of life sciences activities across a number of NASA program offices has made it difficult to conduct research in several important areas, particularly human factors and biospherics. While new coordination efforts are under way, the integration of life sciences efforts across the Agency remains problematic.

- NASA's Life Sciences Division supports diversified programs that could benefit from coordination between the Division and outside organizations. The Division has initiated formal cooperative agreements with the National Institutes of Health and other Federal agencies.

- The increasing importance of foreign space programs has opened up a broad field for potential cooperative projects. These arrangements require international negotiations that are lengthy and involve multiple U.S. agencies.

- The Life Sciences Division has not always been able to create stable relationships with outside scientific groups.

[38]

- Scientists outside the Agency provide a valuable resource to NASA, both as researchers and as advisors to Agency staff.

- Recent program development plans for a balance between external and intramural research, as well as the creation of a new advisory and planning structure, promise desirable change in this area.

- Information concerning life sciences activities is not disseminated as widely as possible and desirable. As a result, many university and industrial researchers find it difficult to secure data on past, current, and future life sciences projects.

Recommendations

- Senior NASA management should support the continuation of recent Division efforts to establish a strong program by:

- Strengthening the Division's role in Agency-wide planning

- Facilitating access to frequent and regular flight opportunities

- Acknowledging the differences between programs of the Life Sciences Division and other NASA program areas

- Indicating to the rest of the Agency that biomedical research relevant to the safe conduct of human space flight is essential to ongoing and future NASA initiatives.

- Senior personnel from the Life Sciences Division should participate in all top-level planning of Agency flight programs.

- NASA should substantially increase the resources for Life Sciences programs to assure implementation of the recommendations given in this report.

- NASA should increase its efforts to expand the numbers of scientists at the Centers and Headquarters and should institute new efforts to provide career development opportunities for existing staff.

- The Life Sciences Division should further its efforts to establish formalized agreements and working groups with other agencies and organizations.

- NASA should provide funds to expand and implement plans to establish Specialized Center of Research (SCOR) units within selected universities, an effort designed to develop young scientists in space life sciences.

- In addition, the Agency should consider the establishment of NASA-supported professorships in space life sciences at selected universities, so that by 1990 one or two internationally recognized bioscientists and clinical investigators can play a significant role in the biomedical research crucial to human space missions of extended duration.

- The Life Sciences Division should generate and maintain a data base through collaborative arrangements with NASA's Scientific and Technical Information Facility and the National Library of Medicine.

Document III-24

Document Title: Announcement of Opportunity, Spacelab Lab Sciences-4, Neurolab, 21 July 1993.

Source: NASA Historical Reference Collection, History Office, NASA Headquarters, Washington, D.C.

This Announcement of Opportunity solicited research proposals for what turned out to be the last Spacelab flight dedicated to life sciences. It is an indication of how far space life sciences had come, both in the sophistication of the areas of science proposed for investigation and in the use of the Shuttle/Spacelab system as a global resource for carrying out life sciences in space. The Neurolab mission was launched on 17 April 1998 and landed on 3 May. Neurolab's 26 experiments targeted one of the most complex and least understood parts of the human body—the nervous system. The primary goals were to conduct basic research in the neurosciences and to expand the understanding of how the nervous system develops and functions in space. Test subjects were crewmembers, rats, mice, crickets, snails, and two kinds of fish.

National Aeronautics and
Space Administration

JULY 21, 1993
AO 93-OLMSA-01

ANNOUNCEMENT OF OPPORTUNITY

Spacelab Life Sciences-4, Neurolab

An Announcement of Opportunity for
The Life and Biomedical Sciences and Applications Division
Flight Research Program

Letters of Intent Due October 1, 1993
Proposals Due December 1, 1993

["Table of Contents" not included]
[2]

ANNOUNCEMENT OF OPPORTUNITY
SPACELAB LIFE SCIENCES-4, NEUROLAB

I. DESCRIPTION OF THE OPPORTUNITY

The National Aeronautics and Space Administration (NASA), in collaboration with its domestic and international partners, announces the opportunity for participation in scientific investigations in the weightless environment of the Space Shuttle. This Announcement of Opportunity is soliciting life sciences investigations in the neuroscience disciplines that will use hardware described in Appendix B [not included]. The specific instructions and regulations governing the proposal format, submission, and evaluation, and the selection of life sciences flight investigations for support by NASA and its partners are defined in this Announcement.

This spacelab [sic] mission, dedicated to the neurosciences, is being carried out as a cooperative effort between NASA and various domestic and international agencies. Domestic partners include several Institutes at the National Institutes of Health including the Division of Research Grants, the National Institute on Aging, the National Institute on Deafness and Other Communication Disorders, and the National Institute of Neurological Disorders and Stroke; the National Science Foundation; and the Office of Naval Research. International partners are the Canadian Space Agency; the French Space Agency, Centre National d'Etudes Spatiales; the German Space Agency, Deutsche Agentur fur Raumfahrtangelegenheiten; the European Space Agency; and the National Space Development Agency of Japan.

The President of the United States declared the 1990's as the "Decade of the Brain" with the primary goal to maximize human potential by advancing and applying scientific knowledge related to the brain and the nervous system. This Spacelab mission, titled "Neurolab," has been established as an essential part of NASA's activities in the "Decade of the Brain." The purpose of this Announcement is to offer the global science community the opportunity to propose distinct and innovative approaches requiring the unique environment of space flight to address the study of neuroscience. The specific areas of neuroscience research being solicited are described in Section III (page 4). Such investigations should be capable of being carried out on a Shuttle/Spacelab flight of two to three weeks' duration. NASA is particularly interested in investigations that use space flight to address fundamental questions in the neurosciences and those that are related to ensuring that a permanent human presence in space can be realized and sustained. NASA intends that investigations selected as a result of this Announcement be accommodated on Neurolab, but other modes of accommodation are possible and will be used if appropriate.

NASA has developed and maintains a number of special facilities and an inventory of standard and specialized Life Sciences Laboratory Equipment (LSLE) to facilitate research in space. International hardware and facilities are also available to be utilized for research in accordance with agreements established with NASA. The crew for Spacelab missions consists of highly trained research scientists as well as career astronauts.

This Announcement does not constitute an obligation on the part of the Government to carry any proposed effort to completion, flight assignment, or actual flight in space. [3] Tentative selection of a proposal for flight does not guarantee a flight assignment. This opportunity is dependent on the availability of funds. Participation in this program is open to all categories of domestic and foreign organizations, including educational institutions, industry, non-profit institutions, NASA Centers, and other Government agencies.

II. BACKGROUND

Space flight affords the unique opportunity to study and characterize basic biological mechanisms in the absence of gravity, one of the fundamental forces that shapes life on Earth. Neurolab provides a laboratory in which to explore the role of gravity on the function and behavior of the nervous system. In addition, space flight presents unique environmental stressors to the nervous system, and study of responses to these stressors will provide new insights into neurologic function and behavior. It is anticipated that the scientific data from Neurolab will result in improvements in the health of people on Earth. Neurolab represents a singular opportunity to conduct studies that will provide a foundation on which to build future space flight research efforts in the brain and behavioral sciences. The overall goals of Neurolab are to:

- Use the unique environment of space flight to study fundamental neurobiological processes
- Increase understanding of the mechanisms responsible for neurologic and behavioral changes that occur in space flight
- Further life sciences goals in support of human space flight
- Apply results from space studies to the health, well-being, and economic benefit of people on Earth.

During the early years of NASA's manned space flight program, efforts in the life sciences were driven by operational medicine and biomedical support of short duration missions such as the Mercury and Gemini flight series. During these missions, no significant problems arose regarding sensory system function. However, during the Apollo missions, a number of astronauts reported mild to severe motion sickness symptoms, and in the early 1970's NASA initiated studies directed at understanding the basic etiology of space motion sickness. In addition, studies were undertaken to develop tests that would predict susceptibility and enhance development of suitable countermeasures.

Over the past two decades, NASA's efforts in the neurosciences have included extensive research directed at understanding neurovestibular and sensorimotor system function (Space Physiology and Medicine, 2nd. edition, Nicogossian, Huntoon, Pool [eds.] Lea & Febinger, 1989; Experimental Brain Research, 62[2], 1986). More recently, it has been realized that attention must be devoted to other physiological and behavioral changes that occur as a result of exposure to space flight. This includes the study of molecular, cellular, and systemic mechanisms involved in changes that occur during the process of adaptation to altered gravity and that may have implications for readapting on return to a gravitational environment.

The Spacelab Life Sciences 4 (SLS-4) Neurolab mission will be the fourth in a series of Spacelab missions sponsored by the Life and Biomedical Sciences and Applications Division that are dedicated to life science space research. The first three missions, SLS-1, SLS-2, and SLS-3, each has its own specific research focus. The SLS-1 (June 1991) and SLS-2 (September 1993) missions have been designed primarily to study the mechanisms involved in the acute cardiovascular and metabolic responses to microgravity and readaptation to Earth. The SLS-3 mission (February 1996) is primarily focused on [4] understanding the effects of microgravity on the musculoskeletal system and on performance. The Neurolab mission, SLS-4, is dedicated to neuroscience research.

The investigations to be selected from proposals in response to this solicitation will be conducted on the SLS-4 Neurolab mission planned for the last quarter of 1997 or the first quarter of 1998. This mission will probably be two to three weeks in duration. The experiments on this mission will be conducted primarily in the Spacelab module with limited use of the Shuttle Middeck. The Spacelab module is a pressurized laboratory facility located in the Shuttle orbiter's cargo bay and attached to the orbiter cabin by a pressurized transfer tunnel. The Spacelab module will be configured to provide the resources required to conduct the selected neuroscience investigations. Implementation of investigations on this mission is limited by the Spacelab/Shuttle environment, available facilities and hardware (Appendix B) [not included], animal housing facilities, and utilization of crewmembers as experimental subjects or operators. These constraints are further defined in Section VI (page 11) [not included].

Potential investigators are urged to become familiar with relevant neuroscience space-related research prior to submitting a formal proposal to NASA. Some references are provided in Section III and an additional partial bibliography of space flight related neuroscience research is presented in Appendix A [not included]. More complete bibliographies can be obtained by writing to:

      Neurolab Program Scientist
      Life and Biomedical Sciences and Applications Division
      Code UL
      NASA Headquarters
      Washington, D.C. 20546

III. ANNOUNCEMENT OBJECTIVES

Proposals submitted in response to this Announcement must address neuroscience questions that require the unique characteristics of space flight. These characteristics include: 1) microgravity and a changing G load during launch, flight, and reentry; 2) increased radiation; 3) altered temporal cues; 4) isolation and confinement; and 5) physiological and psychological changes associated with space flight and exposure to the space craft environment. Studies are being solicited in the following research areas:

- Cellular and Molecular Neurobiology
- Developmental Neurobiology
- Sensory and Motor Systems
- Nervous System Homeostasis and Adaptation
- Behavior, Cognition, and Performance

Proposed investigations should require the unique research environment provided by the Space Shuttle, defined in Section VI (page 11) [not included]. The major candidate hardware that is planned to be available for this mission is included in Appendix B [not included]. The specific objectives include, but are not limited to, the major areas described below.

A. CELLULAR AND MOLECULAR NEUROBIOLOGY

The goal of this research is to provide information about space flight induced changes in nervous system cellular populations and the effects such changes have on information processing and normal nervous system function. The information [5] derived will also provide a better understanding of basic cellular mechanisms on Earth. Various changes in diverse types of cells ranging from bacteria to mammalian cells have been observed during exposure to space flight (In Vitro, 14:165-173, 1978; Aviat. Space Environ. Med., 53:370-374, 1982; Science, 225:228-230, 1984; Appl. Micrograv. Tech., 1:115-122, 1988; Exp. Geront., 26:247-256, 1991). Significant alterations in cell proliferation, cell growth, differentiation, metabolism, membrane properties, electrolyte concentration, and cytoplasmic streaming have been reported. The environmental conditions during space flight may produce a variety of changes in the function, morphology, biochemistry, and metabolism of cells in the nervous system. Alterations in the transcription and translation of neuronal proteins may occur. Another important area of investigation is understanding changes related to ionic balance, membrane structure, and regulation of neurotransmitters, receptors and trophic factors that can affect cell-to-cell interactions and functional activity. Also, examination of possible effects on intracellular transport mechanisms and the ability of cells to survive and regenerate in a microgravity environment are of interest.

Suggested Specific Areas for Space Flight Investigation

- Cellular function, metabolism, and biochemistry
- Gene expression and differentiation, and protein synthesis

- Synthesis, motility, and organelle trafficking
- Cell-cell interactions
- Survival and regeneration

B. DEVELOPMENTAL NEUROBIOLOGY

Research in this area will examine nervous system development in microgravity. The goal is to discern what role gravity plays, and how crucial it is, in normal nervous system development. Only a few studies have been conducted on the effects of space on nervous system development (USSR Space Life Sciences Digest, 14:130-21, 1987; 15:42-44, 1988; 17:31-34, 1988; 21:89-93, 1989). Results from a study on the Russian Cosmos biosatellite demonstrated that the brains of 18-day rat fetuses whose embryonic development from day 13 to day 18 occurred in space flight showed signs of insufficient tissue oxygenation, trends toward delayed cell migration in the cortex, and delayed differentiation of hypothalamic neurosecretory cells (The Physiologist, 28:S81-S82, 1985; Kosmicheskaya Biologiya I Aviakosmicheskaya Meditsina, 21:16-22, 1987). The cell differentiation rate returned to normal during continued embryogenesis after return to 1 G. These preliminary results indicate that exposure to space flight can affect the normal development of the nervous system. Proposed studies could examine the nature and magnitude of any such alterations on cellular, morphological, systems, and functional aspects of the developing nervous system. A main goal of this research would be to study the effects of space flight on the mechanisms and patterns of neuronal birth, differentiation, and death that underlie the formation of the nervous system. Also, the effects of exposure to altered gravity and other features of the space environment on the morphology and physiology of neuronal connectivity, and the functional consequences, is another area of interest. From a systems viewpoint, experiments should examine epigenetic factors, such as neuronal activity and environmental factors during pre- and postnatal development under conditions of altered gravity and space flight.

[6] Suggested Specific Areas for Space Flight Investigation

- Regulation of neurogenesis including neuronal migration and differentiation during embryogenesis
- Process outgrowth, axonal guidance and pathway formation
- Formation and specificity of synapses
- The timing and duration of sensory and motor "critical periods"
- The development of oscillatory events such as circadian rhythms and sleep-wake cycles
- The effect of postnatal exposure to altered gravity on the morphology and physiology of neuronal connectivity
- Functional consequences of space flight induced alterations

C . SENSORY AND MOTOR SYSTEMS

The goal of this research is to better understand the effects of space flight on sensory-motor processing and integration. Sensory perception, sensory-motor transformations, and motor performance are all affected by conditions of space flight (Space Physiology and Medicine, 2nd. edition, Nicogossian, Huntoon, Pool [eds.] Lea & Febinger, 1989; Experimental Brain Research, 62(2), 1986; J. Clin. Pharm., 31:904-910, 1991). Much of this research has been devoted to examining neurovestibular and sensory-motor adaptation to microgravity, readaptation to 1 G, and the underlying mechanisms of space motion sickness. Observations range from cellular alterations to changes in human motor performance during and after flight. For example, changes in various cellular parameters (e.g. striatal muscarinic receptors [Brain research, 593:291-294, 1992], reticular neuron dendritic morphology [The Physiologist, 33:S12-S 15, 1990], number and activity of ventral horn cells [J. of Appl. Physiol., 73 {Suppl.}: 107S-111S, 1992], and synapses of Type II otoconia hair cells [preliminary results, SLS-1 mission]) have been observed in rats exposed to space flight. In all cases these changes are thought to be due to compensatory mechanisms related to altered vestibular input as evidenced by the observation of an increase in gain of vestibular nerve fiber activity in monkeys exposed to microgravity (J. of Appl. Physiol., 73[Suppl.]: 112S-120S, 1992). Functionally, this altered otolith input to the central nervous system may contribute to the changes in vestibulo-ocular reflexes in monkeys (J. of Appl. Physiol., 73[Suppl.]: 121S-131S, 1992) and disrupted postural control in humans observed upon return to Earth (The Amer. J. Otology, 14:9-17, 1993). Studies of interest in this area would involve examination of sensory systems on the single system or multisystem level, in both human and animals, and emphasize research that would serve to advance the understanding of how the nervous system integrates information, carries out sensory-motor transformations, and functions under normal conditions as well as under the special conditions imposed by space travel.

Suggested Specific Areas for Space Flight Investigation

- The structure and function of sensory systems
- The organization and coordination of standing posture, locomotion, flight, swimming, and other volitional movements
- Sensory-motor adaptation to microgravity and readaptation to 1 G
- Sensory-motor mismatch and space motion sickness
- Central cortical and subcortical representation of three-dimensional space
- Multisensory integration and spatial orientation

[7] D. NERVOUS SYSTEM HOMEOSTASIS AND ADAPTATION

The goal of this research is to provide information about space flight induced changes in the homeostasis of the nervous system and of nervous system regulation of other systems. Such changes in homeostatic mechanisms during space flight have been reported (Space Physiology and Medicine, 2nd. edition, Nicogossian, Huntoon, Pool [eds.] Lea & Febinger, 1989; (J. of Appl. Physiol., 73[Suppl.], 1992). Specific findings include observations of autonomic changes, such as altered baroreceptor reflex function, which contributes to post-flight orthostatic dysfunction (J. Clin. Pharm., 31:951-955, 1991), and neuroendocrine changes, such as lowered growth hormone and prolactin secretion in cultured pituitary cells (J. of Appl. Physiol., 73[Suppl.]: 151S-157S, 1992) and diminished mRNA for other hormones in hypothalamic cells (J. of Appl. Physiol., 73[Suppl.]:158S156S, 1992). Research should focus on two areas: (1) Experiments related to effects of space flight on the endocrine, autonomic, and immunological systems and the central nervous system control of these; and (2) Experiments that would evaluate the capability of the nervous system to adapt to changes in the organism or in the external environment caused by space flight.

Suggested Specific Areas for Space Flight Investigation

- Autonomic aspects of nervous system function (e.g. vascular control, intracranial pressure)
- Hypophyseal-pituitary axis, neuro-endocrine and neuro-immune interactions and their consequences for adaptive responses
- Changes in CNS homeostasis and correlated CNS functional changes
- CNS activity (e.g., EEG, evoked potentials) and other dynamic changes in the brain, including their psychological correlates
- Circadian rhythms and sleep cycles
- Consequences of environmental factors (e.g., noise and vibration) on the nervous system

E . BEHAVIOR, COGNITION, AND PERFORMANCE

The goal of this research is to provide information about the effects of space flight on behavior and cognition and on how such changes can affect task performance. Normal motor function, perception, spatial memory, alertness, sleep, and higher cognitive functions of the crew are all of major concern for successful space flight operations. Actual space flight data in this area are limited. Available accounts of behavioral capabilities in space related to perceptual, cognitive, and psychomotor performance and the capacity to do work suggest that with few exceptions decrements in performance have been transitory, occurring early in the mission (J. British Interplanetary Soc., 43:475-488, 1990; USSR Space Life Sciences Digest, 15:58, 1988; 29:52-53, 1991). Many of the reported problems are related to learning new perceptual and motor skills required for functioning in weightlessness and to the spatial disorientation and space sickness that occur early in a mission. The stress associated with space flight (Space Biol. and Aerosp. Med., 17:30-33, 1983; Acta Astronautica, 11:149-153, 1984) and the sleep disturbances, which are common on space missions (Space Physiology and Medicine, 2nd. edition, Nicogossian, Huntoon, Pool [eds.] Lea & Febinger, 1989), may also have an impact on behavior and performance. This research will consist of experiments related to the effects of space flight on cognition, behavior, and human performance.

[8]
Suggested Specific Areas for Space Flight Investigation

- Higher cognitive function, learning, memory, reasoning, and associated physiological and neural mechanisms
- Effects of space flight associated stress on behavior and performance
- Motor performance and coordination
- Alertness, sleep, circadian rhythms, and their effects on performance
- Evaluation of human performance (e.g. crew factors, selection and training, and collective decision making; complex skill acquisition, retention, and transfer; and reaction time and inspection time).
- Small group dynamics
- Perception of time

Document III-25

Document Title: Memorandum from Dr. Harry C. Holloway, Associate Administrator for Life and Microgravity Sciences and Applications, to the Director of Life and Biomedical Sciences and Applications, "Implementation of Revised Life Sciences Peer Review Policy," September 1993.

Source: NASA Historical Reference Collection, History Office, NASA Headquarters, Washington, D.C.

During the 1980s and early 1990s, the NASA life sciences program came under external criticism from the science community regarding the quality of the scientific peer-review process and the management of grants by the Life Sciences Division. In response, a thorough review and revision of policy and process was put into effect. An annual NASA Research Announcement (NRA), reflecting the advice of the National Academy research strategy reports, was put into place. Policies on conflict of interest with respect to peer reviewers were enforced, and the segregation of intramural and extramural budgets were abolished so that scientists from both communities could compete openly and fairly in the same arena.

[no page number]
National Aeronautics and
Space Administration

Washington, DC
20546

TO: UL/Director, Life and Biomedical Sciences and Applications

FROM: U/Associate Administrator for Life and Microgravity Sciences and Applications

SUBJECT: Implementation of Revised Life Sciences Peer Review Policy

1. All NASA intramural life sciences programs will be externally peer reviewed in fiscal year 1994. This will initiate a new review process in which all new intramural life sciences investigations will be initially externally peer reviewed prior to being funded, and all ongoing intramural life sciences investigations will be externally peer reviewed at least every three years. The review process will be conducted in conjunction with a revised, annual Research and Technical Operating Plan (RTOP) process. RTOPs will be developed by NASA Centers in response to requirements set by the NASA Headquarters Life and Biomedical Sciences and Applications Division. The RTOP plan will require written research proposals for every new investigation proposed and progress reports for every on-going investigation.

2. All intramural and extramural research projects and programs, including operationally oriented programs like the Extended Duration Orbiter Medical Program, the Biomedical Monitoring and Countermeasures Program, and the United States-Russian Shuttle-MIR program, whether funded through NASA Announcements of Opportunity (AOs), NASA Research Announcements (NRAs), or RTOPs, will be reviewed to the same standards by qualified scientists not associated with the particular program being reviewed. Peer review panels will be composed primarily of scientists from outside the NASA scientific community. Formation of peer review panels will be coordinated by an extramural, independent support contractor (such as American Institute for Biological Sciences or Universities Space Research Association), which will ensure absence of conflicts of interest of panel members and assure that panel members have no affiliation with, or other conflict of interest regarding the particular program and project being reviewed.
[2]
3. NASA life scientists who are actively involved in research in a specific discipline will not manage extramural research grants in that particular field. Technical and management oversight of both extramural and intramural research will be the responsibility of the appropriate NASA Headquarters program scientists. The management of all intramural and extramural research will be carried out in such a way as to avoid any conflict of interest or the appearance of conflict of interest. Field Center life sciences managers will submit a plan for accomplishment of these directives at their respective Centers by October 1, 1993.

4. You will immediately direct all field installations involved in life sciences research to implement the above policy and assure that it is adhered to. Provide Dr. Earl W. Ferguson, my Special Assistant for Research and Operations, quarterly reports of progress and accomplishments in implementing these requirements for the first three years of its implementation.

[signed]
Harry C. Holloway, M.D.

Document III-26

Document Title: Letter from Robert A. Whitney, Jr., Deputy Surgeon General, to Dr. Harry C. Holloway, Associate Administrator for Life and Microgravity Sciences and Applications, and the attached report, "Special Study Panel Concerning Animal Research," 14 September 1993.

Source: NASA Historical Reference Collection, History Office, NASA Headquarters, Washington, D.C.

The issue of using laboratory animals such as mice, rats, and primates as experimental subjects in space experiments and of having to kill them (not primates) in order to achieve experimental success has always been politically sensitive for an agency like NASA that operates in the public eye. As it anticipated flying a number of rats on the second Space Life Sciences Spacelab mission, NASA asked the Office of the Surgeon General to carry out an independent review of its experimental procedures and justifications.

DEPARTMENT OF HEALTH AND HUMAN SERVICES

Public Health Service
Office of the Surgeon General
Rockville MD 20857

[no page number]

September 14, 1993

Harry C. Holloway, M.D.
Associate Administrator for Life and
    Microgravity Sciences and Applications
National Aeronautics and Space Administration
Washington, D.C. 20546

Dear Dr. Holloway:

As Chairperson of the Special Study Panel Concerning Animal Research, it is my pleasure to forward the enclosed report of our findings and recommendations.

As you requested, we undertook a thorough review of the overall process, justification, and oversight of the experimental protocols requiring animals on the SLS-2 Mission, and assessed the level of sensitivity that exists within NASA and its Field Centers to issues related to conducting animal research in space.

Our panel found the NASA scientific and animal care review process superb. The high level of refinement of the protocols reduces the number of animals required, and the NASA management staff we interviewed were obviously quite sensitive to all of the issues surrounding the use of animals in this type of research.

On behalf of the Panel, I wish to express our thanks for the extensive documentation and background information you provided and for the excellent cooperation we received from NASA staff. If you wish to discuss our findings and recommendations, please give me a call at (301) 443-4000.

> Sincerely,
> [signed]
> Robert A. Whitney, Jr.
> Deputy Surgeon General

Enclosure

[1]

SPECIAL STUDY PANEL CONCERNING ANIMAL RESEARCH

REPORT
September 14, 1993

Introduction

This study panel was formed on August 30, 1993 by Harry C. Holloway, NASA's Associate Administrator for Life and Microgravity Sciences and Applications to provide him with an assessment of the level of sensitivity that exists within NASA and its Field Centers to issues related to conducting animal research in space, particularly to the kind of research planned for the Spacelab Life Sciences 2 (SLS-2) Mission and to recommend ways to improve on NASA's current approach to planning and communications regarding the use of animals in NASA research. Attachment A [not included] to this report lists the panel membership and Attachment B [not included] contains the charge that was provided the panel at its inception.

The panel held two meetings to carry out its charge. The first, on Tuesday, September 7, 1993, focused on clarifying the issues that led to the formation of the panel by Dr. Holloway and on determining the course of action that the panel should take. The second meeting, on Friday, September 10, 1993 was primarily a fact-finding meeting to obtain information on the actual scientific experiments to be carried out on SLS-2, and to discuss the issues related to animal research in general and to research on SLS-2 in particular with the NASA personnel responsible for managing and carrying out the animal research program. Attachment C [not included] lists the attendees at these two meetings. Following the fact-finding portion of the second meeting, the panel met in executive session to consider its findings and develop its recommendations. This report contains those findings and recommendations.

The panel was briefed on the 14 primary scientific experiments on the SLS-2 Mission. Of these, six require laboratory rats:

- Regulation of Blood Volume During Space Flight, C. P. Alfrey, M.D., Baylor College of Medicine;
- Effects of Zero Gravity on Biochemical and Metabolic Properties of Skeletal Muscles in Rats, K. M. Baldwin, Ph.D., University of California at Irvine;
- Bone, Calcium, and Space Flight, E. M. Holton, Ph.D., NASA Ames Research Center;
- [2]Regulation of Erythropoiesis in Rats During Space Flight, A. T. Ichiki, Ph.D., University of Tennessee Medical Center;
- Effects of Microgravity on the Electron Microscopy, Histochemistry, and Protease Activities of Rat Hind Limb Muscles, D. A. Riley, Ph.D., Medical College of Wisconsin; and
- A Study of the Effects of Space Travel on Mammalian Gravity Receptors, M. D. Ross, Ph.D., NASA Ames Research Center.

Four of these experiments (Alfrey, Ichiki, Riley, and Ross) require inflight decapitation and autopsy of a total of five (or six if the crew can maintain an optimistic schedule for completing this activity) rats from among a total of 48 rats onboard. Detailed information was provided to the panel (see Attachment D [not included]) concerning the actual protocols of each of the six experiments involving rats and on the scientific rationale for carrying out these investigations. The protocol provided to the panel describing Dr. Ross' experiment is not an accurate description of the planned SLS-2 activity.

Findings

1. The panel found that the experimental protocols in question have been properly reviewed by appropriate review groups through 15 years of preflight preparation. The most recent scientific body to review these experiments (in 1993) was the independent, external Committee on Space Biology and Medicine of the National Academy of Sciences' Space Studies Board. NASA's animal care and use review of the proposed research is in full compliance with NASA Management Instruction 8910.1, "Care and Use of Animals in the Conduct of NASA Activities," which is based on the Public Health Service Policy on Humane Care and Use of Laboratory Animals. All proposed research activities involving animals adhere to the U.S. Government Principles for the Utilization and Care of Vertebrate Animals Used in Testing Research, and Training.

2. NASA has applied a remarkable level of engineering to assure the well being of the rats involved. A high degree of refinement of animal use has been embedded in the experimental protocols, requiring that only five or six animals undergo euthanasia in space. This appears to be an extraordinarily small number of animals to achieve the inflight requirements of four extensive research projects.

3. The level of sensitivity to proper animal care and use practices among all NASA personnel interviewed by the panel was high. These are [3] personnel directly connected with managing and carrying out the animal research program. NASA Ames Research Center should be commended for having hired an individual with the background and training required to focus on the public affairs issues related to animal experimentation, and for the development of specific public affairs materials and Agency briefings related to raising the level of awareness of NASA management and personnel on this subject.

However, two issues related to the subject of sensitivity are of concern. First, it is clear that NASA has not made full use of the material that it has developed to educate its own employees and the public concerning the wise use of animals in space research. The panel learned that NASA's three background papers (The Use of Laboratory Animals in NASA-Sponsored Research; NASA's Animal Care and Use Policy; and Alternatives to Animal Use in Research) have not been distributed to all NASA personnel associated with projects involving the use of animals. The panel additionally learned that these materials have not been distributed to principal investigators in the NASA program.

Second, additional educational efforts by NASA are required to more fully sensitize all NASA employees, including higher levels of management, to the general public's views surrounding the use of animals in research. It is apparent that NASA's educational efforts to date have not fully sensitized employees to the views of those in the public who are not necessarily opposed to animal research, but who are naive regarding the rigorous requirements NASA imposes prior to animal use in space research.

4. The inflight decapitation of five rats is justified in the proposed protocol by the following statement: "It is necessary that the decapitation procedure be carried out in the absence of anesthesia because there is evidence that anesthetic agents acceptable for use in the Spacelab produce changes in the neural tissue under investigation."

Recommendations

1. NASA should distribute its three fact sheets on animal research to all NASA employees and to those members of the extramural scientific community who participate in NASA's life sciences research program.

2. NASA should augment its educational efforts regarding animal care and use in NASA activities, directing such efforts to all levels of NASA. Particular emphasis should be placed on attempting to increase [4] the awareness of NASA employees of the wide diversity of views held by the general public regarding the use of animals in research. Only through an awareness of these views will it be possible for NASA spokespersons to appreciate that their statements may be interpreted in a context different from that intended.

3. The panel recommends that all press interviews connected or potentially connected with animal research be scheduled through NASA's public affairs office. It further recommends that NASA have a public affairs representative attend any such personal

or telephone interview session to assure that the results of that session are portrayed accurately. This practice has enjoyed widespread success in limiting bias in the reporting of interviews in other Federal agencies.

List of Attachments

A. PANEL MEMBERSHIP

B. CHARGE TO PANEL

C. ATTENDEES AT PANEL MEETINGS

D. MATERIAL PROVIDED TO PANEL

Document III-27

Document Title: Presentation, William Berry, Ames Research Center, to France Cordova, NASA Chief Scientist; Wes Huntress, Associate Administrator for Space Science; Bill Townsend, Deputy Associate Administrator for Space Science; and Dr. Harry Holloway, Associate Administrator for Life and Microgravity Sciences, NASA Headquarters, 29 March 1995.

Source: Archives, Ames Research Center, Moffett Field, California

As NASA conducted a "Zero-Base Review" in 1995, the Ames Research Center perceived itself at risk of being closed or having its activities significantly diminished. Reacting to this situation, Ames managers came up with the idea of focusing the Center's activities on a "Life in the Universe" theme. Dealing with space science, this presentation by Bill Berry of the Ames staff to NASA Headquarters senior staff was influential in getting that focus accepted, but, at the suggestion of Associate Administrator for Space Science Wes Huntress, with the designation of "Astrobiology" rather than "Life in the Universe."

RE-INVENTING AMES SPACE - ALTERNATIVE CONCEPT

COLLABORATION

"The new NASA will of necessity be more reliant on...the private sector than the old NASA."
Excerpt from NASA Federal Laboratory Review (Feb. 1995)

• PROBLEM STATEMENT/SOLUTION

• MISSION

- FUNCTIONAL ANALYSIS [not included]

- APPROACH

- CONCLUSION

[no slide number]

PROBLEM STATEMENT

NASA is anticipating a significant budget decrease in (~ 20%) over the next five years.

The challenge is to preserve a viable Agency science program and to maximize science return from missions.

- Preserve and enhance core competencies, meet current commitments, enable new programs, and improve the quality of products with a downsized workforce and lower total program cost

SOLUTION

- Redefine ARC Space roles to focus on Life in the Universe across NASA Strategic Enterprises to make maximum use of Ames' proven multidisciplinary strengths.

- Streamline bureaucracy (NASA/Center commitment)

- Meet existing commitments, enable new work, improve R&D, and acquire new talent through innovative use of cooperative agreements and outsourcing

- Downsize civil servant workforce in functions that do not require civil servant implementation.

[no slide number]

AMES SPACE MISSION

Lead Center for Life in the Universe

"NASA should develop mechanisms to promote multidisciplinary research between programs of the Scientific Research Enterprise and the MTPE Enterprise." - Excerpt from NASA Federal Laboratory Review (Feb. 1995)

Life in the Universe is the fundamental science theme that unites all the NASA Strategic Enterprises, yet there is no center or HQ element that is responsible for making the necessary linkages among Enterprises and organizations. This is a unique Ames strength that can be amplified at no additional cost to better serve NASA's needs.

- Implement multidisciplinary Life In the Universe program of ground research, technology development, platform aircraft, and space fight missions to explore:

  * The origin, evolution, and distribution of life in the universe;
  * The relationship of gravity and life [sic]
  * The evolution, evaluation, and protection of living planets;
  * The expansion of life from Earth into space.

- Lead Advanced Studies

  NASA must link the strategic plan in terms of accomplishing the NASA Strategic Plan elements. In maintaining global leadership, NASA must also couple the strategic plan with a process that permits new and competing ideas to be developed and evaluated." – Excerpt from NASA Federal Laboratory Review (Feb. 1995)

- Transfer knowledge and technologies across Strategic Enterprises

- Maximize use of NASA ground and flight resources by external community

- Promote the national goals to ensure a scientifically and technically educated citizenry.

- Transfer knowledge and technologies for public sector benefits.

[four slides, "Function Analysis: National Core Functions, R&D, Facilities and Technical Support, and Administration and Management," omitted]

[no slide number]

APPROACH

Support the best people to do the highest quality work in the most efficient manner with the fewest bureaucratic barriers

- Life In the Universe Center located at ARC

- Code S Directorate, Division, Branch, and budget management will be performed by civil servants or by IPAs.

- Life In the Universe cooperative agreement with major university(s) according to the authority of 31 U.S.C. 6301 to 6308.

"There is great appeal in terms of coupling NASA Research Capabilities with the expertise and intellectual resources of a university or consortia of universities...NASA should consider a Space Act Agreement or similar contractual arrangements with a university at one or more of its Centers...The new NASA will of necessity be more reliant on...the private sector than the old NASA." Excerpt from NASA Federal Laboratory Review (Feb. 1995)

"Cooperative agreement is an agreement that provides funds to an educational institution or other non-profit organization to accomplish a public purpose of support or stimulation authorized by Federal Statute. Substantial technical involvement between NASA and the recipient is expected and will be identified in the agreement." NASA Handbook 5800.1C

- Novate all contract work to single prime with gradual phasedown of CS in support functions.

[no slide number]

APPROACH (continued)

- NASA Ames Civil Service R&D workforce to conduct research, development planning, selection, advanced studies in collaboration with universities and industries.

- Civil servants will be maintained and recruited to preserve core competencies necessary to ensure a productive and dynamic space program.

- In implementing national functions, (e.g., lead roles) consortia would be composed of participants across the nation.

- Technology transfer and education programs would be the responsibility of the non-NASA partners with strong NASA volunteer participation.

- CS personnel who perform support functions will not be replaced by CS when they leave the government. These functions will be taken over by the industry/university/consortia partners.

- Dollar savings and quality enhancements result from:

- CS Workforce downsizing via attrition or reassignment into product tasks

- Minimization of unproductive reviews, paperwork, reporting

- More efficient execution of support functions by businesses

- More efficient procurement (less downtime, faster completion of projects)

- Contracted rapid prototyping/fabrication vs. maintenance of in-house staff

- Skill mix adjustments on an "as needed" basis from university/industry specialists rather than permanently employed personnel.

[no slide number]

CONCLUSION

- • This approach will enhance R&D, provide an important integrating function for NASA, meet current commitments, and enable new programs while meeting necessary cost reductions.

  - Life In the Universe provides a focus that unites all of NASA's Enterprises.

  - Cooperative agreement, consortia and single prime contractor approach offers superior cost savings over status quo, transfer of programs to other Centers or privatization of all of Code S, and - -

    * Strengthens NASA science, technology, and flight programs

    * Offers maximum flexibility in skill mix

    * Guarantees an infusion of new people and new ideas

    * Provides opportunities for existing civil service staff to return to the core work

    * Allows downsizing without program terminations

    * Trains the next generation

    * Provides an exciting opportunity to make the national community true hands-on partners in the space program

  - Legal mechanisms are already in place.

[slide, "Backup: Programs," omitted]

Document III-28

Document Title: NASA Policy Directive 8020.7E, "Biological Contamination Control for Outbound and Inbound Planetary Spacecraft," 19 February 1999.

Source: NASA Historical Reference Collection, History Office, NASA Headquarters, Washington, D.C.

This is the most recent NASA policy directive related to measures for avoiding forward and back contamination during solar system exploration.

[no page number]

NASA	Directive: NPD 8020.7E
POLICY	Effective Date: February 19, 1999
DIRECTIVE	Expiration Date: February 19, 2004

Responsible Office: S/Office of Space Science

Subject: Biological Contamination Control for Outbound and Inbound Planetary Spacecraft

1. POLICY

The conduct of scientific investigations of possible extraterrestrial life forms, precursors, and remnants must not be jeopardized. In addition, the Earth must be protected from the potential hazard posed by extraterrestrial matter carried by a spacecraft returning from another planet or other extraterrestrial sources. Therefore, for certain space-mission/target-planet combinations, controls on organic and biological contamination carried by spacecraft shall be imposed in accordance with directives implementing this policy.

2. APPLICABILITY

a. This directive applies to NASA Headquarters and NASA Centers, including Component Facilities, and to NASA contractors where specified by contract.

b. The provisions of this directive cover all space flight missions which may intentionally or unintentionally carry Earth organisms and organic constituents to the planets or other solar system bodies, and any mission employing spacecraft which are intended to return to Earth and/or its biosphere from extraterrestrial targets of exploration.

3. AUTHORITY

42 U.S.C. 2473(c)(1), Section 203(c)(1) of the National Aeronautics and Space Act of 1958, as amended.

4. REFERENCES

a.   Article IX of the Outer Space Treaty of 1967, to which the United States is a party, states in part that "...parties to the Treaty shall pursue studies of outer space, including the Moon and other celestial bodies, and conduct exploration of them so as to avoid their harmful contamination and also adverse changes in the environment of the Earth resulting from the introduction of extraterrestrial matter and, where necessary, shall adopt appropriate measures for this purpose..." ("Treaty on Principles Governing the Activities of States in the Exploration and Use of Outer Space, Including the Moon and Other Celestial Bodies" entered into force October 10, 1967. 18 U.S. Treaties and Other International Agreements at 2410-2498.)

b.   NPG 8020.12x, Planetary Protection Provisions for Robotic Extraterrestrial Missions.

[no page number]
5. RESPONSIBILITY
a.   The Associate Administrator for Space Science, or designee, is responsible for overall administration of NASA's planetary protection policy. This includes the following:

(1) Maintaining the required activities in support of the planetary protection policy at NASA Headquarters.

(2) Assuring that the research and technology activities required to implement the planetary protection policy are conducted.

(3) Monitoring space flight missions as necessary to meet the requirements for planetary protection certification.

b.   The designee for managing and implementing this policy is the Planetary Protection Officer, who is responsible for the following:

(1) Prescribing standards, procedures, and guidelines applicable to all NASA organizations, programs, and activities to achieve the policy objectives of this directive.

(2) Certifying to the Associate Administrator for Space Science and to the Administrator prior to launch; and (in the case of returning spacecraft) prior to the return phase of the mission, prior to the Earth entry, and again prior to approved release of returned materials, that–

(a) All measures have been taken to assure meeting NASA policy objectives as established in this directive and all implementing procedures and guidelines.

(b) The recommendations, as appropriate, of relevant regulatory agencies with respect to planetary protection have been considered, and pertinent statutory requirements have been fulfilled.

(c) The international obligations assessed by the Office of the General Counsel and the Office of External Relations have been met, and international implications have been considered.

(3) Conducting reviews, inspections, and evaluations of plans, facilities, equipment, personnel, procedures, and practices of NASA organizational elements and NASA contractors, as applicable, to discharge the requirements of this directive.

(4) Keeping the Associate Administrator for Space Science informed of developments and taking actions as necessary to achieve conformance with applicable NASA policies, procedures, and guidelines.

c. The Associate Administrator for Space Flight and the Associate Administrator for Life and Microgravity Sciences and Applications, or designees, will ensure that applicable standards and procedures established under this policy, and detailed in subordinate implementing documents, are incorporated into human space flight missions. Any exceptions will be requested and justified to the Administrator through the Asociate [sic] Administrator for Space Science.

d. Program Managers, through their respective Center Director, are responsible for the following:

[no page number]
(1) Meeting the biological and organic contamination control requirements of this directive and its subordinate and implementing documents during the conduct of research, development, test, preflight, and operational activities.

(2) Providing for the conduct of reviews, inspections, and evaluations by the Planetary Protection Officer, pursuant to this directive.

6. DELEGATION OF AUTHORITY

None.

7. MEASUREMENTS

Specific constraints imposed on spacecraft involved in solar system exploration will depend on the nature of the mission and the identity of the target body or bodies. These constraints will take into account current scientific knowledge about the target bodies through recommendations from both internal and external advisory groups, but most notably from the Space Studies Board of the National Academy of Sciences. The most likely constraints on missions of concern will be a requirement to reduce the biological contamination of the spacecraft, coupled with constraints on spacecraft operating procedures, an inventory of organic constituents of the spacecraft and organic samples, and restrictions on the handling and methods by which extraterrestrial samples are returned to Earth. In the majority of missions, there will also be a requirement to document spacecraft flyby operations, spacecraft impact potential, and the location of landings or impact points of spacecraft on planetary surfaces or other bodies. Specific requirements (reviews, documentation, and levels of cleanliness) are detailed in implementing procedures and guidelines, primarily NPG 8020.12x, "Planetary Protection Provisions for Robotic Extraterrestrial Missions," and will be used to measure adherence to this directive.

8. CANCELLATION

NMI 8020.7D dated December 3, 1993.

/s/ Daniel S. Goldin

Administrator

Document III-29

Document Title: Space Studies Board, National Academy of Sciences, "A Strategy for Research in Space Biology and Medicine in the New Century," 28 November 2000.

Source: Space Studies Board, National Academy of Sciences, Washington, D.C.

This report from the Space Studies Board represents the assessment of space life sciences by the scientific community at the turn of the century and its recommendations for future research priorities.

[1]
A Strategy for Research in Space Biology
and Medicine in the New Century

Executive Summary

The core of the National Aeronautics and Space Administration's (NASA's) life sciences research lies in understanding the effects of the space environment on human physiology and on biology in plants and animals. The strategy for achieving that goal as originally enunciated in the 1987 Goldberg report, A Strategy for Space Biology and Medical Science for the 1980s and 1990s,[1] remains generally valid today. However, during the past decade there has been an explosion of new scientific understanding catalyzed by advances in molecular and cell biology and genetics, a substantially increased amount of information from flight experiments, and the approach of new opportunities for long-term space-based research on the International Space Station. A reevaluation of opportunities and priorities for NASA-supported research in the biological and biomedical sciences is therefore desirable.

The strategy outlined in the Goldberg report had two main purposes: "(1) to identify and describe those areas of fundamental scientific investigation in space biology and medicine that are both exciting and important to pursue and (2) to develop the foundation of knowledge and understanding that will make long-term manned space habitation and/or exploration feasible."[2] To achieve these purposes, the Goldberg report identified four major goals of space life sciences:

"1. To describe and understand human adaptation to the space environment and readaptation upon return to earth.

"2. To use the knowledge so obtained to devise procedures that will improve the health, safety, comfort, and performance of the astronauts.

"3. To understand the role that gravity plays in the biological processes of both plants and animals.

"4. To determine if any biological phenomenon that arises in an individual organism or small group of organisms is better studied in space than on earth."[3]

These goals remain valid and form the basis of the present report.

Both the Goldberg report and the 1991 follow-up assessment, Assessment of Programs in Space Biology and Medicine 1991,[4] emphasized basic research and the importance of vigorous ground-based programs aimed at addressing the fundamental mechanisms that underlie observed effects of the space environment on human physiology and other biological processes. The present report strongly reemphasizes that strategy and calls for an integrated, multidisciplinary approach that encompasses all levels of biological organization-the molecule, the cell, the organ system, and the whole organism- and employs the full range of modem experimental approaches from molecular and cellular biology to organismic physiology.

The sections that follow summarize the Committee on Space Biology and Medicine's priorities for NASA-supported research, its recommendations for high-priority research in individual disciplines, and its recommendations for overall priorities for NASA-sponsored research across disciplinary boundaries. The final section outlines significant concerns in the program and policy arena and offers related recommendations.

PRIORITIES FOR RESEARCH

Taking into account budgetary realities and the need for clearly focused programs, the highest priority for NASA-supported research in space biology and medicine in the new century should be given to research meeting one of the following criteria:

1. Research aimed at understanding and ameliorating problems that may limit astronauts' ability to survive and/or function during prolonged spaceflight. Such studies include basic as well as applied research and ground-based investigations as well as flight experiments. NASA programs should focus on aspects of research in which NASA has unique capabilities or that are underemphasized by other agencies.

2. Fundamental biological processes in which gravity is known to playa direct role. As above, programmatic focus should emphasize NASA's capabilities and take into account the funding patterns of other agencies.

A lower priority should be assigned to areas of basic and applied research that are relevant to fields of high priority to NASA but are extensively funded by other agencies, and in which NASA has no obvious unique capability or special niche.

HIGH-PRIORITY DISCIPLINE-SPECIFIC RESEARCH

Because the recommendations for research, and research priorities, in the discipline-specific chapters cover a wide range of fields relevant to space biology and medicine, the committee chose not to reproduce all of those recommendations in full in this executive summary. Instead the committee sought to capture the essence of what is recommended in Chapters 2 through 12, an approach that was best served by condensation, full quotation, or [2] addition of supplemental detail as seemed useful to preserve the intent of the recommendations in their full form and context. The recommendations are numbered only in instances in which the committee considered that there was a clear priority order. [This section (pp. 2-8) not included.]

CROSSCUTTING RESEARCH PRIORITIES

This section summarizes the committee's recommendations for the highest-priority research across the entire spectrum of space life sciences. In the near term, until the research facilities of the International Space Station come online or an additional Spacelab mission is provided, NASA-supported research will necessarily be directed primarily toward ground-based investigations designed to answer fundamental questions and frame critical hypotheses that can later be tested in space. Indeed, as this report emphasizes, understanding the basic mechanisms underlying biological and behavioral responses to spaceflight is essential to designing effective countermeasures and protecting astronaut health and safety both in space and upon return to Earth. For these reasons, the following recommendations for high-priority areas of crosscutting research place emphasis on ground-based studies.

Physiological and Psychological Effects of Spaceflight

Priority should be given to research aimed at ameliorating problems that may limit [9] astronauts' health, safety, or performance during and after long-duration spaceflight. The committee emphasizes that specific priorities may shift to a significant degree depending on the types of missions to be carried out in the future, particularly as related to long-term human exploration of space. For this reason, the recommended areas of research are not given an order of priority.

Loss of Weight-bearing Bone and Muscle

Bone loss and muscle deterioration are among the best-documented deleterious effects caused by spaceflight in humans and animals. Exercise has been only partially successful in preventing muscle weakness and bone loss. Development of effective countermeasures requires advances in several areas of research:

• Research should emphasize studies that provide mechanistic insights into the development of effective countermeasures for preventing bone and muscle deterioration during and after spaceflight.

• Ground-based model systems, such as hindlimb unloading in rodents, should be used to investigate the mechanisms of changes that reproduce in-flight and postflight effects.

• A database on the course of microgravity-related bone loss and its reversibility in humans should be established in preflight, in-flight, and postflight recording of bone mineral density.

• Hormonal profiles should be obtained on humans before, during, and after spaceflight. The relationship between exercise activity levels and protein energy balance in flight should be investigated.

Vestibular Function, the Vestibular Ocular Reflex, and Sensorimotor Integration

During the transitions in gravitational force that occur going into and returning from spaceflight, the vestibular system undergoes changes in activity that can result in debilitating symptoms in astronauts.

• The highest priority should be given to studies designed to determine the basis for the adaptive compensatory mechanisms in the vestibular and sensorimotor systems that operate both on the ground and in space.

• In-flight recordings of signal processing following otolith afferent stimulation should be made to determine how exposure to microgravity affects central and peripheral vestibular function and development.

• Motor learning should be investigated in spaceflight and the results compared with findings obtained in ground-based studies of this process.

Orthostatic Intolerance Upon Return to Earth Gravity

Orthostatic hypotension, present since the very earliest human spaceflights, still affects a high percentage of astronauts returning from spaceflights even of relatively short duration and is an even greater problem for shuttle pilots, who must perform complex reentry maneuvers in an upright, seated position. The problem remains despite the use of extensive antiorthostatic countermeasures by both U.S. and Russian space programs. Studies should focus on determining physiological mechanisms and developing effective countermeasures.

• Current knowledge of the magnitude, time course, and mechanisms of cardiovascular adjustments should be extended to include long-duration exposure to microgravity.

• The specific mechanisms underlying inadequate total peripheral resistance observed during postflight orthostatic stress should be determined.

• Appropriate methods for referencing intrathoracic vascular pressures to systemic pressures in microgravity should be identified and validated, given the observed changes in cardiac and pulmonary volume and compliance.

• Current antiorthostatic countermeasures should be reevaluated to refine those that offer protection and eliminate those that do not. Priority should be given to interventions that may provide simultaneous bone and/or muscle protection.

Radiation Hazards

The biological effects of exposure to radiation in space pose potentially serious health effects for crew members in long-term missions beyond low Earth orbit. High priority is given to the following recommended studies:

• Determine the carcinogenic risks following irradiation by protons and high-atomic-number, high-energy (HZE) particles.

• Determine if exposure to heavy ions at the level that would occur during deep-space missions of long duration poses a risk to the integrity and function of the central nervous system.

• Determine how the selection and design of the space vehicle affect the radiation [10] environment in which the crew has to exist.

• Determine whether combined effects of radiation and stress on the immune system in spaceflight could produce additive or synergistic effects on host defenses.

Physiological Effects of Stress

The immune system interacts closely with the neuroendocrine system. Results indicate a close association between the neuroendocrine status of the host and host defense systems.

• The role that the host response to stressors during spaceflight plays in alterations in host defenses should be determined.

Psychological and Social Issues

The health, well-being, and performance of astronauts on extended missions may be negatively affected by many stressful aspects of the space environment. Mechanisms of response to physiological and psychosocial stressors encountered in spaceflight must be better understood in order to ensure crew safety, health, and productivity.

• Highest priority should be given to interdisciplinary research on the neurobiological (circadian, endocrine) and psychosocial (individual, group, organizational) mechanisms underlying the effects of physical and psychosocial environmental stressors. Cognitive, affective, and psychophysiological measures of behavior and performance should be examined in ground-based analogue settings as well as in flight.

• High priority should be given to evaluation of existing countermeasures (screening and selection, training, monitoring, support) and development of effective new countermeasures.

Fundamental Gravitational Biology

Mechanisms of Graviperception and Gravitropism in Plants

Plants respond to changes in the direction of the gravitational vector by altering the direction of the growth of roots and stems. The gravitropic response requires (1) perception of the gravitational vector by gravisensing cells; (2) intracellular transduction of this information; (3) translocation of the resulting signal to the sites of reaction, i.e., sites of differential growth; and (4) reaction to the signal by the responding cells, i.e., initiation of differential growth.

• Studies of graviperception should concentrate on three problems:

  - The identity of the cells that actually perceive gravity;

  - The intracellular mechanisms by which the direction of the gravity vector is perceived; and

- The threshold value for graviperception (this will require a spaceflight experiment. [sic]

• Studies of gravitropic transduction should focus on the nature of the cellular asymmetry that is set up in a cell that perceives the direction of the gravity vector.

• Studies on the translocation step should concentrate on the nature and mechanism of the translocation of the signals that pass from the site of perception to the site of reaction.

• Studies on the reaction step should focus on the mechanism(s) by which gravitropic signals cause unequal rates of cell elongation, and on the possible effects of gravity on the sensitivity of these cells to the signals.

Mechanisms of Graviperception in Animals

It is known that in several systems sensory stimulation plays a role in the development of the neural connections necessary for normal processing of sensory information. The potential role of gravity in the normal development of the gravity-sensing vestibular system of animals is therefore an important area for ground- and space-based research.

• Ground-based studies should identify the critical periods in vestibular neuron development before initiation of experiments on the effects of microgravity on vestibular development.

• Pre- and postflight functional magnetic resonance imaging (fMRI) studies should be conducted with astronauts to determine the effects of microgravity on neural space maps.

Effects of Spaceflight on Reproduction and Development

To determine whether there are developmental processes that are critically dependent on gravity, organisms should be grown through at least two full generations in space.

• Key model animals should be grown through two life cycles; the highest priority should be given to vertebrate models. If significant developmental effects are detected, control experiments must be performed to determine whether gravity or some other element of the space environment induces these developmental abnormalities.

• [11] An analogous experiment should be carried out with the model plant Arabidopsis thaliana to confirm results obtained on Mir with a preliminary experiment using Brassica rapa.

PROGRAMMATIC AND POLICY ISSUES

Although NASA has responded effectively to many of the programmatic and policy issues raised in the 1987 and 1991 reports,[7,8] significant concerns in the program and policy arena remain unresolved. These concerns focus on issues relating to strategic planning and conduct of space-based research; utilization of the International Space Station (ISS) for life sciences research; mechanisms for promoting integrated and interdisciplinary research; collection of and access to human flight data, specifically; publication of and access to space life sciences research in general; and professional education.

Space-based Research

Development of Advanced Instrumentation and Methodologies

Future life sciences flight experiments on the ISS will depend on the availability of advanced instrumentation to carry out the measurements and analyses required by the research questions and approaches described in this report. In addition, facile data and information transfer between space- and ground-based investigators are crucial.

• NASA should work with the broad life sciences community to identify and catalyze the development of advanced instrumentation and methodologies that will be required for sophisticated space-based research in the coming decade.

• NASA should take advantage of advanced instrumentation developed in other countries.

• The capability for direct, real-time communication between space-based experimenters and principal investigators at their home laboratories should be a high-priority objective for the ISS.

Utilization of the International Space Station for Life Sciences Research

Issues relating to the design and use of the ISS are a major concern of the committee. These issues include (1) changes in the design of the ISS, (2) the diversion of funds intended for scientific facilities and equipment into construction budgets, (3) the adequacy of power and transmission of data to and from Earth, (4) the availability of crew time for research, and (5) an extended hiatus in flight opportunities for life sciences research owing to delays in ISS construction. These issues have alarmed the life sciences communities.

• To better ensure that the ISS will adequately meet the needs of space life sciences researchers, NASA should continue to bring the external user community as well as NASA scientists into the planning and design phases of facility construction.

• NASA should make every effort to mount at least one Spacelab life sciences flight in the period between Neurolab and the completion of ISS facilities.

• NASA should determine whether continuation of shuttle missions for short-term flight experiments after the opening of ISS would be economically and scientifically sound.

Science Policy Issues

Peer Review

The Division of Life Sciences initiated a universal system of peer review in 1994 for all NASA-supported investigators. The new process has the committee's strong support.

• Responsibility for the establishment of peer review panels and for funding decisions should remain a function of the Headquarters Division of Life Sciences.

• NASA should regularly evaluate the composition of scientific review panels to ensure that the feasibility of proposed flight experiments receives appropriate expert evaluation.

Integration of Research Activities

• Principal investigators of projected flight experiments should be brought together with NASA managers and design engineers at the beginning of the planning process to function as an integrated team responsible for all phases of the planning, design, and testing. This integration should continue throughout the life of the project.

• NASA should regularly review and evaluate the NASA Specialized Centers of Research and Training (NSCORT) program to determine whether this mechanism provides the best way to foster interdisciplinary research and increase the scientific value of the life sciences research program.

• NASA should regularly review and evaluate the performance of the National Space Biomedical Research Institute and the impact of its funding on the overall life sciences research [12] budget and program.

Human Flight Data: Collection and Access

The disciplinary chapters of this report repeatedly stress the need for improved, systematic collection of data on astronauts preflight, in space, and postflight.

• NASA should initiate an ISS-based program to collect detailed physiological and psychological data on astronauts before, during, and after flight.

• NASA should make every effort to promote mechanisms for making complete data obtained from studies on astronauts accessible to qualified investigators in a timely manner. Consideration should be given to possible modifications of current policies and practices relating to the confidentiality of human subjects that would ethically ensure astronaut cooperation in a more effective manner.

Publication and Outreach

An essential outcome of scientific research is publication (dissemination of results to the scientific community at large [sic]. The record of peer-reviewed publication, especially of spaceflight experiments, by funded investigators in NASA's life sciences programs needs to be improved, as does the usefulness of the Spaceline Archive to the scientific community.

• NASA should provide funding for data analysis and publication of flight experiments for a sufficient period to ensure analysis of the data and publication of the results.

• NASA should insist on timely dissemination of the results of space life sciences research in peer-reviewed publications. For investigators with previous NASA support, the publication record should be an important criterion for subsequent funding.

• NASA should take as a high priority the completion of data entry into the Spaceline Archive and should ensure that access to the archive is simple and transparent.

Professional Education

NASA should make every effort to ensure the professional training of graduate students and postdoctoral fellows in space and gravitational biology and medicine.

• NASA should take as high priority the support of a small, highly competitive program of postdoctoral fellowships for training in laboratories of NASA-supported investigators in academic and research institutions external to NASA centers.

REFERENCES

1. Space Science Board, National Research Council. 1987. A Strategy for Space Biology and Medical Science for the 1980s and 1990s. National Academy Press, Washington, D.C.

2. Space Science Board, 1987, A Strategy for Space Biology and Medical Science for the 1980s and 1990s, p.xi.

3. Space Science Board, 1987, A Strategy for Space Biology and Medical Science for the 1980s and 1990s, p. 4.

4. Space Studies Board, National Research Council. 1991. Assessment of Programs in Space Biology and Medicine 1991. National Academy Press, Washington, D.C.

5. Space Science Board, National Research Council. 1987. A Strategy for Space Biology and Medical Science for the 1980s and 1990s. National Academy Press, Washington, D.C.

6. Wilson, J.W., Cucinotta, F.A., Shinn, J.L., Kim, M.H., and Badavi, F.F. 1997. Shielding strategies for human space exploration: Introduction. Chapter 1 in Shielding Strategies for Human Space Exploration: A Workshop (John W. Wilson, Jack Miller, and Andrei Konradi, eds.). National Aeronautics and Space Administration.

7. Space Science Board, National Research Council. 1987. A Strategy for Space Biology and Medical Science for the 1980s and 1990s. National Academy Press, Washington, D.C.

8. Space Studies Board, National Research Council. 1991. Assessment of Programs in Space Biology and Medicine 1991. National Academy Press, Washington, D.C.

Chapter Four

The Evolution of Earth Science Research from Space: NASA's Earth Observing System

by John H. McElroy
and
Ray A. Williamson

The detailed observation of Earth from space is one of the notable scientific and technological achievements of the space age. The discipline of Earth observations encompasses scientific research, national security, government operational, and commercial activities. The technologies used in Earth observations have extended from small, handheld cameras to very large Earth-directed intelligence-gathering sensors and telescopes. The latter are the predecessors to the Hubble Space Telescope. The beginnings of these efforts in civilian research are described elsewhere in the *Exploring the Unknown* series.[1] This essay traces the history of one aspect of Earth observations from space, the evolution of NASA's Earth science program between the early 1980s and the mid-1990s, a period of significant growth and change in NASA's efforts to study and understand the many structural elements and interactive processes that make up planet Earth. During that period, NASA restructured its program focus, moving Earth-related research out of the space science program into a distinct, separate program. It is also a period when the political implications of Earth science research become apparent to policymakers.

Early work in Earth observations was often exploratory, descriptive, discipline-specific, and qualitative, as researchers and operators learned the capability of the new tools. By the 1970s, quantitative analyses became increasingly important, and scientists developed sensors incorporating increased geometric and radiometric accuracy and enhanced spectral and spatial resolution. The infusion of techniques and analytical approaches developed for planetary exploration also enriched Earth observations. Large-scale international programs, in which NASA participated, brought new understanding about Earth's interactive biogeochemical systems that linked the land, atmosphere, oceans, and ice cover.

By the beginning of the 1980s, the new quantitative capabilities anchored an ambitious scientific effort soon dubbed the "Mission to Planet Earth" (MTPE). The quantitative capabilities also underpinned a new perspective termed "Earth system science" that sought a new synthesis of the Earth sciences with special emphasis on the interactions and

---

1. Pamela E. Mack and Ray A. Williamson, "Observing the Earth from Space," in John M. Logsdon, ed., *Exploring the Unknown, Selected Documents in the History of the U.S. Civil Space Program, Volume III: Using Space* (Washington, DC: NASA, 1998), pp. 155–384.

feedback mechanisms among the traditional disciplines. Universities offered courses on Earth system science and devoted research laboratories to the subject. The new synthesis offered the possibility for an organized examination of global environmental change, and this effort became the core activity of NASA's Earth science research.

In a very real sense, NASA's Earth Science Enterprise of the early twenty-first century began with the start of NASA's programs in 1958. In that year, NASA's scientists and engineers assumed responsibility for work on TIROS-1, the world's first Earth-observing satellite. The TIROS-1 project began under the Department of Defense (DOD) in 1957 and then was transferred to the new space agency upon NASA's formation. Launched in 1960, TIROS-1 gave the world the first comprehensive images of clouds and other aspects of Earth's weather as seen from the vantage point of Earth orbit.

These images, together with the visually powerful photographs of the whole Earth as seen from the orbit of the Moon, gave the public as well as scientists an entirely new perspective on Earth and sowed the seeds of the perception of our planet as an interactive system of physical and biological subsystems. However, it was not until the early 1980s that this conceptual framework took hold as the basis of a focused research effort.

Between the early 1960s and the development of MTPE in the mid-1980s lay more than two decades of Earth science research from space, focusing on understanding different aspects of the solid Earth, weather and climate, oceans, land, and ice cover. Such research developed the basis for beginning to understand the interactions among these realms. This important research formed the essential groundwork for the scientific concepts that have guided the development and execution of NASA's MTPE and ESE programs. Over about a decade and a half, NASA articulated and began to execute a coordinated program based on the concept of Earth as a deeply interrelated system of systems and the related social and political goal of global habitability. In part because of the potential social and political consequences of the research, that struggle involved not only NASA and NASA-supported scientists, but also the National Research Council, the NASA Advisory Council, the Office of Management and Budget, and Congress.

Three primary themes emerged from an examination of this period in NASA's history: NASA's role in the development of Earth science; the high costs of developing and maintaining the global perspective allowed by satellite systems; and NASA as a focal point for a debate on the pursuit of Earth science and public policy. The history of this period illustrated the important role of scientific advisory committees in setting the agenda for Earth science and in defending that agenda in the political process.

The Early Period

The notion that observations from space would likely further the understanding of how Earth's multiple physical systems function was accepted early. Just how this might work was not known. Hence, during the period from 1960 to 1980, NASA scientists explored the basic feasibility of making useful Earth observations from space. At the outset, it was not evident what could be seen from space, how precisely a particular environmental phenomenon could be measured, or who would find the data sufficiently useful to pay for them with government or private funds. This period also tackled the fundamental question of whether space systems could be made sufficiently reliable for continuing use.[2] In essence, this period first explored and developed the relationships among science, space engineering, and political objectives in Earth science. Space systems were of particular interest for exploring and monitoring Earth's makeup because they provided broad-scale, synoptic data that could be verified by observations conducted in situ or from ground-based instruments.

As the feasibility of making useful observations became established, the research community called for increasingly sophisticated and longer-term measurements. For example, the Landsat series of satellites, which began in 1972 with the relatively low-resolution multispectral sensor (80 meters), was upgraded by the early 1980s with the medium-resolution thematic mapper (30 meters), which carried more spectral bands. Nimbus-7, which was launched in 1978 to gather meteorological and ocean data, was particularly important in stimulating the demand for further measurements.[3] This satellite's launch occurred at the same time as the international scientific community was paying increasing attention to climate change, e.g., as in a major international conference that the U.N.'s World Meteorological Organization (WMO) hosted in 1979.[4]

In many respects, Nimbus-7 was a major influence in shaping the Earth sciences program that followed. The long-lived satellite provided "the first satellite views of the Antarctic ozone hole, the first global view of oceanic primary productivity [ocean color], and early measurements of the Earth's radiation budget and stratospheric chemical species and aerosols."[5] Nimbus-7 also began the systematic production of monthly average sea-ice measurements in the polar regions. Many of the satellite's instruments (Table 1) remained operational for more than a decade and a half, contributing to a growing archive of environmental data.

Table 1. Nimbus Instrumentation	
Instruments	**Functions**
Coastal Zone Color Scanner (CZCS)	The CZCS collected ocean color data from November 1978 to June 1986. CZCS acquired nearly 68,000 images, each covering up to 2 million square kilometers of ocean surface. Imagery from CZCS provided the first view of the distribution and abundance of phytoplankton over the global oceans, and demonstrated the ability to monitor how these patterns change in time and space. (1)

---

2. Abraham Schnapf, ed., *Monitoring Earth's Ocean, Land, and Atmosphere from Space—Sensors, Systems, and Applications, Vol. 97: Progress in Astronautics and Aeronautics* (New York: AIAA, 1985).

3. I. S. Haas and R. Shapiro, "The Nimbus Satellite System: Remote Sensing R&D Platform of the 1970s,"in Schnapf (1985), pp. 71–95; see also H. F. Eden, B. P. Elero, and J. N. Perkins, "Nimbus Satellites: Setting the Stage for Mission to Planet Earth," *EOS, Transactions, American Geophysical Union* 74(29 June 1993): 281–285.

4. World Climate Conference, *Proceedings of the World Climate Conference: A Conference of Experts on Climate and Mankind, Geneva, 12–23 February 1979*, WMO Series, no.537 (Geneva: Secretariat of the World Meteorological Organization).

5. Eden, et al., op. cit.

Earth Radiation Budget (ERB)	The ERB experiment recorded data between 16 November 1978 and 20 June 1980. The principal products are nine years of global albedo, outgoing longwave, and net radiation, plus continuing solar irradiance measurements. Chief uses of the data include studies in regional energy heat budgets, the improvement of climate- and weather-prediction models, interannual climate variations, shortwave bidirectional reflectance from Earth atmosphere scenes, and solar physics. (2)
Limb Infrared Monitor of the Stratosphere (LIMS)	LIMS mapped the vertical profiles of temperature and the concentration of ozone, water vapor, nitrogen dioxide, and nitric acid in the lower to middle stratosphere range, with extension to the stratopause for water vapor and into the lower mesosphere for temperature and ozone. (3)
Stratospheric Aerosol Measurement (SAM-II)	The SAMS II instrument aboard Nimbus-7 obtained polar Arctic and Antarctic aerosol profiles. It measured attenuation of solar radiation by aerosol particles during sunsets and sunrises. (4)
Stratospheric and Mesospheric Sounder (SAMS)	SAMS made global atmospheric measurements of carbon dioxide, water vapor, carbon monoxide, nitrous oxide, methane, and nitric oxide. (5)
Solar Backscatter Ultraviolet and Total Ozone Mapping Spectrometer (SBUV/TOMS)	TOMS measured ozone indirectly by comparing ultraviolet light emitted by the Sun to that scattered from Earth's atmosphere back to the satellite. The TOMS instrument mapped in detail the global ozone distribution, including the Antarctic "ozone hole," which forms September through November of each year. TOMS measured sulfur dioxide released in volcanic eruptions. (6)
Scanning Multichannel Microwave Radiometer (SMMR)	SMMR observed Earth's microwave emission between 6.6 and 37 gigahertz in 10 channels from November 1979 to September 1987. This data set provided an important time series of Earth's microwave signature. With the proper calibration, the measured radiances can be converted into estimates of near-surface wind speed, columnar water vapor, and liquid water of clouds and rain. (7)
Temperature Humidity Infrared Radiometer (THIR)	THIR measured surface and cloud-top temperatures, as well as the water vapor content of the upper atmosphere

1) NASA Goddard Space Flight Center, Satellite Ocean Color, *http://seawifs.gsfc.nasa.gov/SEAWIFS/CZCS_DATA/DOC/CZCS_BACKGROUND.html*, accessed on 23 May 2002.
2) NASA, Nimbus-7 Earth Radiation Budget (ERB) Langley DAAC Project/Campaign Document, *http://charm.larc.nasa.gov/GUIDE/campaign_documents/nimbus7_project.html*, accessed on 23 May 2002.
3) National Space Science Data Center, Limb Infrared Monitor of the Stratosphere, *http://nssdc.gsfc.nasa.gov/database/MasterCatalog?sc=1978-098A&ex=1*, accessed on 23 May 2002.
4) The British Atmospheric Data Center, http://www.badc.rl.ac.uk/data/sam2/, accessed on 28 May 2002.
5) Oxford University, Department of Physics, *http://wwwatm.physics.ox.ac.uk/group/nimbus/radiometers.html*, accessed on 28 May 2002.
6) NASA Goddard Space Flight Center, Total Ozone Mapping Spectrometer, *http://www.gsfc.nasa.gov/gsfc/service/gallery/fact_sheets/earthsci/toms.htm*, accessed on 23 May 2002.
7) Jet Propulsion Laboratory, NIMBUS-7 SMMR Ocean Products: 1979–1984, *http://podaac.jpl.nasa.gov:2031/DATASET_DOCS/smmr_wentz.html*, accessed on 23 May 2002.

The early period of civil Earth observations was dominated by the development of electro-optical imaging sensors, utilizing spatial resolutions ranging from low (1–50 kilometers) to moderate (10–30 meters), with varying spectral characteristics and resolutions.[6] Scientists obtained low spatial resolution imagery from instruments on the TIROS polar-orbiting and Geostationary Operational Environmental Satellites (GOES), and moderate resolution imagery from the Landsat series. They also deployed sounding instruments designed to produce vertical profiles of atmospheric parameters on both polar-orbiting and geostationary weather satellites, but only slowly integrated these data into weather-forecasting models. The TIROS and GOES series of satellites were developed by NASA, with input from the Weather Bureau, and later transferred to operational status within NOAA, which was created in October 1970 by the Weather Bureau and other agencies.[7]

The secret reconnaissance satellites of the U.S. and U.S.S.R. yielded first photographic and later electro-optical data of enhanced spatial resolution, but these data were not offered for most civilian uses until the mid-1990s. First, the federal government released photographic data from the Corona series through 1972.[8] In 2002, NIMA released thousands of frames of imagery from the KH-7 and KH-9 reconnaissance

---

6. Pamela E. Mack and Ray A. Williamson, "Observing the Earth from Space," op. cit.

7. Eileen L. Shea, "A History of NOAA, Being a Compilation of Facts and Figures Regarding the Life and Times of the Original Whole Earth Agency," National Oceanic and Atmospheric Administration, 1987, *http://www.lib.noaa.gov/edocs/noaahistory.html*, accessed May 2003.

8. Albert D. Wheelon, "Corona: The First Reconnaissance Satellite," *Physics Today* (February 1997): 24–30; Philip J. Klass, *Secret Sentries in Space* (New York: Random House, 1971); William E. Burroughs, *Deep Black: Space Espionage and National Security* (New York: Random House, 1986); Dwayne A. Day, John M. Logsdon, and Brian Latell, eds., *Eye in the Sky: The Story of the CORONA Reconnaissance Satellite* (Washington, DC: Smithsonian Institution Press, 1998).

satellites.[9] Sensors aboard the commercial high-resolution satellites, which were first launched in the late 1990s and early 2000s, provided multispectral and panchromatic data of higher resolution than the early reconnaissance satellites.[10]

Through the 1970s and early 1980s, NASA pursued a variety of research programs in cooperation with NOAA and the international community. Among others, these international efforts included the Global Atmospheric Research Program (GARP), the GARP Atlantic Tropical Experiment (GATE), and the Monsoon Experiment (MONEX). The multidisciplinary GARP, which began in the 1960s, was a joint effort between the International Council of Scientific Unions (ICSU) and the WMO, and was the forerunner of the current World Climate Research Program. GATE was conducted in 1974 and was focused, in part, on understanding the interactions between convective activity and large-scale weather systems, and how issues of scale affected these interactions. The 1979 MONEX campaign was part of the Global Weather Experiment (GWE) and was the first comprehensive experiment on the Asian monsoon, focused mainly on meteorological disturbances related to the onset of monsoons and their short-term variability.

In the 1980s, NASA also participated in observing ozone depletion in the upper atmosphere and in determining the cause of that harmful phenomenon. Specifically, it took part in the first field experiment focused on ozone—the National Ozone Experiment (NOZE)—and led the 1987 Airborne Antarctic Experiment (AAOE).

NASA's role in these research efforts helped to establish NASA's scientific credibility in the international community and demonstrated its commitment to working with other national and international agencies. Such research also provided the background data and scientific understanding for the next step in integrating research from all aspects of planet Earth.

Calls for an Integrated Program of Global Observations

In the early 1980s, in response to a growing consensus in the scientific community, the initiative dubbed "global habitability" emerged within NASA. This call for new Earth observations programs was explicitly tied to emerging scientific and public policy concerns related to global change in general, and global climate change in particular. The scientific aspects of these concerns were forcefully articulated in the following three major reports published by NASA and the National Academy of Sciences.

Global Change: Impacts on Habitability

The first of these reports contained the results of a workshop convened by NASA and chaired by scientist Richard Goody. (See Document IV-2.) By the time this workshop was

---

9. Many of these images are available through the National Archives and through the U.S. Geological Center's EROS Data Center in Sioux Falls, SD.

10. John C. Baker, Kevin O'Connell, and Ray A. Williamson, *Commercial Observation Satellites: At the Leading Edge of Global Transparency* (Washington, DC: RAND and ASPRS, 2001).

held, the need to conduct detailed scientific studies of Earth from space was well accepted as the most viable way to obtain global data sets, though questions remained regarding the quality and depth of such observations. Some fifty scientists contributed to this report, which called for a long-term program to accomplish the following:

> . . . to document, to understand, and if possible, to predict long-term global changes that can affect the habitability of the Earth. The major factor contributing to change is human activity. The problem is urgent in the sense that situations are already being created that can take decades to reverse . . . the need is urgent for a commitment to proceed by the federal government.
>
> The program will include studies of the atmosphere, oceans, land, the cryosphere, and the biosphere. On decadal time scales, these regimes and the cycles of physical and chemical entities through them are coupled into a single interlocking system.[11]

Although the Goody report was not the first text to recognize the coupled nature of Earth's biophysical systems,[12] by the early 1980s, scientists had gained sufficient understanding of some critical interactions, such as those between the atmosphere and the oceans, to lead to the outline of a comprehensive research program. Observations to support broad-based research in the Earth sciences had always been a part of NASA's efforts, but the more comprehensive nature of both the desired observations and their underlying scientific foundation had a character quite distinct from the previous efforts of the 1960s and 1970s. The early work tended to be correlative and phenomenological rather than deeply analytical and based on a comprehensive theoretical foundation. The Goody report assembled in one document the principal scientific ideas that led to the eventual formation of NASA's concept of an MTPE. It was also the first major report to recognize explicitly the connection between the scientific enterprise and mankind's interest in maintaining a habitable planet.

The Goody report and the work that went into it raised awareness within the scientific community to a point where it began to be noticed within the broader community of science policy-makers. This awareness also led to a decision within NASA, supported by the White House, to take the issue of global habitability to the United Nations Unispace II Conference, held in Vienna in August 1982,[13] where it was one of seven multilateral initiatives put forward by the U.S. delegation. (See Documents IV-1, IV-3, IV-4, and IV-5.) Global habitability, as proposed by the United States, was an international cooperative research effort to obtain data from space and through other means on changes in the environment that would affect Earth's ability to sustain human life for the long term.

---

11. Jet Propulsion Laboratory (JPL), *Global Change: Impacts on Habitability, A Scientific Basis for Assessment*, 1982, report by the Executive Committee of a workshop held at Woods Hole, Mass., 21–26 June 1982, submitted on behalf of the Executive Committee on 7 July 1982, by Richard Goody, chair.

12. See Charles Mathews, "ERTS-1: Teaching Us a New Way to See," *Astronautics and Aeronautics* 11 (September): 36–40; and Marvin Holter, "Some Thoughts on Earth Observations," in *Symposium Proceedings, Remote Sensing Applied to Energy-Related Problems* (Ann Arbor, MI: Environmental Research Institute of Michigan, 1974).

13. United Nations, *Report of the Second United Nations Conference on the Exploration and Peaceful Uses of Outer Space*, A/CONF.101/10, Vienna, 9–21 August 1982; U.S. Congress, Office of Technology Assessment, *Unispace '82: A Context for International Cooperation and Competition*, OTA-TM-ISC-26 (Washington, DC: U.S. Government Printing Office, March 1983).

However, despite considerable scientific interest in a comprehensive international program, global habitability did not immediately gain the political support within the U.S. that was needed to create a funded international research program.

Toward an International Geosphere-Biosphere Program

Despite the lack of clear political support within the White House for such a program, Goody and his collaborators continued to explore the issue. The following year, the scientific concept of an interdisciplinary approach to Earth science had evolved and strengthened, carrying on and expanding the themes of the Goody report in the report of a National Research Council workshop chaired by Herbert Friedman. (See Documents IV-6 and IV-7.) The following is an excerpt from this report:

The real connections that link the geosphere and biosphere to each other are subtle, complex, and often synergistic; their study transcends the bounds of specialized, scientific disciplines and the scope of limited, national scientific endeavors. . . . If, however, we could launch a cooperative interdisciplinary program in the Earth sciences, on an international scale, we might hope to take a major step toward revealing the physical, chemical, and biological workings of the Sun-Earth system and the mysteries of the origins and survival of life in the biosphere. The concept of an International Geosphere-Biosphere Program (IGBP), as outlined in this report, calls for this sort of bold, "holistic" venture in organized research—the study of whole systems of interdisciplinary science in an effort to understand global changes in the terrestrial environment and its living systems.

Nearly 50 scientists from universities, the government, and the boards of the NRC participated in the formulation of the IGBP, which recommended greater emphasis on international collaboration than did the Goody report. The Goody and Friedman reports reflected a growing consensus in the scientific community for the systematic investigation of Earth as an interacting, complex system. Together, they defined the scientific basis for the MTPE.

The ESSC and the Bretherton Reports

The next influential step in the development of NASA's MTPE was the publication of two reports from the Earth Systems Science Committee (ESSC), formed under the aegis of the NASA Advisory Council, the most senior of the NASA advisory committees. (See Document IV-11.) This committee, chaired by climate scientist Francis Bretherton, was charged with addressing the themes raised by the Goody and Friedman Reports. It recommended a series of specific space missions, together with "an interdisciplinary program of basic Earth system research and in situ measurements to be carried out by NASA, NOAA, the National Science Foundation (NSF), and other federal agencies." It also recommended "an advanced information system to process global data and to facilitate data analysis, data interpretation, and quantitative modeling of Earth system processes."[14]

These two reports, to which more than 230 scientists contributed, laid out a specific road map for what became NASA's MTPE, and its primary component, the Earth Observing System (EOS). The Bretherton reports envisioned a comprehensive design of space missions of varied scope and size, including:

- continued and improved NOAA polar and geostationary satellite programs;
- new operational sensors and platforms;
- coordinated specialized space research missions to examine specific processes;
- centralized Earth data and information system; and
- "Earth Observing System of polar-orbiting platforms planned as part of the U.S. space station program."

The ESSC reports envisioned a broad, inclusive program with the goal of obtaining "a scientific understanding of the entire Earth system on a global scale by describing how its component parts and their interactions have evolved, how they function, and how they may be expected to continue to evolve on all timescales."[15] This remains a daunting task today. Further, Earth system science had the challenge of developing "the capability to predict those changes that will occur in the next decade to century, both naturally and in response to human activity."[16] This vision was seminal in initiating a major new thrust in Earth science, and led to a ten-year, $17-billion program to develop large polar-orbiting platforms that would be launched on the Space Shuttle. It was also a major component in the development of the U.S. Global Change Research Program, a government-wide effort to explore the changes taking place in Earth's climate and other environmental systems.

The report then listed and categorized the measurements required in order to achieve the report's goals. Table 2 summarizes these variables, which shaped the program that evolved from the ESSC reports. The list reveals the very wide range of measurements that were needed (some of which were of uncertain feasibility at the time, and others of which could not realistically be made using space-based observations). This list of variables became the touchstone during each revision of the program during the 1990s.

Equally ambitious was the proposed implementation of the space systems needed to collect the required data. The report proposed using or developing the most advanced technology available for the measurements.

The work of the ESSC portrayed a picture of the Earth sciences of remarkable breadth and great scientific attractiveness. The material from ESSC reports quickly found its way into textbooks and even more quickly into classrooms. Entirely new curricula evolved as the participants in the ESSC returned to their academic institutions and spread the Earth systems view. NASA employed its considerable presentation skills in publishing widely read

---

14. NASA Advisory Council, *Earth System Science, Overview, A Program for Global Change*, report of the Earth System Sciences Committee, 1986a; NASA Advisory Council, *The Crisis in Space and Earth Science, A Time for a New Commitment*, report of the Space and Earth Science Advisory Committee, 1986b.

15. NASA Advisory Council, *Earth System Science: A Closer View*, report of the Earth System Sciences Committee, 1988.

16. Ibid.

documents espousing the new view. Unfortunately, as noted in a later section, the tie to the human exploration program soon proved problematic.

Table 2. Summary of ESSC-Recommended Earth Science Measurements	
External Forcing	
• Solar Irradiance • Ultraviolet Flux • Index of Volcanic Emissions	
Radiatively and Chemically Important Trace Species	
• CO_2 • N_2O • CH_4 • Chlorofluoromethanes • Tropospheric O_3 • CO	• Stratospheric O_3 • Stratospheric H_2O • Stratospheric NO_2 • Stratospheric HNO_3 • Stratospheric HCL • Stratospheric Aerosols
Atmospheric Response Variables	
• Surface Air Temperature • Tropospheric Temperature • Stratospheric Temperature • Surface Pressure • Tropical Winds • Extratropical Winds	• Tropospheric Water Vapor • Precipitation • Components of Earth Radiation Budget • Cloud Amount, Type, Height • Tropospheric Aerosols
Land-Surface Properties	
• Surface-Radiating Temperature • Incident Solar Flux • Snow Cover • Snow Water Equivalent • River Runoff (Volume) • River Runoff (Sediment Loading) • River Runoff (Chemical Constituents)	• Surface Characteristics (Albedo, Roughness, Infrared, and Microwave Emittance) • Index of Land Use Changes (Broad Classification of Vegetation Types) • Index of Vegetation Cover • Index of Surface Wetness • Soil Moisture • Biome Extent, Productivity, and Nutrient Cycling
Ocean Variables	
• Sea-Surface Temperature • Sea-Ice Extent • Sea-Ice Type • Sea-Ice Motion • Ocean Wind Stress • Sea Level	• Incident Solar Flux • Subsurface Circulation • Ocean Chlorophyll • Biogeochemical Fluxes • Ocean CO_2

The Earth Observing System

Even before the publication of the Bretheron reports, NASA's Earth Science and Applications Division began planning for a major program in Earth science research by assembling the Science and Mission Requirements Working Group for the EOS. This twenty-member group, many of whose participants overlapped with the ESSC, developed a two-part report[17] that called for comprehensive observations of Earth according to the following interlocking themes.

1. Research on fundamental processes within traditional disciplines—hydrologic cycle, the biogeochemical cycles, and climate processes.
2. Employ operational capabilities ranging "from detailed in situ and laboratory measurements to the global perspective offered by satellite-based remote sensing."
3. Measure "a number of the parameters of the natural system . . . over time scales of seasons to years, persistent observations of dynamic phenomena . . . to build data records which stretch over a decade or more."
4. Preserve key observations that already have been gathered and provide improved access to them.

This working group developed specific goals for the cycles and processes they wished to characterize (Box A) and derived from them a conceptual research payload consisting of three groups of sensors and a data collection and location system (Box B). This notional payload became the initial baseline of EOS and led to the development of a plan for two large platforms, both in orbit at the same time (one in a morning-crossing Sun-synchronous orbit, the other in an afternoon-crossing orbit). Some instruments were to fly on both platforms; others were to fly on only one.

**BOX A—Earth Science Goals for the 1990s, as Stated by
the EOS Science and Mission Requirements Working Group**

Hydrologic Cycle
- Quantify the processes of precipitation, evaporation, evapotranspiration, and runoff on a global basis.
- Determine what factors control the hydrologic cycle.
- Determine the effects of sea and land ice upon the global hydrologic cycle.
- Quantify the interactions between the vegetation, soil, and topographic characteristics of the land surface and the components of the hydrologic cycle.

Biogeochemical Cycles
- Understand the biogeochemical cycling of carbon, nitrogen, phosphorus, sulfur, and trace metals.

---

17. *Earth Observing System, Science and Mission Requirements Working Group Report, Volume I, Parts 1 and 2*, NASA Technical Memorandum 86129 (Greenbelt, MD: Goddard Space Flight Center, 1984).

- Determine the global distribution of biomass and what controls both its heterogeneous distribution in space and its change over time.
- Determine the global distribution of gross primary production and respiration by autotrophic and heterotrophic organisms, and the annual cycle and year-to-year variation of these processes.
- Determine the transport of sediments and nutrients from the land to inland waters and ocean.
- Quantify the global distribution and transport of tropospheric gases and aerosols, and determine the strengths of their sources and sinks in the ocean, land surface, coastal and inland waters, and upper atmosphere.
- Understand the processes controlling acid precipitation and deposition.

Climatological Processes

- Determine the modes of large-scale and low-frequency variability (month-to-month and year-to-year time scales) of meteorological variables such as wind, pressure, temperature, cloudiness, and precipitation.
- Quantify the large-scale and low-frequency variability of net incoming solar radiation and outgoing long-wave radiation and their relationships to cloudiness.
- Determine the relationships between large-scale and low-frequency variability of meteorological observables and the variability of sea-surface temperatures and current systems.
- Quantify the influences of changes in land-surface evaporation, albedo, and roughness on local and regional climate.
- Assess the influence of sea and land ice cover on global climate.
- Improve and extend knowledge of past climates.
- Determine the role of land biota as sources and/or sinks of carbon dioxide and other radiatively important trace gases.
- Predict climate on a probabilistic basis.

Geophysical Processes

Atmospheric

- Understand the coupling of the chemical, radiative, and dynamic processes of the troposphere, stratosphere, and mesosphere.
- Determine the coupling between the lower and upper atmosphere.
- Improve the quantitative understanding of the variability of atmospheric ozone, including the influence of anthropogenic perturbations.
- Improve the understanding of the mechanisms for the maintenance and variability of atmospheric electric fields.
- Improve the accuracy of deterministic weather forecasting and extend the useful forecast period.

Oceanic

- Measure the mesoscale to large-scale circulation of the ocean and acquire a better understanding of the long-term variability in this circulation.

- Determine the global heat, mass, and momentum coupling between the ocean and atmosphere.
- Understand the processes controlling the dynamics of sea ice and its interaction with the underlying water.
- Determine the upper ocean response to thermal and atmospheric forcing, including the effects of persistent horizontal variability.
- Characterize the exchange processes between surface and deep waters.

Solid Earth
- Determine the global distribution, geometry, and composition of continental rock units.
- Understand the causes of the morphology and structure of the continental crust.
- Understand how episodic processes such as rainfall, runoff, dust storms, and volcanism modify the surface of Earth.
- Understand the dynamics of inland ice sheets.
- Determine the relation between the factors of climate, topography, vegetation, and the geologic substrata and the processes of soil formation and degradation.
- Understand the secular variation in plate velocities.
- Determine the platform, vertical structure, and time variation of mantle convection.
- Measure the global gravity and magnetic fields to reveal with greater accuracy and resolution the structure of the lower crust and upper mantle lithosphere.
- Explain secular variations, including reversals, in Earth's magnetic field.
- Explain secular variations in Earth's long wavelength gravity field in terms of the viscosity structure of the mantle.

SOURCE: NASA (1984)

Box B—Conceptual Research Payload for EOS,
as Described by the EOS Science and Mission Requirements Working Group

1. Data Collection
Advanced Data Collection and Location System (ADCLS)

2. Surface Imaging and Sounding Package (SISP)
Moderate-resolution Imaging Spectrometer (MODIS)
High-resolution Imaging Spectrometer (HIRIS)
High-resolution Multifrequency Mircrowave Radiometer (HMMR)
Laser Atmospheric Sounder and Altimeter (LASA)

3. Sensing with Active Microwaves (SAM)
Synthetic Aperture Radar (SAR)
Radar Altimeter (RA or ALT)
Scatterometer (SCAT)

---

4. Atmospheric Physical and Chemical Monitor (APACM)
   Laser Atmospheric Wind Sounder (LAWS)
   Upper Atmospheric Wind Interferometers
   Tropospheric Composition Monitors

SOURCE: NASA (1984)

Reports of the Committee on Earth Sciences

Concurrent with the NASA studies of the early 1980s, the Space Studies Board (SSB) of the National Research Council (NRC) was conducting its own studies of what types of space-based measurements were needed and how they should be integrated with other Earth science research. The first of these studies (1982) examined research needs for the solid Earth and the oceans;[18] the second (1985) examined the atmosphere and its interactions with the land, oceans, and biota. The second report identified three major goals to accomplish the following.

1. Determine the atmospheric distributions and cycles of mass, energy, momentum, and water vapor and/or chemical constituents important to climate and to the maintenance of life.
2. Understand the physical and chemical dynamics of the atmosphere and its interactions with the land, ice caps, oceans, and biota.
3. Understand the evolution of the atmosphere to its present states and predict its future evolution on time scales less than 100 years, including the effects of anthropogenic and natural perturbations.[19]

These reports strengthened the case for additional Earth science funding and provided the basis for establishing a focused effort within NASA's Space Science and Applications Program.

Mission to Planet Earth

In 1987, NASA released a report authored by former astronaut Dr. Sally Ride, entitled Leadership and America's Future in Space, which, among other programs, suggested that NASA adopt a Mission to Planet Earth (MTPE) as one of its four major themes to take NASA into the twenty-first century.[20] The overall report was an attempt to

---

18. Committee on Earth Sciences, *A Strategy for Earth Science from Space in the 1980's, Part I: Solid Earth and Oceans* (Washington, DC: National Academy Press, 1982).

19. Committee on Earth Sciences, *A Strategy for Earth Science from Space in the 1980's and 1990's, Part II: Atmosphere and Interactions with the Solid Earth, Oceans, and Biota* (Washington, DC: National Academy Press, 1985).

20. Sally Ride, *NASA Leadership and America's Future in Space: A Report to the Administrator*, NASA, August 1987.

establish broad goals for continuing the U.S. leadership in space. The section on MTPE (see Document IV-12) stressed the initiative's "fundamental importance to humanity's future on this planet," noting that it "directly addresses the problems that will be facing humanity in the coming decades."[21] It also stressed the global nature of the science and reminded readers that the very nature of the potential problems faced by humanity demands international cooperation. The Ride report urged NASA to take a clear leadership role in this endeavor.

NASA adopted the term Mission to Planet Earth as the name of its Earth science program, a name that it retained until November 1997, when it was dropped in favor of Earth Science Enterprise. (See Document IV-27.) Then NASA Administrator Dan Goldin considered the latter name more reflective of the program's overall focus.

EOS and the Space Shuttle

The very scope and ambition of EOS, the centerpiece of the early MTPE, became somewhat of a liability in the budgetary process on Capitol Hill. The funding levels that this ambitious program implied could not be sustained, and over the next decade the program was repeatedly reduced in scale, in part because of circumstances within NASA but unrelated to the envisioned EOS scientific program.

For example, the inclusion of the large polar platforms reflected larger national priorities in the early 1980s, which directed that the government's expendable launch vehicles (ELVs) should be phased out in favor of the Space Shuttle. National policy envisioned that the Space Shuttle would become the only launch vehicle for government launches. Although by the time of the Bretherton report, national policy also allowed for the development of private ELVs, no company was willing to risk competing for business with the Space Shuttle, which was offering launch services to private customers as well as supporting government needs.[22] Hence, despite the fact that the Space Shuttle was capable of carrying only a limited payload to polar orbit, it was the only launch vehicle that appeared to be available to the scientists. Further, because of the major investment in the Space Shuttle and NASA's interest in using it as often as possible in order to reduce the cost per flight, proposing a series of polar platforms that could be tended by humans operating from the Space Shuttle seemed a sensible political move at the time.

The assumption that astronauts and payload specialists would be available to trade instruments or repair them as needed, which was taken as a given by NASA officials, affected scientific planning. Any reservations concerning the complexity and reliability of the instrumentation on such large, complex payloads were moderated by the prospect

---

21. Ibid, p. 51.

22. John Logsdon and Craig Reed, "Commercializing Space Transportation," in John M. Logsdon, Roger D. Launius, and Stephen J. Garber, eds., *Exploring the Unknown, Selected Documents in the History of the U.S. Civil Space Program, Vol. 4: Accessing Space* (Washington, DC: NASA SP 4407, 1999), pp. 405–422; Ray A. Williamson, "The U.S.-Europe Technology Gap in Space Transportation: The View from the United States," *Space Policy* (2000).

that onorbit repair and/or replacement would be available, as in the case of the Hubble Space Telescope.[23]

This was despite the fact that most of NASA's ambitious Earth-observation programs could have been pursued more quickly using sensor technology and platforms already available, and which did not require tending by astronauts. Some have argued that the drive to put large, technically challenging instruments on astronaut-serviced platforms actually set back Earth science research, especially for oceanic research.[24]

The subsequent loss of the Space Shuttle *Challenger* on 26 January 1986 abruptly changed the entire outlook for polar-orbiting Earth science platforms. When, in August 1986, President Reagan announced that henceforth the Space Shuttle would be limited to payloads requiring its special capabilities, and that it would not be flown into polar orbit, the science community was faced with a redesign of the EOS platforms. Satellites designed for launch on the Space Shuttle required different configurations and attach points than those designed for ELVs, and could not be configured to fly on both types of vehicles without considerable expense. Further, because the only vehicle capable of carrying the EOS polar platforms to orbit was the very expensive Titan IV,[25] EOS scientists were forced into a conceptual redesign of the entire EOS program, to place the platforms on an intermediate capacity launch vehicle that had not yet been developed.

Fortunately for the EOS program, the Department of Defense also faced the same gap in launch capacity and contracted for the development of the Atlas II ELV, capable of lifting some 4,800 kilograms into low-Earth polar orbit. In December 1999, that vehicle was used to launch Terra, the first satellite derived from the EOS planning for large polar-orbiting platforms.

The Space Studies Board and the Twenty-First Century Study

In the late 1980s, the Space Studies Board, as part of its broad-based study of the direction space science should take in the twenty-first century, formed a task group on earth sciences to study NASA's EOS proposal. It endorsed and extended the EOS proposal using the term Mission to Planet Earth to frame the effort it envisioned.

> "The task group therefore proposes a Mission to Planet Earth to include all the program elements required to understand a planet with an atmosphere,

---

23. For the Hubble Space Telescope, repair and modification has proven expensive but highly successful. However, Hubble orbits 600 kilometers above Earth in an orbit inclined at 23.5 degrees from the equator rather than from polar orbit. Launching into polar orbit requires much more energy (and therefore fuel) and, for the Space Shuttle, would have required a launch from Vandenberg Air Force Base in California. Attempts to construct a Space Shuttle launch facility at Vandenberg were abandoned following the loss of *Challenger*. See Ray A. Williamson, "The Space Shuttle" in John M. Logsdon, Roger D. Launius, and Stephen J. Garber, eds., *Exploring the Unknown, Selected Documents in the History of the U.S. Civil Space Program, Vol. 4: Accessing Space* (Washington, DC: NASA SP 4407, 1999), pp. 161–193.

24. Committee on Earth Sciences, *Earth Observations from Space: History, Promise, and Reality* (Washington, DC: National Academy Press, 1995), p. 32.

25. Estimates of the cost of a Titan IV launch ranged between $180,000 and $250,000 .

hydrosphere, biosphere, solid crust, mantle, and solid-liquid core. The proposal sets forth a concerted and integrated research program on the origin, evolution, and nature of our planet and its place in the solar system."

The study further urged a grand strategy with three primary measurement themes.

1. Global and Synoptic—We have learned that advances in Earth science derive from the synthesis of new ideas that come from global synoptic observations.
2. Long-Term Continuity—The Earth as a system is energetic on many scales, from microseisms to interannual El Niño, to ice ages. However, even if we restrict ourselves to decadal time changes—the human time scale—statistics dictate that measurements over many years will be required before we can make accurate statements about the energetics of the system.
3. Simultaneity—This entails observations of different kinds of processes at the same time. It includes study of the land, ocean, atmosphere, and biota as an interactive system on a global scale. Earth sciences have tended to treat these components as separate disciplines.[26]

The report recognized the poor match between the time periods required for the observations it was urging and the political decision-making process, which tended to shift its focus over relatively short time scales. The report suggested that "there is a need for a national commitment to carrying measurements through the necessary time periods."

Concern Over Mission Delays and Lost Opportunities: The Earth Explorers Study

Although the Space Studies Board's Committee on Earth Science was firmly behind the plan to develop and operate an EOS, many members of the scientific community felt there were important gaps in NASA's approach. For one thing, the EOS platforms were not able to collect all the measurements that Earth scientists wanted. For another, many scientists were unhappy with the lack of flexibility to take new measurements as new knowledge was obtained.

Scientists especially worried that the large platforms would take many years to develop and launch. Not only would the emphasis on the large platforms make it difficult for NASA to respond quickly to scientific findings that suggested a different approach or new types of measurements, or to fill gaps in the collection of important data such as ocean color (see later section, "Small Spacecraft and Data Purchases"), the long lead times for new satellites were making it difficult to attract graduate students or young faculty to the field. Further, managing a large satellite program became more difficult because the operation of every instrument depended on the slowest one in the development queue. Finally, the approval

---

26. Task Group on Earth Sciences, *Space Science in the Twenty-First Century, Imperatives for the Decades 1995 to 2015—Mission to Planet Earth* (Washington, DC: National Academy Press, 1988a).

process for large missions, both within NASA and the Office of Management and Budget, as well as within Congress, became much more complicated and lengthy.

At the same time, technological advances were making it possible to build smaller, cheaper instruments that could be launched on a smaller satellite within a few years of a decision to proceed with a new mission. Hence the CES began to examine specifically the role small satellites might play in furthering the aims of an integrated Earth science research program. Among other things, the CES report suggested the creation of an Earth Explorer series of missions that would permit "the construction of two small missions per year, or one moderate mission every three years."[27]

The 1991 CES Assessment Report

The original configuration proposed in the ESSC activity involved the deployment of an EOS made up of two large platforms, EOS-A and EOS-B, which were descendents of the space station activities. EOS-A was to be launched into a morning orbit, and EOS-B into an afternoon orbit. NASA planned to replace these platforms every five years with identical copies in order to obtain a consistent fifteen-year data set. NASA distilled the variables listed in Table 2 to a key group of twenty-four basic data sets (Table 3) to be obtained from the mission, and NASA committed to obtaining these data for at least fifteen years with consistency and continuity.

By 1991, the Committee on Earth Science was still expressing cautious optimism about NASA's Earth science research plans, feeling that with the expected modernization of NOAA's environmental satellite program, the United States would maintain its substantial international leadership in Earth science research from space. The study noted that progress was still needed in hydrology, land-surface geology and vegetation, and in understanding the gravitational and magnetic fields. It complained that the high costs of obtaining Landsat or SPOT data from the commercial suppliers had severely hampered research in hydrology and land-surface geology and vegetation. The committee also expressed concern over the lack of a synthetic aperture radar satellite and the High-Resolution Imaging Spectrometer (HIRIS), both of which were severely delayed under the EOS program.[28]

The optimism of this report and its support of EOS were soon undercut in part by increasing congressional concern over the size and scope of NASA's Earth science ambitions. The size of the proposed EOS program ($17 billion over a ten-year period) and its connection with studies of global warming caused congressional members already skeptical about the use of Earth science data to support the concerns of the environmental activist community to question this program closely. Further, the George H. W. Bush Administration cut the proposed overall NASA budget, which had been rising. This led to a sharp reduction in total funding for EOS ($11 billion for the same ten-year period), forcing NASA to restructure its Earth science program.

---

27. Committee on Earth Sciences, *Strategy for Earth Explorers in Global Earth Sciences* (Washington, DC: National Academy Press, 1988b).

28. HIRIS was eventually dropped from NASA's list of EOS instrumentation and the United States still has no orbiting civilian synthetic aperture radar system.

Restructuring

NASA gave the task of helping to restructure MTPE to the EOS Engineering Review Committee, chaired by Edward Frieman, then chairman of the National Research Council Board on Sustainable Development. The committee's report (see Document IV-13) cited three main reasons for the study:

- the relatively high costs of EOS and its associated EOS Data and Information System (EOSDIS);
- the long delay the existing EOS structure imposed on the provision of Earth science information of interest to policy-makers; and
- alternate sensors and/or platforms might be preferred on scientific and technical grounds.[29]

This report and the pressures that led to its production resulted in significant changes in NASA's plans for EOS. The committee concluded the following.

- EOS-A and EOS-B (the flagship satellites that were to lead the EOS armada of satellites) can and should be completely reconfigured. Packages of sensors can now be flown in smaller clusters whose choice should be dictated by the scientific priorities set by the findings of the Intergovernmental Panel on Climate Change (IPCC) and the U.S. Global Change Research Program (USGCRP). The resulting program is both resilient and robust.
- The committee believes that [the planned] early missions can be of great benefit in the overall drive to understand global climate change, and the program of precursor missions should be strengthened and integrated into a coherent plan.
- The committee recommends that [the Departments of Energy and Defense], along with the National Oceanic and Atmospheric Administration, be made an integral part of the above coherent plan. Further, launches of small satellites carrying high priority sensors should be carried out by this group by 1995. The plan itself should be put under the aegis of the Committee on Earth and Environmental Sciences of the Federal Coordinating Council on Science, Engineering, and Technology for oversight.
- The committee recommends that NASA reexamine EOSDIS both as to its utility in a redesigned program and on its merits in handling the data and information needs of such a large component of the USGCRP.
- The committee recommends . . . that the overall USGCRP be reexamined with the view of achieving proper balance between the space-based and nonspace-based components. Any needed restructuring should focus primarily on science priorities and timeliness of data availability.

---

29. NASA Advisory Council Report of the Earth Observing System (EOS) Engineering Review Committee, 1991.

The report further urged integrating the objectives of the Earth Probes series of investigative sensors and spacecraft with EOS planning and declared that EOSDIS, the distribution system for EOS data products, should be modified in parallel with changes in EOS. It also urged consideration of a smaller, "intermediate launch vehicle," the Lockheed Martin Titan IIAS, for launching EOS-A and EOS-B. This change would require NASA to lighten the satellite by removing some instruments that were to fly together to accommodate simultaneity of data collection. The committee felt that the need to achieve simultaneity of some measurements could be satisfied by orbiting the necessary sensors in formation on small spacecraft.

Finally, the Frieman Committee noted that a fifteen-year time horizon was unrealistically short and that to follow the sometimes-subtle long-term changes of climate would require a longer commitment. The committee's decision on this point was driven in part by its realization that the world would face "a steadily increasing list of physical and chemical problems in the global environment."

This report led to the first of several efforts to restructure EOS and to shift its focus to an emphasis on global climate change. This change of emphasis meant a reduction or delay in NASA's research on stratospheric chemistry and its effects on ozone depletion, as well as the study of the physics of the solid Earth. It also meant making some compromises in meeting scientific requirements, such as simultaneous measurements and reductions in the quality of some measurements.[30]

The Frieman Committee report and other pressures led NASA to develop a restructuring plan that it delivered to Congress in the spring of 1992. (See Document IV-15.) Congress held several hearings to examine NASA's plans and heard from both supporters and critics of its plans. In general, however, most scientists supported the increased program flexibility that the Frieman Committee report urged, though certain science areas were slighted. (See Document IV-14.)

Rescoping

The restructuring process was only underway for a short time when NASA Administrator Goldin levied an additional reduction of 30 percent in the EOS runout budget through 2000, a funding decrease from $11 billion to $8 billion. The EOS Payload Advisory Panel (see Document IV-17), which had been appointed to recommend the best path for rescoping the program, recommended "reducing the amount of contingency held to handle unexpected problems in instrument development," while also acknowledging that this would lead to a loss in future flexibility in the program.[31] The panel acknowledged that the United States increasingly would have to rely on its international partners to supply some of the desired data. This also meant "relinquishing

30. EOS Investigators Working Group (EOS IWG), *Payload Advisory Panel Recommendations* (Washington, DC: NASA, 1991).

31. EOS Payload Advisory Panel, *Adapting the Earth Observing System to the Projected $8 Billion Budget: Recommendations from the EOS Investigators*, Berrien Moore III and Jeff Dozier, eds. (Washington, DC: NASA, 1992).

to international partners the development of new advanced technologies in laser and active microwave remote sensing."

The resulting rescoping of EOS initially assumed the availability of the Atlas IIAS intermediate launch vehicle for the spacecraft that were to follow EOS-A and EOS-B. In December 1992, NASA allowed industry contracts to study the design of a common spacecraft for these later missions. Yet, shortly after the presentation of these results in the fall of 1993, NASA Administrator Dan Goldin directed a repeat of the studies under the assumption that the spacecraft would be launched on the smaller capacity Delta II launch vehicle. This necessitated the development of smaller, lighter spacecraft and a further reduction of the scientific scope of EOS.

Rebaselining

Congressional pressure to make MTPE more relevant to policy-makers and to increase the timeliness of the program's results to the decision-making process, as well as a realization that the rescoping attempt had left too little reserve in the EOS program, led to another fiscal scrub of the program. In the spring of 1994, NASA began an internal review with three teams (the Project, Core Science, and Review teams), complemented by twenty-six subteams. The results of this effort led both the Payload Advisory Panel and the independent Senior Review Group to conclude that the rescoping effort had added significant risk to the program and affected future science measurements adversely.[32]

Reshaping

Dissatisfaction with aspects of the rebaselined program within both the administration and Congress resulted in NASA undertaking a new set of program studies. At about the same time, the National Science and Technology Council (NSTC) within the White House initiated a broad study of ways in which to improve the efficiency and effectiveness of federal R&D. The Interagency Federal Laboratory Review examined the various laboratories of the Departments of Defense and Energy, and NASA. Administrator Goldin charged the NASA Advisory Council (NAC) with the task of responding for NASA, and NAC convened a NASA Federal Laboratory Review (NFLR) task force for the purpose. That report noted "the task force found little evidence of advanced technology development for and infusion into MTPE." Further, "thorough searches at each of the Centers involved in the MTPE Enterprise led the task force to conclude that, while there is excellent science being pursued within MTPE, there is a lack of definition of scientific milestones and need dates that will provide the national policy process with the necessary

---

32. Angelita Castro Kelly, "The Evolving Earth Observing System (EOS) Mission Operations Concept: Then and Now," *Fourth International Symposium on Space Mission Operations and Ground Data Systems*, ESA-SP-394, vol. 1, 1996, pp.106–113.

information to make decisions in a timely manner."[33] As a result of this study, NASA pursued another internal study, which focused on reducing program costs and examined issues concerned with inserting new technologies into the program after 2000.

In March 1995, NASA also began a study that addressed the need for closer collaboration between NASA and NOAA on MTPE-related issues in order to reduce EOS costs and assist in the transition to operational instruments in NOAA's environmental satellite program. As part of that effort, NASA examined the transition of technology it developed for EOS into the National Polar-Orbiting Environmental Satellite System (NPOESS). NPOESS was a joint effort of the Air Force, NOAA, and NASA to develop and operate a single polar-orbiting environmental system in cooperation with Eumetsat, the European meteorological satellite program.

Overall, these and other studies were focused on lowering program costs by reshaping MTPE with modifications in EOSDIS and redesigning the EOS satellites that were to follow EOS-A and EOS-B. The NPOESS study reflected an increasing realization that one way to obtain a long-term data set was to develop operational instruments returning data of sufficient quality to serve scientific as well as operational needs.

Additional Congressional Reviews

The previous reviews, though informed by congressional concerns, were driven primarily by administration efforts, first within the administration of President H. W. George Bush and then that of President Bill Clinton, to reduce NASA's budget. In part because NASA's Earth science program is the largest single component of the U.S. Global Change Research Program (USGCRP), congressional critics often focused their concerns about the direction of the USGCRP at MTPE, raising questions about the scientific focus and validity of the satellite program that would support USGCRP research.

For example, on 6 April 1995, Congressman Bob Walker, chairman of the House of Representatives Committee on Science, sent a letter to NASA Administrator Daniel Goldin asking a series of questions about MTPE, such as "Is the science that is being conducted as a part of the MTPE/EOS program fully justified on strictly scientific terms? Has it been fully peer-reviewed?" (See Document IV-19.) Separate letters on the same day to the National Academy of Sciences and the National Academy of Engineering requested that the National Research Council (the operating arm of the Academies) Board on Sustainable Development review the scientific progress of MTPE and USGCRP. (See Document IV-20.) Congressman Walker also asked the Academies to assess the role of NASA's EOS within the USGCRP. These letters reflected the views of some members of the Science Committee that NASA's direction may have been verging away from carrying out the best scientific research in favor of supporting particular theories about global warming. The letters also reflected some dissatisfaction within the scientific community that NASA was building a program that had too little flexibility and could not respond quickly and efficiently to new scientific findings that might arise.

---

33. NASA Advisory Council, *NASA Federal Laboratory Review Report*, 28 February 1995, submitted to the White House Office of Science and Technology Policy (OSTP).

The National Research Council referred Congressman Walker's request to the Board on Sustainable Development and its Committee on Global Change Research. In March 1996 testimony before the House of Representatives Committee on Science, Dr. Edward Frieman, chairman of the Board on Sustainable Development, reported on the results of a workshop that the Board and the Committee held in July 1995. This workshop was part of a continuing review of USGCRP and its principal component (in program funding), NASA's EOS.[34] The testimony specifically reported on EOS in the context of the USGCRP and listed several specific recommendations for NASA with respect to EOS.

The National Research Council was generally very supportive of NASA's direction with EOS and its integration with the USGCRP, stating "science is the fundamental basis for the USGCRP and its component projects, and that fundamental basis is scientifically sound."[35] It also stressed that EOS was developed in response to scientific needs and was generally on the right course, but further improvements were needed. Specifically, Frieman's testimony on behalf of the NRC urged that NASA "maintain a science-driven approach to observational and information management technology."[36] Noting that revolutionary spacecraft technologies are likely to reduce future spacecraft costs, Dr. Frieman urged that NASA focus on developing advanced technologies to reduce the costs of continuing observations essential to the needs of the USGCRP. He also pointed out the need for NASA to enhance in situ observation programs, process studies, and large-scale modeling activities in support of its satellite observations. The testimony also addressed the use of data from commercial high-resolution satellites in global change studies, noting that such data can be very useful, but they also have severe limitations in covering many of the data types needed for such studies. In a personal statement, Dr. Frieman said "let me state again my conviction that the USGCRP and NASA's MTPE and EOS are fundamentally sound scientific programs of immense importance to the future welfare of our country and our world."[37]

In 1997, NASA released the results of an internal study,[38] the first of a planned series of MTPE biennial reviews, which underscored NASA's intention to maintain as much flexibility as possible in designing second- and third-generation EOS satellites to take advantage of the latest scientific results and most advanced technologies available. This study was assessed by an independent review panel and continued in the same direction as recommended by the National Academy of Sciences studies. It stressed the need to infuse more new technology into upcoming missions and to use as much off-the-shelf spacecraft and subsystems as possible. The review also supported the NASA Earth System Science Pathfinder program, with its approach of encouraging Principal Investigators to

34. National Research Council, *A Review of the U.S. Global Change Research Program and NASA's Mission to Planet Earth/Earth Observing System* (Washington, DC: National Academy of Sciences, 1995), available at *http://www.gcrio.org/USGCRP/LaJolla/cover.html*

35. Dr. Edward A. Frieman, Chairman, Board on Sustainable Development, National Research Council, statement before the Committee on Science U.S. House of Representatives, 6 March 1996. Dr. Frieman stressed that his testimony also reflected subsequent input from the National Research Council's extensive review process subsequent to the July 1995 workshop.

36. Ibid.

37. Ibid.

38. NASA Office of Mission to Planet Earth, *Mission to Planet Earth Biennial Review* (Washington, DC: NASA, 1997).

assist in developing new missions in tackling new science questions of interest to the global change research community.

EOS Data and Information System

In designing the Earth Observing System, program scientists met the challenging issue of how to collect and distribute the massive amounts of data that the EOS satellites were to generate by developing a concept for a centralized EOS Data and Information System (EOSDIS). EOSDIS had the capability to command and control the EOS satellites and their instruments, and provided archiving, distribution, and information management for all of NASA's Earth science data, including data from in situ studies and non-EOS satellites. EOSDIS was based on an "open" computing architecture that allowed the system to evolve with new computing advances and networking technology. It was a comprehensive system, designed to serve the needs of scientists performing integrated, interdisciplinary studies of Earth's systems.

In the early stages of the development of EOSDIS, all functions were to be carried out at a central location at the Goddard Space Flight Center. Seen from a budget and management standpoint, this arrangement seemed preferable to MTPE managers. However, potential users of EOS data complained that this arrangement limited their input to system development and unnecessarily constrained flexibility as new scientific findings suggested the need to develop new data-processing algorithms or new data products. In a 1989 report, the EOSDIS Science Advisory Panel of the EOS Investigators Working Group (IWG) underscored the importance of establishing a series of "Active Archives," chosen with "(1) demonstrated experience in and institutional commitment to responsibilities of Active Archives, particularly in the area of satellite data processing and archiving; and (2) critical mass of in-house science expertise in the analysis of the type of EOS and other satellite data to be located in the Active Archive."[39]

A 1990 IWG report suggested that NASA should build on existing capabilities within the Earth science community, rather than creating entirely new data centers. "The data panel believes that it is important to build EOSDIS, starting from the real systems of today, into a system capable of handling data from the EOS instruments planned for the end of the decade. The challenge is to gain from experience already bought and paid for—successes as well as failures . . . [and to] select existing sites with expertise in maintaining data collections and scientific use of these products."[40]

Thus in 1990, NASA, in collaboration with other federal agencies and private institutions, created a system of Distributed Active Archive Centers (DAACs), each of which has a specific Earth science focus (Table 4). This structure enables the development of centers of expertise in the use of EOS data and in the integration of these data with other Earth science data sets. NASA required each of the DAACs to identify DAAC User Working Groups to assist in setting data standards and requirements within their disciplines.

---

39. Science Advisory Panel for EOSDIS, *Initial Scientific Assessment of the EOS Data and Information System (EOSDIS)* (Washington DC: NASA, 1989).
   40. EOS IWG, 1990.

Table 4. EOSDIS Distributed Active Archive Centers

- Alaska SAR Facility (ASF), NASA
- GSFC Earth Sciences (GES), Goddard Space Flight Center
- Global Hydrology Resource Center (GHRC), Marshall Space Flight Center
- Land Processes (LP), Langley Research Center
- Atmospheric Sciences Data Center, Langley Research Center
- National Snow and Ice Data Center (NSIDC)
- Oak Ridge National Laboratory (ORNL)
- Physical Oceanography (PO)
- Socioeconomic Data and Applications Center (SEDAC)

The DAACs were set up as part of the plan to develop EOSDIS Version 0 by July 1994. As Version 0 was to be in place prior to the launch of any of the flagship satellites, the data products in Version 0 were developed from satellite data already collected that could serve as prototypes of the data NASA expected to gain from the EOS satellites. This development approach relied upon guidance from DAAC User Working Groups and active participation of scientists who tested the data products and their delivery mechanisms to ensure that the resulting products would meet user needs. Version 0 was declared operational in August 1994.

EOSDIS was seen as an essential component of the EOS program, as this system was the interface between NASA's Earth Science Enterprise and the scientific community that use the data, in large part to support the USGCRP. The importance of this connection is stressed in an April 1992 interim NRC report on EOSDIS. "If EOSDIS fails, so will the Earth Observing System and so may the U.S. Global Change Research Program." (See Document IV-16.) Hence, the recommendations of this report stressed the need to maintain flexibility and reliability of EOSDIS in order to support not only NASA-supported scientists, but also the much broader scientific community involved in USGCRP.

Nevertheless, concerns persisted throughout the development of EOSDIS over the cost and functionality of the data system. In 1994, the NRC issued its final report, which elaborated on the interim report, stressing the need for EOSDIS to achieve wide usage of data from EOSDIS throughout the scientific community.[41]

In 1995, the Congressional General Accounting Office published a report requested by the House Science Committee that commended NASA's efforts to respond to the NRC's 1994 EOSDIS report by making EOSDIS more open, distributed, and flexible, but noted "the significant risks inherent in this large, technically complex project."[42] In particular, the report expressed concern that NASA needed to gain a better knowledge of its projected user base for the EOSDIS data products and to find the appropriate balance between the needs of near- and long-term system development.

---

41. National Research Council, *Panel to Review EOSDIS Plans, Final Report* (Washington, DC: National Academy Press, 1994).

42. U.S. Congress, General Accounting Office, "Earth Observing System: Concentration on Near-term EOSDIS Development May Jeopardize Long-term Success," GAO/T-AIMD-95-103, 1995.

A 1994 study by the Committee on Global Change Research of the Board on Sustainable Development within the NRC resulted in a complementary report, underscoring the need for "the community of researchers and users . . . [to] take the lead in making key decisions. . . . The system must encourage innovation and creativity through broad participation of the scientific, public, and private sectors."[43] The report also recommended that "responsibility for product generation, publication, and user services should be transferred to a federation of partners selected through a competitive process open to all."[44]

NASA accepted this recommendation. The DAACs are now part of the broader Federation of Earth Science Information Partners (ESIPs) that were established by NASA. Partners include government agencies, national laboratories, universities, nonprofit organizations, and commercial businesses, which have collaborated to ensure better development and distribution of scientific data and information.[45]

Small Spacecraft and Data Purchases

Throughout the history of the Earth science program, the apparent commercial value of some Earth observations data has created tensions between some policy-makers who suggest that the private sector should provide Earth science data and many scientists who worry either that commercial objectives will lead to data of much less scientific value or that required data will not be provided at all. The troubled history of the Landsat program illustrated well how difficult it was to find the right balance between satisfying commercial and scientific objectives.[46] Publicly funded science programs generally had to satisfy a variety of national objectives, one of which was to rely on the private sector for systems and data (beyond contracting with commercial firms) whenever there is a commercial potential for contributing to the national economy in doing so.

Such arrangements, even while imbued with laudable social goals, generally complicated the development of new systems, because they often required the development of new institutional arrangements that had not been tried before. The experience with the periodic attempts to provide ocean color data provided an instructive example of the complexities and subsequent delays that the attempt to meet multiple societal objectives may engender in the delivery of useful data for science and applications.

The color of the ocean, as seen from space, indicates its biological activity and potential productivity. Ocean color data reveal the interactions of physical processes with the ocean's biological communities, particularly phytoplankton, which produces organic carbon through the process of photosynthesis. NASA first developed the Coastal Zone Color Scanner (CZCS) for Nimbus-7 (Table 1); this was an instrument designed to collect precise measurements of the intensity of radiation reflected from the ocean in different

43. National Research Council, *A Review of the U.S. Global Change Research Program and NASA's Mission to Planet Earth/Earth Observing System* (Washington, DC: National Academy Press, 1995), p. 23.

44. op. cit, p. 24.

45. See the Federation of Earth Science Information Partners at *http://www.esipfed.org/* for more information about the ESIP federation.

46. See Ray A. Williamson, "The Landsat Legacy: Remote Sensing Policy and the Development of Commercial Remote Sensing," *Photogrammetric Engineering and Remote Sensing* (July 1997), pp. 877–885.

parts of the spectrum. Areas rich in phytoplankton appear green because they contain chlorophyll, the pigment that enables them to convert sunlight and carbon dioxide into organic matter. Areas of high phytoplankton productivity are likely to contain schools of plankton-eating fish.

The data also reveal the interaction of ocean currents with the phytoplankton, imparting further information of value to shipping and fisheries. The success of the CZCS aboard Nimbus whetted scientists' interest in gaining further ocean color data in order to study and understand biophysical processes in more detail. It also suggested that an operational sensor would be of economic value to the Navy (for improved navigation and lower ship operational costs), commercial shipping, and fisheries.

After several attempts to include an ocean color sensor on multisensor satellites, such as the National Oceanic Satellite System (NOSS), the NOAA POES series, the French SPOT satellite, or Landsat 6 (each of which failed in different ways), scientists at NASA Goddard Space Flight Center proposed to fly a sensor called the Sea-viewing Wide Field-of-view Sensor (SeaWiFS) aboard a small satellite in 1990.

During the late 1980s, improvements in materials, circuitry, and satellite design made possible smaller and lighter satellites, thereby reducing the needed size and cost of launch vehicles for some applications. This engineering development led to a broadly based drive to place some instruments on small spacecraft.[47] SeaWiFS was an excellent candidate for such a proposal. A dedicated ocean color sensor that would orbit in a Sun-synchronous polar orbit could cover the entire world at about 1 kilometer of resolution, providing scientifically important data about ocean productivity, ocean currents, and other biophysical ocean phenomena.

The drive to build SeaWiFS also coincided with a private sector interest in providing scientific data on a commercial basis, an interest that gained firm support in Congress. Under pressure from Congress to support commercial purchases of scientific data, NASA issued a Request for Proposal for an ocean color data purchase; Orbital Sciences Corporation won the competition. Orbital offered to build a spacecraft to carry the SeaWiFS sensor if NASA would fund most of the spacecraft and sensor development. In return, Orbital would deliver SeaWiFS data to NASA for its scientific needs and would market the data to shipping companies, fisheries, and other commercial users. In accepting this proposal, NASA in effect became an "anchor tenant" in the deal, allowing Orbital to borrow its share of the development and operational costs from the financial community.

Orbital created Orbimage, Inc. in order to market data from Orbview 2, the satellite that carried the SeaWiFS instrument.[48] In August 1997, more than six years after OSC signed the contract with NASA, Orbimage successfully orbited the SeaWiFS sensor and shortly thereafter began delivering imagery to NASA scientists and to the commercial community. As of the spring of 2003, Orbview 2 was still delivering high-quality ocean color imagery to data users.

---

47. It also led to exaggerated claims that small satellites would revolutionize the satellite world and lead to much lower costs. However, smaller satellites are not necessarily cheaper than larger ones. See U.S. Congress, Office of Technology Assessment, *Affordable Spacecraft: Design and Launch Alternatives*, background paper (Washington, DC: Office of Technology Assessment, 1990), available at *http://www.wws.princeton.edu/~ota/*

48. Orbimage is also developing a series of high-resolution satellites from which the company also plans to market high-resolution data on a commercial basis.

Table 5 summarizes the many steps in the development and deployment of a successor to the CZCS.[49]

Table 5. Ocean Color Measurements	
1978:	The Coastal Zone Color Scanner (CZCS) is launched on Nimbus-7. Its objective was to provide a one-year test of the experimental instrument.
1979:	Planning began for the successor to CZCS.
1979–81:	CZCS included in plans for the National Oceanic Satellite System (NOSS). Instrument cost was estimated at $11.6 million.
1982:	The Marine Experiment (MAREX) specified requirements for the ocean color instrument.
1982–83:	The Ocean Color Instrument (OCI) was studied by Ball Corp. for flight on the NOAA H-I-J POES series. The project was descoped to $7.6 million, and then canceled when NOAA could not fund integration costs.
1984:	NASA EOS Science and Mission Working Group published description of follow-on ocean color measurements for EOS program.
1984–85:	A joint French-U.S. Science Working Group proposed a revised OCI on the SPOT-4 satellite. The cost to U.S. was $14.5 million, with launch and integration costs provided by France. Planned launch date was 1989. NASA cancelled the mission in a cost-cutting move.
1986:	The Bretherton Committee included a CZCS follow-on in its program for Earth system science.
1986:	EOSAT conducted a market survey for commercial OCI data.
1986:	MODIS Instrument Panel issued a report incorporating ocean color measurements in MODIS-T.
1987:	NASA-EOSAT joint Science Working Group specified Sea-viewing Wide Field-of-view Sensor (SeaWiFS) performance; NASA Goddard studied in-house *Explorer*-class spacecraft.
1987:	EOS Science Steering Committee issued its science strategy for EOS, citing need for ocean color measurements.
1988:	NASA Goddard designed the SeaWiFS program, including ground system, data processing and archive, instrument calibration, algorithms and verification, and global science. Instrument was planned for Landsat-6, with an expected launch date of 1991. (When Landsat-6 was launched in 1993, SeaWiFS had been replaced by ballast to balance spacecraft mass.)

49. National Research Council, *Earth Observations from Space: History, Promise, and Reality* (Washington, DC: National Academy Press, 1995); National Research Council, *The Role of Small Satellites in NASA and NOAA Earth Observation Programs* (Washington, DC: National Academy Press, 2000).

1989:	EOSAT withdrew from project, considering it financially unattractive. Santa Barbara Research Center (SBRC), the instrument supplier, continued efforts. Estimated cost was $24 million. SBRC finished instrument design to produce a sensor for a $14-million fixed-price cost. New design was endorsed by science community.
1990:	NASA Goddard studied in-house mission with estimated cost of $56 million using a SeaWiFS/CZCS/MODIS-T engineering model, small *Explorer*-class spacecraft, and Scout launch vehicle.
1990:	Orbital Sciences Corp. (OSC) assumed lead role. NASA proposed to buy data for the research community in request for proposals on December 15.
Mar. 1991:	NASA signed $42-million data purchase contract with OSC, with a specified launch date of 29 August 1993.
April 1992:	NASA Goddard issued research announcement soliciting proposals for science team members. Twenty-seven members are selected from sixty-seven proposals.
1991–92:	NASA restructured EOS program and deleted MODIS-T, added SeaWIFS II data purchase in 1998 (also termed EOS-Color).
Jan. 1993:	First meeting of the SeaWiFS science team.
April 1993:	OSC announced launch delays resulting from stray light problem in sensor.
June 1993:	OSC contract modified for June 1994 launch.
Early 1994:	OSC announced SeaWiFS launch would not be earlier than 29 September 1994; required the Pegasus-XL launcher, which was still under development.
Mid-1994:	NASA rebaselining review recommended termination of EOS-Color, in part due to delays in SeaWiFS.
Mid-1994:	OSC announced launch would not be earlier than December 1994.
January 1995:	OSC requested that NASA and the Air Force rank the seven Pegasus-XL missions they had scheduled to begin in April 1995.
March 1995:	OSC declared that a launch in 1995 was subject to the uncertainties of the Pegasus-XL launch manifest, tentatively set for September 1995.
June 1995:	NASA announced that a launch of SeaWiFS could not occur before November 1995 and was likely to occur no sooner than early 1996.
1 August 1997:	SeaWiFS was successfully launched. Excellent results were being obtained from the mission, and a time series of ocean color measurements has been provided since October 1997. OrbImage continues to sell data and analysis from the system to private sector customers.

Throughout the above period, participants worked hard to obtain the kinds of ocean color data that had first been obtained in 1978. Numerous compromises were sought, and private companies, government agencies, and researchers made numerous attempts to launch a successor to the CZCS first flown on Nimbus-7, using several different funding and institutional mechanisms. The difficulties and delays encountered illustrate the sometimes tortuous process of funding a system that most involved agreed was of value to science, private industry, and, ultimately, the U.S. taxpayer.

International Cooperative Programs and the Committee on Earth Observing Satellites

For both practical and political reasons, international cooperation has long been a thrust of NASA's programs. Indeed, international cooperation is explicitly identified as a NASA focus in its originating legislation.[50] In Earth science, NASA has established bilateral cooperative programs with many individual countries and cooperated within a multilateral framework. A full account of the history of NASA's international Earth science programs is beyond the scope of this chapter. Nevertheless, cooperative research has played an important part in the development and execution of NASA's EOS.

By the mid 1980s, the ability of Canada, Europe, and Japan to field sophisticated Earth observing systems made them especially important potential partners in global Earth science studies. These space programs not only had developed extensive research capabilities, but also had orbited operational environmental space systems designed to contribute to weather observations and forecasts. The proliferation of satellite systems raised the concern that redundant global systems would waste scarce resources. Hence, in 1984, during the Economic Summit of Industrialized Nations, the United States proposed the formation of an international committee to coordinate scientific and operational missions to collect data about Earth and its systems. Other nations agreed and formed the Committee on Earth Observing Satellites (CEOS). (See Documents IV-8, IV-9, and IV-10.)

Members of CEOS are government agencies or international organizations with funding and program responsibilities for satellite observations and data management. Membership also includes Associate Members, agencies or international organizations that have significant program activity (e.g., ground segments) that support the goals of CEOS. Members must agree to provide "nondiscriminatory and full access to data that will be made available to the international community."[51]

CEOS has proved highly effective in coordinating the various global Earth observation programs and in serving as a platform for keeping each country and organization active in Earth observations aware of what other members are contemplating. Its technical committees developed international protocols and standards for data distribution and use.

---

50. The National Aeronautics and Space Act of 1958 calls for "Cooperation by the United States with other nations and groups of nations in work done pursuant to this Act and in the peaceful application of the results, thereof," Public Law 85-568, 72 Stat., 426; See John M. Logsdon, Linda J. Lear, Jannelle Warren-Findley, Ray A. Williamson, and Dwayne A. Day, eds., *Exploring the Unknown, Selected Documents in the History of the U.S. Civil Space Program, Volume I: Organizing for Exploration*, Document II-17 (Washington, DC: NASA, 1995), pp. 334–345.

51. See *http://www.ceos.org/*, accessed January 2003.

One of the major initiatives to grow out of CEOS was the International Global Observing Strategy (IGOS) (see Documents IV-22, 23, and 24), which was first proposed by U.S. delegates to CEOS. U.S. officials originally named their proposed initiative the International Global Observing System. However, after meeting concern from other CEOS members that the term "system" implied developing an expensive new organization, new satellites, and their attendant ground-based infrastructure, the name was changed to the current one, which was more reflective of the goals originally envisioned for the initiative. These goals included the broader involvement of the world's scientific community in global Earth science research and the merging of local and regional in-situ data sets with global satellite data. Accomplishing the former assisted in meeting the latter.

Following formal adoption of the IGOS concept by CEOS at its plenary meeting in 1996, CEOS delegates formed a Strategic Implementation Team to develop the concept in more detail and to select prototype projects within which to test the IGOS concept and make it productive. Data from NASA's EOS and other satellites, from POES and GOES, and the Earth observation satellites of Europe, Japan, and in situ sources have been integrated into the following six prototype demonstration projects focused on contributing to the global public good.

- Global Ocean Data Assimilation Experiment (GODAE)
- Upper Air Measurements
- Long-Term Continuity of Ozone Measurements
- Global Observation of Forest Cover
- Long-Term Ocean Biology Measurements
- Disaster Management Support

In the spring of 2003, these projects were still receiving support from the international research and applications community.

NASA's MTPE and (later) the Earth Science Enterprise program have especially benefited from bilateral projects such as the highly successful TOPEX/Poseidon and Jason missions undertaken with France's Centre Nationale Étude Spatiale (CNES) and the Tropical Rainfall Measuring Mission (TRMM), developed, launched, and operated in cooperation with Japan's NASDA. (See Document IV-26.)

TOPEX/Poseidon, launched in 1992, has provided precise measurements of the changing heights of the world's oceans, crucial information for following El Niño and the Southern Oscillation (ENSO), and predicting their future development. Such information was especially important to alert climate-sensitive industries to future potential variations in weather and climate.[52] The considerable success of this important project led to the development and launch of a follow-on cooperative satellite, Jason-1 (launched in December 2001). Operating the two satellites together has produced a wealth of scientific data about ocean dynamics and valuable information about the progress of ENSO.

---

52. Ray A. Williamson, Henry R. Hertzfeld, Joseph Cordes, and John M. Logsdon, "The Socioeconomic Benefits of Earth Science and Applications Research: Reducing the Risks and Costs of Natural Disasters in the USA," *Space Policy* 18: 57–65, 2002.

TRMM carried five instruments, including the Precipitation Radar (PR), the TRMM Microwave Imager (TMI), and the Visible and Infrared Scanner (VIS), designed to collect precipitation and other data over the tropics, areas that were generally poorly covered by in situ measurements. To improve the repeat cycle of the instruments, the satellite flew in an inclined orbit of 35 degrees, rather than in the more usual Earth observation polar orbit of 90 degrees. TRMM has proved particularly effective in improving the predictive paths of tropical cyclones and in estimating rainfall over the mid latitudes. Data from this satellite were incorporated into ongoing weather forecast models. NASA has been planning a much more ambitious Global Precipitation Mission (GPM) to follow TRMM and provide global precipitation data.

Conclusion

As matters now stand in the Earth science program, the first one of the planned EOS satellites, EOS AM-1 (now dubbed "Terra") is in orbit and performing well. The previous EOS PM-1 has been split into two smaller satellites (AQUA and AURA). Launched on 4 May 2002, AQUA is successfully delivering data to researchers. AURA is scheduled for launch in early 2004. These spacecraft make up the first round of EOS measurements. The second and later rounds of measurements and the means by which they are to be obtained are yet to be defined at this time, but will certainly involve smaller satellites within a constrained program budget.

As an important part of the evolution of the program, NASA also has built and orbited a series of small satellites in order to fill in needs for data not covered by the larger flagship satellites. For example, Quikscat was launched in June 1999 in order to gather important data on surface wind fields over the oceans. Some of these satellites, for example, the New Millennium Program Earth Observing-1 (EO-1) satellite, which carries three advanced imaging instruments, has served as a test bed for new technologies and data-gathering methods.

As currently structured, the program attempts to answer five basic questions about Earth and the influence of human activities upon it.[53]

- How is the global system changing?
- What are the primary forcings of the Earth system?
- How does the Earth system respond to natural and human-induced changes?
- What are the consequences of change in the Earth system for human civilization?
- How well can we predict future changes in the Earth system?

Although the program cannot avoid some measure of controversy because it attempts to collect and analyze data that could have enormous economic and political consequences, it is on a much more settled course than was the case in the mid-1990s.

The history of NASA's MTPE and its successor, the Earth Science Enterprise, is a mixed one. On the one hand, as noted, the effort has produced some very important

---

53. See *http://gaia.hq.nasa.gov/ese_missions/*

scientific results and will contribute to improved predictions of weather and climate, and to a much better understanding of "spaceship Earth." Ultimately, these scientific results could result in better management of Earth's biological systems. However, the program's history also illustrates the extent to which political winds, both those blowing in the narrow politics of science and the broader national political scene, can sway the slender reed of science decisions, rendering the process through which this science has been accomplished tortuous and overly costly.

Since the early 1980s, several billion dollars have been spent on the MTPE and the Earth Science Enterprise, all too much of it on rescoping, rebaselining, and reshaping the program from its original conception in the early and mid-1980s. Though the overhead costs have been high, in many respects the program is now more flexible and capable of responding to new directions in Earth science than the program originally conceived in the early 1980s. By any standard, the Earth Observing System being deployed by NASA is a major contribution to the Earth sciences. The program is not that which was originally proposed, but change was inevitable. The constraints imposed on the program within an agency engaged in the financially even more ambitious undertakings in human spaceflight are severe. Whether the technology-driven cost savings will permit the full goals of the Earth sciences program to be achieved is uncertain and—based on NASA's record over the past two decades—deserving of skepticism. However, if a shortfall results in the Earth sciences program against its stated goals, any disappointment will stem from the lofty goals that were set when the program was established.

Additional Information

The NASA Earth Science Enterprise Web site contains a great deal of useful information not only regarding current plans, but also the history of past missions (*http://www.earth.nasa.gov*). Additional details regarding U.S. and other international instruments and their capabilities can be found on the Committee on Earth Observation Satellites (CEOS) Web site (*http://www.ceos.org*).

Document IV-1

Document Title: Memorandum from Kenneth S. Pedersen, Director of International Affairs, to Associate Administrator for Space Science and Applications, "Global Habitability as a U.S. Initiative at UNISPACE 82," 1 July 1982.

Source: NASA Historical Reference Collection, NASA History Office, NASA Headquarters, Washington, D.C.

This memorandum to Associate Administrator Dr. Burton Edelson describes how NASA might use the opportunity afforded by the United Nations-sponsored Unispace '82 conference to NASA's advantage, demonstrating not only NASA's interest in conducting global-scale Earth studies, but also its interest in international scientific collaboration. The memo reveals insights into NASA's skeptical views about internationally managed programs, a skepticism that the Reagan administration certainly shared.

[stamped "rec'd in Code E Jul 2 1982"]

NASA Routing Slip to:
E/Dr. Edelson
  Keller
  Raney
  Rosendhal
EE/Tilford
[italics indicate handwritten note on routing slip]
I think this paper outlines a positive approach to Unispace on this issue. I would appreciate hearing your views before the 7/7 meeting if possible.

Ken
Kenneth S. Pedersen
7/1/82

****

[no page number]

National Aeronautics and
Space Administration

Washington, D.C.
20546

Reply to attn of: LID-18

TO: E/Associate Administrator
               for Space Science and Applications

FROM: LI-15/Director of International Affairs

SUBJECT: Global Habitability as a U.S. Initiative at UNISPACE 82

Following up on our discussion Monday, I offer some thoughts as to how we ought to play Global Habitability as a UNISPACE initiative and what mode of international collaboration would be preferable from the NASA point of view.

My understanding regarding the Global Habitability program is that it is to be a major thrust involving the clustering of several OSSA programs, with UNISPACE viewed as a target of opportunity to increase the chances of success with OMB. Aside from such strategy considerations, there is also—so far as I have been able to ascertain—genuine NASA program office interest in pursuing such a program on a global scale. A program that is truly global in scope would indeed require international collaboration.

As you know, we have raised the prospect of a UNISPACE Global Habitability initiative with Ambassador Helman of the State Department who is coordinating U.S. preparations for UNISPACE. Helman has discussed the idea at some length with Hans Mark and is enthusiastic about pursuing an initiative, given the successful conclusion of the Woods Hole Summer Study effort, if we get a NASA front office go-ahead.

In considering how best to play Global Habitability at UNISPACE, it appears to me that the program should be cast as a broad scale, highly integrated, timely and relevant NASA scientific research program, into which international participation will be welcomed. There are a variety of means by which international participation culd [sic] be integrated into the program including: AO's; AN's; bilateral agreements; multilateral coordination with programs of other space nations; exchange of space data for ground truth from program-targeted areas (including many LDC's). Our experience with international cooperation has demonstrated that such approaches allow us to maintain control over the scientific integrity and [2] research thrust of the particular program. I anticipate that other countries will be attracted because of the excellent reputation for openness that NASA has built on the basis of international participation in Landsat and other cooperative programs.

There are good reasons to avoid consideration of recasting the program to make it an internationally organized/managed effort. The arguments for confining such a program to research and to single-agency direction in the U.S. would seem to be all the more compelling on the international front. An international management apparatus could be unwieldy and it might serve to restrict and even dilute the science objectives of the program. It would run the risk of treading on the well-established turf of international agencies with already well-defined global environmental objectives. Further, such an apparatus would be subject to the type of political divisions that currently plague so many international programs no matter how meritorious their objectives.

Of course, the program will need to draw upon the unique capabilities of such existing international organizations as UNEP, WMO, FAO, etc. We should be able to characterize our program as contributing to their programs. We can indicate willingness to correlate our research with the research aims of these agencies and to engage in joint projects, in data exchange and in joint efforts to disseminate the results of our research. Within such guidelines, the details of our working with international organizations can be worked out after we present the initiative and are in a better position to work out mutually beneficial arrangements.

A possible scenario for UNISPACE might include the following:

- having Jim Beggs announce Global Habitability as a major U.S. (read NASA) initiative in his speech as U.S. Representative scheduled for August 10;
- featuring of Global Habitability in the U.S. multimedia presentation scheduled for the evening of August 10;
- a possible poster session the following day to elaborate on programmatic aspects and modes of potential collaboration;
- a possible press briefing on Global Habitability and distribution of a package of materials to the press and to UNISPACE delegations;
- inclusion of our plans for the program in the NASA presentation at John Houghton's informal WMO/ICSU meeting on August 11.
- [3] discussion of our plans, as appropriate, with the representatives of other space agencies at UNISPACE.

During all of these presentations, we need to emphasize the NASA lead role and guard against any efforts to create a UN management role for this program.

With regard to the mode of international collaboration for a NASA-managed Global Habitability Program, I am quite pleased with the language in the draft chapter on National/ International Considerations produced last week at Woods Hole. Brent Smith of my staff worked with your people and with a representative of the State Department in drafting the international section. Attached is a copy of the draft International Collaboration subsection which, I feel, underscores many of the points I have made above.

[signature]
Kenneth S. Pedersen

Attachment

cc:
L/Bob Allnutt

****

[4]
4. INTERNATIONAL COLLABORATION

Fundamental to the success of the Global Habitability Research Program will be the participation of individual experimentors [sic], scientific organizations and governmental agencies worldwide. To ensure collaboration that is truly global in scope, NASA will:

- encourage and invite broadscale participation of experimentors [sic] in the development of instrumentation, acquisition of comparative observations to corroborate satellite-obtained information, and in the analysis and interpretation of data. Investigators will be selected through Announcements of Opportunity and Announcement Notices. Unsolicited proposals will be considered. Collaboration will also take place at the scientist-to-scientist level based on agreements with national and international agencies.
- contribute through this program to the ongoing environmental research efforts of the WMO, the FAO, UNEP and other regional and international agencies. NASA will correlate its research with the research aims of these bodies and will engage in data exchange, co-sponsorship of specific projects, and in joint efforts to acquaint national publics and policymakers with results and prognoses.
- [5] consult and coordinate its plans with those of prospective satellite operating agencies to encourage the development of compatible and complementary spaceborne instrumentation and thereby maximize the effectiveness of global measurements. Such coordination should result in cost savings, avoid duplication of effort, and minimize the prospect of incompatible national systems.
- provide for significant program involvement by developing countries in areas such as ground-based observations and the exchange of data that can assist individual countries and regions [to] manage their resources and better deal with changes affecting habitability—e.g., deforestation, desertification, agricultural productivity. Through such collaboration, both NASA and the individual countries will obtain vital information beyond that obtainable using only a single tool of measurement.

Document IV-2

Document Title: Richard Goody, Chairman, Workshop Executive Committee, "Global Change: Impacts on Habitability, A Scientific Basis for Assessment," 7 July 1982.

Source: NASA Historical Reference Collection, NASA History Office, NASA Headquarters, Washington, D.C.

This report, which was requested by NASA, established the initial scientific basis for conducting an integrated study of Earth's systems, incorporating physical, chemical, and biological aspects of the planet, centered on global change. It raised the important question of how human activities might be altering these systems and affecting Earth's habitability.

[title page]

Global Change: Impacts on Habitability
A Scientific Basis for Assessment

A Report by the Executive Committee of a Workshop held at Woods Hole, Massachusetts, June 21-26, 1982

Submitted on behalf of the Executive Committee on July 7, 1982, by Richard Goody
(Chairman)

National Aeronautics and
Space Administration

Jet Propulsion Laboratory
California Institute of Technology
Pasadena, California

[iii]
Executive Summary

The earth is a planet characterized by change, and has entered a unique epoch when one species, the human race, has achieved the ability to alter its environment on a global scale. This report outlines a scientific strategy that would offer a basis for the difficult choices that lie ahead and for the complex decisions that must be made now to protect the integrity of the earth.

NASA could play a central role in this task. The unique perspectives of space observational systems, the ability to manage complex interdisciplinary science programs realized in two decades of planetary exploration, and the overall technical expertise of NASA are essential to the success of the endeavor.

This report is directed to issues that arise on a relatively immediate basis: the time horizons of people now alive. We are concerned with matters that relate to the production of food adequate to the needs of a growing population and with subtle interactions that might affect the planet's ability to sustain and renew the quality of air and water and the integrity of the global chemical cycles essential for life. We approach the task in a spirit of optimism. but with a clear recognition of the magnitude of the challenge.

Needs of the past could be met by the expansion of frontiers, by land clearance, by application of chemical fertilizers and pesticides, by irrigation, and by intensive application of energy resources harvested from the sun by the biosphere in years past. Now, however, humanity is being confronted by the finite dimensions of its world. In the twentieth century, humanity has become a factor in the global cycles of carbon, nitrogen, phosphorus, and sulfur and can affect global and regional air quality and climate; even control of the powerful hydrological cycle is within its grasp. An understanding of the overall system is essential if the human race is to live successfully with global change.

The magnitude and complexity of the scientific issue are prescribed by the choice of time scale. On time scales of a decade or so, the ocean, atmosphere, and biosphere function as an integrated system. Physical, chemical, and biological processes are coupled, and progress in understanding them will require an interdisciplinary science research program of broad scope and content. Such a program would require space observations to provide a global perspective; investigations of specific ecosystems both on the ground and from space to deepen scientific understanding of the overall function of the biosphere; studies of representative estuarine and coastal systems to define the transfer of materials from land to ocean; studies of horizontal and vertical motions in the ocean, recognizing the importance of these motions for climate and their role in regulating the flow of materials to the depths of the ocean; studies of the atmosphere and its chemistry, physics, and motions; and theoretical and laboratory investigations to synthesize new information and to serve as a guide for acquiring new data.

A group of some 50 scientists, including physicists, chemists, and biologists, met recently at Woods Hole to consider these matters and arrived at the following general conclusions. A program of research can be implemented that would respond to the overall concern regarding the future habitability of the planet. Some elements of the program are already in place or can be carried out with a modest redirection of available scientific, administrative, and financial resources. Other elements can be defined now, while some will require further consideration. The investigation of interconnections between land, [iv] ocean, and atmosphere is crucial and the view from space is necessary for integration on a global scale. The task is urgent. Because the human race has reached the point of being able to change the earth significantly on a time scale of mere decades, its ability to mount countermeasures, should it wish to, is restricted in many cases to a time scale that is similar or longer. A commitment by the United States Government is essential to further progress and there is an obvious need for international cooperation. We can see no better use for the mastery of near space than the acquisition of the body of knowledge essential to the future well-being and prosperity of humanity.

[v]

Contents

[vi]

Abstract

This report addresses the feasibility of a major NASA research initiative to document, to understand, and if possible, to predict long-term (5-50 years) global changes that can affect the habitability of the earth. The major factor contributing to change is human activity. The problem is urgent in the sense that situations are already being created that can take decades to reverse. The appropriate science community is available to support such a research program, and the need is urgent for a commitment to proceed by the Federal Government.

The program will involve studies of the atmosphere, oceans, land, the cryosphere, and the biosphere. On decadal time scales, these regimes and the cycles of physical and chemical entities through them are coupled into a single interlocking system. Some part of this system can be studied in a straightforward manner (the atmosphere) and some with great difficulty (the biosphere). A new emphasis for NASA would be to design and carry through studies of the complex interactions. A major effort to involve new scientific talent would be necessary, particularly from biology. New management approaches by the agency would be required.

The program would be large, comparable in size and scope to other major activities of NASA such as the Physics and Astronomy Program and the Planetary Exploration Program. It would require support for at least a decade. The program would involve international collaboration, although the U.S. initiative should be unambiguous and managerially independent. Space systems would be emphasized, but in situ observations would also be required. A large theoretical and laboratory program is essential.

Detailed studies exist of some of the space systems needed to execute this program, e.g., the Ocean Topography Experiment, the upper Atmospheric Research Satellite, and the imaging radar for cryospheric studies. The early funding of such projects and studies of advanced systems is a cornerstone for the ultimate success of the program.

[1]

Global Change: Impacts on Habitability
A Scientific Basis for Assessment

I. Introduction

This report responds to a request by the National Aeronautics and Space Administration for the establishment of a group to discuss the viability of a major research initiative in the area of global habitability, specifically addressed to the question of changes, either natural or of human origin, affecting that habitability. The program should be one of research, which we interpreted to mean a concern with the foundations of knowledge needed for enlightened policy decisions. It should be responsive to human needs. It should be urgent in the sense that it needs to be undertaken now, and not in a few years. The changes in habitability discussed should be of significance within a human life span (50 to 100 years), with less emphasis on short terms of a year or two. And finally, we were asked to consider why this program might be appropriate for a NASA initiative.

Following an organizational meeting in April 1982, a workshop was convened for the week of June 21 through 26. Those present at the workshop are listed in the Appendix.

The membership of the workshop covered an unusually wide range of disciplines. We examined science issues and possible research programs to the extent necessary for responsibly answering the basic question: can a program be formulated that responds to the requirements, that could be rationally developed from the theme of global habitability, and that NASA could successfully carry out? We took the discussion to the point of concluding that the answer to the question is "yes," but not to the point of identifying the optimum programs. That step remains to be taken. This paper presents a selection of scientific issues and program elements developed by the executive committee, without any claim to completeness.

The responsiveness of the program to human needs and the issue of urgency are discussed in Chapter II. Chapter III discusses some scientific issues concerning the problem of global habitability and Chapter IV discusses some of the programs that could lead to substantial increases of knowledge in the foreseeable future. These two chapters together establish the validity of the program and its feasibility. Here, we will briefly cover some general issues that arose during our discussions.

A research program concerned with changes in global habitability would contain some familiar elements and some that are less familiar. The familiar elements involve current activities in atmospheric, oceanic, and land-surface research. The coupling between the ocean and atmosphere for time scales longer than a few years is also a familiar concept, but when we introduce, additionally, biological activity in the sea and on land, the couplings become even more complex. On the decadal time scale the atmosphere, the ocean (the top 500 meters), the biosphere, and the physical properties of the land surface are linked in a fascinating but exasperating symbiosis

[2] Is it possible to design effective programs for investigating a system of such complexity? Some have argued that analytic approaches are usually simplistic and that, on the other hand, the full complexity is beyond existing comprehension. If we were to accept this dichotomy it would be difficult to argue that a program of research appropriate to NASA can be defined. On the contrary, we believe that important aspects of the global problem can usefully be isolated and subsequently integrated into general ecological themes at the appropriate time and space scales. We had no difficulty in identifying major areas in which scientific issues and effective research programs could be for-mulated. A specific example is the nutrient cycle involving nitrogen, carbon, phosphorus, and sulphur. These elements cycle through the atmosphere, ocean, and land biosphere. We have little certain knowledge about them and productive re-search programs can be designed. Some aspects of the cycles, e.g., the carbon dioxide problem, have already emerged as stimulating research areas with significant consequences for public policy.

To break the ecological problem down into manageable elements or, alternatively, to emphasize the complexity of the entire system—these represent different approaches that will continually emerge, because both are valid. Their existence, however, emphasizes the need for the highest quality of scientific leadership for NASA's program. Judgments must be made that require scientific credibility and sophistication from both management and the science community. We make no proposals regarding program management, but scientific leadership and the appropriate program structure within NASA are crucial to the success of the program.

While much needs to be done if the program is to succeed, we believe that there is enough strength in some of its parts and enough potential competence in others for a successful outcome. A sine qua non, however, is a firm national commitment to a research program in global habitability. Given this commitment and the appropriate leadership, the scientific side of the government-science partnership can be counted upon; but it cannot proceed further without an explicit Federal undertaking to carry through the program for a period of at least one decade.

Why should NASA be responsible for this program? The short answer is that NASA can do it and no other Federal agency can. We are proposing a research program that would be global in extent and decadal in time scale. This does not fall fully within the terms of any other single agency. The global character of the problem virtually demands a space capability, and where space systems are not required the problems involve high technology, for which NASA often has appropriate skills. In addition, NASA has no direct operational, service, or regulatory responsibilities, which allows NASA to undertake difficult research programs unimpeded by short-term requirements. NASA has a responsibility for research as part of its mandate under the National Aeronautics and Space Act of 1958. In the pursuit of its responsibilities in space science, NASA has learned how to work with the outside science community as an equal partner. Finally, NASA has experience with interdisciplinary project management of the kind required for this program (Project Viking and the Upper Atmospheric Research Program provide two examples of successful interdisciplinary efforts).

In addition to the appropriateness of the task to the agency, NASA stands to benefit from undertaking it. The agency has developed tools of revolutionary importance for the earth sciences, but the responsibility for the programs has rested with other agencies (the Upper Atmospheric Research Program is the notable exception). This has resulted in split responsibilities that have inhibited the rapid development and utilization of space tools. This contrasts sharply with projects such as Apollo, the Shuttle, and High Energy Astrophysics, for which NASA had overall responsibility, and which were superbly executed. The global habitability program can provide NASA with a mission responsibility in the earth sciences that, properly handled, could lead to outstanding advantages for the earth sciences as a whole.

In summary, our workshop deliberations led us to conclude that a valid program can be developed involving global investigations of changing conditions affecting habitability; and that its development is urgent. We concluded that the long time scales (5 to 50 years) and the global character of the problem define the necessary thrust of the program and that the interactions between land, sea, and air on the one hand, and biology, chemistry, and physical processes on the other, provide some of the crucial aspects of the tasks. We found that a commitment by the Federal Government was essential to further progress and that there are exceptionally difficult questions in science and program management.

It has been suggested from many sides that this program might be announced as a U.S. initiative at Unispace 82 to be held in Vienna in August, 1982. While we are conscious of the great difficulties in effective execution, we nevertheless believe that the step is appropriate for reasons both of feasible science and of good international polity.

[3] **II. Impact**

The human race lives on a planet characterized by change. Change is evident on all space and time scales, from the very large and very long—the hundreds of millions of years associated with rearrangement of the continents—to the very short and relatively local—the day-to-day variations of weather. This is a unique time, when one species, humanity, has developed the ability to alter its environment on the largest (i.e., global) scale and to do so within the lifetime of an individual species member. This report is concerned specifically with changes that may affect the habitability of the earth: the ability of the planet to support communities of plants and animals, to produce adequate supplies of food, and to sustain and renew the quality of air and water and the integrity of the chemical cycles essential for life.

While life has persisted on the earth for more than 3.5 billion years, its course has been threatened many times during the history of the planet. Volcanic eruptions and advancing ice sheets have altered the temperature and circulation patterns of the atmosphere and ocean, leading to great change in local areas. Species have become extinct and others have evolved to take their place. Most species of animals have survived for periods of the order of 10 million years or so and then disappeared, mainly for reasons related to the destruction of habitat. Food, water, air, and climatic conditions changed gradually but inexorably beyond the ability of species to survive.

The human species faces a similar challenge. In contrast to its less successful predecessors, however, humanity has the ability to manage its resources, to plan intelligently for its future, and to preserve the necessary elements of its habitat. On the other hand, if the human race is to be successful in this endeavor it must take steps now to develop the body of knowledge required to permit wise policy choices in the future. The task is urgent.

Humanity has been accustomed in the past to respond to changes as they occurred: to move to the next valley when local resources were impaired or to substitute energy-intensive technology for the natural course of resource renewal. But the next valley is now occupied. Humanity is being confronted increasingly and rapidly with limitations imposed by the finite nature of its habitat. The human race is now a major direct and indirect influence on the chemistry of the atmosphere and on the allocation of resources on land, and is increasingly an influence on the ocean. That its influence can be subtle is illustrated by recent concerns regarding the potential vulnerability of stratospheric ozone.

It has become apparent within the last decade that humanity has developed the ability to alter ozone and thus to change the level of harmful ultraviolet radiation penetrating to the ground. The direct injection into the stratosphere of the exhaust gases of high-flying aircraft, the release of chlorinated gases used as aerosol propellants, as industrial solvents, and as refrigeration-system working fluids, and the creation of complex perturbations within the global nitrogen cycle can all lead to such a change. However, while these activities lead for the most part to a reduction in ozone, they are offset to some extent by thermal disturbances, due to enhanced levels of carbon dioxide, that cause a rise in ozone. Humanity's assessment of its own impact is hampered by its lack of understanding of the underlying physical, chemical, and biological influences regulating ozone in the natural state, and humanity's assessment is forced to proceed in parallel with a research program designed to provide this fundamental body of knowledge. The matter assumes urgency because the gases responsible for changes in ozone—human-made chlorocarbons and biologically formed nitrous oxide—have lifetimes ranging from 50 to 200 years. The self-cleansing function of the atmosphere proceeds slowly and human effects today will persist for centuries.

Carbon is the largest single waste product of modern society. Humanity has added, by burning fossil fuel, over 100 billion tons of carbon to the atmosphere as carbon dioxide since the industrial revolution, with perhaps a quantity of similar magnitude transferred from the biosphere to the atmosphere over this same period as a consequence of land clearance for agriculture. The increase in the burden of atmospheric carbon dioxide is readily detectable. Approximately half of the carbon added to the system remains in the atmosphere and the remainder is presumed to have been taken up by the ocean and transported to the depths of the oceanic abyss.

Massive engineering programs are being initiated to alter the course of major rivers—Soviet plans for three of the ten largest rivers in the world are an example—the consequences of which on ocean physics and therefore on climate would be continental if not global in scale. The increasing pressure on fisheries from humanity and from atmospheric variability has led to decreases and often elimination of traditional protein sources in many parts of the world. The consequences of chemical change can be detected in the coastal waters of all the developed nations. These are no longer isolated problems; taken together, they affect the habitability of that part of the earth most exposed to population pressures.

Changes brought about by human activity must be monitored and assessed and the science community must be prepared to recommend responses when necessary. Changes involve soil erosion, loss of soil organic matter, desertification, deforestation, overgrazing, diversion of freshwater resources, rising levels of air pollutant, and acid rain. On the decadal time scale, land, sea, and atmosphere operate as a coupled system not only in their physical interactions, but also through chemical and biological processes. Water in the upper layers of the ocean has a residence time of a few decades and the cycling time for carbon and nitrogen through the terrestrial biosphere is on the [4] same scale. Physics cannot be separated from chemistry and biology by considering them consecutively. Nor can problems in the ocean be treated in isolation from those on the land or in the atmosphere. The necessary questions relate to the inter-action of land, sea, and atmosphere. Activities on land alter the atmosphere, which in turn affects the physical and chemical dynamics of the ocean. The total effect at decadal time scales depends on the response of the ocean, the flywheel for these interacting components. The habitability of the earth depends not only on the presence of the great mass of water in the ocean, but also on its dynamics. The transport of heat in the ocean helps to determine continental climate; the ocean controls the ex-change of water through the hydrological cycle and determines the long-term balance of carbon and other essential chemicals. To understand this total system, the physical dynamics of the upper layers of the ocean and the exchange from these layers to the atmosphere above and the deeper ocean below must first be understood. Then this knowledge must be used to determine the consequences of change upon the chemistry and biology of the whole system.

This program requires greater knowledge of each of the three components—land, sea, and air, on regional to global scales—but the study of their interactions must particularly be emphasized. One such interaction concerns the direct connection between land and sea. Not only is population increasing rapidly, and not only is this increase most marked at the coastlines of all the continents, but also cumulative human-imposed changes in coastal waters are taking the problems beyond their previous terms of reference.

While it has reached the point of affecting the global system significantly within a few decades, humanity's ability to introduce countermeasures, should it wish to, is on the same time scale or longer. There are inherent dangers that must be anticipated when changes are imposed more rapidly than humanity's ability to achieve control. Steps must be taken now to develop the necessary scientific base. In the past the study of global habitability has been scattered among several independent disciplines, such as atmospheric physics, biology, oceanography, etc. The attempt to isolate problems that are tractable is the nature of scientific investigation. As the knowledge in each discipline has grown, so have the boundaries of investigation. We have now reached the point where the boundaries of each discipline are overlapping, and the next step forward can only come from an interdisciplinary research program. Such a program requires dedicated scientists, sophisticated tools using advanced instruments, computer and satellite technology, and strong managerial leadership.

The ultimate goal of the research program proposed here is that it should provide a scientific basis that can aid in policy decisions. It should be stressed that this cannot be put together in a single predictive model. Only studying the natural environment, understanding its varied processes, and showing that this knowledge can be distilled into the form of several comprehensive models will provide confidence in the utility of models for forecasting future events.

Fortunately, through satellites, computers, and modern communications humanity has the scientific tools to carry out the required research. At the same time the theoretical tools have advanced to the point where the integrated study of the global system is becoming feasible. If both are carried forward, the future may be manageable.

[5] III. Some Science Issues

The problem of global habitability leads directly to some of the most interesting and difficult scientific problems that have emerged in recent decades. In this chapter we consider a few of these in order to illustrate our view that their importance is great and that their intrinsic interest is such as to ensure enthusiastic participation by the research community. We should emphasize, however, that we have made no attempt at completeness.

Global changes that impact the earth's habitability are principally of two kinds: those relating to biological productivity and those affecting human health. Both involve the cycling of energy, water, and essential chemicals through the atmosphere, land, and oceans. The problem is complex and interactive. However, it is possible to identify several crucial scientific questions that must be answered before future changes in habitability can be effectively assessed. It is possible also to formulate a program to address them.

Biological productivity over land is paced by the availability of water, nutrients, and light; in the oceans only the latter two are of concern. Changes in the global rainfall distribution, intensity, and timing are critical for land productivity. The flow of nutrients through the rivers to the coastal zone and their eventual mixing with deep ocean water and similar transfers from land to ocean through the atmosphere are dominant factors for productivity in the oceans. The problem of assessing these changes becomes more complicated when it is realized that natural rainfall patterns over the globe are only marginally predictable by current climate models and are now being perturbed by the introduction of optically active gases, which are changing the temperature regime and impacting the evaporation-condensation cycle. At the same time, other chemicals are being added to the atmosphere, changing the acidity of rainfall, the level of toxic gases, and the amounts of nutrients in the soils, rivers, and eventually on the coastal shelves. They may even modify the intensity of near-ultraviolet radiation reaching the earth's surface. The objective then is to isolate the natural variability of biological productivity, and then to assess the impact of human-induced changes in the global system. What needs to be done to get at the problem?

One area of intensive research has to be the investigation of the global hydrological cycle. Many phases of the hydrological cycle, from evaporation and evapotranspiration through the vertical and horizontal transport of water vapor to cloud processes and ultimately to precipitation, are coupled directly to the circulation of the atmosphere and have important dynamic feedback mechanisms. The circulation is interactively linked to the radiation budget through optically active gases or possible changes in the solar flux and to the transport of minor constituent gases, including those that are chemically and biologically active. It is, therefore, critical that the principal components of the hydrological cycle be thoroughly understood.

Areas of major uncertainties are evapotranspiration over land, evaporation and precipitation over oceans, and cloud formation and precipitation processes. Scientific understanding of the vital role played by vegetation in the transfer of water from soil to air must be improved and quantified. It is necessary to establish climatological distributions of clouds, precipitation over the oceans, and evaporation from the ocean surface; at the present time the last two are unknown by at least a factor of two.

Habitability questions are particularly sensitive to the variations in global climate. The oceans provide the major regulatory mechanism, particularly on the decadal scales where mixing of the upper few hundred meters acts as the flywheel determining the rates of interaction with the atmosphere. The oceanic processes that drive this mixing are partly horizontal and partly vertical, and the quantitative understanding of either is currently inadequate. The horizontal component is related directly to wind stress, fresh-water exchange, and heat budget, and so interacts closely with atmospheric changes. The vertical component regulates the exchange of essential chemicals and organic matter with the very long-term (1,000s of years) deep reservoir; it also determines the exchange of gases, heat, and water through the sea surface. The overall effect on the atmosphere. and hence on climate, is through the net flux of heat and mass through the sea surface, which is closely linked to the underlying transports and is nonuniform over ocean basins and in time. It is necessary to know how changes in ocean circulation alter exchanges with deeper water and how they affect climate.

Another area of almost complete ignorance concerns the interaction between the atmosphere and the land surface. It operates on time scales that range from days to decades and it includes both gradual and abrupt changes. There is, at present, no quantitative understanding of the coupling between the atmospheric and land systems. Climatic models assume constant surface properties (albedo, heat capacity, soil moisture, roughness, etc.). The exchange of radiative energy is governed by surface albedo, atmospheric water content, and clouds. Transfer of latent heat is governed by soil properties and land cover that may be irreversibly altered from grassland to desert or from forest to farmland. Such changes profoundly alter the surface albedo, roughness, and evapotranspiration, which may lead, in turn, to a change in climate and in the chemical composition of the atmosphere.

Another largely unknown factor in the climatic system involves the role of snow and sea ice. Snow has the highest albedo of any material abundant on the earth's surface and its areal extent is more variable than that of any other material. Its presence or absence, therefore, drastically alters the surface radiation budget and, thereby, surface temperature.

[6] Just as snow modifies the albedo of the land, sea ice, which covers up to 10% of the world ocean, drastically modifies the albedo of the surface of the sea. In the "landlocked" north, sea-ice extent doubles in the winter, with the expansion largely occurring in the marginal seas of the subarctic. In the waters surrounding the Antarctic continent the seasonal change in ice extent is much larger. Science lacks a quantitative understanding of decadal change in global ice extent.

Sea ice serves as a deformable heat insulating layer of variable thickness. The heat flux through the cracks in the ice (the leads) is two orders of magnitude larger than through the ice itself. The occurrence of sea ice modifies the seasonal temperature cycle, delaying temperature extremes through the release of its latent heat of freezing in the fall and the uptake of heat for melting in the spring.

Finally, sea ice by its very nature is sensitive to small changes in the atmospheric and oceanic heat fluxes, making it a useful indicator of climatic change. This is particularly true in the southern hemisphere where the movement of the pack is unaffected by blocking land masses (for instance at lat. 60° S, a 1° C change in mean air temperature correlates with a 2.5° latitude decrease in maximum ice extent). A comprehensive program for monitoring the cryosphere can illuminate some of the issues in climate change.

Questions about the availability of the nutrients related to biological productivity again involve chemicals that cycle through the atmosphere, land, rivers, coastal zone, and deep ocean. Removal of chemicals and organic matter across the shelf to the deep ocean is a major pathway in the overall cycles of carbon, nitrogen, and phosphorus. Nutrient addition from rivers is an increasing factor in shelf productivity and it is influenced by human activities. The open ocean contributes significantly to the nutrient balance on the shelf and changes in ocean circulation can also alter this balance. The shelf is a direct source of food for humanity and pressures on this supply are increasing rapidly. The mechanisms and processes that control the transport of these nutrients from terrestrial sources to oceanic "sinks" are known only in the most qualitative way. It is therefore critical to assess the fate of this excess anthropogenic nitrogen to determine its impact on the shelves and coastal zones and ultimately its effect on the rate of fixation of carbon in the open ocean.

At decadal scales ocean dynamics are inhomogeneous within and between basins. There is an intimate relation between ocean circulation and biological production. The scientific focus could be on the longer term consequences of these interactions between physical and biological processes in the oceans. The consequences occur through the exchange of essential or critical elements such as nitrogen and carbon through intermediate depths (300 to 500 meters). These exchange rates can be measured through specific experiments. To generalize these results, however, the processes must be linked to the general circulation and productivity of the upper layers.

How do variations in vertical processes alter the rates of biological and chemical cycles? These changes are driven by mixing over much of the ocean interior. In certain regions, where there is significant net upward or downward movement, the chemical and biological changes are magnified and can play a dominant role in the total exchange: for example, the removal of excess carbon dioxide from the atmosphere probably involves downwelling in the Norwegian Sea. How do changes in upwelling regions affect both regional fish production and global chemical cycles? How are downwelling rates in the high latitudes affected by changes in the general circulation and how, in turn, will these affect the global balance of radiatively important gases? These are some of the questions that need to be addressed on a global scale.

The cycling of carbon, nitrogen, and sulfur through the atmosphere is controlled mainly by photochemical transformations in the troposphere. The sink processes, which include wet and dry deposition and takeup by vegetation, can have important and widespread effects on biospheric productivity, positive in some cases (deposition of sulfur and nitrogen in nutrient states) and negative in others (excessively acidic rainfall with attendant growth reduction and increases in exposure of vegetarian to ozone, SOx, and NOx). Accompanying the major effects of global atmospheric chemistry on biological productivity are other changes in the atmosphere such as visibility degradation.

Although the exact details of the global carbon, nitrogen, and sulfur atmospheric chemical cycles are far from understood, some key issues are clearly defined. The major challenge in understanding both nutrient (carbon, nitrogen, and sulfur) and toxic (ozone, SOx, and NOx) atmospheric chemistry is to determine the concentration distributions of these materials in the atmosphere, understand the relevant atmospheric transformations, and delineate human inputs from natural sources.

There is a need to identify natural and human sources of hydrocarbons and to understand the relevant oxidation mechanisms. Regional to global changes in the hydroxyl and ozone concentrations are very important in determining the pace of photochemical cycles. Tropospheric ozone appears to be increasing not only in polluted air but in "clean air" regions of the troposphere. It is critical to determine the spatial variation in the concentration of tropospheric ozone and other toxic gases remote from industrial sources, especially over agricultural and vegetated regions, so as to isolate the effects on primary productivity.

[7] The concern about human effects on stratospheric ozone has been mentioned earlier and represents an important example of a widely recognized issue impacting global habitability. An extensive effort is underway to improve scientific understanding of the chemical and dynamical influences regulating ozone in both natural and perturbed states. The program recognizes the importance of microbial reactions in mediating the concentrations of gases such as methane, nitrous oxide, and various halocarbons and their effects on ozone.

A major scientific issue related to biological productivity and the carbon cycle is that the total biomass of the earth's surface is unknown to better than a factor of two and that the extent to which the biomass changes in decadal time scales is also not known. Complicating the issue is the fact that human disturbances of natural ecosystems over the last 100 years, including the clearing of forests for agriculture, have changed the vegetation and soil reservoirs by relatively large amounts. Better estimates of both the total biomass on the earth's surface and rate at which it is changing are needed before an understanding can be achieved of how the primary productivity of the earth's surface is being altered by changes in land use, in climate, and in the supply of nutrients. A concerted program using current capabilities for space and ground observation, especially in tropical countries, can be mounted in order to provide information on the extent of biomass change. This program, with related definitions of changing nutrient cycles, climate, and available energies, can help bring about progress in understanding several areas: the global carbon cycle; the interdependence of carbon with the cycles of other nutrients; the recovery time of the earth's biosphere after a major perturbation; and the reasons for the changing productivity of the semiarid regions.

The scientific issues raised here are not all-encompassing. They provide an example of the range of issues that must be addressed in a program of research motivated by the question of habitability. The scientific expertise is available and rapid progress can be expected in most areas. In others, the challenge is more formidable, and success will require imaginative combinations of talents from different disciplines. The program outlined in the next chapter shows the extent to which this may be achieved.

[9] IV. Some Program Elements

A new emphasis that appears in the global habitability program is on the interaction between land, ocean, and atmospheric processes, with biological processes on land and in the sea as a central concern. Programs can be identified that can contribute to the solution of most of the problems posed in Chapter III. Some are already in existence (e.g., weather and climate), but for the most part these are missions of other agencies and are funded at relatively low levels. Some are ready for execution, such as the Upper Atmospheric Research Satellite and the Ocean Topography Experiment (TOPEX), and they must be funded immediately if a global habitability program is to have meaning. Finally, there are programs, particularly in the biological area, that may be feasible but for which even advanced planning does not exist.

We discuss, in this chapter, only a few of the possible programs, with emphasis on those that involve interactions between disciplines. This is where innovative and interesting new approaches are possible, but it should be borne in mind that many of NASA's existing programs are also crucial for the successful execution of a global habitability research program.

A NASA research program cannot concern itself with space systems alone. In each of the following sections, therefore, we indicate the activities required for global observations, for process studies (laboratory research, ground-based measurements), and for theoretical research.

The Atmosphere

NASA already has a well-conceived and productive program in stratospheric chemistry research, mandated by Congress in 1976. The focus of this program is on understanding the stability of the upper-atmospheric ozone layer and on examining changes or threats posed by human activity.

A key new element in the upper-atmosphere program is the Upper Atmospheric Research Satellite (UARS), planned for launch in 1988, which will study the coupling between radiation, chemistry, and dynamics. An integrated program involves sets of global UARS satellite measurements, in situ field measurements utilizing aircraft and balloon measurements, laboratory studies, data analysis, and theoretical work (for details see JPL Publication 78-54, Upper Atmosphere Research Satellite System).

A tropospheric research program, more extensive than that which NASA now has, needs to be developed. The immediate aim would be to provide global data and parameterizations of physical processes for incorporation in numerical models of hydrological and biogeochemical cycles. These models are not ends in themselves: they cannot substitute for careful thought and understanding, but they are helpful in providing guidance to the important measurements and to the required precision of measurement.

One physical process that is inadequately understood is that of cloud formation and its influence on the planetary radiation fields. We have yet to document cloud climatology on a global scale, a task well-suited to earth-orbiting systems. A large program has been proposed to take place from 1983 to 1988 under the auspices of the World Climate Research Program (see International Satellite Cloud Climatology Project (ISCCP): Preliminary Implementative Plan, ICSU/WMO, January 1982). NASA can and should be a major participant in this program. The agency should also undertake detailed cloud and precipitation process studies using aircraft for observation platforms.

Knowledge of tropospheric chemistry lags behind stratospheric chemistry and, at this time, the need is for measurements, from aircraft, over the full tropospheric altitude range and over a range of conditions of the concentration (or flux if possible) of carbon dioxide, carbon monoxide, methane and other hydrocarbons, ozone, ammonia, chlorofluoromethanes, SOx, NOx, hydroxyl, aerosols, and other species. Laboratory and field measurement techniques need to be developed for these studies. Global measurements from space will eventually be required to provide a data base for global tropospheric models.

The Ocean

Global measurements from space are ideally suited for providing improved understanding of the ocean's large-scale, low-frequency circulation, but such measurements are inherently limited to features at or near the sea surface. Of prime importance are measurements made from space of sea-surface elevation and global wind stress, with complementary measurements made within the ocean. A global program combining subsurface measurements with those from satellites can best be accomplished by transmitting in situ data directly to one or more satellites.

The United States has pioneered in the field of ocean observations from space, and NASA must maintain an active program in those aspects relevant to its concerns with global habitability. NASA is ready to proceed with an ocean topographic satellite (TOPEX), which can greatly increase knowledge of the transport of heat and chemical species in the oceans. The funding of this program is an early order of business.

The need to unify space and in situ measurements is particularly pressing in oceanic research. A valuable and distinctive character could be given to the U.S. program by making a major effort to integrate NASA's activities with those of the oceanographic [10] community, with its tried in situ methods. An immediate commitment by NASA to space oceanographic missions is essential if the interest and involvement of U.S. oceanographers is to be increased above its current low level.

Further steps concerned with research into the close coupling between the atmosphere and ocean are under active consideration by many international groups. The United States should participate in their efforts and collaborate in program execution. Current planning by the Joint Scientific Committee of the World Climate Research Program involves:

- A "Cage Experiment" to address the problem of estimating the annual mean surface heat flux to the atmosphere on an ocean-basin scale.
- A "World Ocean Circulation Experiment" to address the problem of estimating the rate of bottom water formation and to determine the dynamics of the slow, large-scale components of the ocean circulation.
- An "Interannual Variability of the Tropical Oceans" program.

These programs fit, to varying degrees, within the objectives of a global habitability program, and all contain major space elements.

An ocean biological productivity program should focus on one major theme: the exchange of major chemicals between near-surface and deeper water. Distinctive regions are the open ocean, centers of upwelling and downwelling, and the coastal zone; each has markedly differing vertical exchange processes. The aim should be to determine physical and chemical conditions and rates of biological processes from in situ measurements, so that the information can be used together with remote sensing data to extend scientific ideas in space and time. Physical programs to measure advection and diffusion should be closely linked with measurements and experiments on dissolved and particulate constituents. A key requirement for assessing the net long-term rates of removal of carbon, nitrogen, and phosphorus from the system is the measurement of sediment and organic matter at the shelf edge prior to removal. New instrumentation for this purpose is needed. To connect the physical changes with net chemical transports, the basic biological dynamics that link these components must be understood. Several focused experiments to study ocean productivity are outlined in the Satellite Ocean Color Science Working Group Report (NASA, 1982). These include river-dominated, wind-upwelling, western-boundary, and open-sea experiments.

The Cryosphere

It is currently possible to measure snow extent using satellite imagery, but it is not possible to obtain by remote sensing other physical data (depth, density, water equivalence, free-water content, age, crystal size, type, etc.). Recent studies of microwave systems have shown promise, and work should be initiated on systems to extend scientific understanding of processes occurring within natural snowpacks.

The situation for sea ice is different in that there are a number of sensor systems that have proven capabilities. The most valuable of these operate in the microwave range and are not limited by either clouds or the absence of light. Of particular value is synthetic aperture radar, which supplies high-spatial-resolution information on ice extent, type, and roughness in a map-like format. Also of value are passive microwave radiometers (low-spatial-resolution information on ice extent and type), radar altimeters (ice extent and topography), and scatterometers (ice extent, type, and surface roughness). In short, the sensors are available; it is their deployment into suitable orbits and the utilization of available data that have lagged behind.

Cryospheric research is an area of central importance to the coupling of the ocean and the atmosphere. Several international and national programs are in progress or under discussion. The role of space observations is crucial, and important contributions are described in documents such as Guideline for the Cryospheric Processes Special Study, (NASA, 1982). And yet, no synthetic aperture radar or other innovative measurement has been authorized by the United States. Again, the ground for a global-habitability thrust is thoroughly prepared.

The Land

We have identified the crucial importance of biology to processes in the atmosphere and the ocean on the decadal time scale; the reaction of global processes upon the biosphere in turn is not only the return path in a fully interacting system, but, in addition, defines the social issue of importance to this study. More than 90% of the global biomass is on or in the land surface but, despite the fact that humanity exists in close contact with it and deploys considerable effort on agricultural and ecological research, there are too few available data for allowing generalizations regarding the global system and no demonstrated concepts upon which a large measurement program can be based.

If resources are devoted, as they should be, to a global habitability program, the greatest single need is to broaden the intellectual base in the area of land biology and to evolve, virtually from first principles, useful measurements addressed to the problems raised in Chapter III.

There is only limited experience to draw on. Landsat images have yielded some interesting details, but there is no substantial reason to suppose that land biomass and its changes (let alone soil biomass) can be reliably measured, on a global scale, with any similar system.

[11] Studies have been made of space systems for measuring soil moisture. There are some reasons for encouragement here, but there is a long path to travel before a credible

program can emerge (JPL Publication 80-57, Joint Microwave and Infrared Studies for Soil Moisture Determination).

Studies of small ecosystems, monocultures, and other activities evolving from chemical studies make continual progress, but they are difficult to relate, with confidence, to global questions. Presumably, enough studies in enough detail would provide some insight, but this only shifts the burden to the need for an intellectual synthesis of extraordinary difficulty.

What achievement, then, can be hoped for in this area? The distribution of land biomass and its natural and human-induced changes are primary targets for investigation. Further efforts to use Landsat data in a quantitative fashion and attempts to define more sophisticated measuring systems should continue.

Studies of new concepts in remote sensing, e.g., to measure soil moisture, should also continue. The existing effort needs to be accelerated. Where the connection with global change can be clearly demonstrated, studies of individual ecosystems should be undertaken. Studies of toxicity can define the significant tropospheric chemical parameters.

Generally speaking, the present choices for a directed program are severely limited, but we believe that they can be increased in number and their quality improved. The key is an able scientific community and the key to that, in turn, is an agency commitment to the task. In the short term there are, fortunately, many related areas in oceanic and atmospheric research where progress towards the overall goal is possible.

The parameterization of surface data to define thermal and moisture fluxes into the atmosphere for global dynamical models is an area for which a feasible program exists. Aircraft and in situ studies can be defined that can determine thermal and transpiration boundary conditions for large ecosystems identifiable from space images. If this is successful, one of the most important links between human activities and climatic change may be understood in quantitative terms.

The Sun

Solar radiation at wavelengths less than 350 nanometers controls the production and destruction of important stratospheric and mesospheric chemical species. Changes in the solar ultraviolet spectral distribution can therefore change the vertical temperature and species profiles in this region of the atmosphere, which may, in turn, change stratospheric circulation patterns.

In the ultraviolet region, intensity variations of several percent have been measured over a period of a single solar rotation. Because of unsatisfactory calibrations in the past, science does not yet have a long-term data set of solar spectral variability. Instrumentation suitable to the task has now been developed for periodic Shuttle flights.

In terms of the total energy from the sun falling on the earth, only recently have direct measurements become available with sufficient accuracy to establish the reality of solar variations. These indicate changes over a few days on the order of a few parts per thousand; small on the scale of phenomena that may affect habitability of the earth.

Changes of a few percent would, however, be important, and a long time series of measurements is required to establish whether they occur. As a minimum, the total solar irradiance should be measured over a full solar magnetic cycle to a precision of 0.05 percent or better.

Interactions

Many of the most interesting and least understood features of the global system concern the interactions between land, sea, air, and life. In this section a few out of many possible research directions concerning interactions are discussed.

For the hydrological cycle, NASA already has several active programs, but the balance between evaporation and precipitation over the oceans is a neglected area. A combination of infrared and microwave sounders with active and passive techniques for determining surface stress offers some promise of yielding data on evaporation. Precipitation can possibly be inferred from cloud forms and from altitudes of cumuliform clouds and from radar reflectivities for stratiform clouds. A directed thrust to determine evaporation and precipitation over the oceans appears to be justified.

The cycling of carbon, nitrogen, and sulfur across land, sea, and air boundaries is a central problem. Some information comes from tropospheric chemical measurements, but the more difficult features require in situ and aircraft measurements. It is necessary to know more about microbiological activity, reservoirs of relevant species in the three media, and transport between them, and it is necessary to integrate this knowledge with coupled numerical models. Many aspects of this problem are feasible now.

Another surface chemical problem concerns the destruction of ozone at the land and sea surface. We have signaled the importance of ground-level ozone but heterogeneous destruction at the surface is poorly understood and a program of research could be designed.

[12]Sensible heat fluxes at the ocean surface are required, together with water fluxes, to define the heat transports in both the ocean and the atmosphere. This important parameter can be obtained from infrared and microwave space soundings under certain conditions, and a more active program should be undertaken.

The problem of desertification is one of interactions. We have discussed the evapotranspiration cycle, and steps that could be taken to further understand it. But what conditions led to irreversible changes on the land, perhaps even to the decay of earlier civilizations? Space tools can, at the least, provide the monitoring of patterns of changing land use. As the physical understanding of desertification increases, observational tools can probably be devised to provide the needed data.

Interactions between land surfaces and the oceans over decadal time scales are changed primarily through runoff. Runoff provides a means for removal of chemical species and organic matter from land and is the major source of fresh water to the oceans. A general survey is required for several select estuarine systems and coastal environs to trace export of various chemicals and organic matter from the terrestrial source to the long-term oceanic sink and to establish a flow data set from which induced changes can be assessed. These are, principally, in situ studies, but attempts must be made to relate to measurements by color scanners and other space observations.

Studies need to be undertaken to understand how changes in biochemical transports and the quantity of fresh water affect coastal productivity. A mixture of space and in situ measurements here is essential to obtain a full representation in both space and time.

Data Sets

Much has been written on the extraordinarily difficult problems involved in handling global data sets so that they are useful for the purpose for which they were gathered. It is often simpler to collect data than to use it. Examples of the destruction of unique information for apparently extraneous reasons are growing. The problem will not look after itself but requires a directed activity of high priority.

The nature of the problem is discussed in Guidelines for the Air-Sea Interaction Special Study (JPL Publication 80-8, 1980), and a strategy is proposed around the following conceptual elements:

- Understanding satellite signals.
- Instrument verification.
- Combining the capabilities of several different remote sensors with remote and in situ capabilities to measure a given variable.
- Utilizing a directly sensed remote measurement as index of a climate change.
- Comprehensive data management to provide an easily usable data base.

Each appears deceptively simple, but requires, in practice, a major effort.

[13] V. National and International Considerations

National Programs

There is no national research program of the same overall scope as that proposed in this report. However, the United States does have scientific programs and operational systems that encompass elements of the proposed program—e.g., the National Climate Program. NASA can work with other agencies and with industrial users through existing relationships in order to make its data and findings available where needed.

The International Aspects

A global habitability program must be international in scope and requires the cooperation of the international science community. No one nation could carry it out alone. These considerations are recognized by scientists and policymakers around the world.

There are many areas of possible collaboration in different geographical regions at different levels of effort. Such areas could, for example, include ground-based measurements of major estuaries, coastal shelves, tropical forests, and oceanic upwelling regions.

The Global Atmospheric Research Program is a good example of an enterprise that has achieved effective international cooperation between large and small nations in its planning, its realization of the required observational system, its resource allocation, and its analysis and interpretation of results.

Other International Programs

The major international programs relevant to the proposed global habitability program fall under the auspices of the United Nations Environmental Program (UNEP), the World Meteorological Organization (WMO), the International Council of Scientific Unions (ICSU), and the United Nations Food and Agriculture Organization (FAO). Although their definition and coordination are carried out by the appropriate international bodies, these programs rely for their execution on national agencies and resources and on scientists in individual countries.

Examples of these programs, all of which are supported by the United States, are:
- The Global Environmental Monitoring Systems, which are concerned with monitoring a variety of environmental factors and are a part of Earthwatch, a program coordinated by UNEP.
- The World Climate Program, with two major subprograms: an Impact Program, which is the responsibility of UNEP and assesses the impact of possible climatic changes on human activities such as agriculture, water supply, etc., and a Research Program, which is the joint responsibility of WMO and ICSU and aims to acquire scientific understanding of the climatic system.
- The Middle Atmosphere Program of ICSU, which aims to observe and understand the middle atmosphere, especially the ozone layer.
- The global agricultural productivity activities of the FAO, which include the regional desert locust project and soil-information and crop-statistic planning-assistance efforts.
- The Consultative Group on International Agricultural Research, which coordinates the research in some fifteen international research centers located in various countries, mostly in the developing world.

International Collaboration

To ensure effective global collaboration, NASA can:
- Encourage and invite participation in the U.S. programs of experimenters from all countries, to develop instrumentation, acquire comparative observations to corroborate satellite-obtained information, and analyze and interpret data.
- Contribute to the environmental research efforts of WMO, FAO, UNEP, and other regional and international agencies. NASA can correlate its research with the research aims of these bodies and can engage in data exchange, cosponsorship of specific projects, and joint efforts to publicize results and prognoses.
- Consult with and coordinate its plans with other satellite- operating agencies.

[15] Appendix

List of workshop attendees, 21-26 June, Woods Hole, Massachusetts

Executive Committee
Richard Goody (Chairman)
Robert Chase (Executive Officer)
Wesley Huntress (Executive Officer)

Moustafa Chahine
Michael McElroy
Ichtiaque Rasool
John Steele
Shelby Tilford (NASA Representative)

Attendees

James Baker
Victor Baker
William Bishop
Harry Blaney III
Daniel Botkin
Geoffrey Briggs
Wallace Broecker
Kenneth Carder
Philip Chandler
Kenneth Coughlin
Paul Crutzen
Robert Duce
Peter Eagleson
James Hansen
Robert Harriss
Howard Hogg
Donald Hornig
John Houghton
Robert Hudson
James Lawless
Robert MacDonald

Robert McNeal
Lynn Margulis
Berrien Moore
Robert Murphy
James Pollack
Ronald Prinn
V. Ramanathan
Mitchell Rambler
William Raney
Charles Robinove
Robert Schiffer
Brent Smith
Gerald Soffen
Vern Soumi [sic]
John Theon
John Walsh
Richard Waring
Robert Watson
Wilford Weeks
Sylvan Wittwer

Document IV-3

Document Title: James M. Beggs, Administrator, NASA, to George A. Keyworth II, Science Advisor to the President, Proposal for "Global Habitability," 14 July 1982.

Source: NASA Historical Reference Collection, NASA History Office, NASA Headquarters, Washington, D.C.

By mid-July 1982, NASA Administrator James. Beggs had agreed to take the global habitability initiative to Unispace '82. This memorandum and attached program description and white paper were NASA's attempt to convince the President's Science Advisor, Dr. George Keyworth, to agree to and support the initiative.

[no pagination]
[stamped "Jul 14, 1982"]

Dr. George A. Keyworth, II
Science Advisor to the President
Old Executive Office Building
Room 360
Washington, DC 20500

Dear Jay:

I would like to advise you of a new NASA Program, known as Global Habitability, which I intend to present as a new U. S. initiative at the UNISPACE Conference the week of August 9 in Vienna, Austria. The initiative primarily consists of inviting the international community to join the U.S. under the leadership of NASA in a cooperative research program to achieve a consolidated approach to understand changes which may affect the habitability of earth. These activities include research on the atmosphere, oceans, land, and biosphere and the requisite space and ground observations to support them. The enclosed paper more fully describes this initiative.

Sincerely,

[signature]
James M. Beggs
Administrator

Enclosure

bcc:
A
AD
E/Edelson
E/Keller
E/Rosendhal   .
E/Thome
AEM

****

July 13, 1982

GLOBAL HABITABILITY PROGRAM

The National Aeronautics and Space Administration is proposing a ten year international cooperative research program to obtain a solid body of knowledge from which policy decisions can be made addressing the questions of change, either natural or of human origin, affecting the habitability of the earth. Called "Global Habitability," the program will focus on biological productivity and the water, chemical and energy cycles which comprise the life support systems of the global habitat. The program will investigate the interactions between land, sea and air on the one hand, and biological, chemical and physical processes on the other. Much of the required effort is currently in process within NASA, other federal agencies and the international community. However, some of the current research would require a change in emphasis and in time additional areas of research will be necessary. The total activities include research on the atmosphere, oceans, land, and biosphere and the requisite space and ground observations to support them.

We live on a planet characterized by change, at a time when man has achieved the ability to alter his environment on a global scale. Within the lifetime of our children the population on the earth is predicted to double. Of concern are the production of food adequate to the needs of this growing population, and the subtle interactions which might affect the planet's ability to sustain and renew the quality of air and water and the integrity of the global chemical cycles essential for life.

Scientists believe that humanity has reached the point of being able to change the Earth significantly on a time scale of mere decades, yet our ability to mount counter measures, should we wish to, is restricted in many cases to a time scale which is similar or longer. An understanding of the overall system is essential if we are to live successfully with global change. Although the time scales for change are decadal, the need for the body of knowledge is in this present decade if enlightened policy decisions are to be made.

Global changes which impact the habitability on the earth are principally of two kinds: those which relate to biological productivity and those which affect human health. Both involve cycles of energy, water, and essential chemicals through the atmosphere, land, and oceans.

Biological productivity over land is paced by the availability of water, nutrients, and light; in the oceans only the latter two concern us. Changes in the global rainfall distribution, intensity and timing are critical for land productivity while the flow of nutrients through the rivers to the coastal zone and their eventual mixing with deep ocean water and similar transfer from land to ocean through the atmosphere are dominant factors for productivity in the oceans. The problem of assessing these changes becomes more complicated when we realize that natural rainfall patterns over the globe are only marginally predictable by current climate models and are now being perturbed by the introduction of optically active gases, changing the temperature regime and impacting the evaporation-condensation cycle. At the same time other chemicals are being added to the atmosphere, changing the acidity of rainfall, the level of toxic gases, and the amounts of nutrients in the soils, rivers, and eventually on the coastal shelves. They may even modify the intensity of near ultraviolet radiation reaching the earth's surface. The objective then is to isolate the natural variability and then to assess the impact of man-induced changes on the global system.

On time scales of a decade, the ocean, atmosphere, and biosphere function as an integrated system. Physical, chemical, and biological processes are coupled, and progress in our understanding will require an interdisciplinary research program of broad scope and content. The program will require space observations to provide global perspective;

investigations of specific ecosystems both on the ground and from space to enhance our appreciation for the overall function of the biosphere; studies of representative estuarine and coastal systems to define the transfer of materials from land to ocean; studies of horizontal and vertical motions in the ocean, recognizing the importance of these motions for climate and their role in regulating the flow of materials to the depths of the ocean; studies of the atmosphere, its chemistry, its physics and its motions; and theoretical and laboratory investigations to synthesize new information serving as a guide to the acquisition of new data.

A global habitability program should be international in scope and be performed with the cooperation of the international science community. There are many areas of possible international collaboration in different geographical regions and at different levels of effort: These include ground-based measurements of major estuaries, coastal shelves, tropical forests, and oceanic upwelling regions, satellite measurements, data analysis and laboratory research. It can thus involve both the large and the small nations.

Consideration is being given to introducing the Global Habitability Program at the UNISPACE 82 Conference in Vienna in August 1982, with the U. S. representative inviting other nations to join in this effort for the benefit of all mankind.

****

UNISPACE WHITE PAPER
July 13, 1982

GLOBAL HABITABILITY PROGRAM

The United States of America, through its National Aeronautics and Space Administration is proposing a ten year international cooperative research program to obtain a solid body of knowledge from which policy decisions can be made addressing the questions of change, either natural or of human origin, affecting the habitability of the earth. Called "Global Habitability," the program will focus on biological productivity and the water, chemical and energy cycles which comprise the life support systems of the global habitat. The program will investigate the interactions between land, sea and air on the one hand, and biological, chemical and physical processes on the other. Much of the required effort is currently in process within NASA, other U.S. federal agencies, and the international community. However, some of the current research would require a change in emphasis, and in time additional areas of research will be necessary. The total activities include research on the atmosphere, oceans, land, and biosphere and the requisite space and ground observations to support them.

We live on a planet characterized by change, at a time when man has achieved the ability to alter his environment on a global scale. Within the lifetime of our children the population on the earth is predicted to double. Of concern are the production of food adequate to the needs of this growing population, and the subtle interactions which might affect the planet's ability to sustain and renew the quality of air and water and the integrity of the global chemical cycles essential for life.

Scientists believe that humanity has reached the point of being able to change the Earth significantly on a time scale of mere decades; yet our ability to mount counter measures, should we wish to, is restricted in many cases to a time scale which is similar or longer. An understanding of the overall system is essential if we are to live successfully with global change. Although the time scales for change are decadal, the need for the body of knowledge is in this present decade if enlightened policy decisions are to be made.

Global changes which impact the habitability on the earth are principally of two kinds: those which relate to biological productivity and those which affect human health. Both involve cycles of energy, water, and essential chemicals through the atmosphere, land, and oceans.

Biological productivity over land is paced by the availability of water, nutrients, and light; in the oceans only the latter two concern us. Changes in the global rainfall distribution, intensity and timing are critical for land productivity while the flow of nutrients through the rivers to the coastal zone and their eventual mixing with deep ocean water and similar transfer from land to ocean through the atmosphere are dominant factors for productivity in the oceans. The problem of assessing these changes becomes more complicated when we realize that natural rainfall patterns over the globe are only marginally predictable by current climate models and are now being perturbed by the introduction of optically active gases, changing the temperature regime and impacting the evaporation-condensation cycle. At the same time other chemicals are being added to the atmosphere, changing the acidity of rainfall, the level of toxic gases, and the amounts of nutrients in the soils, rivers, and eventually on the coastal shelves. They may even modify the intensity of near ultraviolet radiation reaching the earth's surface. The objective then is to isolate the natural variability and then to assess the impact of man-induced changes on the global system.

On time scales of a decade, the ocean, atmosphere, and biosphere function as an integrated system. Physical, chemical, and biological processes are coupled, and progress in our understanding will require an interdisciplinary research program of broad scope and content. The program will require space observations to provide global perspective; investigations of specific ecosystems both on the ground and from space to enhance our appreciation for the overall function of the biosphere; studies of representative estuarine and coastal systems to define the transfer of materials from land to ocean; studies of horizontal and vertical motions in the ocean, recognizing the importance of these motions for climate and their role in regulating the flow of materials to the depths of the ocean; studies of the atmosphere, its chemistry, its physics and its motions; and theoretical and laboratory investigations to synthesize new information serving as a guide to the acquisition of new data.

This program will contribute to the ongoing environmental research efforts of the WMO, the FAO, UNEP and other regional and international agencies. NASA will correlate its research with the research aims of these bodies and will engage in data exchange, co-sponsorship of specific projects, and in joint efforts to acquaint national publics and policymakers with results and prognoses. We suggest the establishment of an international clearing house, perhaps in the form of an existing scientific body, to broadly disseminate the results of the studies.

We wish to encourage and invite broadscale participation of experimenters in the development of instrumentation, the acquisition of comparative observations to corroborate satellite-obtained information, and in the analysis and interpretation of data. Investigators will be selected through Announcements of Opportunity and Announcement Notices. Unsolicited proposals will be considered. Collaboration will also take place at the scientist-to-scientist level based on agreements with national and international agencies. NASA will provide for significant program involvement by developing countries in areas such as ground-based observations and the exchange of data that can assist individual countries and regions to manage their resources and better deal with changes affecting habitability—e.g., deforestation, desertification, agricultural productivity. Through such collaboration, both NASA and the individual countries will obtain vital information beyond that obtainable using only a single tool of measurement.

In addition, NASA will consult and coordinate its plans with those of prospective satellite operating agencies to encourage the development of compatible and complementary spaceborne instrumentation and thereby maximize the effectiveness of global measurements. Such coordination should result in cost savings, avoid duplication of effort, and minimize the prospect of incompatible national systems.

Life has persisted on the Earth for more than 3.5 billion years. The course of life has been threatened many times over the history of the planet. Volcanic eruptions and advancing ice sheets altered the temperature and circulation patterns of the atmosphere and ocean, leading to large changes in local areas. Species became extinct and others evolved taking their place. Most species of animals survived for periods of the order of 10 million years or so, and disappeared mainly for reasons relating to destruction of habitat. Food, water, air and climatic conditions changed gradually but inexorably beyond the limit of the species to survive.

The human species could be facing a similar challenge. In contrast to his less successful predecessors, however, man has the capability to manage his resources, to plan intelligently for the future, and to preserve the necessary elements of his habitat. We can see no better use of man's mastery of near space than the acquisition of the body of knowledge essential for the wise management decisions of the future.

Document IV-4

Document Title: James M. Beggs, NASA Administrator, to George A. Keyworth, Science Advisor to the President, July 1982.

Source: NASA Historical Reference Collection, NASA History Office, NASA Headquarters, Washington, D.C.

Dr. Keyworth apparently responded to the July 7 letter from Mr. Beggs with considerable skepticism about the value of the proposed global habitability program. This letter is NASA's attempt to allay his concerns and prevent the White House from blocking NASA's cooperative initiative.

[no pagination]
[no date]

Honorable George A. Keyworth
Science Advisor to the President
Office of Science and Technology Policy
Executive Office of the President
Washington, DC 20500

Dear Jay:

Thanks for your letter on our Global Habitability initiative for UNISPACE '82. While I can appreciate the concerns you expressed, I believe we are taking appropriate action on each of them.

I understand the sensitivity to the "Global 2000" report and contrary to the pessimism experssed [sic] therein, I believe that mankind has the capability to manage his resources and to plan intelligently for the future. However he must have sound scientific information as a foundation. The Global Habitability program is a serious scientific effort expected to take a decade or so, and its output will be the needed body of knowledge upon which future decisions can be made.

We are well aware of the expertise existing in other government agencies in the research areas related to Global Habitability; indeed NOAA and USGS were involved in the Woods Hole Study which generated the concept. In carrying out the U.S. Program in Global Habitability, many agencies will continue to be involved.

Let me clarify that it is certainly not our intent to turn the Global Habitability effort over to any new or existing international organization. It is our intention that the U.S. program remain entirely under U.S. management and control. At the UNISPACE conference we plan merely to describe our concept and to invite other nations to join this research effort with their own programs, and to share the results. We expect that the only international part of the effort will be some sort of "clearinghouse" for the results of the various national efforts. This might be undertaken by one or more existing groups— either a UN agency such as WMO, UNEP, or FAO, or by international scientific organizations such as ICSU, COSPAR, or IAF. In Vienna, we do not plan for the U.S. to identify which of these organizations might serve this role. We will suggest that an international workshop be held in early 1983 to discuss this concept.

We have now briefed all the U. S. agencies on this concept, and I hope that your concerns are alleviated by the above explanation and our actions.

Sincerely,

[signature]
James M. Beggs
Administrator
(Head U.S. Delegation to UNISPACE '82)

Document IV-5

Document Title: B. I. Edelson, Associate Administrator for Space Science and Applications, NASA, with comments from Hans Mark, Deputy Administrator, NASA, "Global Habitability," 8 November 1982.

Source: NASA Historical Reference Collection, NASA History Office, NASA Headquarters, Washington, D.C.

In this memorandum to a wide variety of officials within NASA, Burton Edelson illustrates his strong support for the global habitability concept and his attempt to turn it into a working program within NASA and with international involvement. As was frequently his habit on others' papers, NASA Deputy Administrator Hans Mark added his comments.

[no page number]
[stamped "Nov 8 1982"]

National Aeronautics and
Space Administration

Washington, D.C.
20546

Reply to Attn of: EL-4

TO: Distribution

FROM: E/Associate Administrator for Space Science and Applications

SUBJECT: Global Habitability

We had a good kickoff for our Global Habitability effort at UNISPACE in Vienna in August, including Mr. Beggs' announcement, Shelby Tilford's and Richard Goody's presentations, and discussions at several separate meetings with international organizations—UNEP, FAO, UNESCO, WMO, and ISCU. What we presented and discussed in Vienna was a "concept." Now we need to turn Global Habitability into a "program." [Italics indicate a handwritten note by Hans Mark] *I fully agree with this point except that we want to make it clear that the "program" we are contemplating may not require new funds.*

As a first task, we should get internal NASA agreement on the program definition, goals, and objectives. I suggest the following:

Global Habitability is a proposed NASA research program to investigate long-term physical, chemical, and biological trends and changes in the earth's environment, including its atmosphere, land masses, and oceans. The program will of necessity involve major international participation and will be coordinated insofar as possible with national research programs and those of international agencies. The program will specifically investigate the effects of natural and human activities on the earth's environment by measuring and modeling important nutrient chemical cycles, and will estimate the future effects on biological productivity and habitability of the earth by man and other species. The program will involve space and suborbital observation, land- and sea-based measurements, laboratory research, and supporting data management technologies over a ten-year or longer period of time. [Italics indicate a handwritten note by Hans Mark] *The program will start by combining existing efforts in these areas before developing new ones.* [In reference to the above paragraph:] *This is an excellent statement.*

As a second step, consistent with the definition and goals established above, we should prepare detailed background material for the Global Habitability program. This would supplement the report of the Goody summer study to include summaries of existing research, analysis of scientific issues, and relations to relevant ongoing programs (climate, weather studies, arctic research, etc.). This should be prepared in document form suitable for international presentation.

Third, we should formulate and document the NASA program. We should identify those ongoing [underlined by Hans Mark with the following comment: "most important"] and planned (i.e., FY 84-88) projects and research tasks which should be part of the program. These should be written into a NASA program plan to include description of the work, schedules, budgets, and responsibility assignments.

[2] Fourth, we should coordinate with other U.S. government agencies, particularly NOAA, NSF, Agriculture, Energy, and Interior, on their ongoing programs in relation to the NASA program and discuss possible collaborations. [Italics indicate a handwritten note by Hans Mark] *Yes.*

Fifth, we should contact and work with other governments and international research organizations we know to be interested in the subject to explore the possibility of parallel research programs or joint programs. [Italics indicate a handwritten note by Hans Mark] *Yes.*

Sixth, and last, we should approach an international body we wish to act as a focal point for the international aspects of the Global Habitability effort. This organization will have several roles; including:

1. sponsoring of international meetings and conferences on Global Habitability,
2. serving as a "clearinghouse" for information on plans and progress of individual participants, and
3. creating and maintaining an archive center for results of the foreign and international programs contributing to the Global Habitability effort.

This body should not itself conduct or direct any research activity nor set goals or standards for national programs. Members of the body might, however, review and analyze the combined results of various contributing programs and report results on these analyses which, in turn, might become contributions to the global program. The

International Council of Scientific Unions (ICSU) seems to be the prime candidate to act as this focal point. [Italics indicate a handwritten note by Hans Mark] *Right.*

We will need schedules for all the activities mentioned above. We should start the first four steps at once; in fact, some are already in motion. As soon as the NASA program is reasonably clear and coordinated with other U.S. agencies, we should contact other countries (step 5) to coordinate with their programs, and select the international organization to serve as a focal point (step 6). We plan to have an international Global Habitability workshop in late spring if possible; if not, early fall (1983).

I propose to act as chairman of a steering committee to formulate the program and move it along through the six steps above. Members of the steering committee will initially consist of NASA representatives. [Italics indicate a handwritten note by Hans Mark] *I agree.*

[3]   When we hit step 4, we will add representatives of other program agencies. Shelby Tilford will be the NASA program manager for Global Habitability. He will appoint and chair a NASA program committee which will be responsible for coordinating all work in OSSA divisions (particularly in EE, EL, EI, and EB) and centers related to Global Habitability, and will be responsible for presenting and reporting all NASA work. Ken Pedersen will coordinate the international aspects of the Global Habitability program.

[signed: "B.I. Edelson"]
B. I. Edelson

cc:
A/Mr. Beggs
AD/Dr. Mark

Distribution:
L/Mr. Allnutt
LI-15/Mr. Pedersen
P/Dr. McDonald
E/Mr. Keller
   Dr. Rosendhal
   Dr. Raney
   Mr. Thome
   Dr. McConnell
EE/Dr. Tilford
EL/Mr. Moore
EB/Dr. Soffen
EI/Mr. Villasenor
EP/Mr. Stanford
ARC/Mr. Syvertson
   Mr. Guastaferro
GSFC/Dr. Hinners
   Dr. Meredith

[Italics indicate a handwritten note at the end of the document]

Nov. 13, 1982

Dear Burt:

   I think that what you are proposing here is very important. We must get some momentum behind this business and I agree with the steps you are proposing.

Best Regards,

Hans.

P.S. Please see my notes.

JPL/Dr. Allen
    Dr. Hibbs
JSC/Mr. Griffin
    Mr. Rice
LaRC/Dr. Hearth
      Mr. Holloway
LeRC/Mr. Stofan
NSTL-ERL/Mr. Hlass
        Mr. Mooneyhan
MSFC/Dr. Lucas
    Dr. Dessler

Document IV-6

Document Title: B. I. Edelson, Associate Administrator for Space Science and Applications, NASA, "Global Habitability," 24 June 1983.

Source: NASA Historical Reference Collection, NASA History Office, NASA Headquarters, Washington, D.C.

This memorandum outlines NASA's activities to support the global habitability concept following the Unispace '82 meeting. Among other things, it notes the creation of the Earth Science and Applications Division within the Office of Space Science and Applications, directed by Dr. Shelby Tilford. It also references the National Academy of Sciences study of the International Geosphere-Biosphere Program, led by Dr. Herbert Friedman. (See Document IV-7.)

[no page number]
[stamped: "Jun 24 1983"]

National Aeronautics and
Space Administration

Washington, D. C.
20546

Reply to Attn of: EE-8

TO: Distribution
FROM: E/Associate Administrator for
          Space Science and Applications
SUBJECT: Global Habitability

    In November 1982, I outlined activities to be undertaken to turn the "Global Habitability concept," which was announced at the UNISPACE Conference in Vienna last

August, into a "Global Habitability program." These outlined activities consisted of six steps—internal NASA agreement on program definition, goals, and objectives; preparation of background material; formulation of the NASA program; coordination with other U.S. government agencies; coordination with other governments and international organizations; and identification of a scientific body to serve as an international focal point. This letter is to report on the progress we've made in each of these six areas since November.

The statement of program definition, goals, and objectives which was identified as NASA's first task, has been revised in light of NASA and external recommendations to read as follows:

Global Habitability is a proposed research program to investigate long-term physical, chemical, and biological trends and changes in the earth's environment, including its atmosphere, land masses, and oceans. The program will specifically investigate the effects of natural and human activities on the earth's environment by measuring and modeling important physical, chemical, and biological processes, and their interactions. The program will involve the acquisition and analysis of space and suborbital observations, land- and sea-based measurements, modeling and laboratory research, and supporting data management technologies over a ten-year or longer period of time. The program will of necessity involve major international participation and will be coordinated, insofar as possible, with national research programs and those of international agencies.

We have made significant progress in our efforts to develop detailed background material to support the Global Habitability program. Four science working groups have been formed to address specific program elements:

- A Global Biology Science Working Group, under the Chairmanship of Dr. Mitchell Rambler of NASA Headquarters, has prepared and [2] published a report entitled "Global Biology Research Program" in January 1983. This report addresses the scientific issues and technical approaches to characterizing biological processes on a global scale.
- A Land Global Habitability Science Working Group, under the Chairmanship of Dr. Sylvan Wittwer of Michigan State University, has drafted a report entitled "Land-related Global Habitability Sciences Research Program." This report deals with the contribution of the land areas of the earth to global processes.
- Dr. Michael McElroy of Harvard University is chairing two working groups which will produce documents relating to physical and chemical aspects of Global Biogeochemical Processes.

These four documents will contribute significantly to providing the scientific foundation for the Global Habitability program. The "Global Biology Research Program" document has been distributed. The remaining documents will be distributed as they are finalized.

The third step outlined in my November 1982 memo was to formulate and document the NASA program. Formulation of the program is now well underway at Headquarters as well as participating field centers. On April 18, 1983, I directed the organizational relocation of OSSA's solid earth research programs to the Environmental Observations Division. I also directed that the Division name be changed to the Earth Science and Applications Division, to better reflect the future focus of Division activities. The creation of the Earth Science and Applications Division, under the direction of Dr. Shelby Tilford, will provide improved coordination of land, oceans, and atmospheric research and will facilitate interdisciplinary studies critical to the Global Habitability program.

Two important Earth Science and Applications Division flight projects, which will support the Global Habitability program, are candidates for Fiscal Year 1985 new starts. The Upper Atmospheric Research Satellite (UARS), which we propose to launch in 1989, will carry sensors for investigating the chemical composition, energetics, and dynamics of the stratosphere and mesosphere. The Ocean Topography Experiment (TOPEX), which would also be launched in 1989, will provide highly precise altimeter data which will contribute to improved understanding of ocean circulation. In addition to these flight projects, augmentations will be requested for Global Habitability-related research and analysis activities in the fiscal year 1985 budgets of both the Earth Science and Applications Division and Life Sciences Division of OSSA.

[3] NASA scientists at Goddard Space Flight Center, Jet Propulsion Laboratory, Ames Research Center, Johnson Space Center, Langley Research Center, and National Space Technology Laboratories continue to participate in working groups and planning committees for Global Habitability-related activities. These centers are currently in the process of redirecting appropriate Earth Science research activities to focus on global processes.

In our efforts to begin domestic coordination of the program, we have met with representatives of other U.S. Government agencies to acquaint them with the goals of the Global Habitability program and to encourage their participation. To date, we have met with Dr. Anthony J. Calio, Deputy Administrator of NOAA; Dr. Dallas Peck, Director, U.S.G.S., Department of Interior; Dr. Orville Bentley, Assistant Secretary for Science and Education, Department of Agriculture; and Dr. Frank Johnson, Associate Director, National Science Foundation. We will continue these discussions as we progress with internal formulation of the NASA program.

Discussions with international organizations and representatives of foreign governments, the fifth step in my November memo, have been limited to discussions of the general Global Habitability concept. Active interest in the concept has been expressed by representatives of the United Kingdom's Natural Environment Research Council and Science and Engineering Research Council, the European Space Agency, the Federal German Government, Saudi Arabian National Center for Science and Technology, World Meteorological Organization, United Nations Environment Programme, Food and Agriculture Organization, and the United Nations University. Discussions with foreign agencies will continue at the conceptual level until the Global Habitability program is formulated.

To implement the sixth and final proposed step, NASA has approached the National Academy of Science (NAS) to coordinate U.S. scientific efforts in Global Habitability, and to represent the U.S. science communities in the international arena through the International Council of Scientific Unions (ICSU). Meetings were held with Dr. Herbert

Friedman of NAS to discuss the government-wide formulation of the effort. During these meetings Dr. Friedman proposed a concept for a broadly-based International Geophysical Biophysical Program (IGBP), of which he proposed that Global Habitability be a part. Dr. Friedman anticipated that IGBP planning and implementation would evolve over approximately a five-year period. The National Research Council (NRC) of NAS will hold a workshop in late July 1983 to formulate and draft the framework of the IGBP. The workshop will encompass NASA's planned research in Global Habitability as well as other geophysical research areas, such as [4] geodynamics, solar-terrestrial interactions, tropospheric chemistry, oceans, and climate. The Workshop results will serve as a basis for possible NAS recommendation of a U.S. national research effort as well as a possible input to the ICSU General Assembly in September 1984.

While we support the concepts and objectives of the IGBP, and feel that such a program may ultimately provide needed scientific coordination for Global Habitability and other international scientific programs, we wish to get started on Global Habitability much sooner than we feel the IGBP will be started. We have, therefore, asked NAS to consider the scientific and societal merit of Global Habitability as an entity separate from the IGBP. We feel that this separate consideration will ensure that all elements of Global Habitability will receive the deserved level of attention from the scientific community. (I might add that we have already received some very favorable publicity from the scientific community on Global Habitability—most notably the attached essay by Lewis Thomas in the June issue of "Discover" magazine.) [omitted]

As you can see, we are progressing in each of the six areas outlined last November. Global Habitability is, however, an ambitious concept which will require continued careful planning and coordination. I intend to ensure that we build upon the progress we've made in the last six months, and will keep you periodically informed of Global Habitability developments.

[signed: "B.I. Edelson"]
B. I. Edelson

Enclosure

cc:
A/Mr. Beggs
AD/Dr. Mark

[5] Distribution:
F/Mr. Terrell
L/Mr. Templeton
LI-15/Mr. Pedersen
P/Dr. McDonald
E/Mr. Keller
　　Dr. Rosendhal
　　Dr. Raney
　　Dr. McConnell

EE/Dr. Tilford
    Mr. Arnold
    Dr. Hudson
    Ms. Phillips
    Mr. Tuyahov
    Mr. Weber
EL/Dr. Briggs
    Ms. Paige
EB/Dr. Soffen
    Dr. Rambler
    Ms. LeSane
EI/Mr. McCoy
EP/Mr. Konkel
ARC/Mr. Syvertson
    Mr. Guastaferro
    Dr. Klein
    Mr. Colin
GSFC/Dr. Hinners
    Dr. Meredith
JPL/Dr. Allen
    Dr. Hibbs
JSC/Mr. Griffin
    Mr. Rice
    Dr. Erickson
LaRC/Dr. Hearth
    Mr. Holloway
LeRC/Mr. Stofan
NSTL/Mr. Hlass
    Mr. Mooneyhan
MSFC/Dr. Lucas
    Mr. Snoddy

Document IV-7

Document Title: Herbert Friedman, Chairman, Commission on Physical Sciences, Mathematics, and Resources, National Research Council, "Toward an International Geosphere-Biosphere Program: A Study of Global Change," July 1983.

Source: NASA Historical Reference Collection, NASA History Office, NASA Headquarters, Washington, D.C.

This document presents the results of a workshop to review the proposed International Geosphere-Biosphere Program. Herbert Friedman of the Naval Research Laboratory was one of the pioneers of the U.S. space program.

[title page]

Toward an International Geosphere-Biosphere Program

A Study of Global Change

Report of a
National Research Council Workshop
Woods Hole, Massachusetts
July 25-29, 1983

Commission on Physical Sciences,
Mathematics, and Resources
National Research Council

[iii]

INTERNATIONAL GEOSPHERE-BIOSPHERE PROGRAM WORKSHOP
Woods Hole, Massachusetts
July 25-29, 1983

Steering Committee
Herbert Friedman, Chairman, Commission on Physical Sciences, Mathematics, and
    Resources, National Research Council, (Study Chairman)
David Johnson, University Corporation for Atmospheric Research
Francis S. Johnson, The University of Texas at Dallas
Thomas F. Malone, Resources for the Future

Participants
Daniel B. Botkin, University of California
Bernard Burke, Massachusetts Institute of Technology
Charles L. Drake, Dartmouth College
Thomas M. Donahue, University of Michigan
John A. Eddy, National Center for Atmospheric Research
James E. Hansen, Goddard Institute for Space Studies
Devrie S. Intriligator, Carmel Research Center
Mark Langseth, Lamont-Doherty Geological Observatory
James McCarthy, Harvard University
Michael B. McElroy, Harvard University
Berrien Moore III, University of New Hampshire
V. Rama Murthy, University of Minnesota
Andrew F. Nagy, University of Michigan
Jack E. Oliver, Cornell University
Ronald G. Prinn, Massachusetts Institute of Technology
David Reichle, Oak Ridge National Laboratory
Roer Revelle, University of California at San Diego

Juan G. Roederer, University of Alaska
Joseph Smagorinsky, Princeton, New Jersey
Verner E. Suomi, University of Wisconsin
[iv]George Woodwell, Marine Biological Laboratory, Woods Hole
Peter Wyllie, California Institute of Technology

Government Participants
Ralph M. de Vries, Office of Science and Technology Policy
Burton I. Edelson, National Aeronautics and Space
    Administration
Bruce Hanshaw, U.3. Geological Survey
Ned A. Ostenso, National Oceanic and Atmospheric
    Administration
Dennis Peacock, National Science Foundation
Mitchell Rambler, National Aeronautics and Space
    Administration
John Schlee, U.S. Geological Survey
Shelby Tilford, National Aeronautics and Space
    Administration
Robert Watson, National Aeronautics and Space
    Administration

National Research Council
Frank Press, President, National Academy of Sciences
Robert M. White, President, National Academy of
    Engineering
Jesse Ausubel, Board on Atmospheric Sciences and Climate
William Easterling, NRC Fellow
Pembroke J. Hart, Geophysics Research Forum
W. Timothy Hushen, Polar Research Board
Raphael G. Kasper, Commission on Physical Sciences,
    Mathematics, and Resources
Nancy Maynard, Board on Ocean Sciences and Policy
John S. Perry, Board on Atmospheric Sciences and Climate
Martha Treichel, Office of International Affairs
Thomas M. Usselman, Committee on Solar Terrrestrial [sic]
Research
[vii]

PREFACE

    The real connections that link the geosphere and biosphere to each other are subtle, complex, and often synergistic; their study transcends the bounds of specialized, scientific disciplines and the scope of limited, national scientific endeavors. For these reasons progress in fundamental areas of ocean-atmosphere interactions, biogeochemical cycles, and solar-terrestrial relationships has come far more slowly than in specialized fields, in spite of the obvious practical importance of such studies. If, however, we could launch a cooperative interdisciplinary program in the earth sciences, on an international scale, we might hope to take a major step toward revealing the physical, chemical, and biological workings of the Sun-Earth system and the mysteries of the origins and survival of life in the biosphere. The concept of an International Geosphere-Biosphere Program (IGBP), as outlined in this report, calls for this sort of bold, "holistic" venture in organized research – the study of whole systems of interdisciplinary science in an effort to understand global changes in the terrestrial environment and its living systems.

The National Research Council IGBP Workshop at Woods Hole in July 1983 considered the major problems for research in five areas that might naturally be coordinated in such a program: the atmosphere, oceans, lithosphere, biosphere, and solar-terrestrial system. A unifying and pervasive theme of the Workshop was global change, on all time scales, from the slow recurrence of the Ice Ages to the shortest transient phenomena. Of pressing importance is the need to understand the often deleterious effects of modern man on natural processes, such as the inevitable climatic impact of carbon dioxide loading of the atmosphere since the industrial revolution. Progress in [viii] understanding global change will require extensive and well-organized observations made over much of the Earth and over a long period of time. The scope of such an effort requires international cooperation and interdisciplinary emphasis.

If we believed – as once in fact we did – that ocean circulation was slow and unchanging, we could hope to establish its fundamental patterns from the collection of random oceanographic observations from all times past. We could mix data taken by Captain Cook on the Endeavour with soundings from the most recent cruise of the Columbus Iselin. But the ocean is constantly changing, and the element of time is all important. Synoptic observations or time-oriented measurements in conjunction with time-dependent models are today indispensable. In place of a Challenger on a lone voyage of exploration we now need a fleet of research vessels taking coordinated observations, accompanied by downlooking spacecraft that can survey entire ocean basins with high spatial resolution on a synoptic basis. All of these must be supported by sophisticated modeling efforts.

If we believed that the Earth was a constant system in which the atmosphere, biosphere, oceans, and lithosphere were unconnected parts, then the traditional scientific fields that study these areas could all proceed at their own pace treating each other's findings as fixed boundary conditions. However, not only is the Earth changing even as we seek to understand it – in ways that involve the interplay of land and sea, of oceans, air, and biosphere – but we cannot even presume that global change will be uniform in space and steady in time. World climate and the acknowledged complexity of climatic change are cases in point. Needed to resolve this complex of change and interplay are coordinated efforts between adjacent scientific disciplines and programs of synoptic observations focused on common, interrelated problems that affect the Earth as a whole. Through this kind of international and interdisciplinary collaboration, we can hope to identify the mechanisms and causes of global change. At the same time we can illuminate areas of particular need and emphasis in each of the related disciplines.

A major challenge to an IGBP will be that of understanding the causes and effects of climate change. Variations in the Earth's climate appear to follow from a long and convolved set of interactions including human and other biological activity, solar radiation, volcanism, [ix] ocean circulation, polar ice and land effects, and the chemistry and dynamics of the atmosphere itself. Also involved on long time scales are the varying gravitational perturbations of the planets and the Moon on the orbit of the Earth and perhaps the varying density of interstellar matter encountered by the Sun in recurrent passages through the spiral arms of the Galaxy. No single factor is clearly dominant. In most cases we must presume, moreover, that any detectable change is the result ultimately of the interplay, feedback, and possible amplification of many causative factors.

The recovery and interpretation of a wide variety of historical, paleontological, and geochemical records provide most of what is known of the history of climate and what insights we have as to the responsivity of the terrestrial environment to internal and external forcing. Tree-ring data, lake deposits and pollen samples, and ocean and polar ice cores are natural diaries that contain information on past meteorological conditions, sea temperatures and ice cover, atmospheric composition, biotic conditions, crustal magnetism, and solar behavior. Such records know no national boundaries, nor is their interpretation the exclusive province of a specific scientific discipline. The critical reading of these data and the ability to relate what is told of global change and interaction are necessary steps toward reliable prediction of future trends in climate and biological response.

Tropospheric chemistry – a field of study that hardly existed a dozen years ago – now looms as ever more important in fixing the conditions for life on Earth. Local atmospheric pollution and the generation of global haze are obvious factors in the habitability of the planet. In addition, the troposphere is both the ultimate source and the ultimate sink for the trace chemistry of the stratosphere, which in turn modulates the amount and spectral distribution of sunlight received at the surface of the Earth. To understand the stratosphere we must know the chemical as well as the dynamical behavior of the global troposphere. Stratospheric ozone is subject to tropospheric chemistry and sensitive to anthropogenic modification. The best-known and most throughly [sic] studied green house gas is carbon dioxide, whose role as a perturber of future surface temperature is a subject of keen international concern. But it is now clear that gaseous leaks from old refrigerators and even the bacteria in the gut of cattle are insidious rivals to carbon dioxide as contributors to global temperature [x] change. Chlorofluoromethanes, nitrous oxide, and methane appear to contribute 70 percent as much to green house warming as does carbon dioxide. Methane in the atmosphere is increasing at more than 1.5 percent (60 x 106 metric tons) per year, and we are completely at a loss to explain how it is conserved.

The power of new technologies for remote sensing of atmospheric, geological, biological, and oceanographic conditions promises to revolutionize our grasp of global conditions and our understanding of global change. In the past, for example, our knowledge of the sea floor was gained from shipbaord [sic] by reflecting sonic waves off the ocean bottom. Along the heavily trafficked sealanes in the North Atlantic, the soundings obtained by many ships combined to provide a reasonably well-filled map. Far less was known of larger areas of the oceans of the world, where commerce was less frequent. SEASAT, following its deployment in 1978, demonstrated the power of global seafloor mapping from the vantage of Earth orbit. Its radar altimeter portrayed the large features of sea mounts and ocean trenches in an immediately obvious way. Surprisingly, however, when noise corrections were made to the same data, an unexpected range of seafloor features of much finer detail was clearly revealed. New features, never before mapped, include underwater volcanoes, tectonic fracture zones of crustal plates, swales of ocean crust with unusual gravitational characteristics, and a smooth underwater plateau as large as the state of California. Fracture zones mapped from the perspective of space can be followed across entire ocean basins, even where they are covered by sediment.

Advantages equally great have come in the past 20 years from orbital sensing of the Earth's biota. Much of the motivation has been to develop means of assessing conditions for the world production of food, fiber, and fuel from renewable biological resources. The focus has been on agricultural crops, forests, and rangeland of economic importance, and the principal tool has been infrared mapping. In recent years, microwave techniques have come to the fore, offering added capability in sensing through overcast and penetrating more deeply into canopies of vegetation.

It is hard to exaggerate the impact of orbital platforms and new technology in the field of solar-terrestrial physics. The ability to make in situ measurements in the vast domain between the middle atmosphere and interplanetary space has revealed the complex topography of near [xi] Earth space and a tangled web of connections that link the lower atmosphere to the Sun. Among the early findings from spacecraft was the unexpected scale and structure of the extended magnetic field of the Earth and the existence of an ordered but inconstant solar wind of charged particles that fills the Sun-Earth space. Orbital measurements of the meteorologically important fluxes of solar radiation are at last providing quantitative data on the perennial question of the influence of solar variations on weather and climate. The cavity radiometer on the Solar Maximum Mission spacecraft, launched in 1980, has made continuous measurements of the total radiative flux from the Sun with a temporal resolution of less than a minute and a short-term accuracy of about 10-5–about 4 orders of magnitude better than what could be done from the ground. Quickly found were depletions of about 0.1 percent lasting for weeks, that result from the masking of the solar disk by sunspots. Future measurements should establish whether a similar variation accompanies the 11-year sunspot cycle.

Ultraviolet imaging from space has literally shown us the Earth in a new light. The aurora now stands revealed in detail that escaped the previous 100 years of observation. Reel-down instruments suspended from balloons provide direct profiles of the chemical composition of the middle atmosphere from tropopause to stratopause; tethered instruments from future satellites promise similar capabilities for characteristics of the lower thermosphere.

These comments sample the content of the discussions that took place at the IGBP Workshop. From the report itself it will be clear that rich opportunities are in store for the earth sciences in the future and that the world scientific community should prepare to meet the challenges boldly.

Herbert Friedman, Chairman
International Geosphere-Biosphere
Program Workshop

[xiii]

CONTENTS

[1]   SUMMARY

The workshop reviewed the major problems for research in the atmosphere, oceans, lithosphere, biosphere, and solar-terrestrial relationships. A unifying theme is global change. Beyond the intellectual drive to understand basic scientific interrelationships is the practical need to gain a better grasp on how to manage the environment and global life-support systems.

A majority conclusion was reached that an International Geosphere-Biosphere Program (IGBP) under the auspices of the International Council of Scientific Unions (ICSU) could provide an effective vehicle for coordinating global measurements from space platforms and ground-based observational networks, exploiting new technologies for observations, implementing improved capabilities for data management, and placing proper emphasis on mathematical modeling with advanced computational facilities.

Unlike the comparatively short-lived International Geophysical Year (IGY), an IGBP must be designed as a long-range, interdisciplinary program. Effective planning over the coming several years could inaugurate a program that would begin to take shape toward the end of this decade and reach maturity in the 1990s.

The Workshop participants urged that such planning begin immediately within national organizations and in all the interested adhering bodies of ICSU. An important target date for a first assessment of plans brought forward from all sources would be the ICSU General Assembly in September 1984.

In implementing an IGBP, careful attention must be paid to all relevant national and international programs already conceived and in various stages of progress. Proper planning should guarantee that these programs are [2] carried forward effectively. The success and timeliness of an IGBP is in large part predicated on gains in understanding as well as a desire for greater interaction between neighboring disciplines starting to emerge from these programs.

INTRODUCTION

[3]   A generation ago researchers concerned with the basic sciences of the Earth joined together in a unifying effort, the International Geophysical Year (IGY). The IGY was an outstanding success, both scientifically and politically. During 1957-1958, coordinated geophysical observations were gathered from pole to pole, and these observations and the research and monitoring that followed have significantly improved our understanding of the physics of the Earth. Thousands of scientists, from 67 nations, shared knowledge, purpose, and tools and created a spirit of cooperation and good will that transcends political differences.

Studies of ecological and biological processes were essentially neglected during the IGY, as were many questions requiring multidisciplinary approaches for their solution. These questions have become increasingly prominent since the IGY. The contemporary concern for the environment has heightened awareness of the need for increased scientific understanding of biogeochemical cycles and of the links between geophysical and biospheric processes. The desire to explore possible future conditions for life on Earth and the role of human activities in shaping those conditions lend urgency to developing the basic scientific knowledge of the geosphere and biosphere. It seems clear that this knowledge will only come from a comprehensive understanding of the historical evolution of the Earth and the dynamics of global change.

Since the IGY, there have been several major international programs, such as the Global Atmospheric Research Program and the International Magnetospheric Study, that included coordinated observations on a global scale. These programs have contributed significantly to our knowledge of geophysics and, in the case of meteorology, [4] our ability to forecast weather. However, the observational programs have generally focused on narrowly defined problem areas rather than on the broad scope of linkages among components of the geosphere and biosphere. Often the programs have been too brief in duration to establish linkages between slowly varying cyclic processes.

Today we find a growing number of current and planned programs, international in scope, that have interdisciplinary breadth. These programs and their national and multinational components are vital to any more broadly based future geosphere-biosphere program. For example, the Scientific Committee on Problems of the Environment (SCOPE) of ICSU has catalyzed pioneering work over the last decade with its program on the biogeochemical cycles of carbon, nitrogen, sulfur, and phosphorus. The theme of biogeochemistry has now been endorsed by a group of U.S. scientists in plans for a NASA program on "Global Habitability." Examples of other current programs close in spirit to an IGBP are the International Lithosphere Program (ILP), the World Climate Research Program (WCRP), and the Study of Travelling Interplanetary

Phenomena (STIP). Each of these, of course, has many important components. For example, the WCRP embraces the planned World Ocean Circulation Experiment (WOCE), whose objective is to describe and understand quantitatively for the first time the large-scale circulations of the world's oceans and their influence on the atmosphere. This objective should be attainable in large part as a result of new remote sensing techniques. WOCE is dependent upon major satellite missions, such as the Topography of the Ocean Experiment (TOPEX) and the Geopotential Radar Mapper (GRM) of the United States and the Earth Resources Satellite (ERS-1) of Europe, as well as surface and subsurface observing platforms. These observational programs are well conceived now and should be carried forward with a sense of urgency.

To understand many geosphere-biosphere phenomena and their interactions, monitoring over long periods will be necessary. For example, to detect and separate anthropogenically caused climate changes from natural ones and to understand the mechanisms involved may require a decade or more of accurate global monitoring of atmospheric and sea temperatures, the amount and distribution of radiatively active gases, vulcanism and aerosols, solar and terrestrial radiation fluxes, and ice masses and sea level. Thus, it becomes important to continue [5] and strengthen existing global observing and monitoring programs (and their national components), such as the programs World Weather Watch (WWW), and perhaps re-evaluate others, like the International Global Ocean Services System (IGOSS) and the Global Environmental Monitoring System (GEMS), in view of the scientific goals of an IGBP.

These and other programs that may relate in conceptual or practical ways to an IGBP are summarized briefly in Appendix E.
[6]

THE POWER OF NEW TECHNOLOGIES

The IGY occurred at the dawn of the Space Age. The 25 years since the IGY have seen the development of remote sensing techniques and space platforms, which have opened up major avenues to investigate important problems in the geosphere-biosphere. Significant strides have been made in describing solar-terrestrial relations and studying the atmosphere through the use of space technology. The new power of this technology in studying the lithosphere, oceans, and the biosphere has been demonstrated, for example, in mapping the oceanfloor and terrestrial ecosystems. New programs are planned to take advantage of these techniques in all the discipline areas; they should be supported preparatory to increased multidisciplinary activity in the 1990s.

Remote sensing is by no means limited to space platforms. Probing the solid Earth by seismic and electromagnetic waves or the oceans by acoustic tomography are important examples of the growing array of ground-based remote-sensing techniques that can unlock the mysteries of the planet Earth.

The potential of high-altitude balloons to explore directly the stratosphere has been enhanced by the demonstration of a "yo-yo" technique to lower and raise a payload repeatedly through a layer about 10 km thick beneath the balloon. A tethered satellite is proposed to facilitate repeated direct measurement of the ionosphere and mesosphere.

Improvements are being made in the quality and reliability of many in situ sensors, platforms for carrying them (e.g., buoys, long-duration balloons), and the use of inexpensive satellite links for data collection and platform location. The cost of some of these systems is [7] within reach of many countries that may wish to be involved in international cooperative research.

The development of communication and computer technologies also opens up new vistas of research not possible 25 years ago. The handling of complex models is one example. Perhaps of equal importance is the opportunity that these technologies present for collecting, processing, archiving, accessing, and exchanging data. Such information will be needed by the many discipline groups internationally whose intellectual efforts must be enlisted if answers are to be obtained to many of the challenging questions raised at the Workshop.

[8]

GOALS AND FINDINGS OF THE WORKSHOP

Invitees to the Workshop were asked to address several specific issues and problems:

1. Need and criteria for an IGBP. Are there significant benefits to be had from increased integration and coordination? Which activities for which reasons (geographical access, satellite coverage, and cost, for example) truly require international cooperation to be successful?

2. Identification of areas in which large measurement programs can serve multiple scientific needs. For example, monitoring of a variety of chemical and radiative processes in the upper atmosphere is essential to understanding both the modification of the ozone layer and the greenhouse effect. Can more efficient use of resources to fulfill global and temporal requirements be made through an IGBP-type program?

3. Identification of critical interdisciplinary problems, interfaces, and linkages. For example, climate variation is related to solar-terrestrial interactions, vulcanism, ocean circulation, and forest and ecosystem processes and is recorded by sedimentary processes.

4. Potential roles of U.S. agencies, for example, the National Oceanic and Atmospheric Administration (NOAA), the National Aeronautics and Space Administration (NASA), the National Science Foundation (NSF), and the Department of Defense in ocean, atmosphere, and solar-terrestrial programs; the U.S. Geological Survey, NSF, and NASA for land surface and lithosphere; the Department of Energy, the Department of Agriculture, NOAA, NASA, and others for biogeochemical cycles.

5. Relationships with existing international programs and projects, such as the Middle Atmosphere Program, the [9] World Climate Research Program, the World Ocean Circulation Experiment, the International Lithosphere Program, and others.

6. Review of relevant papers and plans in the area, including the Space Science Board study Toward a Science of the Biosphere, the NASA Global Habitability plans, the CSTR report on Solar-Terrestrial Research for the 1980's, documents of the World Climate Research Program, including plans for the World Ocean Circulation Experiment, the report of the National Research Council's (NRC's) Carbon Dioxide Assessment Committee, Changing Climate, the NRC's global tropospheric chemistry report, and reports of the Middle Atmosphere Program.

7. How should design of the IGBP be broadened, nationally and internationally, should the concept evoke enthusiastic support at the Workshop?

The participants (Appendix F) represented a broad cross section of the natural-sciences community in the United States. The breadth of the group led to a realization that the study of global change relevant to human welfare must encompass wider time scales and greater areas of investigation than had hitherto been proposed. In particular, it became evident that much information relevant to the future can be gleaned from study of the past. These considerations suggest that investigation of global changes in the geosphere and biosphere should encompass pertinent studies of solar-terrestrial relations, lithospheric processes and their interaction with the atmosphere and oceans, and the history of life as evidenced in the fossil record, as well as the more obvious core of dynamic and radiative interactions among the atmosphere, ocean, and biosphere.

At the same time, questions were raised about the underlying conceptual unity and coherence, as well as practical necessity, of an IGBP. While the overall finding of the NRC group was strongly supportive, the concept remains only broadly defined. The major conclusion of the Workshop was the importance of the topic of global change, and the major recommendation was for further development of programmatic concepts on the national and international levels. The issues and problems listed above must be addressed in more detail by various potential participating individuals and groups to arrive at a consensus both nationally and internationally.

An IGBP must cover the time span of major natural cycles, such as the 11- and 22-year solar magnetic cycles. The characteristic duration of some important processes in the biosphere is even longer. It would be most valuable to design an international program with such intrinsic stability that monitoring parameters of global change could be carried on with necessary fidelity for decades. ICSU could provide the continuity of administrative structure to coordinate an IGBP over the long range.

A report of this Workshop was presented as part of a colloquium on global change organized by ICSU on August 3, 1983, coincident to a meeting of its General Committee in Warsaw. The report was favorably received, and the General Committee recommended global change as a theme for one of the scientific symposia being arranged on the occasion of the ICSU General Assembly in Ottawa, Canada, September 1984. This decision constitutes an invitation to all constituencies of ICSU to develop detailed inputs to a comprehensive discussion meeting at Ottawa. If that discussion leads to ICSU endorsement of a program, several years of intensive planning will need to follow in order to initiate a well-coordinated scientific effort.

In implementing an IGBP, careful attention must be paid to all relevant international programs already conceived and in various stages of progress. Proper planning should guarantee that these programs are carried forward effectively and in due course meshed with moves to accomplish the major goals of an IGBP. With adequate support and proper planning the IGBP could reach full maturity in the 1990s.

Several other desirable characteristics for an IGBP should be mentioned. One is that it should enlist young scientists and help them plan careers around the many exciting problems described here. Another is that universities should be involved. Universities are where much of the most advanced research is performed, and their involvement has implications for the scale of problems to be studied and rates at which results might be obtained. Third, the program should be worldwide, involving both developed and developing countries. Finally, there should be easy, wide access to data and individual freedom of choice for researchers on problems in which to involve themselves.

[11]

THE SCIENTIFIC CHALLENGE

Unraveling the mysteries of the past evolution of the Sun and Earth and predicting the future evolution of this planet and its life forms depend on continued strength to tackle the intellectually challenging problems within each of the disciplines studying the geosphere and biosphere. However, Workshop participants emphasized those scientific problems and opportunities that are multidisciplinary and global in scope, often requiring international cooperation. The scientific challenges were considered from the vantage point of four broad research domains: solar-terrestrial, oceans and atmosphere, the lithosphere, and the biosphere. The following sections summarize problems and opportunities in each area. Appendixes A-D report on the areas in greater detail.

SOLAR-TERRESTRIAL SYSTEMS (see APPENDIX A)

Solar-terrestrial research is concerned with the interactions by which diverse forms of energy generated by the Sun influence the terrestrial environment and with the resulting complex interplay of physical and chemical processes in every element of the system comprising Sun and Earth (Figure 1). The Sun is a variable star. Transient outbursts, such as solar flares, and cyclic changes in solar magnetism, such as the 11- and 22-year variability of sunspots, remain to be understood. The sine and figure of the Sun oscillate in a variety of modes. Understanding the processes causing these variations and the accompanying changes in solar output as they may affect the Earth will be a major preoccupation of solar research for a long time to come.

[14] Magnetospheric perturbations induced by episodic variations in the properties of the solar wind are coupled electrodynamically to the ionosphere, to regions deep in the atmosphere, and to the lithosphere. The processes involved in interactions of the solar wind with the magnetosphere, storage and transfer between regions of the magnetosphere, and coupling to the ionosphere and atmosphere are still poorly understood. Since the atmospheric effects may influence climate change, an understanding of these phenomena may be important to the entire geosphere-biosphere system.

The ionosphere (Figure 2) is a magnetospheric energy and exerts important feedback on the regions of the magnetosphere to which it is linked. Changes in the high-latitude ionosphere alter the global circulation and temperature structure of the thermosphere. The detailed nature of these interactions is not well understood, nor is the question of whether they affect, in a substantial way, the lower atmosphere. Variations in solar short-wavelength radiation and particle emission affect the composition and structure of the middle atmosphere, leading in extreme cases to possibly serious changes in atmospheric ozone. Such effects need to be further analyzed.

Finally, a thorough understanding of the photochemistry of the stratosphere and mesosphere is yet to be attained. Such knowledge would contribute importantly to assessment of the effects of trace gases such as nitrous oxide (formed partly by atmospheric electricity) and fluorocarbons and of variations in solar input on the ozone content of the middle atmosphere. Because of the importance of the ozone ultraviolet screen to the biosphere, these couplings have profound significance.

Key foci in solar-terrestrial research in the next few years include Origins of Plasmas in the Earth's Neighborhood (OPEN, see Appendix E), a program to examine the global flow of energy through the Earth's space environment above the upper atmosphere, and the planned Upper Atmosphere Research Satellite (UARS, see Appendix E), which are designed to improve our understanding of the chemistry and dynamics of the middle atmosphere and the flow of energy through this region.
[15]

OCEANS AND ATMOSPHERE (see APPENDIX B)

Over the last quarter century, our knowledge and understanding of the behavior of the atmosphere and ocean on a global scale have advanced impressively. Operationally oriented observing programs and special research-oriented field programs, in which space technology played a crucial role, have provided new insights into the dynamics and chemistry of the Earth's fluid envelope. This knowledge has provided the basis for the development of numerical models employed in weather prediction and simulation of the dynamics of the oceans and the variability of climate. Similar advances have been made in the understanding of our nearest star and its influence on our planet.

Three critical global multidisciplinary problems are perceived as major foci of attention in ocean-atmosphere research:

1. Dynamic and thermodynamic interactions between components of the climate system (Figure 3). The interactions between oceans, atmosphere, land, biosphere, and cryosphere maintain in delicate balance our climate and the natural productivity vital to life. A scientific strategy for the World Climate Research Program has been developed to address key problems employing emerging technological capabilities for observing the atmosphere, the Earth's surface, and the oceans. An early focus of activity will be the study of interaction between the tropical oceans and the global atmosphere as exemplified by the El Nino/Southern Oscillation phenomenon. In the longer term, the slow circulation and evolution of the world ocean will be addressed through a World Ocean Circulation Experiment.

2. The interactive processes of global chemistry, emphasizing the biogeochemical cycles of substances important to the stable maintenance of life and global climate. Over the past decade, we have come to regard the global atmosphere, oceans, solid Earth, and biota – including man – as an integrated chemical system. It is clear that this system is at present undergoing significant changes in the global and regional cycles of many substances. Many of these changes are coupled to human activity. Major problems include the following:

* Processes regulating the distribution of stratospheric ozone;
* The atmospheric hydrological cycle;
[19] * The tropospheric distribution of oxidants;
* The chemistry of dry and wet deposition;
* The sources and sinks of radiatively important aerosols and gases;
* The ocean's role in biogeochemical cycles in terms of chemical, geological, and biological processes;
* Fluxes between ocean, atmosphere, lithosphere, and biosphere; and
* The role of land surface processes.

3. Sensitivity of the terrestrial system to solar variation. Solar variability on a broad range of time scales can have an impact on many terrestrial processes. Better understanding will require precise measurement of solar energy output and solar ultraviolet radiation. Information on past changes in insolation due to orbital changes should be exploited to understand the sensitivity of climate to solar variations.

Further advances in understanding the oceans and atmosphere will depend critically on the implementation of new technology for sustained data acquisition, processing, archiving, and dissemination to research users.

THE LITHOSPHERE (see APPENDIX C)

Development of the solid Earth has resulted from complex interactions of physical, chemical, and biological processes. The theory of plate tectonics has provided a framework into which we can fit diverse evidence from the solid-Earth sciences and a foundation upon which subsequent investigations of the evolution of the Earth are built.

The plate-tectonics model tells us that thermal convection must take place within the Earth, but we have not yet been able to characterize it. Related to this problem is the characterization of the chemical composition of mantle heterogeneities and their residence times. We do not yet have a clear picture of the various processes at plate boundaries nor quantitative estimates of subduction related recycling of materials and its effects on mantle evolution and basalt genesis (Figure 4). Associated with plate movements and processes are significant changes in paleogeography, paleooceanography, and paleoclimatology, the records of which remain to be fully reconstructed. We are beginning to understand geochemical cycling in the [21] mantle; for example, it is estimated that a volume of water equal to that of the entire ocean circulates through the oceanic crust every 8 million to 10 million years.

Much of what we infer about the early history of the Earth has come from comparative planetology. However, continental platforms and fold belts appear to be unique to the Earth, and except in the near surface, we know little about their structure and can only infer their origin and evolution. They hold the record of most of Earth's history, including that of life. The origin of life, the nature and rates of its evolution, and the nature of major changes and extinctions are among the principal challenges in the Earth sciences. An ocean and an atmosphere must have existed early in Earth history, but the degassing history of the Earth and the quantity of primordial components still being emitted remain to be determined. Activity around geothermal vents in the oceanfloor is offering a new perspective on possible origins of life.

The solid Earth, and especially the sedimentary stratum, on land and on the seafloor, holds the only record of past activity at or near the Earth's surface. This record includes chemical composition, climate variations, past organisms and their evolution, sea-level change, and biogeochemical information about life in the past. Much of this information is relevant to studies of cyclic phenomena and other variations of interest to researchers concerned with the biosphere, the hydrosphere, the atmosphere, and solar-terrestrial relationships. It appears possible that records of solar variability may be preserved in the sediments, that confirmation of hypothesized effects of short-term physical or biochemical events may be found, and that sources and sinks for chemical compounds can be established from the sedimentary record. Significant geological events have resulted from biological activity in the past, and, of course, our supplies of petroleum, coal, phosphate, and other resources are the result of interactions among the lithosphere, atmosphere, hydrosphere, and biosphere.

Many of the techniques and tools used for lithosphere studies are also utilized by or derived from other disciplines. Solar seismology uses methods of terrestrial seismologists. Marine biologists and physical oceanographers use acoustic tools developed by geophysicists, and seismologists are now developing tomographic techniques to study heterogeneity in the Earth's interior similar to those developed by physical oceanographers to [22] study internal variations in the oceans. The TOPEX mission, designed for physical oceanography, also promises important information for solid-Earth scientists, while the GRM mission of the geophysicists will provide information that will enhance the TOPEX data for oceanographers. It is obvious that the interchange of tools and techniques can enhance progress in all areas of investigation of the Earth.

THE BIOSPHERE (see APPENDIX D)

In thinking about global changes in the Earth, we must pay special attention to changes in living organisms – plants, animals, and bacteria – and their interaction with the oceans, the atmosphere, and the lithosphere. On a time scale of the next hundred years, the most important of these interactions will be those between the behavior of living organisms and climatic change.

Changes in the biosphere can bring about climatic change through the global cycles of the "greenhouse gases" – especially carbon dioxide, methane, and nitrous oxides through changes in the regional and planetary albedo and other characteristics of the land surface; and through the effects of land vegetation on evapotranspiration and storage of water and on river runoff. Figure 5 [not included] shows the energy and water balance of the biosphere.

The influence of climatic change on the distribution and abundance of species has long been recognized, but more data are needed to understand better the effects of climatic change on the mass of the biota and on biological productivity. Several examples of feedback between biospheric and climatic change can be described. A diminution in the size of the land biosphere, caused by a rise in atmospheric temperature, will result in an increase in atmospheric CO_2 and hence in further warming. Similarly, because high biological productivity in ocean waters extracts CO_2 from the air and sequesters it in the ocean depths, changes in the intensity of ocean biological production caused by climatic change can result in large changes in atmospheric CO_2, with corresponding climatic effects.

The need for a multidisciplinary approach to study these problems can be illustrated by two questions: (1) How has the concentration of atmospheric carbon dioxide varied during past geologic times, and how will it vary in the future? (2) What is the cause of the present rapid rise in atmospheric methane?

[24] Studies of oxygen isotopes in ocean sediments laid down during the Cretaceous period 60 million to 120 million years ago show that the temperatures of ocean waters at

high latitudes were about 15 degrees warmer than at present. Were these high temperatures brought about by high concentrations of CO_2 in the atmosphere, and, if so, did the high atmospheric CO_2 result from a more rapid vertical ocean circulation, which prevented the sequestering of large quantities of carbon dioxide in the ocean depths?

Measurements of CO_2 in air trapped in ice cores from Greenland and Antarctica show that 10,000 to 20.,000 years ago atmospheric CO_2 was much lower than at present, while 20,000 to 30,000 years ago CO_2 in the air fluctuated from approximately 175 to 300 ppm, apparently over relatively short times. Both of these conditions probably resulted from changes in oceanic biologic production, perhaps related, in the former case, to larger quantities of ocean nitrogen and phosphate, and, in the latter, to fluctuations in ocean biological production, which are poorly understood. The climatic effects of the relatively small variations in insolation resulting from the "Milankovitch" cycles in the Earth's orbital parameters may have been greatly amplified by fluctuations in the sources and sinks of phosphorous and nitrogen in the ocean, their consequences for ocean biological productivity, and, in turn, for atmospheric CO_2 levels. A major uncertainty about future atmospheric CO_2 concerns the question of whether forests will serve as a source or sink for carbon. CO_2 enhances photosynthetic production and reduces water stress on plants, but the significance of this effect in changing the size of the biosphere is conjectural.

Analyses of ice cores show that methane, one of the important greenhouse gases, was relatively constant during the 20,000 years before the seventeenth century and that it has doubled in concentration since that time. its present rate of increase is apparently 1-2 percent per year. Is this increase related to changes in the biological sources of methane (i.e., symbiotic microorganisms in ruminant animals and termites, rice paddies, marshes, and swamps) or in the atmospheric sink? The latter depends on the atmospheric concentration of OH ions. Careful measurements of OH and studies of the size and productivity of methane sources over the next few decades should answer this intriguing question.
[25]

THE NEED FOR AN INTEGRATED APPROACH

From the preceding, it appears clear that a multidisciplinary approach to many problems will be needed. An excellent example involving all four discipline groupings follows.

THE CO_2 QUESTION

It is reasonably well established that the observed continuing anthropogenic increase of atmospheric CO_2 will give rise to tropospheric warming and to shifts in global and regional climate over the next century. A quantitative assessment of the structure and degree of change requires an understanding of the dynamics of climate. The role that the oceans play is paramount in delaying and regionally modifying the climatic response to a given scenario of atmospheric CO_2 increase.

The diminishing reflectivity at the continental snow cover margin provides an important positive feedback, intensifying the high-latitude tropospheric warming.

However, polar sea ice shields the fast-acting atmosphere from the slow-acting, and relatively warmer, sea blow.

The inputs from the geological sciences enter in several ways. Effects of volcanic activity on atmospheric composition is one example – volcanic aerosols may partly counter or mask a greenhouse warming. A second important contribution from geological studies is climatic data for validating climate models in simulating extreme shifts of past climate. This also includes a knowledge of how other configurations of the continents may have influenced past climatic regimes.

Since the atmospheric response is primarily radiatively controlled, the balance of solar radiance and outgoing [26] infrared fluxes in determining the thermal structure of the atmosphere must be better understood. Recent results suggest that the climatic response to a doubling of CO_2, as measured by a surface temperature change, is qualitatively and quantitatively similar to that of a 2 percent increase in the solar constant.

Biogeochemical processes, of course, partially determine the partitioning of increased CO_2 production among the atmosphere, oceans, and biosphere. Determining the net atmospheric loading scenario thus involves an understanding of production, storage, transport, and transformations within and among the active terrestrial media. Changes in other greenhouse gases, such as water vapor, methane, nitrous oxide, and fluorocarbons, further complicate and increase the radiative response. Changes in their concentrations might obscure detection of an actual CO_2-induced signal, and secondary effects may be significant. A CO_2 increase would result in considerable stratospheric cooling, which, in turn, would decrease the rate of destruction of in situ O_3 and, therefore, lead to an increase in its concentration – which promotes still an additional alteration of the temperature and circulation of the stratosphere and affects the biosphere below.

Ultimately, the altered climate of the planet Earth, including the condition of the oceans, will have an impact on the biota – but how?

It is clear from this illustration that the prediction of future states of the regional and global climate resulting from anthropogenic CO_2 sources, and the resulting impact on humanity, will critically depend on a close integration of focused and coordinated research among the relevant disciplines.

A COMPREHENSIVE MONITORING SYSTEM

The interrelationships discussed in the Workshop indicate that many parameters in the geosphere-biosphere need to be monitored on a long-term basis and in a coordinated manner. Indeed, many of the same or similar observations, for example, on solar activity, vulcanism, and carbon fluxes, will be needed by more than one discipline group. In addition, simultaneous observation of physical, chemical, and biological processes, for example, in the oceans, could offer a rich new basis for scientific advance. Some of the needed observational techniques are now available [27] or have been demonstrated; others will be tested in programs under way or planned for the remainder of this decade. This provides the foundation for considering the design of an effective program of sustained observations beginning in the 1990s, which could achieve efficiencies and economies not possible with individual, uncoordinated activities.

COMMON OBSERVATION SYSTEMS

In either long-term monitoring or exploratory research of shorter duration, there are technological opportunities to use common observing platforms (e.g., satellites, buoys) or even sensors (e.g., multispectral imagers, solar and terrestrial radiation detectors, satelliteborne altimeters) to provide data required by several disciplines and studies. Coordination of the system and data specifications and the timing of observations may be feasible in many cases, thus providing increased scientific benefits from a fixed investment.

EXPLORING THE EARTH'S FUTURE FROM ITS PAST

A coordinated program of exploring the lithosphere and cryosphere not only would aid the solid-Earth sciences but would contribute to understanding climate change, biogeochemical cycles and the evolution of life on Earth, and the impact of solar variations. For example, analysis of lake and marine sediments can provide information on the history of climate and sea level, the biological record, and the geomagnetic field, which is also relevant to the past record of cosmic-ray flux.

OTHER COMMON FOCI

If we look at the totality of scientific questions and challenges discussed at the Workshop, a number of additional common thrusts emerge that tie together various disciplines:
  • The dynamics of change: what are the characteristic rates and time scales of various processes - biospheric, solar, atmospheric, oceanic, and geological-and how do they interact?
[29] • Biogeochemical cycles and global chemistry – the interaction with all media and their dynamics, the effects of solar variations, man's impact, the impact of the physical systems on the biological and vice-versa (Figure 6).
  • Development of major integrative models of the geosphere and biosphere – for example, global climate models and models of radiative balance that treat albedo (cryosphere, snow cover, biomass, etc.), volcanic aerosols, and greenhouse gases; global ecosystem models; carbon cycle models; and plate tectonics models.
These thrusts call for multidisciplinary approaches and emphasize the need for international cooperation.

Document IV-8

Document Title: "Terms of Reference of the Committee on Earth Observations Satellites (CEOS)," adopted 25 September 1984.

Source: NASA Historical Reference Collection, NASA History Office, NASA Headquarters, Washington, D.C.

Document IV-9

Document Title: CEOS, "Meeting Minutes," 24–25 September 1984.

Source: NASA Historical Reference Collection, NASA History Office, NASA Headquarters, Washington, D.C.

Document IV-10

Document Title: John H. McElroy, NOAA Assistant Administrator for Environmental Satellite, Data, and Information Services, "Memorandum to CEOS Conference Participants," 19 October 1984.

Source: NASA Historical Reference Collection, NASA History Office, NASA Headquarters, Washington, D.C.

By the late 1970s, with the proliferation of Earth-observing satellite systems, the space-faring nations had begun to recognize the importance of coordinating their satellite programs, not only to avoid unnecessary duplication, but also to capitalize on possible synergies afforded by closer cooperation on weather and climate observations. In September 1984, the United States used the opportunity as hosts of the Economic Summit of Industrialized Nations to propose the creation of the Committee on Earth Observations Satellites (CEOS), which was accepted by the delegates. These documents present the terms of reference for the new organization, the minutes of the meeting, and a transmission memo from Dr. John McElroy, indirectly referring to the rotating chairmanship of the organization by noting that ESA would chair the next meeting.

Document IV-8

[Cover Page]

TERMS OF REFERENCE
OF THE
COMMITTEE ON EARTH OBSERVATIONS SATELLITES

Adopted September 25, 1984

Washington, D.C.
[1]

TERMS OF REFERENCE
OF THE
COMMITTEE ON EARTH OBSERVATIONS SATELLITES

PREAMBLE

Remote sensing from space has evolved from an early period of limited applications satellite programs to a point where distinctions among existing missions result from the technology employed, rather than from the disciplines served in system operations. In the future, a number of international, national, and regional space-borne earth observations systems will operate simultaneously, and support both interdisciplinary and international applications.

The organization of international cooperation in space-borne earth observations systems also is evolving, from mission-specific reviews to the interdisciplinary coordination of multi-mission programs. Beginning with the first Multilateral Meeting on Remote Sensing – held in Ottawa on May 8-9, 1980, and attended by agency representatives from Canada, the European Space Agency, France, India, Japan, and the United States of America – current and potential operators of earth observations s systems have met several times to discuss the means by which mutually beneficial cooperation and coordination could be achieved in both the near and longer term. As a result of these gatherings, the recent past has seen the creation of the Coordination on Land Observation Satellites (CLOS) by agency representatives from France, Japan, and the United States of America in Paris on November 13-14, 1980; the initiation of the Coordination on Ocean Remote Sensing Satellites (CORSS) in Paris on May 10-11, 1982, through the efforts of agency representatives from the European Space Agency and Japan; and the second Multilateral Meeting on Remote Sensing, held in Paris on may 12-13, 1982, and attended by agency representatives from France, Canada, the European Space Agency, India, Japan, and the United States of America.

This framework of initial discussion and cooperation has enhanced the utility of space-borne earth observations data to users worldwide, has encouraged the coordination of program plans among space-borne earth observations system operators, and has fostered international receptivity to, and acceptance of space-borne earth observations system activities and applications.

Consequently, the assembled representatives of international, national, and regional space-borne earth observations systems:

AWARE of the overlap of space-borne earth observation mission objectives and of the interdisciplinary applications of remotely-sensed data;

RECOGNIZING the advantages of ongoing communication and cooperation among space-borne earth observations system operators; and

[2] DESIRING to promote the international growth and potential benefits of space-borne observations of the earth;

have affirmed the value of the activities described above, and have agreed to coordinate informally their current and planned systems for earth observations from space through the organization of a Committee on Earth Observations Satellites (CEOS).

The CEOS will not supersede current or potential agreements by members. Participation in the activities of the CEOS will not be construed as being binding upon space-borne earth observations system operators, or as restricting their right to develop and manage earth observations systems according to their needs.

MEMBERSHIP

International, national, or regional organizations responsible for a space-borne earth observations program currently operating or at least in Phase-B or equivalent of system development, will be eligible for membership in the CEOS. Members must have a continuing activity in space-borne earth observations with responsibility for the overate mission, intended to operate and provide data for some years. Other entities with sensors on board such missions will normally be represented by the entity responsible for the host mission. Initial members of CEOS are Canada Centre for Remote Sensing, Centre National d'Etudes Spatiales (France), European Space Agency, Indian Space Research Organization, Instituto de Pesquisas Espaciais (Brazil), Japan (to be confirmed), U.S. National Aeronautics and Space Administration, and U.S. National Oceanic and Atmospheric Administration. The addition of new members will be with the consensus of current members of the CEOS.

OBJECTIVES

The CEOS will seek to enhance the benefits of space-borne-earth observations for members and the international user community.

The CEOS will serve as a forum for the exchange of technical information to encourage complementarity and compatibility among space-borne earth observations systems that are currently in service or development. Improved complementarity and compatibility will be sought through cooperation in mission planning and the development of compatible data products, services, and applications.

[3]

COOPERATIVE ACTIVITIES

Cooperation in the development and management of remote sensing programs can be of benefit to operators of space-borne earth observations systems and to users of earth observations data. Redundancy among systems and the utility of data can be optimized through the appropriate coordination of complementary and compatible space and ground segments, data management practices and products, and earth observations systems research and development.

CEOS members will exchange technical information on and pursue the potential for coordination of space and ground segments. Such coordination could include discussions on current and future mission parameters, sensor capabilities and intercalibration, and data and telemetry downlink characteristics. In addition, earth observations systems coordination within CEOs could address issues of ground station technical compatibility for back-up satellite tracking, command and control, and sensor and telemetry data reception.

CEOS members will investigate the means for increasing data utility and cost-effectiveness, for both operators and users. CEOS activity could include the coordination of data acquisition, sampling and pre-processing methodologies; the standardization of data formats where appropriate; the increase in compatibility of data archives; and the enhancement of user access to CEOS member data bases, information products, and applications services. CEOS members will seek to assure that the user community is made aware of the satellite programs of members and encourage discussions between the users and the relevant satellite system operators, as necessary.

CEOS members will present their plans for emerging satellite remote sensing technologies and programs and will discuss appropriate approaches for the coordination of future systems. CEO members could address current developments and future directions and opportunities in earth observation from space, including free-flying spacecraft, mission-specific instruments flown on space transportation systems, and the placement of instruments on space platforms.

ORGANIZATION AND PROCEDURES

CEOS will convene at least once every two years in plenary session. The CEOS meeting will be organized and chaired by the host organization. The host organization will provide and distribute minutes of the meeting and will report on any follow-on activities at the next regular meeting. At each meeting of the CEOS, the time, place, and host for the next meeting will be established.

CEOS also may establish, as mutually agreed, and on an ad hoc basis, special temporary Working Groups to investigate specific areas of interest, cooperation and coordination and to report at subsequent plenary meetings. Continuation of each ad hoc Working Group requires confirmation at each plenary session. Conclusions resulting from CEOS plenary sessions, or the findings and recommendations of ad hoc CEOS Working Groups, will be acted upon at the discretion of each CEPS member.

[4] CEOS will replace the Multilateral Meeting on Remote Sensing, the CLOS, and the CORSS. During the development of and action on CEOS activities, the member agencies of CEOS will follow the example of the successful international technical and programmatic cooperation achieved by the Coordination on Geostationary Meteorological Satellites. CEOS members also will consider the issues, concepts, and conclusions arrived at in previous gatherings of the Multilateral Meeting on Remote Sensing, CLOS, and CORSS, and will address current and future activities of space-borne earth observing systems.

CEOS encourages its members to maintain communication as appropriate with other groups and organizations involved in space-borne earth observations activities and applications through the relevant channels within their respective governments. Such groups and organizations include, but are not limited to, the Coordination on Geostationary Meteorological Satellites; the International Polar-Orbiting Meteorological Satellite Group; the Landsat Ground Station Operations Working Group; the Groupement des Operateurs des Stations SPOT; the World Meteorological Organization; the United Nations Committee on the Peaceful Uses of Outer Space; the International Council of Scientific Unions; the Economic Summit of Industrialized Nations; and national, regional, and international remote sensing satellite data archiving, applications, and user organizations.

ADOPTION AND AMENDMENT

These Terms of Reference were adopted at the September 24-25, 1984, meeting of CEOS and may be amended by consensus of the members.

[5]

SUMMARY
REMOTE SENSING TRAINING COORDINATION MEETING

As agreed by the Economic Summit Panel of Experts on Remote Sensing, the United States will host a meeting on coordinating donor activities in remote sensing training for developing countries. The meeting is planned for the Asian Regional Remote Sensing Training Center located at the Asian Institute of Technology (AIT), Bangkok, Thailand.

The date will be during the week of February 4, 1985, and the meeting is scheduled for two days.

Dr. Charles Paul, Forestry and Natural Resources Office, U.S. Agency for International Development, has agreed to assume organizational responsibility for the meeting. Dr. John McElroy, NOAA, will attend as the U.S. representative to the Panel of Remote Sensing Experts. An informal preparatory meeting is taking place October 10, 1984, in Toulouse during the GDTA International Conference on Training for Remote Sensing Users.

For further information on this activity, contact Kenneth Hodgkins, NOAA/NESDIS International Affairs Unit, Washington, DC 20233, telephone 301-763-4586.

Document IV-9

[1]

MINUTES
COMMITTEE ON EARTH OBSERVATIONS SATELLITES
SEPTEMBER 24-25, 1984
WASHINGTON, D.C.

The Committee on Earth Observations Satellites (CEOS) held its first meeting in Washington, D.C. September 24-25, 1984. The group was chaired by the U.S. .National Oceanic and Atmospheric Administration (NOAA) and was convened using the name International Earth Observation Satellite Committee (IEOSC). Attending the meeting were representatives of the Canada Centre for Remote Sensing, Centre National d'Etudes Spatiales (CNES) of France, the European Space Agency (ESA), the Indian Space Research Organization (ISRO), the Instituto de Pesquisas Espaciais (INPE) of Brazil, the Japanese National Space Development Agency (NASDA), U.S. National Oceanic and Atmospheric Administration (NOAA), and U.S. National Aeronautics and Space Administration (NASA), as well as observers from the U.S. Geological Survey and the U.S. Department of State. The attendees are listed in Attachment 1.

Following opening remarks by Dr. John McElroy, Assistant Administrator of NOAA, the group agreed on the name CEOS and that CEOS would replace the previous Multilateral Meetings on Remote Sensing, and the Coordination on Land Observing Satellites (CLOS) and Coordination on Ocean Remote Sensing Satellites (CORSS) groups. The proposed charter, renamed "Terms of Reference", was discussed and revised, and the final agreed document is presented as Attachment 2. The NASDA (Japan) representative requested that formal approval of the Terms of Reference by Japan be rendered upon confirmation from Tokyo.

Presentations were made by each participating agency covering its current and planned earth observation system with presentation materials distributed to all members. Topical discussions covered ocean satellite systems, satellite data products and archives, the role of commercial and government-funded satellite programs, regional user meetings, and training programs. These items are summarized below.

Ocean Satellite Systems-Presentations were made on U.S. plans for coordinating and integrating data from a variety of current and planned satellites of interest to the marine community, with the suggestion that work needs to begin now to ensure an adequate system for handling the rapidly increasing volume of ocean observations expected by the early 1990s. NASA presented its plans for ocean observations from space, as did ESA. Discussion related to ensuring user community involvement and progress made to date on interesting the oceanographic community in satellite data. CEOS members agreed that coordination of planned missions was advantageous to system operators and data users. Specific cooperative arrangements are being addressed in various bilateral discussions between relevant agencies.

[2] Data Issues: There was general agreement on the desirability of increasing coordination and standardization in data management. A Working Group was established to define areas where progress was feasible and to develop a plan for proceeding, mindful of the risk of attacking too large a problem and never making any substantial, practical progress.

Commercial and Government Programs: There was discussion on the proper role for a potential commercial operator of an earth observations system in this or another forum. The Terms of Reference do not exclude the possibility of a commercial organization participating, either directly or as part of a government agency's delegation. The consensus was that there was no immediate case requiring a precise position and that the CAS would continue to discuss this as the situation evolves. All agreed that communication between commercial and government space-borne earth observations systems was desirable. There was not agreement on the formalities or terms under which that communication should take place.

It was agreed that the addition of new members to CEOS can be effected between plenary sessions if there is unanimous agreement among members. If there is not unanimity, the proposed membership will be discussed at the next plenary session.

Regional User Meetings: The discussion centered on the previous Multilateral Meeting sponsorship of regional meetings as a means of increasing interaction between system operators and users throughout the world. The group agreed that while such interaction was desirable, a formal program for regional meetings was unnecessary. There were several action items agreed on regarding increased exchange of information on regional remote sensing organizations, upcoming meetings, and CEOS. (See Action Items 1-3)

Training: Each participant described training activities in its country or agency. There was general agreement on the desirability of coordinating plans for training, and of providing training on-site, in the location of the trainees, rather than bringing the trainees to the location of the training organization. The upcoming international meeting planned for February 1985 in Bangkok to coordinate planned remote sensing training was described, and NOAA will provide participants with further information. (Action Item 3) The results of this meeting will be reported at the next CEOS meeting, and the CEOS will then consider whether CEOS should take further action in training.

[3]   WORKING GROUPS
1. NOAA will chair a Working Group on Data. Each interested participant [Handwritten note: "Dodge/Butler"] will designate a point of contact for this Working Group, and transmit that name to Lisa Shaffer (NOAA) within 6 weeks of the CEOS meeting.
2. ESA will chair a Working Group on Intercalibration and Performance Verification. Each interested participant [Handwritten note: "Schiffer/Thomas"] will designate a point of contact and transmit that name to Guy Duchossois (ESA) within 6 weeks of the CEOS meeting.

                               ACTION ITEMS
1. Each agency will include information on CEOS in its newsletter or other publications as appropriate.
2. Each agency will provide to other members information about relevant regional organizations and upcoming meetings which CEOS members any wish to attend. Agencies will provide information on CAS to relevant regional organizations in which they participate.
3. NOAA will provide information to CEOS members on the planned Bangkok meeting to coordinate plans for remote sensing training.
4. Each agency will provide to other members the results of satellite failure analysis on relevant programs.
5. CCRS will notify members of a planned meeting in Canada on oceanographic satellite data use, and invite representatives of agencies with planned oceanographic satellite programs to participate.

                               NEXT MEETING
    The European Space Agency will host the next meeting of CEOS in April/May 1986 in either Paris or Frascati. Included on the agenda will be presentations on training activities (including a report on the Bangkok meeting, as well as each agency's programs), and the status of commercialization.

                    _____
                           Document IV-10

[no pagination]
[stamped "Oct 19 1984"]

To CEOS Conference Participants

Dear Participant:
    I am pleased to send the minutes of the first meeting of the Committee on Earth Observations Satellites, held September 24 and 25, 1984, in Washington, D.C. Also enclosed are a copy of the agreed Terms of Reference, a complete list of names and addresses of attendees, and information as promised on the remote sensing training meeting scheduled for February 1985 in Bangkok, Thailand.

I again apologize for having to leave before the end of the meeting, due to travel commitments. I understand that the final part of the meeting was at least as useful and productive as the part I attended. The high level of contribution by all attendees confirms my belief that this type of forum is of value to all current and prospective satellite operators, and will ultimately benefit the entire international user community.

If there are any comments or corrections on the minutes or other CEOS documentation, please let Lisa Shaffer know. We look forward to the progress of the two CEOS Working Groups, and to our next meeting to be hosted by ESA.

Best regards,

[signed]
John H. McElroy

Enclosures

Document IV-11

Document Title: Earth System Sciences Committee, NASA Advisory Council, "Earth System Science Overview," May 1986.

Source: NASA Historical Reference Collection, NASA History Office, NASA Headquarters, Washington, D.C.

This document represents the summary results of a committee of scientists chaired by Dr. Francis Bretherton. It was influential in designing the initial framework for what became not only NASA's Mission to Planet Earth (MTPE), but also the U.S. Global Change Research Program. Among other things, the committee's report focused attention on the interlocking, interactive nature of Earth's systems.

[title page]

Earth System Science
Overview

A program for global change

Prepared by the
Earth System Sciences Committee
NASA Advisory Council

National Aeronautics and Space Administration
Washington, DC 20546
May 1986

[inside cover 1]

We, the peoples of the world, face a new responsibility for our global future. Through our economic and technological activity, we are now contributing to significant global changes on the Earth within the span of a few human generations. We have become part of the Earth System and one of the forces for Earth change.

Research holds the key to a deeper understanding of the Earth as an integrated system of interacting components, and of the consequences of global change for humanity. To achieve this understanding, the Earth System Sciences Committee recommends:

OVER THE NEXT DECADE—

- Programs of continuing and operational space observations to extend and to enhance those presently being carried out by the National Oceanic and Atmospheric Administration (NOAA), the National Aeronautics and Space Administration (NASA), and others;
- A coordinated sequence of specialized space research missions for studies of specific Earth System processes, including the Earth Radiation Budget Experiment (ERBE, launched 1984), Laser Geodynamics Satellites (LAGEOS-1, launched 1976, and LAGEOS-2, started 1983), Upper Atmosphere Research Satellite (UARS, started 1982), Navy Remote Ocean Sensing System (N-ROSS, started 1985), Ocean Topography Experiment (TOPEX/ POSEIDON, candidate new start 1987), and Geopotential Research Mission (GRM, candidate new start 1989);
- Pursuit of other observing opportunities to obtain important Earth System data at modest additional cost;
- An interdisciplinary program of basic Earth System research and in situ measurements to be carried out by NASA, NOAA, the National Science Foundation (NSF), and other Federal agencies;
- An advanced information system to process global data and to facilitate data analysis, data interpretation, and quantitative modeling of Earth System processes by the scientific community; and
- A program of instrument development to ready a variety of satellite experiments for implementation in the mid-1990s.

DURING THE SPACE STATION ERA: MID-1990s AND BEYOND—

- An Earth Observing System utilizing polar orbiting platforms planned as part of the U.S. Space Station program and offering opportunities for combined flight of NASA research instruments and NOAA operational payloads, to provide continuous, long-term global Earth observations (candidate NASA new start, 1989);
- Advanced geostationary space platforms to support a new generation of research and operational measurements from geosynchronous orbit; and
- Additional specialized space research missions, including the Rainfall Mission (candidate new start 1991), Magnetic Field Explorer (MFE, candidate new start 1993), Mesosphere-Thermosphere Explorer (MTE, candidate new start 1995), and Gravity Gradiometer Mission (GGM, candidate new start 1997).

BEGINNING AT ONCE—
- Development of new management policies and mechanisms to foster cooperation among NASA, NOAA, NSF, other Federal agencies and commercial firms engaged in Earth System Science and the study of global change; and
- Strengthening of the international agreements and coordination necessary for a truly worldwide study of the Earth.

Each of these steps will contribute to a new framework for Earth studies. By deciding to take these steps now, we can help to ensure that the gifts of the Earth will be preserved and passed on to future generations.

[Inside cover 2]

Preface
    In 1983, the Advisory Council of the National Aeronautics and Space Administration (NASA) established an Earth System Sciences Committee (ESSC) to:
- Review the science of the Earth as an integrated system of interacting components;
- Recommend an implementation strategy for global Earth studies; and
- Define NASA's role in such a program of Earth System Science.

    In charging the Committee with these tasks, the Council emphasized the importance of understanding the Earth as a system, within the context of the solar environment, and of an integrated research program for the study of global change. Both of these objectives are highly relevant to the future habitability of the Earth. In view of the strong collaborative role to be played by the National Oceanic and Atmospheric Administration (NOAA) in any program of Earth System Science, NOAA has requested that it also receive the Committee's recommendations. The present report is thus addressed to NASA and NOAA jointly.
    The Committee began its deliberations in November 1983 with a review of other relevant reports and studies, particularly those of the Space Science Board of the National Academy of Sciences, and a consideration of concurrent approaches to the investigation of global change, such as the Global Habitability study of NASA and the International Geosphere Biosphere study of the National Academy of Sciences (see listing in Appendix A). During 1984, the Committee received the reports of a number of Working Groups formed to assess the status of research programs and opportunities in a variety of specific Earth-science disciplines. These initial meetings and discussions culminated in a June 1985 workshop, during which the Committee achieved a consensus on an implementation strategy for Earth System Science for the next 10-15 years. The present study is believed to mark the first time that such a large and disparate group of Earth scientists, representing broad areas of Earth study, has attempted to define a unified, systematic approach to the scientific investigation of the entire Earth.
    A research strategy for all of Earth science—one that would assign research priorities across all discipline boundaries and among all measurement techniques—is not yet within reach. The Earth System Sciences Committee followed a more restricted approach, first identifying those research areas to which space techniques and observations can make an

outstanding contribution, and then considering the in situ activities and measurements necessary to support and complement the observations from space. Because of the unique benefits of the space perspective, this approach is the one that seems most likely to advance global Earth studies both broadly and rapidly in the years ahead. In the course of its study, the Committee accomplished the following:

- **Identified a unifying scientific strategy** by confirming that the concept of the Earth as an integrated system indeed provides strategic guidance for future Earth-science investigations. [Preface 2] This approach is rapidly gaining the support of scientists from a wide variety of traditional Earth-science disciplines.
- **Defined an integrated program for Earth investigations from space** that (1) builds upon two decades of successful research and operational satellites, and (2) places currently planned near-term missions within the framework of a systematic approach to the study of the Earth as a whole. This near-term space program, complemented by appropriate in situ measurements, furthermore lays the scientific foundation for the recommended global observing program and associated information system in the longer term.
- **Confirmed the importance of new space technology** for the study of the Earth as a system, especially a next generation of large space platforms in polar orbit. Such platforms are currently planned as part of the U.S. Space Station Complex and would be both advantageous and cost-effective for implementing the proposed Earth serving System in the Space Station era.
- **Considered the roles of NASA, NOAA and the National Science Foundation** (NSF) in an integrated Earth System Science research program, calling particular attention to the need for these agencies to collaborate closely in future studies of the Earth System. The important roles of other Federal agencies are also recognized.
- **Identified the issues of management and leadership** as key to the success of these initiatives. United States agencies will need to develop new mechanisms for interacting and carrying out the Federal research program outlined here. Moreover, the success of Earth System Science will also require a collective research effort by the nations of the Earth. Any U.S. program must be part of an effective international collaboration in Earth remote sensing systems and other research activities.

This Overview presents a summary of the current status of Earth System Science and of the implementation strategy recommended by the Committee. A detailed scientific rationale will be published later.

****

[4] Introduction

The Goal of Earth System Science:

To obtain a scientific understanding of the entire Earth System on a global scale by describing how its component parts and their interactions have evolved, how they function, and how they may be expected to continue to evolve on all timescales.

The Challenge to Earth System Science:

To develop the capability to predict those changes that will occur in the next decade to century, both naturally and in response to human activity.

Scientific research continues to yield fundamental new knowledge about the Earth. Studies of the continents, oceans, atmosphere, biosphere, and ice cover over the past thirty years have revealed that these are components of a far more dynamic and complex world than could have been imagined only a few generations ago. These investigations also have delineated, with increasing clarity, the complex interactions among the Earth's components and the profound effects of these interactions upon Earth history and evolution. We can now proceed, for example, to incorporate the global effects of atmospheric wind stress into models of oceanic circulation; to study volcanic activity as a link between convection in the Earth's mantle and worldwide atmospheric properties; and to trace the global carbon cycle through the many transformations of this vital element by terrestrial and ocean biota, atmospheric chemistry, and the weathering of the Earth's solid surface and soils.

Our new knowledge is providing us with deeper insight into the Earth as a system. This insight has set the stage for a more complete and unified approach to its study, Earth System Science.

Complementing our innate curiosity about our planet, the search for practical benefits to improve the quality of human life has long provided a second important motivation for Earth science. Today, human beings in most regions of the globe enjoy greater abundance from the Earth than at any time in our history. Further advances in weather prediction, agriculture and forestry, navigation, and ocean-resource management will accompany a still better understanding of Earth processes.

[5] Now a third and urgent factor spurs the quest for knowledge. The people of the Earth are no longer simple spectators to the drama of Earth evolution but have become active participants on a worldwide scale, contributing to processes of global change that will significantly alter our habitat within a few human generations. In some cases, such as the depletion of the Earth's energy and mineral resources, the effects of human activity are obvious and irreversible. In other cases, such as the alteration of atmospheric chemical composition, the processes of change are more difficult to document, and their consequences harder to foresee. Moreover, the global effects of many human-induced changes cannot readily be distinguished from the results of natural change on the same timescale.

We particularly require a set of Earth observations that will permit us to disentangle the complex interactions among the Earth's components and to document their effects over extended time periods. Such observations will allow us to establish causal relationships among the processes involved and therefore to distinguish between the consequences of human economic and technological activity, on the one hand, and the results of natural change on the other. With this new knowledge, we will then be able to take timely action to ensure an abundant Earth for future generations.

We can begin to meet this challenge today:
  - **Programs of global observations** relevant to a number of Earth System properties have already been carried out with great success. Sensors for future space and in

situ measurements are ready to deploy or in advanced stages of design. The most urgently needed observations can be made through near-term missions and programs that have been thoroughly planned and can now be initiated. The proposed Earth Observing System aboard polar-orbiting platforms now planned as part of the U.S. Space Station Complex appears to provide the most advantageous and cost-effective means of obtaining essential global observations from space from the mid-1990s onwards.

- **Information systems** specifically constructed to process individual sets of global data are already in operation. New developments in computing technology have now made feasible an advanced information system to provide worldwide access to these current data sets, to process the more extensive global data to be obtained in the future, and to facilitate data analysis and interpretation by the scientific community.

- **Existing numerical models** are already contributing to detailed understanding of individual Earth components. Building upon these prototypes, new conceptual and numerical models of the Earth System are now being developed to explore the interactions among the Earth's components and to analyze the global effects of physical, chemical, and biological processes. By furnishing a quantitative understanding of the Earth System, these new models will also provide predictions of the effects of global change on human populations.

- [6] **Federal agencies** are recognizing the need for interdisciplinary research support and interagency cooperation. Moreover, there is a developing consensus on the goals and missions of these agencies in Earth-science research, as exemplified by a recent report of the President's Office of Science and Technology Policy (see Appendix A).

- **A worldwide political awareness** of the necessity for a coordinated, international approach to the global study of the Earth has been created, and cooperative research efforts by many nations across the globe are under way.

If pursued with resolve and commitment, this research program will bring us rewards of knowledge as dramatic, and as relevant to humankind, as any in scientific history. The anticipated achievements of Earth System Science include the following:

- Global measurements: Establishment of the worldwide observations necessary to understand the physical, chemical, and biological processes responsible for Earth evolution on all timescales.

- Documentation of global change: Recording of those changes that will occur in the Earth System over the coming decades.

- Predictions: Use of quantitative models of the Earth System to anticipate future global trends.

- Information base: Assembly of the information essential for effective decision-making to respond to the consequences of global change.

Guided by this new knowledge, the Earth's human societies may wish to consider, for example, modifications in the use of fossil fuels; political, social, and technical planning for the relocation of primary grain-production areas; controls on the disposal of chemical wastes; or redistribution of water in response to drought forecasts.

[7] We have no greater concern than the future of this planet and the life upon it. Exploration of the other planets in the Solar System has confirmed the very special place

of our world among them: the only planet with a biosphere, the only planet with abundant oxygen and liquid water, and the only planet with plate-tectonic processes that renew its surface structure and recycle nutrients essential to life. To preserve it, we must continue to seek a deeper scientific understanding of global Earth processes. Now is the time to meet this challenge through a program of Earth System Science.

[9]
OUR PLANET EARTH

The Earth Sciences

For the present generation of human beings, the continuing search for new knowledge of our planet has been particularly exciting. In an extraordinary burst of research findings over the past 30 years, our view of the solid Earth has been totally transformed. The earlier notion of a static, placid globe has been swept away, replaced by the dynamism and drama of plate tectonics. Enormous sections of the Earth's crust, born at mid-ocean ridges, float upon the convective mantle of the Earth, restlessly jostling against neighboring plates until their ultimate subduction back into the Earth's interior along continental plate boundaries. Patterns of mountain-building, volcanism, and earthquake activity all fit consistently into this new view. Plate tectonics has, for the first time, provided a unified, coherent description of the Earth's crustal features.

The past several decades have also seen remarkable advances in our knowledge of the fluid Earth. The oceans, atmosphere, and ice-covered regions of the planet are now recognized to be closely coupled in shaping the Earth's weather and climate. Research has charted the courses of the world's great ocean currents and revealed the distribution of heat, salt, and nutrients in the ocean interior. Aided by satellite observations of global temperature, moisture, and cloud cover, scientists have constructed numerical models of the atmosphere that have begun to provide reliable predictions of general atmospheric circulation. Studies of the ocean-atmosphere interaction have identified an association between the El Niño ocean-current variation off the South American coast and the Southern Oscillation atmospheric pressure phenomenon that produces effects across the entire tropical Pacific Ocean and beyond. Such investigations are contributing to an initial understanding of the operation of the fluid Earth on a global scale.

The biological Earth is now recognized to exert a major influence on global processes. Ocean biota, for example, have an important effect on climate through net removal of atmospheric carbon dioxide during formation of ocean sediments. Both ocean biota and land ecosystems participate in the global cycles of chemicals essential to life. Furthermore, land biota can also affect climate through their important influence on albedo and water cycling, and through production and emission of various trace gases. All of these findings have established important connections among the components of the planet Earth and thus have emphasized the essential unity of global processes which are only now beginning to be studied systematically.

Science for Practical Benefits

The pursuit of an improved quality of life upon the Earth goes hand in hand with the search for greater scientific understanding of the Earth itself. The application of basic research to human needs is today proceeding more vigorously than ever before.

One of the most important benefits secured to our generation is increasingly accurate global weather prediction. Numerical simulation of atmospheric processes began to become practical in the 1960's with the advent of highspeed computers. These simulations were accompanied by a complementary and dramatic new development: global observations of the Earth's surface and atmosphere from space, beginning with the launch of the first experimental satellite in 1960. The first operational series of polar-orbiting weather satellites began in 1966, and a series of geostationary environmental satellites became operational in 1974. These spacecraft have permitted continuous, global recording of temperature, cloud cover, and other atmospheric variables to supplement an increasingly refined series of measurements made from the ground and within the atmosphere itself. Regional weather forecasts are now based almost entirely upon the predictions of numerical models employing these data.

Studies of the land and the oceans have produced additional benefits for humanity within the past generation. Research into crustal movements and plate tectonics has delineated regions of potential volcanic and earthquake activity and has begun to develop predictors of these events. We have come to understand the origin and distribution of the Earth's vast quantities of petroleum, natural gas, and mineral deposits—particularly since the investigation of environment-specific processes, such as the deposition of metallic-sulfide ores at hydrothermal vents along oceanic spreading centers. Spacecraft observations of ocean color have identified plankton-rich regions and productive time periods important to the aquatic food chain, thus promising more efficient use of our fishing resources. Continued research holds the potential to increase still further the abundant benefits of the Earth.

[10] A New Human Need: Study of Global Change

Human activity is now causing significant changes on a global scale within the span of a few human generations. The burning of fossil fuels, for example, is injecting carbon dioxide into the atmosphere at unprecedented rates. The atmospheric concentration of this gas has increased by nearly 25 percent since the Industrial Revolution, and by over 10 percent since 1958 alone; at this rate it will double within a century. Carbon dioxide is transparent to sunlight entering the atmosphere but blocks the flow of heat radiated outward from the Earth's surface, thus creating a "greenhouse effect" that produces a net warming trend. On the basis of the present rate of increase in atmospheric carbon dioxide, climate models predict an average global increase of at least 2°C in surface temperature during the next century—an increase comparable to that experienced since the last Ice Age 18,000 years ago—together with marked shifts in precipitation patterns. There are also continuing increases in a number of other "greenhouse gases," including methane, chlorofluorocarbons, and tropospheric ozone; although the concentration [11] of these trace species are presently much less than that of carbon dioxide, they are rising much more rapidly. Their effects can also be more pronounced: molecule for molecule, chlorofluorocarbons produce

10,000 times the greenhouse effect of carbon dioxide, in addition to depleting stratospheric ozone.

Moreover, the daily needs of nearly half the world's people for fuel and nourishment are reducing the Earth's vegetation and the productivity of marginal agricultural land. Because of these economic and cultural forces, the extent of the Earth's forest cover has decreased substantially since 1950. Since much of the [12] deforested land is planted to other vegetation, and since substantial afforestation may be occurring at northern midlatitudes, the net effect on carbon-dioxide balance remains unclear, but such changes are almost certain to alter the ecology of the land in a variety of ways. For example, the clearing of tropical forest, often by burning, is reducing the world's greatest reservoir of plant and animal diversity. In marginal agricultural areas, overcropping of the land and uncontrolled animal grazing may be turning productive soil into desert, a major source of dust that in turn can affect atmospheric properties and climate.

All of these human-induced changes are difficult to assess and measure accurately, but it is already evident that they are playing a role in shaping present and future global conditions. Now is the time to document these processes on a global scale and to identify the causal relationships among them, while there is still time to respond effectively.

****

[31]
THE RECOMMENDED PROGRAM

In developing its recommended program, the Earth System Sciences Committee recognized two distinct research eras delineated by the U.S. Space Station development schedule: a current, near-term era, extending over the next decade, that will utilize present satellite capabilities, and a long-term era beginning in the mid-1990's that will draw upon the new capabilities provided by the Space Station. The Committee has also examined the roles of Federal agencies during both of these program periods and placed them in the context of an international effort directed at global Earth studies. Following a presentation of its own budget estimates of the costs of implementing the recommended program, the Committee offers some concluding remarks on Earth System Science.

Priorities for an Implementation Strategy

In determining priorities, the Committee first considered the intrinsic scientific importance of each potential research contribution, particularly its relevance to the Goal and the Challenge of Earth System Science. The relevant reports of the National Academy of Sciences' Space Science Board, such as that of the Committee on Earth Sciences, provided essential guidance for these science-related decisions. Other Academy studies, for example the International Geosphere-Biosphere study, also furnished a valuable scientific perspective.

The Committee next examined the feasibility of proposed program elements in the time periods of interest. The required measurement technology, scientific personnel, and institutional resources must be projected realistically and carry a reasonable assurance of availability. The nature and magnitude of some of the tasks has dictated a careful appraisal of the roles of the Federal agencies engaged in Earth-science studies. Because many of these tasks require satellite observations, the Committee has taken into account the future availability of space observatories. Consideration of the resources and opportunities to be provided by the Space Station program were therefore important to the Committee's conclusions.

Finally, the Committee had to face the constraints, both technical and fiscal, that must inevitably restrict the scope of any national research program, even one as important to our future as Earth System Science. From the perspective of its science strategy, the Committee considered programmatic opportunities to attain the objectives stated in the Academy reports, examining the relevance, readiness, degree of community support, and cost of proposed missions. The Committee has tried to strike a balance between program needs, on the one hand, and a realistic demand on agency resources and capabilities on the other.

In the opinion of the Committee, the sequence, programmatic balance, and – given the high national importance of the Goal and Challenge – the schedule of integrated program recommended here reflect all of these considerations.

The program elements recommended for inclusion in the current, near-term era of research (the next decade) are:

- Continuing and operational space observations;
- Specialized space research missions;
- Other observing opportunities;
- Basic research and in situ observations;
- An advanced information system; and
- Instrument development.

During the second era of research to follow (the mid-1990s and beyond), emphasis will shift to the integrated program of global measurements to be carried out by the proposed Earth Observing System (EOS), as well as NOAA's complement of operational instruments, both of which can utilize new space platforms in polar orbit provided by the Space Station program. These programs will be complemented by ongoing basic research and in situ observations, appropriate specialized space research missions, and observations from new space platforms in geosynchronous orbit.

The Current Era

Continuing and Operational Space Observations

The United States civilian operating Earth-observing satellite systems, together with ground-based calibration and validation programs, furnish continuing global data on the atmosphere, oceans, solid Earth, and important solar and space-environment properties. From these sources, Federal agencies provide operational services critical to the protection of life and property, the national economy, energy development and distribution, and global food supplies.

At the same time, this ongoing measurement program, together with associated in situ investigations, provides a data base that is fundamental to research on the state of the Earth and global change. The Earth System Sciences Committee therefore concurs with the National Academy of Sciences' COSEPUP Panel on Remote Sensing of the Earth in emphasizing that present ongoing and operational satellite measurement systems must be continued [32] and improved as required—to provide accurate, homogeneous, and timely data. The Earth System Science program recommended here is based on the assumption that the operational system now in place will be continued.

NOAA is presently concluding contract negotiations for the next series of Geostationary Operational Environmental Satellite (GOES) spacecraft and payloads and has initiated a procurement through NASA for the continuation of the NOAA polar-orbiting spacecraft series. The GOES initiative will produce a series of Shuttle-launched, three-axis-stabilized spacecraft available for service beginning in 1990. These satellites, GOES-I through M, will offer direct-broadcast capabilities and simultaneous operation of imaging and sounding instruments. A greater number of spectral channels and higher resolution will also be provided. The extended polar-orbiting series, NOAA-K, L, and M, includes an Advanced Microwave Sounding Unit to provide improved soundings in cloud-covered areas and in the stratosphere, together with new information on sea ice, rain rates, and soil moisture. This series will remain in service until NOAA transfers its polar mission to the polar-orbiting platforms planned as part of the Space Station Complex.

Specialized Space Research Missions

Of particular importance in the near term is a carefully constructed sequence of specialized space research missions required for the study of specific Earth System properties and processes. Each is characterized by a choice of orbit and spacecraft design tailored to achieve the particular objectives of the mission. These missions must therefore be flown separately, and in a sequence that yields the optimum scientific return to the Earth System Science program as a whole. They are as follows:

- Earth Radiation Budget Experiment (ERBE). Because of its importance to climate studies, measurement of the Earth's radiation budget has been the objective of many satellite observations since the beginning of the space program. The ERBE program in progress, combining observations from a NASA research experiment (EBBS) in a low-inclination orbit with measurements from the operational NOAA-9 and NOAA-G satellites in polar orbit, is the first to provide these essential characteristics: adequate calibration, wide geographic sampling, broad spectral response, extensive measurements of the angular distribution of reflected and emitted radiation, unbiased diurnal sampling, and high spectral resolution. The projected ERBE observing period is 1985-1989.
- Upper Atmosphere Research Satellite (UARS). Scheduled for a 1989 launch, the approved UARS program is designed to improve understanding of the coupled chemistry and dynamics of the stratosphere and mesosphere, the role of solar radiation in these processes, and the susceptibility of the upper atmosphere to long-term changes in the concentration and distribution with altitude of key atmospheric constituents, particularly ozone. UARS data will be coordinated with results from the Solar Backscatter Ultraviolet (SBUV) spectrometer scheduled to be flown aboard operational meteorological satellites and the Space Shuttle during the UARS mission duration.
- Scatterometer (NSCAT) aboard the Navy Remote Ocean Sensing System (N-ROSS) satellite. The approved N-ROSS program, scheduled for launch in 1991, will carry four sensors: a scatterometer for ocean wind measurements, a

microwave radiometer for measurements of sea-surface temperature, a microwave radiometer to monitor ice extent, and a radar altimeter to measure wave height and to locate oceanic [33] fronts and eddies. NSCAT is planned to provide accurate, global wind-field data over a three-year period that will be of high importance to oceanography and meteorology. The instrument itself will satisfy both the research requirements of the scientific community and the operational requirements of the Navy. In addition to providing NSCAT to the Navy, NASA and NOAA plan to establish a ground data processing system to produce data products, including those of research quality, and to make them available to the oceanographic and meteorological communities.

- Ocean Topography Experiment (TOPEX/POSEIDON). This joint US/France mission, proposed as a 1987 NASA new start, will use radar altimetry to measure the surface topography of the oceans over a period of several years. When combined with appropriate in situ measurements, these observations will permit a determination of the three-dimensional structure of the world's ocean currents. The prime sensor will be a modification of the highly successful 1978 Seasat altimeter providing direct measurement of ocean topography through two-frequency operation. Highly accurate orbital characteristics are to be provided by receivers of the Global Positioning System; laser-tracking retroreflectors will be carried as well. Two experimental French instruments are part of the payload, and launch will be provided by Ariane. The TOPEX/POSEIDON satellite is scheduled to operate during the N-ROSS mission.

- Geopotential Research Mission (GRM). Candidate for a NASA new start in 1989, GRM is designed to measure spatial variations in the Earth's gravity and magnetic field over the entire globe to a resolution of 100 kilometers with unprecedented completeness and accuracy. The mission currently incorporates two low-drag spacecraft in co-planar, 160-km orbits, tracked by Doppler radar to an accuracy of one micrometer per second over their 300-km separation. GRM will yield important new insights into the Earth's remote interior. Measurements of the gravity field will elucidate the pattern and dynamics of thermal convection in the mantle, which drives plate-tectonic motions: observations of the magnetic field and its time variation will constrain models of the geodynamo in the fluid outer core of the Earth. Moreover, the timely flight of GRM is essential to a maximum utilization of data from TOPEX/POSEIDON by providing the geoid to which sea-surface heights are referred for studies of ocean circulation. The mission furthermore has important applications to additional studies of the thermosphere and mantle, and to geodesy.

All of the above missions are either operating, ready to proceed, or in advanced stages of development. Many of the measurements initiated by these missions should be continued as part of a program of long-term global observations from the mid-1990s onward.

Other Observing Opportunities

In addition to the specialized space research missions discussed above, there are opportunities to fly several other instruments, either on NOAA operational satellites, aboard the Space Shuttle, or on other spacecraft. All offer high scientific return at modest cost. Their development could be undertaken either by NASA, by NOAA, through a NASA-NOAA cooperative program, or through international collaboration.

- First among this set is an Ocean Color Imager. The Coastal Zone Color Scanner on the Nimbus 7 satellite, launched in 1978, has already provided a significant start on a long-term data set and has operated well beyond its design lifetime. This data set on biological activity in the world's oceans has a demonstrated utility to the research and ocean-user communities alike. This data stream must be continued with global oceanic coverage and improved flight hardware as soon as possible.
- The surface topography of the continents can be determined by a scanning radar altimeter flown on a series of Shuttle missions. This will furnish a data set of broad applicability in geology, geophysics and hydrology that can facilitate the interpretation of later high-resolution imagery.
- The chemistry of the troposphere is an area of growing emphasis in research and analysis as part of Earth System Science. The only trace chemical constituents of the troposphere currently measurable from space are carbon monoxide and water vapor. Carbon monoxide is indicative both of hydrocarbon oxidation and of the abundance of hydroxyl radicals which control the destruction of a number of other tropospheric gases. A version of the current Space Shuttle instrument, improved to detect carbon monoxide in three layers spanning the full height of the troposphere, would be a good candidate for flight on the NOAA morning satellites.

Basic Research and In situ Observations

A program of basic research is needed to complement and make full utilization of the data from specific missions and projects recommended in this report. In particular, NASA, NOAA, and NSF will all need to expand considerably their basic Earth-science research efforts in order to strengthen ecological studies, fund new multidisciplinary research efforts in support of Earth System Science, and extend research in a number of present Earth-science disciplines, such as tropospheric chemistry.

[36] NASA will require, for example, expanded capabilities to make measurements from non-space platforms, such as aircraft and surface stations, exploiting the latest technology. NOAA will need to improve the ground-based monitoring of sea levels and of long-lived atmospheric constituents from networks of stations and enhance its programs of modeling and research in the application of these data. With respect to spacecraft data, NOAA will need to increase its research on the application of operational satellite observations to Earth System Science problems. NOAA should also actively participate in national and international research projects designed to exploit operational satellite observations for Earth System Science applications, such as programs for cloud and land-surface climatology. Key NSF program enhancements include the establishment of terrestrial ecosystem observatories in which detailed in situ measurements of different biomes (vegetative groupings) can be made on a long-term basis for comparison with global satellite observations, together with studies of ocean circulation and related biogeochemistry.

All three agencies will, in addition, need to support modeling and laboratory studies of Earth System components and their interactions. In shaping programs to attain these objectives, the agencies must furthermore take care to provide for the full participation of the university research community, which will play a pivotal role in the advance of Earth System Science.

Finally, the Committee wishes to stress the potential research importance of Earth System data to be provided by the commercial remote-sensing ventures now beginning operation. The Land Remote Sensing Commercialization Act of 1984 laid down extensive guidelines for the conduct of United States commercial operations; however, it is not yet clear how this act will actually be implemented in detail. The research will, in particular, need access to commercial data for scientific purposes at a price commensurate with the resources of realistic research budgets. Moreover the continuity and quality central of remote-sensing data are of the highest importance to research. These concerns must be met in any plan to transfer research instruments developed for remote sensing to commercial or Federal operational programs.

[37] An Advanced Information System

Of paramount importance to the success of Earth System Science is an advanced information system that will promote productive use of global data. The worldwide space and in situ observations required for a deeper understanding of the Earth System can be utilized only if the research community has effective access to them. The design, development, and management of the requisite information system are tasks that approach, in scope and complexity, the design, development, and operation of space-based observing systems themselves. NASA, NOAA, and NSF will all benefit from such a system and should collaborate in this undertaking. Other interested agencies, such as the U.S. Geological Survey of the Department of the Interior, should participate as well.

The diversity of Earth-data sources mandates an information system of substantial capabilities in which flexibility of use is a key characteristic. There is, to begin with, a wealth of existing Earth System data, scattered among various locations, that could be rapidly applied to research problems if that information were more immediately accessible to the scientific community. Operational data currently processed and used by NOAA also need to be made available to the community through interactive access by remote terminals. In addition, data to be returned by specialized NASA space missions over the next two decades must be processed and distributed widely for scientific analysis. Finally, the information system will need to meet the data-handling requirements of the global system to observe the Earth envisioned for the mid-1990s and beyond. The system must thus permit scientists to obtain and combine data from all of these sources and to carry out detailed analysis of these integrated data on central and local computers. The system should also permit individual research groups to exchange analyses and to develop Earth System models interactively.

Among the more specialized requirements for an advanced information system are the following: the provision of data directories and catalogs, browse capabilities, and full documentation on sensors, missions, algorithms, and data sets; a hierarchical structure, so that active data bases of geophysical and biological properties can be maintained together with archives of more primitive sensor characteristics; the provision of utilities for higher-level data processing; and the linking of local work stations with observing-system control centers, so that qualified users can submit requests for specialized observations rapidly and directly. These features will, moreover, encourage the early formation of a community of interactive users of the system.

Such an information system is clearly a formidable undertaking, but it is essential to the pursuit of Earth System Science. The information system must be designed to accommodate the variety and complexity of the Earth itself, for it will provide our primary means for detecting and examining the processes of Earth evolution—particularly the processes of global change, arising from both natural and human causes, that are of such importance to the future habitability of the Earth. The contents of the information system, and the understanding that they generate, will constitute one of the chief legacies of Earth System Science to future generations.

Instrument Development

Finally, we need to begin, in the near term, a program of instrument development for spacecraft use that will ready a variety of experiments for service by the mid-1990s and beyond. Examples of such instruments are: (1) multichannel imaging spectrometers for study of physical, geochemical, and biological surface properties; (2) synthetic-aperture radars for ice studies, cartography, and surface properties; (3) high-resolution atmospheric sounders incorporating visible, infrared, submillimeter, and microwave channels; (4) laser ranging systems for geodetic measurements; (5) systems for high-precision ocean-floor measurements; (6) laser systems for measuring cloud heights, aerosols, temperature, moisture, chemical composition, and winds; and (7) improved microwave imagers for surface hydrologic studies and precipitation. Such instrument development is being proposed in anticipation of the requirements of the Earth Observing System (discussed below), and prototypes have in several cases already been scheduled for forthcoming Shuttle flights. The concurrent development of advanced instruments for measurements in situ will be needed to support and complement these space initiatives.

The Space Station Era

Research in Earth System Science will change in two fundamental ways beginning in the mid-1990s. First, the near-term program described above will be underway: the flight of specialized space research missions can be expected to narrow the range of variables for further study, and other important elements of the near-term program should be in place. Second, we will have access to new technology, particularly a new generation of advanced observational platforms in space. These two developments will prepare the way for operation of the Earth Observing System.

[38] Earth Observing System/Polar-Orbiting Platforms

By the mid-1990's, we will require a global observing system in space that utilizes highly capable polar-orbiting platforms and returns both research and operational data through the advanced information system described above. The Earth Observing System (EOS) presently under study by NASA, carried out in collaboration with NOAA's operational program, will incorporate both of these essential features. In combination with observations from geosynchronous platforms, several specialized space research missions, and complementary in situ measurements, EOS can provide the extended observations required for a fundamental understanding of the Earth System.

The planned instruments may be divided into three related classes: (1) a group of instruments that images the Earth's surface in the visible, infrared, and microwave regions and sounds the lower atmosphere; (2) a complement of radar instruments that will gather information on the character and structure of the surface; and (3) a group of instruments designed to study the composition and dynamics of the atmosphere and to measure the Earth's energy balance. Also proposed for EOS are a Geodynamics Laser Ranging System, for rapid measurements of crustal deformation over specific tectonic regions, and an Automated Data Collection and Location System to support automated in situ measurement devices.

The phased assembly of the EOS instrument complement has been discussed extensively within the Earth System Sciences Committee and its Working Groups. This scenario of instrument deployment has been found to meet the requirements of Earth System Science as they are anticipated to evolve in the mid-1990s and beyond, and to address the observational goals of the National Academy of Sciences' Research Briefing Panel of 1985 on Earth Remote Sensing:

"To advance our understanding of the causes and effects of global change, we need new observations of the Earth. These measurements must be global and synoptic, they must be long-term, and different processes such as atmospheric winds, ocean currents, and biological productivity must be measured simultaneously . . ."

Accordingly, the Earth System Sciences Committee endorses the planned Earth Observing System as satisfying the requirements of ESSC and its Working Groups, and recommends an EOS new start in 1989. As currently planned, NASA research instruments and NOAA operational instruments will jointly utilize polar-orbiting platforms of the Space Station Complex. Until the present, polar-orbiting satellites have been rather modest, automated devices devoted to a few instruments only, and inaccessible for servicing. The Space Station platforms of the mid-1990s are being designed with Earth System Science requirements in mind and would offer the following advantages:

- Expanded capability for instrument accommodation, power, and data telemetry. Because of the advent of advanced remote-sensing instruments and the need to integrate observations of different kinds, the NASA and NOAA payloads will make greater demands on platform services than can be met by platforms of current design.
- Accessibility for on-orbit servicing, payload augmentation, and instrument replacement by Space Shuttle crews. The periodic refurbishment and replacement of instruments should greatly increase the scientific return of Earth System Science payloads and facilitate the acquisition of long-term, self-consistent data [39] sets. With the assurance of an extended operating lifetime, instruments may also be designed with more advanced features and capabilities than is feasible without serviceability.

The Committee furthermore notes that the Earth Observing System, NOAA operations, and the Space Station Complex are all being planned to include substantial international contributions and cooperation.

Geostationary Platforms

A second step toward a total system for global Earth observations will be provided by advanced platforms in geosynchronous orbit. These offer several fundamental advantages. First, high temporal resolution—limited only by instrument design and cost—

can be brought to bear on the study of rapidly changing, global atmospheric phenomena. In the cases of land and ocean surveys, high temporal resolution helps to minimize data loss resulting from cloud cover and unfavorable atmospheric conditions. Geosynchronous orbit furthermore provides a fixed reference geometry for a given Earth location, facilitating data analysis and interpretation and the study of processes with significant diurnal variations.

Operational geosynchronous satellites, in service since 1974, have carried imager/sounder instruments providing high-resolution visible and infrared images of the Earth. The infrared channels of the sounding instruments have provided temperature and moisture profiles over large areas of the Earth with high frequency. NOAA presently operates two GOES geostationary satellites and should continue to maintain and improve them. Future geosynchronous platforms with increased weight and power capabilities will permit advanced imager/sounder instruments operating in the visible, infrared and microwave spectral regions. The added capability of microwave sounding is not presently available because of the large antenna required for adequate spatial resolution at these high orbital altitudes, but such an advance is being studied as a possible addition to the next generation of NOAA geostationary satellites in the mid-1990s. In addition, a geostationary platform is currently under consideration as a growth element of the U.S. Space Station program. This platform may be expected to extend many of the capabilities and benefits of the Space Station polar platforms to geosynchronous orbit.

Specialized Research Missions

In addition to the specialized research missions recommended for implementation during the current decade and described earlier, several additional missions will be needed in the Space Station era to complement observations carried out by the Earth Observing System.

There is a critical requirement for space measurements of global precipitation. An exploratory mission is needed to test the feasibility of using active and passive microwave data, together with visible and infrared imagery, to derive useful estimates of rainfall amounts and distribution. A low-inclination orbit will permit study of the diurnal cycle of rainfall over the tropics and an assessment of the relationship of heat released into the atmosphere to anomalies in atmospheric circulation.

Also highly desirable in the Space Station era are: (1) a Magnetic Field Explorer (MFE) mission to derive an accurate description of the Earth's magnetic field and its secular variation at the measurement epoch; (2) a Mesosphere-Thermosphere Explorer (MTE) mission to address the chemistry and dynamics of the upper atmosphere, together with its links to the Sun above and the stratosphere below; and (3) a Gravity Gradiometer Mission (GGM) to measure the gradient in the Earth's gravitational field as a complement to the Geopotential Research Mission.

United States Agency Roles

Currently, NASA is responsible for general research and development in civilian satellite technology; NOAA is responsible for operational weather and ocean satellites and for development required to improve these capabilities; NSF is responsible for basic research in all areas of Earth science; and industry is beginning to play a role in land-surface measurements. The Earth System Sciences Committee does not at present see a need for major changes in these basic responsibilities, but it does see a need for more broadly defined roles for the agencies and for a much greater degree of coordination among them. A possible mechanism for fostering such coordination would be a high-level interagency group conducting indepth program reviews and reaching agreement on priorities and implementation. An effective example of this approach is furnished by the Ocean Principals Group, which shapes policy for U.S. oceanographic research.

Role of NASA

NASA must continue its leadership in research from space relevant to Earth System Science. In particular:
- NASA should continue to have the primary responsibility for Earth-sciences research missions from space, including those of broad scientific scope, to study the Earth as an integrated system. NASA should continue to support and foster associated research, including advanced instrumentation development.
- NASA should continue to apply its capabilities in research and technology to the improvement of data transmission, archival, and retrieval techniques for utilization by the Earth-sciences community and by the NOAA operational program.

Role of NOAA

NOAA's role in the Earth sciences must be broadened beyond the present interpretation of its mission in order to meet national needs. The Earth System Sciences Committee urges a strengthening both of the operational satellite program and of the NOAA in situ research program on atmospheric and oceanic processes:
- [41] NOAA should provide the operational services, data-transmission network, and national archives, with an up-to-date interactive capability, for weather, climate, atmospheric chemistry, and oceanographic data in a manner that will support long-term scientific research requirements. Accordingly, NOAA should create formal mechanisms to involve the scientific community in determining and implementing requirements for NOAA's operational space and ground-based systems.
- NOAA should be assigned primary responsibility for conducting a program to obtain, maintain, and make accessible long-term (decadal) data bases that can be used to assess mankind's global impact or potential impact, from civil activities, on the oceans, atmosphere, and land.
- NOAA should conduct research in the atmosphere and oceans, especially applied research, including measurement and diagnostic modeling programs.
- NOAA should continue to have (as provided by the Land Remote Sensing Commercialization Act of 1984) primary responsibility for processing and archiving satellite data for land processes and for providing access to the data by the research community. NOAA should also continue its collaboration with the DoI/USGS in funding, defining, and maintaining an archive of land remote-sensing data for which DoI/USGS has the operational responsibility.

Collaboration Between NASA and NOAA

- NASA and NOAA should continue to investigate the feasibility and practicality of using the polar platforms of the Space Station to support both NASA research needs and NOAA operational needs. NOAA should provide typically 25 percent of the resources on any other operational satellite for support of research instruments and should give greater attention to the calibration and long-term stability of operational instruments in order to support research (as well as operational) needs.
- NASA and NOAA must collaborate in ensuring that satellite data on land processes acquired by NASA are transferred to NOAA as well.

Role of NSF

The role of NSF must be to continue to support studies in basic science and engineering that utilize all types of observations, both in situ and remote. The Committee believes it essential for NSF to view research utilizing satellite and complementary in situ validation and calibration data to be as appropriate for support by NSF as more conventional in situ process studies.

In addition, the Committee hopes that NSF will take an even broader role. Since future satellite programs will give us a new global view, but carry out only part of a given scientific investigation, the satellite missions yield maximum scientific return only if carried out in the context of large-scale field programs. Such programs are exemplified by the Global Atmospheric Research Program (GARP), funded jointly by NASA, NOAA, and NSF, which produced a major improvement in satellite coverage of global weather. Today we are seeing the development of global studies to understand the Southern Oscillation and its manifestation, El Nino (the Tropical Ocean Global Atmosphere Program); large-scale ocean circulation and mixing (the World Ocean Circulation Experiment); global fluxes and transport of material in the ocean; terrestrial ecology; the structure of the Earth's crust; the role of land and ice in paleo-environmental reconstruction and in sea-level changes; and others. All of these will have satellite observations as one of the central elements, and, to be successful, all will require major enhancements by NSF as well as other agencies.

[34] **TABLES 1 A**
OBSERVATIONAL PROGRAMS FOR GLOBAL DATA ACQUISITION:
REPRESENTATIVE EXAMPLES OF APPROVED AND CONTINUING PROGRAMS

Representative Space Programs

Representative Space Programs

Program	Agency/ Status	Objectives
POES: Polar-orbiting Operational Environmental Satellites (e.g., NOAA-7)	NOAA/ Operating	Weather observations
GOES: Geostationary Environmental Satellite System	NOAA/ Operating	Weather observations
DMSP: Defense Meteorological Satellite Program	U.S. Air Forces/ Operating	Weather observations for Department of Defense
METEOSAT: Meteorology Satellite	ESA/Operating	Weather observations
GMS: Geostationary Meteorology Satellite	NASDA (Japan)/ Operating	Weather observations
METEOR-2: Meteorological Satellite-2	USSR/Operating	Weather observations
LANDSAT: Land Remote Sensing Satellite	EOSAT/ Operating	Vegetation, crop, and land-use inventory
LAGEOS-1: Laser Geo-dynamics Satellite-1	NASA/ Operating	Geodynamics, gravity field
ERBE: Earth Radiation Budget Experiment	NASA-NOAA/ Operating	Earth's radiation losses and gains
GEOSAT: Geodesy Satellite	U.S Navy/ Operating	Geodesy, shape of the geoid, ocean and atmospheric properties
GPS: Global Positioning System	U S. Navy-NOAA-NASA-NSF-USGS/ Completion 1989	Geodesy, crustal deformation
SPOT-1: Système Probatoire d'Observation de la Terre-1	France/ Operating	Land use, Earth resources
IRS: Indian Remote Sensing Satellite	Indian/Operating	Earth resources

[34] **TABLES 1 A**
OBSERVATIONAL PROGRAMS FOR GLOBAL DATA ACQUISITION:
REPRESENTATIVE EXAMPLES OF APPROVED AND CONTINUING PROGRAMS

Representative Space Shuttle instruments

Program	Organization/ Status	Objective
ATMOS: Atmospheric Trace Molecules Observed by Spectroscopy	NASA/Current	Atmospheric chemical composition
ACR: Active Cavity Radiometer	NASA/Current	Solar energy output
SUSIM: Solar Ultraviolet Spectral Irradiance Monitor	NASA/Current	Ultraviolet solar observations
SIR: Shuttle Imaging Radar	NASA/Current/ In development	Land-surface observations
MAPS: Measurement of Air Pollution from Shuttle	NASA/Current/ In development	Tropospheric carbon monoxide
SISEX: Shuttle Imaging Spectrometer Experiment	NASA/Planned	Spectral observations of land surfaces
LIDAR: Light Detection and Ranging instrument	NASA/Planned	Surface topography, atmospheric properties

Program	Agency/ Status	Objective
MOS-1: Marine Observation Satellite-1	NASDA (Japan)/Launch 1987	State of sea surface and atmosphere
LAGEOS-2: Laser Geo dynamics Satellite-2	NASA-PSN (Italy)/Launch 1988	Geodynamics, gravity field
SPOT-2: Système Probatoire d'Observation de la Terre-2	France/Launch 1988	Earth remote sensing
UARS: Upper Atmosphere Research Satellite	NASA/Launch 1989	Stratospheric chemistry, dynamics, energy balance
ERS-1: Earth Remote Sensing Satellite-1	ESA/Launch 1990	Imaging of oceans, ice fields, land areas
N-ROSS: Navy Remote Ocean Sensing System	U.S. Navy/ Launch 1991	Ocean topography, surface winds, ice extent
JERS-1: Japan Earth Remote Sensing Satellite-1	NASDA (Japan)/ Launch 1991	Earth resources

[34] **TABLES 1 A**
OBSERVATIONAL PROGRAMS FOR GLOBAL DATA ACQUISITION:
REPRESENTATIVE EXAMPLES OF APPROVED AND CONTINUING PROGRAMS

Representative International Programs for Measurements *In Situ*

Program	Organization/ Status	Objective
GEMS: Global Environment Monitoring System	UNEP/Begun 1974	Monitoring of global environment
World Ozone Program	WMO-NASA-UNEP/Operating	Atmospheric composition
Crustal Dynamics Project	NASA-23 nations/Begun 1979	Tectonic plate movement and deformation
Man and the Biosphere	UNESCO/ Operating	Ecological studies
International Biosphere Reserves	UN/Operating	Long-term ecological studies
ISCCP: International Satellite Cloud Climatology Project (World Climate Research Program)	WMO-ICSU/ Begun 1983	Measure interaction of clouds and radiation
ISLSCP: International Satellite Land Surface Climatology Protect (World Climate Research Program)	WMO-ICSU/ Begun 1985	Measure interactions of land-surface processes with climate
TOGA: Tropical Ocean Global Atmosphere Program (World Climate Research Program)	WMO-ICSU/ Begun 1985	Variability of global interannual climate events
GRID: Global Resource Information Database	UNEP/Begun 1985	Information on global resources

[35] **TABLES 1 B**

OBSERVATIONAL PROGRAMS FOR GLOBAL DATA ACQUISITION:
REPRESENTATIVE EXAMPLES OF PROPOSED FUTURE PROGRAMS

Representative Space Programs

Program	Agency/ Status	Objectives
TOPEX/POSEIDON: Ocean Topography Experiment	NASA-CNES (France)/Start 1987, Launch 1991.	Ocean surface topography
POES: Polar-orbiting Operational Environmental Satellite system—follow-on missions (NOAA K,L,M)	NOAA/ Planned	Advanced capabilities for weather observations
GOES, Geostationary Operational Environmental Satellite system—follow-on missions (e.g., GOES-Next)	NOAA/ Planned	Advanced capabilities for weather observations
RADARSAT—Canadian Radar Satellite	Canada/Start 1986, Launch 1991	Studies of arctic ice, ocean studies, Earth resources
MOS-2: Marine Observation Satellite-2	NASDA (Japan)/ Launch about 1990	Passive and active microwave sensing
GRM: Geopotential Research Mission	NASA/Start 1989, Launch 1992	Measure global geoid and magnetic field

[35] **TABLES 1 B**
OBSERVATIONAL PROGRAMS FOR GLOBAL DATA ACQUISITION:
REPRESENTATIVE EXAMPLES OF PROPOSED FUTURE PROGRAMS

Individual instruments for long-term global observations

Program	Agency/ Status	Objective
OCI: Ocean Color Imager	NASA-NOAA/ Planned	Ocean biological productivity
ERB: Earth Radiation Budget instrument	NASA/ Planned	Earth radiation budget on synoptic and planetary scales
Carbon-Monoxide Monitor	NASA/Planned	Monitor tropospheric carbon monoxide
Total Ozone Monitor	NASA/Planned	Monitor global ozone
GLRS: Geodynamics Laser Ranging System	NASA/Planned	Crustal deformations over specific tectonic areas
Laser Ranger	NASA/Planned	Continental motions
Scanning radar altimeter	NASA/Planned	Continental topography

Program	Agency/ Status	Objective
Eos: Earth Observing System/Polar-Orbiting Platforms. NASA-NOAA program:	NASA-NOAA/ NASA Start 1989, Launch 1994	Long-term global Earth observations
NASA research payloads	NASA/Planned	Surface imaging, sounding of lower atmosphere, measurements of surface character and structure; atmospheric measurements; Earth radiation budget; data collection and location of remote measurement devices
NOAA operational payloads	NOAA/ Planned	Weather observations and atmospheric composition; observations of ocean and ice surfaces; land surface imaging; Earth radiation budget; data collection and location of remote measurement devices; detection and location of emergency beacons; monitoring of space environment

[35] **TABLES 1 B**
OBSERVATIONAL PROGRAMS FOR GLOBAL DATA ACQUISITION:
REPRESENTATIVE EXAMPLES OF PROPOSED FUTURE PROGRAMS

Representative Space Programs

Program	Agency/Status	Objective
European Polar-Orbiting Platform (Columbus)	ESA/Planned	Long-term comprehensive research, operational, and commercial Earth observations
Rainfall mission	NASA/Start 1991, Launch 1994	Tropical precipitation measurements
MFE: Magnetic Field Explorer	NASA/Start 1993, Launch 1996	Secular variability of Earth's mgnetic field
MTE: Mesosphere-Thermosphere Explorer	NASA/Start 1995, Launch 1998	Chemistry and dynamics of upper atmosphere
GGM: Gravity Gradiometer Mission	NASA/Start 1997, Launch 2000	Gradient in Earth's gravitational field

[35] **TABLES 1 B**
OBSERVATIONAL PROGRAMS FOR GLOBAL DATA ACQUISITION:
REPRESENTATIVE EXAMPLES OF PROPOSED FUTURE PROGRAMS

Representative International Programs for
Measurements *In situ*

Program	Organization/ Status	Objective
WOCE: World Ocean Circulation Experiment (World Climate Research Program)	WMO-ICSU-IOC- NSF-NASA-NOAA/ 1987 enhancement	Detailed understanding of ocean circulation
IGBP: International Geosphere-Biosphere Program (Global Change)	ICSU/Proposed	Study of global change on timescale of decades to centuries
GOFS: Global Ocean Flux Study	NSF-NCAA-NASA/ Enhancement	Production and fate of biogenic materials in the global ocean
GTCP: Global Tropospheric Chemistry Program	NSF-NASA- NOAA/ Enhancement	Tropospheric chemistry and its links to biota
Ocean Ridge Crest Processes	NSF-USGS-NOAA/ Enhancement	Chemistry and biology of deep-sea thermal vents, plate motions, crustal generation
Sensing of the Solid Earth	NSF-USGS-DoD- NASA/Enhancement	Large-scale mantle convection, studies of continental lithosphere
Ecosystem Dynamics	NSF/Enhancement	Studies of long-term ecosystems biogeochemical cycles
Greenland Sea Project	ISCU/Planned	Atmosphere - sea ice - ocean dynamics

[40] **TABLE 2**

REPRESENTATIVE EXAMPLES OF PROPOSED SATELLITE MEASUREMENTS*

Measurement	Implementation: Current Era	Implementation: Space Station Era
Solar energy output	ERBE, UARS	Eos
Ice extent, dynamics	DMSP, N-ROSS, ERS-1, JERS-1	Eos, DMSP, RADARSAT
Weather and climate: physical parameters	POES, GOES, DMSP, MOS-1, N-ROSS, ERS-1, JERS-1, (WWW)	POES, GOES, DMSP, MOS-2, Eos, RADARSAT, (WWW)
Stratospheric ozone chemistry & dynamics	UARS, POES	Eos
Tropospheric Chemistry	CO Monitor	Eos
Ocean surface winds & ocean currents	N-ROSS, TOPEX/ POSEIDON, ERS-1, GRM, MOS-1, GEOSAT, (TOGA), (WOCE)	MOS-2, Eos, (TOGA), (WOCE)
Ocean spectral reflectivity, ocean productivity	OCI, (GOFS)	Eos
Precipitation, rainfall rates	Concept and technique development	Rainfall mission over tropics, Eos, GOES
Surface spectral reflectivity, land-surface biology, continental geology	LANDSAT, Shuttle instruments, SPOT, (ISLSCP)	Eos, EOSAT, SPOT

Measurement	Implementation: Current Era	Implementation: Space Station Era
Geopotential field & mantle circulation	GRM, (Global Digital Seismic Network)	(Global Digital Seismic Network)
Continental topography	Scanning radar altimeter	Eos
Magnetic field	GRM	MFE
Vegetation cover	LANDSAT, SPOT, JERS-1	Eos
Crustal deformation and plate tectonics	LAGEOS-1, LAGEOS-2, GPS, Laser Ranger, Shuttle instruments, (VLBI)	GLRS, Eos, GPS, LAGEOS-1, LAGEOS-2, (VLBI)
Land-surface energy and moisture budgets	Concept and technique development	Eos
Biome extent and productivity	Concept and technique development	Eos
Winds, especially in tropics	GOES, Concept and technique development	Eos

*Programs of complementary measurements *in situ* appear in parentheses, e.g., (WOCE).

Document Title: Sally K. Ride, "Mission to Planet Earth," in NASA Leadership and America's Future in Space: A Report to the Administrator, August 1987.

Source: NASA Historical Reference Collection, NASA History Office, NASA Headquarters, Washington, D.C.

This influential discussion is excerpted from NASA Leadership and America's Future in Space: A Report to the Administrator. *This report was prepared by former astronaut Dr. Sally K. Ride, and it provided a strong impetus within and beyond NASA for instituting a focused NASA program to conduct comprehensive studies of Earth as a planet, much as NASA had explored other planets in the solar system.*

MISSION TO PLANET EARTH

Mission to Planet Earth is an initiative to understand our home planet, how forces shape and affect its environment, how that environment is changing, and how those changes will affect us. The goal of this initiative is to obtain a comprehensive scientific understanding of the entire Earth System, by describing how its various components function, how they interact, and how they may be expected to evolve on all time scales.

The challenge is to develop a fundamental understanding of the Earth System, and of the consequences of changes to that system, in order to eventually develop the capability to predict changes that might occur—either naturally, or as a result of human activity.

Background

With the launch of the first experimental satellites in the 1960s, NASA pioneered the remote sensing of Earth from space. Over the past two decades, the scientific community has concluded that Earth is in a process of global change, and scientists now believe that it is necessary to study Earth as a synergistic system. As stated in the Earth System Sciences Committee report cited earlier, "Global observations, new space technology, and quantitative models have now given us the capability to probe the complex, interactive processes of Earth evolution and global change." Interactive physical, chemical, and biological processes connect the oceans., continents, atmosphere, and biosphere of Earth in a complex way. Oceans, ice-covered regions, and the atmosphere are closely linked and shape Earth's climate; volcanism links inner Earth with the atmosphere; and biological activity significantly contributes to the cycling of chemicals (e. g., carbon, oxygen, and carbon dioxide) important to life. And now it is clear that human activity also has a major impact on the evolution of the Earth System.

Global-scale changes of uncertain impact, ranging from an increase in the atmospheric warming gases, carbon dioxide and methane, to a hole in the ozone layer over the Antarctic, to important variations in vegetation covers and in coastlines, have already been observed with existing measurement capabilities. The potentially major consequences, either detrimental or beneficial, suggest an urgent need to understand these variations.

We currently lack the ability to foresee changes in the Earth System, and their subsequent effects on the planet's physical, economic, and social climate. But that could change; this initiative would revolutionize our ability to characterize our home planet, and would be the first step toward developing predictive models of the global environment.

Strategy and Scenario

The guiding principle behind this initiative is to adopt an integrated approach to observing Earth. The observations from various sensors on platforms and satellites will be coordinated to perform global surveys and also to perform detailed observations of specific phenomena.

Mission to Planet Earth proposes:
1. To establish and maintain a global observational system in space, which would include experiments and free-flying platforms, in polar, low-inclination, and geostationary orbits, and which would perform integrated, long-term measurements.

2. To use the data from these satellites along with in-situ information and numerical modeling to document, understand, and eventually predict global change.

As illustrated in Figure 5, the global observational system would include a suite of nine orbiting platforms:

Four sun-synchronous polar platforms: two provided by the United States and one each provided by the European Space Agency (ESA) and the Japanese National Space Development Agency (NASDA). The first platform would be launched in 1994 and all four platforms would be in orbit by 1997. These platforms would provide global polar coverage with morning and afternoon crossing times.

Five geostationary platforms: three provided by the U.S. and one each by ESA and NASDA. These platforms would all be launched and deployed between 1996 and 2000.

["Figure 5. Mission to Planet Earth" image not included]

Low-inclination, low-altitude payloads would also be included in the system. The Earth Radiation Budget Experiment satellite, launched from the Space Shuttle in 1984, and the synthetic aperture radar sensors, SIR-A and SIR-B, flown on the Shuttle in 1981 and 1984, are the types of experiments that would fall into this category. Another example would be a proposed Space Station-attached payload designed to obtain coverage of tropical rainfall with sampling at all local times.

The integrated system would measure, the full complement of the planet's characteristics, including: global cloud cover, vegetation cover, and ice cover; global rainfall and moisture; ocean chlorophyll content and ocean topography; motions and deformations of Earth's tectonic plates; and atmospheric concentration of gases such as carbon dioxide, methane, and ozone.

Space-based observations would also be coordinated with ground-based experiments and the data from all observations would be integrated by an essential component of this initiative: a versatile, state-of-the-art information management system. This tool is critical to data analysis and numerical modeling, and would enable the integration of all observational data and the development of diagnostic and predictive Earth System models.

This global observational system would be designed to operate for decades, serviced either by astronauts or robotic systems to ensure long life and to provide the continuing data collection, integration, and analysis required by this initiative.

Because of its international and interdisciplinary nature, the Mission to Planet Earth requires the strong support and involvement of other U.S. government agencies (particularly the National Science Foundation and the National Oceanic and Atmospheric Administration) and of our international partners. The roles of the various Federal agencies have been examined in detail by the Earth System Sciences Committee. NASA's responsibilities would include the information management system and platforms and experiments described previously. Most important, NASA would also provide the supporting technology, space transportation, space support services, and much of the scientific leadership.

Technology, Transportation, and Orbital Facilities

This initiative requires advances in technology to enhance observations, to handle and deliver the enormous quantities of data, and to ensure a long operating life. Sophisticated sensors and information systems must be designed and developed, and advances must be made in automation and robotics (whether platform servicing is performed by astronauts or robotic systems).

To achieve its full scope, this initiative requires the operational support of Earth-to-orbit and space transportation systems to accommodate the launching of polar and geostationary platforms. This does not represent a large number of additional launches, but it does require the capability to launch large payloads to polar orbit; Titan IVs would be used to accomplish this. Since the envisioned geostationary platforms would be lifted to low-Earth orbit, assembled at the Space Station, and then lifted to geosynchronous orbit with a space transfer vehicle well-developed orbital facilities are essential. By the late 1990s, the Space Station must be able to support on-orbit assembly, and a space transfer vehicle must exist.

SUMMARY

NASA, with its technical and scientific expertise, is uniquely suited to lead Mission to Planet Earth. Only from Earth orbit can we gain the perspective necessary to observe the Earth System and the interaction of its components on a global scale. We now understand what to observe, and how to observe it. While we do not yet know how the data will piece together, the resulting Earth System models, developed and refined over years of study, are the important products of this initiative, and would establish NASA as a responsive agency ready to meet the challenge of a genuine time-critical need. Championing this initiative would establish the United States at the forefront of a world recognized need to understand our changing planet.

Document IV-13

Document Title: Edward Frieman, Chairman, Earth Observing System Engineering Review Committee, NASA, "Report of the Earth Observing System Engineering Review Committee," September 1991.

Source: NASA Historical Reference Collection, NASA History Office, NASA Headquarters, Washington, D.C.

This report presents the results of a committee formed by NASA to review its plans for providing the satellite platforms and sensors needed to carry out a comprehensive study of Planet Earth. This document was instrumental in helping NASA redefine its program in light of a substantial Bush administration cut in NASA's future budget for MTPE. Particularly significant was this report's emphasis on the use of smaller satellite platforms than NASA had planned, as well as the consideration of formation flying to solve some of the issues of simultaneity of observations that scientists felt were crucial to deeper understanding of Earth's systems. This report specifically examines NASA's program in light of the needs of the broader U.S. Global Change Research Program.

[title page]

**REPORT OF THE
EARTH OBSERVING SYSTEM (EOS)
ENGINEERING REVIEW COMMITTEE**

SEPTEMBER, 1991

Chairman: Edward Frieman

Members: D. James Baker
Peter Banks
Greg Canavan
Richard Goody
Veerabdhadran Ramanathan

Warren Washington
Albert Wheelon

[1]
EARTH OBSERVING SYSTEM (EOS)
ENGINEERING REVIEW COMMITTEE

TABLE OF CONTENTS

****

[5]**1. BACKGROUND AND PERSPECTIVES**

In early 1991, NASA established the Earth Observing System (EOS) Engineering Review Committee and asked it to provide advice regarding implementation of EOS, including size of platforms, instrument configuration, and launch requirements and sequencing. The Committee met in public meetings twice, in May at NASA Headquarters in Washington, D.C., and in July at Scripps Institution of Oceanography in La Jolla, California, for a total of eight days. The formal charges to the Committee and the members are listed in the appendices.

During its meetings the Committee heard briefings from NASA personnel, representatives of other agencies, individuals, and industry. The Committee considered a series of NASA-generated alternative options for implementing the EOS program. The Committee also considered options for building and flying relevant instruments presented by representatives of the Departments of Energy and Defense, by individuals from industry, and by interested parties who responded to a published notice.

To carry out its charge, the Committee heard information prepared by NASA on technical and scientific tradeoffs of alternative approaches both to the EOS instruments and to the flight of the instruments. The Committee also received limited information on costs and budgets. The Committee considered alternative instruments and platform launch sequences with the objective of meeting the scientific objectives of understanding climate change and global warming and accelerating the most critical measurements. The Committee also discussed certain issues relating to cost and to scientific and technical issues associated with using high-altitude, unmanned aircraft missions as complements to current ground and space-based missions.

The Committee carried out its deliberations in the context of the needs of an overall observing system as identified by the Intergovernmental Panel on Climate Change (IPCC) and the U.S. Global Change Research Program (USGCRP). In its analysis, the Committee recognized the need for beginning as soon as possible the measurements by those instruments that most directly relate to data for understanding climate change and global warming.

The Committee found that its consideration of implementation of the Earth Observing System led inevitably to consideration of two other closely linked elements of NASA's Mission to Planet Earth: Earth Probes and the Earth Observing System Data and Information System (EOSDIS). The Committee believes that the Earth Probes, providing as they do critical near-term measurements for global change, must be considered as an integral and high

[6]priority part of the EOS process. For the data system, the Committee believes that EOSDIS must scale as the space program itself scales; if there is major change in one, there must be major change in the other.

In its first meeting, the Committee was presented with NASA's original technical and budget plan based on two series of large platforms for a 15-year program. The Committee was impressed with the scientific scope and technical advances proposed, but noted at that time that budget limitations might dictate change. Even before the second meeting, overall budget constraints began to loom.

In its second meeting, the Committee was faced with two new realities. First, the budget scenario provided by the Senate for FY1992 and the reduction of the total budget to the year 2000 from approximately $16 billion to $11 billion clearly makes it impossible to meet the scope and timing of the EOS program as originally conceived. Second, new launch vehicle options now available allow intermediate size platforms to be considered for the EOS instruments.

Moreover, the Committee discovered during its deliberations that a better understanding of the simultaneity needs of the EOS instruments has relaxed the requirements for collocation [sic] of all EOS instruments on one large spacecraft as originally proposed. We have found that a smaller platform could accommodate those groupings of instruments that must fly on a single platform. We have also found that it is fully feasible to fly platforms in close enough formation to meet any additional simultaneity needs. It is thus now possible to group sensors with natural simultaneity requirements which in themselves address the priority needs of climate research.

The set of new constraints and options led the Committee to ask NASA to provide a series of options for launch and instruments that could meet the priorities of the IPCC and the USGCRP, fit within the estimated budget envelope, and that would fly the most essential instruments as early as possible. The Committee hoped to find a flight option that was at once fully responsive to the needs of the global change research program and resilient, noting that resilience flows from thinking about the consequences of budget cuts and technical failures as well as budget enhancements and successful flights. NASA provided the Committee with a scientific framework and a series of options which the Committee has carefully considered.

In the discussion that follows, the Committee presents its view of EOS science and its consideration of implementation options. The discussion is followed with a list of findings and recommendations.

****

[20]4. FINDINGS AND RECOMMENDATIONS

4.1. Program Design

BACKGROUND

The Committee finds that the scientific planning for EOS is sound and commends NASA for its hard work and long effort in bringing this important issue of Earth remote sensing to the forefront of the climate debate.

The Committee finds that the EOS program is faced with two new realities. First, the budget scenario provided by the Senate for FY1992 and the reduction of the total budget to the year 2000 clearly make it impossible to meet the scope and timing of the EOS program as originally conceived. Second, new launch vehicle options now available allow intermediate size platforms to be considered for the EOS instruments.

SIMULTANEITY

The Committee finds that a mixture of intermediate and small satellites could carry the EOS sensors while maintaining science and budget resilience. The Committee emphasizes that the requirement for measurement simultaneity does not require that large groups of instruments be co-located on single large spacecraft requiring a Titan launch. All the instruments that must be on the same platform can fit on an intermediate-sized spacecraft. Additional simultaneity required can be satisfied by flying spacecraft in formation. The sensors that measure related science should be flown during the same time interval. The current scenario of launch vehicles indicates that the appropriate mix of rockets will be available at the right time.

ADVANTAGES OF INTERMEDIATE AND SCR SATELLITES

The Committee finds that deploying EOS instruments on several satellites is preferable to assembling them on a single platform. This concept provides quicker development and flight since fewer components are required, and protection against launch vehicle failure and early instrument demise since replacement satellites and replacement instruments can be flown quicker and more cheaply. Resilience against unexpected budget changes is also provided since smaller sets of instruments can be added or delayed with less consequence for the full program. Since each set of instruments to fly can have a narrower focus, there is easier optimization of science needs. Finally, there is quicker technical evaluation and innovation since instruments can be tried without jeopardizing the full program.

[21]LAUNCH OPTIONS

The Committee finds that a variety of launch vehicles will soon be available for launching EOS payload combinations into appropriate orbits: Titan 4, Atlas IIAS and others, Delta II, Pegasus, Titan II, and possibly Taurus. The Committee believes that the flight of EOS instruments on intermediate and smaller platforms launched with Atlas and Delta class and smaller rocket boosters is a better option. This strategy provides the EOS program with enhanced budgetary resiliency, more technical options for the EOS instruments, and the possibility of greater scientific return. The Committee also finds that a separation of goals into long-term monitoring (climate satellites) and shorter-term, detailed process studies is scientifically sound and also increases the program's responsiveness to scientific needs.

The Committee recognizes that some science is lost from the original $16 billion program, since some of the original instruments will be delayed, and some may never fly. But the intermediate and smaller satellites do offer the possibility of earlier launch of critical instruments. Moreover, the more flexible option now proposed will provide quicker technical innovation that may make up at least partly for the loss of some of the original instruments.

INSTRUMENT GROUPINGS

The Committee recommends that NASA plans for individual EOS missions be organized principally around prioritized USGCRP science goals. In contrast to the previous approach using large platforms, the Committee believes that by concentrating upon the flight of crucial instruments on intermediate and smaller spacecraft, it should be possible to more rapidly acquire key climate and Earth process information needed for the USGCRP.

The Committee finds that the NASA option which puts the instruments on three basic spacecraft launched by Atlas IIAS or smaller rockets demonstrates that a significant part of the mission needs can be met. More importantly, it can be accommodated under the funding profile established by the U.S. Senate Appropriations Committee. It might be possible to divide one or more of the three spacecraft into smaller satellites, depending on the instruments to be flown and their orbit priorities. The set of instruments listed for EOS-A and the set for EOS-B should no longer be thought of as separate packages, but should be considered together as a set. The Committee recommends that NASA use the Atlas IIAS concept discussed here as a proof of concept, and explore the details of instrument selection and flight with the relevant scientific communities.

[22] The other part of the EOS needs might well be met with the technical talents that could be brought to bear by DOE and DOD. These agencies have skills in this general area, and are interested in being involved. A climate mission for monitoring changes in forcing functions and radiation budget is one example that might be of interest here.

The possibility of using common platforms for the two or more of the missions should be explored. As additional funds become available, it will be important to consider scientific needs before simply adding more of the same type of satellite in the same orbit. Other orbits, such as lower sun-synchronous, non-sun-synchronous, elliptical, and geostationary should be considered.

The Livermore concept of a "Brilliant Eyes" version of EOS involving multiple satellites flying at 400 km altitude and carrying an array of sensors was presented to the Committee. Some of the sensors involve new technology in the early stages of development. The Committee saw that many of the techniques proposed for this program could be relevant if they are adequately developed. For example, the lower altitude allows significant reductions (i.e., ten times or more) in the weight of mainline instruments, but at the expense of other complications. It is clearly desirable to fly lidar and SAR missions in lower altitude orbits. The Committee recommends that support for instrument technical development be supported at a level adequate to demonstrate that accurate calibrated measurements can be made. This development should be carried out as quickly as possible. The overall concept, however, is viewed as being in too premature a state at this time to seriously recommend that it replace the EOS system as proposed here.

The Committee also noted that a vigorous industry has arisen to build and launch small satellites rapidly. However, these spacecraft cannot carry the large mainline instruments that are central to the EOS mission. If the weight and especially volume of those instruments can be reduced significantly through the efforts of NASA, DOE and DOD, these small satellites could play an important role as EOS evolves. It would also be useful to explore increasing the volume of the fairings on the small launchers to increase their capability to carry EOS instruments.

EARTH PROBES

Establishment of the Earth Probes program is a major achievement in recognizing the importance of Earth measurements and providing a line of missions devoted to process studies and technical demonstration. The missions now in the Earth Probes line provide essential information during the coming decade for the U.S. Global Change Research Program. The Earth [23] Probes line is not merely a set of precursor missions to EOS, but an essential continuing element of that program. Other agencies such as DOE and DOD may be able to make important contributions to Earth Probes, especially in the area of new technology for lidars and SARs. The Committee suggests that NASA include the Earth Probes line in the general concept of EOS by using the term "Early EOS Missions".

LEAD TIMES

A procedure must be identified to allow EOS spacecraft and instrument combinations to be built and flown in less than the nine years experienced by virtually all NASA scientific spacecraft programs in recent times. SDI space experiments of substantial complexity were repeatedly mounted in less than two years, often working through NASA mechanisms. This may require streamlining of NASA procurement processes.

STRATEGIC PLAN

NASA's strategic planning for EOS needs sharpening, especially in the light of the new realities. This process could be based on the discussions held between NASA and the Committee. As the program evolves and changes, it is important that the strategy be comprehensive and clear. Changes stemming from either scientific, technological, or financial reconsideration of the mission must be understood and thoroughly documented. The full scientific community, including those from other federal agencies, must be involved in the development of the strategy. The Committee recommends that NASA refine its plan, based on the documentation that was prepared for the review.

4.2. Partnerships

DOD and DOE

Just as NASA now works with its non-U.S. partners, it should work with the Departments of Defense and Energy. The Defense Department has contributed importantly through its sustained support of technology for SDI and reconnaissance. However, the Committee believes that a more vital role can be identified for DOD if it is willing to commit to talent, funding, special procurement authority, and especially data collected from its own space systems. For example, DOD needs for wide-area surveillance could address improved imagers along the lines of the proposed HIRIS instrument. DOD experience in synthetic aperture radar and lidar could provide early development of a multi-frequency radar instrument for study of vegetation and soil moisture.

[24] There is enormous scientific and engineering talent in the three national security laboratories of the Energy Department and its contractors and scientific collaborators. Some of this talent should be applied to the EOS program. An adequate budget should be provided beginning in FY92 that enables them to develop sensors and to concentrate on Earth Probes and other early space missions that will contribute directly to understanding global change. They should concentrate on reducing the weight and size of the mainline EOS instruments. They should have a role in developing laser technology for direct measurement of winds (e.g., LAWS), ice-sheet altimetry (e.g., GLRS), and water vapor lidar systems.

The Committee recommends that instrument development by DOD and DOE should be carried out as quickly as possible. In the view of the Committee, by using the full technical capabilities of DOD, DOE and NASA, the U.S. government could strive to have a pre-EOS measurement capability (e.g., climate monitoring) in space by 1995. The U.S. government should use all of its resources to validate this new technology in time to support its test by 1995.

The DOE laboratories can and should make a very strong contribution to ground measuring systems and in developing instruments for remotely piloted aircraft for making precise measurements of atmospheric processes. The talents and resources of the two agencies together could provide useful impetus in flying an early climate mission for monitoring changes in forcing functions and radiation budget.

The Committee recommends that DOD and DOE be encouraged to be involved in the U.S. Global Change Observing System and that they have adequate funds for such within their own budgets.

NOAA

NOAA has the mission for monitoring the Earth, a key element of global change. In the long-term, it will not be possible for NASA to sustain a monitoring program. Therefore, the Committee recommends that provision should be made early for flight of pre-operational instruments, transfer of such instruments to operational programs when ready, and full involvement of NOAA in EOSDIS planning and execution. NOAA must budget adequately for such monitoring and data activities required by USGCRP.

[25]4.3. EOSDIS

The EOS Data and Information System is the largest data and information system contemplated by this nation. The Committee highly commends NASA for its effort to date in developing EOSDIS.

In light of the need for getting data sets out to the community now, the Committee supports the emphasis placed on Version 0. The Committee recommends that Version 0 of the program should have high priority and possibly be strengthened, The Pathfinder data sets offer an opportunity to address key questions of global change and to provide prototypes for the generation and distribution of data sets in the EOS era. An element of the Pathfinder activities which could receive greater emphasis is the validation of data sets.

We caution, however, that EOSDIS should not develop a life of its own, independent of the degree of the experimental and observational program. It must be flexibly planned to fit with programs at different levels of activity. We further urge that cooperation with international and national agencies be emphasized in order to increase overall effectiveness and to provide hedges against funding vagaries. Later versions of the system must be paced by the actual flight of instruments, which is slower under the new option of multiple flights of spacecraft.

The Committee also has various concerns about the evolving plans for EOSDIS, including its high cost, the flexibility of its development, and the current lack of simulation models supporting the adopted system architecture and hypothetical user groups. With these considerations in mind, the Committee encourages NASA to proceed slowly with plans to put implementation in the hands of a single industrial contractor until more is learned from Version 0. A more flexible approach that could be considered would be based on a combination of national laboratories and university participation. An important aspect of this approach could be the day-to-day involvement of the scientific staff of the laboratories in the joint role of EOS investigators and EOSDIS developers, a feature that would be lacking in the industrial contract approach. We recommend that NASA carry out a review by technical experts to establish the best way to proceed.

4.4 A Coordinated Effort

The Committee regards EOS as an essential part of a total U.S. system in which NASA, NOAA (with its operational environmental satellites), DOE, and DOD are partners in a national and international effort. Furthermore, the Committee strongly affirms that the NASA EOS remote sensing program must be considered as a vital segment of an extensive observing system [26] that includes in situ and other types of measurements systems that will provide an integrated observing system. In this regard, it is important to regard EOS as a process for gathering global change data rather than as a platform for carrying such instruments. EOS is a process for successively improving our understanding of human impact on the Earth and refining the questions to be answered.

The Committee recommends that the CEES, with strong NASA participation and advice from a national science community, develop and carry through a U.S. Global Change Observing Program based on the following space-based elements:
- A series of science-driven process studies using small and intermediate-size space systems, remotely piloted aircraft, in situ and ground-based programs.
- A series of satellites aimed at monitoring key climate variables designed to establish a baseline and warning capability and to fill critical data gaps.
- New remote sensing technology developments.
- A major effort on a comprehensive Earth Science Data System including EOSDIS that is proportional to the sensors flown and the scientific data collected.

The Committee notes that many of the technical objectives of the NASA program require complex instruments and new methods of attack, and the Committee has seen that there is a wealth of ideas involving a variety of new space launch vehicles, instruments, and other methodologies which could make important contributions to the USGCRP over the next 15 years.

The Committee recommends broadened participation by all relevant federal agencies in the space measurements associated with the USGCRP. All capable and interested agencies should contribute to the execution of this plan. The plan must be put together in such a way that its robustness and resilience to funding vagaries is self-evident.

The new option proposed here must be implemented in the context of the IPCC and USGCRP priorities. In consideration of the total USGCRP, careful attention must be paid to the way in which the dominant space-based activity is planned and executed compared to the in situ component. All of the components of the USGCRP must be adequately funded. The overall program should be examined by the CEES, with oversight by the Office of Science and Technology Policy (OSTP) and OMB.

[27] **4.6 Infrastructure**

The extent of the manpower requirement is difficult to define at the present time. EOS, as originally planned, would increase the stream of environmental data by at least two orders of magnitude. NASA has made a good start in supporting infrastructure needs with its global change fellowship program. We believe that U.S. and international programs are in the long term dependent on substantial increases in our scientific infrastructure.

The Committee recommends that the CEES consult with the NAS to prepare a plan for the increase of scientific manpower and infrastructure for global change research and submit a budget for implementation in FY1994.

Document IV-14

Document Title: D. James Baker, Joint Oceanographic Institutions Incorporated, "Testimony to Congress," 26 February 1992.

Source: NASA Historical Reference Collection, NASA History Office, NASA Headquarters, Washington, D.C.

In addition commenting on NASA's restructuring plan for the Earth Observing System (EOS), this document provides a summary description of the relationship between the MTPE and other national and international Earth observation programs. This document is particularly noteworthy for viewing MTPE in a broad national and international context. Dr. Baker in 1993 became the Administrator of the National Oceanographic and Atmospheric Administration.

[title page]

Testimony to Congress

Senate Commerce, Science and Transportation Committee
Subcommittee on Science, Technology and Space

Hearing on Mission to Planet Earth
and NASA's Plan to Restructure the
Earth Observing System

February 26, 1992

by D. James Baker
Joint Oceanographic Institutions Incorporated
1755 Massachusetts Avenue N.W.
Washington, D.C. 20036

[1]

Thank you, Senator Gore, for the opportunity to present some comments about NASA's Earth Observing System and Mission to Planet Earth.

The successful launches last year of the European Space Agency's Earth Resources Satellite-1 (ERS-1) in April, a NASA ozone monitor on a Soviet Meteor spacecraft in May, and NASA's Upper Atmosphere Research Satellite in September marked the beginning of a new era of remote sensing of the Earth; a true Mission to Planet Earth. After two decades of development of sensors, platforms, and data relay and archive systems, NASA and its sister space agencies around the world have begun a comprehensive series of missions for the next decade and beyond. The recent revelations of ozone depletion in mid-latitudes is due to the foresight of Shelby Tilford and his colleagues, the administration, and the Congress, all of whom together pushed for UARS many years before measurements of stratospheric constituent was [sic] recognized by the public as a major global problem.

Mission To Planet Earth will provide a broad range of data on solar and Earth processes. The data will be fundamental to the major studies of global change now being undertaken by the scientific community as part of the World Climate Research Program and the International Geosphere-Biosphere Program. The Mission is the major part of the U.S. Global Change Research Program, the U.S. contribution to these international efforts. The results of UARS are the first major achievement of that program, and show the wisdom of support of this new technology.

The satellite programs are not cheap. In fiscal year 1992, the total U.S. Global Change Research budget is close to $1 billion, with the NASA effort accounting for almost three-quarters of that amount. By the end of the century, the total cost for Mission to Planet Earth is expected to be more than $10 billion, the largest unmanned mission ever proposed by any space agency. With such expenditures, it is essential that we have a consensus on the best ways to proceed with such a program.

Today, NASA's definition of Mission to Planet Earth includes three space-based program elements, a comprehensive data and information system, ground-based research, and a community of scientists performing research with the data acquired. The three space-based elements to be operated by NASA and its international partners include a set of near-term missions with specific disciplinary goals called Earth Probes, a broad measuring program called the Earth Observing System that will begin in the late 1990s, and a Geostationary Platforms program that will start after the turn of the century. The full system will provide a constellation of satellites in a variety of orbits around the Earth. The program also includes shuttle flights of instruments for test and short-term (a few days) measurements. Mission to Planet Earth does not make all the measurements of the Earth that could be carried out, but together with the operational satellite system, it will [2] make major contributions to most of the areas of Earth science known to be important today.

The goals of Mission to Planet Earth are consistent with the goals and priorities of the World Climate Research Program, which focuses on the physical aspects of climate, and the International Geosphere-Biosphere Program, which focuses on the chemical and biological issues. The time scales have been chosen so that the program focuses on the issues of climate prediction most closely associated with global warming, a subject of much public and political concern. The priorities for this aspect of the program have been set by the Intergovernmental Panel on Climate Change (IPCC) which has coordinated an assessment of the state of understanding of climate change and provided a consensus viewpoint on next steps.

Precursor Missions and Operational Programs

In the near-term the stage will be set for comprehensive measurements by a series of precursor missions that are the culmination of planning that began in the late 1970s. These near-term missions include those that have already been planned for a long time and now are ready to fly in the next two or three years, a set of missions under a new line at NASA called Earth Probes, and missions flown by non-U.S. space agencies. The near-term missions will provide a transition from today's limited duration research missions to the long-term and comprehensive Earth Observing System.

The Upper Atmosphere Research Satellite, the design for which started in 1978, was launched in September 1991 to help answer these and other questions. UARS carries ten instruments that are aimed respectively at the sun, through the limb of the Earth's atmosphere, and directly down at the Earth. Data from UARS will be used to study energy input and loss, global photochemistry, the dynamics of the upper atmosphere, and the coupling between the upper and lower atmosphere. Total Ozone Mapping Spectrometers, which have already given more than a decade of information on ozone flying on NASA's Nimbus-7, will fly on several spacecraft in the decade of the 1990s. Several flights of the Shuttle Solar Backscatter Ultraviolet (SSBUV) Experiment are also planned during the decade for ozone measurements. A Global Ozone Monitoring Experiment (GOME) is planned for ERS-2, the follow-on to ESA's ERS-1, to fly in 1994.

As part of Mission to Planet Earth, NASA has arranged the Atmospheric Laboratory for Applications and Science (ATLAS) program. This program involves experiments flown on the Shuttle for about a week on an annual basis for about a decade to gather data throughout the Sun's 11-year activity cycle. ATLAS-1 experiments will study the chemical makeup of the atmosphere between approximately 15 and 600 km [3] above Earth's surface. The instruments will also measure the total energy contained in sunlight and how that energy varies, investigate how Earth's magnetic and electric fields and atmosphere influence one another, and examine sources of ultraviolet light in the Universe. Radiation from the sun and clouds in the atmosphere are key elements of the climate system. Up to 1990 there was a set of instruments flying which made up the Earth Radiation Budget Experiment (ERBE). At that time, budget cuts forced the U.S. to discontinue radiation measurements on the NOAA satellites, but fortunately France and the Soviet Union began designing a scanning radiometer for measuring the components of the radiation budget. The instrument, called Scanner Radiatsionnogo Balansa (ScaRaB), is now in a prototype phase for test this year. Fully calibrated instruments are scheduled for flight in 1993 aboard two Soviet Meteor spacecraft to provide appropriately phased measurements with coverage of the diurnal cycle with four observations per 24-hour period.

Discussions are now being held in the U.S. about the possibility of flying small radiation instruments based on the developments of the Strategic Defense Initiative. These instruments, called "Brilliant Eyes" could provide an interim set of measurements complementing the ScaRaB data before the more comprehensive data that will be collected by the Earth Observing System starting in the late 1990s. A set of measurements of aerosols in the stratosphere and troposphere and radiation is also under discussion as a small mission ("climsat") that might take place in the mid-1990s.

Precipitation is also part of the global energy budget. Direct measurement of precipitation, particularly over the oceans, has always been difficult. In the mid-1990s, the Tropical Rainfall Measurement Mission (TRMM), a joint U.S./Japanese initiative, will monitor rainfall with active and passive microwave instruments, together with visible and infrared to derive rainfall amount and distribution between 35 degrees north and south latitude. Radiation will be monitored by this mission with some of the same instruments that are planned for flight on the Earth Observing System. TRMM data will be used directly in climate models that are critical to understanding global change. However, TRMM will not measure precipitation outside the tropics and subtropics, it will have relatively large sampling errors over land, and it will have a limited lifetime. These problems will be addressed with the global measurements and long lifetime proposed for the instruments of the Earth Observing System.

In the 1990s, several ocean-related satellite missions are either flying or planned for flight. The European Space Agency's ERS-1 was launched in early 1991 carrying instruments for measuring sea surface temperature, sea surface height, waves and winds, and ice concentration and extent. In mid-1992, a precision altimeter mission will be [4] flown jointly by the U.S. and France, called TOPEX/POSEIDON. A follow-on to the Navy's Geosat altimetry mission is planned for 1995, but there are no precision altimetry missions currently scheduled for the late 1990s. The next flight of a radar scatterometer is planned for 1996; NASA will have an instrument on the Japanese Advanced Earth Observation Satellite (ADEOS). After ADEOS, the next scatterometers will be carried by ERS-2 and EOS.

In 1993, an ocean color mission (SeaStar) is planned as a joint venture between NASA and a private company, Orbital Sciences Incorporated. In 1996, another ocean color instrument is scheduled for flight by the Japanese on ADEOS. In the late 1990s, an instrument for measuring ocean color is planned as part of NASA's Earth Observing System. Ocean color data near coasts is also provided by the land sensing systems Landsat and SPOT.

ESA's ERS-1 carries and ERS-2 will carry a synthetic aperture radar, as will Japan's ERS-1 and Canada's Radarsat. Thus the radar coverage of sea state and ice concentration will be good during the 1990s. The USSR Almaz satellite, launched this year, also has a radar system. These instruments, with their high spatial resolution (on the order of tens of meters) and their ability to see through clouds are also valuable for measurements on land. Radarsat, planned for launch in late 1994 or early 1995, will carry a synthetic aperture radar designed for ice measurements. Stereoscopic SAR imagery will also point out geological structures and help identify potential mining sites. Radarsat will also monitor and map renewable resources for the agricultural and forestry industries.

All of these SAR measurements are single-frequency. For the future, multifrequency SARs are now being tested by aircraft and shuttle flight. These SARs will allow measurements of the amount of vegetation at the land surface and of soil moisture, two quantities that are not easily measurable with current techniques. The SAR system planned as part of the Earth Observing System will be multi-frequency.

Geological and geophysical processes also vitally affect life on Earth. Volcanic eruptions, earthquakes, landslides, uplift and subsidence and associated hazards in coastal areas are all of interest and importance. Today, the European Space Agency, France, India, the U.S., the USSR, and Japan all operate satellites that monitor processes occurring at the land surface. France's Satellite Pour l'Observation de la Terre (SPOT), which began measurements in 1986, provides stereoscopic views of the land surface at high resolution. India and Japan operate systems with resolution similar to Landsat.

All of the current systems are undergoing upgrades in terms of more wavelengths and higher spatial resolution. For example, Landsat 7 will probably be an exact copy of Landsat 6, but upgrades or an entirely new system to help meet DOD needs are also under [5] consideration. The new Thematic Mapper for Landsat 6 (to be launched in 1992) will provide more detailed spectrum coverage than Landsats 4 and 5, which use 7 spectral bands and have 30 meter resolution. All of these systems have one major drawback, however, and that is lack of time resolution. The spatial detail is good, but it takes about a month to get full global coverage. Thus rapid changes in time are not adequately monitored by the existing systems. This is a drawback that can be addressed with the multi-satellite program proposed for EOS.

Measurements of the Earth's gravity field are under discussion for a mission for the late 1990s called ARISTOTELES (Applications and Research Involving Space Technologies Observing the Earth's Field from Low Earth Orbiting Satellites). This mission, which involves flying a proof mass inside a satellite and then tracking its detailed movements as it is affected by the gravity field, is being planned jointly between the U.S. and ESA. A magnetic field measurement is also being planned as a joint U.S./French mission, that would take place before EOS. The magnetic measurements are expected to be continued as part of EOS.

Precise satellite geodetic measurements, providing information on crustal deformation, continental drift and plate tectonics, Earth and ocean tides, and changes in the Earth's geopotential have been carried out since 1976 in a joint project between the U.S. and Italy called LAGEOS (Laser Geodynamics Satellite). LAGEOS is a large heavy satellite covered with laser retroreflectors in a circular orbit 6,000 km from Earth. Accuracies of a few centimeters per year of continental drift have been obtained from LAGEOS. A second such satellite is planned for launch in late 1992 by the Shuttle.

The Earth Observing System

Recognizing the need to take the next steps toward a long-term comprehensive space measurement system and the long lead times necessary for such planning, NASA and the scientific community began planning the next stage of Mission to Planet Earth, the Earth Observing System (EOS), in the early 1980s. The initial conception of EOS was that it would pull together the many strands of disciplinary measurements into one long-term program that was as comprehensive as possible. In this way the scientific community would benefit by having long-term data, and NASA would have a focused program that could be supported as a single unit. This initial general conception has held up; by consolidating its program, NASA was able to show the importance of the program and achieved formal approval from the administration and Congress in 1989 to go ahead with what I have noted is the largest unmanned space mission to be implemented.

The plans for EOS are to put into space the next generation of instruments for remote sensing, starting in 1998, for flight of a period of at least 15 years. This time [6] period, to be achieved by using missions end-to-end, is more than three times the normal span of a single mission. This initial phase covers the time over which major environmental change can occur. For example, in 15 years we can expect to see several atmospheric biennial oscillations, three to five El Niños, and an entire solar cycle. Although in the long-term we will need a commitment to such measurements for an indefinite period, the agreement to fund a series of satellites to make such measurements is an important new step for Earth sciences.

The concept of developing the next generation of instruments follows the recommendations of a series of National Research Council reports to extend the spectral coverage to more wavelengths, more finely divide the spectrum, and to provide higher spatial resolution. The set of instruments proposed for EOS includes a variety of visible, infrared, and microwave imagers and sounders, active radar altimeters and scatterometers, radiation and chemical sensors, particle and aerosol detectors, and geodetic positioning and ranging. In addition, new techniques are being developed for direct measurements of, for example, winds by laser systems (lidar) and water vapor by differential lidar systems.

The EOS program involves NOAA and non-U.S. agencies as well. The European Space Agency plans a series of Polar Orbiting Earth observation Missions (POEM) with two series, one (the M series) focused on meteorology, ocean and ice processes and the other (the N series) with a focus on land resources and related atmospheric processes. The Japanese plan at least two platforms, one in polar orbit and one in an inclined orbit for that time period. Starting in the early 21st century, NOAA satellites will be part of the program.

EOS is designed to provide broad and high-resolution spectral and spatial coverage. One particular strength of EOS is that it is designed to provide for the first time a long-term, simultaneous set of measurements of the same phenomena on ever-increasing spatial scales (local, regional, and global) that can be used to integrate and extrapolate our understanding of ecological and hydrological processes.

One of the important concepts of EOS is that of simultaneity. The remote sensing community has learned that missions flying one or two instruments can do much, but that for quantitative interpretation of measurements, it is good if instruments can point at the same point on Earth as measurements are made. This can be done by flying the instruments on the same spacecraft or by flying spacecraft in formation. The need for flying several instruments simultaneously has been justified by many different studies that show the advantage of having instruments look down through the same column of air, or look at the same piece of ocean or land surface at the same time. For example, data from surface imagers must be corrected for atmospheric moisture. Merging sequential data [7] streams can place technical and financial burdens on ground-based data management and computational systems.

The original NASA design for EOS was to have two series of large satellites, EOS-A and EOS-B. Each series would have three satellites, each with a lifetime of 5 years, so that there would be an overlapping 15-year time series of data, with the EOS-A series starting in 1998, and the EOS-B series about 2 1/2 years later. The satellites were planned to be virtually identical, so that cost savings could be made by duplicating spacecraft. The arguments for a large satellite were based largely on the need for flying instruments simultaneously. Additional arguments for the large spacecraft included cost and risk reduction.

But large satellites also bring disadvantages. With all the instruments on one satellite, there is a real risk of losing them all with a launch failure or a satellite system failure. All the instruments must fly at the same altitude and orbit, thus forcing compromises for the sensors. For example, biological processes and radiation studies must include the daily (or diurnal variations) of the signals. A sun-synchronous orbit does not permit measurements of diurnal signals, and therefore these processes cannot be measured well with the planned EOS orbit. The timing of the launch is dependent on the last instrument to get ready; if there is an instrument problem, then the whole satellite is delayed (in fact, the launch date of the first EOS satellite has already been moved to 1998 from an original announcement of 1994). Trying to put many instruments on one platform forces compromises in viewing angle and direction. Instruments may interact in ways that are not understood, and vibration from one may affect another. Finally, with a single large system, the only way to respond to budget cuts is to delay the program. For Earth measurements, long delays are not acceptable.

For all these reasons, and despite general agreement on the need for long-term comprehensive measurements, the scientific community has been far from unanimous about the implementation of the program on large platforms each with many instruments. In the past year and a half, in recognition of the need for agreement on implementation for such a large and important program, the Office of Science and Technology Policy and the National Space Council asked first the National Research Council and then NASA to establish review panels to examine the issues of implementation.

The first review in 1990 by a National Research Council Panel concluded that the arguments for simultaneity made the case for one large satellite for EOS to carry the instruments that must fly together along with some others. At the same time the NRC Panel strongly urged that the other instruments be divided up among several intermediate and smaller satellites, whose size and orbit should be decided on scientific issues. The NRC Panel urged NASA to develop a contingency and backup plan in case of technical [8] or budgetary problems.

In 1991, a number of new realities impinged on the EOS planning. The Congress cut the EOS request drastically for FY 1992 and 1993, and cut the total funds for the program requested to the year 2000 by one-third. The only way that NASA could meet these budget cuts was either to slip the launch and hope to get more funds later, or to plan to fly the large satellite without some of the instruments that were planned. Neither one of these scenarios was acceptable either to the scientific community that supported the program or to Congress and the National Space Council.

Fortunately, new technical developments had occurred. For unrelated military reasons, the Air Force had decided to re-open the Western Test Range for support of the Atlas IIAS launch vehicle, an intermediate vehicle. Thus the way was open for NASA to design to this size rocket. At the same time, a number of studies were carried out by industry, DOD and DOE on formation flying of satellites to provide simultaneous measurements. These studies showed that it was indeed possible to fly satellites close enough together and to control their position well enough so that accurate simultaneous measurements could be made. Thus the simultaneity requirement could also be met by flying smaller satellites.

In mid-1991, an Engineering Review Panel was appointed by the National Space Council and NASA to review these new developments and make recommendations on implementation to NASA. The Panel confirmed the new developments and worked with NASA to come up with scenarios involving intermediate and smaller satellites that could both meet the scientific requirements and the budget constraints. The Panel suggested redesign of the platform, reassessment of the data system, and use of the Atlas IIAS (or equivalent) launch vehicle. They also recommended that the instrument technology being developed at the Departments of Defense and Energy be incorporated into the planning for EOS.

As a consequence of the review, the plans for the large satellite have been dropped. NASA has now developed an implementation plan consistent with those recommendations. The current plan involves a morning platform focused on energy budget, surface imaging, and lower atmospheric chemistry to fly in 1998, and an afternoon platform with similar instruments and including an atmospheric sounder. Smaller satellites focusing on ocean color, aerosols and clouds, and geodetic and topography measurements are proposed in the 2000 time period. A platform with focus on surface winds, atmospheric dynamics and chemistry, and aerosols and clouds is planned for 2002. These successive payloads correspond roughly to the priorities as established by the international global change research programs: climate with emphasis on energy, radiation, and water, oceans and tropospheric chemistry, and solar radiation and land.

This plan does provide some budget resilience by allowing smaller packages to fly earlier. However, some science is lost because the overall budget is smaller. Moreover, there will be a major budget increase required between 1994 and 1995 to ensure flight of the satellites on schedule. It is clear that adding funds to this program would improve the schedule for the delivery of instruments and spacecraft. Establishing a stable and adequate funding profile would allow for long-term planning and thus help to improve technical and schedule performance.

It is interesting to note that other space agencies are also looking into the possibility of flying small satellites for earth observation. The European and Japanese platforms originally proposed to contribute to EOS have been downsized. The Canadian Space Agency is also exploring the use of a series of small satellites to collect data complementary to other Earth observation programs. Along these lines, consideration is being given to piggy-backing remote sensing instruments on other systems, particularly communications satellites. Motorola is planning a 77-satellite constellation in low Earth orbit for communications. It might be possible to put small Earth-sensing instruments derived from Brilliant Eyes technology aboard each of these satellites, providing a different kind of view of Planet Earth. Discussions on this possibility are now going on.

The Earth Observing System Data and Information System

The amount of data coming down from satellites and being collected by Earthbound sensors is increasing rapidly. By the time of the late 1990s, with the many satellites in place that we have discussed above, there will be at least one terabyte per day of raw data being collected. This amount, 10^{12} pieces of information per day, is equal to the information in the Library of Congress per day that must be processed, archived, and made available for distribution. This is an enormous task, the largest data and information task that any government agency has yet faced.

In recognition of the magnitude of the problem, NASA has put a major emphasis on the EOS Data and Information System. EOSDIS is planned to acquire and maximize the utility of a comprehensive, global, 15-year data set. To show the commitment of NASA to the data system, NASA will allocate only 40% of the EOS funding to spacecraft hardware and 60% to ground-based activities, including EOSDIS and related science. This contrasts with the usual mission, where about 70% goes to spacecraft hardware and 30% to ground-based activities.

The development of EOSDIS has been initiated by building on the existing infrastructure within the research community. The first stage, called Version Zero, is bringing together existing satellite data sets including the Advanced Very High Resolution Radiometer (AVHRR), the TIROS Operational Vertical Sounder (TOVS), and the data [10] from the Geostationary Operational Environmental Satellite (GOES) as "Pathfinder Data Sets" into a prototype data system. From this prototype, NASA, working with the other agencies in the U.S. government that have had experience with working with large data sets (primarily the defense community), will develop an evolving data and information system that will serve the communities interested. As far as long-term responsibility is concerned, NOAA will archive and distribute the atmosphere and ocean data; USGS will handle the land-related data. The architecture will be open and distributed so that it can evolve with advances in computing and networking technology. Access to the raw data from EOS will be provided to operational agencies for forecasting and assessment purposes. NASA will provide for the commercial distribution of data on a nondiscriminatory basis to all other users. The advisory committees have suggested that the development of EOSDIS be commensurate with the development of the space hardware system, and NASA has agreed to proceed in that way.

A good example of an early data system now being used is WetNet. This is a five-year pilot program using the Special Sensor Microwave Imager (SSMI) data from DMSP in a remote interactive computer network for science investigators from NASA, NOAA, the university community and the private sector. The SSMI measurements are being used to retrieve global fields of precipitation, snow cover, oceanic wind speed, water vapor, and cloud water. WetNet ensures the timely production of these fields, and fosters an environment where these algorithms can be tested and new algorithms (for example, vegetation mapping) can be tested. The WetNet data analysis and archive system is based primarily on the Man-Computer Interactive Data Access System (McIDAS) developed at the Space Science and Engineering Center at the University of Wisconsin at Madison.

Issues for the Future

In this discussion, one must understand that NASA is not the only space agency involved. In the U.S., the National Oceanic and Atmospheric Administration (NOAA) is responsible for the operational weather satellites called the Polar Operational Environmental Satellites (POES) and the Geostationary Operational Environmental Satellites (GOES). The Air Force is responsible for the Defense Meteorological Satellite Program (DMSP) that meets DOD needs. Outside the U.S., the European Space Agency and the space agencies of the USSR and Japan operate operational weather satellites that provide continuing and useful data. These agencies and those of Canada, China, France, Germany, India, and Italy either operate or participate with other countries in operating a variety of Earth remote sensing missions. The number of countries is increasing each year, for example, Brazil and South Africa are expected to enter the field of remote sensing soon.

Other U.S. agencies besides NASA, notably the Departments of Energy and [11] Defense, have important capabilities in remote sensing developed for national security reasons. The Energy Department has operated a system for detecting nuclear explosions. As a consequence, both of these agencies have technical and scientific expertise in satellite systems. The agencies are expected to be more involved in satellite measurements in the future, bringing additional resources of human talent and funds to bear on measurements of global change.

For the future, the issues that will face Mission to Planet Earth are those that face all large scientific experiments: can the funding be sustained in the face of other budget pressures, will there be a large enough scientific community to carry out the program, and will the program deliver what it promises? At the present time, it is my view that the funding levels provided by Congress this year probably can be sustained provided that NASA and its constituencies provide a reconfigured program that meets the recommendations of the Engineering Review Committee. The reconfigured program will have the support of a much larger portion of the Earth science community, which will help with Congress and the administration. It won't be easy in the face of other NASA budget pressures for the shuttle and space station, but it is possible.

As far as a user community is concerned, it is clear that new scientists, trained in the study and use of remote sensing data, will be required. But our educational system is delivering fewer scientists in general rather than more, and those scientists tend not to go into Earth science. It may be that the long-term nature of Mission to Planet Earth is an advantage here: the rapid increases of funds now being put into science education may bear fruit in time for a new generation of Earth scientists to be ready for the data when it comes down from EOS and its sister systems in earnest. The Earth science community needs to ensure that such a new generation is in fact nurtured.

Will the program deliver what it promises? This is the most difficult question of all. Quantitative knowledge and improved skill of prediction of a complex system like the Earth comes slowly, as witness the slow improvement of weather prediction. Therefore it is unlikely that there will be, in the near term, new understanding that will have a major impact on policy determination. We could get lucky—discovery of the deepening ozone hole led almost immediately to governmental policy decisions to reduce the amount of chlorofluorocarbons emitted into the atmosphere. But on the whole, we have to anticipate that understanding will come slowly and that early results are not necessarily going to provide spectacular new information for policy makers. But even as the understanding comes slowly, the data that will be provided by a comprehensive observing system provides the ability to make assessments of the state of the environment now. These assessments in my view are well worth the price of a global observing system, and when we add the understanding that will be gained, we have a system well worth the cost.

[12] There is another problem which in my view is the most serious facing us for the establishment of long-term measurements. That problem is the transition from these research measurements to an operational system that will provide the data we need into the indefinite future. NASA and its sister agencies have been remarkably successful in getting attention and funding for the elements of Mission to Planet Earth, but at the same time in the U.S. the civil operational system has been starved. The operational environmental satellite system operated by NOAA, which must be the backbone of any long-term U.S. contribution to a global climate observing system, is showing the strain of budget neglect. Without a robust operational system, we will not have the long-term data sets we need.

There are two ways in which this problem makes itself known: first by the transition of new instruments from research to operations, and second by the modernization of the existing operational programs. Both areas have shown real difficulty in recent years. As far as instruments are concerned, very few instruments have made a smooth transition from research missions to operations. One of the successful examples is the microwave sensor for ice and snow that is now flying on the Defense Meteorological Satellite System. An altimeter and scatterometer, both having demonstrated their worth, are flying on the ERS-1 and ERS-2, but no commitment has been made for flight beyond those satellites. The Earth Radiation Budget scanner instrument (ERB), the Coastal Zone Color Scanner (CZCS), and the Total Ozone Mapping Spectrometer (TOMS) have all proved their maturity and the measurements they make are extremely important. But none of these has been selected for flight on an operational satellite. In an ideal world, these instruments would now be flying on NOAA operational satellites, but budget problems have precluded this.

The second problem comes from modernization of the existing operational programs. The new geostationary satellites (Geostationary Operational Environmental Satellites (GOES)) ordered by NOAA from NASA in 1986 have run into well-publicized trouble and cost overruns. A large part of this problem is due to the fact that NASA cut the funding for research and development for operational satellites, and the funds were never restored in NOAA's budget. Cost-cutting led to cutting corners in the early stages of design, and as a consequence the program for replacement was too rushed. The problems we see today can be traced directly back to the lack of early funds. It appears that the program is now back on track and that working satellites will eventually be produced. But if the old satellites die before the new ones are launched, the U.S. will lose its ability to monitor broad weather patterns across the country. To alleviate that problem, the U.S. has made an agreement with the European Space Agency to move one of their satellites, Meteosat 3, farther west to cover the Atlantic. The U.S., originally the pioneer in these areas, now finds itself in a position where it must turn for help to other countries.

[13] We have seen a related problem in the funding for remote sensing of land: faulty reasoning about the possibility of commercializing land remote sensing data led the U.S.

to reduce government funding for the program and transfer it to private industry. But the market has not proved to be adequate to sustain the system: data sales can come close to meeting operational costs, but the costs of developing and launching new satellites is well beyond what can be covered with the profits from data sales. As I mentioned at the beginning of the article, the military has now recognized the importance of Landsat data, and may be willing to help its funding. But something must be done, otherwise we would lose this important part of our operational program.

Let me conclude with a point about lead times for technology development. Compared to ground-based measurements, this new space-based technology not only is expensive, but also requires long development periods. Scientific and operational user communities are used to rapid turn-arounds of experiments and data. But they have had to change their style of working in order to incorporate these periods of many years between conception of an idea and flight of an instrument. As a consequence, space technology is still not well integrated with in situ programs; achieving coordination between space agencies and other agencies has been difficult. The recent work of the CEES investigating the possibilities of using small satellites with relatively fast turn-around times is a good step in the right direction.

The above points to a real problem of coordination in our government. The operational systems the U.S. has, including Landsat, the NOAA operational polar and geostationary satellites, the Department of Defense Meteorological Satellite Program, and EOS are not integrated in any coherent way: each agency tends to go its own way. The classified technical developments useful for remote sensing do not get transitioned into the civil side. This problem of an uncoordinated, piecemeal approach has been recognized by the Department of Commerce Inspector General in a report on coordination of remote sensing across the government, and it is a problem that should be taken up by the National Space Council, chaired by the Vice President and the Office of Management and Budget. If it is not, then we will lose the opportunity to apply this technology for a common good.

Thank you for the opportunity to testify.

Document IV-15

Document Title: NASA, "Report to Congress on the Restructuring of the Earth Observing System," 9 March 1992.

Source: NASA Historical Reference Collection, NASA History Office, NASA Headquarters, Washington, D.C.

This report details NASA's plans for restructuring the EOS, as requested by the House of Representatives Subcommittee on VA, HUD, and Independent Agencies. This reports how NASA expected to reduce the EOS budget from $16 billion to $11 billion through 2000. It also describes the role of the Earth Observing System Data and Information System (EOSDIS) and how NASA's program would interact with those of other agencies.

[title page]

REPORT TO CONGRESS
ON
THE RESTRUCTURING OF THE EARTH OBSERVING SYSTEM

Submitted by:
THE NATIONAL AERONAUTICS AND SPACE ADMINISTRATION
March 9, 1992

Content:

1. The Science Objectives of EOS
2. New Spacecraft Configurations
3. The Impact of the Restructured EOS on Understanding the Climate
4. The Impact of EOS on Adaptation and Mitigation Strategies
5. EOS Data and Information System
6. Interdisciplinary Investigations
7. Missions in Advance of EOS
8. The Role of the Departments of Energy and Defense
9. Summary of the Mission to Planet Earth
10. Funding Requirements

[1]

THE RESTRUCTURING OF THE EARTH OBSERVING SYSTEM

The following is a report to the Congress on the restructuring of the Earth Observing System program, the centerpiece of NASA's Mission to Planet Earth. This report is submitted pursuant to the requirements stated in the Conference Report accompanying HR 2519, the VA, HUD, Independent Agencies Appropriations.

The Earth Observing System (EOS) is designed to provide comprehensive, long-term observations from space of changes that are occurring on the Earth from natural and human causes so that we can have a sound scientific basis for policy decisions to protect our future. EOS will provide the continuation and extension of observations of the Earth that are now being undertaken as part of Mission to Planet Earth.

As is directed by the Committees on Appropriations, EOS is being restructured for three principal reasons:

- To focus the science objectives of EOS on the most important problem of global change – global climate change.
- To increase the resilience and flexibility of EOS by flying the instruments on multiple smaller platforms rather than a series of large platforms.
- To reduce the cost of EOS, including spacecraft, instruments, data system and science, from $16 billion to $11 billion through FY 2000.

The process by which NASA has restructured EOS to meet these constraints has been thorough, involving reviews by an external committee, evaluation by the scientists who will use the EOS data, and systematic engineering studies of spacecraft configurations and launch options. In July 1991 NASA presented the constraints to the EOS Engineering Review Committee chaired by Dr. E. Frieman (the Frieman Committee), and discussed with them options for how the EOS spacecraft could be reconfigured. The Frieman Committee in its report endorsed these options as a "proof of concept" for an EOS which contains a "favorable measure of resiliency." In August 1991 NASA discussed payload options at the Seattle meeting of the Investigator Working Group of all EOS investigators, and in October, conducted a formal review by [2] the EOS Payload Advisory Panel. This committee is comprised of the EOS interdisciplinary investigators and is formally charged with recommending EOS payloads to NASA. Concurrently, extensive engineering studies were conducted by the Goddard Space Flight Center to determine the most effective spacecraft configuration so that the instruments can be accommodated on smaller platforms. In December 1991 the NASA Administrator conducted a thorough review of and approved the restructured EOS program.

The final payload configurations for the restructured EOS satisfy the Congressional constraints – they focus on climate change, are launched on multiple smaller platforms, and have a total cost of $11 billion through FY 2000. The final payloads are very similar to those endorsed by the Frieman Committee and wholly consistent with the Committee's recommendations. These payloads satisfy the recommendations of the EOS Payload Advisory Panel with the caveat that some of the instruments will fly later than recommended due to the constraints on the budget.

The National Space Council, chaired by the Vice President, is currently developing a National Space Policy Directive covering the space-based elements of the U.S. Global Change Research Program. This document will direct NASA to implement the restructured EOS program as part of an overall space-based global change observation system.

1. THE SCIENCE OBJECTIVES OF EOS: The purpose of the restructured Earth Observing System, is to determine the extent, causes, and regional consequences of global climate change. The extent – the change in average temperature and the time scale over which it will occur – is presently unknown. The causes can be either natural or human induced. It is the latter, however, that must be understood if we are to alter human behavior and avoid the climate changes which are most detrimental. The regional consequences – changes in precipitation patterns, length of growing seasons, severity of storms, sea level, etc. – must be understood if we are to determine which aspects of climate change will be detrimental and how we should adapt to those changes which cannot be avoided.

The behavior and evolution of the climate system is determined by complex processes within and interactions among the atmosphere, the oceans, the biology, and the snow and ice systems of the Earth. Scientists and policy-makers have documented in recent years which of these processes and interactions are most important to understand. In particular, the Intergovernmental Panel on Climate Change, the scientific priorities of the US Global Change Research Program, and the EPA report "Policy Options for Stabilizing Global Climate" are all in agreement that if we are to understand the climate and what the human species is doing to it, we must develop a quantitative understanding of seven key issues:

- The Role of Clouds, Radiation, Water Vapor, and Precipitation.
- The Productivity of the Oceans, their Circulation, and Air-Sea Exchange.
- The Sources and Sinks of Greenhouse Gases, and their Atmospheric Transformations.
- Changes in Land Use, Land Cover, Primary Productivity, and the Water Cycle.
- The Role of the Polar Ice Sheets, and Sea Level.
- The Coupling of Ozone Chemistry with Climate and the Biosphere.
- The Role of Volcanoes in Climate Change.

[3] The instruments that will fly in the restructured EOS program have been selected to address these key scientific issues associated with climate change. The original EOS program addressed a broader range of global change issues, including studies of stratospheric chemistry and its controlling influence on ozone depletion, and aspects of solid Earth physics. The original EOS program included a total of 30 selected instruments. By focusing on climate change, the required instruments are reduced to 17 that need to fly before 2002; 6 that are deferred, and 7 that are deselected.

2. NEW SPACECRAFT CONFIGURATIONS: In the original EOS program the instruments were to be flown on a series of large platforms, each containing about 15 instruments and launched on a Titan-IV with solid motor upgrades. There are certain advantages to the large platform approach:

(i) The total cost for flying a set instruments on a large platform is less than the total cost of flying that same set of instruments on multiple smaller platforms.
(ii) It is easier on a large platform to accommodate those instruments which need to make simultaneous observations of the Earth. The sensitivity of the EOS instruments is such that they generally rely on simultaneous measurements from other instruments, through the same column of air, in order to make atmospheric corrections and/or to pursue their science objectives.

However, there are also major disadvantages to the large platform approach:

(i) The large platform with multiple expensive instruments is inflexible to budget shortfalls or overruns. The only solution to budget problems is to slip the launch of the platform and all its instruments, thereby delaying the acquisition of EOS data.
(ii) The large platform with multiple instruments is inflexible to technical problems. If one major instrument experiences development delays, the launch of the entire platform will slip.
(iii) The orbit of a large platform cannot be optimized for each of the multiple instruments it will carry. Instruments can have different requirements for orbital altitude and inclination and for the time of day at which the spacecraft crosses over points on the Earth.

(iv) A launch failure, which is the greatest single risk to space missions, will result in a greater catastrophic loss to a large platform mission than to smaller platform missions.

When the EOS program was designed, prior to FY 1991, the only available launch vehicle for EOS spacecraft was a Titan-IV. EOS spacecraft are launched from the [4] Western Test Range at Vandenberg Air Force Base in California into sun-synchronous, polar orbit. Such orbits provide both global coverage and allow the spacecraft to pass over points on the Earth at the same time each day, and thus observe with constant sunlight conditions. The launch vehicles available from the west coast were expected to be only a Delta-class, which is too small to accommodate most of the required payload configurations for EOS, or the large Titan-IV-class, which can easily accommodate the large platform, with 15 instruments. Accordingly, EOS was designed in the large platform configuration.

In the spring of 1991, the Air Force announced that it would seek funds to construct facilities at Vandenberg to launch intermediate launch vehicles, in the Atlas IIAS-class. Such vehicles can accommodate payloads which are approximately one-third those of the Titan-IV, and which are large enough to include appropriate groupings of EOS instruments. Thus it is now possible, whereas it was not earlier, to reconfigure the EOS payloads to fly on multiple smaller platforms.

The Frieman Committee, and most of the other advisers that NASA has consulted on EOS, advocated that it is appropriate to fly the EOS instruments on multiple platforms. This approach has advantages which are the reverse of the disadvantages of the large platform approach. It is flexible in being able to accommodate budget shortfalls and overruns, as well as technical difficulties. The optimum orbits can be chosen for each of the instruments. Finally, it is more resilient to catastrophic loss from a single launch failure.

NASA, with input from the Frieman Committee, and detailed recommendations from the EOS Payload Advisory Panel, has reconfigured EOS to fly the 17 instruments required for global climate change studies on (i) three intermediate spacecraft that can be launched on Atlas IIAS-class vehicles, (ii) one smaller spacecraft to be launched on a Delta-class vehicle, and (iii) two small spacecraft to be launched on Pegasus-class vehicles. The names of the spacecraft, their launch dates and vehicles, and their purposes are as follows:

- EOS-AM (June 1998, Atlas IIAS-class) – Characterization of the terrestrial surface; clouds, aerosols and radiation balance.
- COLOR (1998, Pegasus-class) – Oceanic biomass and productivity.
- AERO (2000, Pegasus-class) – Atmospheric aerosols.
- EOS-PM (2000, Atlas IIAS-class) – Clouds, precipitation, and radiative balance; terrestrial snow and sea ice; sea-surface temperature and ocean productivity.
- ALT (2002, Delta-class) – Ocean circulation and ice sheet mass balance.
- CHEM (2002, Atlas IIAS-class) – Atmospheric chemical species and their transformations; ocean surface stress.

[5] The instrument configurations and the observations they will make are shown pictorially in Figure 1. (The acronym for each of the instruments is defined in the Appendix to this report.) The impact of these missions on our understanding of the climate system of the Earth is discussed in the next section.

The orbits for EOS-AM and EOS-PM are both sun-synchronous, polar, but with different crossing times. The EOS-AM spacecraft primarily observes surface features and thus prefers a morning crossing time when cloud cover is a minimum. In contrast, EOS-PM includes an instrument which is a candidate, next-generation atmospheric sounder for deployment on future NOAA operational satellites. It prefers an afternoon crossing time which is most useful for contributing data for meteorological forecasting. Moreover, both EOS-AM and EOS-PM observe characteristics of the terrestrial surface and of the atmosphere. By having measurements at two different times of the day it is possible to study diurnal variations in these features. The orbits for COLOR, ALT, and CHEM are also sun-synchronous, polar, but the optimum orbit for AERO is 57 degree inclination.

As can be seen in Figure 1, certain instruments are flown on more than one spacecraft. The MODIS-N instrument, which is a moderate resolution spectrometer capable of observing both the surface and the atmosphere, is included on both EOS-AM and EOS-PM because MODIS-N has critical simultaneity requirements with the other instruments on each of these spacecraft. That is, the observations from MODIS-N seen simultaneously through the same column of air are required to interpret the data from these other instruments. By including MODIS-N on each of these spacecraft it is unnecessary to do complicated formation flying, and the EOS-AM and EOS-PM spacecraft can each be flown in its optimum orbit. Moreover, MODIS-N is the central instrument on EOS, and with it flying on two separate spacecraft the program has important redundancy. The CERES instrument, which measures the energy budget of the Earth, is flown on both EOS-AM and EOS-PM to give measurements at two different times of the day. The SAGE-III instrument, which measures atmospheric aerosols, is flown on AERO and CHEM to provide measurements from two different orbits, 57-degree inclination and polar, respectively.

The launch of the EOS-AM spacecraft is in June 1998, six months earlier than the planned launch of the first large platform in the original EOS program. By reducing the size of the payload it is possible to launch earlier. The launch dates of the remaining EOS spacecraft will occur over the next four years, through the year 2002. Every effort will be made to hold the launch date of EOS-AM in June 1998, and then EOS-PM in 2000. The program has flexibility in that limited budget shortfalls or overruns can be accommodated by minor slips in the launch dates of the COLOR, AERO, ALT and/or CHEM spacecraft.

EOS is to be a long-term program, providing continuous observations of the causes of global climate change. The principal EOS spacecraft, EOS-AM, EOS-PM, etc., will [6] thus be repeated twice on five-year centers for at least fifteen year coverage. However, the payloads on the follow-on EOS spacecraft could change, depending on the evolution of scientific understanding of global change and the development of technology. For example, the ASTER instrument on the first EOS-AM spacecraft could be replaced by the high resolution spectrometer, HIRIS, on the second EOS-AM spacecraft. Actual decisions on instruments to fly on follow-on spacecraft do not need to be made for some years. However, technology development efforts need to continue to insure that subsequent generation instruments are available when needed.

3. THE IMPACT OF THE RESTRUCTURED EOS ON UNDERSTANDING THE
CLIMATE: Scientists agree that the climate system of the Earth is governed by a series of
interacting components which are illustrated in Figure 2. External inputs to the climate
come from human activities, shown on the right of the figure, and natural inputs such as
volcanoes or changes in solar output, shown on the left. The internal components of the
climate system – the components of the atmosphere, the oceans, and the biology – are
then coupled together through a series of interactions denoted by the arrows in the
figure. Interactions involving primarily physical and chemical processes are included in
the gray box labeled "Physical Climate System." Interactions that also involve mainly the
chemical and biological processes of the Earth are included in the box labeled
"Biogeochemical Systems."

Our knowledge of the individual components which govern the climate system of the
Earth is represented by large-scale numerical models which should (i) include all the
relevant physical, chemical, and/or biological processes that determines the behavior of
each component and its interactions with other components, (ii) include comprehensive
data on the current state of the component, and (iii) allow us to predict the future of the
component. However, the accuracy and completeness of current models of the
components of the climate system are inadequate, and also vary widely depending on the
component being modeled. The lack of appropriate observational data is the principal
impediment to improving models of the components and to coupling these models
together into a comprehensive model of the whole climate system.

Consider a hypothetical scale for the accuracy and completeness of numerical
models of the Earth in which current models for a five-day regional weather forecast are
a "10". On this conceptual scale, then, current climate models that focus upon
atmospheric dynamics, the so-called general circulation models (GCM's), rate
approximately a "6". There is similarity in the physics that is important in five-day weather
forecast models and in global atmosphere models. However, GCM's have time scales of
weeks to years and are more strongly influenced by the poorly understood effects of
clouds and aerosols which can cause heating or cooling of the atmosphere and surface.
In addition, the couplings within the climate system, for instance between the ocean and
the atmosphere, are more complex over the inherently longer time scales of the climate
system.

[7] On this same conceptual scale, the accuracy and completeness of current global
models for ocean circulation rate approximately a "4"; global models of ocean circulation
that include the calculation of marine biological activity are just now being developed.
Global models of terrestrial ecosystem that treat the rate at which they exchange carbon
and water with the atmosphere rate perhaps a "2" on this scale; global terrestrial ecosystem
models that treat the effects of climate change and feedbacks to the climate change are
also just now being developed. These models, particularly those involving the biology,
suffer from lack of data to determine the relevant processes which govern their behavior,
as well as from inadequate data to determine the current state of the component.

As is shown in Figure 2, the components of the climate system are coupled to each
other. Thus, an adequate, comprehensive model for the climate system requires adequate
and comprehensive models for each of the components, and verification of the method
for coupling the components together. Even if we improve the accuracy and completeness

of models of the atmosphere, their utility to predict the future of the climate is severely limited unless we also improve the accuracy and completeness of the models of both the oceans and the terrestrial components. Furthermore, the regional consequences of climate change, on sea levels and agriculture, cannot be assessed without accurate models of the oceans and the biology of the Earth. However, this does not mean that all models need to reach the highest level of sophistication or that they need to reach the same level of completeness.

It is unrealistic, and unnecessary, to attempt to develop a global atmosphere model that will provide specific forecasts on a daily basis that is analogous to five-day weather forecast models. We are, after all, looking only for changes in the climate by region, and by month or by season, on which to base policy. Thus, on our scale where weather forecast models are a "10", a realistic goal for global atmosphere models is an "8".

Our goals for models of the oceans and the biology of the Earth can be less demanding since many of the controlling processes operate on longer time scales and therefore may be treated initially by relatively simpler models. In addition, advances of 50-200 % in these less well modeled subsystems will yield significant contributions toward the development of integrated climate system models. Continuing with the simple scale concept, a realistic goal for models of both the oceans and the biology of the Earth is a "6", though the treatment of the physical circulation of the oceans will need perhaps to reach a level more consistent with the general circulation models of the global atmosphere (e.g. an "8"). Clearly, models of the terrestrial biological systems, which we currently rate as a "2", have the furthest to improve; however, initial improvements in these more data-sparse components will be relatively rapid and, as mentioned, very valuable.

The EOS spacecraft are designed to provide the observations that will allow us to make the necessary improvements in the models of the Earth's climate system, and to monitor its long-term evolution. These observations need to be supplemented with [8] extensive, coordinated ground and aircraft measurements, which are to be funded outside of the EOS program and in many cases by other Federal agencies or our international partners. The color-coding on the figure describing the climate system (Figure 2) indicates which of the EOS spacecraft will study each of the components. The actual observations that each of the EOS spacecraft will make, and the instruments that will make them, are shown pictorially in Figure 1 in section 2.

The EOS-AM spacecraft (shown in green in the figure describing the climate system), and its accompanying small mission COLOR (shown in blue) will provide the observations that will allow the required improvements in models of the biology of the Earth. These spacecraft will study terrestrial ecosystems, marine biogeochemistry, land use, soil, terrestrial energy/moisture, tropospheric chemistry, etc. Improvements in models of the biology require observations of the surface of the Earth and its vegetation in many different wavelengths of radiation. This quantitative information then provides the basis for understanding the processes which govern the biology. EOS-AM, in particular, is a major improvement in wavelength coverage over any mission that will precede it, and thus should provide the required improvements in models of the component of the climate system that are the least developed.

The EOS-PM spacecraft (shown in red in the figure), and its accompanying small mission AERO (shown in yellow) will provide the observations that will allow the required improvements in models of the dynamics of the atmosphere of the Earth. These spacecraft will study atmospheric physics/dynamics, global moisture, etc. The EOS-PM and AERO instruments have improved spectral resolution over preceding missions, and thus should produce the required quantitative understanding of the hydrological cycle and other aspects of the energy balance in the atmosphere.

The observations from the EOS-AM and EOS-PM spacecraft also complement each other. Spacecraft which observe the surface of the Earth in general observe the atmosphere as well, in part to make needed atmospheric corrections to their measurements. Conversely, spacecraft which observe the atmosphere can see the surface. Thus, the EOS-AM spacecraft can also observe aspects of the atmosphere, such as clouds and their effects on the Earth's radiation balance, and the EOS-PM spacecraft, aspects of the surface. These measurements come at different times of the day – EOS-AM in the morning and EOS-PM in the afternoon – and provide an opportunity to study variations in the atmosphere and on the surface during the daily heating and cooling cycles.

Models of the oceans of the Earth should improve in the next few years with observations of global circulation from the US/France TOPEX/Poseidon mission, which will be launched in 1992, and from observations of air-sea interactions from the Japanese ADEOS mission in 1995, which includes the flight of two US instruments. The time scales for phenomena in the oceans, however, are quite long, and thus the TOPEX/Poseidon and ADEOS missions, which are of limited duration, will not provide all the observations needed to make the required improvements in ocean models. The EOS mission will thus continue the TOPEX/Poseidon and ADEOS measurements, [9] at the same spatial and spectral resolution, from the ALT and CHEM spacecraft, respectively.

The chemistry of the lower atmosphere, or troposphere, is determined by emissions from the biology of the Earth and from human activities. However, the chemistry of the troposphere is difficult to study from space with current instrumentation. The CHEM spacecraft, which is the last of the first series of EOS spacecraft, is planned to carry an advanced instrument that will for the first time provide detailed measurements of the chemistry of the troposphere. These measurements will provide important information of the transformations of "greenhouse" gases in the troposphere, and an essential check on the models of the biology and atmosphere that are developed using data from earlier EOS spacecraft.

Models of ozone depletion, and the governing processes in the stratosphere, should improve substantially in the next few years with observations from the Upper Atmosphere Research Satellite (UARS). This problem is important to climate change studies in that ozone depletion in the lower stratosphere can affect the heating or cooling of the troposphere. Thus, the CHEM spacecraft, which launches essentially ten years after UARS, will also carry instruments to check the accuracy of models of stratospheric chemistry.

The launch sequence for the EOS spacecraft is determined by technical, programmatic, and scientific reasons. The scientific reasons provide a compelling logic. The first two spacecraft, EOS-AM and COLOR, will provide the data to improve the models of biology of the Earth – the models most in need of improvement. The second two, EOS-PM and AERO, will provide the data needed to improve the models of the dynamics and energy balance of the atmosphere, which is of primary influence on the climate. The EOS-AM and EOS-PM spacecraft, will work together to observe the daily heating and cooling cycles of the surface and atmosphere. The ALT and CHEM missions are last in the sequence and will provide an important continuation of ocean data which is started in advance of EOS. Moreover, CHEM, at the end of the sequence, can provide an opportunity to fly instruments to study tropospheric chemistry, which are not currently available, as well as an opportunity to monitor the chemistry controlling ozone ten years after UARS.

Models of the climate system should attain the required near-term improvements with observations from the first sequence of EOS spacecraft (EOS-AM, COLOR, EOS-PM, AERO, ALT, and CHEM), and from related ground and aircraft measurements. However, it is important to note that the timescales for processes which govern the climate system of the Earth can extend for several decades. Moreover, to check the accuracy of the models, and to look for changes in the external inputs to the climate system, from human activity and from natural variations, it is important to continue the studies of the components of the climate system. Thus, the sequence of EOS spacecraft is planned to be repeated to provide fifteen year coverage of the climate system.

[10] In summary, models of the climate system of the Earth – of the atmosphere, the oceans, and the biology – all need to be improved if we are to determine the extent, causes, and especially the regional consequences of climate change. The EOS-AM and COLOR spacecraft will provide the data need to improve the models of the biology. The EOS-PM and AERO spacecraft will provide the data to improve the models of the dynamics and energy balance of the atmosphere. The EOS-AM and EOS-PM spacecraft will work together to provide data on daily variations of both the surface and the atmosphere. And the ALT and CHEM spacecraft will extend and improve the models of the ocean and atmospheric chemistry.

4. THE IMPACT OF EOS ON ADAPTATION AND MITIGATION STRATEGIES: Strategies to deal with climate change fall in two categories: adaptation and mitigation strategies. Adaptation strategies provide options for adjusting human behavior to deal with the impacts of climate change. These strategies come into effect primarily after the climate change has occurred. Mitigation strategies provide options for avoiding the impacts of human activity on the climate. They come into effect prior to the full onset of the climate change and can slow its pace.

Observations from the EOS spacecraft, and from related ground-based studies, will provide the scientific underpinning for adaptation and mitigation strategies in two ways:

- First, individual EOS spacecraft will monitor (i) changes in the components of the climate system to which we may have to adapt, and (ii) the inputs of human activity into the climate system, which we may have to mitigate.
- Second, the entire suite of EOS spacecraft, the data they will produce, and the models that will result (See Section 3) will determine (i) the future behavior of the components of the climate system that affect human activity, thereby providing the basis for comprehensive adaptation strategies; and (ii) the impact of individual aspects of human behavior on the entire climate system, thereby providing the basis for comprehensive mitigation strategies.

Consider first the components of the climate system which may change, requiring our adaptation. As can be seen by the arrows in Figure 2, which show the linkages between the components of the climate system, two components have a direct impact on human activity: the Atmospheric Physics/Dynamics component, which describes the actual climate change, and the Terrestrial Ecosystems component. Listed in Figure 3 are aspects of human activity which are sensitive to changes in climate and/or terrestrial ecosystems: farming, managed forests and grasslands, human migration, political tranquility, natural landscapes, and tourism and recreation. In addition, there are components of the climate system which cause secondary changes to which human behavior or nature will be sensitive: changes in the Ocean Dynamics component can affect coastal settlements and structures; the Terrestrial Energy/Moisture component affects water resources; and the Marine Biogeochemistry [11] component affects marine ecosystems. The list of sensitive aspects of human activity and nature was prepared by the National Research Council study, Policy Implications of Greenhouse Warming.

Individual EOS spacecraft will monitor changes in the components of the climate system that affect human activity. The color coding in Figure 3 is the same as in Figure 2 and illustrates which EOS spacecraft monitors which component. The EOS-AM spacecraft will monitor changes in Terrestrial Ecosystems. The EOS-PM spacecraft, its accompanying small mission AERO, and aspects of the CHEM mission will monitor changes in Atmospheric Physics/Dynamics, and thus the climate. The EOS-AM and EOS-PM spacecraft, which observe the Earth at different times of the day, will work together to provide observations of daily variations on the surface and in the atmosphere, and, in particular, will monitor the Terrestrial Energy/Moisture component. The ALT and CHEM spacecraft will monitor the Ocean Dynamics component, and the COLOR spacecraft, Marine Biogeochemistry.

Monitoring the components of the climate system which affect human activity will provide information on trends and on changes that are of immediate concern. This information by itself, however, will not allow us to determine the future behavior of each of the components of the climate system that affects us, and thus will not permit us to develop long-term plans for how we will adapt to change. Rather, determining the future behavior of components of the climate system requires that we understand the climate system as a whole since the components are coupled together. For example, as is illustrated in Figure 2, the future of the Terrestrial Ecosystems component depends on the behavior of the Global Moisture, Soil, and Tropospheric Chemistry components, and in turn on other components. Predicting the future behavior of each of the components that affect human activity will be done by the entire suite of EOS spacecraft and related ground-based studies, working together to provide the data on which to build the coupled models for biology, oceans and atmosphere of the Earth described in the previous section (Section 3).

Consider next the human activities that can affect the climate system, which we may chose to mitigate. As illustrated by the arrows in Figure 2, such human activities fall in three categories: the emissions of pollutants into the troposphere or lower atmosphere, e.g. chlorofluorocarbons (CFC's); the emissions of carbon dioxide into the troposphere from the burning of fossil fuels and deforestation; and changes in land use patterns.

Individual EOS spacecraft will monitor the inputs to the climate system from these human activities. As is illustrated by the color-codes in Figure 3, the EOS-AM spacecraft will monitor changes in land use patterns. The CHEM spacecraft will monitor changes in carbon dioxide concentrations and the combination of CHEM and EOS-AM will monitor the emissions of pollutants into the atmosphere.

Monitoring the inputs of human activity will give us a measure of the magnitude of the influence we are having on the climate system. However, to develop comprehensive mitigation strategies it is necessary to trace the impact of specific human activities [12] throughout the climate system, determine which are most detrimental, and then consider the economic and social consequences of changing these activities. For example, the emissions of pollutants and carbon dioxide into the atmosphere effects the Tropospheric Chemistry component. In turn, as is illustrated in Figure 2, Tropospheric Chemistry affects Terrestrial Ecosystems and, through a more complicated chain, Atmospheric Physics/Dynamics, both of which impact back on human activities. Tracing the effects of specific human activities throughout the climate system to determine which are most detrimental will be done by the entire suite of EOS spacecraft, their related ground-based studies, and the models of the components of the climate system that will result.

The time scales during which adaptation strategies need to be developed is relatively long; the impact of any potential climate change is not expected for decades. Thus, the full EOS system of spacecraft and related ground-based studies can be the principal source of data on changes on the Earth for developing adaptation strategies. In contrast, the time scale for developing mitigation strategies needs to be shorter if we are to avoid detrimental impacts of climate change. Here, we are in the position of having to develop knowledge of the climate system and mitigation strategies concurrently throughout the Mission to Planet Earth. There are prudent, economically sound steps that can be taken today, in the absence of certain knowledge of climate change, to serve as an "insurance" policy against unwelcome changes. The early missions of the Mission to Planet Earth program will reveal additional steps that should be taken. Knowledge of the climate system sufficient to develop comprehensive mitigation strategies will be determined by the full observations of EOS, the related ground-based studies, and the models they generate.

5. EOS DATA AND INFORMATION SYSTEM (EOSDIS): EOSDIS is designed to serve the needs of the entire Mission to Planet Earth program. It will improve, archive, and distribute data from the missions that fly in advance of EOS. It will process, archive and distribute data from, and will control the EOS spacecraft.

EOSDIS is designed to be an evolutionary system. It is to be developed with an open architecture, which will allow the easy insertion of new technology that comes on the market, either software or hardware, and will enable the system to respond to changing user requirements. Indeed, the management approach for EOSDIS is to (i) build the system in increments, (ii) try each increment out on the users, and (iii) make appropriate adjustments at each stage. This approach will require a highly interactive arrangement with the contractor who will develop the core of the EOSDIS system. The contractor will be tasked to perform prototyping activities, develop the system in successive versions, each version responding to the lessons of the previous version, and make appropriate modifications to respond to changing user needs. EOSDIS is unprecedented in scope and accessibility. It is larger and more comprehensive than any previous civilian system; it is more interactive than any other large data system. We need to learn as we develop EOSDIS, so that the final system that results is truly responsive to the needs of all the users of EOS data.

[13]   The first major activity in EOSDIS, known as Version 0, is to develop an experience base for handling large Earth science data sets by reworking existing data sets into more user-friendly formats. This effort, of course, has the added advantage that it will take existing data sets that have been underutilized and ensure that the information they contain on key issues of global change is more widely available to researchers. Examples of data sets that are being reworked as part of Version 0 include: Advanced Very High Resolution Radiometer measurements of sea surface temperature and vegetation from the NOAA polar orbiting satellites, Special Sensor Microwave/Imager measurements of water vapor and precipitation from the Defense Meteorological Satellites, and Multispectral Scanner land cover data from Landsat.

The subsequent development of EOSDIS will be phased to meet the requirements of missions that are to fly in advance of EOS, and then of EOS itself. For example, the required computing capability and institutional support will be brought on-line only as is needed to handle the data. Thus, the restructuring of EOS, which results in delays in some of the observations from what was expected in the original program (e.g. the flight of EOS-PM occurs later than the original large platform), permits the EOSDIS that will be required by the year 2000 to be smaller than initially envisioned. This adjustment can be made without altering the basic architecture, or the evolutionary design of EOSDIS. However, the budgets for EOSDIS have been reduced substantially, and scale approximately with the reduced cost of the space hardware for EOS.

To validate the approach that is being pursued for EOSDIS, NASA has requested that the National Research Council convene a panel of experts to review our plans, including the validation of its engineering and technical underpinning, and to assess whether current plans provide sufficient resiliency to be adaptable to changing requirements. The report is due in the summer of 1992.

The EOS data will be processed, archived and distributed through Distributed Active Archive Centers, or DAAC's, which are located at major research centers around the country. The DAAC's and their data holdings will be:

- Goddard Space Flight Center (Greenbelt, MD.)- Climate Change, Atmospheric Chemistry, and Biogeochemistry.
- EROS Data Center (Sioux Fall, SD) - Land Surfaces.
- Jet Propulsion Laboratory (Pasadena, CA.) - Physical Oceanography.
- Langley Research Center (Hampton, VA) - Radiation Budget, Clouds and Aerosols and Tropospheric Chemistry.
- Marshall Space Flight Center (Huntsville, AL.) - Hydrological Cycle and Hydrodynamics.
- National Snow and Ice Data Center (Boulder, CO.) - Cryosphere.
- Alaska SAR Facility (Fairbanks, AK.) - Sea Ice, Polar Processes Imagery.

[14] Each of these DAAC's has tasks to perform in the Version 0 effort and will receive a

small amount of institutional support to do so. However, the DAAC's will be activated to process EOS data only as needed. The EOS-AM spacecraft data will be processed at the DAAC's at the Goddard Space Flight Center, the EROS Data Center, and the Langley Research Center. Thus, these DAAC's will be equipped with the appropriate computer facilities and staffing levels in advance of the launch of EOS-AM. The remaining DAAC's will be equipped later, in time to process the data sets for which they are primarily responsible.

In addition, the EOS data system includes affiliated DAAC's which are designed to serve the needs of specific research and policy-making communities. The most significant affiliated DAAC is the Consortium for International Earth Science Information Network (CIESIN), located in Michigan, which has the responsibility to use the EOS data products to support socioeconomic studies of the human dimension of global change. CIESIN will also serve as the interface with the non-traditional user community such as educators and policy makers.

Additional DAAC's may also be established to take advantage of unique research capabilities at national laboratories. The EOS data system and its DAAC's are expected to serve a worldwide user community that could approach 10,000 scientists.

The EOS data will be acquired through the TDRSS satellite network, and transmitted to the ground through the White Sands, NM ground station. The data will then be transmitted to the NASA facility in Fairmont, WV., where it will be archived in raw form and distributed to the DAAC's for processing and archiving of the data products. The Fairmont facility will also coordinate the Independent Verification and Validation (IV&V) of the EOS data system. Data for potential future operational instruments will also be directly downlinked to operational ground receiving terminals.

6. INTERDISCIPLINARY INVESTIGATIONS: As part of the EOS program, NASA has established 29 Interdisciplinary Teams, whose job it is to develop a quantitative understanding of the processes which govern the climate system of the Earth, and to incorporate this knowledge into predictive models which can serve as the basis for policy decisions. These Interdisciplinary Teams are being established at 12 US universities, 7 NASA facilities, and 9 international research facilities and universities.

7. MISSIONS IN ADVANCE OF EOS: In advance of EOS some 16 free-flying satellites will be launched to make observations of global change on the Earth. These missions are to be conducted by the US and our international partners. In addition, three major Space Lab series will be launched on the Space Shuttle to study global change, and there will be continuous, routine observations from the NOAA and military weather satellites. This extensive period of observations is known as Phase 1 of the Mission to [15] Planet Earth.

It is important to note that all of the instruments that will fly on the EOS spacecraft have less capable versions, and in some cases identical versions, that will fly on these Phase 1 missions. Thus, Mission to Planet Earth is an evolutionary program. It is providing important information on global change today, will provide more such information throughout Phase 1, and will evolve into more comprehensive and detailed observations when the EOS spacecraft fly.

8. THE ROLE OF THE DEPARTMENTS OF ENERGY (DOE) AND DEFENSE (DOD)- As noted by the Frieman Committee, the DOE and DOD can make important contributions to the Mission to Planet Earth. In particular, the DOE may be able to provide additional spacecraft to Phase 1 of the Mission to Planet Earth. Although there are a large number of missions that fly during Phase 1, several review committees have noted that there is a discontinuity in Earth radiation budget observations, between NASA's measurements on the EBBS and NOAA 9/10 spacecraft, and NASA's measurements on the US/Japan TRMM mission and subsequent EOS spacecraft. The DOE is considering providing one or more small spacecraft which can provide these observations. To be useful the observations must commence no later than 1995, since TRMM flies in 1997, and their calibration must be well understood to help provide continuity between the earlier and future observations.

Further, the DOE and DOD may be able to assist in the development of new technologies for use in subsequent generations of EOS instruments. The restructured EOS does not include a Laser Atmospheric Wind Sounder, an advanced Synthetic Aperture Radar, or a light-weight, high-resolution imaging spectrometer. These instruments will be important for future global change studies. It would be very helpful for DOE and DOD to execute technology demonstration projects that would lead to reliable, more compact, and less expensive versions of these and other advanced instruments.

9. SUMMARY OF THE MISSION TO PLANET EARTH: The structure of the Mission to Planet Earth – the Phase 1 missions, the EOS spacecraft, the DAAC's, the interdisciplinary investigations, and the science and policy objectives – is illustrated in Figure 4, and summarized as follows:

1. During Phase 1 of Mission to Planet Earth, which begins now and continues through 1998, a large number of satellites will be launched by NASA, other Federal Agencies, and our international partners to monitor the climate and study specific processes which may control its evolution.
2. Beginning in 1998, the Phase 1 satellites will be replaced by a constellation of intermediate and small satellites as part of the Earth Observing System, or EOS. The instruments on the EOS spacecraft are more capable and can provide more comprehensive long-term observations than those from the Phase 1 spacecraft.
[16]   3. All of the data on the interactions and processes which control the climate will be processed and archived in Distributed Active Archive Centers, or DAAC's.
4. The data in the DAAC's will be made readily available to all researchers who study global change. In particular, it will be the the [sic] basis for studies by the Interdisciplinary Teams that NASA has established.
5. These Interdisciplinary Teams, along with other researchers, are charged with developing a quantitative understanding of the seven key science issues (listed in section 2) and incorporating this knowledge into predictive models that will determine the extent, causes, and regional consequences of global climate change.

10. FUNDING REQUIREMENTS: The President's budget estimates for fiscal year 1994 through 1997 reflect a government-wide policy to present a nominal freeze in budget

authority in all outyears for domestic discretionary spending programs. This serves to emphasize that in FY 1994 and subsequent years, each proposal for increased funding will have to compete on its merits" against all other domestic discretionary programs, and decisions made accordingly within the caps and overall fiscal constraints. The outyear funding estimates shown in this report represent preliminary program planning estimates that reflect a continuation and extension of the programs proposed in the FY 1993 budget. These programs may not be funded at these levels, however, given competition with other priorities and fiscal constraints. Each program will compete each year on the merits.

The restructured EOS program, including the spacecraft (EOS-AM, EOS-PM, COLOR, ALT, and CHEM), EOSDIS, and the supporting science has been designed to accommodate the $11 billion funding cap through FY 2000, as directed by Congress. Indeed, the design philosophy has been: First, to eliminate all instruments that are not essential to study global climate change. Second, to insure that there is no unnecessary duplication of observations with international missions studying global climate change. Third, to provide an adequate data and information system, and supporting science. Fourth, to launch the first EOS spacecraft, EOS-AM, as early as is programmatically feasible. And fifth, to launch all other EOS spacecraft as early as permitted by the funding profile associated with the $11 billion cap.

The EOS program has four basic components: the space hardware – spacecraft and instruments, the EOSDIS data system, the supporting science, and mission operations. The funding profiles for each component, through FY 2000, is shown in Figure 5. The profiles total to $11 billion. The profiles have relatively modest growth from FY 1992 to FY 1993, in compliance with Congressional direction, and then increase substantially from FY 1994 to FY 1995. The funding profiles level, increasing only at the rate of inflation after FY 1997. This level funding profile is assumed to [17] continue after FY 2000 to complete the development of the ALT and CHEM spacecraft, to continue, through repeated launches of EOS spacecraft, the long-term monitoring of the climate, and to continue data reduction, distribution, and analysis.

The total cost of the space hardware component, through FY 2000, is approximately $6 billion. The ground segment – EOSDIS and science – totals to $5 billion. This proportion of ground to flight segments is very high by the standards of any previous NASA program. It reflects the strong commitment of NASA to use EOSDIS to improve, archive and distribute the data from spacecraft in Phase 1 of the Mission to Planet Earth, thereby substantially increasing the understanding of global change that will result from these missions. Developing an operational EOSDIS in advance of the launch of the EOS-AM spacecraft will also insure the immediate return of useful information from all the EOS spacecraft. After FY 2000, the proportion of ground to space segments increases; by end of the full 15-year operational life of EOS, the total cost of the ground segment will be approximately 60% of the overall mission cost.

The space hardware costs are broken out by individual spacecraft mission (e.g. EOSAM, EOS-PM, etc.) in Figure 6. The cost estimates used for the instruments that were to have flown on the original large EOS platform and the EOS-AM spacecraft bus are well understood since both elements are under contract (although the spacecraft contract will be modified to reflect the current downsized configuration). Cost estimates for subsequent spacecraft and for the instruments that are not currently under development are preliminary, based on derivation from the current EOS-AM definition and development estimates and analogy with previous flight project experience. These estimates will be updated before beginning development of these spacecraft and instruments.

Most of the space hardware costs in FY 1993 and FY 1994 are associated with development of the EOS-AM spacecraft and instruments, and the beginning of the development of EOS-PM. The growth in the budget in FY 1995 is essential to maintain the schedule of EOS-AM and EOS-PM and to begin the full development of AERO, CHEM, and ALT. It should be noted that the small COLOR mission is not shown in this figure. NASA assumes that it can acquire the data from COLOR as a data purchase only, in the same manner as it is acquiring the data from the SeaStar mission that will fly during Phase 1. Accordingly, the costs for COLOR are bookkept in EOSDIS. The category called "Other" in the figure includes the costs of a copy of the CERES instrument which is to fly on the US/Japan TRMM mission in 1997 and begin measurements of the Earth's radiation budget prior to EOS. The "Other" category also includes a wedge to fund additional instruments for flights of opportunity, such as to provide measurements of solar luminosity, or to study other aspects of global change that may become important in the coming years. [18] $1.1 billion, Japan - $1.8 billion, Canada - $0.35, and NOAA/DOD/DOE - $2.5 billion. (Costs for European, Japanese and Canadian missions are estimates only).

The costs for the science portion of EOS are shown in Figure 7. Included here are the costs for the Interdisciplinary Teams, the costs for scientists to develop the algorithms to process EOS data, and the costs for both of their computer facilities. In the NASA budget, the costs for personnel and computer facilities to do algorithm development are bookkept with the costs for the spacecraft for which the algorithms will be developed. NASA is committed to having the algorithms for each EOS spacecraft fully developed prior to its launch, so that there will be no delay in the use of the data. The costs for algorithm development thus increase beginning in FY 1995, in preparation for the EOS-AM launch.

Figure 8 illustrates the costs associated with EOSDIS. Funds for Version 0 support NASA's efforts to take existing Earth science data sets and make them more available and useful to researchers. This effort will both improve our understanding of global change and provide an experience base on which to develop the full EOSDIS. Each of the DAAC's will participate in Version 0 and is provided a small level of institutional support to do so. Funds for Version 1, etc. support the establishment of the EOSDIS capability – its basic architecture and associated networking. As in the case of algorithm development, NASA is committed to have the appropriate level of EOSDIS capability on line prior to the launch of EOS-AM. Funding is also provided for the Independent Verification and Validation activities and the in-house support for managing the data system at the Goddard Space Flight Center. Finally, funding is provided to establish the data processing capability at selected DAAC's, as is needed. The first DAAC's to be brought on line to process EOS-AM data are the Goddard Space Flight Center, the EROS Data Center, and the Langley Research Center. The remaining DAAC's will be activated as they are needed to process data from subsequent EOS spacecraft.

Cost comparisons with the original EOS program are difficult since the restructuring has been such a fundamental change. It is interesting to note, however, that to fly most of the instruments on the original large platform now requires three spacecraft – EOS-AM, EOS-PM, and CHEM – in the restructured program. The total cost of these three intermediate satellites is of course larger than the cost of the original large platform, in part because the program is spread over more years and paid for now in inflated dollars, and in part because the economy of scale of the large platform is lost. However, as can be seen in Figure 9, if we compare the cost profile required to launch the large platform and its instruments in late 1998 with the cost profile available for EOS-AM, EOS-PM, and CHEM, we see that the large platform required substantially more up-front funding. Equivalently, with limited funds available in FY 1993 and FY 1994, the large platform could not be launched until the year 2000, or later. The multiple platform approach has the clear advantage that it can begin the crucial EOS measurements early, with EOS-AM in June 1998, with only a modest growth in funds required in the early years.

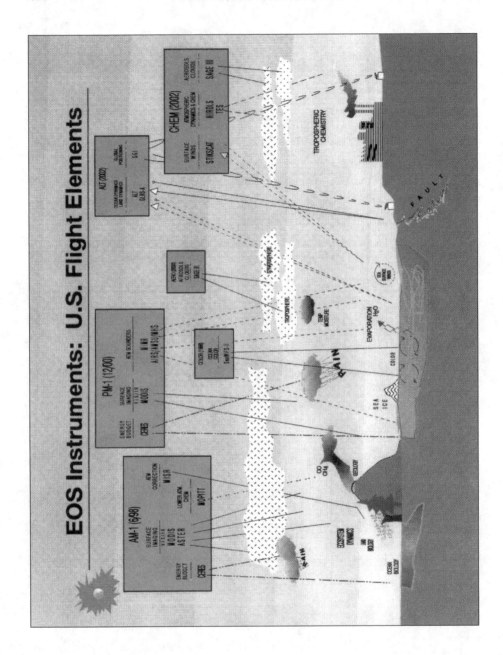

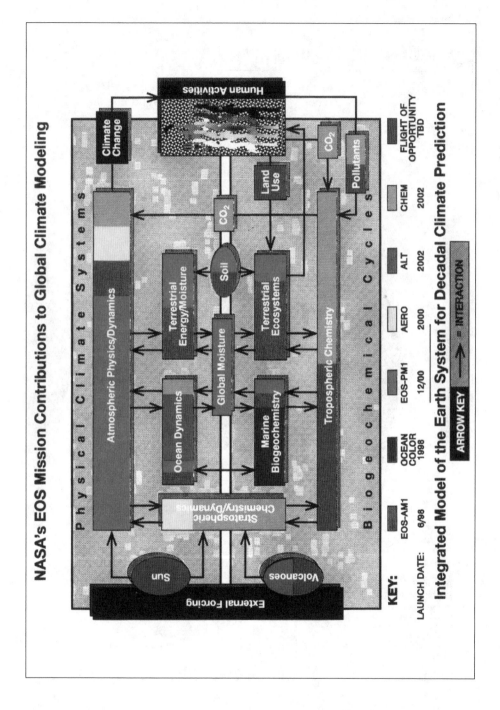

NASA's EOS Mission Contributions to Global Climate Modeling

Integrated Model of the Earth System for Decadal Climate Prediction

Mission to Planet Earth

Purpose	Scientific Issues	Interdisciplinary Investigations

Determine Extent, Causes, and Regional Consequences of Global Climate Change

The Role of Clouds, Radiation, Water Vapor, and Precipitation

The Productivity of the Oceans, their Circulation, and Air-Sea Exchange

The Sources and Sinks of Greenhouse Gases, and their Atmospheric Transformations

Changes in Land Use, Land Cover, Primary Productivity, and the Water Cycle

The Role of the Polar Ice Sheets and Sea Level

The Coupling of Ozone Chemistry with Climate and the Biosphere

The Role of Volcanoes in Climate Change

LaRC—Radiation and Clouds
Goddard Institute for Space Studies—Interannual Climate Variability
NCAR—Climate Modeling
MRI (Japan)—Atmosphere/Ocean/Land Interactions
GSFC—Atmosphere/Ocean/Land and 4-D Data Assimilation
Penn. State U./MSFC—Water Cycle
GSFC—Global Water and Energy Cycle
BMRC (Australia)—Atmospheric Modeling
U. of Washington—Physical Climate over Oceans
U. of Texas—Geodynamics
JPL—Air-Sea Interaction
Woods Hole Ocean Inst.—Biogeochemical Fluxes over Oceans
CSIRO (Australia)—Physical and Biological Oceanography
Chilworth Research (U.K.)—Oceans
Oregon State U.—Physical and Biological Oceanography
Cornell U.—Tectonic/Climatic Dynamics
CCRS (Canada)—Northern Hemisphere Terrestrial Biosphere
Colorado State U.—Carbon Budget in Grasslands
U. of New Hampshire—Biogeochemical Cycles
GSFC—Biosphere-Atmosphere Interactions
INPE (Brazil)—Amazonia
U. of California at Santa Barbara—Snow Hydrology and Chemistry
LERTS (France)—Climatic Processes in Arid/Semi-Arid Lands
U. of Washington—Oceans-Atmosphere Interactions
CCRS (Canada)—Cryospheric Monitoring in Canada
GSFC—Middle Atmosphere Chemistry and Dynamics
LaRC—Radiative/Chemical/Dynamical Processes in the Atmosphere
U. of Cambridge (U.K.)—Middle Atmosphere and Thermosphere
U. of Hawaii—Volcanism and Climate

KEY

Earth Probes Mission
1 Includes U.S. TOMS Instrument
2 Launch by NASA
3 Includes U.S. TOMS and NSCAT (Sea Surface Winds) Instruments
4 Proposed

Distributed Active Archive Centers	Earth Observing System Spacecraft	Phase I Spacecraft

		NASA
		UARS
		  Upper Atmosphere
		TOPEX/Poseidon
		(with France)
		  Ocean Circulation
		LAGEOS
		  Lower Geodynamics
		ATLAS Series
		  Atmos/Solar Effects
		LITE Series
		  Atmospheric Aerosols
		SRI Series
		  Surface Radar Images
		TOMS
		  Ozone Mapping
		SeaWiFS/SeaStar
		  Ocean Color
		TRMM
		  Clouds, Hydrology, Rain

• LaRC
  Radiation
  Budget
  Aerosols
  Tropospheric
  Chemistry
 	
• MSFC
  Hydrology and
  Hydrodynamics
 	
• GSFC
  Upper
  Atmosphere
  Dynamics
  Global
  Biosphere
  Geophysics
 	
• JPL
  Ocean Circulation,
  Air/Sea Interaction
 	
• EROS Data Center
  Land Process and
  Imagery
 	
• Alaska SAR Facility
  Sea Ice and Polar
  Processes Imagery
 	
• National Snow and
  Ice Data Center
  Cryosphere, Snow
  and Ice Data
  Product of Level 2
  and Above
	• EOS-AM
	- Clouds, Aerosols, and
	  Radiative Balance
	- Characterization of
	  Terrestrial Surface
	
	• EOS-PM
	- Clouds, Precipitation,
	  and Radiative Balance
	- Terrestrial Snow and Sea
	  Ice
	- Sea-Surface Temperature
	  and Ocean Productivity
	
	• EOS-COLOR
	- Oceanic Biomass and
	  Productivity
	
	• EOS-AERO
	- Atmospheric Aerosols
	
	• EOS-ALT
	- Ocean Circulation
	- Ice Sheet Mass Balance
	
	•EOS-CHEM
	- Atmospheric Chemical
	  Species and their
	  Transformations
	- Ocean Surface Stress
		Other U.S. (Data)
		POES (NOAA)
		  Global Environment
		GOES (NOAA)
		  Global Environment
		DMSP (DOD)
		  Hydrology and Sea Ice
		Geosat (DOD)
		  Ocean Topography and Sea Ice
		Landsat-6 (EOSAT)
		Landsat-7 (NASA/ DOD)
		  Surface Images
		Earth Radiation Budget
		Measurement (DOE)[4]
		
		International (Data)
		Meteor-3 (Russia)[1]
		  Ozone/Radiation Budget
		ERS-1/2 (ESA)
		  Global Environment
		JERS-1 (Japan)
		  Sea Ice Characteristics
		Radarsat (Canada)[2]
		  Sea Ice Characteristics
		ADEOS (Japan)[3]
		Sea Surface/Atmosphere

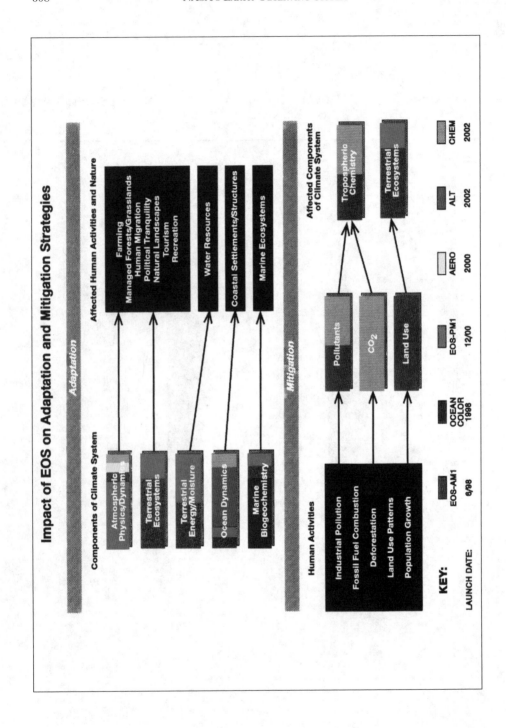

Impact of EOS on Adaptation and Mitigation Strategies

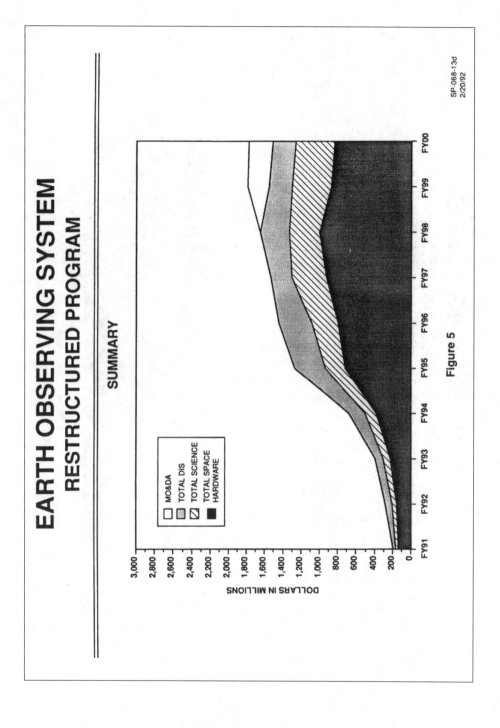

EARTH OBSERVING SYSTEM
RESTRUCTURED PROGRAM

SUMMARY

Figure 5

SP-068-13d
2/20/92

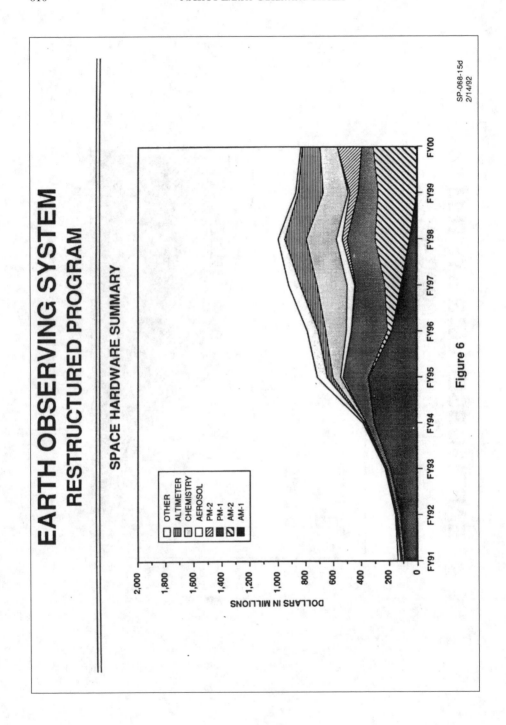

SP-068-15d
2/14/92

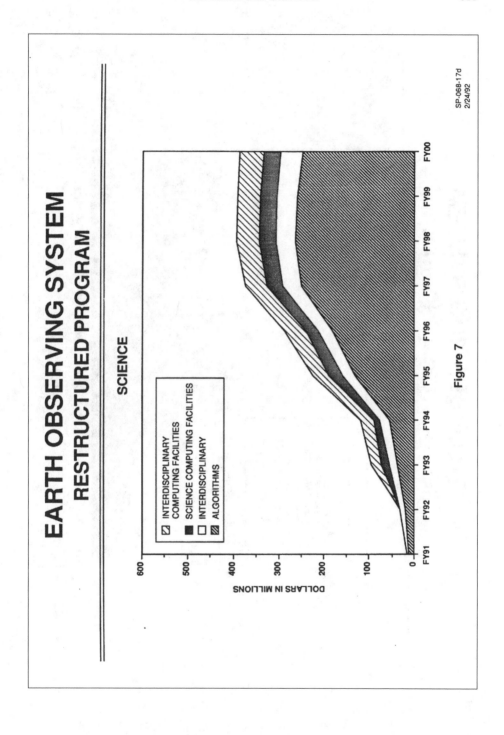

EARTH OBSERVING SYSTEM
RESTRUCTURED PROGRAM

SCIENCE

Figure 7

SP-068-17d
2/24/92

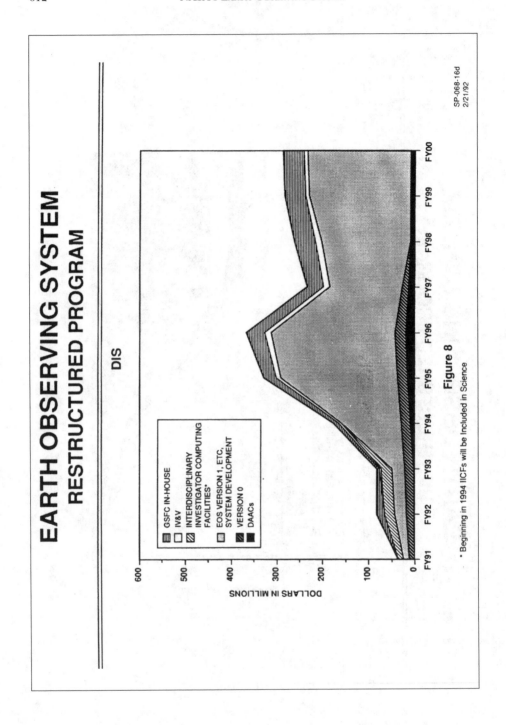

Figure 8

* Beginning in 1994 IICFs will be Included in Science

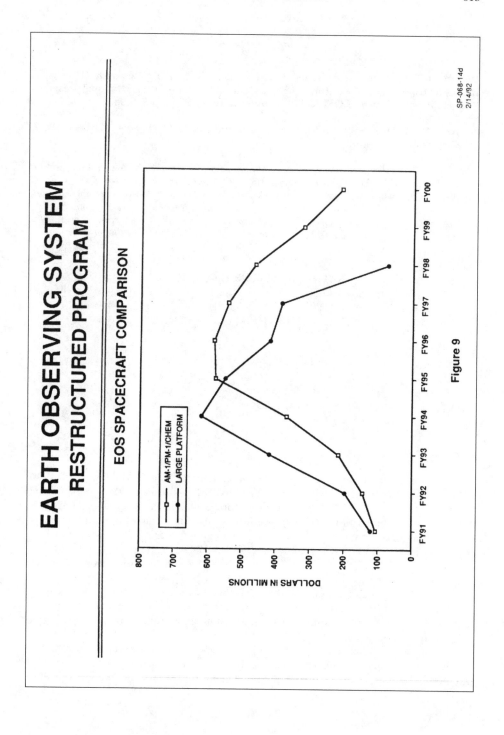

EARTH OBSERVING SYSTEM
RESTRUCTURED PROGRAM

EOS SPACECRAFT COMPARISON

Figure 9

SP-068-14d
2/14/92

Document IV-16

Document Title: Letter from Frank Press, Chairman, National Research Council, to Daniel S. Goldin, Administrator, NASA, 9 April 1992, transmitting National Research Council, Panel to Review EOSDIS Plans, "Interim Report," 9 April 1992.

Source: NASA Historical Reference Collection, NASA History Office, NASA Headquarters, Washington, D.C.

This document includes 1) a letter of transmittal from Frank Press to the new NASA Administrator (Goldin had become NASA Administrator on 1 April 1992) and 2) the interim report of the National Research Council panel convened to examine NASA's plans for the EOSDIS. The interim report, though not complete at the time of delivery, illustrates the scientific community's concern that EOSDIS must provide support to the U.S. Global Change Research Program, that it must serve a large number of users, and that it must maintain flexibility to evolve as technology and scientific needs evolve.

[no page number]

NATIONAL RESEARCH COUNCIL

2101 CONSTITUTION AVENUE WASHINGTON, D. C. 20418
OFFICE OF THE CHAIRMAN April 9, 1992

Mr. Daniel S. Goldin
Administrator
National Aeronautics and
    Space Administration
400 Maryland Avenue SW, Room 7137
Washington DC 20546

Dear Administrator Goldin:

Enclosed is an interim report by the National Research Council on NASA's plans for EOSDIS as well as a transmittal letter from the Chair of the Panel that prepared this report. As you know, EOSDIS is a very complex program, and the demands on the Panel that prepared this interim report were extraordinary–in understanding the program, in coping with a demanding schedule, and in reaching judgements. At the same time, my colleagues and I appreciate the importance of EOSDIS. To quote from the attached report: "If EOSDIS fails, so will EOS, and so may the U.S. Global Change Research Program."

It was against such an understanding that the National Research Council accepted this task, believing that we are obliged to assist the government, even when the time is short, the amount of information to be marshaled great, and the imperative to provide judgements urgent.

I believe the Panel that prepared this report has done an exceptional job, ably assisted by the people of NASA. At the same time, the judgements as well as the limits of this interim report should be clear. While the Panel supports the schedule for procuring a contractor for the EOSDIS Core System, it finds major shortcomings in the actual plans for EOSDIS, and provides substantial recommendations for implementing the program that the Panel believes will help ensure its success. Therefore, this report cannot be construed as an endorsement of NASA's current plans for EOSDIS, but rather a substantial critique of flaws, which, if addressed, will in the Panel's judgement help ensure a strong and responsive program over the long term. The Panel believes that the terms of the contract as stated in the Request for Proposal are sufficiently flexible to accommodate its recommendations.

The limits of the report should also be plain. It is an interim report, provided in response to requests from NASA and other interested parties for an early alert as to the Panel's views of [no page number] EOSDIS plans. The Panel's final report this August will offer detailed analyses for these interim judgements, and will also respond directly to the specific issues as posed in the Terms of Reference for this task.

I look forward to your comments on this interim report. And the Panel looks forward to a discussion with NASA officials involved in EOSDIS planning on this report and any further issues to be considered in preparing the final report. We are arranging for your colleagues at NASA with responsibility for the EOSDIS Project to be briefed by the Panel next week, and intend to release it publicly on April 17th.

Yours sincerely,

[signed]
Frank Press
Chairman

cc: L. Fisk, J. Alexander, S. Tilford, D. Butler-NASA
      A. Bromley, K. Erb-Office of Science and Technology Policy
      J. Hezir, J. Fellows-Office of Management and Budget
      S. Harrison-National Space Council
      R. Corell-National Science Foundation

Congressman George Brown, R. Byerly, P. Cunniffe-House of Representatives Committee on Science, Space, and Technology

Senator Albert Gore, S. Palmer-Senate Subcommitee on Science, Technology, and Space

****
[1]**National Research Council**

Panel to Review EOSDIS Plans

Interim Report

This interim report identifies several issues regarding NASA's plans for developing the Earth Observing System Data and Information System (EOSDIS) and offers a number of recommendations that NASA should consider as it proceeds with procuring a contractor to build the system. This report does not respond in detail to the items in the terms of reference–that will be the subject of the panel's final report. Given the short time available for the panel's initial assessment, it has not been able to pursue the issues it identified to the depth it would like. The panel hopes, nevertheless, that NASA will find its interim conclusions and recommendations useful in the negotiations that will take place with the selected contractor to define the ongoing work plans for the EOSDIS Project.

The appendices of this report include NASA's letter of request for this study, the terms of reference for the task, a list of the members of the panel and brief biographies, the work done and the meetings held to enable the panel to write this interim report, a brief description of EOSDIS for readers not familiar with the Project, and a brief description of the U.S. Global Change Research Program and its objectives.

The panel was selected to have the competencies demanded by its charge–in understanding the needs of those who will use EOSDIS (including both EOS and non-EOS investigators), in the computer science and technology underlying EOSDIS, in the creation and implementation of large data systems, and in the recent history of large space-based data systems. The fact that the procurement for the EOSDIS Core System was concurrent with the panel's work required extreme care to avoid either the reality or perception of conflict of interest. Thus, in addition to following the National Research Council's standard procedures for dealing with bias and conflict of interest, the panel–and those who provided it information and briefings–took pains to consider only publicly available information. The panel, to the best of its knowledge, has not been provided with nor has it considered any proprietary information related to the procurement.

[2]
OBJECTIVES AND MAJOR FINDINGS

In combination with other programs of the U.S. Global Change Research Program, the Earth Observing System (EOS) is intended to reduce the current uncertainties about global climate change. Its Data and Information System (EOSDIS) is essential to the success of EOS. If EOSDIS fails, so will the Earth Observing System and so may the U.S. Global Change Research Program. The panel has been told repeatedly by responsible government officials that EOS is critical to the larger, global change program–one involving many agencies of government, and other national and international participants–and that EOSDIS offers a unique opportunity to begin building a national, and eventually, international, information system for global change research.

To achieve these aspirations, EOSDIS will have to evolve to meet the changing needs of global change research over the next two decades and beyond. The panel believes that the recommendations offered in this report are necessary to ensure that growth and evolution. Specifically, the panel offers its judgments in terms of the following objectives it believes essential to the success of EOSDIS:

- EOSDIS must facilitate the integration of data related to the aims of the U.S. Global Change Research Program. Without this integration, the multidisciplinary and interdisciplinary research objectives of the U.S. Global Change Research Program will not be achieved. The EOSDIS program must be structured and managed to facilitate interactions with the other agencies involved in the U.S. Global Change Research Program so that existing data and future data collected by NASA and by other national and international organizations–using research and operational satellites as well as in situ sources–are available to all global change research scientists.

• EOSDIS must serve a large and broad set of users to facilitate the aims of the U.S. Global Change Research Program in supporting a community concerned with understanding the earth as a system. To serve that larger community, EOSDIS must provide its information [3] in a manner that is simple, transparent, and inexpensive; it also must assure availability of its data to both the earth science community and the larger scientific community. EOSDIS must ensure that service to current users–including those involved with Version 0–will not be interrupted as the development of the system proceeds, and that Version 1 and subsequent versions will be implemented as soon as possible to meet the needs of the users, both in the EOS program and in the larger U.S. Global Change Research Program.

• EOSDIS, as it evolves, must maintain the flexibility to build rapidly on relevant advances in computer science and technology, including those in databases, scalable mass storage, software engineering, and networks. Doing so means that EOSDIS should not only take advantage of new developments, but also should become a force for change in the underlying science and technology where its own needs will promote state-of-the-art developments. Flexibility also requires organizational and management structures and processes that can respond to evolving requirements and implement the means for meeting them.

• EOSDIS needs substantive user participation in the design and development of the system, including involvement in the decisions on data acquisition and archiving, standard or ad hoc product generation, and interfaces that directly affect science users.

• The structure of the EOSDIS management organization and the attention it gives to the project should reflect the importance of the program in terms of its role as one of the major and most costly programs NASA has ever undertaken as well as its central role in the U.S. Global Change Research Program.

The EOS program was recently restructured from a mission consisting of two large, orbiting platforms containing a total of 30 instruments to a series of six smaller spacecraft containing a total of 20 instruments. The amount of data expected to be collected from EOS, however, has decreased only slightly: from 330 gigabytes/day to 240 gigabytes/day. The estimate for the total amount of processed data (from the EOS spacecraft and the other missions and instruments that will be flown) that will be managed by EOSDIS changed from 1300 gigabytes/day to about 1100 gigabytes/day, a reduction of only 15 percent. Furthermore, the [4]EOSDIS Plans capabilities of the EOSDIS System are tied to the existence of the seven Distributed Active Archive Centers (DAACs) and the data they contain, more than to the flight rates. Although the panel will certainly examine this issue further for the final report, it appears that the recent restructuring of the EOS flight program has had little effect on the requirements for EOSDIS and thus does not affect the preliminary conclusions of this interim report.

In general, the panel does not see any serious risk to the EOSDIS program due to unavailable or inadequate technology. The panel believes that the prototyping plans of the EOSDIS Project Office, to be implemented after the contractor is selected, should be accelerated in order to assure that Version 1 is completed in accord with design objectives.

There are risks, however, in two aspects of the planning for EOSDIS. One area of risk derives from the scale and pace of changes in computer and data management technology that can be expected over the long-term life of the program, and from the great diversity of users who must interface with EOSDIS. NASA needs to focus immediate attention on planning how EOSDIS will evolve to continue to be a useful system as the scientific needs and the technology change over time.

Another area of risk concerns the management structure of EOSDIS. EOSDIS is an exceptionally large and complicated project that will cost several billion dollars, involve thousands of people, and continue for many years. The management will involve a complex mix of government, contractors, and a scientific community that is diverse and spread around the world. Each has an important role to play, and each will interact in a variety of ways with the other elements. In its recommendations in this interim report the panel has attempted to provide a number of mechanisms and approaches that it believes will help define these roles and interactions.

NASA, of course, must have the ultimate responsibility for implementing EOSDIS. To do so effectively, however, NASA should first ensure proper internal management attention and should use its own earth and computer science personnel, who can contribute significantly to the successful design of the system. Secondly, NASA needs to bring the scientific user community into the project as a partner, rather than regarding users simply as customers. Finally, NASA must accept the leadership role necessary to provide the essential unity among the user community (including other federal agencies and international participants), DAAC elements (management and scientific), and contractors. The complexity of this project demands that a structure be developed to ensure that all interests are properly integrated into the design of EOSDIS.

[5] The panel believes that NASA can proceed prudently with the procurement process for EOSDIS, provided the agency builds in the flexibility to make the adjustments necessary to ensure the success of the project. The conclusions and recommendations offered in this interim report can help NASA to incorporate that flexibility into work plans during the contract negotiations that will soon take place. This flexibility could be accommodated within the scope of the current procurement as long as it is planned ahead of final contract negotiations and the contract terms are compatible with this approach. The panel believes that its recommendations should not materially affect the EOSDIS schedule and that they can be implemented in work plans resulting from the pending contract negotiations. It is important to all users that EOSDIS implementation proceed as closely as possible to the planned schedule.

The panel has divided its assessment into three parts: user interactions, EOSDIS architecture, and EOSDIS management. The recommendations for each area offer actions that NASA should consider in order to meet the objectives for EOSDIS described above without halting the current procurement. The panel also recognizes that requirements may change over time and that NASA may have to adjust its work plans over the life of the project.

In order to be of service to NASA during this important stage of negotiating with the selected contractor, the panel believes that it is necessary to provide this advice now, in this interim report. The final report will expand on the issues discussed in this interim report and will respond in detail to the terms of reference.

CONCLUSIONS AND RECOMMENDATIONS

The following are the panel's judgments concerning the user interaction, architecture, and management issues that it believes must be addressed if EOSDIS is to meet the objectives integral to its success. In each instance, the panel points to strengths and weaknesses in the program, and offers recommendations.

USER INTERACTIONS

Strengths

[6] NASA has stated its intention to incorporate user feedback throughout EOSDIS development and evolution. The panel applauds this approach. The ability of EOSDIS to serve the broad spectrum of users will be the final measure of EOSDIS success. In this context, it should be acknowledged that NASA has led other agencies in developing the Global Change Master Directory, which will be a comprehensive description of all global change data sets. The panel also commends NASA for its plan to share software code and toolkits with users who wish to import them for their own systems.

Panel Concerns

In its review, the panel has identified several areas in which an augmentation or strengthening of critical user interactions could substantially improve the likelihood for success of the EOSDIS program. Areas of concern are NASA's Science Data Plan, links with other agencies, use of Pathfinder data sets, treatment of operational and historical data, long-term archiving, involvement of nontraditional communities, and ability to provide customized data sets.

Science Data Plan. Version 0 science data requirements are being compiled into a Science Data Plan by the EOSDIS Project through regular interactions with the user community. The intent is to solicit regular review of these requirements from the science community to make certain that evolving needs are adequately reflected in the EOSDIS Project planning. Care must be taken to ensure that the Science Data Plan continues to emphasize the links between global change research objectives and the acquisition of individual data sets. A clearer picture of base-level requirements can be achieved by a continuing assessment of science objectives, existing holdings that might meet the objectives, and requirements for future data streams.

The panel recommends that the Science Data Plan identify the links between global change research objectives and existing and planned data sets.

Interagency Links. The research priorities of the U.S. Global Change Research Program cut across the missions of individual federal agencies. The distribution of current holdings as well as data to be acquired underscores the need for interagency interoperability and cooperation. NASA has been an active participant in interagency efforts for the U.S. Global Change Research Program through a variety of working groups, and is currently a full partner in developing a tri-agency (NASA, NOAA, USGS) data and information implementation plan, of which EOSDIS [7] is a critical component. The panel endorses the efforts of these agencies to work cooperatively.

The Global Change Master Directory is an excellent first step in helping users to identify relevant data sets for global change research. A similar effort is needed in achieving interoperability for access to the data. Success will require both technical developments and leadership in order to integrate and provide broad access to disparate data types currently distributed throughout the agencies. The panel believes that NASA is the logical agency to initiate this step in the context of EOSDIS. Moreover, EOSDIS will be much more effective in broadening its user base if it serves as the vehicle for integrating data.

> **The panel recommends that NASA expand its efforts to increase interagency links by assuming an active leadership role among the agencies in achieving interoperability not only at the level of the Global Change Master Directory, but also at the level of providing access to the actual data.**

Pathfinder Data Sets. Prototyping has been a routine component of EOSDIS planning and Version 0 implementation by the Project Office. NASA has been successful in establishing prototype earth science data systems that are currently acquiring, processing, distributing, and archiving pre-EOS data. Lessons from such prototyping activities can identify problems associated with the manipulation and distribution of extremely large data sets.

Pathfinder data sets provide an early means to evaluate the handling of large data sets, the development of products, and the distribution of data and products. NASA and NOAA are cooperating in a Pathfinder data program for selected satellite data. This program will be extremely valuable to the U.S. Global Change Research Program and to the prototyping of various functions of the overall data and information system.

> **The panel recommends that NASA develop ways to integrate the efforts of existing data centers and centers of data supported by NSF, DOE, and USGS with the NOAA/NASA Pathfinder activities. Further, the Pathfinder data program now under way should be accelerated.**

Operational and Historical Data. Data from past and currently operating satellites already are being provided to several DAACs. NASA has shown considerable foresight in recognizing the importance of data streams from NASA, NOAA, DOD, and foreign satellites in establishing [8]long-term data sets for global change research. Although the EOSDIS Request for Proposal addresses data management of NASA's EOS platform instruments as well as NASA's commitment to maintaining data sets acquired by pre-EOS sensors, the panel wishes to emphasize the need for the accessibility of non-EOS instrument data streams to EOSDIS users.

The panel believes that the full benefit of EOSDIS to the U.S. Global Change Research Program will not be realized until an effort similar to that for EOS data is undertaken to manage the immense collection of historical data related to global change research already collected through operational observing systems. This collection includes the routine data from the space-based and surface-based observing systems of NOAA and DOD, as well as the routine and special data collected by USGS, USDA, EPA, DOE, NSF, and the Census Bureau. Integration, interpretation, and synthesis of such data, as part of a modern data and information system for long-term operational measurement, are critical to the goals of the U.S. Global Change Research Program and the interpretation of EOS measurements.

The panel recommends several ways to address the issue of integrating the operational and research data from other agencies into EOSDIS:

- NASA should articulate a plan for incorporating operational and non-EOS instrument data streams into EOSDIS. Where EOS and non-EOS instruments have similar functions, NASA should develop a strategy to enhance the use of both data streams. This strategy should also include consideration of cross-calibration between basic radiometric data and higher-level products of an EOS instrument with a non-EOS Instrument.
- To test the interoperability of EOSDIS and to integrate the critical longterm operational data that now exist at Affiliated Data Centers into a global change data and information system, NASA should perform a full-function test of the EOSDIS architecture and software on some of the Affiliated Data Centers, in particular, centers with holdings (such as long-term satellite or in situ data records) critical to the U.S. Global Change Research Program and to the synthesis and interpretation of data from EOS instruments.
- NASA should articulate its policy on how Affiliated Data Centers will move up through the different levels of interoperability that are specified for linkage with EOSDIS.

[9]**Long-Term Archiving**. Long-term archiving of EOS data is an issue that has not been addressed. Long-term commitment to maintaining data collected as part of EOSDIS is a critical component of the U.S. Global Change Research Program. NASA, in its response to questions from the panel, correctly pointed out that the issue of maintaining long-term archives is one that must be addressed by all participating federal agencies. Without a concrete plan and agency coordination for establishing permanent data archives, however, the overall objectives of EOS, and, therefore, of the U.S. Global Change Research Program, are jeopardized. As in the case of increasing interagency links, the panel believes that NASA can provide the leadership in addressing this need.

The panel recommends that NASA develop an adequate plan and technology for long-term data archiving in conjunction with the other federal agencies participating in the U.S. Global Change Research Program.

Involvement of Nontraditional Communities. NASA has identified ways for broadening the user community and providing information about EOSDIS to those unfamiliar with the system through professional journals and newsletters. Such publications may be adequate for reaching users in certain disciplines but may be ineffective for those in other fields, particularly in the nonphysical sciences. For example, one of the science priorities identified in the U.S. Global Change Research Program is to assess the human dimensions of global change. A detailed plan for involving potential user communities beyond the traditional disciplines associated with the earth and environmental sciences has not been clearly delineated for the panel.

Many approaches could be taken to encourage users from nontraditional communities (e.g., legal, educational, political, and social). A useful approach could include the distribution of sample products that would allow users to become familiar with the various types of data sets available and to judge whether those data would be helpful to their research.

> **The panel recommends that NASA take an active role in facilitating access to EOSDIS by other, nontraditional disciplines through a program that includes representatives from those disciplines in NASA's user advisory groups and develops products useful to them.**

[10] **Customized Data Sets**. NASA clearly recognizes the importance of involving the user community in the development of EOSDIS. An approach to encourage active user participation is to provide customized data integration and synthesis of various products. The availability of software tools that conform to standards in an open architecture environment would facilitate participation by active users. For example, these tools might enable a user to assemble a customized set of specific time- and/or space-averaged data that could not otherwise be assembled without the user having to develop new software.

> **The panel recommends that NASA encourage broad user participation by providing greater opportunities to create customized data sets.**

EOSDIS ARCHITECTURE

Strengths

The panel in its several lengthy discussions with EOSDIS technical staff was impressed by the staffs competence and motivation. The staff has devised a process for designing the EOSDIS Core System that would rely on open systems, including multiple levels of interoperability for both users and the DAACs as well as the ability to handle evolving international standards. These two approaches–use of an open system and adoption of standards even though they will change over the lifetime of EOSDIS–will strengthen the program.

The Project plans to deliver EOSDIS in incremental stages (via Versions 1 to 6 and Data Product Levels 0 to 6) that are expected to provide the flexibility necessary to meet user needs, to respond to budget uncertainties over the next decade, and to adjust to EOS flight schedules.

Panel Concerns

Design Control. Any large software system requires design criteria that are set by project management and articulated clearly and precisely throughout the project hierarchy. This is particularly true for EOSDIS because of four reasons: (1) the unprecedented size of the system's [11] storage and processing capacity; (2) the extraordinary heterogeneity of both user computation systems and user requirements; (3) the large variation in scale of both the mass stores and the granules of data to be simultaneously managed; and (4) the high degree of evolution expected in the system. The combination of these factors will make the design, implementation, and evolutionary control of the system a substantial architectural challenge.

Although NASA has assured the panel that EOSDIS will serve the needs of global change researchers, the EOSDIS Core System Statement of Work and the Functional and Performance Requirements documents of the Request for Proposal seem to be based on the management of data holdings resident with or owned by NASA or the DAACs and the created data products related to those holdings. It is entirely likely that data and/or data archives that are not within the exclusive purview of NASA or the DAACs will need to be made accessible to users through EOSDIS, without changing ownership of the data or the autonomy of the data repository. In anticipation of the need for accessibility, EOSDIS software should be built in the form of modular components with open, configuration-controlled interfaces so that other national and international agencies will be able to link with the system and provide products and services to the broader global change research community.

The panel believes that responsibility for the design criteria and for their enforcement to guide the system architecture must reside with the government. The government must assure that the contractor's detailed architecture and implementation decisions follow the directions given by the government system architects.

The panel recommends that NASA produce a clear, concise statement of the design criteria for EOSDIS that focuses on facilitating global change research and that NASA communicate these criteria throughout the Project hierarchy.

The panel recommends that NASA strengthen its internal system architecture team by acquiring additional experienced people and that it give them the responsibility, authority, and budget to ensure that the design criteria are met as the system design and implementation proceed. A technical project of the magnitude and complexity of EOSDIS should have the very best system architecture team possible. NASA should make every effort to acquire such talent.

[12] **Logically Distributed System**. The research that will be possible through the resources provided by EOSDIS is difficult to characterize at present. Some research will focus on narrow disciplinary questions, while other work will be interdisciplinary. Since we cannot, indeed should not, attempt to specify the future directions that earth science research will take, EOSDIS must be flexible enough to respond to a wide variety of approaches. Furthermore, EOSDIS will be only a part, albeit a major one, of the efforts directed at managing data and information for global change research.

The EOSDIS development plan provides for centralized control over the specification and implementation of the system. Each DAAC will implement an Information Management System that will be centrally developed by a single contractor. Although a centralized system is desirable for the management, operation, and control of the satellite and its instruments, the data will be distributed and dispersed among geographically separate and discipline-specific DAACs. Achieving the proper balance between the common elements that should be developed centrally and those that should be developed in a distributed fashion is critical to the success of the overall U.S. Global Change Research Program. At present, it appears as though the EOSDIS development plan is too heavily oriented toward a centralized approach.

THE PANEL RECOMMENDS THAT THE EOSDIS PROJECT ADAPT ITS DEVELOPMENT PLAN TO ENSURE A MORE LOGICALLY DISTRIBUTED SYSTEM, INCLUDING:

- **Designing EOSDIS so that all users (EOS and non-EOS investigators, DAACs, other data centers) can easily build selectively on top of EOSDIS components. EOSDIS should not constrain local implementation of diverse functions by users and DAACs. The development plan should reflect a philosophy that it is "easy to interact with EOSDIS" with minimum loss of autonomy. EOSDIS must be able to tolerate different versions of functionality and partial sharing of the components and toolkits it exports.**
- **Identifying those areas of interdisciplinary research that will require special interfaces among discipline-specific products and formats. The Project should specify the interfaces, build prototypes, and run simulations to exercise them, permitting users to evaluate them prior to developing final specifications and proceeding to full implementation. A contractor team that resides at each DAAC and works closely with the DAAC as well as the contractor's "central core" team should facilitate the development of these prototypes.**

[13] This type of distributed development can be accomplished within the scope of the current procurement as long as it is planned **ahead of final contract negotiation, and contract terms are compatible with this approach.**

Incremental Prototyping. The current EOSDIS development plan closely ties the availability of the distributed archive and product generation functions to the EOS flight schedule. There is much work that should be done, however, prior to the first scheduled launch of EOS instruments in 1998 to strengthen prototyping efforts already under way. For example, there are both existing archives and data expected from pre-EOS satellites that will be invaluable to the U.S. Global Change Research Program. Although the EOSDIS Project team has initiated the early prototyping effort for Version 0, more can and should be done to benefit current global change research and to enhance user feedback for final system design.

> The panel recommends that EOSDIS Project management extend its incremental development plan so that all user interfaces, all toolkits, and the end-to-end network system are:

- Specified in detail early in the development of Version 1 and prototyped or simulated sufficiently, and

- Evaluated in depth by users and DAACs prior to full implementation in Version 1. This will require a system network simulation and sufficient testing tools for users to assess and validate the specified functionality.

Usability Evaluation. Prudent practice in the design of complex data management systems ordinarily includes a means of measuring the usability of the data. To the extent possible, such measures should be quantitative. Early evaluation exercises should be designed to measure ease of use, quality of interface specifications; and convenience of interoperability of heterogeneous system components. These exercises should ensure that individual users and data archivers can acquire piecemeal both functional capabilities and data sets. It is also prudent practice to involve independent judgment by having this evaluation performed by a group other than those responsible for developing the system.

> **The panel recommends a usability evaluation program starting as soon as possible that involves:**

- **Selecting key functions, interfaces, and system behavior attributes for evaluation;**
- **Defining a set of metrics and expected values of those metrics for each parameter to be evaluated;**
- **Creating prototypes, simulations, and test suites to stress aspects of usability;**
- **Using the evaluations to guide final specification of system components; and**
- **Implementing this program so that most of the evaluation and validation is done by groups other than the prime contractor.**

EOSDIS MANAGEMENT

Strengths

NASA is to be commended for developing the plans for EOS as its flagship for U.S. participation in global climate change research. NASA and the EOS Project are further to be commended for their dedication to producing an adequate data system for EOS and for its user community. The unprecedented level of funding allocated for EOSDIS and the high level of planned contingency funding are evidence of the commitment NASA has made to this important national research effort. The panel is impressed with the degree of dedication and commitment of the EOSDIS Project team. The team is working diligently and competently toward both prototyping key system and subsystem capabilities and planning for the procurement of the full EOSDIS system.

[15] *Panel Concerns*

Visibility and Management Attention. Although EOSDIS appears to receive substantial attention from management at NASA Headquarters, in the panel's view, EOSDIS lacks the attention of senior management at the Goddard Space Flight Center. The EOS Project is the largest single development effort the Goddard Center has undertaken. Even without the flight hardware components, EOSDIS by itself probably satisfies that description. EOSDIS is an extremely complex interdisciplinary science project and must integrate the most advanced data and system technologies. EOSDIS also contains both the flight operations segment and the ground data system. The fact that schedules overlap and that the prime contractor probably will use different groups of personnel to implement these two very different elements will amplify the government's oversight and management challenge. Yet the panel has heard substantial evidence that from the management standpoint, EOS and EOSDIS are treated like ordinary projects within the Goddard Center. For example, the Project Manager for EOSDIS is two management levels down within the Flight Operations Directorate, which is only one of ten directorates at the Goddard Center. In addition, the Project Office is quite small for the task at hand, with plans for only 45 government employees when fully staffed. This small core of dedicated staff provides inadequate programmatic and managerial depth and expertise in the development of large, distributed data systems and in computer science and technology.

Given the preeminent position of EOS and EOSDIS in the U.S. Global Change Research Program, the panel believes that it is essential to increase the level of management visibility of the Project and the size and skills of the Project staff. In addition to learning from other government agencies that have had experience in the development and operation of large distributed data handling systems, NASA could, as needed, add to the Project experienced systems development personnel from other parts of the government.

The panel suggests that greater flexibility in defining success criteria and in using the process for setting award fees for direct feedback from the Project Manager to senior-level contractor management would help to assure that the contractor will do an outstanding job on EOSDIS. The panel commends NASA for including users in its performance board for contract evaluation and urges the active participation of users in setting award fees.

The panel recommends that the EOSDIS Project Manager have higher management visibility within Goddard Space Flight Center. The staff authorizations and skills should be sized to the scope and complexity of the Project. Further, the Project [16]could augment its staff with experienced personnel from other parts of the government in addition to NASA.

The panel recommends that the EOSDIS Project use the award fee process to best advantage through greater differentiation of success and failure criteria for evaluating contractor performance and by involving users in determining award fees.

Scientific Involvement at Goddard Space Flight Center. The Goddard Center's in-house earth scientists have a very limited role in the management and operations aspects of the EOSDIS Project. Although NASA has established a variety of science advisory and data working groups, such groups cannot replace the continuing and even daily involvement of the external scientific community and the Goddard Center staff to ensure that the eventual system is responsive to user needs.

Likewise, the nation's computer science community currently has very limited involvement in the Project, despite the fact that EOSDIS, to be successful, must implement the latest advances in scientific data management technology and, in some cases, stimulate the development of new technologies. The development of EOSDIS would benefit from substantive use of expertise in systems design and exploitation of information processing technology. Because underlying technologies, such as storage density, processor speeds, and transmission rates, are doubling roughly every three years, EOSDIS must be able to exploit rapidly expanding capabilities during its lifetime of a generation or more.

EOSDIS will also stretch the limits of what can be done by a mammoth database management system shared by a very diverse and demanding user community. Certainly, many of the underlying technologies such as storage will evolve on their own. Other technologies, however, will have to be encouraged, such as large-scale data management, visualization, and integration of heterogeneous information. Possible ways to stimulate technology include establishing an intramural computer science research capability comparable to those in other sciences, supporting and using the external computer science community, and using DAACs to establish formal and informal links with the computer science research community in their neighboring universities.

The panel recommends that NASA involve Goddard Space Flight Center earth scientists to a greater degree in the management and operations of EOSDIS and also involve computer scientists both inside and outside of NASA to explore research and [17] technology in those areas where EOSDIS will stress the state of the art in science and technology and where EOSDIS will evolve most rapidly.

DAAC Involvement. The DAACs are not well integrated into the EOSDIS management structure, particularly during the development phase. The DAAC managers do not have well-defined authority or accountability in building EOSDIS. DAACs should be involved early, in contrast to the current plan, in which their primary role appears to be to operate the hardware and software at their sites after delivery, and to deliver data products to users.

There should be mechanisms for feedback on scientific utility and operational effectiveness from the individual DAACs and associated archive centers to the central Project since the DAACs will be the primary sites for user interaction. There should be a coherent overall development, management, and science advisory structure that includes the DAACs. The panel understands that DAAC managers and scientists are involved in advisory roles. Advisory roles, however, are not sufficient for developing capabilities for and at the DAACs.

Overall, the centralized management of the design and implementation of EOSDIS functions at each DAAC is not conducive to active user involvement and responsiveness to changing technology. What is needed is a structure that strengthens the local role of each DAAC beyond the present DAAC advisory group and thus enhances the responsiveness of each DAAC in meeting the needs of its user community, gives the DAAC some control over its destiny, and yet ensures that an interoperable system is developed to meet the requirements of EOSDIS.

The panel recommends that NASA create, at each DAAC, a Development Team of full-time staff and active science users to address DAAC and user concerns. These teams should evaluate EOSDIS planning and implementation, including architecture, DAAC interface definitions, and other deliverables essential to ensuring that the DAACs will be responsive to user needs and that the EOSDIS system will be interoperable. In accomplishing these tasks, the teams should monitor the contractor's activities on behalf of user communities and prepare test data sets to verify system interfaces. Each DAAC Development Team should validate that DAAC's operational capability to use the evolving EOSDIS system as each of the program releases is implemented. Finally, NASA should provide the DAACs with modest funding to respond to specific user needs so that the DAACs will be able to parallel the evolution of the user community's ability to manipulate, integrate, and model data.

****

[A7]
Appendix C: Panel Members and Biographies

PANEL TO REVIEW EOSDIS PLANS

Charles A. Zraket, Chair, Center for Science and International Afairs, Harvard University
D. James Baker, Joint Oceanographic Institutions Incorporated
Kenneth I. Daugherty, Defense Mapping Agency
Richard E. Hallgren, American Meteorological Society
John E. Hopcroft, Cornell University
Kenneth C. Jezek, Ohio State University
Anita K. Jones, University of Virginia
Thomas R. Karl, National Oceanic and Atmospheric Administration
Ethan J. Schreier, Space Telescope Science Institute
Gael F. Squibb, Harvard-Smithsonian Center for Astrophysics
Jeffrey D. Ullman, Stanford University

Staff

Richard C. Hart, Space Studies Board
Monica B. Krueger, Computer Science and Telecommunications Board
Norman Metzger, Commission on Physical Sciences, Mathematics, and Applications
Lorraine W. Wolf, Board on Earth Sciences and Resources

Document IV-17

Document Title: Berrien Moore III, Institute for the Study of Earth, Oceans and Space, University of New Hampshire, and Jeff Dozier, Center for Remote Sensing and

Environmental Optics, University of California, Santa Barbara, Executive Summary, "Adapting the Earth Observing System to the Projected $8 Billion Budget: Recommendations from the EOS Investigators," 14 October 1992.

Source: NASA Historical Reference Collection, NASA History Office, NASA Headquarters, Washington, D.C.

This document, the results of the EOS Payload Advisory Panel deliberations, focused on making the EOS program fit into the procrustean bed of $8 billion over ten years; the program had recently been reduced to $11 billion from $17 billion. This so-called rescoping, or descoping, necessitated, among other things, a redesign of most of the EOS platforms to fit on a Delta launch vehicle, rather than the larger Atlas IIAS. The panel's recommendations include several reductions in the complexity of EOS instruments and a subsequent reduction in the quality of possible scientific observations.

[title page]

Adapting the Earth Observing System to the Projected $8 Billion Budget: Recommendations from the EOS Investigators

Edited by

Berrien Moore III
Institute for the Study of Earth, Oceans and Space
University of New Hampshire
Durham, NH
and

Jeff Dozier
Center for Remote Sensing and Environmental Optics
University of California
Santa Barbara, CA 93106-3060

October 14,1992

****

[1]

1   Executive Summary: Synopsis of Recommendations for the $8 Billion Program

We believe that a properly structured $8 billion funding profile through the rest of this decade is enough to design and put in place the initial components of the Earth Observing System, NASA's major contribution to the Global Change Research Program, whose purpose is to study and understand natural and anthropogenic changes in the Earth System. The EOS program will include an integrated space-based observing system, creation of a global data base of crucial measurements that span a 15-year period, development of better predictive models so that plausible changes can be understood, and a comprehensive data and information system that fosters synergistic interactions between observations and models and enables and encourages interdisciplinary research.

We note, however, that the descope of EOS to $8 billion requires difficult tradeoffs to maximize science in a cost driven program. One key choice is the amount of contingency held to handle unexpected problems in instrument development and changing science requirements driving the specifications for the instruments and data system. This contingency must be balanced against the savings that would result from complete elimination of instruments and their associated scientific information.

We favor reducing this contingency and therefore accepting a loss in future EOS flexibility. The reduction last year from $17 billion to $11 billion has already reduced EOS to a minimum set of instruments to pursue the focused objective of global climate change. In this reduction the measurement capabilities of the remaining instruments were significantly reduced and further reductions are not reasonable.

The increase in risk associated with this reduction in contingency is implicitly mitigated because EOS is a long-term measurement program, with instruments flown on five-year intervals. Consequently, instrument development problems or changes in science specifications could be handled in the next versions of the instruments. The first copies of some instruments may be deficient. For example, instruments with detector arrays may have some failed detectors at launch, or instrument noise may be greater than specification. Some problems can be fixed in later data processing; others will require correction in later versions of the instruments. Some level of resilience and flexibility, however, must be maintained to allow EOS to be carried out under normal (expected) levels of uncertainty in the budget and also to allow for the necessary technology developments that benefit US. competitiveness.

At $8 billion, EOS must depend increasingly on our European and Japanese partners. Failure to accomplish planned international cooperation on ADEOS, POEM, TRMM, and their follow-on missions will leave gaping holes in the international Earth Observing System. We note that in the $8 billion program the U.S. is relinquishing to international partners the development of new advanced technologies in laser and active microwave remote sensing. Finally, any further budget cuts will require wholesale elimination of information critical to understanding global climate change.

In developing our recommendations we considered carefully the proposals of the Red/Blue Team. Their recommendations are carefully constructed and we applaud their efforts. Generally, the Payload Advisory Panel concurs with the Red/Blue Team. There are, however, important differences. We summarize the differences by highlighting some of the Panel's recommendations and contrasting them to those of the Red/Blue Team. Our recommendations are discussed more fully in the major sections that follow this one. The major recommendations that deal with program additions (Sections 1.1-1.4) are discussed roughly in priority order. The [2] recommendation (Section 1.6) about the Wide Band Data Collection System may lead to an additional deletion and cost saving and is considered high priority.

1.1 High-Resolution Imaging Spectrometer

- The Panel recommends that the current HIRIS science investigation continue as planned through its projected completion in 1994. This effort will provide the foundation for imaging spectrometry of canopy chemistry, and other applications in the Earth sciences, whether such measurements remain in EOS, migrate to Landsat 8, or find some other venue.

The Red/Blue Team recommended cancellation of HIRIS. The Payload Advisory Panel concurs that the original HIRIS is too expensive for an $8 billion program; however, we believe that a solid case exists for an imagining [sic] spectrometer with spectral resolution and spectral coverage of HIRIS, but with lower spatial resolution. Our recommendation seeks to learn more fully the basis and strength for that case. In the body of this Report, we make several specific recommendations about imaging spectrometry. We do not recommend that development of the currently envisioned HIRIS II instrument be accelerated, but we do recommend a modest study of a new imaging spectrometer to help estimate future costs of a lower-spatial resolution instrument, while the science requirements for canopy chemistry are better defined and while the use of imaging spectrometer data in other important Earth science applications is further explored.

1.2 Instruments for Stratospheric Chemistry and Dynamics

- The Payload Advisory Panel supports the Red/Blue Team proposal to fly the EOSCHEMISTRY package in about 2002. EOS-CHEMISTRY with HIRDLS, MLS, SAGE III, and SOLSTICE satisfies all minimum requirements. If SAFIRE is substituted for MLS, the mission satisfies all minimum requirements except for ClO measurement. A timely selection between MLS and SAFIRE should be done; the
- We recommend that two SAGE flights be carried out by the year 2000. Continuity of the data record will be more useful if later flights can overlap previous ones. We also reiterate the need to have simultaneous flights with both polar and inclined orbits to achieve global spatial distributions. Because of its enhanced capabilities, the Panel recommends flights of SAGE III rather than SAGE II as soon as possible.

Global stratospheric measurements of temperature, winds, aerosols and clouds, long lived trace gasses, some radical and most reservoir species are needed, because these quantities can vary strongly both spatially and seasonally. Long-term, high-precision, continuous monitoring of ozone, temperature and some reservoir gasses is needed for global trends. Measurements and mapping at fine spatial resolution are required to examine the mixing process in the polar vortex, troposphere-stratosphere exchange, and to resolve localized stratospheric synoptic scale events that can produce important localized regions of heterogeneous chemical reactions and ozone depletion.

ESA's POEM-ENV package, likely to fly in 1998, may include GOMOS, MIPAS and SCIAMACHY and will measure many of the required species, but it lacks capability to measure [3] the key radical OH, measures ClO only under ozone hole conditions where the concentration exceeds 1 ppb, and does not measure key reservoir species in the chlorine and bromine families (HCl and HBr). None of the POEM-ENV instruments provide the high horizontal resolution data available from HIRDIS.

The ESA measurements will contribute to our understanding of stratospheric processes and to monitoring global trends, and the flight of ESA's POEM-ENV in about 1998 followed by the flight of EOS-CHEMISTRY post 2000 would provide a valuable time series of many important stratospheric variables. Currently we have insufficient information about the sensitivity and precision of the constituent measurements and the risk associated with the ESA instruments. As a consequence, the Payload Advisory Panel plans to invite the principal investigators of MIPAS, GOMOS, and SCIAMACHY to address the panel about the space heritage and risk associated with the instruments, and the sensitivity and precision of the measurements of temperature, aerosols and constituents that will be possible.

Finally, the Payload Advisory Panel iterates the recommendation made previously by itself and by the Atmospheres Panel. We recognize that this additional flight of SAGE is above and beyond the Red/Blue Team recommendation.

1.3 Measurements of Tropospheric Aerosol

- We recommend that the proposed MISR polarization measurement on the EOS AM-1 be included expeditiously by the EOS Project to further the capability of deriving global distributions of aerosol properties from space.

Tropospheric aerosols have been posed as a possible paradigm for understanding many of the dominant discrepancies between patterns of temperature increase that are measured versus inferred from climate models for global warming. EOS observations can make important progress to the resolution of this issue by global mapping of tropospheric aerosol opacities on a time scale of a few days out to interannual.

The EOS aerosol workshop in December 1991 recommended an assessment of the capability of EOSP under realistic conditions through simulation and field ground truth measurements to provide unique additional information through its polarization capability. A preliminary assessment is currently being carried out by the EOSP team which, if successful, supports the inclusion, as recommended by the Red/Blue Team, of EOSP on the second EOS-AM platform.

The MISR team proposes to add a polarization measurement on the EOS AM-1 platform through a relatively minor enhancement of its MISR instrument. The data would be used by the EOSP team to refine and validate their algorithms for deriving global fields of aerosol opacities using polarimetry.

1.4 Solar Irradiance Monitoring

- We recommend that plans be made for prompt flight (within the next 3-4 years) of solar monitoring from a small satellite or a flight of opportunity. A flight would need to occur within the next several years if it is to have a good chance of connecting with UARS. The method for continuation of solar monitoring after this first "gap filler" should be determined soon, i.e., within the next year or so.

[4]

- Solar spectral variability is an important aspect of solar variability for climate purposes. Climate forcing due to changes of solar irradiance depends on the spectrum of the changes. To a large degree, the arguments about the need for overlap of successive instruments applies to monitoring of the spectrum, as well as the total solar irradiance. The length of existing record at risk due to a potential gap in monitoring is much less for the spectral radiance. We recommend that plans for prompt flight of continued solar monitoring include SOLSTICE as well as ACRIM.

Solar monitoring of irradiance variability is crucial for issues of long-term climate change. Satellite monitoring during the past decade confirms the existence of significant total solar irradiance variations, of about 0.1%, and it is important to know whether there are larger variations on longer time scales. Measurement of solar irradiance change relies on instrumental precisions [sic]. Consequently, temporal overlap of the instruments is required. The Shuttle ACRIM calibration is aimed at this problem, but it is not proven that this experiment will eliminate the need for overlapping instruments to provide an adequate long-term record. Moreover, of the solar irradiance instruments currently flying the one with the potential for longest life is probably ACRIM II on UARS; however, it is unlikely that its lifetime will be more that [sic] a few years.

The first identified flight following the Red/Blue Team recommendation of ACRIM (and SOLSTICE) is on CHEMISTRY 2002, which implies a large gap in solar monitoring.

1.5 Descoping or Failure to Fund Major Instruments: AIRS, MODIS, LAWS, EOS SAR

1.5.1 Descoping of the Atmospheric Infrared Sounder (AIRS)

- We support the Red/Blue Team's recommendation to reduce AIRS from two spectrometers to a single spectrometer. We caution, however, that the modifications eliminate some important measurements of cloud emissivity and water vapor at fine vertical resolution near the surface. AIRS can still provide temperature and humidity profiles at the accuracy needed to improve climate modeling and numerical weather prediction.

AIRS has been reduced from two spectrometers to a single spectrometer, which effectively cuts the spectral coverage by half. Elimination of full spectral coverage will possibly reduce the accuracy of the spectral calibration in some IR channels. In addition, the signal-to-noise requirements and the resulting NEΔT, which had been relaxed previously to 0.35°C in the 15 μm region and 0.2°C in the 4 μm region must now revert to the original requirements of nearly 0.1°C throughout the spectrum. This results from the reduction of detector elements that previously provided the required accuracy and redundancy by co-addition of their signals. The end result is an intrinsically less reliable focal plane with more demanding requirements on optics, detectors, filters, and signal-to-noise ratios to achieve the goals of AIRS for EOS and NOAA.

The loss of AIRS full spectral coverage will result in gaps for determining the infrared spectral emissivity of clouds and the surface. We have also lost the opportunity to test and verify a new concept for determining humidity (from daytime observation in the 3.4 μm region) with a vertical resolution of just a few hundred meters above the surface, and we have relinquished the capability to map globally the horizontal distribution of several important trace atmospheric gasses.

[5] With strict adherence to the original signal-to-noise requirements, we believe that AIRS can still achieve its basic science goal to provide temperature and humidity profiles with the same accuracy and resolution originally specified.

1.5.2   Descoping of the Moderate-Resolution Imaging Spectroradiometer (MODIS)

- Because of the pivotal role of the MODIS instrument in supporting other instruments and in providing key products for several land, ocean, and atmosphere studies, we urge that Project and Program proceed carefully before instituting any further reductions in the specifications and capability of MODIS.

The original MODIS was descoped in the previous restructuring by eliminating the tilting instrument, MODIS-T. The current instrument is being further reviewed and contingency is being cut. Among the suite of descope options being considered are detector performance, band-to-band and focal plane-to-focal plane registration, and in-flight calibration. The MODIS Science Team is considering the implications of the suggested descope.

1.5.3   Laser Atmospheric Wind Sounder (LAWS)

- We encourage NASA to develop interagency and international partnerships, involving the LAWS team, that would lead to achieving measurements of the tropospheric wind field.

The LAWS instrument will provide critical information on the tropospheric wind field. However, its flight requires both a separate platform and additional funding.

1.5.4   Multipolarization, Multifrequency Synthetic Aperture Radar (EOS SAR)

- We encourage NASA to develop interagency and international partnerships to design and build a multifrequency, multipolarization SAR that will address the broad science objectives of global climate change.

EOS SAR is required to measure globally biomass, soil moisture, snow accumulation, and polar ice dynamics. The proposed EOS SAR instrument has been descoped to provide the minimal capability to measure key parameters in ecosystem dynamics, hydrology, solid Earth, and cryospheric science.

1.6   Wide Band Data Collection System (WBDCS)

- At the next Payload Advisory Panel Meeting, the WBDCS Team should be prepared to justify the inclusion of the WBDCS on an EOS platform in the context of other EOS priorities.

The Payload Advisory Panel recognized the potential value for the direct broadcast of geophysical data using a satellite relay system. Earthquake monitoring, tsunami warning, snow-fall data, and real-time alert of volcanic eruptions are some of the potential fields that [6] would benefit from such a system. The Red/Blue team recommends the WBDCS for flight on PM-1. As currently defined, however, the Panel felt unable to endorse the flight of the WBDCS on any specific EOS platform without additional information on data rate requirements and the necessary tracking capabilities of the ground stations.

1.7   Scatterometer Data for EOS

- The Payload Advisory Panel reaffirms the necessity of flying an NSCAT-class scatterometer throughout the EOS time frame. Specifically, the Panel encourages continued discussions between NASA and NASDA for the flight of NSCAT-2 on the ADEOS-II mission, scheduled for launch in 1999. Such a flight, and its follow-ons, assures continuity of the time series of ocean wind measurements begun by NSCAT/ADEOS-I and will provide a unique data set through simultaneous scatterometer, microwave radiometer, and ocean color measurements.

Scatterometer measurements of wind velocity (both speed and direction) are crucial for studies of the ocean's role in climate variability, wind-forced upper ocean circulation and heat transport, regional and basin-wide air-sea interaction, marine meteorology, and ocean productivity. In the operational arena, scatterometer measurements of surface wind velocity can be assimilated into regional and global atmospheric forecast/analysis systems to yield improved weather forecasts.

Because climatically important oceanic and air-sea interaction processes occur over a wide range of temporal and spatial scales, measurements of key dynamic variables must be obtained frequently and with high resolution extending over long periods and having extensive coverage. The sampling characteristics of a two-swath, NSCAT-class scatterometer in the ADEOS orbit allow coverage of more than 95% of the global oceans every two days with 50 km resolution. NSCAT-2 will be a near-copy of NSCAT (in terms of frequency, measurement technology, sampling, and ground processing); launch on ADEOS-II (before the planned end of the ADEOS-I mission) will ensure a continuous, multi-year data set of consistent wind velocity measurements.

1.8   Satellite Radar Altimeter

- We recommend that NASA proceed immediately to identify and secure funding to proceed with a joint U.S./France TOPEX Follow-On mission to launch in 1998.

- As an alternative, the concept of moving the EOS-ALTIMETRY mission forward to near 1998 should be examined. The possibility of combining TOPEX follow-on and EOS-ALTIMETRY along with the option for flying the GLAS in this period offer the possibility of cost savings in the overall cost profile of the Mission to Planet Earth.

The importance of altimeter data in developing an understanding of ocean circulation has led to the TOPEX/Poseidon mission. The accuracy of the TOPEX sea-surface height measurement is not matched by any current or previous satellite altimeter missions. To evaluate changes in global ocean circulation patterns and global mean sea-level over decadal time scales requires that the EOS-ALTIMETRY missions preserve the height measurement accuracy of the TOPEX/Poseidon system.

[7] Although the currently considered concept of a CNES provided solid state altimeter and a DORIS tracking system for POD provide a promising approach, it is important that the capabilities of the CNES package meet TOPEX/Poseidon standards. The Panel will continue to monitor the TOPEX follow-on mission.

The most pressing issue concerns funding for TOPEX follow-on. To meet a 1998 launch date, TOPEX follow-on would need to begin Phase C/D in FY 1995. This means that efforts must begin in early 1993 to identify funding. Since this mission should be funded under the Earth Probe line, it is critical that the Earth Probe line in FY95 be maintained at the FY94 year plus inflation. Such a mission would have the broad support of the oceanographic community, is critical to a number of global change objectives, and should be of relatively low cost, given the joint France/U.S. involvement.

The Red/Blue Team recommendation supported only an ALTIMETRY mission in 2002 and did not address or recommend a more timely oceanographic mission.

1.9   International Instruments on EOS Platforms

1.9.1   ASTER

- ASTER should be flown, to provide subpixel variability of surface temperature, mineral composition, surface topography, and three-dimensional mapping of volcanic plumes.
- ASTER should be flown out of phase with Landsat 7 for eight-day interleaved coverage.
- Cooperation with NASDA, whereby we fly ASTER on EOS AM-1 and they fly NSCAT on ADEOS and CERES and LIS on TRMM, is crucial to the EOS mission's dynamical observations.

Given the possibility that Landsat 7 will fly in 1997 or 1998, the role and cost of ASTER must be carefully examined. The Panel discussed the issues associated with ASTER and concluded that Landsat 7 is complimentary to ASTER. It does not replace the need for ASTER.

1.9.2   MIMR

- Given the possibility that new microwave radiometers may be flown by the U.S. Department of Defense, ESA, and NASDA in 1998-2000, we need to consider the desirability of overlap of so many similar instruments.

ESA is considering building MIMR for both EOS PM-1 and POEM AM spacecraft. A decision on this plan may be made at the European Ministers meeting in November. This issue needs to be revisited at the next Payload Advisory Panel Meeting.

1.10   EOS Data and Information System

1.10.1   EOS Data Products

- The Red/Blue Team has started the process of refining the list of science data products to be provided in EOSDIS. However the EOS investigators cannot relinquish responsibility [8] for this task. Hence the science panels and instrument teams of the EOS Investigators' Working Group must systematically develop the list of core data products, including science requirements, algorithm heritage, alternative approaches, and intermediate products.
- The EOS Project must work with the appropriate EOS investigators to better estimate the data system loads associated with each product, and consider whether the data product should be produced routinely on [sic] only on demand, and whether the coded algorithm could be distributed instead of the calculated data product.

Evaluation of the relationship between the cost of EOSDIS and the list of high-priority data products requires analysis of the science, algorithms, products, and the associated requirements on EOSDIS for systems engineering, processing, archive, and distribution. The EOS IWG recognizes that it must play an active role in the definition and development of EOS data products.

1.10.2   Transition from Version 0 to Version 1

- The EOS Project must work with the science community in the development of the transition from EOSDIS Version 0 to Version 1 to ensure that necessary services are maintained and that required capabilities are added in an orderly manner. The transition must be examined in the context of the types of science problems and information system technology that will be available in the mid-1990s.

The release of Version 1 is scheduled to occur 2 years after the release of Version 0. The EOS Project must adequately plan for the transition between these two efforts, to ensure that the Version 0 capabilities evolve smoothly into Version 1. The ECS contractor must study the designs, results, and experience from the Version 0 effort and assess the feasibility of using Version 0 products in the development of Version 1. The specific steps in the transition—determined by the EOS Project, the EOS contractor, and the EOSDIS Advisory Panel—must be based on these assessments, experience at the DAACs, science user feedback, technical factors, and cost.

1.10.3   EOSDIS User Model

- The EOS Project must develop a user model—numbers of users and their characteristics—that is based on investigators' proposed work with EOS instruments, on existing scientific data production systems, and on processing scenarios and benchmarks.
- The EOSDIS IV & V (Independent Verification and Validation) contract must support the EOS Project's effort to examine the system from the scientific users' viewpoints. It must not be merely a requirements-tracing exercise.

Establishment of users' requirements for EOSDIS is difficult, because they will change as the interaction between the Earth and information science communities improves and as scientists gain experience in using the tools developed by the ECS contractor. To evaluate the effectiveness of EOSDIS from the scientists' perspectives, end-to-end scenarios and benchmarks will be needed that use representative data sets to address science issues, to evaluate the functional capabilities of EOSDIS and measure the performance of the system.

[9] 1.10.4   Effects of Budget Reductions on EOSDIS

- The EOS IWG (through its EOSDIS Advisory Panel) must examine the architecture, design, and assumptions of the newly selected contractor, and analyze the cost sensitivity of the system's attributes. The IWG can then assess where costs might be reduced.

The significant steps in meeting the 30% reduction include reduction of the suite of data products available at launch to about 100 Level 2 products and about 100 additional Level 3 and 4 products and deferred migration of existing data sets into Version 0 (in cases were the data are available through an existing operational system), and reduction in contingency.

1.10.5   Other Procurements: EDOS, ECOM, IV&V

- NASA should design flexibility into EOSDIS to support network data delivery via networks whenever economically feasible and plan for the insertion of NREN technology in EOSDIS when it is operationally available.
- The IV&V contract must provide specific analysis and testing functions appropriate for the evolutionary development of EOSDIS.

Late this fall, NASA plans to begin other procurements related to EOSDIS:

EROS, EOS Data and Operations System, to bring data from the EOS satellites and deliver Level 0 data to ECS;

ECOM, EOS Communications, to establish networks, especially for operational delivery of spacecraft data;

IV&V, Independent Verification and Validation.

We are particularly worried about networks, both for connecting investigators to the DAACs and for interconnection among investigators. We no longer view physical media as the "normal" mode of delivering data to scientists.

We recommend that the IV&V contract provide specific analysis and testing functions appropriate for the evolutionary development of EOSDIS, to assure that the design of the system fully and correctly implements the requirements, that evolutionary changes are implemented consistently and correctly to best meet the scientists' needs, and that costly redesigns are avoided.

1.10.6   Data Assimilation in EOSDIS

• The EOS IWG and the broader science community must evaluate the scientific requirements for assimilated data available through EOSDIS, so that the processing loads can be accommodated.

The computational requirements for data assimilation (production of level 4 data products) are huge. It is now not possible to define the computing requirements precisely, because of the needs to define the demand for assimilated data products and their quality specifications. All [10] these needs are evolving, and many of them are rapidly changing disciplines of research and development.

The research-quality data sets that are needed for NASA Earth science applications require a level of internal physical and chemical consistency that is far beyond that achieved in present day data products used for numerical weather prediction. The data sets are expected to be used for problems with time scales of years to decades, instead of hours to days. The data sets will be used for problems far more difficult than the problem of prediction. Furthermore, many more types of data (e.g., constituent, land-surface, and oceanic) will be assimilated.

Document IV-18

Document Title: "NASA Management Instruction 1102.16," 28 June 1993.

Source: NASA Historical Reference Collection, NASA History Office, NASA Headquarters, Washington, D.C.

This document describes in brief the roles and responsibilities of the NASA Associate Administrator of the Office of Mission to Planet Earth. Of interest is that the associate administrator also had responsibility for the institutional management of the Goddard Space Flight Center in Greenbelt, Maryland.

[1] NASA   NMI 1102 .16
MANAGEMENT   Effective Date   June 28, 1993

INSTRUCTION   Expiration Date   June 28, 1997

Responsible Office: Y/Office of Mission to Planet Earth
SUBJECT: ROLE AND RESPONSIBILITY - ASSOCIATE ADMINISTRATOR
           FOR MISSION TO PLANET EARTH (AA/MTPE)

1. OBJECTIVES OF POSITION
    a. Provide overall planning, directing, executing, and evaluating of NASA Mission to
        Planet Earth Program.
    b. Provide institutional management of the Goddard Space Flight Center (GSFC).

2. ORGANIZATIONAL SETTING
    a. The Office of Mission to Planet Earth is a Program Office, and the AA/MTPE
        reports to the Administrator.
    b. The basic organization of the office of Mission to Planet Earth is shown on
        Attachment A.
    c. The Office of Mission to Planet Earth will support and actively encourage cultural
        diversity in the composition and development of its workforce.

3. RESPONSIBILITIES
    The AA/MTPE has the following responsibilities and is delegated authority to take
    such action as necessary to carry out assigned functions within such limitations as
    may be established by the Administrator, applicable laws, and regulations. In the
    execution of these responsibilities, the AA/MTPE will coordinate, as appropriate,
    with other NASA program and functional offices and Field Installations.
    a. Programmatic Responsibilities
        (1) Develops the Mission to Planet Earth programs within NASA that are directed
            toward obtaining a scientific understanding of the entire Earth system on a global
            scale by describing how its component parts and their interactions have evolved,
            how they function, and how they may be expected to continue to evolve on all
            time scales. These programs include the following areas:

[2]   (a) The Flight Systems Division is responsible   for the development of all
                MTPE flight systems, including the Earth Observing System (EOS), Earth
                probes, Space Shuttle and aircraft payloads: development of National
                Oceanic and Atmospheric Administration (NOAA) operational flight systems
                on a reimbursable basis; and serves as the liaison to the office of Space Flight
                for Space Transportation System (STS) matters, Office of Space Science for
                Expendable Launch Vehicle (ELV) matters, and office of Life and
                Microgravity Sciences and Applications for flight systems interfaces.
            (b) The Operations, Data and Information Systems Division is responsible for the
                Management of mission operations, data systems development, applications
                and data management for MTPE programs, including the nonflight elements
                of EOS, and serves as the liaison to the Office of Space Communications.
            (c) The Science Division is responsible for the management of all science
                research efforts and the definition of flight programs related to MTPE and

serves as liaison to the Office of Aeronautics for investigating the environmental impact of aviation. The research and analysis program will include process studies, modeling, data analysis, assessment activities, and the aircraft support program.

(d) The Technology Innovation and Advanced Planning Division is responsible for providing technology infusion in flight systems and instruments and will serve as liaison to the office of Advanced Concepts and Technology, the Office of Safety and Mission Assurance, and other Government agencies involved in relevant advanced technology development.

[3]   (2) Manages the MTPE programs and policies, including the planning, developing, advocating, monitoring, and overseeing to ensure completion within approved resources and schedule, subject to final review by the Administrator.

(3) Includes the following primary line responsibility:

(a) Detailed scientific and technical definitions of specific flight and ground-based programs, including development of and commitment to, Agency cost and schedule plans for their accomplishment.

(b) Presentation, advocacy, and defense of these programs and related policies within NASA, the Administration, and to Congress.

(c) General management of these programs at the responsible Field Installation to ensure that scientific, technical, schedule, and resource commitments are met.

(4) Coordinates the MTPE programs with other NASA organizations and other NASA Associate Administrators to ensure the optimum use of resources and facilities to accomplish the goals of the Agency.

(5) Maintains relationships with universities, the scientific community, industry, other U.S. and foreign Government agencies, and international organizations. Selects MTPE flight investigations and investigators.

(6) Coordinates the MTPE programs, facilities, and functions with the Office of Safety and Mission Assurance to ensure safety, reliability, maintainability, and quality assurance.

b. Institutional Management Responsibilities. The overall institutional management of GSFC is the responsibility of the AA/MTPE. Major organizational changes and key personnel assignments that have a direct impact on programmatic responsibilities of other Officials-in-Charge of Headquarters Offices will be coordinated with the affected officials. In view of the major involvement in operations at GSFC [4] on the part of the Office of Space Communications (Code O) and the office of Space Science (Code S), the AA/MTPE will establish a mechanism to involve these Associate Administrators in the institutional management of GSFC sufficiently to ensure the representation of their interests.

4.   LINE OF SUCCESSION

Those officials authorized to act for the AA/MTPE are designated in Attachment B. [omitted]

[Signed by Daniel S. Goldin]
Administrator

ATTACHMENTS:
A.   Organization Chart.
B.   Authority to Act, for the Associate Administrator for
     Mission to Planet Earth.

DISTRIBUTION:
SDL 1

Document IV-19

Document Title: Letter from Representative Robert Walker, Chairman of the House Committee on Science, to NASA Administrator Daniel Goldin, 6 April 1995.

Source: NASA Historical Reference Collection, NASA History Office, NASA Headquarters, Washington, D.C.

Oversight of the Executive Branch is an important responsibility of Congress. This letter from the chairman of the House Committee on Science refers to a similar letter written on the same day to the presidents of the National Academies of Sciences and Engineering, requesting a study of NASA's EOS. (See Document IV-20.) This letter not only requests information from NASA on the status of NASA's EOS, but it also raises specific concerns regarding the overall direction of the program.

April 6, 1995

The Honorable Dan Goldin
Administrator
National Aeronautics and Space Administration
Washington, D.C. 20546-0001

Dear Mr. Administrator:

The Committee on Science is rapidly approaching the time when we will be forced to make a series of important decisions regarding the funding for programs within the National Aeronautics and Space Administration. One of the issues that has caused me some concern is the scientific and technical justification for the Earth Observing System which is a part of the Mission to Planet Earth program.

I am asking the National Academy of Sciences and Engineering to undertake an independent, and much broader look at the United States Global Research Program and as a part of that study they will also be looking at MTPE/EOS.

However, as the Administrator of NASA I would appreciate it if you would initiate an immediate review of MTPE/EOS to respond to the following concerns:

1. Is the science that is being conducted as a part of the MTPE/EOS program fully justified on strictly scientific terms? Has it been fully peer-reviewed?

2. In the planning process for implementation of MTPE/EOS are there active measures being taken to fully ensure that we are continually re-evaluating both scientific and technical options? Is new scientific knowledge being introduced to ensure that science breakthroughs are utilized?

[2]
The Honorable Dan Goldin
April 6, 1995
Page 2

Are new technologies being considered and integrated such as miniaturization or inclusion of appropriate sensors on platforms of opportunity such as Iridium?

3. What assurances are there that you have done an adequate job in thoroughly scrubbing all elements of the program to ensure that there are no additional CIESIN type undertakings included in MTPE/EOS?

4. Is the current program adequately scoped to ensure that the data generated by MTPE/EOS will be appropriately processed, and will all of that data be properly archived and made fully available for use by the scientific community?

Based on the current legislative schedule it would be particularly helpful to the Committee if your findings could be made available to us no later than late summer of this year. If there are any questions concerning this request, or if there is a need for additional guidance or information please feel free to call on Mr. David Clement, Chief of Staff, at 202-225-8772.

Cordially,
[signed]
Robert S. Walker
Chairman

Document IV-20

Document Title: Congressman Robert Walker, Chairman of the House Committee on Science, to Dr. Bruce Alberts, President, National Academy of Sciences and Dr. Robert White, President, National Academy of Engineering, 6 April 1995.

Source: NASA Historical Reference Collection, NASA History Office, NASA Headquarters, Washington, D.C.

The National Academy of Science and the National Academy of Engineering had a major role in designing NASA's MTPE and its EOS. They also had, and maintain, a role in the design of the U.S. Global Change Research Program (USGCRP). This letter requests a review of both the USGCRP and MTPE. In particular, in light of the major part NASA's satellite systems play in gathering data for the USGCRP, this letter requests an assessment of NASA's role, through the MTPE and the EOS, in the USGCRP.

April 6, 1995

Dr. Bruce Alberts,
President, National Academy of Sciences
Dr. Robert White,
President, National Academy of Engineering
2101 Constitution Avenue
Washington, DC 20418

Gentlemen,

One of the issues the Committee on Science will examine this year is the U.S Global Change Research Program. I am therefore requesting a study of that program by the National Research Council's Board on Sustainable Development. A review of the program in this time frame is also called for under Public Law 101-606, the Global Change Research Act of 1990. I suggest that these efforts be merged.

It is my understanding that the National Research Council's Board on Sustainable Development has responsibility for continuing scientific and engineering technical guidance to the Federal Government on the programs and plans of the USGCRP. In my capacity as Chairman of the Committee on Science I have similar oversight jurisdiction for the USGCRP in the United States House of Representatives.

Over the years I have had a strong interest in this interagency scientific program and have had a specific interest in its major space based observational component, the National Aeronautics and Space Administration program titled Mission to Planet Earth, and the subprogram entitled the Earth Observing System. It is apparent that the USGCRP could make significant contributions to the improvements of scientific understanding of the Earth system and I believe that it could provide critical information to help us understand the fundamental nature of climate change.

[2]
Dr. Alberts & Dr. White
April 6, 1995
Page 2

In response to a number of conversations I have had with senior members of the scientific community I firmly believe that the success of long-term programs such as the USGCRP and smaller programs such as MTPE must be based on continual evolution in response to new scientific insights and now [sic], and expanded opportunities provided by technological and scientific innovation. In addition, programs such as USGCRP and MTPE can, and should be responsive to new challenges posed by both national and international changes in public policy.

Much has changed on both the national and international scenes since these programs were first proposed. Needless to say, one of the most significant changes is the current critical review of all government programs in view of budget issues and our own national security. As the Committee moves toward making basic decisions on the future of our National investment in science and technology I would request that the Chair of the Board of Sustainable Development be directed to organize a comprehensive review of the United States Global Change Research Program and the Mission to Planet Earth/Earth Observing System programs within the National Aeronautics and Space Administration. This review will serve as the major review of these programs called for under Public Law 101-606, the Global Change Research Act of 1990.

Because of the large magnitude of the NASA effort in the USGCRP it is suggested that this effort be supported by and coordinated with the NASA Administrator. I particularly want to note that the linkage and integration between the NASA efforts and the remainder of the USGCRP are absolutely essential and critical to the success of the larger program. However, it is also important that there be active coordination with all other key Federal participants such as the Subcommittee on Global Change of CENR.

As Chairman of the Committee I am particularly interested in receiving:

1. A review of the scientific progress to date and an assessment of the implications of those scientific insights for future USGCRP and MTPE/EOS activities;

2. An assessment of the current observational strategy for the USGCRP, with particular attention to the role of NASA's Earth Observing System. This assessment should include consideration of any

[3]
Dr. Alberts & Dr. White
April 6, 1995
Page 3

adjustments that might be necessary to respond to: (i) recent improvements in scientific understanding, (ii) the needs of the general user community, and, (iii) the requirement for the EOSDIS to develop into an operational system routinely accessed by various users versus the need to rapidly incorporate new and emerging technologies.

3. An assessment of USGCRP and MTPE/EOS plans for data and information management in the context of efficiently addressing: (i) the needs of the scientific community, (ii) the rapid evolution of data and information management technologies, and (iii) opportunities to provide new scientific information products to support economic development and public policy needs, and

4. An assessment of where efficiencies are most likely to be found in terms of: (i) convergence of systems and activities across agencies, (ii) integration of space based and in situ programs, (iii) increased reliance on international partners, and, (iv) utilization of commercial capacity or capability or the privatization of EOSDIS.

Based on the current schedule of the Committee and the normal budget cycle of the Federal programs we are charged with overseeing, your initial findings and any guidance should be received by the Committee not later than September of this year, and I would strongly suggest that any possible preliminary findings that might be available at earlier dates would be extremely helpful and deeply appreciated.

I hope that you will be able to respond favorably to this request and if there are any questions or clarifications required please do not hesitate to contact the Committee Chief of Staff, Mr. David D. Clement, at 202-225-8772.

Cordially,
[signed]
Robert S. Walker
Chairman

Document IV-21

Document Title: NASA Technical Memorandum 4679, "Spaceborne Synthetic Aperture Radar: Current Status and Future Directions," Report to the Committee on Earth

Sciences, Space Studies Board, National Research Council, April 1995.

Source: NASA Historical Reference Collection, NASA History Office, NASA Headquarters, Washington, D.C.

This memorandum on the promise of synthetic aperture radar for supporting Earth science research was prepared to support the National Research Council's review of satellite synthetic aperture radar technology research within the United States.

[title page]

NASA Technical Memorandum 4679
National Aeronautics and Space Administration
Scientific and Technical Information Office

April 1995

Spaceborne Synthetic Aperture Radar:
Current Status and Future Directions

A Report to the Committee on Earth Sciences
Space Studies Board
National Research Council

D. L. Evans,	Editor
J. Apel	E. Kasischke
R. Arvidson	F. Li
R. Bindschadler	J. Melack
F. Carsey	B. Minster
J. Dozier	P. Mouginis-Mark
K. Jezek	J. van Zyl

****

[iii]

Preface

In June 1994 NASA's Office of Mission to Planet Earth (OMTPE) identified the need for a broadly scoped review of the nation's civilian spaceborne Synthetic Aperture Radar (SAR) program, and requested the Committee on Earth Studies (CES) of the Space Studies Board (SSB) of the National Research Council (NRC) to undertake such a review. The Board was charged with answering the following questions:

(1) Is multiparameter SAR the optimum spaceborne approach to characterize the critical geophysical parameters identified by the interdisciplinary Spaceborne Imaging Radar-C, X-Band Synthetic Aperture Radar (SIR-C/X-SAR) Earth science

team, and if not, are the data products nevertheless of credible utility in Earth science? For example:

- How well can SAR be used to estimate biomass, characterize vegetation type, characterize forest stand maturity, clear-cut, or regrowth?
- How well can SAR be used to characterize snow-water equivalent, distinguish between new and refrozen ice, or estimate ice volume?
- How well can SAR characterize oceanographic features and parameters such as internal waves, oil slicks, wave direction, and air-sea interaction?
- How well can SAR characterize soil moisture?
8 How well can SAR characterize geologic features such as rock type/composition, and surface texture?

(2) With respect to all of the above questions, how important are the multiple wavelength, multi-polarizing, variable-incidence angle capabilities to these characterizations?

(3) What is the potential of spaceborne radar interferometry in topographic mapping and surface change monitoring connected with natural hazards?

(4) What is the complementary nature of a spaceborne radar interferometry project to monitoring crustal strain and deploying dense arrays of Global Positioning System (GPS) receivers in selected areas of seismic hazard?

(5) What are the priorities in SAR technology development which are critical not only to NASA's maintaining leadership in spaceborne SAR technology, but to providing societally relevant geophysical parameters?

(6) What is the priority of SAR science in the context of the overall national and international Earth observing areas?

(7) With the answers to these questions as backdrop, how might the international space program community make the best use of its resources while satisfying individual programmatic requirements through joint planning and cooperation in future SAR flight projects?

(8) What would an appropriate role be for NASA in such an international SAR program? For example:
[iv]
- What would an appropriate role be for the NASA Earth Observing System Data and Information System (EOSDIS) in such a program?
- How could the SIR-C/X-SAR or the EOS SAR Science Team expertise be used?

This report was prepared as background material through a series of discipline-oriented workshops held throughout the fall of 1994. Science Discipline Panel workshops and meetings, held in Late October and early November focused on Questions 1-4. The Technology Panel addressed Question 5 during two separate meetings on November 15 and 29. An Interagency panel formed to help address Questions 6-8 will report separately. The Chairmen of the Discipline Panels met on December 5, 1994 to discuss the report format and their key findings. Because of the broad interest in this subject, attempts have been made to solicit input and comments on this material from as broad a community as time allowed, including the international community. However, this should only be considered a "snapshot" of a rapidly evolving field, as new results are reported on a continuing basis.

****

[1-1]

1-Executive Summary

INTRODUCTION

This report provides a context in which questions put forth by NASA's Office of Mission to Planet Earth (OMTPE) regarding the next steps in spaceborne synthetic aperture radar (SAR) science and technology can be addressed. It summarizes the state-of-the-art in theory, experimental design, technology, data analysis and utilization of SAR data for studies of the Earth, and describes potential new applications.

This report is divided into five science chapters and a technology assessment. The science chapters are Ecology, Hydrology, Marine Science and Applications, Ice Sheets and Glaciers, and Solid Earth Sciences and Topography. Each Chapter outlines key science questions in the context of Mission to Planet Earth that can be addressed with SAR data. In addition, the chapters summarize the value of existing SAR data and currently planned SAR systems, and identify gaps in observational capabilities that need to be filled to address the scientific questions. Both demonstrated and potential capabilities are described, with appropriate references cited. Both NASA and non-NASA sources of SAR data are included, and the importance of multiple wavelengths, multiple polarizations, and variable incidence angles are substantiated for each measurement. A summary of sensors is included as an Appendix. Cases where SAR provides complementary data to other (non-SAR) measurement techniques are also described.

The chapter on technology assessment outlines SAR technology development which is critical not only to NASA's providing societally relevant geophysical parameters, but to maintaining competitiveness in SAR technology, and promoting economic development.

RECOMMENDATIONS

SAR data provide unique information about the health of the planet and its biodiversity, as well as critical data for natural hazards and resource assessments. Interferometric measurement capabilities uniquely provided by SAR are required to generate global topographic maps, to monitor surface topographic change, and to monitor glacier ice velocity and ocean features. Multiparameter SAR data are crucial for accurate land cover classification, measuring above-ground woody plant biomass, delineation of wetland inundation, measurement of snow and soil moisture, characterization of oil slicks, and monitoring of sea ice thickness.

The suite of spaceborne SAR systems and programs currently envisioned by the international community provides an important framework for addressing key science issues and applications. However, additional activities, and interferometric/multiparameter measurement capabilities are required for long-term environmental monitoring and commercial applications.

This report recommends NASA take an aggressive leadership role in an international SAR program to meet these needs. Specific near-term steps should be to:

(1) **Establish interagency and international SAR science teams**. These teams would be funded to exploit data from both NASA and non-NASA sources, and would be charged with development and testing/validation of new applications, both scientific and commercial in nature.

[1-2]

(2) **Initiate an advanced technology effort**. The initial focus of such a program should be on lowering the costs of the operational elements and exploiting the functionality of SAR. The NASA airborne radar (AIRSAR) which flies on the NASA DC-8 should be the focus of this activity.

(3) **Design an evolvable flight program.** The long-term objective of such a program is an operational interferometric, multiparameter spaceborne SAR for long-term environmental monitoring and commercial applications. Immediate steps toward this goal are to initiate an interferometric spaceborne mission, and to continue multiparameter measurements through additional flights of the Spaceborne Imaging Radar-C, X-Band Synthetic Aperture Radar (SIR-C/X-SAR) as a free-flyer or on the Space Shuttle.

Document IV-22

Document Title: Office of Science and Technology Policy Concept Paper, "The Global Observing System," 18 April 1995.

Source: NASA Historical Reference Collection, NASA History Office, NASA Headquarters, Washington, D.C.

Document IV-23

Document Title: Task Force on Observations and Data Management, "Concept for an Integrated Global Observing Strategy," 9 February 1996.

Source: NASA Historical Reference Collection, NASA History Office, NASA Headquarters, Washington, D.C.

Document IV-24

Document Title: Letter from Robert T. Watson, Associate Director for Environment and Co-Chair, Committee on Environment and Natural Resources, to Dr. Charles F. Kennel, Associate Director for Mission to Planet Earth, NASA, 22 March 1996.

Source: NASA Historical Reference Collection, NASA History Office, NASA Headquarters, Washington, D.C.

This set of documents illustrates the initial discussions over the development of an Integrated Global Observing Strategy. As Document IV-22 shows, the concept began as an International Global Observing System, which connoted to some a new entity, requiring significant new investment. The concept soon evolved into an International Global Observing Strategy. The key feature of the concept is that scientists should attempt to find ways to integrate the many different forms of data that scientists are collecting throughout the globe into data products that reflect the condition of the Earth and changes taking place in Earth systems. After about a year of discussion, the Global Observing Strategy was accepted by the CEOS and other stakeholders in the global Earth observations community, and a CEOS committee formed to find ways to implement it.

Document IV-22

[cover letter, dated July 17, 1995]
MEMORANDUM TO: CENR Task Force on Observations and Data Management
                       Co-Chairs
FROM: Robert T. Watson
D. James Baker
SUBJECT: International Global Observing System

In January of this year, the Office of Science and Technology Policy formed an ad hoc interagency group to develop a concept paper on an international Global Observing System (GOS) for Earth observations. The conclusion of this group, contained in the attached paper, is that creation of such a system has many potential benefits and is worthy of further study.

There has been a significant expansion of Earth observations funding and capability over the last decade. The United States, Japan, and several European and Asian nations have extensive space-based operational and research Earth observation programs which will extend into the future. These same nations and many others support ground based Earth observation sites and networks around the world. Although much has been accomplished, current and planned measurements still fall short of meeting the requirements of the major international science programs, including the World Climate Research Program and the International Geosphere-Biosphere Programme. In spite of the continuing need to increase both the quantity and quality of Earth system measurements, further budget expansion in this area is very unlikely.

Improving the effectiveness of international cooperation holds significant promise for continuing to improve global-scale Earth observations, even without significant increases in available funding. While international coordination of these efforts is generally accepted as necessary and desirable, and, in some cases, already well established as practice, there is considerable opportunity for improvement. Commitment to full and open data exchange and creation of a mutually agreed framework for the definition, planning and conduct of such observations could lead to a more efficient division of labor among implementing agencies around the world.

As the co-chairs of the Committee on Environment and Natural Resources Research, we request the Task Force to define a strawman international GOS framework and create a plan for organizing. U.S. civil Earth observations efforts in this context. An important element of this plan should be improving the US support for, and thus the effectiveness of, relevant existing international planning mechanisms, such as the United Nations Global Climate, Ocean, and Terrestrial Observing Systems and the space agency Earth Observations International Coordination Working Group. The draft plan should be submitted to us no later than November 1, 1995.

Attachment

cc: CENR Subcommittee Chairs

[1]   THE GLOBAL OBSERVING SYSTEM

Introduction

Understanding the processes which govern the Earth system is essential in enabling us to safeguard, maintain and improve this system for our use and that of our descendants. The United States is committed to a broad range of programs which observe the Earth system and use these observations to model, predict, and identify trends in the future behavior of the system. We all – individuals, families, companies, academic research institutions and governments – use the results of these efforts to adjust our social and economic patterns to improve the quality of our lives and to improve our economic situation.

For decades the United States has supported observations of the atmosphere, land, and ocean. Such observations are conducted routinely across the United States, in the overlying atmosphere, in our coastal waters, in many other countries and in the deep ocean. These in situ observations have been complemented in recent years by remote sensing observations from aircraft and, increasingly, from space. In addition, new observational capabilities have been developed within basic research programs.

Observational data is processed, analyzed, and provides input for the preparation of products for our everyday use and our edification. For example, NOAA's National Weather Service provides a wide range of prediction services including forecasts of weather and climate; NOAA's National Ocean Service provides forecasts of ocean and coastal tides and currents; and the Department of Agriculture provides crop forecasts. Observational data is also essential for research conducted by university and Federal laboratory scientists on the origin of the planet, the impact of ocean circulation on the atmosphere, the formation of storms, and the development of icebergs.

Advances in scientific theory, when combined with improvements in observing systems and great increases in computer capacity, now enable scientists to model the Earth's atmosphere, land, and ocean processes on a regional and global scale with steadily improving reliability. However, to conduct such modeling most effectively requires the development of comprehensive regional and global databases which, in turn, require us to improve and expand our observations of the planet, both geographically and over time.

These new applications require us to obtain data for a wide variety of parameters. For example, in the atmosphere it is important to measure, among other parameters, the temperature, chemical composition, and cloud cover throughout the atmospheric column. At the land and ocean surface, we need to observe such variables as winds, precipitation, incident radiation, precipitation, vegetation and ice cover. In the oceanic water column, observations of temperature, pressure, salinity, mixed-layer depth, and currents are all important.

Needs and Opportunities for Improved Coordination

As operational and scientific communities come to rely on these new observing systems, there is a need to identify and compile the observational requirements of the various user communities, both national and international (focusing on commonalties among them), and outline the observational capabilities necessary to fulfill these requirements, in order to assure that all key measurements are being made and that unnecessary duplication is avoided. As these requirements and other needs for observational data are identified, it is essential that they are [2] brought to the attention of national and international agencies with responsibilities for observational systems.

This process includes considering ways in which existing capabilities can be improved and/or expanded to meet these requirements, while avoiding building separate new systems, and evaluating complementary capabilities of remote sensing and in situ observational systems to obtain needed data and information.

Scientists in the United States, working together with their counterparts in other countries, have begun planning for new observing systems directed at the Earth's atmosphere, ocean and land. These are called the Global Climate, Ocean, and Terrestrial Observing Systems (GCOS, GOOS, and GTOS, respectively), all of which will provide for both in situ and remote sensing observations.

It is essential to move towards planning, development, and implementation of these systems both nationally and internationally within a global, multidisplinary (sic) approach which minimizes duplications, redundancies, and inefficiencies among separate observing systems and more closely links those communities that generate requirements with those that operate the observing systems. The goal of such an approach is a comprehensive Global Observing System (GOS).

A Global Observing System

The Global Observing System (GOS) would be an internationally coordinated system of mutually funded experimental and operational space-based and in situ data acquisition, archive and distribution systems and programs for Earth observations and environmental monitoring. The GOS would be created through the mutual agreement of participating countries to contribute discrete elements of the overall system within an agreed comprehensive framework. Such a system is necessary in order to efficiently acquire continuous, global, well-calibrated quantitative data about the Earth for applications in meteorology, oceanography, geology and geophysics, natural disaster reduction, terrestrial ecology, biology, atmospheric science, and climate change research.

Integration of existing and potential capabilities into a coherent worldwide observing system will require: 1) integrating and prioritizing requirements from multiple user communities; 2) providing systematic requirements-capabilities analysis; 3) providing a comprehensive framework for Earth observations in which members can commit to providing specific discrete elements; and 4) providing coordination within and between the satellite and in situ observing system communities.

An early requirement in GOS implementation is to define an appropriate scope or focus for the system which serves as a basis to prioritize or limit the types of measurements made by the system. This scope must cut across operational and research users, but must also provide a paradigm for prioritizing required observations within the comprehensive, and overwhelming, list of all measurements that are significant for overall Earth system studies.

[3] Given the potential difficulty in prioritizing a comprehensive set of observational requirements, member countries may decide to develop the GOS through an incremental approach which ultimately evolves into a comprehensive system (e.g. initially focusing on climate, land cover, ocean color, etc.). The GOS could focus initially on a limited set of basic measurements (including many of those currently being made on a routine basis), which could be expanded as the system developed and countries become comfortable with the process.

Physical Infrastructure

The GOS would build on the existing operational systems, which currently include satellites provided by the U.S., Europe, and Japan, and in situ observation systems from most countries of the world, the data from which is voluntarily exchanged through systems and procedures developed in the WMO and IOC. This global constellation of operational geostationary and polar orbiting satellites is the backbone of the GOS space component. Current and planned research missions, such as UARS, Topex/Poseidon, ERS, JERS, ADEOS, Radarsat, EOS, Envisat, TRMM, Meteor, and others are expected to demonstrate new capabilities that could provide the basis for future operational observations. Extensive in situ data collection activities take place around the world such as WWW, GAW, UNEP GEMS and GRID, and IOC GLOSS, IGOSS, and IODE. To support the essential ongoing research in Earth system science, all useful data must be calibrated, documented, retained and available for retrospective analysis as well as for routine operational uses. The GOS should thus also include a distributed data and information system to collect, process, maintain, and distribute the data and products to the full user community.

Some physical elements will require redundancy and assurance of uninterrupted service to support operational requirements. Other elements may be operated for discrete periods of time as demonstrations of technology or to support one-time research campaigns. Others may attempt to provide continuous service but, to reduce costs, the system may accept the risk of possible temporary interruptions in case of failures. For certain climate variables, temporary losses may not be critical. A set of operating principles should be established at the beginning and accepted by all participants as a condition of their incorporation into the system.

Role of the Private Sector

While governments undertake mutual commitments to provide discrete elements of the overall system, direct involvement of private, commercial entities, should be encouraged when governments determine that their needs can best be met through the private sector. Ultimate responsibility for the system should remain with governments, as the system is intended to meet public needs for basic information about the Earth and its environment.

Functions

Efficiently collecting Earth observations data for multiple uses demands work in three key areas. First, coordination of various user groups with varying, yet sometimes overlapping, data requirements will be necessary in order to promote the formation of integrated requirements for environmental observations. This will help agencies to respond more effectively to the needs of the science and operational communities.

Second, there must be a commitment by countries participating in GOS to meet observational requirements on a long-term and continuous basis, and to coordinate with each other in doing so. Finally, a commitment to collectively meet integrated observational requirements will require coordination of various system operators (both in situ and space-based) in order to minimize [4] duplication and allow assessment of in situ vs. space-based approaches. It should be recognized in such an effort that in some cased [sic] redundant systems or a multiplicity of approaches may be necessary (e.g. use of redundant systems in remote, harsh environments).

Some aspects of these three essential functions for the GOS are currently being done by existing groups or mechanisms. These groups should be incorporated into the GOS approach to the greatest extent possible. A prime requirement of the GOS is creating a mechanism to better translate the program recommendations of these groups (e.g. GCOS, GTOS, GOOS) into national activities.

Approach

Once consensus can be reached among international partners on the fundamental need for a GOS and a concept of the system, there are at least two distinct approaches to GOS implementation. The first way envisages the participating countries acting in concert under an agreed strategy or framework through existing groups (or modified existing groups). This approach would involve initial informal discussions nationally and internationally which could result in more formal meetings or statements. The goal of this approach would be to reach international consensus on the general framework of the GOS and to pursue implementation of the system through selected groups (while focusing attention away from ineffectual groups or mechanisms).

Alternatively, after the initial steps leading to consensus among participating countries, GOS members could decide to conclude a more formal agreement to ensure that the system is implemented and operated consistent with its intended purposes. A major challenge in Earth observations is persuading countries to develop and modify their long-term plans and investments jointly, thus requiring these countries to share responsibility with others for the collection of key data sets. Present uncertainties about the commitments of countries to execute their national plans results in some national systems that are duplicative on an international scale. In order to generate effective interdependence (and the consequent cost savings), an intergovernmental coordinating mechanism could be part of the implementation approach of the GOS. A formal commitment to the long-term provision of space-based and in situ monitoring capabilities would provide more formal support for these existing informal and non-binding coordination groups.

Such an agreement could suffice as a guide for national programs which would codify goals and operating principles, and designate existing mechanisms through which members could coordinate programs.

If participating countries believe that there is a need for a mechanism through which countries make specific commitments to undertake certain observations, coordinate programs, and resolve issues, they may decide to approach the GOS implementation through an international agreement with an associated oversight structure that would have sufficient authority to take actions needed to achieve efficient operations and high quality services. This mechanism would include senior level representatives of governments providing the funding and/or other resources for the system and would provide the formal authority and ultimate decision-making. This oversight mechanism in which commitments are made could be an integrating element across existing groups such as GCOS, GTOS, GOOS, and CEOS (which could continue to perform many of the assessment, planning, and coordination functions).

[5] Next Steps

The U.S. should take the lead in defining the concept, structure and implementation steps for the GOS. While national and international implementation should proceed in parallel, many aspects of the international implementation of GOS (e.g. consensus building) do not depend on the specific national elements finally endorsed as GOS contributions (i.e. the U.S. component to the GOS). The following are proposed national and international steps in the process of establishing the GOS.

• OSTP will brief the CENR Task Force on Observations and Data Management late April 1995; GOS concept paper, approved by principals of the informal OSTP drafting group, will be circulated for wider interagency review.

• In May/June 1995, OSTP will host an informal meeting of key international scientists and policy-level individuals in order to discuss the GOS concept, structure and implementation steps.

• By fall of 1995, the CENR Task Force should develop a comprehensive strategy for GOS, reflecting the input from informal national and international discussions.

  • The CENR Task Force should arrange a workshop to discuss the GOS concept and implementation plans with wider U.S. community (e.g. scientists, academia, operational users, private sector). Involvement of the National Academy of Science should be encouraged.

  • Strategy should have firm recommendations on structure, role of existing mechanisms and groups, and identification of any new agreements or mechanisms required for GOS.

• As the U.S. Secretariat for GCOS, GTOS, and GOOS, the CENR Task Force should designate representatives to meet with the heads of those organizations and solicit their views on the GOS. Discussion of the GOS concept should be placed on the agenda for upcoming meetings of the Joint Scientific and Technical Committees of the three systems as well as meetings of the sponsoring organizations (e.g. UNEP, WMO, IOC, ICSU).

• In addition to the above-listed specific implementation steps, efforts should be made to obtain the endorsement of the GOS concept by the following international groups by the end of 1995:

– Committee on Earth Observation Satellites Plenary
– International Group of Funding Agencies
– Economic Summit of Industrialized Nations (G-7)

In the context of this paper, the term "Global Observing System" is used to define a concept for an integrating framework of space-based and in situ observations and should not be confused with the existing WMO Global Observing System.

Document IV-23

[cover letter pg. 1]
[stamped: "RECEIVED, Feb 20 1996, NOAA/NESDIS/IA]
Reply to Attn of: YM

Dr. D. James Baker
Co-Chair, Committee on Environment
 and Natural Resources
U.S. Department of Commerce
National Oceanic and Atmospheric
 Administration
Washington, DC 20230

Dear Dr. Baker:

In your memorandum of July 17, 1995, you and Dr. Watson requested the Task Force on Observations and Data Management (TFODM) create a plan for organizing U.S. civil Earth observations in the context of an integrated global observing system. Enclosed is the "Concept for an Integrated Global Observing Strategy" developed in response to your request.

This document is the product of an interagency effort within the TFODM, led by Jim Rasmussen as the chair of the TFODM's Integrated Observing System Working Group. It proposes an overall aim for an Integrated Global Observing Strategy (IGOS), a rationale for integrated global observations, and guidelines for U.S. and international implementation. In this document, we have exchanged the term "strategy" for "system" to reflect that we are essentially proposing a framework in which national governments and international organizations can commit to providing observation systems that are elements of an internationally accepted observing strategy. The document concludes with a schedule of next steps.

In our roles as TFODM Chairman and Vice-Chairman, we have already met with some of our counterparts internationally and received generally positive responses to this initiative. Additionally, we have briefed the Board on Sustainable Development of the National Academy of Sciences and have had initial discussions concerning their potential role in reviewing the IGOS concept as it evolves.

[2] We are continuing to work on scientific and technical definition of possible IGOS components, and look forward to working with you in the continued national and international development of the IGOS.

Sincerely,

[signed]   [signed]
Charles F. Kennel   Robert S. Winokur
Chairman, Task Force on   Vice-Chairman, Task Force on
    Observations and Data Management   Observations and Data Management

Enclosure
cc: CENR Subcommittee Chairs

[An identical letter was sent to Dr. Robert T. Watson, Co-Chair, CENR.]

[1]   CONCEPT FOR AN INTEGRATED GLOBAL OBSERVING STRATEGY
            Committee on Environment and Natural Resources
            Task Force on Observations and Data Management

FOREWORD

The environment, including physical climate and managed or natural ecosystems, affects every aspect of society and economy. The environment is often viewed as both a resource and a risk factor. Minimizing environmental risks while sustaining maximum environmental resources requires understanding the processes that govern the Earth system. The diversity and complexity of the Earth system has led to the development of diverse observing systems. These systems may be broadly defined as either long-term continuous observing systems or time-limited experimental observing programs. They are focused on specific needs for a variety of applications and scientific disciplines, including meteorology and climatology, oceanography, geology and geophysics, marine and terrestrial ecology and biodiversity, atmospheric chemistry, human health and geography.

The concept of an Integrated Global Observing Strategy (IGOS) has arisen from a realization that the integration of existing and new observing capabilities into a coherent system, or family of systems, would most efficiently serve the needs of society. A reliable cost effective source of global data would constitute a valuable resource for a variety of important applications such as understanding and predicting environmental stresses, planning the allocation of energy resources, and assessing agricultural productivity. To accomplish this goal, the IGOS will address three important tasks. The first is coordinating the recommendations of user groups, with different yet sometimes overlapping data requirements, to promote the implementation of multi-purpose observing and data systems. Since environmental monitoring often spans national boundaries, the second task will be to ensure coordinated national commitments to the long-term implementation of the various components of a coherent global system. Lastly, a similar coordination effort will be required of operating agencies, including both space-based and in situ observing systems and their associated data systems, in order to minimize duplication and enhance the complementary nature of each component. These three tasks constitute the purpose of IGOS.

This paper is intended as a U.S. contribution to the international effort to organize systematic programs for global observation of atmospheric, oceanic and terrestrial variables. Recognizing the extensive worldwide effort which is addressing the issue of global climate change and its consequences, it takes into account the international plans that have been developed for overlapping Global Climate, Ocean and Terrestrial Observing Systems, and the programmatic plans of operating and/or funding agencies of many nations. In this perspective, it is particularly important to design a single integrative framework that could lead to effective [2] task-sharing, agreements among these agencies. The strategy in IGOS underlines its intended role in coordinating the mechanism for setting priorities.

It is also essential that the proposed scientific and technical programs be explicitly and clearly linked to services deliverable to society, with emphasis on short-term applications as well as longer-term needs to grow more food, enhance public health, safeguard transport systems, maintain clean air and water, promote economic development and expand human knowledge. This is the objective and vision the IGOS concept is aiming to capture.

1. INTRODUCTION

Understanding of the global Earth system, the atmosphere, land and oceans, has progressed to the point where we can use this knowledge to predict future climatic variations and change with some degree of reliability, or at least to assess the probability of natural hazards, and thus adjust our daily practice and long-term plans in order to improve the quality of our lives and our economic situation. **Global observations of a wide range of environmental parameters are needed to characterize the current state of the Earth system, identify significant trends and predict future changes, as they affect economic activities and the quality of life.**

A vast amount of environmental data is already being acquired and exchanged for a variety of applications from air quality control to tide predictions, from the assessment of the frequency of seismic hazards to weather forecasting. Existing operational data acquisition, analysis and distribution systems, such as the World Meteorological Organization's (WMO) World Weather Watch (WWW), already constitute an essential but far from complete source of information for scientific studies and assessments of the global Earth system and climate. Many existing surface-based observing systems are focused on national territories, leaving major gaps over the "global commons" such as the open ocean, and thus falling short of providing an adequate determination of the global environment and its characteristics such as climate.

Scientists from many countries are working together to plan new observing systems directed at the Earth's atmosphere, ocean and land. These are called the Global Climate, Ocean and Terrestrial Observing Systems, all of which will provide both in situ and remote sensing observations. The Global Climate Observing System (GCOS) is intended to acquire systematic lone-term observations to monitor and characterize aspects of climate in the atmosphere and ocean and over land. In addition to climate-related oceanic observations required for GCOS, the Global Ocean Observing System (GOOS) is a long-term program to acquire oceanic observations to monitor living marine resources, the coastal zone environment and the health of the ocean, as well as to continue providing marine meteorological and operational ocean observations. IGOS involves only the climate module of GOOS. The Global Terrestrial Observing System (GTOS) is being designed to document changes in land use and terrestrial ecosystems in support of sustainable development.

It is essential try move toward planning, development and implementation of these system, both nationally and internationally, within a global multi-disciplinary approach which minimizes duplications, redundancies and inefficiencies among separate observing systems and more closely links thee communities that generate requirements with those that [3] operate observing systems and deliver useful products. While international coordination of these efforts is generally accepted as desirable and, in some cases, is already well established as normal practice, there is considerable opportunity for improvement. In particular, commitment to full and open data exchange and the formulation of a mutually agreed framework for the planning and implementation of such observation programs would lead to more effective sharing of tasks among the implementing agencies around the world.

The goal of the Integrated Global Observing Strategy (IGOS) is to create an international mechanism for consultations among responsible implementing agencies in order to translate the program recommendations of GCOS, GOOS, GTOS and other relevant scientific bodies into nationally-funded activities fitting within a common framework. The overall aim of this framework is to: Acquire information of global Earth system phenomena needed for assessment of changes in the global environment, prediction of climatic variations and long-term climate change, strategic management of renewable natural resources and guidance of long-term investments.

2. RATIONALE FOR INTEGRATED GLOBAL OBSERVATIONS

Much of what is currently known about the Earth's climate, environment and natural resources is actually based on the analysis of data acquired for applications, such as weather and river flow forecasting, crop predictions, the establishment of tide tables or the assessment of environmental impacts on health. However, these existing data sources do not fully meet the requirements for characterizing or monitoring global Earth system processes, which can be distinctly different from requirements for regional or global operational applications.

Understanding Earth system mechanisms **requires long-term consistent global observations.** Existing observing systems have often been developed independently, each focused on a specific application and/or geographical domain. The Integrated Global Observing Strategy, on the other hand, is conceived from the outset to characterize coupled Earth system processes on a global scale, as required to understand and predict global environmental phenomena.

Earth system processes involve a considerably **wider range of parameters** than are routinely observed for existing applications. Examples of the measurements that will be needed include chemical agents that affect the production and destruction of ozone both in the stratosphere and troposphere, solar irradiance and the concentrations of greenhouse gases and aerosols that are basic climate forcing factors, observations of the thermal structure and circulation of the world oceans, and soil moisture and vegetative state that affect plant growth, evaporation, precipitation, and land use.

Furthermore, adequate provisions need to be made for **long-term consistency** of the data, based on stable absolute calibration standards and uniform processing schemes. Or the capability for re-analyzing archived records using a uniform procedure at a later time.

Finally an Integrated Global Observing Strategy must **assure adequate sampling of climate variability and the environment**, on the relevant time-scales. [4] This implies systematic (but not necessarily continuous) observation over long periods of time and ties to historical data bases, as well as resolving (at least in a statistical sense) rapidly changing environmental processes such as those linked to the diurnal and seasonal cycles.

The Integrated Global Observing Strategy described here would reach existing operational observing systems and provide essential required data on basic environmental parameters. with appropriate consistency in time and uniformity in space, for both applications and scientific studies. On the other hand, **the Integrated Global Observing Strategy is not a substitute for research observation programs**, like intensive field studies, oceanographic observation campaigns using research vessels, or one-of-a-kind experimental satellite missions, that are indispensable in acquiring a deeper scientific insight into Earth system processes.

3.0 GUIDELINES FOR IMPLEMENTATION

A prime requirement for the IGOS is the creation of a mechanism to better translate into national contributions the recommendations of international planning groups for the GCOS, GOOS and GTOS. The IGOS would be established through the mutual agreement of participating nations to contribute discrete elements of the system, within an internationally agreed framework. While effective international coordination already exists in some instances, there is considerable opportunity for improvement. The formulation of a mutually agreed framework for the planning and conduct of relevant national observations and international data exchange can lead to more efficient sharing of tasks among implementing agencies around the world.

The IGOS would include and **build upon existing operational observing and data systems**, which currently include environmental satellites and numerous in situ observation systems or networks maintained by all countries of the world for a variety of applications, as well as procedures developed for the voluntary exchange of environmental data. The IGOS would concentrate on additions and improvements, and more effective utilization of existing observing and data systems in order to provide enhanced observations of basic environmental parameters, and ensure adequate global coverage.

The IGOS is meant to **foster and ensure consistent long-term records** of basic environmental data for a diversity of practical applications and multi-disciplinary scientific studies, and would encourage, but not explicitly organize, one-of-a-kind field observation campaigns or experimental satellite projects specifically intended for scientific investigation of particular processes or phenomena.

The activities included in the IGOS are expected to be contributed by national governments to serve long-term general interests of all nations of the world: For this reason, the U.S. should advocate that data acquired should be distributed in accordance with the principle of **full and open international exchange of environmental data and products**, adopted by several intergovernmental bodies and other national and international entities.

It is important that the global data sets derived in the concept of IGOS are linked to the more regional and local observations networks and data sets, so that these embedded data can be interpreted in the wider global context.

[5] While governments would undertake mutual commitments to provide the elements of an overall system, the involvement of commercial entities would be encouraged when governments determine that requirements can best be met through the private sector. Ultimate responsibility for the system would remain with governments, as the system is intended to meet public needs for basic information about the Earth and its environment.

3.1 U.S. IMPLEMENTATION

In July 1995 the Committee on the Environment and Natural Resources (CENR) assigned the Task Force on Observations and Data Management (TFODM) the responsibility to define a conceptual global observing system framework and create a plan for organizing U.S. civil Earth observations in this context. The Task Force had earlier established a Secretariat and three working groups (Integrated Observing Systems, User Needs, and Data Management) to act as the catalyst for coordinating the global, national and regional observing and data system requirements and plans of the CENR Subcommittees and U.S. federal agencies, and serve as the primary CENR interface for international consultations on planning and implementation of a global observing system and its related data management system. This includes coordination with the planning efforts underway by the international scientific and intergovernmental agencies (e.g. WMO, Intergovernmental Oceanographic Commission [IOC], the United Nations Environment Programme [UNEP] and the International Council of Scientific Unions [ICSU]), as well as by the space agencies (through the Committee on Earth Observation Satellites [CEOS]). Coordination will be undertaken with the established international scientific research and observations programs (e.g., World Climate Research Programme [WCRP], International Geosphere and Biosphere Programme [IGBP], GCOS, GODS, and GTOS) to synthesize and to assess broad user needs, and to analyze the opportunities for minimizing unnecessary redundancies and filling gaps. The U.S. planning process is outlined in Fig. 1.

Fig. 1 US PLANNING PROCESS FOR IGOS

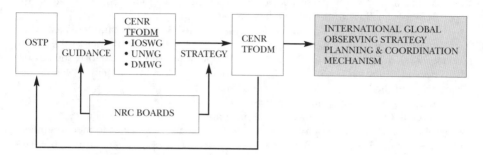

[6] This process will require Task Force initiatives aimed at bilateral and multilateral international coordination of global observations and data management plans and proposed programs, including specific missions, sensors, and related in situ and ground-based measurements in order to seek consensus on mutually supportive initiatives; and will provide a means for coordination of implementation plans with involved U.S. federal agencies and the scientific community. Overall scientific oversight and guidance will be solicited through interactions with the National Research Council's (NRC) Board on Sustainable Development (BSD), and its Committee on Global Change Research (CGCR), in consultation with the Space Studies Board (SSB), the Ocean Science Board (OSB). and the Board on Atmosphere Sciences and Climate (BASC). The responsibilities of the respective Task Force Working Groups and its Secretariat are outlined in sections 3.1.1 through 3.1.4.

3.1.1 USER NEEDS WORKING GROUP

The User Needs Working Group (UNWG) will coordinate and synthesize the needs (requirements) of U.S. users of global, regional, state and local scale observations and data. These user needs will originate from the CENR subcommittees, the federal agencies and other sources such as the Interagency Working Group on Sustainable Development Indicators (IWGSDI) and the Federal Geographic Data Committee. Specific UNWG responsibilities will include:
• working with the individual CENR subcommittees to develop, refine, and prioritize coordinated user need statements for global. regional, state and local observations and data;
• identifying the time and space sampling, coverage, accuracy and continuity requirements, and data availability for key environmental indicators;
• assessing the adequacy of existing and planned observing and data sources; developing recommendations for remedial actions including budget initiatives and direct support by the Task Force;
• advising on the evolving needs of the broad user community for environmental and natural resources data and information.

3.1.2 DATA MANAGEMENT WORKING GROUP

The Data Management Working Group (DMWG) will coordinate CENR efforts to develop and use federal data and information systems and standards to promote environmental and natural resources data system inter-operability. Included within the specific responsibilities of the DMWG will be:

• coordination of the data and information systems, standards, and policies being implemented and used by participating agencies for CENR programs, such as the Global Change Data and Information System, while cooperating with other efforts such as the Government Information Locator System, the National Environmental Data Index program, and the Federal Geographic Data Committee;

[7]• international coordination of data and information systems, services, and policies to support CENR requirements for international data and information exchange;

• coordination of the development of data and information products to meet the needs identified by the User Needs Working Group; specific tasks assigned to it by the Task Force, particularly for advice and support in assessing and meeting the data and information management requirements of the Task Force;

• coordination with the other Task Force working groups and with the Committee on Geophysical and Environmental Data (CGED) of the National Academy of Sciences.

3.1.3 INTEGRATED OBSERVING SYSTEM WORKING GROUP

The Integrated Observing System Working Group (IOSWG) will coordinate the Federal agency planning for satellite and ground-based observing program elements to meet the requirements of the CENR Subcommittees, as synthesized by the User Needs Working Group and other scientific advisory bodies. Specific responsibilities include:

• taking the lead in the planning and interagency coordination for U.S. participation in the Integrated Global Observing Strategy;

• working with the Task Force's Secretariat, federal agency planning offices for GCOS, GOOS, and GTOS, and the NRC Boards and their various sub-groups to coordinate the planning for satellite, ground-based and in situ components of the international effort;

• identifying with the UNWG observational requirements that are not being met and recommending possible corrective actions;

• addressing specific observing systems issues assigned by the Task Force Chair, including the calibration of instruments and continuity of observations.

3.1.4 SECRETARIAT

A Secretariat, established by the Task Force and representative of the interests of Task Force member agencies, will support the activities of the Task Force (and its Working Groups). It will operate in accordance with a work plan approved by the Task Force and will fulfill other responsibilities as assigned. In carrying out these duties, the Secretariat may establish ad hoc interagency panels with limited objectives. Acting at the direction of the Task Force and in accordance with the strategy outlined in the Office of Science and Technology Policy (OSTP) concept paper on the Global Observing System, the Secretariat will facilitate the CENR interface for international consultations on scientific planning and implementation of the Integrated Global Observing Strategy. This includes coordination with the planning efforts underway by intergovernmental organizations (e.g., WMO, IOC and UNEP), as well as by the space agencies (through CEOS) and with the international scientific programs (WCRP and IGBP). The U.S. position on these matters will be developed and relined through deliberations of the CENR Task Force Working Group, and approved by the Task Force itself.

[8] 3.2 INTERNATIONAL IMPLEMENTATION

The strategy for the international implementation of a truly integrated global observing system will require a very high level mechanism to ensure that:

• The overall strategy is science-driven and emphasizes the delivery of useful products as its goal.

• Priorities are set and commitments are made by each, participating nation or mulfti-national consortium (e.g. regional satellite organization) so that a practical, effective and coordinated program evolves.

• The specific requirements emanating from more narrowly focused planning groups (e.g. GCOS, GOOS, GTOS, IGBP, WCRP, etc.) are considered in an integrated context, and resource commitments are coordinated in such a way as to optimize the implementation of the systems and to ensure that critical elements of the global system are indeed implemented through international cooperation.

It is envisaged that U.S. representation to this "high level integrating mechanism" would be arranged under auspices of the OSTP, thus ensuring inter-agency cooperation within the broad US CENR community. Likewise OSTP would seek to arrange participation from other nations in this mechanism at all levels, as an effective IGOS would require some degree of partnership among implementing agencies and governments. One possible framework for initiating this mechanism is the Group of Seven Industrialized Nations (G-7). The mechanism would be informal at first, but evolve toward a standing body with a secretariat supporting it as soon as international agreement on the structure and function of this body is agreed. The global community of nations could be included in this effort through participation of specific intergovernmental organizations such as WMO, UNESCO/IOC and UNEP among others. International implementation approaches will be addressed by CEOS and the in situ community in two international workshops on IGOS planned during 1996.

An initial step in the strategy toward developing the international implementation of an integrated global observing capability will be to reach international agreement on the overall operational and research rationale for the global system and to identify, in priority order, the elements required to be implemented and their general sequence. It is proposed that NRC's Boards be engaged to provide the scientific guidance needed to develop the initial U.S. recommendations and proposed initial contribution to this important first step (see section 3.1). ICSU could be asked to organize the international review and adoption of the overall scientific rationale.

The high level integrating mechanism would also define and coordinate the specific national and multinational contributions to the overall integrated system by obtaining national commitments toward implementing specific elements according to an agreed schedule. These commitments would address specific requirements of GCOS, GOOS, and GTOS, but have the added dimension of treating the global environment in an integrated way and meeting the requirements through sharing the responsibilities among participating nations and by addressing, in a synthesized way, broad global environmental interests. This would avoid implementing limited and potentially duplicative systems that in might be designed and deployed for more narrowly defined objectives.

The international implementation strategy must be designed to be flexible and evolutionary. It will have to be inclusive in order to take advantage of the scientific planning and disciplinary strength of the GCOS, GOOS, GTOS and other research (e.g. IGBP and USGCRP) and operational programs and systems (e.g. WMO/WWW). It must have authority to set priorities and put in place a scheme so that resources are committed to address these priorities effectively.

The functioning of the IGOS is illustrated in Fig 2. The planning and coordination mechanism at the international level would be reflected at the national level through the national representation to CEOS, GOOS, GCOS, GTOS, etc., the implementing agencies (e.g. NASA, NOAA, NSF, DOE, DOI, EPA etc.) and the scientific oversight (NRC) as proposed in section 3.1 above.

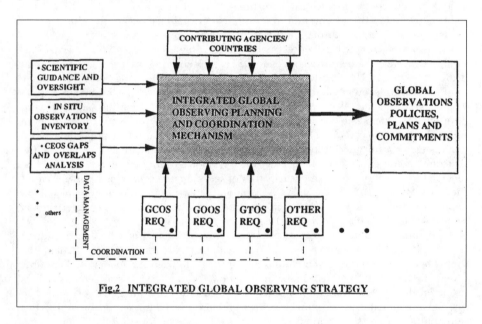

Fig.2 INTEGRATED GLOBAL OBSERVING STRATEGY

3.3. SCHEDULE
The provisional schedule for IGOS planning and implementation is:
- July '95   Task Force on Observations and Data Management assigned IGOS responsibilities
- Oct '95   Informal international discussions with CEOS and GCOS
- Oct '95   Request to NRC/BSD for oversight of IGOS scientific planning
- Feb '96   Preliminary IGOS Conceptual Plan submission to CENR
- Spring '96 IGOS Scientific Rationale Review through NRC Board on Sustainable Development (BSD) and other NRC Bodies as appropriate

- Mar '96   First international workshop on IGOS strategic planning (CEOS lead; Seattle, WA. March 27-28, 1996)
- TBD.   Second international workshop on IGOS strategic planning (GCOS/GOOS/GTOS, IGFA and others; Geneva)
- TBD   Draft proposal for international implementation framework for IGOS
- Feb'97   Symposium on Integrated Observing Systems, American Meteorological Society Annual Meeting, Long Beach, CA, February 2-7, 1997

Document IV-24

[no pagination]

EXECUTIVE OFFICE OF THE PRESIDENT
OFFICE OF SCIENCE AND TECHNOLOGY POLICY

March 22, 1996

Dr. Charles F. Kennel
Associate Administrator
 for Mission to Planet Earth
National Aeronautics and Space Administration
Washington, DC 20546

Dear Charlie:

The purpose of this letter is to endorse your effort to develop an internationally agreed strategy for integrated global Earth observations. I have discussed this issue with Dr. D. James Baker, my co-chair on the Committee on Environment and Natural Resources (CENR) of the National Science and Technology Council (NSTC), and we both agree that this initiative should be pursued. The CENR Task Force on Observations and Data Management chaired by you and Dr. Robert Winokur of the National Oceanic and Atmospheric Administration should continue to coordinate U.S. agency participation in this activity.

It is clear that future development of U.S. observing systems, both space-based and in situ, must be well coordinated with complimentary efforts undertaken by other nations, building on the excellent record of international cooperation achieved to date in the study of the Earth's environment. Improving the integration of the planning and implementation of observing systems in response to user requirements should benefit both funding agencies and users.

I think the appropriate next step is to begin informal discussions with our international colleagues about their perspectives on an integrated strategy. The concept paper drafted by your Task Force provides a useful overview of many of the issues implicit in devising such a strategy. It should serve as an excellent starting point for international discussions.

I have requested Mr. Peter Backlund to continue staffing this issue for the NSTC.

Sincerely,
[signed]
Robert T. Watson
Associate Director for Environment
and
Co-Chair, Committee on Environment and Natural Resources

CC: NOAA/Dr. Baker
    NOAA/Dr. Winokur

Document IV-25

Document Title: Rear Admiral J. J. Dantone, Jr., U.S. Navy, Memorandum for Under Secretary of Defense (Acquisition and Technology), "Shuttle Radar Topographic Mission with Attached Memorandum of Understanding Between NASA and the Defense Mapping Agency," 8 July 1996.

Source: NASA Historical Reference Collection, NASA History Office, NASA Headquarters, Washington, D.C.

NASA and the Department of Defense have always cooperated on programs of mutual interest. This document reproduces the memorandum that specified the terms under which the Defense Mapping Agency and NASA were to cooperate in the production of a high-quality digital terrain elevation model. The project involved using the C-band microwave Shuttle synthetic aperture radar (SAR) modified with a boom extended horizontally from the Shuttle payload bay to make possible an interferometric SAR, capable of modeling the surface of Earth with high precision. The system also included a cooperative mission with the German space agency that used an X-band antenna.

[no page number]
[stamped "8 Jul 1996"]

DEFENSE MAPPING AGENCY

MEMORANDUM FOR UNDER SECRETARY OF DEFENSE (ACQUISITION AND TECHNOLOGY)

SUBJECT: Shuttle Radar Topographic Mission (SRTM)

1. The Defense Mapping Agency (DMA) and National Aeronautics and Space Administration (NASA) have completed negotiations on a Memorandum of Understanding for the cooperative flight of the SRTM on the NASA Space Shuttle. This mission will provide DMA with terrain height data from which we will generate near-global Digital Terrain Elevation Data (DTED) in support of military requirements.

2.   SRTM will be a reflight, scheduled for May 2000 on the NASA Space Shuttle, of the Spaceborne Imaging Radar/C-Band Synthetic Aperture Radar (SAR) hardware, modified with a boom and additional antenna to operate as an interferometer. The instrument will be capable of collecting near-global, contiguous SAR and Interferometric SAR (IFSAR) data covering all land areas between approximately 60° North and 56° South latitude that can be processed to generate terrain height data with approximately 16 meters absolute vertical accuracy at 1 second or arc latitude and longitude. DMA will receive Interferometric Terrain Height Data and C-Band SAR data in the form of orthorectified image strips. The terrain height data generated from the mission will satisfy the elevation requirements for the DMA product DTED Level 2.

3.   DMA is very pleased with this initiative to obtain data in support of Department of Defense requirements for world terrain elevation data. We appreciate the strong support that you and the senior leadership of the Department have given this endeavor.

4.   Enclosed is a signed copy of the Memorandum of Understanding.

Enclosure a/s   [signature]
                                                 J. J. DANTONE, Jr.
                                                 RADM, USN

cc:
USD(C)
DUSD(SPACE)
ASD(C31)
Chairman, JROC
Joint Staff (J-5)

REPLY TO THE FOLLOWING: [hand-checked]
Headquarters
8613 Lee Highway
Fairfax, Virginia 22031-2137

                          ****

[no page number]

                    MEMORANDUM OF UNDERSTANDING
                             BETWEEN
          THE NATIONAL AERONAUTICS AND SPACE ADMINISTRATION
                               AND
                  THE DEFENSE MAPPING AGENCY
                 FOR A COOPERATIVE FLIGHT OF THE
                SHUTTLE RADAR TOPOGRAPHY MISSION

ARTICLE I: GENERAL

1.   The National Aeronautics and Space Administration (hereinafter referred to as NASA) and the Defense Mapping Agency (hereinafter referred to as DMA) desire to extend fruitful cooperation of previous joint projects and affirm their mutual interest in carrying out further cooperative activities for Earth sciences. The Shuttle Radar Topography Mission (SRTM) will consist of the Spaceborne Imaging Radar-C (SIR-C) and, following negotiations with the German (DARA) and Italian (ASI) Space Agencies, the X-Band Synthetic Aperture Radar (X-SAR). Both instruments will be modified to operate as interferometers. The concept for SRTM has been developed in conjunction with scientists and engineers from NASA, DMA and other Department of Defense (hereinafter referred.to as DOD) agencies. SIR-C/X-S AR were flown on NASA's SRL-1 and SRL-2 missions in April and October 1994.
2.   The primary objective of SRTM will be to measure the topographic surface of the Earth. SRTM will produce terrain height data for use by the civilian and defense community as provided herein.

ARTICLE II: PURPOSE

This Memorandum of Understanding (MOU) defines the terms and conditions under which NASA and DMA agree to cooperate on the SRTM mission. SRTM is described in the October 11, 1995 SRTM cost review package (See Attachment 1). Specifically, this MOU describes the managerial, technical, and operational interfaces between NASA and DMA that will be necessary to ensure continuation of and compatibility between their respective activities.

ARTICLE III: SCOPE

SRTM will be a reflight, on the NASA Space Shuttle, of the SIR-C/X-SAR hardware, modified with a boom and additional antenna to operate as an interferometer. The instrument will be capable of collecting near-global, contiguous SAR and Interferometric SAR (IFSAR) [2]data covering all land areas between approximately 60° north and 54° south latitude that can be processed to generate terrain height data with approximately 16 meter absolute vertical accuracy at 1 second of arc latitude and longitude. The terrain data generated from the mission will satisfy the terrain height data set specifications, referred to as Interferometric Terrain Height Data-2 (ITHD-2), described in Attachment 2.

ARTICLE IV: AUTHORITY

1.   The authority for NASA entering into this MOU is Section 203(c) of the National Aeronautics and Space Act of 1958, as amended (42 U.S.C. 2473(c)).

2.   The authority for DMA entering into this MOU is 32 Code of Federal Regulations (C.F.R) Part 399.
ARTICLE V: NASA RESPONSIBILITIES

1.   To implement this cooperative program, NASA will, in accordance with the SRTM Implementation Plan to be developed, carry out the following responsibilities:

a) Provide the existing SIR-C/X-SAR hardware in functional condition, as well as any infrastructure necessary for a reflight of such;

(b) Design the SRTM system in cooperation with DMA, secure or fabricate the necessary additional components, and modify the existing SIR-C hardware and software to operate as an interferometer;

(c) Integrate and test the SRTM system and prepare for launching by the Space Shuttle;

(d) Design and implement necessary ground systems for mission operations, SAR and IFSAR data processing, and generating the registered C-band scansar digital elevation model (DEM) data sets;

(e) Negotiate with DARA and ASI to provide for the upgrade, delivery, and integration of the interferometric X-SAR instrument, along with the necessary ground systems, mission operations and data processing support, and exchange of data consistent with Article VIII of this agreement and Article IX of the NASA/DARA SIR-C/X-SAR MOU as amended;

(f) Launch the SRTM system in late FY 2000 or earlier on the Space Shuttle in accordance with the schedule as defined in the SRTM Implementation Plan;

[3] (g) Continue to examine and consult with DMA on alternatives for an earlier launch in either FY 1998 or FY 1999 consistent with Attachment 3 (letters between the Undersecretary of Defense for Acquisition and Technology and the Administrator of NASA);

(h) Operate the SRTM system, acquire, calibrate and process the SAR and IFSAR data for each ascending and descending pass over land areas, and generate a strip of DEM data sets in accordance with the schedule defined in the SRTM Implementation Plan under development;

(i) Produce a mosaic of the strips of DEM data sets from each ascending and descending orbit and use them to generate an ITHD-2 data set for DMA;

(j) Provide to DMA the C-Band SAR data from ascending and descending orbits into orthorectified image strips that can be made into a mosaic (image data will be provided in National Image Transfer Format-2 (NITF-2)); and,

(k) Consistent with Article VIII, archive and distribute data from SRTM:

ARTICLE VI: DMA RESPONSIBILITIES

1. To implement this cooperative program, DMA will, in accordance with the SRTM Implementation Plan to be developed, carry out the following responsibilities:

(a) Function as a liaison to the DOD community for SRTM;

(b) Represent DOD by defining the data processing priorities;

(c) Provide technical guidance to NASA for preparation and delivery of data sets to DMA;

(d) Provide technical review of SRTM, and appropriate representation at NASA SRTM design reviews in accordance with the schedule outlined in the SRTM Implementation Plan;

(e) Provide the Jet Propulsion Laboratory (JPL) with the best available digital topographic data base that will facilitate the verification of SRTM terrain height data during the processing;

(f) Function as liaison to solicit support for SRTM-related joint research with the Advanced Research Project Agency (ARPA) in the use of High Performance Computing Centers (HPCC) for generation of SRTM global data sets free of usage fees to NASA;

[4] (g) Validate the terrain height data from NASA and populate the terrain height data base for DMA accordingly;

(h) Produce a 3 second of arc latitude and longitude data set from the ITHD-2 data set; and,

(i) Provide to NASA copies of the ITHD-2 data set and any data set generated from it for distribution. Notwithstanding the provisions of Article VIII, NASA shall not be restricted from agency access to and use of any data set derived from SRTM. Consistent with Article VIII, provide copies of the ITHD-2 data set and any data set generated from it to other parties.

ARTICLE VII: POINTS OF CONTACT [omitted]

ARTICLE VIII: DATA RIGHTS AND RELEASE

1. It is the intent of the parties that no data from SRTM will be classified.

2. For purposes of Titles 5 and 10 of the United States Code, the raw phase history information, ITHD-2, and strip DEM with 30 meter spatial resolution outside the United States are under the control of DOD.

3.   NASA retains the right to distribute raw (amplitude) and processed image data obtained from SRTM. The data will be archived in the appropriate NASA data center and made available to all users without restriction and at no more than the cost of fulfilling the user request.

4.   Terrain height data greater than or equal to three seconds of arc latitude and longitude generated from SRTM may be released and distributed by NASA and DMA without restrictions.

5.   Raw phase history information, terrain height data of 1 second of arc latitude and longitude (i.e., ITHD-2 data sets), or strip DEM with 30 meter spatial resolution, over the United States may be released and distributed by NASA and DMA without restrictions.

6.   Release of raw phase history information, terrain height data of 1 second of arc latitude and longitude (i.e., ITHD-2 data sets), or strip DEM data sets with 30 meter spatial resolution outside the United States will be in accordance with guidelines developed and agreed jointly by DMA and NASA, provided nothing in these guidelines shall prevent DOD from releasing this data for national security or security assistance purposes. The Parties will consult at least annually to review these guidelines.

[7]
ARTICLE X: LIABILITY

1.   The purpose of this clause is to establish a cross-waiver of liability between the Parties and their related entities, in the interest of encouraging participation in the exploration exploitation, and use of outer space. This cross-waiver of liability shall be broadly construed to achieve this objective.

2. As used in this cross-waiver, the term:

   (a) "Party" means a person or entity that signs this agreement;
   (b) "Related Entity" means:

      (i)   a contractor or subcontractor or a Party at any tier;
      (ii) a user or customer of a Party and any tier; or
      (iii) a contractor or subcontractor of a user or customer of a Party at any tier.

   "Contractor" and "Subcontractors" include suppliers of any kind;

   (c) "Damage" means:

      (i)   bodily injury to, or other impairment of health of, or death of, any person;
      (ii) damage to, loss of, or loss of use of any property;
      (iii) loss of revenue or profits; or
      (iv) other direct, indirect, or consequential damage;

(d) "Payload" means any property to be flown or used on or in the Shuttle; and

(e) For the purposes of this Agreement, "Protected Space Operations" means all launch vehicle and payload activities on Earth, in outer space, or in transit between Earth and outer space done in implementation of this Agreement. Protected Space Operations begins at the signature of this Agreement and ends when all activities done in implementation of this Agreement are completed. It includes, but is not limited to:

(i) research, design, development, test, manufacture, assembly, integration, operation, or use of: the Space Shuttle, transfer vehicles, payloads, related support equipment, and facilities and services.

(ii) all activities related to ground support, test, training, simulation, or guidance and control equipment and related facilities or services. "Protected Space Operations" excludes activities on Earth which are conducted on return from space to develop further a payload's product or process for use other than for Shuttle-related activities necessary to complete implementation of this Agreement.

[8] 3.   (a) Each Party agrees to a cross-waiver of liability pursuant to which each Party waives all clams against any of the entities or persons listed in sub-paragraphs (i) through (iv) of this section based on damage arising out of Protected Space Operations. This cross-waiver shall apply only if the person, entity, or property causing the damage is involved in protected space causing the damage is involved in protected space [sic] operations under a NASA agreement for Shuttle services and the person, entity, or property damaged is damaged by virtue of its involvement in protected space operations under a NASA agreement for Shuttle services. The cross-waiver shall apply to any claims for damage, whatever the legal basis for such claims, including but not limited to delict and tort (including negligence of every degree and kind) and contract, against:

(i) another Party;

(ii) any party who has signed a NASA agreement that includes Shuttle services;

(iii) a related entity of another Party;

(iv) the employees of any of the entities identified in sub-paragraphs (i) and (iii) of this section.

(b) In addition, each Party shall extend the cross-waiver of liability as set forth in sub-paragraph (a) of this section to its own related entities by requiring them, by contract or otherwise, to agree to waive all claims against the entities or persons identified in paragraphs (a)(i) through (a)(iv) of this section.

(c) Notwithstanding the other provisions of this section, this cross-waiver of liability shall not be applicable to:

(i) claims between a Party and its own related entity or between its own related entities;

(ii) claims made by a natural person, his/her estate, survivors, or subrogees for injury or death of such natural person, except when a subrogee is one of the Parties;

(iii) claims for damage caused by willful misconduct;

(iv) intellectual property claims;

(v) contract claims between the Parties based on the express contractual provisions of this Agreement;

(vi) claims for damage based on a failure of the Parties of their related entities to flow down the cross-waiver.

(e) Nothing in this section shall be construed to create the basis for a claim or suit where none would otherwise exist.

[9] ARTICLE XI: SETTLEMENT OF DISPUTES

NASA and DMA agree that any dispute as to the interpretation or implementation of this Agreement shall be resolved through consultation between the points of contact identified in Article 7. Any dispute not resolved at this level shall first be referred to the NASA Associate for Mission to Planet Earth and the Director of DMA for resolution. Any dispute which cannot be resolved at this level shall be referred to the NASA Deputy Administrator and the DOD Undersecretary of Defense for Acquisition and Technology for settlement.

ARTICLE XII: ENTRY INTO FORCE, DURATION AND TERMINATION

This agreement shall enter into force upon the last signature and shall remain in force for eight years after the launch of SRTM. The agreement may be amended at any time by mutual written agreement and may be terminated by either of the consenting parties after ninety (90) days notification of intent to terminate.

For DMA: For NASA:

[signature]   [signature]
J. J. Dantone, Jr.   W. F. Townsend
Rear Admiral, USN   Deputy Associate Administrator for
                                        Mission to Planet Earth (Programs)

Date [handwritten "8 Jul 96"]   Date [handwritten "6/7/96"]

Document IV-26

Document Title: NASA, "Memorandum of Understanding Between the National Aeronautics and Space Administration of the United States of America and the National Space Development Agency of Japan for Joint Development of the Tropical Rainfall Measuring Mission," 20 October 1995.

Source: NASA Historical Reference Collection, NASA History Office, NASA Headquarters, Washington, D.C.

The success of the Tropical Rainfall Measuring Mission (TRMM) results in large part from the partnership between NASA and NASDA in the design and execution of this mission, which was designed to gather detailed observations of rainfall in the tropics, regions that are generally poorly covered by Earthbound observations of precipitation. NASA provided four instruments, including the Microwave Imager for the spacecraft, and NASDA provided one, the Precipitation Radar. NASDA launched TRMM on 27 November 1997 from Tanegashima Space Center aboard an HII rocket. In addition to specifying the contributions of both parties to the agreement and various legal arrangements, this document specifies the data exchange policies that govern the partnership. (See Appendix.)

[title page]

Memorandum of Understanding

between the

National Aeronautics and Space Administration

of the United States of America

and the

National Space Development Agency of Japan

For Joint Development of the

Tropical Rainfall Measuring Mission

[2]

CONTENTS

[3]

Preamble

The National Aeronautics and Space Administration of the United States of America (hereinafter referred to as "NASA") and the National Space Development Agency of Japan (hereinafter referred to as "NASDA"), together hereinafter referred to as "the Parties",

Recognizing that tropical rainfall is essential to the distribution of water throughout the Earth system, and over two-thirds of the worldwide precipitation occurs in the tropics, releasing the energy that helps to power the global atmospheric circulation which shapes both weather and climate;

Recognizing that tropical rainfall also plays a key role in the sporadic "El Niño" climate anomalies that trigger floods and droughts around the world, and measuring tropical rainfall from the Earth's surface is difficult because of its high variability, and moreover, surface observations are not feasible over the vast ocean and jungle regions of the tropics;

Noting that advances in technology now make it possible to obtain these essential measurements from space;

Considering that in 1972, the first imaging microwave radiometer orbits on NASA's Nimbus-5 gave evidence that instantaneous rainfall rates could be measured from space, and in the early 1980's, the Communications Research Laboratory of Japan (hereinafter referred to as "CRL") developed an airborne microwave radar and radiometer system to investigate the interference of rain in satellite-to-ground communications;

Recalling that in 1986, the Tropical Rainfall Measuring Mission (TRMM) Science Steering Group, a team of experts in atmospheric, oceanic, and remote sensing sciences, began investigating the scientific justification and implementation process of a satellite mission to study systematically tropical rainfall, and in early 1987, NASA, NASDA, and CRL instituted a study of the feasibility of implementing TRMM as a Joint space project;

Considering that the resulting three-year TRMM mission is a part of a systematic, integrated program designed to increase the extent and accuracy of rainfall and latent heat measurements and provide strides in weather and climate research;
Noting that the goals of TRMM are as follows:

(1) to advance the Earth System Science objective of understanding the global energy and water cycle by means of providing distributions of rainfall and inferred heat over the global tropics; [4]

(2) to understand the mechanisms through which tropical rainfall and its variability influence global circulation and to improve our ability to model these processes in order to predict global circulation and rainfall variability at monthly and longer time scales; and

(3) to evaluate a space-based system for rainfall measurement; and

Noting that NASA and NASDA will share data from TRMM for research, operational and other uses under the terms of the International Earth Observing system (IEOS) Data Exchange Principles (DEP) contained in the Appendix to this MOU.

Considering the above mentioned circumstances, have agreed as follows:

Article I - Purpose

The Purpose of this Memorandum of Understanding (hereinafter referred to as the "MOU") is to establish the terms and conditions under which NASA and NASDA will cooperate in the joint development, launch, operations and use of the Tropical Rainfall Measuring Mission (hereinafter referred to as "TRMM") for peaceful purposes.

Article II - Mission Description and Participation

1. The primary objective of the TRMM Project is to measure the distribution and variability of tropical rainfall and latent heat releases on a monthly basis for three years to advance the scientific understanding of the global energy and water cycles.

2. Accordingly, an observatory (hereinafter referred to as the "TRMM Observatory"), consisting of a satellite to be provided by NASA, carrying four NASA-provided instruments and one NASDA-provided instrument, is planned for launch in 1997. The TRMM Observatory shall be launched using an H-II launch vehicle provided by NASDA for injection into a near-circular orbit at a nominal initial altitude of 380 kilometers with an inclination of 35 degrees. Launch shall be from the Yoshinobu Launch Complex of NASDA's Tanegashima Space Center located on the Tanegashima Island of Japan (hereinafter referred to as the "launch site").

3. After an initial checkout period and a reduction in altitude to 350 kilometers, the TRMM Observatory is planned to be operated for three years at a nominal altitude of 350 km. The TRMM Observatory shall be operated by NASA utilizing its Tracking and Data Relay Satellite System (TDRSS) for command, control, data acquisition, and routine tracking. Science data received at the NASA Goddard Space Flight Center (GSFC) shall be processed by the TRMM Science Data and Information System (TSDIS) and the Earth Observing System Data and Information System (EOSDIS). Science data will also be sent to the NASDA Earth Observation Center and processed by the TRMM Data Processing System and Earth Observation Information [5] System (EOIS). The NASDA-provided Communications and Broadcasting Engineering Test Satellite (COMETS) shall be included in an experimental manner to relay mutually selected mission data from the TRMM Observatory to the Japanese ground station located at Tsukuba, Japan.

Article III - NASA Responsibilities

To implement this cooperative program, NASA shall use its best efforts to carry out the following responsibilities:

1. Development of Instruments

    Design, develop, produce, calibrate and space-qualify the following rainfall instruments which, together with the NASDA-provided Precipitation Radar (PR) are primary to the TRMM mission success criteria:

    a. TRMM Microwave Imager (TMI)
    b. Visible Infrared Scanner (VIRS)

    In addition, design, develop, produce, calibrate and space-qualify the following EOS instruments, which are Flights of Opportunity for the NASA EOS Program

    c. Clouds and Earth's Radiant Energy System (CERES)
    d. Lightning Imaging Sensor (LIS)

    These four instruments together will be referred to as the "NASA-provided instruments." The TMI and VIRS together with the PR will be referred to as the "rainfall instruments" and data obtained from these instruments will referred to as the "rainfall data."

2.   Development of and Integration of Instruments with the TRMM Spacecraft

    a.   Design, develop, produce, and space-qualify the TRMM spacecraft;

    b.   Integrate the NASA-provided instruments and the NASDA-provided Precipitation Radar (PR) with the TRMM spacecraft to create the TRMM Observatory;

    c.   Provide NASDA with a spacecraft interface simulator designed to verify the PR-to-spacecraft interfaces with regard to command and data handling software, optical signals, bus hardware, and the non-optical functions;

    d.   Provide NASDA with a TRMM spacecraft-to-PR drill template and a set of test kinematic mounts; [6]

    e.   Obtain all necessary national and international radio frequency approvals for TRMM for U.S. territory and in-orbit operations at the appropriate time;

    f.   Provide NASDA with a NASA User Test System designed to show TRMM/COMETS system compatibility with regard to command and telemetry data; and

    g.   Provide necessary support to NASDA for the import of test pyrotechnic devices into the United States.

3.   Shipping and Launch Site Activities

    a.   Provide all shipping and handling of the TRMM Observatory until the TRMM Observatory is handed over to NASDA for H-II integration, and package and ship the TRMM Observatory, including the test equipment needed at launch, to the launch site;

    b.   Perform all required post-shipping tests of the TRMM Observatory at the launch site;

    c.   Perform the radio-frequency (RF) end-to-end test at the launch site for command and telemetry flow;

    d.   Support NASDA in its application for the Reaction Control System (RCS) special loading license, as necessary;

    e.   Provide the hydrazine fueling and pressurization of the Reaction Control System (RCS) of the TRMM Observatory;

    f.   Support a mutually-agreed-upon integration and test program for the TRMM Observatory, with the H-II launch vehicle;

    g.   Determine TRMM Observatory readiness for launch;

h. Maintain a launch team at the launch site to perform all observatory-related launch activities; and

i. Pack and ship all materials to be returned from the launch site to GSFC.

[7]
4. Operation after the Separation of the TRMM Observatory from the Launch Vehicle

a. Control the TRMM Observatory after separation from the launch vehicle, which includes control and monitoring of all instruments including the PR;

b. Utilize and operate the TDRSS and GSFC institutional support facilities for communications, tracking, telemetry and command links to the TRMM Observatory;

c. Utilize and operate the Mission Operations Center at GSFC to support the operation of the TRMM Observatory, and to assure successful mission control;

d. Perform an initial on-orbit check-out of the TRMM Observatory with NASDA supporting the planning and data analysis for the PR checkout;

e. Plan and conduct the operation of TRMM including the PR with the support of NASDA for PR planning and PR engineering data analysis;

f. Plan and conduct calibration and validation to verify performance of NASA-provided instruments;

g. Operate the TRMM Observatory as required to perform mutually-agreed-upon telemetry, tracking and command and science data experiments designed to demonstrate successfully the capability of NASDA's COMETS data relay satellite and meet Space Network Interoperability Panel (SNIP) demonstration testing objectives; and

h. Operate the TRMM Observatory after the 3 year mission life for possible extended duration. Termination of the operation and the disposal of the TRMM Observatory shall be decided with agreement from NASDA.

5. Data System

a. Design, develop, and operate the NASA TSDIS to capture, produce, distribute, and archive all mission data in accordance with the International Earth Observing System (IEOS) Data Exchange Principles (DEP) in the Appendix;

b. Provide all PR housekeeping data and other data necessary to check the PR status, as mutually agreed;

c. Receive TRMM mission data at White Sands Complex (WSC) and produce TRMM Level Zero Processed (LZP) data; [8]

d. Provide LZP data, NASDA-requested Quick Look (QL) data, NASA-provided instrument calibration data and the other necessary data to perform data processing at NASDA, as mutually agreed;

e. Provide the higher level products other than LZP data to NASDA when available, as mutually agreed;

f. Provide ground truth data to NASDA, as mutually agreed;

g. Provide NASDA with NASA algorithms used to generate standard products, as mutually agreed;

h. Provide NASA's catalogue data to NASDA, as mutually agreed; and

i. Establish the network interfaces mutually agreed with NASDA for providing some of mission data, mission operation information through EOSDIS-EOIS communication lines.

6. Safety

NASA shall comply with NASDA's safety requirements and safety instructions, with assistance from NASDA.

7. Other

a. Develop and maintain instrument-to-spacecraft interface control specifications, mission specifications, Operation and Interface Specification (OIS), and flight operation plans, as detailed in the TRMM Implementation Plan (TRMM IP);

b. Hold meetings and reviews periodically in the United States of America, as required and mutually agreed to carry out the responsibilities set forth by this MOU;

c. Exchange personnel and information with NASDA, as needed, to address engineering, operations, and science issues; and

d. Consult with NASDA on arrangements related to radio-frequency interference issues.

[9]

Article IV – NASDA Responsibilities

To implement this cooperative project, NASDA shall use its best efforts to carry out the following responsibilities:

1.   Development of Instrument

    a.   Design, develop, produce, calibrate, and space-qualify the PR;

    b.   Provide all shipping and handling of the PR, from Japan to the integration site at GSFC, including the test equipment required for integration and testing at GSFC;

    c.   Perform all post-shipping tests of the PR at the integration site (GSFC), as mutually agreed;

    d.   Support a mutually-agreed-upon integration and test program of the PR with the TRMM Observatory; and

    e.   Pack and ship the PR ground support equipment (GSE) to be returned from GSFC to Japan.

2.   Development of Launch Vehicle and Launch Site Facilities

    a.   Design, develop, produce and space-qualify the H-II launch vehicle including the Payload Attach Fitting (PAF);

    b.   Deliver a test PAF to the GSFC for NASA to perform structural test and integration test of the TRMM Observatory;

    c.   Deliver a flight PAF to GSFC to perform interface testing of the H-II and the TRMM Observatory and transport the flight PAF to the launch site for integration with the TRMM Observatory;

    d.   With the support of NASA, provide six test pyrotechnic devices; and

    e.   Provide the facilities required at the launch site to perform all launch operations, as mutually agreed.

3.   Shipping and Launch Site Activities

    a.   Provide the launch vehicle including the launch vehicle test equipment required at the launch site; [10]

    b.   With NASA support, integrate the TRMM Observatory with the H-II launch vehicle;

    c.   Perform launch operations of the launch vehicle at the launch site;

    d.   Provide NASA with Self Contained Atmosphere Protection Equipment (SCAPE) suits for hazardous operations, as mutually agreed;

    e.   Provide necessary support to NASA for transportation and launch site activities in Japan related to the TRMM Observatory, as mutually agreed;

f. Provide real-time launch operation information to the designated NASA data pick-up point at a mutually-agreed-upon point, as mutually agreed;

g. Support the RF end-to-end test, including obtaining all necessary radio frequency approvals required by the Government of Japan;

h. Apply for Reaction Control System (RCS) special loading license with support from NASA and support NASA in the RCS hydrazine fueling and pressurization, as mutually agreed;

i. Determine when all launch systems are ready for launch; and

j. Launch the TRMM Observatory to the mutually-agreed-upon insertion orbit and deploy the TRMM Observatory from the launch vehicle.

4. Operation after the Separation of the TRMM Observatory

a. Support planning and data analysis for PR initial on-orbit check-out;

b. Integrate PR operation requests and support NASA's planning and conducting of PR operations including data analysis to verify proper instrument operations;

c. Utilize and operate the COMETS Data Relay Satellite and Tsukuba Space Center/NASDA to perform mutually-agreed-upon TRMM/COMETS data relay capability experiments; and

d. Plan and conduct calibration and validation to verify performance of the PR. [11]

5. Data Systems

a. Design, develop, and operate the NASDA TRMM data system to capture, produce, distribute and archive all PR mission data and other sensor mission data, as mutually agreed, in accordance with the IEOS DEP in the Appendix;

b. Provide the processed data of PR and other sensor products to NASA, as mutually agreed, when available;

c. Provide NASA with the PR calibration data and ground truth data, as mutually agreed;

d. Provide NASA with NASDA algorithms used to generate standard products, as mutually agreed;

e. Provide NASDA's catalogue data to NASA, as mutually agreed; and

f. Establish the network interfaces mutually agreed with NASA for providing the mission operation information through EOSDIS-EOIS communication lines.

6. Safety

NASDA is responsible for the overall safety control of the launch, including the TRMM launch operation at Tanegashima Space Center performed by NASA. NASDA shall assist NASA in its efforts to comply with Japanese safety regulations.

7. Other

    a. Develop and maintain spacecraft-to-launch vehicle Interface Control Specifications (ICS), launch operation ICS, TRMM/COMETS ICS, a TRMM/COMETS experiment plan and procedure, and PR Instrument Operation Documents as detailed in the TRMM IP;

    b. Hold meetings and reviews periodically in Japan, as required and mutually agreed, to carry out the responsibilities set forth by this MOU;

    c. Exchange personnel and information with NASA, as needed, to address engineering, operations, and science issues; and

    d. Provide necessary support to NASA on arrangements related to radio-frequency interference issues.

["Article V – Program and Project Management" section omitted]
["Article VI – Scientific Investigations" section omitted]
["Article VII – Funding and Limits of Obligation" section omitted]

[15]

Article VIII - Data Distribution

1. NASA and NASDA shall share all TRMM data and make such data available to other users for research, operational and other uses under the terms of the IEOS DEP (contained in the Appendix to this MOU). The shared data shall include all products from the NASA-provided instruments, the NASDA-provided instrument, and ground truth data used to validate the TRMM products.

2. For the purposes of the DEP, NASA and NASDA are joint Data Providing Agencies (DPAs) for the TRMM instrument data and agree that such data shall be distributed to all users without restriction, in a manner consistent with applicable laws and regulations.

3. NASA and NASDA shall each create and maintain a catalogue of TRMM data acquired and processed at their facilities, exchange this information by means of standardized formats, and make it freely available to users. NASA and NASDA shall establish interoperative capability between their respective catalogue systems.

4. The analyzed results from TRMM shall be made available to the general scientific community through publication in appropriate journals or presentations at scientific

conferences as soon as possible and consistent with standard scientific practices. The publication shall indicate thereon, as appropriate, that the publication is based on results obtained from the joint NASA-NASDA TRMM mission. Each Party shall take necessary measures that NASA, NASDA or their related researchers shall provide a copy of reports and/or publications to the other Party freely. And in the event such reports or publications are copyrighted, both Parties shall take necessary steps that NASA and NASDA shall be granted a royalty free right under the copyright to reproduce, use, and distribute such copyrighted work for their purpose by the copyright holder.

["Article IX – Coping with Anomaly in Development Phase and Schedule Impacts" section omitted]
["Article X – Coping with Anomaly In-Flight" section omitted]
["Article XI – Personnel Accommodation" section omitted]
["Article XII – Necessary Equipment" section omitted]
["Article XIII – Transfer of Technical Data and Goods" section omitted]
["Article XII – Necessary Equipment' section omitted]
["Article XIII – Transfer of Technical Data and Goods' section omitted]
["Article XIV – Inventions and Patent Rights" section omitted]
["Article XV – Customs and Taxes" section omitted]
["Article XVI – Public Information" section omitted]
["Article XVII – Liability" section omitted]
["Article XVIII – Registration" section omitted]
["Article XIX – Settlement of Disputes" section omitted]
["Article XX – Implementing Arrangements" section omitted]
["Article XXI – Amendment" section omitted]
["Article XXII - Entry into Force, Duration and Termination" section omitted]
[24]

Appendix

IEOS DATA EXCHANGE PRINCIPLES

The Data Exchange Principles contained in this document establish the basis on which the Agencies listed below (hereinafter referenced as the "Agencies") will share the data from the International Earth Observing System (IEOS) among themselves and make such data available to other users. These Agencies are the four Agencies who are responsible for the Earth Observation programmes of the Space Station partners and who will act as Delegations with respect to implementation of the Principles, along with the operational organisations closely related to them. The Agencies are: the European Space Agency (ESA) along with the European Organization for the Exploitation of Meteorological Satellites (EUMETSAT) [to be confirmed]; the United States National Aeronautics and Space Administration (NASA) along with the United States National Oceanic and Atmospheric Administration (NOAA); the Japanese Science and Technology Agency (STA) along with the National Space Development Agency of Japan (NASDA), the Ministry of International Trade and Industry of Japan (MITI), the Japan Environment Agency (JEA), and the Japan Meteorological Agency (JMA); and the Canadian Space Agency (CSA).

The IEOS is currently composed of the following platforms and their corresponding Earth Observation instruments which are listed in the IEOS Implementation Plan: the NASA Earth Observing System (EOS), beginning with EOS-AM1; the ENVISAT-1 element of the ESA Polar Orbit Earth Observation Mission (POEM) programme; the NOAA Polar-orbiting Operational Environmental Satellites (POES), beginning with NOAA-N; the Japanese Earth Observing System (JEOS) beginning with the Advanced Earth Observing Satellite (ADEOS); and the NASA/Japanese Tropical Rainfall Measuring Mission (TRMM). The IEOS Agencies will endeavor to include future Earth Observation missions, as appropriate, within the IEOS framework, including application of these Data Exchange Principles.

Any Agency may propose an addition to the IEOS. With the unanimous agreement of all Agencies, a new element may be added to the IEOS and its provider may become an Agency for purposes of these Data Exchange Principles.

The following Principles address the criteria of access and utilisation of data from the above platforms. Modalities of implementation will be agreed by the parties in the IEOS Implementation Plan. Detailed Terms and Conditions for the practical execution of these Principles will be documented in the IEOS Implementation Plan and agreed by the Agencies. The definitions attached to these Data Exchange Principles are an integral part of them, and will be referred to for the correct implementation of all arrangements and cooperative activities carried out in the IEOS. [25]

1. All IEOS Data will be available for peaceful purposes to all users on a non-discriminatory basis and in a timely manner.

2. There will be no period of exclusive data use. Where the need to provide validated data is recognized, any initial period of exclusive data use should be limited and explicitly defined. The goal should be release of data in some preliminary form within three months after the start of routine reception of instrument data.

3. All IEOS Data will be available for the use of each of the Agencies and its designated users at the lowest possible cost for non-commercial use in the following categories: Research, Applications, and Operational Use for the Public Benefit.

4. Agencies which designate users for Research Use and for Applications Use will do so through an Announcement of Opportunity or similar process. The designation will include a definition of the data to be provided. Research Users shall be required to submit their results for publication in the scientific literature and Applications Users shall be required to publish their results in a technical report and both shall be required to provide their results to the designating Agency and to the Data Providing Agency.

5. Any of the Agencies may designate national users of the respective countries or Member States of the Agencies as it deems appropriate to be given access to all IEOS data at the lowest possible cost for Non-commercial Operational Use for the Public Benefit, provided the designating Agency assumes responsibility for ensuring that all the terms and conditions for data use are met. This use will have to be reported to the

Data Providing Agency on the basis of commonly agreed criteria including type, usage, and final destination of the data. Designation of users outside the national territory of the Agencies or their member states (e.g., international organisations and agencies in non-participating countries) for Non-commercial Operational Use for the Public Benefit will be done only with the agreement of the Data Providing Agency.

6. For purposes other than 3 above, the specified data will be made available in accordance with terms and conditions to be established by the Data Providing Agency.

7. Each Data Providing Agency will fulfill the data requests of the other Agencies and their designated users to the maximum extent possible. In the event that these data requests exceed the Data Providing Agency's capacity, the Data Providing Agency and the designating Agency will pursue alternative arrangements to fulfill such requests.

8. All data required by the Agencies and their designated users will be made available on condition that the recipient agrees to applicable intellectual property rights terms and conditions and/or proprietary rights consistent with these Data Exchange Principles, and ensures that the data shall not be distributed to non-designated parties, nor used in ways other than those for which the data were provided, without the written consent of the Data Providing Agency. [26]

9. Any of the Agencies may delegate some of its functions to other entities; in which case, such Agency will remain responsible for ensuring compliance with these Data Exchange Principles.

10. Agencies will harmonise criteria and priorities for data acquisition, archiving, and purging, in consultation with other relevant organisations.

Definitions

The following Definitions apply in the context of these Principles:

Applications Use of data is a limited proof of concept study toward: 1) the solution of an applied program to demonstrate the utility of the data; or 2) the demonstration of the operational use of the data.

Data refers to original Earth observation sensor output and higher level products created from it by the Data Providing Agency as part of the standard set of products.

Data Providing Agency is the Agency which has primary responsibility for the distribution of data from a particular instrument or is the owner of such data. The Data Providing Agency will be defined in agreements between the operator of the platform carrying the instrument and the instrument provider should the two be different.

Lowest Possible Cost for designated users is no more than the additional cost of resources, above the cost of the normal planned data system operations required to fill a specific user request. These costs may include media, labour, expenses for operating and maintaining equipment, as well as delivery charges for mail or electronic transmission.

The above costs should not include non-recurring costs such as research, development, and space segment capital cost. However, it may include a reasonable amount towards additional capital cost of data provision.

Non-commercial Use is the utilisation of data to provide a service for the public benefit as distinguished from conferring an economic advantage on a particular user or group of users.

Non-commercial Operational Use for the Public Benefit is the utilisation of data to provide a regular service for the public benefit as distinguished from conferring an economic advantage on a particular user or group of users. An example is the use of data to carry out a mandate of environmental observation and prediction. These activities can be carried out by national or international agencies or other entities designated by these agencies to support their public benefit mandate. Such a user may be requested by the Data Providing Agency and/or the designating Agency to provide a periodic status report back to them. [27]

Non-discriminatory Basis means that all users in a clearly defined data use category can obtain data on the same terms and conditions, and the categories are defined in such a way that all potential users will be included in categories with access to the data.

Research Use of data is utilisation of data in a study or investigation which aims to establish facts or principles.

Document IV-27

Document Title: Letter from W. F. Townsend, Acting Associate Administrator for Mission to Planet Earth, to Daniel S. Goldin, NASA Administrator, "Name Change," 24 November 1997.

Source: NASA Historical Reference Collection, NASA History Office, NASA Headquarters, Washington, D.C.

In late 1997, as part of a general reorganization of NASA's offices, NASA Administrator Daniel Goldin directed the Office of Mission to Planet Earth to change its name to the Earth Science Enterprise. This memo codifies that change, but it also raises other possible names for the office in an effort to make the name more descriptive of its true focus. Mr. Goldin chose not to change his mind. The memo illustrates the difficulty of describing in a program title the breadth of NASA's Earth science research.

[no page number]
[Stamped Nov 24 1997]

TO: A/Administrator
FROM: Y/Acting Associate Administrator for Mission to Planet Earth
SUBJECT: Name Change

We are preparing all the necessary material to formalize changing the name of the Enterprise from Mission to Planet Earth to Earth Science, as you requested in the October 30, 1997 Senior Management Council Meeting. However, before we submit the necessary NASA Program Directive change for your signature, I wanted to make sure that we take maximum advantage of this opportunity to select a new name that is more descriptive of what we really do. In that regard I want to call to your attention other possibilities that you should at least consider. We have identified several candidates, including pros and cons for each. The candidates and key considerations follow. In order that we can move expeditiously to complete the name change, please check your preference.

[CHECKED] **Earth Science**. Your choice clearly has the virtue of brevity as well as parallelism with Space Science. However, most of the public associates Earth Science with geology, so this title will not readily communicate the breadth of our research. It also does not distinguish the applications potential that is clearly different from Space Science.

_____ **Earth Science & Applications**. This option acknowledges directly that there are near-term practical uses of the information we generate. It may cause confusion with the European Space Agency (ESA). This is my personal preference.

_____ **Earth System Science**. This option more accurately conveys the breadth of our research, though it requires some explanation. It also provides some distinction from other research agencies with less breadth. It does not recognize our applications efforts.

_____ **Earth System Science & Applications**. Somewhat longer, but not out of line with the titles of two of the other three Enterprises.

_____**Other**

We have polled our advisory committee and conclude that any of these alternatives are workable.

[2] With either your indication that you prefer to stick with Earth Science or a new preference, we will move immediately to make the office documentation changes required by NHB 1101.3, and begin using the new name in all our publications and correspondence.

[signature]
W. F. Townsend

Concurrence: [signature]   [handwritten] 12/11/97
               Daniel S. Goldin   Date
               Administrator

cc:
Z/Mr. A. Ladwig
Y/Mr. M. Mann
YF/Mr. M. Luther
YM/Mr. D. Norton
YS/Dr. N. Maynard

Biographical Appendix

A

Loren W. Acton (1936–   ) earned a B.A. degree from Montana State University in 1959 and a Ph.D. from the University of Colorado in 1965. Trained as a solar physicist, Dr. Acton was a Payload Specialist aboard STS-51F/Spacelab 2. STS-51F carried 13 major experiments in astronomy, astrophysics, and life sciences. In 1985, he was the Senior Staff Scientist with the Space Sciences Laboratory, Lockheed Research Laboratory in Palo Alto, California. See "Acton, Loren W.," biographical file, NASA Historical Reference Collection, NASA History Office, NASA Headquarters, Washington, DC.

Bruce Michael Alberts (1938–   ) earned a B.A. in biomedical sciences from Harvard College in 1960 and a Ph.D. in biophysics from Harvard University in 1965. Since 1993, he has served as the President of the National Academy of Sciences and as Chairman of the National Research Council. See *Who's Who in America, 2000*, 54th Ed., New Providence, NJ: Marquis Who's Who, 1999.

B

D. James Baker (1937–   ) served as Administrator of the National Oceanographic and Atmospheric Administration in the Department of Commerce from 1993 to 2001. See *Who's Who in America, 1996*, New Providence, NJ: Marquis Who's Who, 1996.

Bodo Bartocha (1928–   ) earned his B.S. in 1951, M.S. in 1953, and Ph.D. (inorganic chemistry) in 1956. He has served as an adjunct professor in the Arizona Research Laboratory of the University of Arizona since 1993. Previous positions include: Head, Propellant Branch, U.S. Naval Ordnance Laboratory, 1958 to 1961; Director of Research and Acting Director, Development, Research, and Development Management, Propellant Plant, 1961 to 1964; Deputy Chief Scientist, International Research and Development Management, Office of Naval Research, 1964 to 1966; Assistant Technology Director, Advanced Planning and Development, Naval Ordnance Station, 1966 to1967; Staff Associate and Deputy Head, Office of Planning and Policy Studies, Policy Analysis, National Science Foundation, 1967 to1970; Deputy Executive Secretary, Executive Council, Management, 1970 to1971; Executive Assistant to the Assistant Director, National and International Programs, 1971; and Fellow, American Association for the Advancement of Science. See "Bartocha, Bodo," in *American Men and Women of Science, 1998–99*, 20th Ed., New Providence, NJ: R. R. Bower, 1998.

Jacques Maurice Beckers (1934–   ) earned his Ph.D. from the University of Utrecht, The Netherlands, in 1969, and immigrated to the United States in 1962. He currently holds a position with the National Solar Observatory, Sacramento Peak, in Sunspot, New Mexico. Other positions include: Senior Researcher at the National Solar Observatory in Tucson, Arizona, where he served as Director from 1993 to 1998; Astrophysicist with the European Solar Observatory, Germany, 1988 to 1993; Director, Advanced Development Program, National Optical Astronomy Observatory, 1984 to 1988; and Director, Multiple Mirror Telescope Observatory, 1979 to 1984. He is a member of the Norwegian Academy of Science and the Royal Netherlands Academy of Science. See "Beckers, Jacques Maurice," in *Who's Who in America, 1956*, 10th Ed., New Providence, NJ: Marquis Who's Who, 1955.

James M. Beggs (1926–   ) was nominated by President Reagan on 1 June 1981 to become the sixth Administrator of the National Aeronautics and Space Administration. Beggs took his oath of office on 10 July 1981. Prior to his appointment as NASA Administrator, he had been Executive Vice President and a Director of General Dynamics Corporation, St. Louis, Missouri. He served with NASA from 1968 to 1969 as Associate Administrator, Office of Advanced Research and Technology. From 1969 to 1973, he was Undersecretary of Transportation. He went to Summa Corporation, Los Angeles, California, as Managing Director of Operations and joined General Dynamics in January 1974. Before NASA he also had been with Westinghouse Electric Corporation, in Sharon, Pennsylvania, and Baltimore, Maryland, for 13 years. His resignation from NASA was effective 25 February 1986. Since leaving NASA, Mr. Beggs has worked as a consultant. See "Beggs, James M.," biographical file, NASA Historical Reference Collection, NASA History Office, NASA Headquarters, Washington, DC.

Orville George Bentley (1918–   ) earned a B.S. from South Dakota State College in 1942, an M.S. in biochemistry from the University of Wisconsin in 1947, and a Ph.D. from the University of Wisconsin in 1950. Dr. Bentley served as Assistant Secretary of Agriculture for Science and Education from 1982 to 1989. See *Who's Who in America, 2000*, 54th Ed., New Providence, NJ: Marquis Who's Who, 1999.

Lloyd V. Berkner (1905–1967) was involved in most of the early spaceflight activities of the United States. Trained as an electrical engineer, he was interested in atmospheric propagation of radio waves, but after World War II became a scientific entrepreneur of the first magnitude. He was heavily involved in the planning for and execution of the International Geophysical Year in 1957 and 1958, and served in a variety of positions in Washington, DC, where he could influence the course of science policy. See "Berkner, Lloyd V.," biographical file, NASA Historical Reference Collection, NASA History Office, NASA Headquarters, Washington, DC.

Warren Walt Berning (1920–   ) is a physicist by training but also an expert in meteorology. He is currently a Senior Scientist at the Physical Science Laboratory of New Mexico State University. In 1947, he started a career with the Ballistic Research Laboratory at Aberdeen Proving Ground as a Meteorologist, and stayed until 1950. From 1950 through 1967, he worked as a Physicist at Aberdeen. See "Berning, Warren Walt," in *American Men and Women of Science, 1998–99*, 20th Ed., New Providence, NJ: R. R. Bower, 1998.

Charles A. Berry earned a B.A. degree from the University of California Berkeley in 1945, an M.D. from the University of California Medical School in San Francisco in 1947, and an M.A. in public health from the Harvard School of Public Health in 1956. He began working with NASA in 1958 when he participated in the selection and flight monitoring of the Mercury 7 astronauts. From 1963 to 1971, Berry served as Director of Medical Research and Operations at the Johnson Space Center. In 1971, Dr. Berry became NASA Director for Life Sciences at NASA Headquarters, where he remained until retirement in 1973. See "Berry, Charles A.," biographical file, NASA Historical Reference Collection, NASA History Office, NASA Headquarters, Washington, DC.

J. David Bohlin (1939–   ) has been the Senior Scientific Program Executive at NASA Headquarters since 1996. Prior to his current position, he served as Chief Scientist of the Space Physics Division from 1990 through 1995 and as Chief of Solar Physics from 1978 through 1990. He started his career in 1970 with the Naval Research Laboratory as an Astrophysicist of Solar Physics. See "Bohlin, David J.," *in American Men and Women of Science, 1998–99*, 20th Ed., New Providence, NJ: R. R. Bower, 1998.

George E. Brown, Jr., (1920–1999) (D-CA) served in the House of Representatives from 3 January 1963 to 3 January 1971, and then again from 3 January 1973 to the present. He chaired the House Committee on Science, Space, and Technology for a number of years, and currently is its ranking minority member. See *Biographical Directory of the American Congress, 1774–1996*, Alexandria, VA: CQ Staff Directories, Inc., 1997.

C

Anthony J. Calio (1936–   ) earned a B.A. in physics from the University of Pennsylvania in 1953. Calio also did graduate work in physics at the University of Pennsylvania and the Carnegie Institute of Technology. Dr. Calio joined NASA in 1963 and began work at the Electronics Research Center in Cambridge, Massachusetts. Dr. Calio became Director of Science and Applications at the Johnson Space Center in 1969. From 30 November 1975 to 1 October 1977, Dr. Calio served as the Deputy Associate Administrator for the Office of Space Science at NASA Headquarters. On 1 October 1977, Dr. Calio became NASA's Associate Administrator for Space Science and Applications, where he remained until becoming Deputy Administrator for the National Oceanic and Atmospheric Administration in 1981. In 1985, Calio became Administrator of the National Oceanic and Atmospheric Administration, where he remained until 1987, when he departed to work in private industry. See "Calio, Anthony J.," biographical file, NASA Historical Reference Collection, NASA History Office, NASA Headquarters, Washington, DC.

Alastair Graham Walter Cameron (1925–   ) is an astrophysicist/educator. He received his B.S. and Ph.D. degrees in Canada before coming to the U.S. in 1959, and became a naturalized citizen in 1963. He was an assistant professor of physics at the University of Iowa, in Ames, from 1952 to 1954, and served as Senior Scientist at the Goddard Institute of Space Studies in New York from 1961 to 1966. Since 1997, he has served as the Donald

H. Menzel Professor of Astrophysics at Harvard University. He is the recipient of the 1983 Distinguished Public Service Medal from NASA, the 1988 J. Lawrence Smith Medal from the National Academy of Sciences, the 1989 Harry H. Hess Medal, the 1994 International Astronomical Union (IUA) Meteorological Society Leonard Medal, and was the 1997 American Astronomical Society (AAS) Russell Lecturer. Professional memberships include the National Academy of Sciences, the American Association for the Advancement of Science, the American Geophysical Union, the American Astronomical Society, and the International Astronomical Union. See "Cameron, Alastair Graham Walter," in *Who's Who in America, 2000*, 54th Ed., New Providence, NJ: Marquis Who's Who, 1999.

Richard Charles Canfield (1937–) started his career as a visiting scientist at the High Altitude Observatory of the National Center for Atmospheric Research. He left the center in 1969. At the Sacramento Peak Observatory he worked as an Astrophysicist from 1970 to 1976, an Associate Resident Physicist from 1976 to 1980, and a Resident Physicist from 1980 to 1985. He has been an Astronomer at the University of Hawaii since 1985. Professional memberships include the American Astronomical Society and the International Astronomical Union. See "Canfield, Richard Charles," in *American Men and Women of Science, 1998–99*, 20th Ed., New Providence, NJ: R. R. Bower, 1998.

James (Jimmy) Carter (1924–) was the thirty-ninth President of the United States, from 1977 to 1981. He was a naval officer and a businessman before entering politics. A member of the Georgia State Legislature from 1962 to 1966, he was Governor of Georgia from 1971 to 1975. See "Carter, Jimmy," biographical file, NASA Historical Reference Collection, NASA History Office, NASA Headquarters, Washington, DC.

Talbot Albert Chubb (1923–) is an authority in geophysics and astrophysics. From 1959 through 1981, he headed the Upper Air Physics branch of the Naval Research Laboratory. He has been the President of Research Systems, Incorporated, since 1981. See "Chubb, Talbot Albert," in *American Men and Women of Science, 1998–99*, 20th Ed., New Providence, NJ: R. R. Bower, 1998.

John F. Clark (1920–) was an Electrical Engineer at the Naval Research Laboratory from 1942 to 1958. His experiences at the laboratory varied from rocket experiments to measuring the electrical properties of Earth's atmosphere. He is an authority in atmospheric and space sciences and applications, and joined NASA in 1958 as Chief of the Ionospheric Physics Program Office. From 1963 to 1965, he served as Deputy Associate Administrator for Space Sciences and Applications at Headquarters. From 1962 to 1965, he concurrently served as Chairman of NASA's Space Science Steering Committee. He was Director of the Goddard Space Flight Center from 1965 until he retired in 1976. See "Clark, John F.," biographical file, NASA Historical Reference Collection, NASA History Office, NASA Headquarters, Washington, DC.

France Cordova (1947–) earned a B.A. in English from Stanford University and a Ph.D. in physics from the California Institute of Technology. From 1979 to 1989, Cordova served as Staff Scientist in the Earth and Space Science Division at the Los Alamos National Laboratory in New Mexico. She subsequently became Deputy Group Leader of the Space Astronomy and Astrophysics Group at Los Alamos. Cordova left Los Alamos to become a professor at Pennsylvania State University and later became head of the Astronomy and Physics Department. She was on extended leave from her post while serving as NASA Chief Scientist from 1993 to 1996. See "Cordova, Dr. France," biographical file, NASA Historical Reference Collection, NASA History Office, NASA Headquarters, Washington, DC.

D

Joseph John Dantone, Jr., (1942–) earned a B.S. from the United States Naval Academy in 1964 and an M.S. in aerospace engineering and material management from the Naval Postgraduate School in 1971. He served as Deputy Director of the National Reconnaissance Office from 1994 to 1996. He also served as Director of the Defense Mapping Agency from 1996 to 1997 and as Director of the National Imagery and Mapping Agency from 1997 to 1998. See *Who's Who in America, 2000*, 54th Ed., New Providence, NJ: Marquis Who's Who, 1999.

Robert James Davis (1929–) is an authority on stellar astronomy and is currently an astrophysicist at the Smithsonian Astrophysics Observatory. At Harvard University, he earned a B.A. in 1951, an M.A. in 1956, and a

Ph.D. (in astronomy) in 1960. See "Davis, Robert James," in *American Men and Women of Science, 1998–99*, 20th Ed., New Providence, NJ: R. R. Bower, 1998.

William Gould Dow (1896–2000) was an emeritus professor of electrical engineering and computer science at the University of Michigan. During his tenure, he was responsible for creating and organizing thirteen laboratories and research units, including space physics research, plasma engineering, and Cooley electronics. He also served on the panel of scientists who helped form NASA in the late 1950s. See "Dow, William Gould," biographical file, NASA Historical Reference Collection, NASA History Office, NASA Headquarters, Washington, DC.

Hugh Latimer Dryden (1898–1965) was the Director of the National Advisory Committee for Aeronautics (NACA) from 1947 until the creation of the National Aeronautics and Space Administration (NASA) in 1958. He was named Deputy Administrator of the new aerospace Agency, created in response to the Sputnik crisis. Before NACA, he was Associate Director of the National Bureau of Standards, where he had served since 1918 in scientific research. Influenced by Dr. Joseph S. Ames, who for many years was Chairman of the NACA and a pioneer in aerodynamics, Dryden undertook a study of fluid dynamics at the Bureau of Standards while continuing his courses at the Johns Hopkins University Graduate School. His laboratory work was accepted by the university, and he received the degree of doctor of philosophy in 1919. He served as the NASA Deputy Administrator until his death on 2 December 1965. For further information on Hugh Dryden, see Michael Gorn, *Hugh L. Dryden's Careeer in Aviation and Space* (Washington, DC: Monographs in Aerospace History, No. 5, 1996) or Richard K. Smith, *The Hugh L. Dryden Papers, 1898–1965*, Baltimore, MD: The Johns Hopkins University Library, 1974.

Richard B. Dunn (1927–) earned a B.M.E. from the University of Minnesota in 1949, an M.S. in 1950, and a Ph.D. from Harvard University in 1961. He is an authority on astronomy and mechanical engineering, and currently is a Physicist at the National Solar Observatory, Sunspot, New Mexico. See "Dunn, Richard B.," in *American Men and Women of Science, 1998–99*, 20th Ed., New Providence, NJ: R. R. Bower, 1998.

Andrea K. Dupree (1939–) is currently the Associate Director of the Center for Astrophysics at Harvard University. She has served as the Senior Astrophysicist for the Smithsonian Astrophysics Observatory since 1979. She received her B.A. from Wellesley College in 1960 and her Ph.D. in astronomy from Harvard in 1968. She was the 1972 Phillips Lecturer for Haverford College and was awarded the 1973 Bart J. Bok Prize by Harvard University. Her professional experiences include research fellow for the Harvard College Observatory; Research Associate; Senior Research Associate for astronomy and astrophysics; member of the Space and Earth Science Advisory Board, NASA; member, Space Science Board, National Academy of Sciences; lecturer, Harvard University; and member, executive committee, American Association for the Advancement of Science. While at the observatory, she produced a report entitled "Interaction Between Solar Physics and Astrophysics." Professional memberships include the American Astronomical Society, Vice President in 1988; the International Astronomical Union (IAU); the Committee on Space Research; and the American Association for the Advancement of Science. See "Dupree, Andrea K.," in *American Men and Women of Science, 1998–99*, 20th Ed., New Providence, NJ: R. R. Bower, 1998.

E

John Allen Eddy (1931–) has served as the Senior Scientist and Director of the Office of Interdisciplinary Earth Studies, National Center of Atmospheric Research, since 1986. He received his B.S. degree from the U.S. Naval Academy in 1953 and his Ph.D. (astrogeophysics) from the University of Colorado in 1962. Other positions held include Physicist, National Center for Atmospheric Research; professor adjunct, University of Colorado; and Research Associate, Harvard-Smithsonian Center for Astrophysics. Professional memberships include the American Astronomical Society; the American Association for the Advancement of Science; the American Geophysical Union; Sigma Xi; and the International Astronomical Union. See "Eddy, John Allen," in *American Men and Women of Science, 1998–99*, 20th Ed., New Providence, NJ: R. R. Bower, 1998.

Burton I. Edelson (1926–2002) was NASA's Associate Administrator for Space Science and Applications between 1982 and 1988. He earned his B.S. from the U.S. Naval Academy in 1947 and served for twenty years in the service. He then returned to school and received a Ph.D. from the University of California at San Diego in 1969. Thereafter he worked with the Communications Satellite Corporation for fourteen years before coming to NASA. See "Edelson, Burt I.," biographical file, NASA Historical Reference Collection, NASA History Office, NASA Headquarters, Washington, DC.

Krafft A. Ehricke (1917–1984) was an aeronautical engineer and physicist, and one of the members of the German V-2 rocket team that surrendered to the United States Army in the closing days of World War II brought to the United States as a nucleus of America's postwar rocket development program. In 1947, he worked for the Army's missile program under Dr. Wernher von Braun. From 1956 to 1965, he worked for the Convair division of the General Dynamics Corporation, where he had a major design and management role in the development of the Atlas and Centaur rockets. He later served as the Chief Scientific Adviser to the space division of Rockwell International Corporation, the builder of the Apollo manned spacecraft and the Space Shuttle. See "Ehricke, Krafft A.," biographical file, NASA Historical Reference Collection, NASA History Office, NASA Headquarters, Washington, DC.

F

Lennard A. Fisk was the Chief Scientist of NASA prior to his retirement in 1993. Before this appointment, he served as the Associate Administrator for Space Science and Applications, and was responsible for planning and directing NASA programs, including the institutional management of the Goddard Space Flight Center in Greenbelt, Maryland, and the Jet Propulsion Laboratory in Pasadena, California. From 1971 to 1977, he was an Astrophysicist at the NASA Goddard Space Flight Center. Prior to NASA, he was the National Academy of Sciences Postdoctoral Research Fellow at Goddard from September 1969 to June 1971. He also had served as Vice President for Research and Financial Affairs at the University of New Hampshire. See "Fisk, Lennard A.," biographical file, NASA Historical Reference Collection, NASA History Office, NASA Headquarters, Washington, DC.

James C. Fletcher (1919–1991) was the Administrator (1971–1977) of the National Aeronautics and Space Administration who gained the approval of the Nixon Administration on 5 January 1972 to develop the Space Shuttle as the follow-on human space flight effort of the Agency. He also served as NASA Administrator a second time (1986–1989), following the loss of the Space Shuttle Challenger on 28 January 1986. Fletcher received an undergraduate degree in physics from Columbia University and a doctorate in physics from the California Institute of Technology. After holding research and teaching positions at Harvard and Princeton universities, he joined Hughes Aircraft in 1948, and later worked at the Guided Missile Division of the Ramo-Wooldridge Corporation. In 1958, Fletcher co-founded the Space Electronics Corporation in Glendale, California. He was later named Systems Vice President of the Aerojet General Corporation in Sacramento, California. In 1964, he became President of the University of Utah, a position he held until he was named NASA Administrator in 1971. Dr. Fletcher died at his home in suburban Washington on 22 December 1991. See "Fletcher, James C.," biographical file, NASA Historical Reference Collection, NASA History Office, NASA Headquarters, Washington, DC.

David J. Forrest is a research associate professor in the Institute of Earth, Oceans, and Space (EOS), and the Space Science Center at the University of New Hampshire. He received his Ph.D. in physics from UNH in 1969. His thesis work involved the design, fabrication, and flight of three different balloon experiments. Since then he has been involved in at least six balloon experiments, some as the project scientist and some as a senior mentor for young scientists. In 1971, he became the Instrument Scientist for the Gamma Ray Experiment on NASA's OSO-7 mission. In mid-1970, he designed, and was the Experiment Scientist and the UNH technical manager for, the Gamma Ray Spectrometer (GRS) for NASA's Solar Maximum Mission. The GRS experiment was delivered on time and within budget, and was the first of six instruments delivered and integrated onto the SMM spacecraft. The GRS operated without a single failure for nearly 10 years, from launch in 1980 until the SMM reentry in late 1989. See "Forrest, David J.," at *http://www.catsat.sr.unh.edu/personnel/dforrest.html* on the Internet.

Lawrence William Fredrick (1927–   ) is an eminent part of the faculty at the University of Virginia, Charlottesville. His tenure with the university spans four decades. From 1965 to 1995, he taught astronomy. In 1995, he became a research professor. He was a member of the American Astronomical Society from 1969 through 1980. A published writer with expertise in the field of astronomy, he received his B.A. (1952) and M.A. (1954) from Swarthmore College, and his Ph.D. from the University of Pennsylvania (1959). See "Fredrick, Lawrence William," in *Who's Who in America, 1956*, 10th Ed., New Providence, NJ: Marquis Who's Who, 1955.

Herbert Friedman (1916–2000) earned his Ph.D. in physics from the Johns Hopkins University in 1940. He conducted his first experiments in rocket astronomy with a V-2 rocket in 1949. He performed hundreds of experiments including the traced solar cycle variations of x-rays and ultraviolet radiations from the Sun, and the measured ultraviolet fluxes of early-type stars. Dr. Friedman received the National Medal of Science, the Nation's highest honor for scientific achievement, as well as other numerous awards and merits. His scientific and technical contributions include 50 patents and about 300 published papers. He served on many science advisory committees, including the President's Science Advisory Committee, the General Advisory Committee of the Atomic Energy Commission, and the Space Science Board of the National Academy of Sciences. See "Friedman, Herbert," biographical file, NASA Historical Reference Collection, NASA History Office, NASA Headquarters, Washington, DC.

Edward Allan Frieman (1926–   ) earned a B.S. from Columbia University in 1946, an M.S. in physics from Columbia University in 1948, and a Ph.D. in physics from the Polytechnic Institute of Brooklyn in 1952. An educator by profession, he served as Vice Chairman of the White House Science Council from 1981 to 1989. He also served on the Defense Science Board from 1984 to1990. In 1992, he became Vice President of the Space Policy Advisory Board. He also served as Chairman of the NASA Earth Observing Systems Engineering Review from 1991 to 1992. See *Who's Who in America, 2000*, 54th Ed., New Providence, NJ: Marquis Who's Who, 1999.

G

Riccardo Giacconi (1932–   ) became head of the Hubble Space Telescope Science Institute in 1981 and served through launch in 1990 on STS-31. Prior to Hubble, he served as Associate Director of the High Energy Astrophysics Division at the Harvard University Smithsonian Center for Astrophysics. A pioneer in the field of x-ray astronomy, he led the team that launched UHURU, the first x-ray satellite, in 1970. See "Giacconi, Riccardo," biographical file, NASA Historical Reference Collection, NASA History Office, NASA Headquarters, Washington, DC.

Peter A. Gilman received his B.A. from Harvard College (1962) and his S.M. (1964) and Ph.D. (meteorology, 1966) from the Massachusetts Institute of Technology. His professional experiences include: assistant professor of astrogeophysics and lecturer at the University of Colorado; adjunct professor of astrophysics, planetary, and atmospheric science; member, Solar and Space Physics Committee, Space Science Board, National Academy of Sciences; Chairman, Advanced Study Program, Head, Solar Variability Section, and Director, High Altitude Observatory; and Senior Scientist, National Center for Atmospheric Research. Professional memberships include the American Association for the Advancement of Science, the American Meteorological Society, the American Astronomical Society, the American Geophysical Union, and the International Astronomical Union. See "Gilman, Peter A.," in *American Men and Women of Science, 1998–99*, 20th Ed., New Providence, NJ: R. R. Bower, 1998.

Harold Glaser (1924–   ) is an authority of theoretical physics. Early in his career he worked as a Senior Physicist in the Applied Physics Laboratory at Johns Hopkins University (from 1952 through 1954). From 1954 through 1957, he was Head of the Theoretical Analysis Section of the System Analysis Branch of the Naval Research Laboratory. He joined NASA in 1966 as the Deputy Chief of Solar Physics in the Office of Space Science and Applications. From 1975 through 1980, he was the Director of NASA's Terrestrial Programs. See "Glaser, Harold," in *American Men and Women of Science, 1998–99*, 20th Ed., New Providence, NJ: R. R. Bower, 1998.

T. Keith Glennan (1905–1995) was the first Administrator of the National Aeronautics and Space Administration, formally established on 1 October 1958, under the National Aeronautics and Space Act of 1958. Within a short time after NASA was formally operational, Glennan incorporated several organizations involved in space exploration projects from other federal agencies to ensure that a viable scientific program of space exploration could be reasonably conducted over the long term. A resident of Reston, Virginia, for twenty years, after his retirement in the late 1980s he moved to Mitchellville, Maryland, and died there on 11 April 1995. See "Glennan, T. Keith," biographical file, NASA Historical Reference Collection, NASA History Office, NASA Headquarters, Washington, DC.

Harry J. Goett (1910–) earned a degree in physics from Holy Cross College in 1931 and another in aeronautical engineering from NYU in 1933. After holding a number of engineering posts with private firms, he became a Project Engineer at Langley Aeronautical Laboratory in 1936. He later moved to Ames Aeronautical Laboratory and was Chief of the Full-Scale and Flight Research Division from 1948 to 1959. He was Director of the Goddard Space Flight Center from 1959 to 1965 and then became a Special Assistant to NASA Administrator James E. Webb. Later he was Director for Plans and Programs at Philco's Western Development Labs in California. A position with Ford Aerospace and Communications ended with his retirement. See "Goett, Harry J.," biographical file, NASA Historical Reference Collection, NASA History Office, NASA Headquarters, Washington, DC.

Leo Goldberg (1913–1987) was the Director of the Kitt Peak National Observatory from 1971 to 1977. He served as a professor of astronomy and as Director of the observatory at the University of Michigan and Harvard University from 1948 to 1971. A former president of the International Astronomical Union and the American Astronomical Society, he received three degrees, including his Ph.D., from Harvard University. See "Goldberg, Leo," biographical file, NASA Historical Reference Collection, NASA History Office, NASA Headquarters, Washington, DC.

Daniel S. Goldin (1940–) initiated a revolution to transform America's aeronautics and space program. To date, he is NASA's longest-serving Administrator. Before coming to NASA, Goldin was Vice President and General Manager of the TRW Space and Technology Group in Redondo Beach, California. During his twenty-five-year career at TRW, Goldin led projects for America's defense and conceptualized and managed production of advanced communication spacecraft, space technologies, and scientific instruments. He began his career at NASA's Lewis Research Center, Cleveland, Ohio, in 1962 and worked on electric propulsion systems for human interplanetary travel. See "Goldin, Daniel S.," Administrator file, NASA Historical Reference Collection, NASA History Office, NASA Headquarters, Washington, DC.

Richard Mead Goody (1921–) earned a Ph.D. from Cambridge University in 1949. Dr. Mead worked as a professor of applied sciences at Harvard University from 1958 to 1991. He also served as Director of the Blue Hill Observatory from 1958 to 1970. See *Who's Who in America, 2000*, 54th Ed., New Providence, NJ: Marquis Who's Who, 1999.

Albert Gore (1948–) was the forty-fifth Vice President of the United States. He was elected as President Clinton's Vice President in 1992 after serving in the House of Representatives (D-TN) from 1977 to 1985 and the Senate from 1985 to 1993. He graduated with a degree in government, with honors, from Harvard College in 1969 and attended Vanderbilt Law School after serving in the Vietnam War. As Vice President, he worked with Russian Prime Minister Chernomyrdin on the Gore-Chernomyrdin Commission to establish an agreement for Russian participation in the International Space Station. In 2000, he received the Democratic nomination for President of the United States. See "Gore, Albert," biographical file and "Gore, Albert," space statements file, NASA Historical Reference Collection, NASA History Office, NASA Headquarters, Washington, DC.

Charles F. Green received his B.A. and M.A. degrees from the University of Kansas, and in 1915 joined the University of Illinois as a graduate assistant in mathematics. World War I interrupted his work, and he enlisted in the Air Corps, serving overseas as a test pilot. Upon his return, he received his Ph.D. from the University of Illinois and remained on the staff until joining General Electric in Schenectady, New York, in 1929. He was among the group of experts sent to Europe early in 1945 to investigate engineering achievements of the Axis powers. When he returned, he brought with him information on the Germans' progress in guided missiles and jet aircraft, which he obtained by visiting their military, industrial, and research centers. See "Green, Charles F.," biographical file, NASA Historical Reference Collection, NASA History Office, NASA Headquarters, Washington, DC.

Milton Greenberg (1918–) served as the Deputy Director of the United States Air Force Geophysics Research Directorate from 1950 through 1954. In 1954, he became the Director and served until 1958. In 1957, he participated in the Rocket and Satellite Research Panel. And from 1958 through 1986, he was President and Chairman of GCA Corporation. See "Greenberg, Milton," in *American Men and Women of Science, 1998–99*, 20th Ed., New Providence, NJ: R. R. Bower, 1998.

Jesse Leonard Greenstein (1909–   ) received his A.B., A.M., and Ph.D. from Harvard University. He is an astrophysicist and has served as a Lee A. Dubridge Emeritus Professor of Astrophysics since 1980. Other professional experiences include staff member, Hale Observatory; professional astrophysicist; fellow, National Research Council, Yerkes Observatory, University of Chicago; and research associate, McDonald Observatory, Chicago and Texas. Professional memberships include the National Academy of Sciences, the American Astronomical Society, the American Philosophical Society, the Royal Astronomical Society, and the Royal Belgian Academy of Science. In 1972, he was chairman of the review board panel of the National Academy of Sciences report on "Astronomy and Astrophysics for the 1970s." See "Greenstein, Jesse Leonard," in *American Men and Women of Science, 1998–99*, 20th Ed., New Providence, NJ: R. R. Bower, 1998.

H

Fred Theodore Haddock, Jr., (1919–   ) is an emeritus professor at the University of Michigan. He received his B.S. (1941) and M.S. (1950) degrees from the University of Maryland, and his honorary D.Sc. degree from Southwestern, at Memphis, and Ripon College. He is an expert in radio astronomy. He started his career in 1941 with the U.S. Naval Research Laboratory as a Physicist and Electronic Scientist. He has been associated with NASA since the early 1960s, as a member of the Advisory Committee of Planetary and Interplanetary Science, Office of Space Science; the Ad Hoc Working Group on Apollo Science Experiment and Training, 1962 to 1963; the NASA Headquarters Astronomy Subcommittee, 1967 to 1969; consultant to the Office of Space Science and Applications since 1970; the NASA Astronomy Missions Board, the Radio Astronomy Panel, 1968 to 1971; the Ad Hoc Committee on National Astronomy Space Observance, 1970; the NASA Outer Planets Grand Tour Mission; and the Radio Astronomy Team since 1971. He is a member of the American Astronomical Society, Vice President 1961 to 1963; a fellow with the Institute of Aeronautics and Astronautics; a fellow with the Royal Astronomical Society; and a member of Sigma Xi. See "Haddock, Frederick Theodore, Jr.," in *American Men and Women of Science, 1998–99*, 20th Ed., New Providence, NJ: R. R. Bower, 1998.

John W. Harvey (1941–   ) was one of eight finalists selected to be Payload Specialists aboard STS-51F to perform experiments on Spacelab 2. He graduated from the University of Colorado and was a scientist at the Kitt Peak National Observatory in Tucson, Arizona. See "Harvey, John W.," biographical file, NASA Historical Reference Collection, NASA History Office, NASA Headquarters, Washington, DC.

Karl G. Henize (1926–1993) received a B.A. degree in mathematics in 1947 and an M.A. degree in astronomy in 1948 from the University of Virginia, and his Ph.D. (astronomy) in 1954 from the University of Michigan. NASA selected him as a Scientist-Astronaut in August 1967. He completed the initial academic training and the fifty-three-week jet pilot training program at Vance Air Force Base, Oklahoma. He was a member of the astronaut support crew for the Apollo 15 mission and for the Skylab 2, 3, and 4 missions, and Mission Specialist for the ASSESS-2 Spacelab simulation mission in 1977. He has logged 2,300 hours of flying time in jet aircraft. He was also a Mission Specialist on the Spacelab-2 mission (STS 51-F), which launched from Kennedy Space Center, Florida, on 29 July 1985. In 1986, he accepted a position as Senior Scientist in NASA's Office of Space Science. See "Henize, Karl G.," biographical file, NASA Historical Reference Collection, NASA History Office, NASA Headquarters, Washington, DC.

Harry H. Hess (1906–1969) was one of the Senior Scientists involved in analyzing the lunar samples returned to Earth by Project Apollo. Blair Professor of Geology at Princeton University, he was Chair of the Space Science Board of the National Academy of Sciences during the Apollo era. See "Hess, Harry H.," biographical file, NASA Historical Reference Collection, NASA History Office, NASA Headquarters, Washington, DC.

Noel W. Hinners (1935–   ) was trained in geochemistry and geology at Rutgers University, California Institute of Technology, and Princeton University. He began his career in 1963 with Bellcomm, Inc., working on the Apollo program, and came to NASA Headquarters in 1972 as the Deputy Director of lunar programs in the Office of Space Science. From 1974 to 1979, he was NASA Associate Administrator for Space Science. He also served, from 1979 to 1982, as Director of the Smithsonian Institution's National Air and Space Museum, and he was Director of the Goddard Space Flight Center in Greenbelt, Maryland between 1982 and 1987. He then became Associate Deputy Administrator of NASA. He left the Agency in 1989 to join the Martin Marietta Corporation as Vice President of Strategic Planning. See "Hinners, Noel W.," biographical file, NASA Historical Reference Collection, NASA History Office, NASA Headquarters, Washington, DC.

Harry C. Holloway served as NASA's Associate Administrator for Life and Microgravity Sciences and Applications from 1993–1996, while on temporary assignment to NASA from the School of Medicine at the Uniformed Services University of the Health Sciences in Bethesda, Maryland. Dr. Holloway also served NASA in the capacities of chairman of the NASA aerospace medicine advisory committee and member of the U.S./U.S.S.R. joint working group on space biology and medicine. See "Holloway, Harry C.," biographical file, NASA Historical Reference Collection, NASA History Office, NASA Headquarters, Washington, DC.

Thomas Edward Holzer (1944–   ) was a member of the ad hoc committee on the role of theory in space science from 1980 through 1982. He also served on the NASA-sponsored panel on solar terrestrial theory from 1978 through 1979. In 1989, he was chairman of a heliospheric SR&T review panel sponsored by the Space Physics Division of NASA and served concurrently as a Senior Scientist at the High Altitude Observatory of the National Center of Atmospheric Research and adjunct professor of astrophysics and planetary and atmospheric science at the University of Colorado. See "Holzer, Thomas Edward," in *American Men and Women of Science, 1998–99*, 20th Ed., New Providence, NJ: R. R. Bower, 1998.

Robert Franklin Howard (1932–   ), with his colleague Harold Zirin, attempted to build a 65-centimeter Solar Optical Telescope (SOT) in the mid-1960s. Among his many positions, he served as Director of the National Solar Observatory, Tucson, Arizona, from 1984 to 1988. From 1988 to 1998, he was an astronomer of the observatory and later assumed the status of astronomer emeritus. See "Howard, Robert Franklin," in *Who's Who in America, 1956*, 10th Ed., New Providence, NJ: Marquis Who's Who, 1955.

Karl George Hufbauer (1937–   ) is a science historian. He started his career in 1966 at the University of California, Irvine, as an assistant professor, and then a professor. From 1984 through 1990, he worked for NASA Headquarters as a contract historian. He is the author of *Exploring the Sun, 1991*, which won the Emme prize in 1993. See "Hufbauer, Karl George," in *Who's Who in America, 2000*, 54th Ed., New Providence, NJ: Marquis Who's Who, 1999.

James W. Humphreys earned a B.S. in chemical engineering from the Virginia Military Institute in Lexington, Virginia. He subsequently received an M.D. from the Medical College of Virginia, Richmond in 1939 and an M.S. in surgery from the University of Colorado in 1951. From 1967 to 1970, Dr. Humphreys served as NASA's Director of Space Medicine before his appointment as NASA Director of Life Sciences in the Office of Manned Space Flight later in 1970. See "Humphreys, James W.," biographical file, NASA Historical Reference Collection, NASA History Office, NASA Headquarters, Washington, DC.

Arthur James Hundhausen (1936–   ) started his professional experience in 1964 in the theoretical division of Los Alamos Science Laboratory. He has concurrently served as a Staff Scientist at the High Altitude Observatory and a lecturer at the University of Colorado, since 1971. In 1976, he was a participant in the National Academy of Sciences "Solar Physics Study." See "Hundhausen, Arthur James," in *American Men and Women of Science, 1998–99*, 20th Ed., New Providence, NJ: R. R. Bower, 1998.

Wesley Huntress (1942–   ) earned a B.S. in chemistry from Brown University in 1964 and a Ph.D in chemical physics from Stanford University in 1968. Huntress subsequently began work on the science staff at California Institute of Technology's Jet Propulsion Laboratory (JPL). He left JPL and came to NASA where he served as Assistant to the Director of the Earth Sciences and Applications Division, from 1988 to 1990. Dr. Huntress directed the Solar System Exploration Division from 1990 to 1992. In 1993, he became Associate Administrator for Space Science, where he remained until joining the Carnegie Institution of Washington staff in 1998. See "Huntress, Wesley," biographical file, NASA Historical Reference Collection, NASA History Office, NASA Headquarters, Washington, DC.

J

Roy Jackson (1919–   ) earned a B.A. degree in mechanical engineering with an aeronautical option from Stanford University in 1941. He served as section head in the wind tunnel at NASA Ames Research Center during World War II while in the Navy. Jackson worked in various posts at Northrop from 1953 to 1970. Jackson served as NASA Associate Administrator for Aeronautics and Space Technology from 1970 to 1973. After leaving NASA, Jackson again worked at Northrop as the Vice President of Program Management. See "Jackson, Roy," biographical file, NASA Historical Reference Collection, NASA History Office, NASA Headquarters, Washington, DC.

Kenneth A. Janes is a professor of astronomy in the College of Arts and Sciences at Boston University. He received his B.A. degree from Harvard College, his M.S. degree from San Diego State University, and his Ph.D. from Yale University. In 1976, he was the Study Director of the "Solar Physics Study" of the National Academy of Sciences. See "Janes, Kenneth A." at *http://www.bu.edu/htbin/webph/query.pl?id=X4876&ret*

Robert T. Jones designed swept-back wings in 1944 while working as a scientist for the National Advisory Committee for Aeronautics. Pivoting them back created less wind resistance, allowing for supersonic speeds with the same engine power. Virtually every commercial and military jet uses the design today. He also invented the oblique wing, a new design that is thought to improve fuel economy and increase airspeed. See "Jones, R. T.," biographical file, NASA Historical Reference Collection, NASA History Office, NASA Headquarters, Washington, DC.

Walton L. Jones, Jr. (1918–   ) earned a B.S. degree in 1939 and an M.D. in 1942 from Emory University. Dr. Jones came to NASA in 1964 on detail and remained until his retirement from the Navy in 1966. Before NASA, he served as Director of the Aviation and Medicine Technical Division of the Navy Bureau of Medicine and Surgery, and as Aerospace Medical Advisor to the Bureau of Aeronautics. Dr. Jones served as Director of NASA's Biotechnology and Human Research Division, later became Deputy Director for Life Sciences, and then Director of the Office of Occupational Medicine. See "Jones, Walton L., Jr.," biographical file, NASA Historical Reference Collection, NASA History Office, NASA Headquarters, Washington, DC.

K

W. W. Kellogg (1917–   ) was a meteorologist with the Rand Corporation between 1947 and 1959. He has held a senior position with the National Center for Atmospheric Research, Boulder, Colorado, since 1959. See *Who's Who in America, 2000*, 54th Ed., New Providence, NJ: Marquis Who's Who, 1999.

Charles F. Kennel (1939–   ) earned a B.A. from Harvard College in 1959 and a Ph.D. in astrophysical sciences from Princeton University in 1964. Before coming to NASA, Dr. Kennel was a tenured member of the University of California, Los Angeles, Department of Physics since 1967. He served as NASA Associate Administrator for Mission to Planet Earth from 1993–1996. In 1998, Dr. Kennel was appointed to the NASA Advisory Council (NAC). In 2001, he became Chair of the NAC. See "Kennel, Charles F.," biographical file, NASA Historical Reference Collection, NASA History Office, NASA Headquarters, Washington, DC.

George A. Keyworth II (1939–   ) was Director of the Office of Science and Technology Policy and Science Advisor to President Ronald Reagan from 1981 to 1986. Formerly the head of the Los Alamos Scientific Laboratory, Keyworth earned his Ph.D. in nuclear physics from Duke University in 1968. He began work at Los Alamos after graduation and remained there until 1981. See "Keyworth, George Albert II," *1986 Current Biography Yearbook*, pp. 265–68.

Mukul Ranjan Kundu (1930–   ) is a scholar of solar and stellar radio physics and supernova remnants. He received his B.S., with honors, from the University of Paris. He is a member of the editorial board of solar physics, received the 1978 U.S. Senior Scientist Award from the Humbolt Foundation, and was the 1989 American Physics Society Fellow. He also is a member of the American Astronomical Society, the American Geophysical Union, and the International Astronomical Union of Radio Science. A physics and astronomy educator, he is a professor at the University of Maryland, College Park. The author of *Solar Radio Astronomy, 1965*, he was named the National Academy of Sciences Fellow in 1967, 1974, 1975, and 1986. In 1976, he participated in the National Academy of Sciences "Solar Physics Study." See "Kundu, Mukul Ranjan," in *Who's Who in America, 2000*, 54th Ed., New Providence, NJ: Marquis Who's Who, 1999.

James Edward Kupperian, Jr., (1925–1984) earned his bachelor's degree in naval architecture, from the Webb Institute, his master's degree from the University of Delaware, and his doctorate in physics from the University of North Carolina. He was a Physicist with the Naval Research Laboratory before joining NASA in 1959. From 1959 to 1970, he was Chief of the Astrophysics Branch at Goddard Space Flight Center. At Goddard he conceived and planned the observatory series of spacecraft, including the Orbiting Astronomical Observatory (OAO). He was the Project Scientist for the OAO. See "Kupperian, James Edward, Jr.," in *Who's Who in America, 1980–1981*, 41st Ed., New Providence, NJ: Marquis Who's Who, 1980.

L

Louis John Lanzerotti (1938–   ), physicist, received his B.S. from the University of Illinois, 1960; and his M.A., 1963, and his Ph.D., 1965, from Harvard University. A postdoctoral fellow with Lucent Technologies Bell Labs, he also served as a member of the technical staff of AT&T Bell Labs. His professional experiences include: member, Polar Research Board, National Research Council (NRC), 1982 to 1991; chairman, NRC's Space Studies Board, 1988 to1994; member, Ocean Studies Board, 1995 to 1999; member, NASA Physical Science Committee, 1975 to 1979; chairman, Space and Earth Advisory Committee; member, Advisory Committee on the Future of the U.S. Space Program, 1990; and member, Vice President's Space Policy Advisory Board, 1992 to 1993. He coauthored *Diffusion in Radiation Belts* and co-edited two books on space physics. He is the recipient of the Antarctic Service Medal, 1979, and NASA's Distinguished Public Service Award, 1988 and 1994. He is a fellow of the American Institute of Aeronautics and Astronautics, the Institute of Electrical Electronics Engineers, the American Physics Society, the American Geophysical Union, and the American Association for the Advancement of Science. He is a member of the National Academy of Engineering, the International Academy of Astronautics, and the Woods Hole Oceanographic Institution. See "Lanzerotti, Louis John," in *Who's Who in America, 2000*, 54th Ed., New Providence, NJ: Marquis Who's Who, 1999.

Cecil Eldon Leith, Jr., (1923–   ) is a physicist by education and career. He earned his B.A., in 1943, and his Ph.D., in 1957, at the University of California, Berkeley. He started his career in 1946 as an Experimental Physicist at the Lawrence Radiation Laboratory, Berkeley, and transferred to the Lawrence Radiation Laboratory in Livermore, California, as a Theoretical Physicist. From 1968 through 1983, he worked as a Senior Scientist at the National Center for Atmospheric Research, Boulder, Colorado, and participated in the National Academy of Sciences "Solar Physics Study." A fellow of the American Physics Society, he received the Meisinger Award in 1967 and the Rossby Research Medal in 1982, while a fellow with the American Meteorological Society. See "Leith, Cecil Eldon, Jr.," in *Who's Who in America, 2000*, 54th Ed., New Providence, NJ: Marquis Who's Who, 1999.

William Liller (1927–   ) is a scholar of archaeoastronomy. He earned his B.A. from Harvard College in 1949, and his M.A. in 1950 and his Ph.D. (1953, astronomy) from the University of Michigan. In 1952, he was an assistant at the McMath-Hulbert Observatory, University of Michigan. From 1953 to 1960, he was an instructor, then an assistant professor of astronomy at the University of Michigan. He has been the Robert Wheeler Willson Professor of Applied Astronomy at Harvard University since 1970. Professional memberships include the American Astronomical Society, the Royal Astronomical Society of Canada, fellow with the American Association for the Advancement of Science, the American Academy of Arts and Sciences, the International Astronomical Union, and the British Astronomical Association. See "Liller, William," in *American Men and Women of Science, 1998–99*, 20th Ed., New Providence, NJ: R. R. Bower, 1998.

Robert Peichung Lin (1942–   ) is an academician in the fields of solar and space plasma physics and high-energy astrophysics. He earned his B.S. from the California Institute of Technology (1962) and his Ph.D. (physics) from the University of California at Berkeley (1967). He started his career in 1967 as an assistant resident physicist at the University of California, and then he served as a resident physicist, from 1979 to 1988. He participated in the National Academy of Sciences "Solar Physics Study" in 1976. From 1979 to 1988, he was resident physicist. He has been the Associate Director of the Space Science Laboratory at the University of

California, Berkeley, since 1992. See "Lin, Robert Peichung," in *American Men and Women of Science, 1998–99*, 20th Ed., New Providence, NJ: R. R. Bower, 1998.

Richard Emery Lingenfelter (1934–) received his B.A. from the University of California, Los Angeles, in 1956. He has been a Research Physicist at the University of California, San Diego, Center for Astrophysics and Space Sciences, since 1979. His concentrations are astrophysics and cosmic ray physics. He was a professor in residence at the University of California, Los Angeles, in the Department of Geophysics and Space Physics from 1969 to 1979. Concurrently, he worked in the Department of Astronomy from 1974 to 1979. See "Lingenfelter, Richard Emery," in *American Men and Women of Science, 1998–99*, 20th Ed., New Providence, NJ: R. R. Bower, 1998.

Jeffrey L. Linsky (1941–) earned his B.S., 1963, from the Massachusetts Institute of Technology, and his M.A., 1965, and Ph.D. (astronomy), 1968, from Harvard University. He is an authority on space and solar physics, and has taught at the University of Colorado in the Department of Astronomy, Planetary, and Atmospheric Science as an adjunct professor since 1979. In tandem, he was a member of the Joint Institute of Laboratory Astrophysics from 1968 to 1971. See "Linsky, Jeffrey L.," in *American Men and Women of Science, 1998–99*, 20th Ed., New Providence, NJ: R. R. Bower, 1998.

George M. Low (1926–1984), a native of Vienna, Austria, came to the United States in 1940 and received an aeronautical engineering degree from Rensselaer Polytechnic Institute in 1948 and a master of science in the same field in 1950. He joined the NACA in 1949 and specialized in experimental and theoretical research in several fields at Lewis Flight Propulsion Laboratory. He became Chief of Manned Spaceflight at NASA Headquarters in 1958. In 1960, Low chaired a special committee that formulated the original plans for the Apollo lunar landings. In 1964, he became Deputy Director of the Manned Spacecraft Center in Houston, the forerunner of the Johnson Space Center. He became Deputy Administrator of NASA in 1969 and served as Acting Administrator from 1970 to 1971. He retired from NASA in 1976 to become President of Rensselaer, a position he held until his death. In 1990, NASA renamed its quality and excellence award after him. See "Low, G. M.," Deputy Administrator files, NASA Historical Reference Collection, NASA History Office, NASA Headquarters, Washington, DC.

Alan M. Lovelace (1929–) was born in St. Petersburg, Florida, and was educated at the University of Florida in Gainesville. He received a B.S. degree in chemistry in 1951, an M.S. degree in organic chemistry in 1952, and a Ph.D. degree in organic chemistry in 1954. Shortly after the Korean conflict, he served in the United States Air Force from 1954 to 1956. He began his government career as a scientist at the Air Force Materials Laboratory (AFML), Wright-Patterson Air Force Base, Dayton, Ohio. In January 1964, he was named Chief Scientist of the Air Force Materials Laboratory and was named Director in 1967. In October 1972, he became the Director of Science and Technology for the Air Force Systems Command at Headquarters, Andrews Air Force Base, Maryland. In September 1973, he became the Principal Deputy to the Assistant Secretary of the Air Force for research and development. One year later, he left the Department of Defense to become the Associate Administrator of the NASA Office of Aeronautics and Space Technology. He became Deputy Administrator after the departure of George Low in June 1976. He retired from NASA in July 1981 and accepted a position as Corporate Vice President of Science and Engineering with the General Dynamics Corporation at St. Louis, Missouri. See "Lovelace, Alan M.," Deputy Administrator files, NASA Historical Reference Collection, NASA History Office, NASA Headquarters, Washington, DC.

M

Robert Moffat MacQueen (1938–) received his B.S. from Rhodes College in 1960 and his Ph.D. from Johns Hopkins University in 1968. He was a Senior Research Scientist at the National Center for Atmospheric Research in Boulder, Colorado, from 1967 to 1990. In the early to mid-1970s he was the principal investigator on NASA's Apollo and Skylab programs, and from 1976 to 1979 he was the principal investigator for NASA's Solar Maximum Mission. From 1973 to 1976, he was a member of the Committee on Space Astronomy of the National Academy of Sciences; from 1977 to 1979 he was a member of the Committee on Space Physics; and from 1983 to 1986 he was a member of the Space Science Board. See "MacQueen, Robert Moffat," in *Who's Who in America, 2000*, 54th Ed., New Providence, NJ: Marquis Who's Who, 1999.

Hans Mark (1929–) became NASA Deputy Administrator in July 1981. He had previously served as Secretary of the Air Force from July 1979 until February 1981 and as Undersecretary of the Air Force from 1977 to 1979. In February 1969, Mark became Director of NASA's Ames Research Center, Mountain View, California, where he managed research and applications efforts in aeronautics, space science, life science, and space technology. He received a Ph.D. in physics from the Massachusetts Institute of Technology in 1954. Born in Mannheim, Germany, he came to the United States in 1940 and became a citizen in 1945. Upon leaving NASA, he became Chancellor of the University of Texas-Austin. See "Mark, Hans," Deputy Administrator files, NASA Historical Reference Collection, NASA History Office, NASA Headquarters, Washington, DC.

Franklin D. Martin is a native of China Grove, North Carolina. He received a B.A. in physics and math from Pfieffer College, Misenheimer, North Carolina, in 1966 and his Ph.D. in physics from the University of Tennessee in 1971. He began his career with NASA in 1974. Before joining NASA, he served as a Physicist with the Naval Oceanographic Office and as an Aerospace Engineer for Lockheed. He was Assistant Administrator for the Office of Exploration from 1988 to 1989. Prior to that appointment, in 1986 he was named Deputy Associate Administrator for the Space Station Office. From 1983 through 1986, Dr. Martin was at NASA Goddard Space Flight Center as the Director of the Office of Space and Earth Sciences. See "Martin, Franklin D.," biographical file, NASA Historical Reference Collection, NASA History Office, NASA Headquarters, Washington, DC.

Daniel G. Mazur (1916–1984) earned his degree in electrical engineering from Worchester Polytechnic Institute in Worchester, Massachusetts. He began his federal career with the Philadelphia Navy Yard and transferred to Washington, DC, in 1946. Prior to joining NASA, as one of the first employees in 1958, he was at the Naval Research Laboratory. In 1964, he received NASA's Medal for Exceptional Scientific Achievement for his work on communications satellites. He retired in 1973 as Associate Deputy Director for Engineering of the Goddard Space Flight Center. See "Mazur, Daniel G.," biographical file, NASA Historical Reference Collection, NASA History Office, NASA Headquarters, Washington, DC.

Frank Bethune McDonald (1925–) received his B.S. degree from Duke University (1948), and his M.S. (1951) and Ph.D. (1955) degrees from the University of Minnesota. Professional memberships include the National Academy of Sciences, the American Geophysical Union, and the American Physics Society. He was the Head of the Fields and Particles Branch at NASA Goddard Space Flight Center from 1959 through 1970. From 1964 to 1989 he was, concurrently, Project Scientist for explorer satellites and high-energy astronomy observatories, and served in 1976 as a participant in the National Academy of Sciences "Solar Physics Study," and as the Chief of the Laboratory of High Energy Astrophysics. See "McDonald, Frank Bethune," in *American Men and Women of Science, 1998–99*, 20th Ed., New Providence, NJ: R. R. Bower, 1998.

John H. McElroy earned a B.S.E.E. from the University of Texas at Austin, an M.E.E. from the Catholic University of America, and a Ph.D. from the Catholic University of America. McElroy began his NASA career in 1966 at Goddard Space Flight Center, where he directed research on laser communication systems, tracking and radiometry, and advanced satellite communications technology. Dr. McElroy also served as Deputy Director of Goddard Space Flight Center from 1980 to 1982. See "McElroy, Dr. John H.," biographical file, NASA Historical Reference Collection, NASA History Office, NASA Headquarters, Washington, DC.

David Morrison came to NASA on an Intergovernmental Personnel Act (IPA) appointment from the University of Hawaii and was named to the position of Acting Deputy Associate Administrator for Space Science in 1981. A noted astronomer with a wealth of experience in space science programs, he was appointed Chief of the Space Division at NASA Ames Research Center in 1994. See "Morrison, David," biographical file, NASA Historical Reference Collection, NASA History Office, NASA Headquarters, Washington, DC.

Robert H. Moser (1923–) earned a B.S. from Loyola University in 1944 and an M.D. from Georgetown University in 1948. He served as Flight Controller for Project Mercury from 1959 to 1962. He was also a consulting member of the Project Gemini medical evaluation team from 1962 to 1966. He consulted for the Apollo program from 1967 to 1973. From 1978 to 1982, he served as chairman of the life sciences advisory committee. He was a member of the NASA Advisory Council from 1983 to 1988. He was a member of the advisory committee for the NASA Space Station program from 1988 to 1993. See *Who's Who in America, 2000*, 54th Ed., New Providence, NJ: Marquis Who's Who, 1999.

George E. Mueller (1918–   ) was Associate Administrator for the Office of Manned Space Flight at NASA Headquarters, 1963 to 1969, where he was responsible for the completion of project Apollo and for the initial development of the Space Shuttle. He moved to the General Dynamics Corporation as Senior Vice President in 1969 and remained until 1971, became President of Systems Development Corporation (1971 to 1980), and later Chairman and CEO (1981 to 1983). See "Mueller, George E.," biographical file, NASA Historical Reference Collection, NASA History Office, NASA Headquarters, Washington, DC.

Thomas A. Mutch (1931–1980) earned a Ph.D. in geology from Princeton University in 1960. Previously a professor of geological sciences at Brown University, he led the Viking spacecraft's imaging "science team" from 1969 to 1977. He became the NASA Associate Administrator for Space Science in 1979 and died in a 1980 climbing accident in the Himalayas. See "Mutch, Thomas A.," biographical file, NASA Historical Reference Collection, NASA History Office, NASA Headquarters, Washington, DC.

Dale D. Myers (1922–   ) served as NASA Deputy Administrator from 1986 until 1989. He had been Undersecretary of the U.S. Department of Energy from 1977 to 1979. From 1974 to 1977, he was Vice President, Rockwell International, and President, North American Aircraft Group, El Segundo, California. He was also the Associate Administrator for Manned Space Flight at NASA from 1970 to 1974. From 1969 to 1970, Mr. Myers served as Vice President/Program Manager, Space Shuttle Program, Rockwell International. He was Vice President and Program Manager, Apollo Command/Service Module Program, North American-Rockwell from 1964 to 1969. After leaving NASA in 1989, Myers returned to private industry. See "Myers, Dale D.," NASA Deputy Administrator Files, NASA Historical Reference Collection, NASA History Office, NASA Headquarters, Washington, DC.

N

John E. Naugle (1923–   ) was trained as a physicist at the University of Minnesota. He began his career studying cosmic rays by launching balloons to high altitudes. In 1959, he joined NASA Goddard Space Flight Center, Greenbelt, Maryland, where he developed projects to study the magnetosphere. In 1960, he was named Manager of NASA's Fields and Particles research program. He later served as NASA's Associate Administrator for the Office of Space Science and as the Agency's Chief Scientist before his retirement in 1981. See "Naugle, John E.," *First Among Equals: The Selection of NASA Space Science Experiments*, Washington, DC: NASA SP-4215, 1991.

Homer E. Newell (1915–1983) earned his Ph.D. in mathematics at the University of Wisconsin in 1940, and then he served as a theoretical physicist and mathematician at the Naval Research Laboratory from 1944 to 1958. Concurrently, he was Science Program Coordinator for Project Vanguard and Acting Supervisor of the Atmosphere and Astrophysics Division. In 1958, he transferred to NASA to assume responsibility for planning and development of the new space science program. He soon became Deputy Director of Space Flight programs. In 1961, he became Director of the Office of Space Science, and in 1963 he was named Associate Administrator for the Office of Space Science and Applications. Over the course of his career, he became an internationally known authority in the field of atmospheric and space sciences. He is the author of numerous scientific articles and seven books, including *Beyond the Atmosphere: Early Years of Space Science* (Washington, DC: NASA SP-4211, 1980). He retired from NASA in 1973. See "Newell, Homer E.," biographical file, NASA Historical Reference Collection, NASA History Office, NASA Headquarters, Washington, DC.

Gordon Allen Newkirk, Jr., (1928–1985) earned his Ph.D. at the University of Michigan, with a concentration in astrophysics. He began his career at Upper Air Research Observatory in 1953. In 1955, he joined the senior research staff of the High Altitude Observatory. He remained there for three decades and, concurrently, from 1961 to 1965, was an adjunct professor of the Department of Astrogeophysics at the University of Colorado. From 1965 to 1976—by this time he had become Director of the High Altitude Observatory—he taught physics and astrophysics. See "Newkirk, Gordon Allen, Jr.," in *Who's Who in America, 1955*, 9th Ed., New Providence, NJ: Marquis Who's Who, 1954.

M. H. Nichols was on the Rocket and Satellite Research Panel in 1957 and worked at Palmer Physical Laboratory at Princeton University. See "Nichols, M. H.," at *http://www.hq.nasa.gov/office/pao/History/SP-4211/appen-b.htm*

Arnauld E. T. Nicogossian (1936–) earned an M.S. degree in aerospace medicine from Ohio State University, an M.D. from Teheran University, and trained in internal medicine at Mt. Sinai Hospital in New York. Dr. Nicogossian started at NASA in 1972 at the Johnson Space Center, where he conducted biomedical research and was the Crew Surgeon for the Apollo Soyuz Test Project. He came to Headquarters in 1976 and established an operational medicine and space medicine research program for the Space Shuttle and Space Station era programs. He also served as Chief Health and Medical Officer. In 1993, Dr. Nicogossian became Deputy Associate Administrator for Space Flight Activities, Life and Microgravity Sciences and Applications, and later became Associate Administrator for Life and Microgravity Sciences. Dr. Nicogossian also served as a Senior Advisor to NASA Administrator O'Keefe before retiring in January 2003. See "Nicogossian, Arnauld E. T.," biographical file, NASA Historical Reference Collection, NASA History Office, NASA Headquarters, Washington, DC.

Robert Wilson Noyes (1934–) received his B.A. from Haverford College (1957) and his Ph.D. from the California Institute of Technology in 1963. A prominent figure in the science community, his concentrations are solar and stellar physics. He began his current career as a physicist for the Smithsonian Astrophysical Observatory in 1962. Concurrently he was a lecturer of astronomy at Harvard University from 1962 until 1973 and Associate Director for the Center for Astrophysics from 1973 until 1980. He participated in the National Academy of Sciences "Solar Physics Study." His professional memberships include: the American Astronomical Society and the International Astronomical Union. See "Noyes, Robert Wilson" in *American Men and Women of Science, 1998–99*, 20th Ed., New Providence, NJ: R. R. Bower, 1998.

O

Hugh Odishaw (1916–1984) became Assistant to the Director of the National Bureau of Standards from 1946 to 1954, served as Executive Director of the U.S. National Committee for the International Geophysical Year from 1954 to 1965, and became the Executive Secretary of the Division of Physical Sciences in the National Academy of Sciences, 1966 to 1972. See "Odishaw, Hugh," biographical file, NASA Historical Reference Collection, NASA History Office, NASA Headquarters, Washington, DC.

Goetz K. H. Oertel (1934–) was educated in Germany but received his Ph.D. from the University of Maryland in 1963. He started his career with NASA at Langley Research Center in 1963 as an Aerospace Engineer. From 1968 to 1975, he was NASA's Chief of Solar Physics. From 1975 to 1983, he was the Director of the Defense and Civilian Nuclear Waste Programs at the U.S. Department of Energy. Professional memberships include the American Physics Society, the American Astronomical Society, the International Astronomical Union, the Cosmos Club, and Sigma Xi. See "Oertel, Goetz K. H.," in *Who's Who in America, 2000*, 54th Ed., New Providence, NJ: Marquis Who's Who, 1999.

Bernard M. Oliver (1916–1995) earned a B.A. in electrical engineering from Stanford University in 1935, an M.S. in electrical engineering from the California Institute of Technology in 1936, and a Ph.D. in electrical engineering from the California Institute of Technology in 1940. Dr. Oliver codirected a NASA Ames Research Center Cyclops summer study, which examined using radio telescopes to search for extraterrestrial life. Dr. Oliver served as a Senior Manager for the NASA SETI effort, until congressional cancellation in 1993. Dr. Oliver was a key player in finding private funding to continue the effort. He remained a Senior Scientist for the privatized project, called Phoenix, until his death in 1995. See "Oliver, Bernard M.," biographical file, NASA Historical Reference Collection, NASA History Office, NASA Headquarters, Washington, DC.

Frank Quimby Orrall (1925–) earned his B.S. from the University of Massachusetts in 1950, and his M.A. in 1953 and Ph.D. (astronomy) in 1955 from Harvard University. His professional career began with a position at the College Observatory of Harvard University in 1955 that ended in 1956. From 1956 to 1964, he was a Solar Physicist at the Sacramento Peak Observatory. He participated in the 1976 "Solar Physics Study" of the National Academy of Sciences. He also served as a professor of physics at the University of Hawaii. See "Orrall, Frank Quimby," in *American Men and Women of Science, 1998–99*, 20th Ed., New Providence, NJ: R. R. Bower, 1998.

P

Thomas O. Paine (1921–1992) graduated from Brown University in 1942 with a B.A. degree in engineering. He received his M.S. (1947) and Ph.D. (1949, physical metallurgy) degrees from Stanford University, and received honorary degrees from Brown University, Clarkson College of Technology, Nebraska Wesleyan University, the University of New Brunswick (Canada), Oklahoma City University, and Worcester Polytechnic Institute. He was appointed Deputy Administrator of NASA on 31 January 1968. When James E. Webb retired on 8 October 1968, Paine was named Acting Administrator of NASA. His nomination as NASA's third Administrator, on 5 March 1969, was confirmed by the Senate on 20 March 1969. During his administration, the first seven Apollo manned missions were flown, and a total of twenty astronauts orbited the Earth; fourteen astronauts traveled to the Moon; and four walked upon the surface of the Moon. He resigned from NASA on 15 September 1970, returned to the General Electric Company in New York City as Vice President and Group Executive, Power Generation Group, where he remained until 1976. In 1985, the White House chose him as Chair of a National Commission on Space to prepare a report on the future of space exploration. He was a tireless spokesman, for fifteen years after leaving NASA, for an expansive view of what should be done in space. The Paine Commission took most of a year to prepare its report, largely because it solicited public input in hearings throughout the United States. The Commission report, Pioneering the Space Frontier, was published in May 1986. It espoused "a pioneering mission for 21st-century America . . . to lead the exploration and development of the space frontier, advancing science, technology, and enterprise, and building institutions and systems that make accessible vast new resources and support human settlements beyond Earth orbit, from the highlands of the Moon to the plains of Mars." The report also contained a "Declaration for Space" that included a rationale for exploring and settling the solar system, and outlined a long-range space program for the United States. He died at his home in Los Angeles, California, on 4 May 1992. See "Paine, Thomas O.," Administrator file, NASA Historical Reference Collection, NASA History Office, NASA Headquarters, Washington, DC, and Roger D. Launius, "NASA and the Decision to Build the Space Shuttle, 1969–1972," *The Historian 57* (Autumn 1994): 17–34.

Eugene Newman Parker (1927–) earned his B.S. degree from the University of Michigan in 1948, his Ph.D. degree from the California Institute of Technology in 1951, and his D.S. degree from Michigan State University in 1975. He developed the theory of the origin of the dipole magnetic field of Earth and the theoretical basis for x-ray emissions from the Sun and the stars. He served concurrently at the University of Chicago as professor of physics, from 1962 to 1995, and professor of astronomy and astrophysics, from 1967 to 1995, and was also the Chairman of the "Solar Physics Study" of the National Academy of Sciences. In 1995, he achieved status as professor emeritus. Professional memberships include: the National Academy of Sciences (H. K. Arctowski Award, 1969); the U.S. Medal of Science Award, 1989; the American Astronautical Society (Henry Norris Russess Lecturer, 1969); the George Ellery Hale Award, 1978; the American Geophysical Union (John Adam Fleming Award, 1968); and the William Bowie Medal, 1990. See "Parker, Eugene Newman," in *Who's Who in America, 2000,* 54th Ed., New Providence, NJ: Marquis Who's Who, 1999.

William Charles Parkinson (1918–) earned his B.S.E. degree from the University of Michigan in 1940, his M.S. in 1941, and his Ph.D. in 1948. He has been in the Physics Department at the University of Michigan since 1947, and he was named professor emeritus of physics in 1988. He served as a member of the subcommittee on nuclear structure of the Nuclear Regulatory Commission from 1959 to 1968, and he supported the advisory panel of the National Science Foundation on physics from 1966 to 1969. From 1969 to 1970, he was a member on the nuclear physics subpanel on management and costs of nuclear programs. See "Parkinson, William Charles," in *Who's Who in America, 2000,* 54th Ed., New Providence, NJ: Marquis Who's Who, 1999.

Dallas Lynn Peck (1929–) earned a B.S. from California Institute of Technology in 1951, an M.S. from California Institute of Technology in 1953, and a Ph.D. from Harvard University in 1960. A geologist by profession, Dr. Peck worked for the United States Geological Survey from 1954 to 1995. He served as Chief Geologist in the Office of Geochemistry and Geophysics from 1977 to 1981, and as Director of the office from 1981 to 1993. He also served as a member of the Lunar Sample Review Board from 1970 to 1971. See *Who's Who in America, 2000,* 54th Ed., New Providence, NJ: Marquis Who's Who, 1999.

Kenneth S. Pedersen (1939–) served in numerous government agencies—Office of Equal Opportunity, Department of Commerce, Atomic Energy Commission, and Nuclear Regulatory Commission—prior to coming to NASA in 1982 as Director of International Affairs. In 1988, Pedersen was appointed as NASA Associate Administrator for External Relations, serving until 1990, when he left NASA to accept an academic appointment at Georgetown University. See "Pedersen, Kenneth S.," biographical file, NASA Historical Reference Collection, NASA History Office, NASA Headquarters, Washington, DC.

Charles J. Pellerin, Jr., has a B.S. in physics from Drexel University and a Ph.D. in physics from Catholic University of America. He is a longtime NASA official who began his career at the Goddard Space Flight Center as he was completing his Ph.D. in physics in 1974. The next year he moved to NASA Headquarters where he managed the development and integration of scientific instrumentation for flight on the Space Shuttle. In 1983, he was named Director of Astrophysics in NASA's Office of Space Science and Applications, and in 1992 was appointed Deputy Associate Administrator for Safety and Mission Quality, and served in that position until 1994. See "Pellerin, Charles J., Jr.," biographical file, NASA Historical Reference Collection, NASA History Office, NASA Headquarters, Washington, DC.

Laurence E. Peterson (1931–   ) is a professor of physics at the University of California, where he started his career in 1962 as a resident physicist. He received his B.S. in 1954 and his Ph.D. in 1960 from the University of Minnesota. Dr. Peterson has served on the NASA Space Science Steering Committee since 1964; has been the Director of the Center for Astrophysics and Space Science, University of California, since 1988; was a fellow with the National Science Foundation, in 1958 and 1959; was a member of the Guggenheim Foundation from 1973 to 1974; and is a fellow of the American Physics Society. He is a member of the American Institute of Aeronautics and Astronautics, and the International Astronomical Union. He was a member of the panel of participants in the National Academy of Sciences "Solar Physics Study." See "Peterson, Laurence E.," in *Who's Who in America, 2000*, 54th Ed., New Providence, NJ: Marquis Who's Who, 1999.

Leonard Richard Piasechi worked for the Reynolds Metals Company as a Research Engineer from 1953 to 1956, an Engineer Group Supervisor from 1956 to 1960, Chief of the Solid Propellant Rockets Section from 1960 to 1963, Deputy Division Chief from 1963 to 1964, and Chief of the Propulsion Section and Voyager Propulsion from 1964 to 1965. See "Piasechi, Leonard Richard," in *American Men and Women of Science, 1997–98*, 19th Ed., New Providence, NJ: R. R. Bower, 1997.

Richard W. Porter attended the University of Kansas and received his B.S. in 1934. He was awarded a Ph.D. by Yale University in 1937. As an electrical engineer, he worked on missile programs with the General Electric Company before working Earth sciences programs at the National Academy of Sciences. In 1964, he was the Academy's delegate to the Committee on Space Research (COSPAR). See assorted government officials biographical file, NASA Historical Reference Collection, NASA History Office, NASA Headquarters, Washington, DC.

Frank Press (1924–   ) served as President Carter's Science Advisor and Director of the Office of Science and Technology Policy from 1977 to 1981. Upon leaving this post, he was elected nineteenth President of the National Academy of Sciences. Press earned his Ph.D. in geophysics from Columbia University and has earned twenty-eight additional honorary doctorates. See "Press, Frank," biographical file, NASA Historical Reference Collection, NASA History Office, NASA Headquarters, Washington, DC.

J. DeWitt Purcell (1912–1986) was a Physicist with the Naval Research Laboratory for nearly three decades before retiring in 1975. As an authority on optics, spectroscopy, and solar astronomy, he authored hundreds of technical papers. Additionally, he did optics work for the Skylab space project. See "Purcell, J. D.," in the "Obituaries" of *The Washington Post*, section C, page 08, in the 9 July 1986 issue.

R

Ronald Reagan (1911–2004) was elected President of the United States in 1980, assumed office in January 1981, and served until 1989. He was President when the maiden flight of the Space Shuttle was launched in 1981. In 1984, he mandated the construction of an orbital space station. Reagan declared "America has always been greatest when we dared to be great. We can reach for greatness again. We can follow our dreams to distant stars, living and working in space for peaceful, economic, and scientific gain. Tonight I am directing NASA to develop a permanently manned space station and to do it within a decade." See "Reagan, Ronald," biographical file, NASA Historical Reference Collection, NASA History Office, NASA Headquarters, Washington, DC, and Sylvia D. Fries, "2001 to 1994: Political Environment and the Design of NASA's Space Station System," *Technology and Culture*, Vol. 29, No.3 (July 1988): 568–93.

Edmond Morden Reeves (1934–) earned his B.S. in 1956, his M.S. in 1957, and his Ph.D. in 1959 from Western Ontario University, Canada. He has been the Director of NASA's Flight Systems Division, Office of Life and Microgravity Science and Applications, since 1993. From 1961 to 1978, he was a research physicist and lecturer of astronomy at Harvard University, and from 1978 to 1982 he was the Senior Research Associate of the Harvard College Observatory. He also served, from 1968 to 1971, on NASA's Solar Physics Panel and the Astronomy Missions Board. See "Reeves, Edmond Morden," in *American Men and Women of Science, 1998–99*, 20th Ed., New Providence, NJ: R. R. Bower, 1998.

William A. Rense (1914–) earned his B.S. (1935) from Case Western University, and his M.S. (1937) and his Ph.D. (1939, physics) from Ohio State University. He is a scholar of space physics. He began his career at the University of Colorado, Boulder, in 1949, as an associate professor of physics, and became professor emeritus in 1980. His professional experiences include: being a physics instructor at Louisiana State University, 1939 to 1940; an assistant professor at the University of Miami, 1940; visiting assistant professor at Rutgers University, 1941 to 1942; and assistant professor/associate professor at Louisiana State University, 1943 to 1949. From 1956 to 1978, he was Codirector of the Laboratory of Atmospheric and Space Physics. He is a member of the American Geophysical Union, the American Physics Society, Sigma Xi, and the American Astronomical Society. See "Rense, William A.," in *American Men and Women of Science, 1998–99*, 20th Ed., New Providence, NJ: R. R. Bower, 1998.

Sally K. Ride (1951–) was the first American woman to fly in space. She was chosen as an astronaut in 1978 and served as a Mission Specialist for STS-7 (1983) and STS 41-G (1984). Ride was also a member of the Presidential Commission on the Space Shuttle Challenger Accident in 1986, and she chaired a NASA taskforce that prepared a report from 1986 to 1987 on the future of the civilian space program entitled *Leadership and America's Future in Space* (Washington, DC: U.S. Government Printing Office, 1987). She resigned from NASA in 1987 to join the Center for International Security and Arms Control at Stanford University. She left Stanford in 1989 to assume the directorship of the California Space Institute, part of the University of California at San Diego. Ride also served as a member of the Columbia Accident Investigation Board. See "Ride, Sally K.," biographical folder, NASA Historical Reference Collection, NASA History Office, NASA Headquarters, Washington, DC.

Walter Orr Roberts (1915–1990) was an astronomer at the University of Colorado's High Altitude Observatory. He was also instrumental in the creation of the National Center for Atmospheric Research in 1960 and directed the program on food, climate, and the world's future for the Aspen Institute for Humanistic Studies, 1974 to 1981. He was heavily involved in the debate over "nuclear winter" and the possibility of the "Greenhouse Effect" on Earth in the 1980s. See "Roberts, Walter Orr," *1990 Current Biography Yearbook*, p. 660.

Nancy G. Roman (1925–) received her B.A. from Swarthmore College in 1946 and her Ph.D. in astronomy from the University of Chicago in 1949. In 1955, she took a position as an astronomer with the Radio Astronomy Branch of the Naval Research Laboratory. She was the head of the Microwave Spectroscopy Section when, in 1959, she left NRL to become the head of the Observational Astronomy Program at NASA. She remained at NASA, serving as Chief of the Astronomy and Relativity Programs in the Office of Space Science until her retirement in 1979. See "Roman, Nancy G.," biographical file, NASA Historical Reference Collection, NASA History Office, NASA Headquarters, Washington, DC.

Milton W. Rosen (1915–), an electrical engineer by training, joined the staff of the Naval Research Laboratory in 1940, where he worked on guidance systems for missiles during World War II. From 1947 to 1955, he was in charge of Viking rocket development. He was Technical Director of project Vanguard, the scientific Earth satellite program, until he joined NASA in October 1958 as Director of Launch Vehicles and Propulsion in the Office of Manned Space Flight. In 1963, he became Senior Scientist for NASA's Deputy Associate Administrator for Space Science (engineering). In 1974, he retired from NASA to become Executive Secretary of the Space Science Board at the National Academy of Sciences. See "Rosen, Milton W.," biographical file, NASA Historical

Reference Collection, NASA History Office, NASA Headquarters, Washington, DC; and Milton W. Rosen, *The Viking Rocket Story*, New York: Harper, 1955.

David Maurice Rust (1939–) was a research fellow of solar physics at the National Center of Atmospheric Research while finishing his Ph.D. studies in 1966. From 1966 to 1968, he was a Carnegie Fellow of Astrophysics at the Mt. Wilson and Palomar Observatory. From 1968 to 1974, he was an Astrophysicist with the Sacramento Peak Observatory, Air Force Cambridge Research Laboratories. He later became the Senior Staff Scientist at American Science and Engineering, Incorporated, and remained in that position until 1983. While at American, he was invited to participate in the "Solar Physics Study" sponsored by the National Academy of Sciences. He has been in his current position as Physicist with the Applied Physics Laboratory, Johns Hopkins University, since 1983. His professional memberships include the American Geophysical Union, the American Astronomical Society, and the International Astronomical Union. See "Rust, David Maurice," in *American Men and Women of Science, 1998–99*, 20th Ed., New Providence, NJ: R. R. Bower, 1998.

S

William R. Schindler (1928–) was head of the Delta Rocket Program at Goddard Space Flight Center, where Echo, Telstar, Relay, and Syncom communication satellites were placed into orbit, as well as the Tiros weather satellites, several Explorer scientific satellites, an orbiting solar observatory, and the Ariel satellite. He received his first training in rocketry as a member of the Vanguard project team in late 1957. He joined NASA at its beginning in 1958 and became the Delta project Technical Director. See "Schindler, William R.," biographical file, NASA Historical Reference Collection, NASA History Office, NASA Headquarters, Washington, DC.

Robert C. Seamans, Jr., (1918–) was born on 30 October 1918 in Salem, Massachusetts. He attended Lenox School, and he earned a B.S. degree in engineering at Harvard College in 1939; an M.S. degree in aeronautics at Massachusetts Institute of Technology (MIT) in 1942; and a D.S. degree in instrumentation from MIT in 1951. Dr. Seamans also received the following honorary degrees: Doctor of Science from Rollins College (1962) and from New York University (1967); and Doctor of Engineering from Norwich Academy (1971), from Notre Dame (1974), and from Rensselaer Polytechnic Institute (RPI) in 1974. In 1960, Dr. Seamans joined NASA as Associate Administrator. In 1965, he became Deputy Administrator, retaining many of the general management-type responsibilities of his previous title and also serving as Acting Administrator. During his years at NASA, he worked closely with the Department of Defense in research and engineering programs and served as cochairman of the Astronautics Coordinating Board. Through these associations, NASA was kept aware of military developments and technical needs of the Department of Defense, and Dr. Seamans was able to advise that agency of NASA activities which had application to national security. For further information on Robert C. Seamans, Jr., see his autobiography, *Aiming at Targets*, NASA SP-4106, 1996.

Albert F. Siepert earned a B.A. from Bradley University in 1936. Siepert began his NASA career as Director of Administration at NASA Headquarters in 1958. He later served as Deputy Director of the NASA launch operations center at Kennedy Space Center. See "Siepert, Albert F.," speech file, NASA Historical Reference Collection, NASA History Office, NASA Headquarters, Washington, DC.

Henry J. Smith (1928–) received his Ph.D. degree from Harvard University in 1955. He was named Deputy Director of the Physics and Astronomy Program, Office of Space Science and Applications (OSSA), in 1963. An astronomer by education, he was the Director of Harvard Observatory's Boyden Station, South Africa, from 1952 to 1954, and Director of the Harvard Observatory solar project in New Mexico, from 1955 to 1962. Prior to NASA, he was Chief of the Sun-Earth Relations Section of the National Bureau of Standards Central Radio Propagation Division in Boulder, Colorado. See "Smith, Henry J.," biographical file, NASA Historical Reference Collection, NASA History Office, NASA Headquarters, Washington, DC.

Kurt R. Stehling (1919–1997) was head of the Launch Vehicle Division of Project Vanguard at the Naval Research Laboratory and transferred with the program to NASA. In the early years of NASA, he was Senior Scientist on the staff of the NASA Administrator. In the mid-1960s, he was Vice President of Electro-Optical Systems Corporation, and in 1971 he became a Senior Aerospace and Technology Advisor on Undersea Technology at the National Oceanic and Atmospheric Administration. See "Stehling, Kurt R.," biographical file,

NASA Historical Reference Collection, NASA History Office, NASA Headquarters, Washington, DC.

Homer J. Stewart (1915–) earned his doctorate in aeronautics from Caltech in 1940, joining the faculty there two years before that. In 1939, he participated in pioneering rocket research with other Caltech engineers and scientists, including Frank Malina, in the foothills of Pasadena. Their efforts resulted in the creation of the Jet Propulsion Laboratory (JPL), and Stewart maintained his interest in rocketry there. He was involved in developing the first American satellite, Explorer I, in 1958. In that year, on leave from Caltech, he became Director of NASA's Program Planning and Evaluation Office, and returned to Caltech in 1960 to a variety of positions, including Chief of the Advanced Studies Office at JPL, from 1963 to 1967, and professor of aeronautics. See "Stewart, Homer J.," biographical file, NASA Historical Reference Collection, NASA History Office, NASA Headquarters, Washington, DC; and Clayton R. Koppes, JPL, *American Space Program: A History of the Jet Propulsion Laboratory*, New Haven, CT: Yale University Press, 1982, pp. 23, 32, 44, 47, 79–80, 82.

Andrew J. Stofan (1935–) began his career with NASA in 1958 as a Research Engineer at the Lewis Research Center. Throughout his 30 years at NASA, he held numerous managerial and administrative positions. He was the Associate Administrator for the Space Station Office from 30 June 1986 until his retirement on 1 April 1988. See "Stofan, Andrew J.," biographical file, NASA Historical Reference Collection, NASA History Office, NASA Headquarters, Washington, DC.

Morton J. Stoller (1917–1963) was a leading figure in the Nation's weather and communications satellite program. He joined the National Advisory Committee for Aeronautics in 1939 as an Electrical Engineer at the Langley Aeronautical Laboratory, and in 1958 he became NASA's Chief of Space Science in the Office of the Assistant Director for Space Science. In early 1960, he was named Assistant Director for the Satellite and Sounding Rocket program, in the Office of Space Flight, and in 1962 he was named Director of the Office of Applications. See "Stoller, Morton J.," biographical file, NASA Historical Reference Collection, NASA History Office, NASA Headquarters, Washington, DC.

William G. Stroud (1923–) earned a B.A. degree from Pennsylvania State University and graduate degrees in physics from the University of Chicago and Princeton University. In 1959, before joining NASA, he was Chief of the Astro-Instrumentation Branch of the Astro-Electronics Division at the U.S. Army Signal Research and Development Laboratory, Fort Monmouth, New Jersey. In 1960, he was Chief of Meteorology in the Satellite Applications Systems Division of the Goddard Space Flight Center and Project Manager for Tiros I, the meteorological satellite launched 1 April 1960. He became Chief of the Aeronomy and Meteorology Division at Goddard, and then Special Assistant to the Director of Flight Projects there. See "Stroud, William G.," biographical file, NASA Historical Reference Collection, NASA History Office, NASA Headquarters, Washington, DC.

Hubertus Strughold (1898–1986) served as Director of the Physiological Institute of the University of Heidelberg from 1946 through 1947. Later in 1947, he joined the staff of the U.S. Air Force School of Aviation Medicine, Randolph Field, Texas. In 1949, he was named Chief of the newly founded Department of Space Medicine. In 1951, he received the title of Professor of Aviation Medicine from the United States Air Force's Air University, who named him the First Professor of Space Medicine. In 1958, he was appointed Advisor for Research, USAF Aerospace Medical Center, Brooks Air Force Base, Texas, and was named Chairman of the Advanced Studies Group of the Center in 1960. In early 1962, he was named Chief Scientist of the Aerospace Medical Division (AFSC). See "Strughold, Hubertus," biographical file, NASA Historical Reference Collection, NASA History Office, NASA Headquarters, Washington, DC.

Ernst Stuhlinger (1913–) was a physicist who earned his Ph.D. at the University of Tbingen in 1936 and continued research into cosmic rays and nuclear physics until 1941, while serving as an assistant professor at the Berlin Institute of Technology. He spent two years as an enlisted man in the German army on the Russian Front before being assigned to the rocket development center at Peenemünde, Germany. There he worked principally on guidance and control of rockets. After World War II, he came to the United States as part of Project Paperclip and worked with Wernher von Braun at Fort Bliss, Texas, and then at the Redstone Arsenal in Huntsville, Alabama. He transferred to Marshall Space Flight Center in 1960, was Director of the Space Science Lab from until 1968, and was Associate Director for Science from 1968 to 1975. Then he retired and became an adjunct professor and senior research scientist with the University of Alabama, at Huntsville. He directed early planning for lunar exploration and the Apollo telescope mount, which flew on Skylab, and produced a wealth of scientific information about the Sun. He also was responsible for the early planning on the High-Energy Astronomy Observatory and contributed to the initial phases of the space telescope project. His work included studies of electric propulsion and scientific payloads for the Space Shuttle. See "Stuhlinger, Ernst," biographical file, NASA Historical Reference Collection, NASA History Office, NASA Headquarters, Washington, DC.

Peter Andrew Sturrock (1924–) earned his B.A. in 1945, his M.A. in 1948, and his Ph.D. (math) in 1951 from Cambridge University. He began his professional career as the Harwell Senior Fellow, Atomic Energy Research Establishment, England, in 1951. Other positions held include: fellow, St. John's College, 1952 to 1955; research associate for microwaves, Stanford University, 1955 to 1958; and the Ford Fellow for Plasma Physics, European Organization for Nuclear Research, Switzerland. See "Sturrock, Peter Andrew," in *American Men and Women of Science, 1998–99*, 20th Ed., New Providence, NJ: R. R. Bower, 1998.

T

William B. Taylor (1925–) graduated from the U.S. Military Academy at West Point in 1945 and rose to the rank of Major as an engineer in the Army Corps of Engineers, in support of the Nuclear Power Program. He was a Technical Operations Officer involved in atomic weapons tests and operations as part of the Manhattan Project from 1946 to 1952. In 1951, he received an M.S. degree from Johns Hopkins University. He retired in 1954 and joined NASA as a Systems Engineer in 1962. He was the Assistant Director for Engineering Studies in the Office of Manned Space Flight. See "Taylor, William B.," biographical file, NASA Historical Reference Collection, NASA History Office, NASA Headquarters, Washington, DC.

Adrienne F. Timothy supported the Apollo Telescope Mount (ATM) Reduction and Analysis System at the American Science and Engineering (AS&E) Corporation from 1971 to 1974. She began her career with NASA in 1974 as a Staff Scientist for Solar Physics, Physics, and Astronomy programs in the Office of Space Science. In 1975, she became Chief of the Solar Physics Branch, where she planned and directed a national program of space science research in the solar physics discipline. In 1977, she became Program Manager for advanced programs and technology in NASA's Solar Terrestrial Division of the Office of Space Science. Then in June of 1978 she became Assistant Associate Administrator for Space Science. See "Timothy, Adrienne F.," biographical file, NASA Historical Reference Collection, NASA History Office, NASA Headquarters, Washington, DC.

Richard Tousey (1909–1997) received his Ph.D. in physics from Harvard in 1933 and an honorary D.S. degree in 1961 from Tufts University, where he was a research instructor from 1935 to 1941. In 1941, he joined the Naval Research Laboratory (NRL) as a Space Scientist and stayed until his retirement in 1978. He was credited with pioneering NRL's rocket spectroscopy research and was the principal investigator for four successful solar experiments carried out by astronauts aboard the Skylab space station in 1973 and 1974. In 1990, he received the George Ellery Hale Prize from the American Astronomical Society for his work. Other honors include: the Henry Draper Medal of the National Academy of Sciences, the Frederick Ives Medal of the Optical Society of America, the E.O. Hulbert Award of the Naval Research Laboratory, the Progress Medal of the Photographic Society of America, the Prix Ancel of the Societe Francaise de Photographie, and the Eddington Medal of the Royal Astronomical Society. He was the Henry Norris Russell Lecturer of the American Astronomical Society and the Darwin Lecturer of the Royal Astronomical Society. See "Tousey, Richard," biographical file, NASA Historical Reference Collection, NASA History Office, NASA Headquarters, Washington, DC.

John W. Townsend, Jr., (1924–) was the Deputy Director of the Goddard Space Center from 1965 to1968 and Director from 1987 to 1990. Before NASA he worked at the Naval Research Laboratory from 1949 to 1958 and was Branch Head from 1955 to 1958. He supported several positions in various scientific fields, including Deputy Administrator of the Environment Sciences Services Administration (1968 to 1970), Associate Administrator of the National Oceanic and Atmospheric Administration (1970 to 1977), and President of the Fairchild Space and Electronics Company. He also had roles in the International Academy of Astronautics, NASA Advisory Council, National Academy of Engineering, National Research Council Space Applications Board, and the Office of Technology Assessment Advisory Board Panel on International Cooperation and Competition in Civilian Space

Activities. He earned his B.A., M.A, and Sc.D. from Williams College. See "Townsend, John W., Jr.," biographical file, NASA Historical Reference Collection, NASA History Office, NASA Headquarters, Washington, DC.

U

Roger K. Ulrich (1942–   ) earned his B.S. degree, 1963, and his Ph.D. degree, 1968, from the University of California, Berkeley. He was active in the study of the solar interior and used theoretical helioseismology, analysis of the solar neutrino problem, and solar atmosphere dynamics and observational methods that included the operation of the 150-foot tower on Mt. Wilson and participation in two space helioseismology experiments on the NASA/ESA SOHO mission. The 150-foot tower project provided access to observe facilities with a unique long-termed digital database of solar activity that extended over two solar cycles. He used the Mt. Wilson facility to make the first observation and identification of Alfven waves on the solar surface. Professional memberships include the American Astronomical Society and the International Astronomical Union. See "Ulrich, Roger K.," in *Who's Who in America, 1980–81*, 41st Ed., V. 2, and at *http://www.physics.ucla.edu//people/faculty_members/ulrich.html*

V

Wernher von Braun (1912–1977) was the leader of what has been called the "rocket team" that developed the German V-2 ballistic missile in World War II. At the conclusion of the war, von Braun and some of his chief assistants—as part of a military operation called Project Paperclip—came to America and were installed at Fort Bliss in El Paso, Texas, to work on rocket development and use the V-2 for high-altitude research. They used launch facilities at the nearby White Sands Proving Ground in New Mexico. In 1950, his team moved to the Redstone Arsenal near Huntsville, Alabama, to concentrate on the development of a new missile for the Army. They built the Army's Jupiter Ballistic missile, and before that the Redstone, used by NASA to launch the first Mercury capsules. The story of von Braun and the "rocket team" has been told many times. See, as examples, David H. DeVorkin, *Science with a Vengeance: How the Military Created the U.S. Space Sciences After World War II*, New York: Springer-Verlag, 1992; Frederick I. Ordway III and Mitchell R. Sharpe, *The Rocket Team*, New York: Thomas Y. Crowell, 1979; Erik Bergaust, *Wernher von Braun*, Washington, DC: National Space Institute, 1976.

W

Arthur Bertram Cuthbert Walker, Jr., (1936–2001) earned his B.S. from Case Institute of Technology in 1957, and his M.S. and Ph.D. (physics) from the University of Illinois, respectively, in 1958 and 1962. He was a noted Astronomer and Space Physicist at Stanford University from 1965 until 2001. His professional experiences include: member of the technical staff at the Space Physics Lab of the Aerospace Corporation, from 1965 to 1968, where he was promoted from Staff Scientist to Senior Staff Scientist in 1970; Director of the Space Astronomy Project in 1972 and 1973; and Associate Dean of Graduate Studies at Stanford University in 1975. His professional memberships include: Sigma Xi, the American Physics Society, the American Geophysical Union, the American Astronomical Union, the American Astronomical Society, and the International Astronomical Union. See "Walker, Arthur Bertram Cuthbert, Jr.," in *American Men and Women of Science, 1998–99*, 20th Ed., New Providence, NJ: R. R. Bower, 1998.

Robert S. Walker (1942–   ) (R-PA) earned a B.S. in education from Millersville University and an M.A. in political science from the University of Delaware. Congressman Walker was a high school teacher and a congressional aide prior to being elected to Congress. He served from 1977 to 1997 in the House, and he was elected Chairman of the House Science Committee in 1995. After his retirement from Congress, Walker served as Chairman of the Commission on the Future of the U.S. Aerospace Industry. See "Walker, Robert S.," biographical file, NASA Historical Reference Collection, NASA History Office, NASA Headquarters, Washington, DC.

Robert T. Watson (1922–   ) earned a B.A. from DePauw University in 1943 and a Ph.D. from Massachusetts Institute of Technology in 1951. Dr. Watson worked for the Department of Commerce from 1971 to 1990. See *Who's Who in America, 2000*, 54th Ed., New Providence, NJ: Marquis Who's Who, 1999.

James E. Webb (1906–1992) was NASA's Administrator from 1961 to 1968. Previously, he had been an aide to a congressman in New Deal, Washington, an aide to Washington lawyer Max O. Gardner, and a business executive with the Sperry Corporation and the Kerr-McGee Oil Company. He was the Director of the Bureau of the Budget between 1946 and 1950, and Undersecretary of State from 1950 to 1952. See W. Henry Lambright, *Powering Apollo: James E. Webb of NASA*, Baltimore, MD: Johns Hopkins University Press, 1995.

Robert M. White (1923–   ) served as Head of the U.S. Weather Bureau and the Environmental Science Services Administration in the 1960s, as Administrator of the National Oceanic and Atmospheric Administration in the 1970s, and as Head of the National Academy of Engineering in the late 1980s. See "White, Robert M.," biographical file, NASA Historical Reference Collection, NASA History Office, NASA Headquarters, Washington, DC.

Edward Lawrence (Larry) Winn, Jr., (1919–   ) (R-KS) earned his B.A. from the University of Kansas in 1941. Before his congressional seat, he was with a radio station in Missouri for two years, spent two years with North American Aviation, worked two years as a private builder, has been Vice President of the Winn-Rau Corporation since 1950, was the director of the National Association of Home Builders for 14 years, and is past president of the Home Builders Association of Kansas. A Republican, Winn was elected to the 90th Congress, 2 January 1967, and served his State until 3 January 1985. He was a ranking representative from Kansas on the House Science and Technology Committee and did not seek reelection to the 99th Congress. See "Winn, Larry, Jr.," biographical file, NASA Historical Reference Collection, NASA History Office, NASA Headquarters, Washington, DC.

Y

Herbert F. York (1923–   ) has been associated with scientific research in support of national defense since World War II. He was Director of the Livermore Radiation Laboratory for the University of California before moving to the Department of Defense in March 1958 as Chief Scientist of the Advanced Research Projects Agency. As a result of a Department of Defense reorganization, he became the Director of Research and Engineering in December 1958. He supported the position, the third-ranking civilian office after the Secretary and Deputy Secretary of Defense, until 1961. He then moved to the University of California, San Diego, as chancellor and professor of physics. He also served as a member of the President's Science Advisory Committee under both Eisenhower and Johnson, and was later Chief Negotiator for the comprehensive test ban during the Carter Administration. See "York, Dr. Herbert F.," biographical file, NASA Historical Reference Collection, NASA History Office, NASA Headquarters, Washington, DC; and Herbert F. York, *Making Weapons, Talking Peace: A Physicist's Odyssey from Hiroshima to Geneva*, New York: Basic Books, 1987.

Z

Harold Zirin (1929–   ) earned and received his B.A. in 1950, his M.A. in 1951, and his Ph.D. (astrophysics) in 1953 from Harvard University. His professional career began at the Rand Corporation in 1952. From 1953 to 1955, he was an instructor of astronomy at Harvard University. He was a senior researcher at the High Altitude Observatory, of the University of Colorado, from 1955 to 1964. With his colleague Robert Howard, Zirin attempted to build a 65-centimeter Solar Optical Telescope (SOT) in the mid-1960s. Since 1964, Dr. Zirin has been a professor of astrophysics at the California Institute of Technology (CIT). See "Zirin, Harold," in *American Men and Women of Science, 1998–99*, 20th Ed., New Providence, NJ: R. R. Bower, 1998.

Index

A

B

C

E

F

O

Q

R

S

T

U

V

The NASA History Series

Reference Works, NASA SP-4000:

Grimwood, James M. *Project Mercury: A Chronology.* NASA SP-4001, 1963.

Grimwood, James M., and C. Barton Hacker, with Peter J. Vorzimmer. *Project Gemini Technology and Operations: A Chronology.* NASA SP-4002, 1969.

Link, Mae Mills. *Space Medicine in Project Mercury.* NASA SP-4003, 1965.

Astronautics and Aeronautics, 1963: Chronology of Science, Technology, and Policy. NASA SP-4004, 1964.

Astronautics and Aeronautics, 1964: Chronology of Science, Technology, and Policy. NASA SP-4005, 1965.

Astronautics and Aeronautics, 1965: Chronology of Science, Technology, and Policy. NASA SP-4006, 1966.

Astronautics and Aeronautics, 1966: Chronology of Science, Technology, and Policy. NASA SP-4007, 1967.

Astronautics and Aeronautics, 1967: Chronology of Science, Technology, and Policy. NASA SP-4008, 1968.

Ertel, Ivan D., and Mary Louise Morse. *The Apollo Spacecraft: A Chronology, Volume I, Through November 7, 1962.* NASA SP-4009, 1969.

Morse, Mary Louise, and Jean Kernahan Bays. *The Apollo Spacecraft: A Chronology, Volume II, November 8, 1962– September 30, 1964.* NASA SP-4009, 1973.

Brooks, Courtney G., and Ivan D. Ertel. *The Apollo Spacecraft: A Chronology, Volume III, October 1, 1964–January 20, 1966.* NASA SP-4009, 1973.

Ertel, Ivan D., and Roland W. Newkirk, with Courtney G. Brooks. *The Apollo Spacecraft: A Chronology, Volume IV, January 21, 1966–July 13, 1974.* NASA SP-4009, 1978.

Astronautics and Aeronautics, 1968: Chronology of Science, Technology, and Policy. NASA SP-4010, 1969.

Newkirk, Roland W., and Ivan D. Ertel, with Courtney G. Brooks. *Skylab: A Chronology.* NASA SP-4011, 1977.

Van Nimmen, Jane, and Leonard C. Bruno, with Robert L. Rosholt. *NASA Historical Data Book, Volume I: NASA Resources, 1958–1968.* NASA SP-4012, 1976, rep. ed. 1988.

Ezell, Linda Neuman. *NASA Historical Data Book, Volume II: Programs and Projects, 1958–1968.* NASA SP-4012, 1988.

Ezell, Linda Neuman. *NASA Historical Data Book, Volume III: Programs and Projects, 1969–1978.* NASA SP-4012, 1988.

Gawdiak, Ihor Y., with Helen Fedor, compilers. *NASA Historical Data Book, Volume IV: NASA Resources, 1969–1978.* NASA SP-4012, 1994.

Rumerman, Judy A., compiler. *NASA Historical Data Book, 1979–1988: Volume V, NASA Launch Systems, Space Transportation, Human Spaceflight, and Space Science.* NASA SP-4012, 1999.

Rumerman, Judy A., compiler. *NASA Historical Data Book, Volume VI: NASA Space Applications, Aeronautics and Space Research and Technology, Tracking and Data Acquisition/Space Operations, Commercial Programs, and Resources, 1979–1988.* NASA SP-2000-4012, 2000.

Astronautics and Aeronautics, 1969: Chronology of Science, Technology, and Policy. NASA SP-4014, 1970.

Astronautics and Aeronautics, 1970: Chronology of Science, Technology, and Policy. NASA SP-4015, 1972.

Astronautics and Aeronautics, 1971: Chronology of Science, Technology, and Policy. NASA SP-4016, 1972.

Astronautics and Aeronautics, 1972: Chronology of Science, Technology, and Policy. NASA SP-4017, 1974.

Astronautics and Aeronautics, 1973: Chronology of Science, Technology, and Policy. NASA SP-4018, 1975.

Astronautics and Aeronautics, 1974: Chronology of Science, Technology, and Policy. NASA SP-4019, 1977.

Astronautics and Aeronautics, 1975: Chronology of Science, Technology, and Policy. NASA SP-4020, 1979.

Astronautics and Aeronautics, 1976: Chronology of Science, Technology, and Policy. NASA SP-4021, 1984.

Astronautics and Aeronautics, 1977: Chronology of Science, Technology, and Policy. NASA SP-4022, 1986.

Astronautics and Aeronautics, 1978: Chronology of Science, Technology, and Policy. NASA SP-4023, 1986.

Astronautics and Aeronautics, 1979–1984: Chronology of Science, Technology, and Policy. NASA SP-4024, 1988.

Astronautics and Aeronautics, 1985: Chronology of Science, Technology, and Policy. NASA SP-4025, 1990.

Noordung, Hermann. *The Problem of Space Travel: The Rocket Motor.* Edited by Ernst Stuhlinger and J. D. Hunley, with Jennifer Garland. NASA SP-4026, 1995.

Astronautics and Aeronautics, 1986–1990: A Chronology. NASA SP-4027, 1997.

Astronautics and Aeronautics, 1990–1995: A Chronology. NASA SP-2000-4028, 2000.

Management Histories, NASA SP-4100:

Rosholt, Robert L. *An Administrative History of NASA, 1958–1963.* NASA SP-4101, 1966.

Levine, Arnold S. *Managing NASA in the Apollo Era.* NASA SP-4102, 1982.

Roland, Alex. *Model Research: The National Advisory Committee for Aeronautics, 1915–1958.* NASA SP-4103, 1985.

Fries, Sylvia D. *NASA Engineers and the Age of Apollo.* NASA SP-4104, 1992.

Glennan, T. Keith. *The Birth of NASA: The Diary of T. Keith Glennan.* J. D. Hunley, editor. NASA SP-4105, 1993.

Seamans, Robert C., Jr. *Aiming at Targets: The Autobiography of Robert C. Seamans, Jr.* NASA SP-4106, 1996.

Garber, Stephen J., editor. *Looking Backward, Looking Forward: Forty Years of U.S. Human Spaceflight Symposium.* NASA SP-2002-4107, 2002.

Project Histories, NASA SP-4200:

Swenson, Loyd S., Jr., James M. Grimwood, and Charles C. Alexander. *This New Ocean: A History of Project Mercury.* NASA SP-4201, 1966; rep. ed. 1998.

Green, Constance McLaughlin, and Milton Lomask. *Vanguard: A History.* NASA SP-4202, 1970; rep. ed. Smithsonian Institution Press, 1971.

Hacker, Barton C., and James M. Grimwood. *On the Shoulders of Titans: A History of Project Gemini.* NASA SP-4203, 1977.

Benson, Charles D., and William Barnaby Faherty. Moonport: *A History of Apollo Launch Facilities and Operations.* NASA SP-4204, 1978.

Brooks, Courtney G., James M. Grimwood, and Loyd S. Swenson, Jr. *Chariots for Apollo: A History of Manned Lunar Spacecraft.* NASA SP-4205, 1979.

Bilstein, Roger E. *Stages to Saturn: A Technological History of the Apollo/Saturn Launch Vehicles.* NASA SP-4206, 1980, rep. ed. 1997.

SP-4207 not published.

Compton, W. David, and Charles D. Benson. *Living and Working in Space: A History of Skylab.* NASA SP-4208, 1983.

Ezell, Edward Clinton, and Linda Neuman Ezell. *The Partnership: A History of the Apollo-Soyuz Test Project.* NASA SP-4209, 1978.

Hall, R. Cargill. *Lunar Impact: A History of Project Ranger.* NASA SP-4210, 1977.

Newell, Homer E. *Beyond the Atmosphere: Early Years of Space Science.* NASA SP-4211, 1980.

Ezell, Edward Clinton, and Linda Neuman Ezell. *On Mars: Exploration of the Red Planet, 1958–1978.* NASA SP-4212, 1984.

Pitts, John A. *The Human Factor: Biomedicine in the Manned Space Program to 1980.* NASA SP-4213, 1985.

Compton, W. David. *Where No Man Has Gone Before: A History of Apollo Lunar Exploration Missions.* NASA SP-4214, 1989.

Naugle, John E. *First Among Equals: The Selection of NASA Space Science Experiments.* NASA SP-4215, 1991.

Wallace, Lane E. *Airborne Trailblazer: Two Decades with NASA Langley's Boeing 737 Flying Laboratory.* NASA SP-4216, 1994.

Butrica, Andrew J., editor. *Beyond the Ionosphere: Fifty Years of Satellite Communication.* NASA SP-4217, 1997.

Butrica, Andrew J. *To See the Unseen: A History of Planetary Radar Astronomy.* NASA SP-4218, 1996.

Mack, Pamela E., editor. *From Engineering Science to Big Science: The NACA and NASA Collier Trophy Research Project Winners.* NASA SP-4219, 1998.

Reed, R. Dale, with Darlene Lister. *Wingless Flight: The Lifting Body Story.* NASA SP-4220, 1997.

Heppenheimer, T. A. *The Space Shuttle Decision: NASA's Search for a Reusable Space Vehicle.* NASA SP-4221, 1999.

Hunley, J. D., editor. *Toward Mach 2: The Douglas D-558 Program.* NASA SP-4222, 1999.

Swanson, Glen E., editor. *"Before this Decade Is Out . . .": Personal Reflections on the Apollo Program.* NASA SP-4223, 1999.

Tomayko, James E. *Computers Take Flight: A History of NASA's Pioneering Digital Fly-by-Wire Project.* NASA SP-2000-4224, 2000.

744

Morgan, Clay. *Shuttle-Mir: The U.S. and Russia Share History's Highest Stage*. NASA SP-2001-4225, 2001.

Leary, William M. *"We Freeze to Please": A History of NASA's Icing Research Tunnel and the Quest for Flight Safety*. NASA SP-2002-4226, 2002.

Mudgway, Douglas J. *Uplink-Downlink: A History of the Deep Space Network 1957–1997*. NASA SP-2001-4227, 2001.

Center Histories, NASA SP-4300:

Rosenthal, Alfred. *Venture into Space: Early Years of Goddard Space Flight Center*. NASA SP-4301, 1985.

Hartman, Edwin P. *Adventures in Research: A History of Ames Research Center, 1940–1965*. NASA SP-4302, 1970.

Hallion, Richard P. *On the Frontier: Flight Research at Dryden, 1946–1981*. NASA SP-4303, 1984.

Muenger, Elizabeth A. *Searching the Horizon: A History of Ames Research Center, 1940–1976*. NASA SP-4304, 1985.

Hansen, James R. *Engineer in Charge: A History of the Langley Aeronautical Laboratory, 1917–1958*. NASA SP-4305, 1987.

Dawson, Virginia P. *Engines and Innovation: Lewis Laboratory and American Propulsion Technology*. NASA SP-4306, 1991.

Dethloff, Henry C. *"Suddenly Tomorrow Came . . .": A History of the Johnson Space Center*. NASA SP-4307, 1993.

Hansen, James R. *Spaceflight Revolution: NASA Langley Research Center from Sputnik to Apollo*. NASA SP-4308, 1995.

Wallace, Lane E. *Flights of Discovery: 50 Years at the NASA Dryden Flight Research Center*. NASA SP-4309, 1996.

Herring, Mack R. *Way Station to Space: A History of the John C. Stennis Space Center*. NASA SP-4310, 1997.

Wallace, Harold D., Jr. *Wallops Station and the Creation of the American Space Program*. NASA SP-4311, 1997.

Wallace, Lane E. *Dreams, Hopes, Realities: NASA's Goddard Space Flight Center, The First Forty Years*. NASA SP-4312, 1999.

Dunar, Andrew J., and Stephen P. Waring. *Power to Explore: A History of the Marshall Space Flight Center*. NASA SP-4313, 1999.

Bugos, Glenn E. *Atmosphere of Freedom: Sixty Years at the NASA Ames Research Center*. NASA SP-2000-4314, 2000.

General Histories, NASA SP-4400:

Corliss, William R. *NASA Sounding Rockets, 1958–1968: A Historical Summary*. NASA SP-4401, 1971.

Wells, Helen T., Susan H. Whiteley, and Carrie Karegeannes. *Origins of NASA Names*. NASA SP-4402, 1976.

Anderson, Frank W., Jr. *Orders of Magnitude: A History of NACA and NASA, 1915–1980*. NASA SP-4403, 1981.

Sloop, John L. *Liquid Hydrogen as a Propulsion Fuel, 1945–1959*. NASA SP-4404, 1978.

Roland, Alex. *A Spacefaring People: Perspectives on Early Spaceflight*. NASA SP-4405, 1985.

Bilstein, Roger E. *Orders of Magnitude: A History of the NACA and NASA, 1915–1990*. NASA SP-4406, 1989.

Logsdon, John M., editor, with Linda J. Lear, Jannelle Warren-Findley, Ray A. Williamson, and Dwayne A. Day. *Exploring the Unknown: Selected Documents in the History of the U.S. Civil Space Program, Volume I, Organizing for Exploration.* NASA SP-4407, 1995.

Logsdon, John M., editor, with Dwayne A. Day and Roger D. Launius. *Exploring the Unknown: Selected Documents in the History of the U.S. Civil Space Program, Volume II, Relations with Other Organizations.* NASA SP-4407, 1996.

Logsdon, John M., editor, with Roger D. Launius, David H. Onkst, and Stephen J. Garber. *Exploring the Unknown: Selected Documents in the History of the U.S. Civil Space Program, Volume III, Using Space.* NASA SP-4407, 1998.

Logsdon, John M., general editor, with Ray A. Williamson, Roger D. Launius, Russell J. Acker, Stephen J. Garber, and Jonathan L. Friedman. *Exploring the Unknown: Selected Documents in the History of the U.S. Civil Space Program, Volume IV, Accessing Space.* NASA SP-4407, 1999.

Logsdon, John M., general editor, with Amy Paige Snyder, Roger D. Launius, Stephen J. Garber, and Regan Anne Newport. *Exploring the Unknown: Selected Documents in the History of the U.S. Civil Space Program, Volume V, Exploring the Cosmos.* NASA SP-2001-4407, 2001.

Siddiqi, Asif A. *Challenge to Apollo: The Soviet Union and the Space Race, 1945–1974.* NASA SP-2000-4408, 2000.

Monographs in Aerospace History, NASA SP-4500:

Launius, Roger D. and Aaron K. Gillette, compilers, *Toward a History of the Space Shuttle: An Annotated Bibliography.* Monograph in Aerospace History, No. 1, 1992.

Launius, Roger D., and J. D. Hunley, compilers, *An Annotated Bibliography of the Apollo Program.* Monograph in Aerospace History, No. 2, 1994.

Launius, Roger D. *Apollo: A Retrospective Analysis.* Monograph in Aerospace History, No. 3, 1994.

Hansen, James R. *Enchanted Rendezvous: John C. Houbolt and the Genesis of the Lunar-Orbit Rendezvous Concept.* Monograph in Aerospace History, No. 4, 1995.

Gorn, Michael H. Hugh L. *Dryden's Career in Aviation and Space.* Monograph in Aerospace History, No. 5, 1996.

Powers, Sheryll Goecke. *Women in Flight Research at NASA Dryden Flight Research Center, from 1946 to 1995.* Monograph in Aerospace History, No. 6, 1997.

Portree, David S. F. and Robert C. Trevino. *Walking to Olympus: An EVA Chronology.* Monograph in Aerospace History, No. 7, 1997.

Logsdon, John M., moderator. *Legislative Origins of the National Aeronautics and Space Act of 1958: Proceedings of an Oral History Workshop.* Monograph in Aerospace History, No. 8, 1998.

Rumerman, Judy A., compiler, *U.S. Human Spaceflight, A Record of Achievement 1961–1998.* Monograph in Aerospace History, No. 9, 1998.

Portree, David S. F. *NASA's Origins and the Dawn of the Space Age.* Monograph in Aerospace History, No. 10, 1998.

Logsdon, John M. *Together in Orbit: The Origins of International Cooperation in the Space Station.* Monograph in Aerospace History, No. 11, 1998.

Phillips, W. Hewitt. *Journey in Aeronautical Research: A Career at NASA Langley Research Center.* Monograph in Aerospace History, No. 12, 1998.

Braslow, Albert L. *A History of Suction-Type Laminar-Flow Control with Emphasis on Flight Research.* Monograph in Aerospace History, No. 13, 1999.

Logsdon, John M., moderator. *Managing the Moon Program: Lessons Learned From Apollo.* Monograph in Aerospace History, No. 14, 1999.

Perminov, V. G. *The Difficult Road to Mars: A Brief History of Mars Exploration in the Soviet Union.* Monograph in Aerospace History, No. 15, 1999.

Tucker, Tom. *Touchdown: The Development of Propulsion Controlled Aircraft at NASA Dryden.* Monograph in Aerospace History, No. 16, 1999.

Maisel, Martin D., Demo J. Giulianetti, and Daniel C. Dugan. *The History of the XV-15 Tilt Rotor Research Aircraft: From Concept to Flight.* NASA SP-2000-4517, 2000.

Jenkins, Dennis R. *Hypersonics Before the Shuttle: A Concise History of the X-15 Research Airplane.* NASA SP-2000-4518, 2000.

Chambers, Joseph R. *Partners in Freedom: Contributions of the Langley Research Center to U.S. Military Aircraft in the 1990s.* NASA SP-2000-4519, 2000.

Waltman, Gene L. *Black Magic and Gremlins: Analog Flight Simulations at NASA's Flight Research Center.* NASA SP-2000-4520, 2000.

Portree, David S. F. *Humans to Mars: Fifty Years of Mission Planning, 1950–2000.* NASA SP-2001-4521, 2001.

Thompson, Milton O., with J. D. Hunley. *Flight Research: Problems Encountered and What They Should Teach Us.* NASA SP-2000-4522, 2000.

Tucker, Tom. *The Eclipse Project.* NASA SP-2000-4523, 2000.

Siddiqi, Asif A. *Deep Space Chronicle: A Chronology of Deep Space and Planetary Probes, 1958–2000.* NASA SP-2002-4524, 2002.

Merlin, Peter W. *Mach 3+: NASA/USAF YF-12 Flight Research, 1969–1979.* NASA SP-2001-4525, 2001.

Anderson, Seth B. *Memoirs of an Aeronautical Engineer—Flight Tests at Ames Research Center: 1940–1970.* NASA SP-2002-4526, 2002.

Renstrom, Arthur G. *Wilbur and Orville Wright: A Bibliography Commemorating the One-Hundredth Anniversary of the First Powered Flight on December 17, 1903.* NASA SP-2002-4527, 2002.

No monograph 28.

Chambers, Joseph R. *Concept to Reality: Contributions of the NASA Langley Research Center to U.S. Civil Aircraft of the 1990s.* SP-2003-4529, 2003.

Peebles, Curtis, editor. *The Spoken Word: Recollections of Dryden History, The Early Years.* SP-2003-4530, 2003.

Jenkins, Dennis R., Tony Landis, and Jay Miller. *American X-Vehicles: An Inventory—X-1 to X-50.* SP-2003-4531, 2003.